14

CRM SERIES

Configuration Spaces

Geometry, Combinatorics and Topology

edited by

A. Bjorner, F. Cohen, C. De Concini,
C. Procesi and M. Salvetti

ISBN 978-88-7642-430-4
ISBN 978-88-7642-431-1 (eBook)

Contents

Authors' affiliations

A. ADEM – Department of Mathematics, University of British Columbia, Vancouver BC V6T 1Z2, Canada
adem@math.ubc.ca

M. AMRAM – Shamoon College of Engineering, Bialik/Basel Sts., 84100 Beer-Sheva, Israel
and
Department of Mathematics, Bar-Ilan University, 52900 Ramat-Gan, Israel
meiravt@sce.ac.il, meirav@macs.biu.ac.il

D. ARINKIN – Department of Mathematics, University of North Carolina at Chapel Hill Chapel Hill, NC 27599-3250, USA
arinkin@email.unc.edu

A. BAHRI – Department of Mathematics, Rider University, Lawrenceville, NJ 08648, U.S.A.
bahri@rider.edu

E. ARTAL BARTOLO – Departamento de Matemáticas, IUMA, Universidad de Zaragoza, C. Pedro Cerbuna 12, 50009 Zaragoza, Spain
artal@unizar.es

M. BENDERSKY – Department of Mathematics, CUNY, East 695 Park Avenue New York, NY 10065, U.S.A.
mbenders@xena.hunter.cuny.edu

N. BERLINE – Ecole Polytechnique, Centre de Mathematiques Laurent Schwartz, 91128 Palaiseau Cedex, France
nicole.berline@math.polytechnique.fr

C.-F. BÖDIGHEIMER – Mathematisches Institut, Universität Bonn, Endenicher Allee 60, 53115 Bonn, Germany
boedigheimer@math.uni-bonn.de

F. CALLEGARO – Scuola Normale Superiore, Piazza dei Cavalieri, 7, 56126 Pisa, Italia
f.callegaro@sns.it

J. I. COGOLLUDO-AGUSTÍN – Departamento de Matemáticas, IUMA, Universidad de Zaragoza, C. Pedro Cerbuna 12, 50009 Zaragoza, Spain
jicogo@unizar.es

D. C. COHEN – Department of Mathematics, Louisiana State University, Baton Rouge, Louisiana 70803, USA
cohen@math.lsu.edu

F. R. COHEN – UR Mathematics, 915 Hylan Building, University of Rochester, RC Box 270138, Rochester, NY 14627, USA
cohf@math.rochester.edu

A. DIMCA – Laboratoire J.A. Dieudonné, UMR du CNRS 6621, Université de Nice Sophia Antipolis, Parc Valrose, 06108 Nice Cedex 02, France
dimca@unice.fr

M. FALK – Department of Mathematics and Statistics, Northern Arizona University, Flagstaff, Arizona 86011-5717, USA
michael.falk@nau.edu

D. GARBER – Department of Applied Mathematics, Faculty of Sciences, Holon Institute of Technology, 52 Golomb St., PO Box 305, 58102 Holon, Israel
garber@hit.ac.il

S. GITLER – Department of Mathematics, Cinvestav, San Pedro Zacatenco, Mexico, D.F. CP 07360 Apartado Postal 14-740, Mexico
samuel.gitler@gmail.com

C. GIUSTI – Mathematics Department, Willamette University, 219 Ford Hall, 900 State Street Salem, Oregon 97301, USA
cgiusti@willamette.edu

E. GODELLE – Université de Caen, Laboratoire LMNO, UMR 6139 du CNRS, Campus II, 14032 Caen cedex, France
eddy.godelle@unicaen.fr

J. M. GÓMEZ – Department of Mathematics, Johns Hopkins University, Baltimore, MD 21218, USA
jgomez@math.jhu.edu

A. HENDERSON – School of Mathematics and Statistics, University of Sydney NSW 2006, Australia
anthony.henderson@sydney.edu.au

H. KAMIYA – School of Economics, Nagoya University, Nagoya, 464-8601, Japan
kamiya@soec.nagoya-u.ac.jp

T. Kohno – IPMU, Graduate School of Mathematical Sciences, The University of Tokyo, 3-8-1 Komaba, Meguro-ku, Tokyo 153-8914, Japan
kohno@ms.u-tokyo.ac.jp

G.LEHRER – School of Mathematics and Statistics F07, University of Sydney, NSW 2006, Australia
gustav.lehrer@sydney.edu.au

A. LEVIN – State University - Higher School of Economics, Department of Mathematics, 20 Myasnitskaya Ulitsa, Moscow, 101000, Russia
and
Laboratory of Algebraic Geometry, GU-HSE, 7 Vavilova Street, Moscow, 117312, Russia
alevin@hse.ru

A. LIBGOBER – Department of Mathematics, University of Illinois, 851 S. Morgan Str., Chicago, IL 60607
libgober@math.uic.edu

I. MARIN – Université Paris Diderot 175 rue du Chevaleret 75013 Paris
marin@math.jussieu.fr

D. MORONI – Istituto di Scienza e Tecnologia dell'Informazione "Alessandro Faedo" (ISTI), Area della Ricerca CNR di Pisa, Via G. Moruzzi, 1, 56124 Pisa, Italia
davide.moroni@isti.cnr.it

L. PARIS – Université de Bourgogne, Institut de Mathématiques de Bourgogne, UMR 5584 du CNRS, B.P. 47870, 21078 Dijon cedex, France
lparis@u-bourgogne.fr

R. RANDELL – Department of Mathematics, University of Iowa, Iowa City, Iowa 52242, USA
richard-randell@uiowa.edu

M. SALVETTI – Departimento di Matematica, Università di Pisa, Largo B. Pontecorvo, 5, 56127 Pisa, Italia
salvetti@dm.unipi.it

J. SHARESHIAN – Department of Mathematics, Washington University, St. Louis, MO 63130, USA
shareshi@math.wustl.edu

D. SINHA – Mathematics Department, University of Oregon, Eugene, OR 97403, USA
dps@math.uoregon.edu

A. I. SUCIU – Department of Mathematics, Northeastern University, Boston, MA 02115, USA
a.suciu@neu.edu

A. TAKEMURA – Department of Mathematical Informatics, University of Tokyo, Bunkyo Tokyo, 113-0033, Japan
takemura@stat.u-tokyo.ac.jp

M. TEICHER – Department of Mathematics, Bar-Ilan University, 52900 Ramat-Gan, Israel
teicher@macs.biu.ac.il

H. TERAO – Department of Mathematics, Hokkaido University, Sapporo, 060-0810, Japan
hterao00@za3.so-net.ne.jp

U. TILLMANN – Mathematical Institute, Oxford University, 24-29 St Giles, Oxford OX1 3LB, United Kingdom
tillmann@maths.ox.ac.uk

A. VARCHENKO – Department of Mathematics, University of North Carolina at Chapel Hill, Chapel Hill, NC 27599-3250, USA
anv@email.unc.edu

M. VERGNE – Universite Paris 7 Denis Diderot, Institut Mathématique de Jussieu, 175 rue du Chevaleret - 75013 Paris, France
vergne@math.jussieu.fr

A. VILLA – Dipartimento di Matematica "Giuseppe Peano", Università di Torino, Via Carlo Alberto, 10, 10123 Torino, Italia
andrea.villa@unito.it

M. L. WACHS – Department of Mathematics, University of Miami, Coral Gables, FL 33124, USA
wachs@math.miami.edu

M. YOSHINAGA – Department of Mathematics, Hokkaido University, North 10, West 8, Kita-ku, Sapporo, 060-0810, Japan
mhyo@math.kyoto-u.ac.jp

Introduction

The present volume arises from the intensive research period "Configuration spaces: Geometry, Combinatorics and Topology" which took place at the Centro di Ricerca Matematica Ennio De Giorgi, Pisa, in May-June 2010.

The program was very intense and included two conferences, nine minicourses and two weekly seminars one of which organized by Filippo Callegaro and run by the younger participants.

The period covered a large number of different topics all centering around the notion of configuration spaces. These included among others:

1. Study of local systems on the complements of a hyperplane arrangements. Characteristic and resonance varieties. Cohomology and monodromy computations. Study of the fundamental group of complements of arrangements. Hodge theoretical aspects.
2. Qualitative and quantitative problems related to the study of partition functions. Relations with the theory of Box Splines and applications to index theory. Enumeration of lattice points in rational polytopes. Relations with toric geometry.
3. Invariants of braids and knots. Topological quantum field theory. Applications to low dimensional topology. Construction of representations of braid groups and related groups.
4. Applications of the theory of hyperplane arrangements and toric arrangements to the study of the combinatorics of matroids and generalizations.
5. Homotopy theory aspects of the study of configuration spaces and moduli spaces. Relations between cohomology of braid and mapping class groups.
6. Combinatorial aspects of the theory of Coxeter and Artin groups. Cohomological computations for both abelian and non abelian local systems.

7. Toric geometry. Moment angle complexes and applications. Geometry and topology of real and complex toric varieties.

Many of these topics were covered by the minicourses whose list we include here:

Fred Cohen: Moment-angle complexes, their stable structure and cohomology.

Eduard Looijenga: Aspects of the KZ system.

Stefan Papadima, Alex Suciu: Cohomology jumping loci and homological finiteness properties.

Claudio Procesi: Splines and partition functions

Dev Sinha: Hopf rings in topology and algebra.

Alexander Varchenko: The quantum integrable model of an arrangement of hyperplanes.

Michele Vergne: Remarks on Box splines.

Sergey Yuzvinsky: A short introduction to arrangements of hyperplanes

Sergey Yuzvinsky: Resonance varieties for arrangements and their relations to combinatorics and algebraic geometry

We feel that the papers appearing in this volume very well represent the spirit and the topics of the research period.

We very warmly thank both the authors of these papers and the other participants who very much contributed to the success of this activity.

Finally we wish to thank the staff of Centro di Ricerca Ennio De Giorgi and of the Scientific Section of Edizioni della Scuola Normale for a very efficient job both in running our period and in editing the present volume.

Pisa, July 2012

A. Bjorner, F. Cohen, C. De Concini,
C. Procesi and M. Salvetti

On the structure of spaces of commuting elements in compact Lie groups

Alejandro Adem* and José Manuel Gómez

Abstract. In this note we study topological invariants of the spaces of homomorphisms $\mathrm{Hom}(\pi, G)$, where π is a finitely generated abelian group and G is a compact Lie group arising as an arbitrary finite product of the classical groups $SU(r)$, $U(q)$ and $Sp(k)$.

1 Introduction

Let $\mathcal{P}$ denote the class of compact Lie groups arising as arbitrary finite products of the classical groups $SU(r)$, $U(q)$ and $Sp(k)$. In this article we use methods from algebraic topology to study the spaces of homomorphisms $\mathrm{Hom}(\pi, G)$ where π denotes a finitely generated abelian group and $G \in \mathcal{P}$. Our main interest is the computation of invariants associated to these spaces such as their cohomology and stable homotopy type, as well as their equivariant K-theory with respect to the natural conjugation action. The natural quotient space under this action is the space of representations $\mathrm{Rep}(\pi, G)$, which can be identified with the moduli space of isomorphism classes of flat connections on principal G–bundles over M, where M is a compact connected manifold with $\pi_1(M) = \pi$. Thus our results provide insight into these geometric invariants in the important case when $\pi_1(M)$ is a finitely generated abelian group.

Our starting point is the observation (see [3]) that when $G \in \mathcal{P}$ and π is a finitely generated abelian group, the conjugation action of G on the space of homomorphisms $\mathrm{Hom}(\pi, G)$ satisfies the following property: for every element $x \in \mathrm{Hom}(\pi, G)$ the isotropy subgroup G_x is connected and of maximal rank. This property plays a central part in our analysis. Indeed, let $T \subset G$ be a maximal torus; in general if a compact Lie group

*Partially supported by NSERC.

G acts on a compact space X with connected maximal rank isotropy subgroups then there is an associated action of W on the fixed–point set X^T and many properties of the space X are determined by the action of W on X^T (see [3, 8]). For our examples this means that a detailed understanding of the W-action on the subspace $\mathrm{Hom}(\pi, G)^T = \mathrm{Hom}(\pi, T)$ can be used to describe key homotopy–theoretic invariants for the original space of homomorphisms.

This approach can be used for example to obtain an explicit description of the number of path–connected components in $\mathrm{Hom}(\pi, G)$. Indeed we show that if $\pi = \mathbb{Z}^n \oplus A$, where A is a finite abelian group, then the number of path–connected components in $\mathrm{Hom}(\pi, G)$ equals the number of distinct orbits for the action of W on $\mathrm{Hom}(A, T)$

In [1] a stable splitting for the spaces of commuting n-tuples in G, $\mathrm{Hom}(\mathbb{Z}^n, G)$, was derived for any Lie group G that is a closed subgroup of $GL_n(\mathbb{C})$. Here we show that this splitting can be generalized to the spaces of homomorphisms $\mathrm{Hom}(\pi, G)$ when $G \in \mathcal{P}$ and π is any finitely generated abelian group. This is done by constructing a stable splitting on $\mathrm{Hom}(\pi, G)^T = \mathrm{Hom}(\pi, T)$ and proving that this splitting lifts to the space $\mathrm{Hom}(\pi, G)$. Suppose that $\pi = \mathbb{Z}/(q_1) \oplus \cdots \oplus \mathbb{Z}/(q_n)$, where $n \geq 0$ and $q_1, \ldots, q_n$ are integers. Here we allow some of the q_i's to be 0 and in that case $\mathbb{Z}/(0) = \mathbb{Z}$. Thus $\mathrm{Hom}(\pi, G)$ can be seen as the subspace of G^n consisting of those commuting n-tuples $(x_1, \ldots, x_n)$ such that $x_i^{q_i} = 1_G$ for all $1 \leq i \leq n$. For $1 \leq r \leq n$ let $J_{n,r}$ denote the set of all sequences of the form $\mathfrak{m} := \{1 \leq m_1 < \cdots < m_r \leq n\}$. Given such a sequence $\mathfrak{m}$ let $P_{\mathfrak{m}}(\pi) := \mathbb{Z}/(q_{m_1}) \oplus \cdots \oplus \mathbb{Z}/(q_{m_r})$ be a quotient of π. Let $S_1(P_{\mathfrak{m}}(\pi), G)$ be the subspace of $\mathrm{Hom}(P_{\mathfrak{m}}(\pi), G)$ consisting of those r-tuples $(x_{m_1}, \ldots, x_{m_r})$ in $\mathrm{Hom}(P_{\mathfrak{m}}(\pi), G)$ for which at least one of the x_{m_i}'s is equal to 1_G.

Theorem 1.1. *Suppose that $G \in \mathcal{P}$ and that π is a finitely generated abelian group. Then there is a G-equivariant homotopy equivalence*

$$\Theta : \Sigma\,\mathrm{Hom}(\pi, G) \to \bigvee_{1 \leq r \leq n} \Sigma \left(\bigvee_{\mathfrak{m} \in J_{n,r}} \mathrm{Hom}(P_{\mathfrak{m}}(\pi), G)/S_1(P_{\mathfrak{m}}(\pi), G) \right).$$

In Section 4 we determine the homotopy type of the stable factors appearing in the previous theorem for certain particular cases. In particular we determine the stable homotopy type of $\mathrm{Hom}(\pi, SU(2))$ for any finitely generated abelian group.

Suppose now that G is any compact Lie group. The fundamental group of the spaces of homomorphisms of the form $\mathrm{Hom}(\mathbb{Z}^n, G)$ was computed in [7]. Let $\mathbb{1} \in \mathrm{Hom}(\mathbb{Z}^n, G)$ be the trivial representation. If $\mathbb{1}$ is chosen as the base point, then by [7, Theorem1.1] there is a natural isomorphism $\pi_1(\mathrm{Hom}(\mathbb{Z}^n, G)) \cong (\pi_1(G))^n$. Here we show that the methods applied in [7] can be used to compute $\pi_1(\mathrm{Hom}(\pi, G))$ for any choice of base point if we further require that $G \in \mathcal{P}$ and that π is a finitely generated abelian group. Write π in the form $\pi = \mathbb{Z}^n \oplus A$, with A a finite abelian group. Then the space of homomorphisms $\mathrm{Hom}(\pi, G)$ can naturally be identified as a subspace of the product $\mathrm{Hom}(\mathbb{Z}^n, G) \times \mathrm{Hom}(A, G)$. Given $f \in \mathrm{Hom}(A, T)$ let

$$\mathbb{1}_f := \mathbb{1} \times f \in \mathrm{Hom}(\pi, G) \subset \mathrm{Hom}(\mathbb{Z}^n, G) \times \mathrm{Hom}(A, G).$$

Every path–connected component in $\mathrm{Hom}(\pi, G)$ contains some $\mathbb{1}_f$ and thus it suffices to consider the elements of the form $\mathbb{1}_f$ as base points in $\mathrm{Hom}(\pi, G)$. With this in mind we have the following.

Theorem 1.2. *Let $\pi = \mathbb{Z}^n \oplus A$, with A a finite abelian group and let $G \in \mathcal{P}$. Suppose $f \in \mathrm{Hom}(A, T)$ and take $\mathbb{1}_f$ as the base point of $\mathrm{Hom}(\pi, G)$. Then there is a natural isomorphism $\pi_1(\mathrm{Hom}(\pi, G)) \cong (\pi_1(G_f))^n$ where $G_f = Z_G(f)$ is the subgroup of elements in G commuting with $f(x)$ for all $x \in A$.*

In Section 6 we study the equivariant K-theory of the spaces of homomorphisms $\mathrm{Hom}(\pi, G)$ with respect to the conjugation action by G. When π is a finite group, then $\mathrm{Hom}(\pi, G)$ is the disjoint union of homogeneous spaces of the form G/H where H is a maximal rank subgroup. Using this it is easy to see that $K^*_G(\mathrm{Hom}(\pi, G))$ is a free module over the representation ring of rank $|\mathrm{Hom}(\pi, T)|$. This result can be generalized for finitely generated abelian groups of rank 1 in the following way.

Theorem 1.3. *Suppose that $G \in \mathcal{P}$ is simply connected and of rank r. Let $\pi = \mathbb{Z} \oplus A$ where A is a finite abelian group. Then $K^*_G(\mathrm{Hom}(\pi, G))$ is a free $R(G)$-module of rank $2^r \cdot |\mathrm{Hom}(A, T)|$.*

It turns out that $K^*_G(\mathrm{Hom}(\pi,G))$ is not always free as a module over $R(G)$. In fact, as was pointed out in [3], the $R(SU(2))$-module $K^*_{SU(2)}(\mathrm{Hom}(\mathbb{Z}^2, SU(2)))$ is not free. However, $K^*_{SU(2)}(\mathrm{Hom}(\mathbb{Z}^2, SU(2))) \otimes \mathbb{Q}$ turns out to be free as a module over $R(SU(2)) \otimes \mathbb{Q}$. The next theorem shows that a similar result holds for all the spaces of homomorphisms that we consider here.

Theorem 1.4. *Suppose that $G \in \mathcal{P}$ is of rank r and that π is a finitely generated abelian group written in the form $\pi = \mathbb{Z}^n \oplus A$, where A is a finite abelian group. Then $K^*_G(\mathrm{Hom}(\pi, G)) \otimes \mathbb{Q}$ is a free module over $R(G) \otimes \mathbb{Q}$ of rank $2^{nr} \cdot |\mathrm{Hom}(A, T)|$.*

The layout of this article is as follows. In Section 2 some general properties of the spaces of homomorphisms $\mathrm{Hom}(\pi, G)$ are determined. In Section 3 we study the cohomology groups with rational coefficients of these spaces. In Section 4 Theorem 1.1 is proved and some explicit examples are computed. In Section 5 the fundamental group of the spaces $\mathrm{Hom}(\pi, G)$ are computed for any choice of base point. Finally, in Section 6 we study the problem of computing $K_G^*(\mathrm{Hom}(\pi, G))$, where G acts by conjugation on $\mathrm{Hom}(\pi, G)$.

Both authors would like to thank the Centro di Ricerca Matematica Ennio De Giorgi at the Scuola Normale Superiore in Pisa for inviting them to participate in the program on Configuration Spaces: Geometry, Combinatorics and Topology during the spring of 2010.

2 Preliminaries on spaces of commuting elements

Let π be a finitely generated discrete group and G a Lie group. Consider the set of homomorphisms from π to G, $\mathrm{Hom}(\pi, G)$. This set can be given a topology as a subspace of a finite product of copies of G in the following way. Fix a set of generators $e_1, \ldots, e_n$ of π and let F_n be the free group on n-letters. By mapping the generators of F_n onto the different e_i's we obtain a surjective homomorphism $F_n \to \pi$. This surjection induces an inclusion of sets $\mathrm{Hom}(\pi, G) \hookrightarrow \mathrm{Hom}(F_n, G) \cong G^n$. This way $\mathrm{Hom}(\pi, G)$ can be given the subspace topology. It is easy to see that this topology is independent of the generators chosen for π. In case π happens to be abelian, then any map $F_n \to \pi$ factors through $F_n \to \mathbb{Z}^n \to \pi$ yielding an inclusion of spaces $\mathrm{Hom}(\pi, G) \hookrightarrow \mathrm{Hom}(\mathbb{Z}^n, G) \hookrightarrow G^n$. Thus the space of homomorphisms $\mathrm{Hom}(\pi, G)$ can be seen as a subspace of the space of commuting n-tuples in G, $\mathrm{Hom}(\mathbb{Z}^n, G)$.

In this note we collect some facts about these spaces of homomorphisms in the particular case that π is a finitely generated abelian group and G belongs to a suitable family of Lie groups. We are mainly interested in the following family of Lie groups.

Definition 2.1. Let $\mathcal{P}$ denote the collection of all compact Lie groups arising as finite cartesian products of the groups $SU(r)$, $U(q)$ and $Sp(k)$.

Whenever G belongs to the family $\mathcal{P}$ the space of homomorphisms $\mathrm{Hom}(\pi, G)$ satisfies the following crucial condition as we prove below in Proposition 2.3.

Definition 2.2. Let X be a G-space. The action of G on X is said to have connected maximal rank isotropy subgroups if for every $x \in X$, the isotropy group G_x is a connected subgroup of maximal rank; that is, for every $x \in X$ we can find a maximal torus T_x in G such that $T_x \subset G_x$.

Proposition 2.3. *Suppose that π is a finitely generated abelian group and $G \in \mathcal{P}$. Then the conjugation action of G on* $\mathrm{Hom}(\pi, G)$ *has connected maximal rank isotropy subgroups.*

Proof. Choose generators $e_1, \ldots, e_n$ of π. As pointed out above we can use these generators to obtain an inclusion of G-spaces $\mathrm{Hom}(\pi, G) \hookrightarrow \mathrm{Hom}(\mathbb{Z}^n, G)$. Given this inclusion it suffices to show that the conjugation action of G on $\mathrm{Hom}(\mathbb{Z}^n, G)$ has connected maximal rank isotropy groups. In [3, Example 2.4] it was proven that the action of G on $\mathrm{Hom}(\mathbb{Z}^n, G)$ has connected maximal rank isotropy subgroups if and only if $\mathrm{Hom}(\mathbb{Z}^{n+1}, G)$ is path–connected. The proposition follows by noting that $\mathrm{Hom}(\mathbb{Z}^k, G)$ is path–connected for all $k \geq 0$ whenever $G \in \mathcal{P}$. □

Suppose that a compact Lie group G acts on a space X with connected maximal rank isotropy subgroups. Choose a maximal torus T in G and let W be the Weyl group. By passing to the level of T-fixed points, the action of G on X induces an action of the Weyl group W on X^T. Many properties of the action of G on X are determined by the action of W on X^T as explained in [8] and in some situations the former is completely determined by the latter up to isomorphism. For example, we can use this approach to produce G-CW complex structures on the spaces of homomorphisms as is proved next.

Corollary 2.4. *Suppose that π is a finitely generated abelian group and $G \in \mathcal{P}$. Then* $\mathrm{Hom}(\pi, G)$ *with the conjugation action has the structure of a G-CW complex.*

Proof. Since π is a finitely generated abelian group it can be written in the form $\pi = \mathbb{Z}^n \oplus A$, where A is a finite abelian group. Let $X := \mathrm{Hom}(\pi, G)$ with the conjugation action of G. Note that $X^T = \mathrm{Hom}(\pi, G)^T = T^n \times \mathrm{Hom}(A, T)$. Since $\mathrm{Hom}(A, T)$ is a discrete set, it follows that X^T has the structure of a smooth manifold on which W acts smoothly. In particular, by [9, Theorem 1] it follows that X^T has the structure of a W-CW complex. Since the conjugation action of G on X has connected maximal rank isotropy subgroups then by [3, Theorem 2.2] it follows that this W-CW complex structure on X^T induces a G-CW complex on X. □

This approach can also be used to determine explicitly the structure of these spaces of homomorphisms whenever π is a finite abelian group.

Proposition 2.5. *Suppose that π is a finite abelian group and $G \in \mathcal{P}$. Then there is a G-equivariant homeomorphism*

$$\Phi : \mathrm{Hom}(\pi, G) \to \coprod_{[f] \in \mathrm{Hom}(\pi, T)/W} G/G_f.$$

Here $[f]$ runs through a system of representatives of the W-orbits in $\mathrm{Hom}(\pi, T)$ *and each G_f is a maximal rank subgroup with* $W(G_f) = W_f$.

Proof. Consider the G-space $X := \mathrm{Hom}(\pi, G)$. Note that $X^T = \mathrm{Hom}(\pi, T)$ is a discrete set endowed with an action of W. By decomposing X^T into the different W-orbits we obtain a W-equivariant homeomorphism

$$X^T \cong \bigsqcup_{[f]\in \mathrm{Hom}(\pi,T)/W} W/W_f.$$

Here $[f]$ runs through a set of representatives for the action of W on $\mathrm{Hom}(\pi, T)$. For each $f \in \mathrm{Hom}(\pi, T)$ let G_f denote the subgroup of elements in G commuting with $f(x)$ for all $x \in \pi$. This group is a maximal rank subgroup in G as $T \subset G_f$. Moreover, by [8, Theorem 1.1] it follows that $W(G_f) = W_f$. Also note that if we let G act on the left on the homogeneous space G/G_f then $(G/G_f)^T = W/W_f$. Let

$$Y = \bigsqcup_{[f]\in \mathrm{Hom}(\pi,T)/W} G/G_f.$$

The left action of G on Y has maximal rank isotropy and there is a W-equivariant homeomorphism $\phi : X^T \to Y^T$. By [8, Theorem 2.1] there is a unique G-equivariant extension $\Phi : X \to Y$ of ϕ and this map is in fact a homeomorphism. □

3 Rational cohomology and path–connected components

In this section we explore the set of path connected components and the rational cohomology groups of the spaces of homomorphisms $\mathrm{Hom}(\pi, G)$.

Suppose that G is a compact connected Lie group and let T be a maximal torus in G. Assume that G acts on a space X of the homotopy type of a G-CW complex with maximal rank isotropy subgroups. Consider the continuous map

$$\begin{aligned} \phi : G \times X^T &\to X \\ (g, x) &\mapsto gx. \end{aligned}$$

Since G acts on X with maximal rank isotropy subgroups for every $x \in X$ we can find a maximal torus T_x in G such that $T_x \subset G_x$. As every pair of maximal tori in G are conjugate it follows that for every $x \in X$ we can find some $g \in G$ such that $gx \in X^T$. This shows that ϕ is a surjective map. The normalizer of T in G, $N_G(T)$ acts on the right on $G \times X^T$ by

$(g, x) \cdot n = (gn, n^{-1}x)$ and the map ϕ is invariant under this action. Thus ϕ descends to a surjective map

$$\begin{aligned} \varphi : G \times_{N_G(T)} X^T = G/T \times_W X^T &\to X \\ [g, x] &\mapsto gx \end{aligned}$$

The map φ is not injective in general. Indeed, as was proven in [4], given $x \in X$ there is a homeomorphism $\varphi^{-1}(x) \cong G_x^0/N_{G_x^0}(T)$, where G_x^0 denotes the path–connected component of G_x containing the identity element. Let $\mathbb{F}$ be a field with characteristic relatively prime to $|W|$. Then as observed in [4] the space $G_x^0/N_{G_x^0}(T)$ has $\mathbb{F}$–acyclic cohomology. The Vietoris-Begle theorem shows that φ induces an isomorphism in cohomology with $\mathbb{F}$-coefficients. As a consequence we obtain the following proposition (first proved in [4]).

Proposition 3.1. *Suppose that G is a compact connected Lie group acting on a spaces X with maximal rank isotropy subgroups. If $\mathbb{F}$ is a field with characteristic relatively prime to $|W|$ then*

$$H^*(X; \mathbb{F}) \cong H^*(G/T \times_W X^T; \mathbb{F}) \cong H^*(G/T \times X^T; \mathbb{F})^W.$$

Remark 3.2. Suppose that G acts on X with *connected* maximal rank isotropy groups. As pointed out above the map φ is not injective in general since $\varphi^{-1}(x) \cong G_x^0/N_{G_x^0}(T)$ for $x \in X$. Under the given hypothesis we have $G_x^0 = G_x$. By [8, Theorem 1.1] the assignment $(H) \mapsto (WH)$ defines a one to one correspondence between the set of conjugacy classes of isotropy subgroups of the action of G on X and the set of conjugacy classes of isotropy subgroups of the action of W on X^T. Thus the different isotropy subgroups of the action of W on X^T determine how far the map φ is from being injective. In particular, if W acts freely on X^T then φ is a continuous bijection and thus a homeomorphism if for example X^T is compact.

Suppose now that $G \in \mathcal{P}$ and let π be a finitely generated abelian group. By Proposition 2.3 the conjugation action of G on $\mathrm{Hom}(\pi, G)$ has connected maximal rank isotropy subgroups. In this case $\mathrm{Hom}(\pi, G)^T = \mathrm{Hom}(\pi, T)$. As a consequence of the previous result the following is obtained.

Corollary 3.3. *Suppose that $G \in \mathcal{P}$ and let π be a finitely generated abelian group. Then there is an isomorphism $H^*(\mathrm{Hom}(\pi, G); \mathbb{Q}) \cong H^*(G/T \times \mathrm{Hom}(\pi, T); \mathbb{Q})^W$.*

As an application of Corollary 3.3 the following can be derived.

Corollary 3.4. *Suppose that $G \in \mathcal{P}$ and let π be a finitely generated abelian group written in the form $\pi = \mathbb{Z}^n \oplus A$. Then the number of path–connected components in* $\mathrm{Hom}(\pi, G)$ *equals the number of different orbits of the action of W on* $\mathrm{Hom}(A, T)$

4 Stable splittings

In this section we show that the fat wedge filtration on a finite product of copies of G induces a natural filtration on the spaces of homomorphisms $\mathrm{Hom}(\pi, G)$. It turns out that this filtration splits stably after one suspension whenever π is a finitely generated abelian group and $G \in \mathcal{P}$.

Suppose that π is a finitely generated abelian group. Using the fundamental theorem of finitely generated abelian groups π can be written in the form

$$\pi = \mathbb{Z}/(q_1) \oplus \cdots \oplus \mathbb{Z}/(q_n),$$

where $n \geq 0$ and $q_1, \ldots, q_n$ are integers. Here we allow some of the q_i's to be 0 and in that case $\mathbb{Z}/(0) = \mathbb{Z}$. This way we can see $\mathrm{Hom}(\pi, G)$ as the subspace of G^n consisting of those commuting n-tuples $(x_1, \ldots, x_n)$ such that $x_i^{q_i} = 1_G$ for all $1 \leq i \leq n$. The fat wedge filtration on G^n induces a natural filtration on the space of homomorphisms $\mathrm{Hom}(\pi, G)$. To be more precise, for each $1 \leq j \leq n$ let

$$\begin{aligned} S_j(\pi,G) =& \{(x_1, \ldots, x_n) \in \mathrm{Hom}(\pi, G) \subset G^n \mid x_i = 1_G \\ & \text{for at least } j \text{ of the } x_i\text{'s}\}. \end{aligned}$$

This way we obtain a filtration of $\mathrm{Hom}(\pi, G)$

$$\begin{aligned} \{(1_G, ..., 1_G)\} &= S_n(\pi, G) \subset S_{n-1}(\pi, G) \subset \cdots \subset S_0(\pi, G) \\ &= \mathrm{Hom}(\pi, G). \end{aligned} \tag{4.1}$$

Note that each $S_j(\pi, G)$ is invariant under the conjugation action of G. In particular each $S_j(\pi, G)$ can be seen as a G-space that has connected maximal rank isotropy subgroups. On the level of the T-fixed points the filtration (4.1) induces a filtration of $\mathrm{Hom}(\pi, G)^T$

$$\begin{aligned} \{(1_G, ..., 1_G)\} &= S_n(\pi, G)^T \subset S_{n-1}(\pi, G)^T \subset \cdots \subset S_0(\pi, G)^T \\ &= \mathrm{Hom}(\pi, G)^T. \end{aligned} \tag{4.2}$$

For each $1 \leq i \leq n$ consider $\mathrm{Hom}(\mathbb{Z}/q_i, T) = \{t \in T \mid t^{q_i} = 1\}$. Note that each $\mathrm{Hom}(\mathbb{Z}/q_i, T)$ is a space endowed with the action of W. Whenever $q_i = 0$ we have $\mathrm{Hom}(\mathbb{Z}/q_i, T) = T$ and if $q_i \neq 0$ then $\mathrm{Hom}(\mathbb{Z}/q_i, T)$ is a discrete set. Since T is abelian it follows that

$$\mathrm{Hom}(\pi, G)^T = \mathrm{Hom}(\pi, T) = \mathrm{Hom}(\mathbb{Z}/q_1, T) \times \cdots \times \mathrm{Hom}(\mathbb{Z}/q_n, T).$$

Moreover, the filtration (4.2) is precisely the fat wedge filtration of $\mathrm{Hom}(\pi, G)^T$ where we identify $\mathrm{Hom}(\pi, G)^T$ with the above product. It is well known that the fat wedge filtration on a product of spaces splits stably after one suspension. More precisely, for each $0 \leq j \leq n-1$ we can find a continuous map

$$r_j : \Sigma S_j(\pi, G)^T \to \Sigma S_{j+1}(\pi, G)^T$$

in such a way that there is a homotopy h_j between $r_j \circ \Sigma(i_j)$ and $1_{\Sigma(S_{j+1}(\pi,G)^T)}$. Here

$$i_j : S_{j+1}(\pi, G)^T \to S_j(\pi, G)^T$$

denotes the inclusion map. Moreover, both the map r_j and the homotopy h_j can be arranged in such a way that they are W-equivariant. The W-action that we have in sight is the diagonal action of W on the product $\mathrm{Hom}(\mathbb{Z}/q_1, T) \times \cdots \times \mathrm{Hom}(\mathbb{Z}/q_n, T)$. Consider the action of G on $\Sigma \mathrm{Hom}(\pi, G)$ with G acting trivially on the suspension component. This action has connected maximal rank isotropy subgroups and $(\Sigma \mathrm{Hom}(\pi, G))^T = \Sigma \mathrm{Hom}(\pi, T)$. By [8, Theorem 2.1] we can find a unique G-equivariant extension

$$R_j : \Sigma S_j(\pi, G) \to \Sigma S_{j+1}(\pi, G)$$

of r_j and a unique G-equivariant homotopy H_j between $R_j \circ \Sigma(I_j)$ and $1_{\Sigma(S_{j+1}(\pi,G))}$ extending h_j. Here $I_j : S_{j+1}(\pi, G) \to S_j(\pi, G)$ as before denotes the inclusion map.

Let $J_{n,r}$ denote the set of all sequences of the form $\mathfrak{m} := \{1 \leq m_1 < \cdots < m_r \leq n\}$. Note that $J_{n,r}$ contains precisely $\binom{n}{r}$ elements. Given such a sequence $\mathfrak{m}$, there is an associated abelian group $P_{\mathfrak{m}}(\pi) := \mathbb{Z}/(q_{m_1}) \oplus \cdots \oplus \mathbb{Z}/(q_{m_r})$ obtained as a quotient of π and also a G-equivariant projection map

$$\begin{aligned} P_{\mathfrak{m}} : \mathrm{Hom}(\pi, G) &\to \mathrm{Hom}(P_{\mathfrak{m}}(\pi), G) \\ (x_1, \ldots, x_n) &\mapsto (x_{m_1}, \ldots, x_{m_r}). \end{aligned}$$

The above can be used to prove the following theorem.

Theorem 4.1. *Suppose that $G \in \mathcal{P}$ and that π is a finitely generated abelian group. Then there is a G-equivariant homotopy equivalence*

$$\Theta : \Sigma \mathrm{Hom}(\pi, G) \to \bigvee_{1 \leq r \leq n} \Sigma \left(\bigvee_{\mathfrak{m} \in J_{n,r}} \mathrm{Hom}(P_{\mathfrak{m}}(\pi), G)/S_1(P_{\mathfrak{m}}(\pi), G) \right).$$

Proof. Note that each $S_j(\pi, G)^T$ has the homotopy type of a W-CW complex and this implies that each $S_j(\pi, G)$ has the homotopy type of a G-CW complex by [3, Theorem 2.2]. The different maps R_j and the homotopies H_j induce a G-equivariant homotopy equivalence

$$\begin{aligned}\Sigma\,\mathrm{Hom}(\pi, G) &\simeq \bigvee_{0\leq r\leq n-1} \Sigma S_r(\pi, G)/S_{r+1}(\pi, G)\\ &= \bigvee_{1\leq r\leq n} \Sigma S_{n-r}(\pi, G)/S_{n-r+1}(\pi, G).\end{aligned}$$

To finish the theorem we will show that for each $1 \leq r \leq n$ there is a G-equivariant homotopy equivalence

$$S_{n-r}(\pi, G)/S_{n-r+1}(\pi, G) \simeq \bigvee_{\mathfrak{m}\in J_{n,r}} \mathrm{Hom}(P_{\mathfrak{m}}(\pi), G)/S_1(P_{\mathfrak{m}}(\pi), G).$$

To see this note that the different projection maps $\{P_{\mathfrak{m}}\}_{\mathfrak{m}\in J_{n,r}}$ can be assembled to obtain a G-map

$$\begin{aligned}\eta : \mathrm{Hom}(\pi, G) &\to \prod_{\mathfrak{m}\in J_{n,r}} \mathrm{Hom}(P_{\mathfrak{m}}(\pi), G)/S_1(P_{\mathfrak{m}}(\pi), G)\\ (x_1, ..., x_n) &\mapsto \{\overline{P}_{\mathfrak{m}}(x_1, ..., x_n)\}_{\mathfrak{m}\in J_{n,r}}.\end{aligned}$$

The map η sends $S_{n-r}(\pi,G)$ onto $\bigvee_{\mathfrak{m}\in J_{n,r}} \mathrm{Hom}(P_{\mathfrak{m}}(\pi),G)/S_1(P_{\mathfrak{m}}(\pi),G)$ and $S_{n-r+1}(\pi, G)$ is mapped onto the base point. It is easy to see that η induces a G-equivariant homeomorphism

$$S_{n-r}(\pi, G)/S_{n-r+1}(\pi, G) \cong \bigvee_{\mathfrak{m}\in J_{n,r}} \mathrm{Hom}(P_{\mathfrak{m}}(\pi), G)/S_1(P_{\mathfrak{m}}(\pi), G)$$

and the theorem follows. □

Remark 4.2. A case of particular importance in the previous theorem is $\pi = \mathbb{Z}^n$. In this case $\mathrm{Hom}(\mathbb{Z}^n, G)$ is precisely the space of commuting ordered n-tuples in G. The previous theorem provides a simple proof for the stable equivalence provided in [1] for the spaces $\mathrm{Hom}(\mathbb{Z}^n, G)$ whenever $G \in \mathcal{P}$.

Example 4.3. Suppose that $\pi = \mathbb{Z}^n$. Let $1 \leq r \leq n$. For any $\mathfrak{m} \in J_{n,r}$ we have

$$\mathrm{Hom}(P_{\mathfrak{m}}(\pi), G)/S_1(P_{\mathfrak{m}}(\pi), G) \cong \mathrm{Hom}(\mathbb{Z}^r, G)/S_r^1(G),$$

where $S_r^1(G) \subset \mathrm{Hom}(\mathbb{Z}^n, G)$ is the subspace of those commuting n-tuples $(x_1, \ldots, x_n)$ with at least one of the x_i equal to 1_G. These stable factors were identified independently in [2, 5] and [6] in the particular case where $G = SU(2)$. Let $n\lambda_2$ denote the the Whitney sum of n-copies of the canonical vector bundle over $\mathbb{R}P^2$ and let s_n denote its zero section. Then

$$\mathrm{Hom}(\mathbb{Z}^n, SU(2))/S_n^1(SU(2)) \cong \begin{cases} \mathbb{S}^3 & \text{if } n = 1, \\ (\mathbb{R}P^2)^{n\lambda_2}/s_n(\mathbb{R}P^2) & \text{if } n \geq 2. \end{cases}$$

Example 4.4. Suppose now that $\pi = \mathbb{Z}/(q_1) \oplus \cdots \oplus \mathbb{Z}/(q_n)$ is any finitely generated abelian group and $G = SU(2)$. Let T be the maximal torus consisting of 2×2 diagonal matrices with entries in $\mathbb{S}^1$ and determinant 1. In this case $W = \mathbb{Z}/2$ acts by permuting the diagonal entries of elements in T. Next we determine the stable factors of the form

$$\mathrm{Hom}(P_{\mathfrak{m}}(\pi), SU(2))/S_1(P_{\mathfrak{m}}(\pi), SU(2)),$$

where $\mathfrak{m} = \{1 \leq m_1 < \cdots < m_r \leq n\}$ is fixed. We consider the following cases.

- Suppose $P_{\mathfrak{m}}(\pi)$ is a finite group so that $q_{m_i} \neq 0$ for all $1 \leq i \leq r$. Assume further that at least one of the q_{m_i}'s is odd. By Proposition 2.5 there is a homeomorphism

$$\mathrm{Hom}(P_{\mathfrak{m}}(\pi), SU(2)) \cong \bigsqcup_{[f]\in \mathrm{Hom}(P_{\mathfrak{m}}(\pi),T)/W} G/G_f.$$

Here $[f]$ runs through all the W-orbits in $\mathrm{Hom}(P_{\mathfrak{m}}(\pi), T)$. In this case

$$G/G_f = G/T \cong \mathbb{S}^2$$

for all orbits corresponding to elements f for which W_f is trivial. On the other hand, when f is fixed by W the corresponding orbit is $G/G_f = G/G = *$. Since we are assuming that one of the q_{m_i}'s is odd, then every $f \in \mathrm{Hom}(P_{\mathfrak{m}}(\pi), T)$ corresponding to r-tuples $(x_{m_1}, \ldots, x_{m_r})$ in $\mathrm{Hom}(P_{\mathfrak{m}}(\pi), SU(2))$ with $x_{m_i} \neq 1_G$ for all i satisfies $W_f = 1$. This shows that

$$\mathrm{Hom}(P_{\mathfrak{m}}(\pi), SU(2))/S_1(P_{\mathfrak{m}}(\pi), SU(2)) \cong \left(\bigsqcup_{A(\mathfrak{m},\pi)} \mathbb{S}^2\right)_+.$$

Here $A(\mathfrak{m}, \pi)$ is the number of W-orbits in $\mathrm{Hom}(P_{\mathfrak{m}}(\pi), T)$ corresponding to r-tuples that don't contain the element 1. This number is precisely

$$A(\mathfrak{m}, \pi) = \frac{1}{2}(q_{m_1} - 1)\cdots(q_{m_r} - 1).$$

• Suppose now that $q_{m_i} \neq 0$ is even for all $1 \leq i \leq r$. In this case we have two possibilities for the W-orbits in $\mathrm{Hom}(P_{\mathfrak{m}}(\pi), T)$. If $[f]$ represents the orbit $[(-1, \ldots, -1)]$ then $W_f = W$ and the corresponding orbit is $G/G_f = *$. For all other orbits $[f] \in \mathrm{Hom}(P_{\mathfrak{m}}(\pi), T)/W$ corresponding to r-tuples $(x_{m_1}, \ldots, x_{m_r})$ in $\mathrm{Hom}(P_{\mathfrak{m}}(\pi), SU(2))$ with $x_{m_i} \neq 1_G$ for all i we have $W_f = 1$ and as before $G/G_f \cong \mathbb{S}^2$. This shows that

$$\mathrm{Hom}(P_{\mathfrak{m}}(\pi), SU(2))/S_1(P_{\mathfrak{m}}(\pi), SU(2)) \cong \left(\bigsqcup_{A(\mathfrak{m},\pi)} \mathbb{S}^2 \right) \sqcup \mathbb{S}^0,$$

where now $A(\mathfrak{m}, \pi)$ is the number of W-orbits in $\mathrm{Hom}(P_{\mathfrak{m}}(\pi), T)$ corresponding to r-tuples in $\mathrm{Hom}(P_{\mathfrak{m}}(\pi), T)$ that don't contain the element 1 and that are different from $(-1, \ldots, -1)$. This number is precisely

$$A(\mathfrak{m}, \pi) = \frac{1}{2} \left((q_{m_1} - 1) \cdots (q_{m_r} - 1) - 1 \right).$$

• We now consider the case where $q_{m_i} = 0$ for some $1 \leq i \leq r$. If $q_{m_i} = 0$ for all $1 \leq i \leq r$ then

$$\begin{aligned} &\mathrm{Hom}(P_{\mathfrak{m}}(\pi), SU(2))/S_1(P_{\mathfrak{m}}(\pi), SU(2)) \\ &= \mathrm{Hom}(\mathbb{Z}^r, SU(2))/S_r^1(SU(2)) \end{aligned}$$

and these stable factors are as in Example 4.3. Suppose then that $q_{m_i} \neq 0$ for some i. For simplicity and without loss of generality we may assume that

$$P_{\mathfrak{m}}(\pi) = \mathbb{Z}^k \oplus \mathbb{Z}/(q_{m_{k+1}}) \oplus \cdots \oplus \mathbb{Z}/(q_{m_r})$$

for some $1 \leq k < r$ and $q_{m_i} \neq 0$ for $k+1 \leq i \leq r$. Since the inclusion map $S_1(P_{\mathfrak{m}}(\pi), SU(2)) \hookrightarrow \mathrm{Hom}(P_{\mathfrak{m}}(\pi), G)$ is a cofibration, we have

$$\begin{aligned} &\mathrm{Hom}(P_{\mathfrak{m}}(\pi), SU(2))/S_1(P_{\mathfrak{m}}(\pi), SU(2)) \\ &\cong (\mathrm{Hom}(P_{\mathfrak{m}}(\pi), SU(2)) \setminus S_1(P_{\mathfrak{m}}(\pi), SU(2)))^+. \end{aligned}$$

Here if X is a locally compact space then X^+ denotes its one point compactification. Consider the map

$$\begin{aligned} \varphi_{\mathfrak{m}} : G/T \times_W \mathrm{Hom}(P_{\mathfrak{m}}(\pi), T) &\to \mathrm{Hom}(P_{\mathfrak{m}}(\pi), G) \\ (g, (t_{m_1}, \ldots, t_{m_r})) &\mapsto (g t_{m_1} g^{-1}, \ldots, g t_{m_r} g^{-1}). \end{aligned}$$

This map is surjective as the action of G on $\mathrm{Hom}(P_{\mathfrak{m}}(\pi), G)$ has connected maximal rank isotropy. Moreover

$$\varphi_{\mathfrak{m}}(g, (t_{m_1}, \dots, t_{m_r})) \in S_1(P_{\mathfrak{m}}(\pi), SU(2))$$

if and only if $t_{m_i} = 1$ for some $1 \leq i \leq r$. Let $Q(P_{\mathfrak{m}}(\pi), T)$ denote the subset of

$$\mathrm{Hom}(\mathbb{Z}/(q_{m_{k+1}}) \oplus \cdots \oplus \mathbb{Z}/(q_{m_r}), T)$$

consisting of those $(r-k)$-tuples $(t_{m_{k+1}}, \dots t_{m_r})$ in T such that $t_{m_i} \neq 1$ for $k+1 \leq i \leq r$. Using the Cayley map as in [2, Section 7] we can find a W-equivariant homeomorphism

$$T \setminus \{1\} \cong \mathfrak{t}.$$

Here $\mathfrak{t}$ denotes the Lie algebra of T. Using this identification and the restriction of the map $\varphi_{\mathfrak{m}}$, we obtain a surjective map

$$\psi_{\mathfrak{m}}: G/T \times_W (\mathfrak{t}^k \times Q(P_{\mathfrak{m}}(\pi), T)) \to \mathrm{Hom}(P_{\mathfrak{m}}(\pi), G) \setminus S_1(P_{\mathfrak{m}}(\pi), SU(2))$$

Moreover, $\psi_{\mathfrak{m}}$ is injective except where the action of W on $\mathfrak{t}^k \times Q(P_{\mathfrak{m}}(\pi), T)$ is not free. We need to consider two cases.

- Suppose first that that q_{m_i} is odd for some $k+1 \leq i \leq r$. In that case W acts freely on $\mathfrak{t}^k \times Q(P_{\mathfrak{m}}(\pi), T)$ and we have a W-equivariant homeomorphism

  $$\mathfrak{t}^k \times Q(P_{\mathfrak{m}}(\pi), T) \cong \bigsqcup_{A(\mathfrak{m},\pi)} \mathfrak{t}^k \times W.$$

  Here $A(\mathfrak{m}, \pi)$ is the number of W-orbits in $Q(P_{\mathfrak{m}}(\pi), T)$. This number is precisely

  $$A(\mathfrak{m}, \pi) = \frac{1}{2}(q_{m_{k+1}} - 1) \cdots (q_{m_r} - 1).$$

  Therefore

  $$\begin{aligned}(G/T \times_W (\mathfrak{t}^k \times Q(P_{\mathfrak{m}}(\pi), T)))^+ &\cong \left(\bigsqcup_{A(\mathfrak{m},\pi)} G/T \times_W (\mathfrak{t}^k \times W) \right)^+ \\ &\cong \bigvee_{A(\mathfrak{m},\pi)} (G/T \times \mathfrak{t}^k)^+.\end{aligned}$$

  Note that $G/T = \mathbb{S}^2$ and thus

  $$(G/T \times \mathfrak{t}^k)^+ \cong \Sigma^k(\mathbb{S}^2_+) \cong \mathbb{S}^{k+2} \vee \mathbb{S}^k.$$

In this case the map $\psi_{\mathfrak{m}}$ is a homeomorphism as the action of W on $\mathfrak{t}^k \times Q(P_{\mathfrak{m}}(\pi), T)$ is free. This shows that if q_{m_i} is odd for some $k+1 \leq i \leq r$ then

$$\begin{aligned}&\mathrm{Hom}(P_{\mathfrak{m}}(\pi), SU(2))/S_1(P_{\mathfrak{m}}(\pi), SU(2))\\ &\cong \left(\bigvee_{A(\mathfrak{m},\pi)} \mathbb{S}^{k+2}\right) \vee \left(\bigvee_{A(\mathfrak{m},\pi)} \mathbb{S}^{k}\right).\end{aligned}$$

- Suppose now that $P_{\mathfrak{m}}(\pi) = \mathbb{Z}^k \oplus \mathbb{Z}/(q_{m_{k+1}}) \cdots \oplus \mathbb{Z}/(q_{m_r})$ and that q_{m_i} is even for every $k+1 \leq i \leq r$. In this case we have two kinds of elements in $Q(P_{\mathfrak{m}}(\pi), T)$. On the one hand we have the $(r-k)$-tuple $(-1, \ldots, -1)$ on which W acts trivially. For all other elements in $Q(P_{\mathfrak{m}}(\pi), T)$ the action of W is free. This shows that there is a W-equivariant homeomorphism

$$\mathfrak{t}^k \times Q(P_{\mathfrak{m}}(\pi), T) \cong \mathfrak{t}^k \sqcup \left(\bigsqcup_{A(\mathfrak{m},\pi)} \mathfrak{t}^k \times W\right).$$

Here $A(\mathfrak{m}, \pi)$ denotes the number of W-orbits in $Q(P_{\mathfrak{m}}(\pi), T)$ different from the trivial orbit $[(-1, \ldots, -1)]$. This number is precisely

$$A(\mathfrak{m}, \pi) = \frac{1}{2}\left((q_{m_1} - 1) \cdots (q_{m_r} - 1) - 1\right).$$

The map $\psi_{\mathfrak{m}}$ is no longer injective. Note that $G/T \times_W \mathfrak{t}^k$ is the Whitney sum of k-copies of the canonical vector bundle over $\mathbb{R}P^2$ and ψ maps the zero section onto the n-tuple $(-1, \ldots, -1)$. On the other hand, the restriction of $\psi_{\mathfrak{m}}$ onto the factor

$$G/T \times_W \left(\bigsqcup_{A(\mathfrak{m},\pi)} \mathfrak{t}^k \times W\right) \cong \bigsqcup_{A(\mathfrak{m},\pi)} G/T \times \mathfrak{t}^k$$

is injective. This shows that if q_{m_i} is even for every $k+1 \leq i \leq r$ then

$$\begin{aligned}&\mathrm{Hom}(P_{\mathfrak{m}}(\pi), SU(2))/S_1(P_{\mathfrak{m}}(\pi), SU(2))\\ &\cong (\mathbb{R}P^2)^{k\lambda_2}/s_k(\mathbb{R}P^2) \vee \left(\bigvee_{A(\mathfrak{m},\pi)} \mathbb{S}^{k+2}\right) \vee \left(\bigvee_{A(\mathfrak{m},\pi)} \mathbb{S}^{k}\right).\end{aligned}$$

We can use the above for example to establish the stable homotopy type of the space of homomorphisms $\mathrm{Hom}(\mathbb{Z}^2 \oplus \mathbb{Z}/(2) \oplus \mathbb{Z}/(3), SU(2))$. In this case we have that after one suspension $\mathrm{Hom}(\mathbb{Z}^2 \oplus \mathbb{Z}/(2) \oplus \mathbb{Z}/(3), SU(2))$ is homotopy equivalent to

$$\left(\bigvee^{2} (\mathbb{R}P^2)^{2\lambda_2}/s_2(\mathbb{R}P^2)\right) \vee \left(\bigvee^{2} \mathbb{S}^4\right) \vee \left(\bigvee^{8} \mathbb{S}^3\right) \vee \left(\bigvee^{2} \mathbb{S}^2\right) \\ \vee \left(\bigvee^{2} \mathbb{S}^2_+\right) \vee \left(\bigvee^{4} \mathbb{S}^1\right) \vee \mathbb{S}^0.$$

Example 4.5. Suppose that $\pi = \mathbb{Z} \oplus A$, where A is a finite abelian group. Choose $G \in \mathcal{P}$ and assume that A is such that the action of W on $\mathrm{Hom}(A, T) \setminus \{1\}$ is free. Since W fixes the trivial homomorphism $1 \in \mathrm{Hom}(A, T)$, then the decomposition of $\mathrm{Hom}(A, T)$ into W-orbits shows that in particular $|W|$ divides $(|\mathrm{Hom}(A, T)| - 1)$ under this assumption.

We will show that in this case

$$\Sigma \,\mathrm{Hom}(\pi, G) \simeq \Sigma \left(\bigvee_k T\right) \vee \Sigma \left(\bigvee_k G/T \wedge T\right) \vee \Sigma G \vee \left(\bigsqcup_k G/T\right)_+$$

Here $k := (|\mathrm{Hom}(A, T)| - 1)/|W|$ is the number of distinct W-orbits on the set $\mathrm{Hom}(A, T)$ that are different from the one corresponding to the trivial homomorphism.

Indeed, using Theorem 4.1 we obtain a homotopy equivalence

$$\Sigma \,\mathrm{Hom}(\pi, G) \simeq \Sigma \,\mathrm{Hom}(\pi, G)/S_1(\pi, G) \\ \vee \Sigma \,\mathrm{Hom}(\mathbb{Z}, G)/S_1(\mathbb{Z}, G) \vee \Sigma \,\mathrm{Hom}(A, G)/S_1(A, G).$$

Trivially $\mathrm{Hom}(\mathbb{Z}, G)/S_1(\mathbb{Z}, G) = G$. Also, since A is a finite abelian group then by Proposition 2.5 we have

$$\mathrm{Hom}(A, G) \cong \bigsqcup_{[f] \in \mathrm{Hom}(A,T)/W} G/G_f.$$

Here $[f]$ runs through all the W-orbits in the finite set $\mathrm{Hom}(A, T)$ and G_f is a maximal rank subgroup such that $W(G_f) = W_f$. In $\mathrm{Hom}(A, T)$ we have two different kinds of orbits. On the one hand, we have the orbit corresponding to the trivial homomorphism in $\mathrm{Hom}(A, T)$. For this orbit we have $W_f = W$ and $G_f = G$. The assumptions on A imply that for all other orbits in $\mathrm{Hom}(A, T)/W$ we have $W_f = 1$ and thus $G_f = T$. This shows that

$$\mathrm{Hom}(A, G)/S_1(A, G) = \left(\bigsqcup_k G/T\right)_+.$$

We now determine the stable factor $\mathrm{Hom}(\pi, G)/S_1(\pi, G)$. For this consider the map

$$\varphi : G/T \times_W \mathrm{Hom}(\pi, T) \to \mathrm{Hom}(\pi, G).$$

Since the action of G on $\mathrm{Hom}(\pi, G)$ has maximal rank isotropy subgroups φ is surjective. Moreover, the restriction of φ induces a surjective map

$$\varphi_| : G/T \times_W ((T \setminus \{1\}) \times (\mathrm{Hom}(A, T) \setminus \{1\})) \to \mathrm{Hom}(\pi, G) \setminus S_1(\pi, G).$$

Since the action of W on $\mathrm{Hom}(A, T) \setminus \{1\}$ is free we have that this restriction map is a homeomorphism. Also

$$G/T \times_W ((T \setminus \{1\}) \times (\mathrm{Hom}(A, T) \setminus \{1\})) \cong \bigsqcup_k G/T \times (T \setminus \{1\}).$$

This shows that

$$\mathrm{Hom}(\pi, G)/S_1(\pi, G) \cong \bigvee_k (G/T \times (T \setminus \{1\}))^+$$

Note that

$$(G/T \times (T \setminus \{1\}))^+ \cong (G/T \times T)/(G/T \times \{1\}).$$

and it is easy to see that there is a homotopy equivalence

$$\Sigma((G/T \times T)/(G/T \times \{1\})) \simeq \Sigma T \vee \Sigma G/T \wedge T.$$

This shows that

$$\Sigma\, \mathrm{Hom}(\pi, G)/S_1(\pi, G) \cong \Sigma\left(\bigvee_k T\right) \vee \Sigma\left(\bigvee_k G/T \wedge T\right)$$

proving the claim.

5 Fundamental group

In this section we study the fundamental group of the spaces of homomorphisms $\mathrm{Hom}(\pi, G)$ under different choices of base point.

Suppose first that $\pi = \mathbb{Z}^n$ and that G is a compact Lie group. Let $\mathbb{1} \in \mathrm{Hom}(\mathbb{Z}^n, G)$ be the trivial representation. If we give $\mathrm{Hom}(\mathbb{Z}^n, G)$ the base point $\mathbb{1}$ then by [7, Theorem 1.1] there is a natural isomorphism $\pi_1(\mathrm{Hom}(\mathbb{Z}^n, G)) \cong (\pi_1(G))^n$. The tools applied in [7] can be used to generalize this result to the class of spaces of homomorphisms $\mathrm{Hom}(\pi, G)$. Here we need to assume that π is a finitely generated abelian group and that G is a Lie group in the class $\mathcal{P}$. Under these assumptions [7, Theorem 1.1] can be generalized for any choice of base point in $\mathrm{Hom}(\pi, G)$. Write π in the form $\pi = \mathbb{Z}^n \oplus A$, where A is a finite abelian group. Suppose first that $n = 0$ and thus π is a finite group. In this case by Proposition 2.5 there is a homeomorphism

$$\Phi : \mathrm{Hom}(\pi, G) \to \bigsqcup_{[f]\in \mathrm{Hom}(\pi,T)/W} G/G_f,$$

where each G_f is a maximal rank isotropy subgroup with $W(G_f) = W_f$. For each maximal rank subgroup $H \subset G$ we have $\pi_1(G/H) = 1$. It follows that $\pi_1(\mathrm{Hom}(\pi, G)) = 1$ for any choice of base point in this case. This handles the case of finite groups. Suppose then that $n \geq 1$. Let $T \subset G$ be a maximal torus. Note that $\mathrm{Hom}(\pi, G)^T = \mathrm{Hom}(\pi, T)$ and since T is abelian we have $\mathrm{Hom}(\pi, T) = \mathrm{Hom}(\mathbb{Z}^n, T) \times \mathrm{Hom}(A, T) = T^n \times \mathrm{Hom}(A, T)$. Choose $f \in \mathrm{Hom}(A, T)$ and let $\mathbb{1} \in \mathrm{Hom}(\mathbb{Z}^n, T)$ denote the trivial representation. Let

$$\mathbb{1}_f := \mathbb{1} \times f \in \mathrm{Hom}(\mathbb{Z}^n, T) \times \mathrm{Hom}(A, T) = \mathrm{Hom}(\pi, T) \hookrightarrow \mathrm{Hom}(\mathbb{Z}^n, G)$$

and denote by $\mathrm{Hom}(\pi, G)_{\mathbb{1}_f}$ the path-connected component of $\mathrm{Hom}(\pi, G)$ containing $\mathbb{1}_f$. It is easy to see that

$$\mathrm{Hom}(\pi, G) = \bigsqcup_{[f]\in \mathrm{Hom}(A,T)/W} \mathrm{Hom}(\pi, G)_{\mathbb{1}_f}.$$

Since the fundamental group does not depend, up to isomorphism, on the base point chosen on a path–connected space, it suffices to compute $\pi_1(\mathrm{Hom}(\pi, G)_{\mathbb{1}_f})$ for any $f \in \mathrm{Hom}(A, T)$, where $\mathrm{Hom}(\pi, G)_{\mathbb{1}_f}$ is given the base point $\mathbb{1}_f$. Fix $f \in \mathrm{Hom}(A, T)$. Note that $\mathrm{Hom}(\pi, G)_{\mathbb{1}_f}$ is invariant under the conjugation action of G and this action has connected maximal isotropy subgroups. In this case $\mathrm{Hom}(\pi, G)^T_{\mathbb{1}_f} \cong T^n \times W/W_f$. As pointed out in Section 3 we have a surjective map

$$\varphi : G/T \times_W \mathrm{Hom}(\pi, G)^T_{\mathbb{1}_f} \cong G/T \times_{W_f} T^n \to \mathrm{Hom}(\pi, G)_{\mathbb{1}_f}$$

that has connected fibers. As observed before this map is not injective in general; however, there is a large set on which this has this desirable property. Let $\mathcal{F}(\pi, f)$ be the subset of $\mathrm{Hom}(\pi, G)^T_{\mathbb{1}_f}$ on which W acts freely. Then the restriction of φ

$$\varphi_| : G/T \times_W \mathcal{F}(\pi, f) \to \varphi(G/T \times_W \mathcal{F}(\pi, f)) \subset \mathrm{Hom}(\pi, G)_{\mathbb{1}_f}$$

is a homeomorphism onto its image. Assume further that G is not a torus. Then under this additional assumption the complement of $G/T \times_W \mathcal{F}(\pi, f)$ is an analytic subspace of $G/T \times_W (T^n \times W/W_f)$ of co-dimension at least 2. Indeed, if G is not a torus then G/T is a smooth manifold of dimension at least 2 and $\mathrm{Hom}(\pi, G)^T_{\mathbb{1}_f} \setminus \mathcal{F}(\pi, f)$ submanifold of $\mathrm{Hom}(\pi, G)^T_{\mathbb{1}_f}$ of co-dimension at least 1. This proves the claim. Note that when G is a torus then W is trivial, $\mathcal{F}(\pi, f) = \mathrm{Hom}(\pi, G)^T_{\mathbb{1}_f}$ and the map φ is a homeomorphism.

Following [7] we have the following definition.

Definition 5.1. Define $\mathcal{H}^r_f$ to be the image of $G/T \times_W \mathcal{F}(\pi, f)$ under the map φ. We refer to $\mathcal{H}^r_f$ as the regular part of $\mathrm{Hom}(\pi, G)_{\mathbb{1}_f}$. Also define $\mathcal{H}^s_f := \mathrm{Hom}(\pi, G)_{\mathbb{1}_f} \setminus \mathcal{H}^r_f$. We refer to $\mathcal{H}^s_f$ as the singular part of $\mathrm{Hom}(\pi, G)_{\mathbb{1}_f}$.

Note that $\mathrm{Hom}(\pi, G)_{\mathbb{1}_f}$ is a compact real analytic space and since $\mathcal{H}^s_f$ is the image of the compact analytic space $(G/T \times_W T^n \times W/W_f) \setminus (G/T \times_W \mathcal{F}(\pi, f))$ under the analytic map φ, it follows that $\mathcal{H}^s_f$ is a compact analytic subspace of $\mathrm{Hom}(\pi, G)_{\mathbb{1}_f}$. As a consequence of the Whitney stratification theorem it follows that $\mathrm{Hom}(\pi, G)_{\mathbb{1}_f}$ can be given the structure of a simplicial complex in such a way that $\mathcal{H}^s_f$ is a subcomplex. Note that when G is a torus $\mathcal{H}^s_f$ is empty. On the other hand, if G is not a torus then using the fact that the complement of $G/T \times_W \mathcal{F}(\pi, f)$ is an analytic subspace of $G/T \times_W (T^n \times W/W_f)$ of co-dimension at least 2 and an argument similar to the one provided in [7, Lemma 2.4] the following lemma is obtained for any $G \in \mathcal{P}$.

Lemma 5.2. *The space* $\mathrm{Hom}(\pi, G)_{\mathbb{1}_f}$ *is a compact simplicial complex and the singular part* $\mathcal{H}^s_f$ *is a subcomplex. Also,* $\mathcal{H}^s_f$ *is nowhere dense and does not disconnect connected open subsets of* $\mathrm{Hom}(\pi, G)_{\mathbb{1}_f}$.

The previous lemma can be used as a first step for the computation of $\pi_1(\mathrm{Hom}(\pi, G)_{\mathbb{1}_f})$. Indeed, suppose that X is a compact simplicial complex and that $Y \subset X$ is a subcomplex. Assume that $X \setminus Y$ is dense and that Y does not separate any connected open set in X. If $x_0 \in X \setminus Y$ is the base point then by [7, Lemma 2.5] the inclusion map $i : X \setminus Y \to X$ induces a surjective homomorphism $i_* : \pi_1(X \setminus Y, x_0) \to \pi_1(X, x_0)$. This can be applied in our situation. Choose $x_0 \in \mathcal{H}^r_f$ as the base point. Using Lemma 5.2 we obtain that the inclusion $i : \mathcal{H}^r_f \hookrightarrow \mathrm{Hom}(\pi, G)_{\mathbb{1}_f}$ induces a surjective homomorphism $i_* : \pi_1(\mathcal{H}^r_f) \to \pi_1(\mathrm{Hom}(\pi, G)_{\mathbb{1}_f})$. The same argument shows that the inclusion map

$$i : G/T \times_W \mathcal{F}(\pi, f) \hookrightarrow G/T \times_W \mathrm{Hom}(\pi, G)^T_{\mathbb{1}_f}$$

induces a surjective homomorphism

$$i_* : \pi_1(G/T \times_W \mathcal{F}(\pi, f)) \to \pi_1(G/T \times_W \operatorname{Hom}(\pi, G)^T_{\mathbb{1}_f}).$$

Since $\varphi_| : G/T \times_W \mathcal{F}(\pi, f) \to \mathcal{H}^r$ is a homeomorphism and the fundamental group of a connected space does not depend on the choice of base point, up homeomorphism, we obtain the following proposition (compare [7, Corollary 2.6]).

Proposition 5.3. *Suppose that $G \in \mathcal{P}$ and that π is a finitely generated abelian group. Then the map*

$$\varphi : G/T \times_W \operatorname{Hom}(\pi, G)^T_{\mathbb{1}_f} \to \operatorname{Hom}(\pi, G)_{\mathbb{1}_f}$$

is π_1-surjective.

Note that

$$G/T \times_W \operatorname{Hom}(\pi, G)^T_{\mathbb{1}_f} \cong G/T \times_{W_f} T^n.$$

Since W_f acts freely on G/T the projection map p induces a fibration sequence

$$T^n \to G/T \times_{W_f} T^n \xrightarrow{p} (G/T)/W_f \cong G/N_{G_f}(T).$$

The tail of the homotopy long exact sequence associated to this fibration is the exact sequence

$$\pi_1(T^n) \to \pi_1(G/T \times_{W_f} T^n) \to \pi_1(G/N_{G_f}(T)) \to 1. \tag{5.1}$$

Note that $\mathbb{1} \in T^n$ is a fixed point of W_f. Therefore the map

$$\begin{aligned} s : G/N_{G_f}(T) &\to G/T \times_{W_f} T^n \\ [g] &\mapsto [g \times \mathbb{1}] \end{aligned}$$

is a section of p and in particular the sequence (5.1) splits. This proves that $\pi_1(G/T \times_{W_f} T^n)$ is generated by $\pi_1(T^n)$ and $s_*(\pi_1(G/N_{G_f}(T)))$. Next we prove the following lemma.

Lemma 5.4. *If $\alpha : [0,1] \to G/N_{G_f}(T)$ is a loop then $\varphi \circ s \circ \alpha$ is homotopic to the the trivial loop in $\operatorname{Hom}(\pi, G)_{\mathbb{1}_f}$. Therefore $s_*(\pi_1(G/N_{G_f}(T))) \subset \operatorname{Ker}(\varphi_*)$.*

Proof. Let $\alpha : [0,1] \to G/N_{G_f}(T)$ be a loop. Note that $\mathrm{Hom}(\pi, G)_{\mathbb{1}_f}$ can be seen as a subspace of $\mathrm{Hom}(\mathbb{Z}^n, G) \times \mathrm{Hom}(A, G)$. Under this identification $\beta := \varphi \circ s \circ \alpha$ is the loop in $\mathrm{Hom}(\pi, G)_{\mathbb{1}_f}$ given by

$$\begin{aligned}\beta := \varphi \circ s \circ \alpha : [0,1] \to \mathrm{Hom}(\pi, G)_{\mathbb{1}_f} &\subset \mathrm{Hom}(\mathbb{Z}^n, G) \times \mathrm{Hom}(A, G)\\ t &\mapsto (\mathbb{1}, \alpha(t) f \alpha(t)^{-1}).\end{aligned}$$

Let $G_f = Z_G(f)$ be the subspace of elements in G commuting with $f(x)$ for all $x \in A$ and $G \cdot f$ the space of elements in $\mathrm{Hom}(A, G)$ conjugated to f. Then

$$\beta : [0,1] \to \{\mathbb{1}\} \times G \cdot f \subset \mathrm{Hom}(\pi, G)_{\mathbb{1}_f}.$$

There is a homeomorphism $G \cdot f \cong G/G_f$ and G_f is a maximal rank subgroup in G as $T \subset G_f$. In particular the homogeneous space G/G_f is simply connected. The simply connectedness of $G \cdot f$ shows that up to homotopy β is the trivial loop in $\mathrm{Hom}(\pi, G)_{\mathbb{1}_f}$ proving the lemma. □

The previous lemma together with Proposition 5.3 and the fact that $\pi_1(G/T \times_{W_f} T^n)$ is generated by $\pi_1(T^n)$ and $s_*(\pi_1(G/N_{G_f}(T)))$ show that the map

$$\begin{aligned}\sigma_f : T^n &\to \mathrm{Hom}(\pi, G)_f \subset \mathrm{Hom}(\mathbb{Z}^n, G) \times \mathrm{Hom}(A, G)\\ (t_1, \ldots, t_n) &\mapsto (t_1, \ldots, t_n) \times \{f\}\end{aligned}$$

is π_1-surjective. On the other hand, the inclusion $T \subset G_f$ shows that $T^n \subset \mathrm{Hom}(\mathbb{Z}^n, G_f)$ and there is a commutative diagram

$$\begin{array}{ccc} T^n & \xrightarrow{\sigma_f} & \mathrm{Hom}(\pi, G)_{\mathbb{1}_f}. \\ \downarrow & \nearrow_{i_f} & \\ \mathrm{Hom}(\mathbb{Z}^n, G_f) & & \end{array} \tag{5.2}$$

In the previous commutative diagram i_f denotes the map

$$\begin{aligned}i_f : \mathrm{Hom}(\mathbb{Z}^n, G_f) &\to \mathrm{Hom}(\pi, G)_f \subset \mathrm{Hom}(\mathbb{Z}^n, G) \times \mathrm{Hom}(A, G)\\ (x_1, \ldots x_n) &\mapsto (x_1, \ldots x_n) \times \{f\}.\end{aligned}$$

The inclusion map $T \subset G_f$ is π_1-surjective and by [7, Theorem 1.1] the map induced by the inclusion $\pi_1(\mathrm{Hom}(\mathbb{Z}^n, G_f)) \to \pi_1(G_f^n) \cong (\pi_1(G_f))^n$ is an isomorphism. This proves that the map $\pi_1(T^n) \to \pi_1(\mathrm{Hom}(\mathbb{Z}^n, G_f))$ is surjective. Using the commutativity of diagram (5.2) and the fact that σ_f is π_1-surjective we obtain the following corollary.

Corollary 5.5. *Suppose that π is a finitely generated abelian . Then the map*

$$i_f : \mathrm{Hom}(\mathbb{Z}^n, G_f) \to \mathrm{Hom}(\pi, G)_{\mathbb{1}_f}$$

described above is π_1-surjective.

We are now ready to prove the following theorem which is the main theorem of this section.

Theorem 5.6. *Let $\pi = \mathbb{Z}^n \oplus A$, with A a finite abelian group and $G \in \mathcal{P}$. Let $f \in \mathrm{Hom}(A, T)$ and let $\mathbb{1} := \mathbb{1} \times f \in \mathrm{Hom}(\pi, G)$ be the base point of $\mathrm{Hom}(\pi, G)$. Then there is a natural isomorphism $\pi_1(\mathrm{Hom}(\pi, G)) \cong (\pi_1(G_f))^n$, where $G_f = Z_G(f)$ is the subgroup of elements in G commuting with $f(x)$ for all $x \in A$.*

Proof. Suppose first that π is a finite group and thus $n = 0$. Then as proved above $\pi_1(\mathrm{Hom}(\pi, G)) = 1$ for any choice of base point and the theorem is true in this case. Suppose then that $n \geq 1$. By Corollary 5.5 the map i_f is π_1-surjective. We now show that in fact $(i_f)_* : \pi_1(\mathrm{Hom}(\mathbb{Z}^n, G_f)) \to \pi_1(\mathrm{Hom}(\pi, G)_{\mathbb{1}_f})$ is an isomorphism. This together with the isomorphism $\pi_1(\mathrm{Hom}(\mathbb{Z}^n, G_f)) \cong (\pi_1(G_f))^n$ provided by [7, Theorem 1.1] proves the theorem.

To start note that G_f is such that $\pi_1(G_f)$ is torsion free. Therefore we can write $\pi_1(G_f) = \mathbb{Z}^a$ for some integer a and the map

$$(i_f)_* : \pi_1(\mathrm{Hom}(\mathbb{Z}^n, G_f)) \cong \mathbb{Z}^{na} \to \pi_1(\mathrm{Hom}(\pi, G)_{\mathbb{1}_f})$$

is a surjection. This shows in particular that $\pi_1(\mathrm{Hom}(\pi, G)_{\mathbb{1}_f})$ is abelian and of rank at most na. We are going to show that in fact

$$r := \mathrm{rank}_{\mathbb{Z}}(\pi_1(\mathrm{Hom}(\pi, G)_{\mathbb{1}_f})) = na.$$

The only way this is possible is that $(i_f)_*$ is an isomorphism, proving the theorem. We now verify this. The universal coefficient theorem together with the Hurewicz theorem provide an isomorphism of $\mathbb{Q}$-vector spaces

$$H^1(\mathrm{Hom}(\pi, G)_{\mathbb{1}_f}; \mathbb{Q}) \cong \mathbb{Q}^r, \tag{5.3}$$

$$H^1(\mathrm{Hom}(\mathbb{Z}^n, G_f); \mathbb{Q}) \cong \mathbb{Q}^{na}. \tag{5.4}$$

On the other hand, since the conjugation action of G on $\mathrm{Hom}(\pi, G)_{\mathbb{1}_f}$ has connected maximal rank isotropy subgroups, then by Theorem 3.1 there is an isomorphism

$$H^*(\mathrm{Hom}(\pi, G)_{\mathbb{1}_f}; \mathbb{Q}) \cong H^*(G/T \times_W (\mathrm{Hom}(\pi, G)_{\mathbb{1}_f})^T; \mathbb{Q}).$$

In this case

$$G/T \times_W (\mathrm{Hom}(\pi, G)_{\mathbb{1}_f})^T \cong G/T \times_{W_f} T^n.$$

In particular

$$H^1(\mathrm{Hom}(\pi, G)_{\mathbb{1}_f}; \mathbb{Q}) \cong H^1(G/T \times T^n; \mathbb{Q})^{W_f} \cong H^1(T^n; \mathbb{Q})^{W_f}. \quad (5.5)$$

The second isomorphism follows from the fact that $H^1(G/T; \mathbb{Q}) = 0$ as G/T is simply connected. On the other hand, by [8, Theorem 1.1] it follows $W(G_f) = W_f$. The conjugation action of G_f on $\mathrm{Hom}(\mathbb{Z}^n, G_f)$ also has maximal rank isotropy subgroups. The same argument as above yields the following isomorphisms

$$H^1(\mathrm{Hom}(\mathbb{Z}^n, G_f); \mathbb{Q}) \cong H^1(G_f/T \times T^n; \mathbb{Q})^{W_f} \cong H^1(T^n; \mathbb{Q})^{W_f}. \quad (5.6)$$

Equations (5.5) and (5.6) show that there is an isomorphism of $\mathbb{Q}$-vector spaces

$$H^1(\mathrm{Hom}(\mathbb{Z}^n, G_f); \mathbb{Q}) \cong H^1(\mathrm{Hom}(\pi, G)_{\mathbb{1}_f}; \mathbb{Q})$$

and thus $n = ra$ by (5.3). □

6 Equivariant K-theory

In this section we study the G-equivariant K-theory of the spaces of homomorphisms $\mathrm{Hom}(\pi, G)$. We divide our study according to the nature of the group π. From now on we fix T a maximal torus in G and let W be the associated Weyl group.

6.1 Finite abelian groups

We first consider the case where π is a finite abelian group.

Fix a finite abelian group π and $G \in \mathcal{P}$. By Proposition 2.5 there is a G-equivariant homeomorphism

$$\Phi : \mathrm{Hom}(\pi, G) \to \bigsqcup_{[f] \in \mathrm{Hom}(\pi, T)/W} G/G_f.$$

Given a subgroup $H \subset G$ we have

$$K_G^q(G/H) \cong \begin{cases} R(H) & \text{if } q \text{ is even,} \\ 0 & \text{if } q \text{ is odd.} \end{cases}$$

By [10, Theorem 1] it follows that if $H \subset G$ is a subgroup of maximal rank then $R(H)$ is a free module over $R(G)$ of rank $|W|/|WH|$. As a corollary of this the following is obtained.

Corollary 6.1. *Let $G \in \mathcal{P}$ and π be a finite abelian group. Then $K_G^0(\mathrm{Hom}(\pi, G))$ is a free $R(G)$-module of rank $|\mathrm{Hom}(\pi, T)|$ and $K_G^1(\mathrm{Hom}(\pi, G)) = 0$.*

Proof. Using Proposition 2.5 and the above we have $K_G^1(\mathrm{Hom}(\pi, G)) = 0$ and

$$K_G^0(\mathrm{Hom}(\pi, G)) \cong \bigoplus_{[f] \in \mathrm{Hom}(\pi, T)/W} R(G_f).$$

Each $R(G_f)$ is a free module over $R(G)$ of rank $|W|/|W_f|$. Note that $W(G_f) = W_f$, where W_f denotes the isotropy subgroup at f, under the action of W on the finite set $\mathrm{Hom}(\pi, T)$. The partition of $\mathrm{Hom}(\pi, T)$ into the different W-orbits provides the identity

$$|\mathrm{Hom}(\pi, T)| = \sum_{[f] \in \mathrm{Hom}(\pi, T)/W} |W|/|W_f|.$$

This proves that $K_G^0(\mathrm{Hom}(\pi, G))$ is free as a module over $R(G)$ of rank $|\mathrm{Hom}(\pi, T)|$. □

Remark 6.2. The previous corollary is not true in general if $G \notin \mathcal{P}$. For example, it can be seen that $K^*_{PU(3)}(\mathrm{Hom}((\mathbb{Z}/(3))^2, PU(3)))$ is not free as a module over $R(PU(3))$.

6.2 Abelian groups of rank one

We now consider the case where π is a finitely generated abelian group of rank one. Thus we can write π in the form $\pi = \mathbb{Z} \oplus A$ where A is a finite abelian group.

Suppose that X is a G-CW complex. The skeleton filtration of X induces a multiplicative spectral sequence (see [11]) with

$$E_2^{p,q} = H_G^p(X; \mathcal{K}_G^q) \Longrightarrow K_G^{p+q}(X). \tag{6.1}$$

The E_2-term of this spectral sequence is the Bredon cohomology of X with respect to the coefficient system $\mathcal{K}_G^q$ defined by $G/H \mapsto K_G^q(G/H)$.

Suppose that $G \in \mathcal{P}$ and that $\pi = \mathbb{Z} \oplus A$, where A is a finite abelian group. Then Corollary 2.4 gives $X := \mathrm{Hom}(\pi, G)$ the structure of a G-CW complex and we can use the previous spectral sequence to compute $K_G^*(\mathrm{Hom}(\pi, G))$. In [3, Theorem 1.6] a criterion for the collapse of the spectral sequence (6.1) without extension problems was provided. This criterion can be used in this case to compute the structure of $K_G^*(\mathrm{Hom}(\pi, G))$ as a module over $R(G)$. Let Φ be the root system associated to (G, T). Fix a subset Φ^+ of positive roots of Φ and let

$\Delta = \{\alpha_1, \ldots, \alpha_r\}$ be an ordering of the corresponding set of simple roots. Suppose that $W_i \subset W$ is a reflection subgroup. Let Φ_i be the corresponding root system and Φ_i^+ the corresponding positive roots. Define

$$W_i^\ell := \{w \in W \mid w(\Phi_i^+) \subset \Phi^+\}.$$

The set W_i^ℓ forms a system of representatives of the left cosets in W/W_i by [12, Lemma 2.5]. In a precise way, this means that any element $w \in W$ can be factored in a unique way in the form $w = ux$ with $u \in W_i^\ell$ and $x \in W_i$. In order to apply the criterion provided in [3, Theorem 1.6] we must verify the hypothesis required there. In particular we need to show that X^T has the structure of a W-CW complex in such a way that there is a CW-subcomplex K of X^T such that for every element $x \in X^T$ there is a unique $w \in W$ such that $wx \in K$. We construct such a subcomplex next. To start note that $X^T = \mathrm{Hom}(\pi, G)^T = T \times \mathrm{Hom}(A, T)$ and $\mathrm{Hom}(A, T)$ is a discrete set endowed with an action of W. If we assume that G is simply connected then as pointed out in [3, Section 6.1] the (closed) alcoves in T provide a structure of a W-CW complex in T in such a way that $K(\Delta)$, the alcove determined by Δ, is a sub CW-complex of T such that any element in T has a unique representative in $K(\Delta)$ under the W-action. Moreover, for each cell σ in $K(\Delta)$, the isotropy subgroup W_σ is a reflection subgroup of the form W_I for some $I \subset \Delta$. Here W_I denotes the reflection subgroup generated by the reflections of the form s_α for $\alpha \in I$. This can be used to produce a sub CW-complex in $\mathrm{Hom}(\pi, G)^T$ satisfying similar properties in the following way. Let $f_1, \ldots, f_m$ be a set of representatives for the W-orbits in the discrete set $\mathrm{Hom}(A, T)$. We can choose each f_i in such a way that the isotropy subgroup W_{f_i} is a reflection subgroup of W of the form W_{I_i} for some $I_i \subset \Delta$. For each $1 \le i \le m$ let $W_{f_i}^\ell$ be a system of minimal length representatives of W/W_{f_i} as defined above. Define

$$L(\Delta) := \bigcup_{i=1}^{m} \bigcup_{u \in W_{f_i}^\ell} \left(u^{-1}K(\Delta) \times \{f_i\}\right) \subset T \times \mathrm{Hom}(A, T) = X^T.$$

Defined in this way $L(\Delta) \subset X^T$ is a sub CW-complex. We now show that $L(\Delta)$ is such that for every element $x \in X^T$ there is a unique $w \in W$ such that $wx \in L(\Delta)$. To see this, since $K(\Delta) \subset T$ satisfies this property, it suffices to see that for any i and any $v_1, v_2 \in W$ there are unique $v \in W$ and $u \in W_{f_i}^\ell$ such that $v_1 K(\Delta) \times v_2 f_i = v(u^{-1}K(\Delta) \times f_i)$. Indeed, suppose that $v_1, v_2 \in W$. Using the defining property of $W_{f_i}^\ell$ we can find unique $u \in W_{f_i}^\ell$ and $x \in W_{f_i}$ such that $v_1^{-1}v_2 = ux$. Let $v = v_1 u$. Then $v_1 = vu^{-1}$ and in particular $v_1 K(\Delta) = vu^{-1}K(\Delta)$.

Also, $x = v^{-1}v_2 \in W_{f_i}$ and thus $v_2 f_i = v f_i$. This shows that $v \in W$ and $u \in W^{\ell}_{f_i}$ are the unique elements such that $v_1 K(\Delta) \times v_2 f_i = v(u^{-1}K(\Delta) \times f_i)$. On the other hand, note that $H^*(X^T;\mathbb{Z})$ is torsion–free and of rank $2^r \cdot |\mathrm{Hom}(A,T)|$, where r denotes the rank of the Lie group G. Also note that the isotropy subgroups of the action of W on $\mathrm{Hom}(\pi,G)^T = \mathrm{Hom}(\pi,T)$ are of the form W_I, with $I \subset \Delta$.

The above work shows that the conditions of [3, Theorem 1.6] are satisfied yielding the next theorem.

Theorem 6.3. *Suppose that $G \in \mathcal{P}$ is simply connected and of rank r. Let $\pi = \mathbb{Z} \oplus A$ where A is a finite abelian group. Then $K^*_G(\mathrm{Hom}(\pi,G))$ is a free $R(G)$-module of rank $2^r \cdot |\mathrm{Hom}(A,T)|$.*

Remark 6.4. Combining Corollary 6.1 and Theorem 6.3 it follows that $K^*_G(\mathrm{Hom}(\pi,G))$ is free as a module over $R(G)$ whenever π is a finitely generated abelian group of rank less or equal to 1 and $G \in \mathcal{P}$ is simply connected. As already pointed out in [3] this result does not extend to all finitely generated abelian groups π as $K^*_{SU(2)}(\mathrm{Hom}(\mathbb{Z}^2, SU(2)))$ contains torsion as a $R(SU(2))$-module.

However, if we tensor with the rational numbers the previous result does extend to the family of finitely generated abelian groups and all Lie groups $G \in \mathcal{P}$. This is done next.

6.3 Finitely generated abelian groups

We show that $K^*_G(\mathrm{Hom}(\pi,G)) \otimes \mathbb{Q}$ is a free $R(G) \otimes \mathbb{Q}$-module for all finitely generated abelian groups π and all Lie groups $G \in \mathcal{P}$.

Let G be a compact Lie group with $\pi_1(G)$ torsion–free act on a compact space X with connected maximal rank isotropy. If we further assume that X^T has the homotopy type of a W-CW complex then by [3, Theorem 1.1] $K^*_G(X) \otimes \mathbb{Q}$ is a free module over $R(G) \otimes \mathbb{Q}$ of rank equal to $\sum_{i\geq 0} \mathrm{rank}_{\mathbb{Q}} H^i(X^T;\mathbb{Q})$. This theorem can be applied in our situation. Let π be a finitely generated abelian group and $G \in \mathcal{P}$. Let $X := \mathrm{Hom}(\pi,G)$. Then the conjugation action of G on X has connected maximal rank isotropy subgroups and X has the homotopy type of a G-CW complex by Proposition 2.3 and Corollary 2.4. Note that $H^*(\mathrm{Hom}(\pi,G)^T;\mathbb{Q})$ is a $\mathbb{Q}$-vector space of rank $2^{nr} \cdot |\mathrm{Hom}(A,T)|$, where r is the rank of G. This proves that the hypotheses of [3, Theorem 1.1] are satisfied in this case yielding the following.

Theorem 6.5. *Suppose that $G \in \mathcal{P}$ is of rank r and that π is a finitely generated abelian group written in the form $\pi = \mathbb{Z}^n \oplus A$, where A is a finite abelian group. Then $K^*_G(\mathrm{Hom}(\pi,G)) \otimes \mathbb{Q}$ is a free module over $R(G) \otimes \mathbb{Q}$ of rank $2^{nr} \cdot |\mathrm{Hom}(A,T)|$.*

References

[1] A. ADEM and F. R. COHEN, *Commuting elements and spaces of homomorphisms*, Math. Ann. **338** (2007), 587–626.

[2] A. ADEM, F. R. COHEN and J. M. GÓMEZ, *Stable splittings, spaces of representations and almost commuting elements in Lie groups*, Math. Proc. Camb. Phil. Soc. **149** (2010) 455–490.

[3] A. ADEM and J. M. GÓMEZ, *Equivariant K-theory of compact Lie group actions with maximal rank isotropy*, Journal of Topology (2012); doi:10.1112/jtopol/jts009

[4] T. BAIRD, *Cohomology of the space of commuting n-tuples in a compact Lie group*, Algebraic and Geometric Topology **7** (2007) 737–754.

[5] T. BAIRD, L. C. JEFFREY and P. SELICK, *The space of commuting n-tuples in $SU(2)$*, Illinois J. Math., to appear.

[6] M. C. CRABB, *Spaces of commuting elements in $SU(2)$*, Proc. Edinb. Math. Soc. **54** (2011), 67–75.

[7] J. M. GÓMEZ, A. PETTET and J. SOUTO, *On the fundamental group of* $\mathrm{Hom}(\mathbb{Z}^k, G)$, Math. Zeitschrift **271** (2012), 33–44.

[8] V. HAUSCHILD, *Compact Lie group actions with isotropy subgroups of maximal rank*, Manuscripta Math. **34** (1981), no. 2-3, 355–379.

[9] S. ILLMAN, *Smooth equivariant triangulations of G-manifolds for G a finite group*, Math. Ann. **233** (1978), no. 3, 199–220.

[10] H. PITTIE, *Homogeneous vector bundles on homogeneous spaces* Topology **11** (1972), 199–203.

[11] G. SEGAL, *Equivariant K-theory*, Inst. Hautes Études Sci. Publ. Math. **34** (1968), 129–151.

[12] R. STEINBERG, *On a theorem of Pittie*, Topology **14** (1975), 173–177.

On the fundamental group of the complement of two real tangent conics and an arbitrary number of real tangent lines

Meirav Amram[1,2], David Garber[2] and Mina Teicher[3]

Abstract. We compute the simplified presentations of the fundamental groups of the complements of the family of real conic-line arrangements with up to two conics which are tangent to each other at two points, with an arbitrary number of tangent lines to both conics. All the resulting groups turn out to be big.

1 Introduction

Line arrangements as simply as they are, carry many open questions around them (including topological and combinatorial questions), which are slowly solved. In this paper, we go up to conic-line arrangements, where the parallel questions about them are even less understood.

We compute here a simplified presentation of the fundamental group of complements of a family of conic-line arrangements. The only known results so far in this direction are [4, 5].

In general, the fundamental group of complements of plane curves is an important topological invariant with many different applications. Unfortunately, it is hard to compute and to understand.

This invariant was used by Chisini [9], Kulikov [17, 18] and Kulikov-Teicher [19] in order to distinguish between connected components of the moduli space of smooth projective surfaces, see also [11].

Moreover, Zariski-Lefschetz hyperplane section theorem (see [20]) stated that

$$\pi_1(\mathbb{P}^N - S) \cong \pi_1(H - (H \cap S)),$$

[1] Partially supported by the Emmy Noether Institute Fellowship (by the Minerva Foundation of Germany).

[2] Partially supported by a grant from the Ministry of Science, Culture and Sport, Israel and the Russian Foundation for Basic research, the Russian Federation.

[3] The research was partially supported by the Oswald Veblen Fund and by the Minerva Foundation of Germany.

where S is an hypersurface and H is a generic 2-plane. Since $H \cap S$ is a plane curve, we can compute $\pi_1(\mathbb{P}^N - S)$ in an easier way, by computing the fundamental group of the complement of a plane curve.

A different direction for the need of fundamental groups' computations in general context is for getting more examples of Zariski pairs [30, 31]. A pair of plane curves is called *a Zariski pair* if the two curves have the same combinatorics (*i.e.* the same singular points and the same arrangement of irreducible components), but their complements are not homeomorphic. Some examples of Zariski pairs can be found at [6–8, 10, 12, 26–29].

Let $T_{n,m}$ be the family of real conic-line arrangements in $\mathbb{CP}^2$ with two conics, which are tangent to each other at two points, and with an arbitrary number of tangent lines to each one of the conics. Moreover, we assume that no line is passing through the tangency points between the two conics, and no three lines are intersected in a multiple point. The main result of this paper is the simplified presentation of the fundamental group $\pi_1(\mathbb{CP}^2 - T_{n,m})$, which is achieved by an inductive computation of the braid monodromy factorization of the curve:

Theorem 1.1. *Let Q_1, Q_2 be two real tangent conics in $\mathbb{CP}^2$ and let $\{L_i\}_{i=1}^n$ and $\{L'_j\}_{j=1}^m$ be n and m real lines which are tangent to Q_2 and Q_1 respectively, see Figure* 1.1. *Moreover, we assume that no line is passing through the tangency points between the two conics, and no three lines are intersected in a multiple point. Denote:*

$$T_{n,m} = Q_1 \cup Q_2 \cup \left(\bigcup_{i=1}^{n} L_i \right) \cup \left(\bigcup_{j=1}^{m} L'_j \right).$$

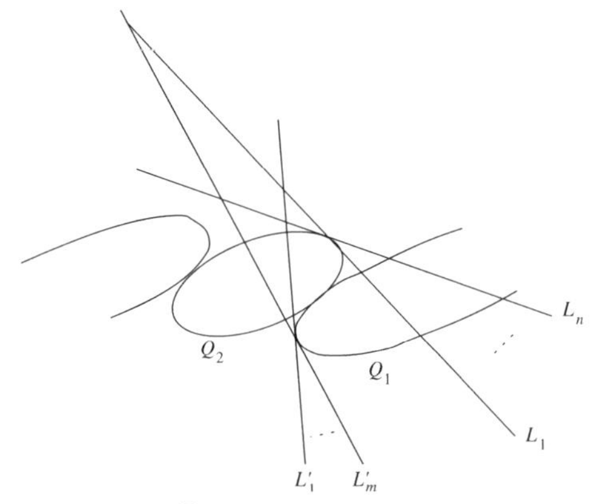

Figure 1.1. The arrangement $T_{n,m}$.

Then the fundamental group $\pi_1(\mathbb{CP}^2 - T_{n,m})$ is generated by the generators x_2 (related to Q_1), x_5 (related to Q_2), $x_6, \dots, x_{n+4}$ (related to $L_2, \dots, L_n$), $x_{n+5}, \dots, x_{n+m+4}$ (related to $L'_1, \dots, L'_m$). This group admits the following relations:

1. $x_{n+5}^{-1} x_5 x_{n+5} = x_{n+6}^{-1} \cdots x_{n+m+4}^{-1} \cdot x_5 \cdot x_{n+m+4} \cdots x_{n+6}$,
2. $(x_2 x_i)^2 = (x_i x_2)^2$, *where* $i = 5, 6, \dots, n+4$,
3. $(x_5 x_i)^2 = (x_i x_5)^2$, *where* $n+5 \le i \le n+m+4$,
4. $[x_i, x_j] = e$, *where* $6 \le i \le n+4$ *and* $j = 5, n+5, \dots, n+m+4$,
5. $[x_i, x_2 x_5 x_2^{-1}]$, *where* $6 \le i \le n+4$,
6. $[x_2^{-1} x_i x_2, x_j] = e$, *where* $6 \le i < j \le n+4$,
7. $[x_2, x_i] = e$, *where* $n+6 \le i \le n+m+4$,
8. $[x_5 x_2 x_5^{-1}, x_i] = e$, *where* $n+5 \le i \le n+m+4$,
9. $[x_{n+5}, x_i] = e$, *where* $i = 2, n+6, \dots, n+m+4$,
10. $[x_5^{-1} x_i x_5, x_j] = e$, *where* $n+6 \le i < j \le n+m+4$.

These arrangements may appear as a branch curve of a generic projection to $\mathbb{CP}^2$ of a surface of general type (see for example [15]).

Remark 1.2. Note that any real conic-line arrangement consists of two real tangent conics in $\mathbb{CP}^2$ (which are tangent to each other at two points) and $n+m$ real lines which are tangent to the conics (where no line is passing through the tangency points between the two conics, and no three lines are intersected in a multiple point), can be transferred to the arrangement presented in Figure 1.1 by translations, rotations and equisingular deformations. These actions do preserve the fundamental group of the complement of the arrangement (see *e.g.* [14]).

A group is called *big* if it contains a free subgroup, generated by two or more generators. By the above computations, we have the following corollary:

Corollary 1.3. *The fundamental group $\pi_1(\mathbb{CP}^2 - T_{n,m})$ is big.*

Algorithmically, this paper uses local computations (local braid monodromies and their induced relations, see also [1]), braid monodromy techniques of Moishezon-Teicher (see [13,21–25]), the van Kampen Theorem (see [16]) and some group simplifications and calculations for having a simplified presentation of fundamental groups (see also [2,3]).

The new idea we are using in this paper for computing the presentation of the fundamental group in the general setting is the construction of the BMF of the general curve in an inductive way. Explicitly, we construct the BMF of the conic-line arrangement $T_{n,m}$ by generalizing the behavior of the BMFs of the conic-line arrangements $T_{1,m}$, $T_{2,m}$ and $T_{3,m}$.

Moreover, the BMFs of the conic-line arrangements $T_{1,m}$, $T_{2,m}$ and $T_{3,m}$ themselves are deduced by studying the corresponding BMFs of the family $T_{i,1}$, $T_{i,2}$ and $T_{i,3}$ for $i \in \{1, 2, 3\}$.

The paper is organized as follows. In Section 2, we compute the braid monodromy factorization of the arrangement $T_{n,m}$. In Section 3, we compute the simplified presentation of the fundamental group by the van Kampen theorem and group simplifications.

ACKNOWLEDGEMENTS. We would like to thank Mutsuo Oka for correcting a wrong assumption in the first version of this paper. We also wish to thank an anonymous referee for useful suggestions.

2 Braid monodromy factorizations

In this section, we compute the braid monodromy factorizations (BMFs, see [19]) of the arrangement $T_{n,m}$ for any n, m. Note that we dealt with the arrangements $T_{1,1}$ and $T_{2,0}$ in [1]. We start with some notations.

Notation 2.1. Let z_{ij} (respectively $\overline{z}_{ij}$) be a path below (respectively above) the axis, which connects points i and j. We denote by Z_{ij} (respectively $\overline{Z}_{ij}$) the counterclockwise half-twist of i and j along z_{ij} (respectively $\overline{z}_{ij}$).

We denote $Z^2_{i,jj'} = Z^2_{ij'} \cdot Z^2_{ij}$ and $Z^2_{ii',jj'} = Z^2_{i',jj'} \cdot Z^2_{i,jj'}$.

Conjugation of braids is denoted by $a^b := b^{-1}ab$.

Example 2.2. The path in Figure 2.1(a) is constructed as follows: take a path z_{34} and conjugate it by the full-twist $\overline{Z}^2_{13}$. Its corresponding half-twist is $Z_{34}^{\overline{Z}^2_{13}}$. The path in Figure 2.1(b) is constructed as follows: take again z_{34} and conjugate it first by Z^2_{23} and then by Z^2_{13}. Its corresponding half-twist is $Z_{34}^{Z^2_{23} Z^2_{13}}$.

Figure 2.1. Examples of conjugated braids.

Now, we compute the BMF of $T_{n,m}$:

Proposition 2.3. *Let Q_1, Q_2 be two real tangent conics in $\mathbb{CP}^2$. Consider two sets of n and m real lines, such that n lines are tangent to Q_2,* *and m lines are tangent to Q_1 (see Figure 1.1). Define:*

$$T_{n,m} = Q_1 \cup Q_2 \cup \left(\bigcup_{i=1}^{n} L_i \right) \cup \left(\bigcup_{j=1}^{m} L'_j \right).$$

Then:

$$\begin{aligned}
\Delta^2_{T_{n,m}} = {} & Z_{23} \cdot (Z_{23})^{Z_{12}^{-2} \prod_{i=6}^{n+4} \overline{Z}_{2i}^2} \cdot (Z_{45})^{\overline{Z}_{24}^2 Z_{5,n+5}^2} \cdot \tilde{Z}_{45} \cdot \overline{Z}_{24}^4 \cdot (Z_{34}^4)^{Z_{13}^2} \\
& \cdot Z_{13}^4 \cdot \prod_{i=6}^{n+4} \overline{Z}_{2i}^4 \cdot \prod_{i=n+5}^{n+m+4} (Z_{5i}^4)^{Z_{5,n+5}^2} \cdot (Z_{15}^2)^{Z_{13}^2} \\
& \cdot \prod_{i=6}^{n+4} \left[(Z_{4i}^2)^{Z_{45}^2} \cdot (Z_{5i}^2)^{Z_{15}^2 Z_{13}^2} \right] \cdot \prod_{i=6}^{n+4} (Z_{1i}^2)^{Z_{15}^2 Z_{13}^2} \\
& \cdot \prod_{6 \leq i < j \leq n+4} (\overline{Z}_{ij}^2)^{\overline{Z}_{2i}^2} \cdot (Z_{2,n+5}^2)^{Z_{24}^2 Z_{23}^2} \cdot (Z_{3,n+5}^2)^{Z_{13}^2} \\
& \cdot \prod_{i=n+6}^{n+m+4} \left[(\overline{Z}_{2i}^2)^{(\prod_{j=n+4}^{6} \overline{Z}_{2j}^{-2})} \cdot \tilde{Z}_{3i}^2 \right] \cdot \prod_{i=n+6}^{n+m+4} Z_{n+5,i}^2 \\
& \cdot \prod_{n+6 \leq i < j \leq n+m+4} (Z_{ij}^2)^{Z_{5i}^{-2} Z_{5,n+5}^2} \cdot \\
& \cdot \prod_{i=n+5}^{n+m+4} (Z_{1i}^2)^{Z_{1,n+5}^2 Z_{13}^2} \cdot \prod_{i=6}^{n+4} \left[\prod_{j=n+5}^{n+m+4} (Z_{ij}^2)^{(\prod_{k=n+5}^{j} Z_{kj}^{-2})} \right].
\end{aligned}$$

The skeletons of $\tilde{Z}_{45}$ and $\tilde{Z}_{3i}^2$ appear in Figure 2.2.

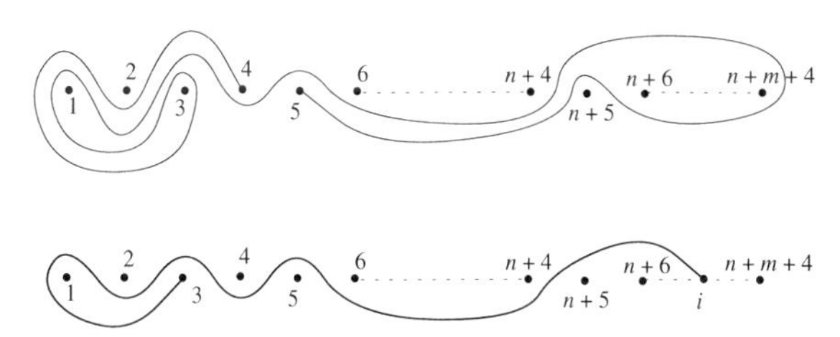

Figure 2.2. Skeletons for the BMF of $T_{n,m}$.

Proof. We first construct the BMF of $T_{1,m}$ by generalizing the BMF of $T_{1,1}$, $T_{1,2}$, $T_{1,3}$ in the following way: while adding m lines, indeed we have more intersection points between the lines and the conics, and between the lines themselves. One can easily check that the braid monodromies of the new intersection points have similar structure to the structure of the braid monodromies of the points appearing already in the arrangements $T_{1,1}$, $T_{1,2}$, $T_{1,3}$, and in that way, we can write the BMF of the general arrangement $T_{1,m}$:

$$\Delta^2_{T_{1,m}} = Z_{23} \cdot (Z_{23})^{Z_{12}^{-2}} \cdot (Z_{45})^{\overline{Z}_{24}^2 Z_{56}^2} \cdot \tilde{Z}'_{45} \cdot \overline{Z}_{24}^4 \cdot (Z_{34}^4)^{Z_{13}^2} \cdot Z_{13}^4 \cdot \prod_{i=6}^{m+5} (Z_{5i}^4)^{Z_{56}^2}$$
$$\cdot (Z_{15}^2)^{Z_{13}^2} \cdot (Z_{26}^2)^{Z_{24}^2 Z_{23}^2} \cdot (Z_{36}^2)^{Z_{13}^2} \cdot \left(\prod_{i=7}^{m+5} \left[\overline{Z}_{2i}^2 \cdot \tilde{Z}'^2_{3i} \right] \right) \cdot \prod_{i=7}^{m+5} Z_{6i}^2$$
$$\cdot \prod_{7 \le i < j \le m+5} (Z_{ij}^2)^{Z_{5i}^{-2} Z_{56}^2} \cdot \prod_{i=6}^{m+5} (Z_{1i}^2)^{Z_{16}^2 Z_{13}^2}.$$

Then we proceed in the same manner to compute the BMF of $T_{2,m}$ and $T_{3,m}$, as presented in the following table. The table is constructed as follows: in each row we write the factors which are related to the same type of singularities in the different arrangements.

	$\Delta^2_{T_{1,m}}$	$\Delta^2_{T_{2,m}}$	$\Delta^2_{T_{3,m}}$
(1)	Z_{23}	Z_{23}	Z_{23}
(2)	$(Z_{23})^{Z_{12}^{-2}}$	$(Z_{23})^{Z_{12}^{-2} \overline{Z}_{26}^2}$	$(Z_{23})^{Z_{12}^{-2} \overline{Z}_{26}^2 \overline{Z}_{27}^2}$
(3)	$(Z_{45})^{\overline{Z}_{24}^2 Z_{56}^2}$	$(Z_{45})^{\overline{Z}_{24}^2 Z_{57}^2}$	$(Z_{45})^{\overline{Z}_{24}^2 Z_{58}^2}$
(4)	$\tilde{Z}'_{45}$	$\tilde{Z}''_{45}$	$\tilde{Z}'''_{45}$
(5)	$\overline{Z}_{24}^4 \cdot (Z_{34}^4)^{Z_{13}^2}$	$\overline{Z}_{24}^4 \cdot (Z_{34}^4)^{Z_{13}^2}$	$\overline{Z}_{24}^4 \cdot (Z_{34}^4)^{Z_{13}^2}$
(6)	Z_{13}^4	$Z_{13}^4 \cdot \overline{Z}_{26}^4$	$Z_{13}^4 \cdot \prod\limits_{i=6,7} \overline{Z}_{2i}^4$
(7)	$\prod\limits_{i=6}^{m+5} (Z_{5i}^4)^{Z_{56}^2}$	$\prod\limits_{i=7}^{m+6} (Z_{5i}^4)^{Z_{57}^2}$	$\prod\limits_{i=8}^{m+7} (Z_{5i}^4)^{Z_{58}^2}$
(8)	$(Z_{15}^2)^{Z_{13}^2}$	$(Z_{15}^2)^{Z_{13}^2} \cdot (Z_{46}^2)^{Z_{45}^2} \cdot (Z_{56}^2)^{Z_{15}^2 Z_{13}^2}$	$(Z_{15}^2)^{Z_{13}^2} \cdot \prod\limits_{i=6}^{7} \left[(Z_{4i}^2)^{Z_{45}^2} \cdot (Z_{5i}^2)^{Z_{15}^2 Z_{13}^2} \right]$
(9)	–	$(Z_{16}^2)^{Z_{15}^2 Z_{13}^2}$	$\prod\limits_{i=6}^{7} (Z_{1i}^2)^{Z_{15}^2 Z_{13}^2} \cdot (Z_{67}^2)^{\overline{Z}_{26}^2}$
(10)	$(Z_{26}^2)^{Z_{24}^2 Z_{23}^2} \cdot (Z_{36}^2)^{Z_{13}^2} \cdot \prod\limits_{i=7}^{m+5} \left[\overline{Z}_{2i}^2 \cdot \tilde{Z}'^2_{3i} \right]$	$(Z_{27}^2)^{Z_{24}^2 Z_{23}^2} \cdot (Z_{37}^2)^{Z_{13}^2} \cdot \prod\limits_{i=8}^{m+6} \left[(\overline{Z}_{2i}^2)^{\overline{Z}_{26}^{-2}} \cdot \tilde{Z}''^2_{3i} \right]$	$(Z_{28}^2)^{Z_{24}^2 Z_{23}^2} \cdot (Z_{38}^2)^{Z_{13}^2} \cdot \prod\limits_{i=9}^{m+7} \left[(\overline{Z}_{2i}^2)^{\overline{Z}_{27}^{-2} \overline{Z}_{26}^{-2}} \cdot \tilde{Z}'''^2_{3i} \right]$
(11)	$\prod\limits_{i=7}^{m+5} Z_{6i}^2 \cdot \prod\limits_{7 \le i < j \le m+5} (Z_{ij}^2)^{Z_{5i}^{-2} Z_{56}^2}$	$\prod\limits_{i=8}^{m+6} Z_{7i}^2 \cdot \prod\limits_{8 \le i < j \le m+6} (Z_{ij}^2)^{Z_{5i}^{-2} Z_{57}^2}$	$\prod\limits_{i=9}^{m+7} Z_{8i}^2 \cdot \prod\limits_{9 \le i < j \le m+7} (Z_{ij}^2)^{Z_{5i}^{-2} Z_{58}^2}$
(12)	$\prod\limits_{i=6}^{m+5} (Z_{1i}^2)^{Z_{16}^2 Z_{13}^2}$	$\prod\limits_{i=7}^{m+6} (Z_{1i}^2)^{Z_{17}^2 Z_{13}^2} \cdot \prod\limits_{j=7}^{m+6} (Z_{6j}^2)^{(\prod_{k=7}^{j} Z_{kj}^{-2})}$	$\prod\limits_{i=8}^{m+7} (Z_{1i}^2)^{Z_{18}^2 Z_{13}^2} \cdot \prod\limits_{i=6}^{7} \left[\prod\limits_{j=8}^{m+7} (Z_{ij}^2)^{(\prod_{k=8}^{j} Z_{kj}^{-2})} \right]$

The skeletons of $\tilde{Z}'_{45}$, $\tilde{Z}''_{45}$, $\tilde{Z}'''_{45}$ and $\tilde{Z}'^2_{3i}$, $\tilde{Z}''^2_{3i}$, $\tilde{Z}'''^2_{3i}$ appear in Figures 2.3 and 2.4 respectively.

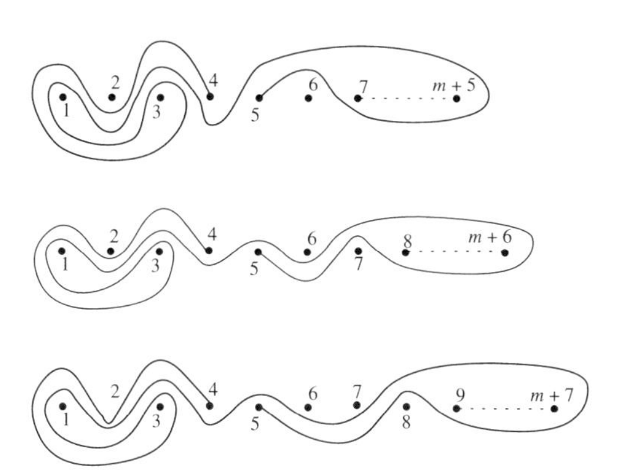

Figure 2.3. The skeletons of $\tilde{Z}'_{45}$, $\tilde{Z}''_{45}$, $\tilde{Z}'''_{45}$.

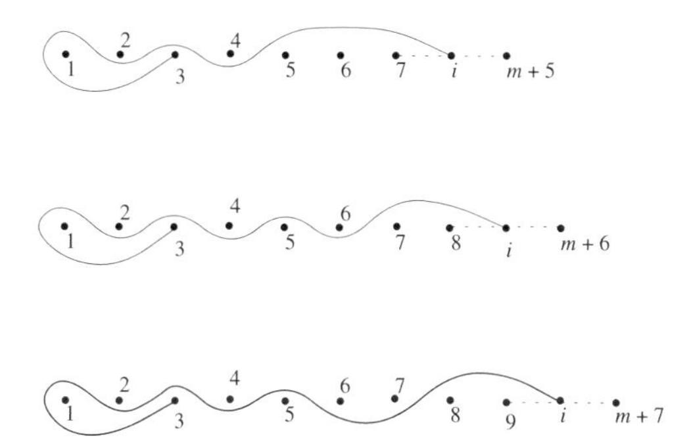

Figure 2.4. The skeletons of $\tilde{Z}'^2_{3i}$, $\tilde{Z}''^2_{3i}$, $\tilde{Z}'''^2_{3i}$.

In the above table, rows (1) and (2) correspond to the branch points of Q_1. Rows (3) and (4) correspond to the branch points of Q_2. Row (5) corresponds to the two tangency points between Q_1 and Q_2. Row (6) corresponds to the tangency points between Q_1 and the lines which are tangent to it. Row (7) corresponds to the tangency points between Q_2 and the lines which are tangent to it. Row (8) corresponds to the intersection points between the lines $\{L_i\}_{i=1}^n$ and the conic Q_2. Row (9) corresponds to the intersection points between the lines $\{L_i\}_{i=1}^n$ themselves. Row (10) corresponds to the intersection points between the lines $\{L'_j\}_{j=1}^m$ with Q_1. Row (11) corresponds to the intersection points between the lines $\{L'_j\}_{j=1}^m$ themselves. Row (12) corresponds to the intersection points between the lines $\{L_i\}_{i=1}^n$ and $\{L'_j\}_{j=1}^m$.

In the next step, we generalize the factorization for any n, *i.e.* we compute $\Delta^2_{T_{n,m}}$. In the following table, each row includes the general form of the corresponding row in the previous table:

	$\Delta^2_{T_{n,m}}$
(1)	Z_{23}
(2)	$(Z_{23})^{Z_{12}^{-2} \prod_{i=6}^{n+4} \overline{Z}_{2i}^2}$
(3)	$(Z_{45})^{\overline{Z}_{24}^2 Z_{5,n+5}^2}$
(4)	$\tilde{Z}_{45}$
(5)	$\overline{Z}_{24}^4 \cdot (Z_{34}^4)^{Z_{13}^2}$
(6)	$Z_{13}^4 \cdot \prod_{i=6}^{n+4} \overline{Z}_{2i}^4$
(7)	$\prod_{i=n+5}^{n+m+4} (Z_{5i}^4)^{Z_{5,n+5}^2}$
(8)	$(Z_{15}^2)^{Z_{13}^2} \cdot \prod_{i=6}^{n+4} \left[(Z_{4i}^2)^{Z_{45}^2} \cdot (Z_{5i}^2)^{Z_{15}^2 Z_{13}^2} \right]$
(9)	$\prod_{i=6}^{n+4} (Z_{1i}^2)^{Z_{15}^2 Z_{13}^2} \cdot \prod_{6 \leq i < j \leq n+4} (\overline{Z}_{ij}^2)^{\overline{Z}_{2i}^2}$
(10)	$(Z_{2,n+5}^2)^{Z_{24}^2 Z_{23}^2} \cdot (Z_{3,n+5}^2)^{Z_{13}^2} \cdot \prod_{i=n+6}^{n+m+4} \left[(\overline{Z}_{2i}^2)^{(\prod_{j=n+4}^{6} \overline{Z}_{2j}^{-2})} \cdot \tilde{Z}_{3i}^2 \right]$
(11)	$\prod_{i=n+6}^{n+m+4} Z_{n+5,i}^2 \cdot \prod_{n+6 \leq i < j \leq n+m+4} (Z_{ij}^2)^{Z_{5i}^{-2} Z_{5,n+5}^2}$
(12)	$\prod_{i=n+5}^{n+m+4} (Z_{1i}^2)^{Z_{1,n+5}^2 Z_{13}^2} \cdot \prod_{i=6}^{n+4} \left[\prod_{j=n+5}^{n+m+4} (Z_{ij}^2)^{(\prod_{k=n+5}^{j} Z_{kj}^{-2})} \right]$

The skeletons of $\tilde{Z}_{45}$ and $\tilde{Z}_{3i}^2$ appear in Figure 2.2. □

3 The computation of the fundamental groups

In this section, we compute the simplified presentation of the fundamental group of the complement of the general arrangement $T_{n,m}$, $n, m \geq 0$.

Proof of Theorem 1.1. Applying the van Kampen theorem on the BMF of $T_{n,m}$ (computed in Proposition 2.3), we get the following presentation for $\pi_1(\mathbb{CP}^2 - T_{n,m})$:
Generators: $\{x_1, x_2, x_3, x_4, x_5, x_6, x_7, \ldots, x_{n+4}, x_{n+5}, \ldots, x_{n+m+4}\}$.
Relations:

1. $x_2 = x_3$
2. $x_3 = x_1^{-1} x_3^{-1} x_4^{-1} x_5^{-1} \cdot x_{n+4} x_{n+3} \cdots x_3 x_2 x_3^{-1} \cdots x_{n+3}^{-1} x_{n+4}^{-1} \cdot x_5 x_4 x_3 x_1$
3. $x_4 x_3 x_2 x_3^{-1} x_4 x_3 x_2^{-1} x_3^{-1} x_4^{-1} = x_{n+5} x_5 x_{n+5}^{-1}$
4. $x_3 x_1 x_3^{-1} x_1^{-1} x_3^{-1} x_4 x_3 x_1 x_3 x_1^{-1} x_3^{-1} = x_5^{-1} \cdot x_{n+5}^{-1} x_{n+6}^{-1} \cdots x_{n+m+4}^{-1} \cdot x_{n+5} x_5 x_{n+5}^{-1} \cdot x_{n+m+4} \cdots x_{n+6} x_{n+5} \cdot x_5$
5. $(x_3 x_2 x_3^{-1} x_4)^2 = (x_4 x_3 x_2 x_3^{-1})^2$
6. $(x_3 x_1 x_3 x_1^{-1} x_3^{-1} x_4)^2 = (x_4 x_3 x_1 x_3 x_1^{-1} x_3^{-1})^2$
7. $(x_1 x_3)^2 = (x_3 x_1)^2$
8. $(x_2 \cdot x_3^{-1} x_4^{-1} \cdots x_{i-1}^{-1} x_i x_{i-1} \cdots x_4 x_3)^2 = (x_3^{-1} x_4^{-1} \cdots x_{i-1}^{-1} x_i x_{i-1} \cdots x_4 x_3 \cdot x_2)^2$, where $6 \leq i \leq n+4$
9. $(x_5 x_{n+5})^2 = (x_{n+5} x_5)^2$
10. $(x_5 x_{n+5}^{-1} x_i x_{n+5})^2 = (x_{n+5}^{-1} x_i x_{n+5} x_5)^2$, where $n+6 \leq i \leq n+m+4$
11. $[x_3 x_1 x_3^{-1}, x_5] = e$

12. $[x_5x_4x_5^{-1}, x_i] = e$, where $6 \le i \le n+4$
13. $[x_5x_3x_1x_3^{-1}x_5x_3x_1^{-1}x_3^{-1}x_5^{-1}, x_i] = e$, where $6 \le i \le n+4$
14. $[x_5x_3x_1x_3^{-1}x_5^{-1}, x_i] = e$, where $6 \le i \le n+4$
15. $[x_{j-1}x_{j-2}\cdots x_4x_3x_2x_3^{-1}x_4^{-1}\cdots x_{i-2}^{-1}x_{i-1}^{-1}x_ix_{i-1}x_{i-2}\cdots x_4x_3x_2^{-1}x_3^{-1}x_4^{-1}\cdots x_{j-2}^{-1}x_{j-1}^{-1}, x_j] = e$, where $6 \le i < j \le n+4$
16. $[x_4x_3x_2x_3^{-1}x_4^{-1}, x_{n+5}] = e$
17. $[x_3x_1x_3x_1^{-1}x_3^{-1}, x_{n+5}] = e$
18. $[x_5x_4x_3x_2x_3^{-1}x_4^{-1}x_5^{-1}, x_{n+5}^{-1}x_{n+6}^{-1}\cdots x_{i-2}^{-1}x_{i-1}^{-1}x_ix_{i-1}x_{i-2}\cdots x_{n+6}x_{n+5}] = e$, where $n+6 \le i \le n+m+4$
19. $[x_5x_3x_1x_3x_1^{-1}x_3^{-1}x_5^{-1}, x_{n+5}^{-1}x_{n+6}^{-1}\cdots x_{i-2}^{-1}x_{i-1}^{-1}x_ix_{i-1}x_{i-2}\cdots x_{n+6}x_{n+5}] = e$, where $n+6 \le i \le n+m+4$
20. $[x_{n+5}, x_i] = e$, where $n+6 \le i \le n+m+4$
21. $[x_{n+5}x_5^{-1}x_{n+5}^{-1}x_ix_{n+5}x_5x_{n+5}^{-1}, x_j] = e$, where $n+6 \le i < j \le n+m+4$
22. $[x_3x_1x_3^{-1}, x_{n+5}^{-1}x_ix_{n+5}] = e$, where $n+5 \le i \le n+m+4$
23. $[x_i, x_{n+5}^{-1}x_{n+6}^{-1}\cdots x_{j-1}^{-1}x_jx_{j-1}\cdots x_{n+6}x_{n+5}] = e$, where $6 \le i \le n+4$ and $n+5 \le j \le n+m+4$
24. $x_{n+m+4}x_{n+m+3}\cdots x_4x_3x_2x_1 = e$. (Projective relation).

Now, we simplify this presentation. Since $x_2 = x_3$ and $(x_1x_3)^2 = (x_3x_1)^2$ (by relations (1) and (7)), we derive a simpler presentation as follows (we combine relations (5) and (7) into one set):

Generators: $\{x_1, x_2, x_4, x_5, x_6, x_7, \ldots, x_{n+4}, x_{n+5}, \ldots, x_{n+m+4}\}$.
Relations:

1. $x_1^{-1}x_2x_1 = x_4^{-1}x_5^{-1} \cdot x_{n+4}x_{n+3}\cdots x_4x_2x_4^{-1}\cdots x_{n+3}^{-1}x_{n+4}^{-1} \cdot x_5x_4$
2. $x_4x_2x_4x_2^{-1}x_4^{-1} = x_{n+5}x_5x_{n+5}^{-1}$
3. $x_1^{-1}x_2^{-1}x_1x_4x_1^{-1}x_2x_1 = x_5^{-1} \cdot x_{n+5}^{-1}x_{n+6}^{-1}\cdots x_{n+m+4}^{-1} \cdot x_{n+5}x_5x_{n+5}^{-1} \cdot x_{n+m+4}\cdots x_{n+6}x_{n+5} \cdot x_5$
4. $(x_2x_i)^2 = (x_ix_2)^2$, where $i = 1, 4$
5. $(x_1^{-1}x_2x_1x_4)^2 = (x_4x_1^{-1}x_2x_1)^2$
6. $(x_4^{-1}\cdots x_{i-1}^{-1}x_ix_{i-1}\cdots x_4x_2)^2 = (x_2^{-1}x_4^{-1}\cdots x_{i-1}^{-1}x_ix_{i-1}\cdots x_4x_2^2)^2$, where $6 \le i \le n+4$
7. $(x_5x_{n+5})^2 = (x_{n+5}x_5)^2$
8. $(x_5x_{n+5}^{-1}x_ix_{n+5})^2 = (x_{n+5}^{-1}x_ix_{n+5}x_5)^2$, where $n+6 \le i \le n+m+4$
9. $[x_2x_1x_2^{-1}, x_5] = e$
10. $[x_5x_4x_5^{-1}, x_i] = e$, where $6 \le i \le n+4$
11. $[x_5x_2x_1x_2^{-1}x_5x_2x_1^{-1}x_2^{-1}x_5^{-1}, x_i] = e$, where $6 \le i \le n+4$
12. $[x_5x_2x_1x_2^{-1}x_5^{-1}, x_i] = e$, where $6 \le i \le n+4$
13. $[x_{j-1}x_{j-2}\cdots x_4x_2x_4^{-1}\cdots x_{i-2}^{-1}x_{i-1}^{-1}x_ix_{i-1}x_{i-2}\cdots x_4x_2^{-1}x_4^{-1}\cdots x_{j-2}^{-1}x_{j-1}^{-1}, x_j] = e$, where $6 \le i < j \le n+4$
14. $[x_4x_2x_4^{-1}, x_{n+5}] = e$
15. $[x_1^{-1}x_2x_1, x_{n+5}] = e$
16. $[x_5x_4x_2x_4^{-1}x_5^{-1}, x_{n+5}^{-1}x_{n+6}^{-1}\cdots x_{i-2}^{-1}x_{i-1}^{-1}x_ix_{i-1}x_{i-2}\cdots x_{n+6}x_{n+5}] = e$, where $n+6 \le i \le n+m+4$
17. $[x_5x_1^{-1}x_2x_1x_5^{-1}, x_{n+5}^{-1}x_{n+6}^{-1}\cdots x_{i-2}^{-1}x_{i-1}^{-1}x_ix_{i-1}x_{i-2}\cdots x_{n+6}x_{n+5}] = e$, where $n+6 \le i \le n+m+4$
18. $[x_{n+5}, x_i] = e$, where $n+6 \le i \le n+m+4$
19. $[x_{n+5}x_5^{-1}x_{n+5}^{-1}x_ix_{n+5}x_5x_{n+5}^{-1}, x_j] = e$, where $n+6 \le i < j \le n+m+4$
20. $[x_2x_1x_2^{-1}, x_{n+5}^{-1}x_ix_{n+5}] = e$, where $n+5 \le i \le n+m+4$
21. $[x_i, x_{n+5}^{-1}x_{n+6}^{-1}\cdots x_{j-1}^{-1}x_jx_{j-1}\cdots x_{n+6}x_{n+5}] = e$, where $6 \le i \le n+4$ and $n+5 \le j \le n+m+4$
22. $x_{n+m+4}x_{n+m+3}\cdots x_5x_4x_2^2x_1 = e$.

Now we apply relations (4) and (18) to get a much convenient presentation (we also combine relations (7) and (8) into one set):

1. $x_1^{-1}x_2x_1 = x_4^{-1}x_5^{-1} \cdot x_{n+4}x_{n+3}\cdots x_4x_2x_4^{-1}\cdots x_{n+3}^{-1}x_{n+4}^{-1} \cdot x_5x_4$
2. $x_2^{-1}x_4x_2 = x_{n+5}x_5x_{n+5}^{-1}$
3. $x_1^{-1}x_2^{-1}x_1x_4x_1^{-1}x_2x_1 = x_5^{-1} \cdot x_{n+6}^{-1}\cdots x_{n+m+4}^{-1} \cdot x_5 \cdot x_{n+m+4}\cdots x_{n+6} \cdot x_5$

4. $(x_2x_i)^2 = (x_ix_2)^2$, where $i = 1, 4$
5. $(x_1^{-1}x_2x_1x_4)^2 = (x_4x_1^{-1}x_2x_1)^2$
6. $(x_4^{-1} \cdots x_{i-1}^{-1}x_ix_{i-1} \cdots x_4x_2)^2 = (x_2^{-1}x_4^{-1} \cdots x_{i-1}^{-1}x_ix_{i-1} \cdots x_4x_2^2)^2$, where $6 \leq i \leq n+4$
7. $(x_5x_i)^2 = (x_ix_5)^2$, where $n+5 \leq i \leq n+m+4$
8. $[x_2x_1x_2^{-1}, x_5] = e$
9. $[x_5x_4x_5^{-1}, x_i] = e$, where $6 \leq i \leq n+4$
10. $[x_5x_2x_1x_2^{-1}x_5x_2x_1^{-1}x_2^{-1}x_5^{-1}, x_i] = e$, where $6 \leq i \leq n+4$
11. $[x_5x_2x_1x_2^{-1}x_5^{-1}, x_i] = e$, where $6 \leq i \leq n+4$
12. $[x_{j-1}x_{j-2} \cdots x_4x_2x_4^{-1} \cdots x_{i-2}^{-1}x_{i-1}^{-1}x_ix_{i-1}x_{i-2} \cdots x_4x_2^{-1}x_4^{-1} \cdots x_{j-2}^{-1}x_{j-1}^{-1}, x_j] = e$, where $6 \leq i < j \leq n+4$
13. $[x_4x_2x_4^{-1}, x_{n+5}] = e$
14. $[x_1^{-1}x_2x_1, x_{n+5}] = e$
15. $[x_5x_4x_2x_4^{-1}x_5^{-1}, x_{n+6}^{-1} \cdots x_{i-2}^{-1}x_{i-1}^{-1}x_ix_{i-1}x_{i-2} \cdots x_{n+6}] = e$, where $n+6 \leq i \leq n+m+4$
16. $[x_5x_1^{-1}x_2x_1x_5^{-1}, x_{n+6}^{-1} \cdots x_{i-2}^{-1}x_{i-1}^{-1}x_ix_{i-1}x_{i-2} \cdots x_{n+6}] = e$, where $n+6 \leq i \leq n+m+4$
17. $[x_{n+5}, x_i] = e$, where $n+6 \leq i \leq n+m+4$
18. $[x_5^{-1}x_ix_5, x_j] = e$, where $n+6 \leq i < j \leq n+m+4$
19. $[x_2x_1x_2^{-1}, x_i] = e$, where $n+5 \leq i \leq n+m+4$
20. $[x_i, x_{n+6}^{-1} \cdots x_{j-1}^{-1}x_jx_{j-1} \cdots x_{n+6}] = e$, where $6 \leq i \leq n+4$ and $n+5 \leq j \leq n+m+4$
21. $x_{n+m+4}x_{n+m+3} \cdots x_5x_4x_2^2x_1 = e.$

By relations (9) we get the following simplified relations (we combine relations (8), (11), (19) into one set):

1. $x_1^{-1}x_2x_1 = x_4^{-1}x_5^{-1} \cdot x_{n+4}x_{n+3} \cdots x_4x_2x_4^{-1} \cdots x_{n+3}^{-1}x_{n+4}^{-1} \cdot x_5x_4$
2. $x_2^{-1}x_4x_2 = x_{n+5}x_5x_{n+5}^{-1}$
3. $x_1^{-1}x_2^{-1}x_1x_4x_1^{-1}x_2x_1 = x_5^{-1} \cdot x_{n+6}^{-1} \cdots x_{n+m+4}^{-1} \cdot x_5 \cdot x_{n+m+4} \cdots x_{n+6} \cdot x_5$
4. $(x_2x_i)^2 = (x_ix_2)^2$, where $i = 1, 4$
5. $(x_1^{-1}x_2x_1x_4)^2 = (x_4x_1^{-1}x_2x_1)^2$
6. $(x_4^{-1} \cdots x_{i-1}^{-1}x_ix_{i-1} \cdots x_4x_2)^2 = (x_2^{-1}x_4^{-1} \cdots x_{i-1}^{-1}x_ix_{i-1} \cdots x_4x_2^2)^2$, where $6 \leq i \leq n+4$
7. $(x_5x_i)^2 = (x_ix_5)^2$, where $n+5 \leq i \leq n+m+4$
8. $[x_2x_1x_2^{-1}, x_i] = e$, where $5 \leq i \leq n+m+4$
9. $[x_5x_4x_5^{-1}, x_i] = e$, where $6 \leq i \leq n+4$
10. $[x_5, x_i] = e$, where $6 \leq i \leq n+4$
11. $[x_{j-1}x_{j-2} \cdots x_4x_2x_4^{-1} \cdots x_{i-2}^{-1}x_{i-1}^{-1}x_ix_{i-1}x_{i-2} \cdots x_4x_2^{-1}x_4^{-1} \cdots x_{j-2}^{-1}x_{j-1}^{-1}, x_j] = e$, where $6 \leq i < j \leq n+4$
12. $[x_4x_2x_4^{-1}, x_{n+5}] = e$
13. $[x_1^{-1}x_2x_1, x_{n+5}] = e$
14. $[x_5x_4x_2x_4^{-1}x_5^{-1}, x_{n+6}^{-1} \cdots x_{i-2}^{-1}x_{i-1}^{-1}x_ix_{i-1}x_{i-2} \cdots x_{n+6}] = e$, where $n+6 \leq i \leq n+m+4$
15. $[x_5x_1^{-1}x_2x_1x_5^{-1}, x_{n+6}^{-1} \cdots x_{i-2}^{-1}x_{i-1}^{-1}x_ix_{i-1}x_{i-2} \cdots x_{n+6}] = e$, where $n+6 \leq i \leq n+m+4$
16. $[x_{n+5}, x_i] = e$, where $n+6 \leq i \leq n+m+4$
17. $[x_5^{-1}x_ix_5, x_j] = e$, where $n+6 \leq i < j \leq n+m+4$
18. $[x_i, x_{n+6}^{-1} \cdots x_{j-1}^{-1}x_jx_{j-1} \cdots x_{n+6}] = e$, where $6 \leq i \leq n+4$ and $n+5 \leq j \leq n+m+4$
19. $x_{n+m+4}x_{n+m+3} \cdots x_5x_4x_2^2x_1 = e.$

By relations (10), relations (9) become $[x_4, x_i] = e$, where $6 \leq i \leq n+4$. These relations enable us to proceed in simplification:

1. $x_1^{-1}x_2x_1 = x_{n+4}x_{n+3} \cdots x_6x_2x_6^{-1} \cdots x_{n+3}^{-1}x_{n+4}^{-1}$
2. $x_2^{-1}x_4x_2 = x_{n+5}x_5x_{n+5}^{-1}$
3. $x_1^{-1}x_2^{-1}x_1x_4x_1^{-1}x_2x_1 = x_5^{-1} \cdot x_{n+6}^{-1} \cdots x_{n+m+4}^{-1} \cdot x_5 \cdot x_{n+m+4} \cdots x_{n+6} \cdot x_5$
4. $(x_2x_i)^2 = (x_ix_2)^2$, where $i = 1, 4$
5. $(x_1^{-1}x_2x_1x_4)^2 = (x_4x_1^{-1}x_2x_1)^2$

6. $(x_6^{-1} \cdots x_{i-1}^{-1} x_i x_{i-1} \cdots x_6 x_2)^2 = (x_2^{-1} x_6^{-1} \cdots x_{i-1}^{-1} x_i x_{i-1} \cdots x_6 x_2^2)^2$, where $6 \leq i \leq n+4$
7. $(x_5 x_i)^2 = (x_i x_5)^2$, where $n+5 \leq i \leq n+m+4$
8. $[x_2 x_1 x_2^{-1}, x_i] = e$, where $5 \leq i \leq n+m+4$
9. $[x_4, x_i] = e$, where $6 \leq i \leq n+4$
10. $[x_5, x_i] = e$, where $6 \leq i \leq n+4$
11. $[x_{j-1} x_{j-2} \cdots x_6 x_2 x_6^{-1} \cdots x_{i-2}^{-1} x_{i-1}^{-1} x_i x_{i-1} x_{i-2} \cdots x_6 x_2^{-1} x_6^{-1} \cdots x_{j-2}^{-1} x_{j-1}^{-1}, x_j] = e$, where $6 \leq i < j \leq n+4$
12. $[x_4 x_2 x_4^{-1}, x_{n+5}] = e$
13. $[x_1^{-1} x_2 x_1, x_{n+5}] = e$
14. $[x_5 x_4 x_2 x_4^{-1} x_5^{-1}, x_{n+6}^{-1} \cdots x_{i-2}^{-1} x_{i-1}^{-1} x_i x_{i-1} x_{i-2} \cdots x_{n+6}] = e$, where $n+6 \leq i \leq n+m+4$
15. $[x_5 x_1^{-1} x_2 x_1 x_5^{-1}, x_{n+6}^{-1} \cdots x_{i-2}^{-1} x_{i-1}^{-1} x_i x_{i-1} x_{i-2} \cdots x_{n+6}] = e$, where $n+6 \leq i \leq n+m+4$
16. $[x_{n+5}, x_i] = e$, where $n+6 \leq i \leq n+m+4$
17. $[x_5^{-1} x_i x_5, x_j] = e$, where $n+6 \leq i < j \leq n+m+4$
18. $[x_i, x_{n+6}^{-1} \cdots x_{j-1}^{-1} x_j x_{j-1} \cdots x_{n+6}] = e$, where $6 \leq i \leq n+4$ and $n+5 \leq j \leq n+m+4$
19. $x_{n+m+4} x_{n+m+3} \cdots x_5 x_4 x_2^2 x_1 = e.$

We treat relations (18). Substituting $i = 6$ and $j = n+5$, we get $[x_6, x_{n+5}] = e$. Substituting $i = 6$ and $j = n+6$, we get $[x_6, x_{n+6}] = e$. If we take $i = 6$ and $j = n+7$, we get $[x_6, x_{n+6}^{-1} x_{n+7} x_{n+6}] = e$, which is $[x_6, x_{n+7}] = e$. Continuing this process by substituting $i = 6$ and $n+8 \leq j \leq n+m+4$, we get $[x_6, x_j] = e$ for $n+8 \leq j \leq n+m+4$. Now we substitute $i = 7$ and $n+5 \leq j \leq n+m+4$. From these substitutions we get: $[x_7, x_j] = e$ for $n+5 \leq j \leq n+m+4$. In the same manner, we get $[x_i, x_j] = e$ for $8 \leq i \leq n+4$ and $n+5 \leq j \leq n+m+4$. This enables us to simplify the above list of relations (we combine relations (9), (10) and (18) into one set):

1. $x_1^{-1} x_2 x_1 = x_{n+4} x_{n+3} \cdots x_6 x_2 x_6^{-1} \cdots x_{n+3}^{-1} x_{n+4}^{-1}$
2. $x_2^{-1} x_4 x_2 = x_{n+5} x_5 x_{n+5}^{-1}$
3. $x_1^{-1} x_2^{-1} x_1 x_4 x_1^{-1} x_2 x_1 = x_5^{-1} \cdot x_{n+6}^{-1} \cdots x_{n+m+4}^{-1} \cdot x_5 \cdot x_{n+m+4} \cdots x_{n+6} \cdot x_5$
4. $(x_2 x_i)^2 = (x_i x_2)^2$, where $i = 1, 4$
5. $(x_1^{-1} x_2 x_1 x_4)^2 = (x_4 x_1^{-1} x_2 x_1)^2$
6. $(x_6^{-1} \cdots x_{i-1}^{-1} x_i x_{i-1} \cdots x_6 x_2)^2 = (x_2^{-1} x_6^{-1} \cdots x_{i-1}^{-1} x_i x_{i-1} \cdots x_6 x_2^2)^2$, where $6 \leq i \leq n+4$
7. $(x_5 x_i)^2 = (x_i x_5)^2$, where $n+5 \leq i \leq n+m+4$
8. $[x_2 x_1 x_2^{-1}, x_i] = e$, where $5 \leq i \leq n+m+4$
9. $[x_i, x_j] = e$, where $6 \leq i \leq n+4$ and $j = 4, 5, n+5, \ldots, n+m+4$
10. $[x_{j-1} x_{j-2} \cdots x_6 x_2 x_6^{-1} \cdots x_{i-2}^{-1} x_{i-1}^{-1} x_i x_{i-1} x_{i-2} \cdots x_6 x_2^{-1} x_6^{-1} \cdots x_{j-2}^{-1} x_{j-1}^{-1}, x_j] = e$, where $6 \leq i < j \leq n+4$
11. $[x_4 x_2 x_4^{-1}, x_{n+5}] = e$
12. $[x_1^{-1} x_2 x_1, x_{n+5}] = e$
13. $[x_5 x_4 x_2 x_4^{-1} x_5^{-1}, x_{n+6}^{-1} \cdots x_{i-2}^{-1} x_{i-1}^{-1} x_i x_{i-1} x_{i-2} \cdots x_{n+6}] = e$, where $n+6 \leq i \leq n+m+4$
14. $[x_5 x_1^{-1} x_2 x_1 x_5^{-1}, x_{n+6}^{-1} \cdots x_{i-2}^{-1} x_{i-1}^{-1} x_i x_{i-1} x_{i-2} \cdots x_{n+6}] = e$, where $n+6 \leq i \leq n+m+4$
15. $[x_{n+5}, x_i] = e$, where $n+6 \leq i \leq n+m+4$
16. $[x_5^{-1} x_i x_5, x_j] = e$, where $n+6 \leq i < j \leq n+m+4$
17. $x_{n+m+4} x_{n+m+3} \cdots x_5 x_4 x_2^2 x_1 = e.$

Now we simplify relations (6) and (10). We start by substituting $i = 6$ in relations (6) to get $(x_2 x_6)^2 = (x_6 x_2)^2$. We continue with $i = 6$, $j = 7$ in relations (10) to get $[x_2^{-1} x_6 x_2, x_7] = e$ (using $(x_2 x_6)^2 = (x_6 x_2)^2$). Using these two relations, we simplify relations (6) for $i = 7$: $(x_2 x_6^{-1} x_7 x_6)^2 =$

$(x_6^{-1}x_7x_6x_2)^2$. We add the expression $x_2^{-1}x_2 = e$ in four locations as follows:

$$x_6x_2x_6^{-1}(x_2^{-1}x_2)x_7x_6x_2x_6^{-1}(x_2^{-1}x_2)x_7$$
$$= x_7x_6x_2x_6^{-1}(x_2^{-1}x_2)x_7x_6x_2x_6^{-1}(x_2^{-1}x_2).$$

Since $(x_2x_6)^2 = (x_6x_2)^2$, we can rewrite the relation as:

$$x_2^{-1}x_6^{-1}x_2x_6x_2x_7x_2^{-1}x_6^{-1}x_2x_6x_2x_7 = x_7x_2^{-1}x_6^{-1}x_2x_6x_2x_7x_2^{-1}x_6^{-1}x_2x_6x_2.$$

We use the relation $[x_2^{-1}x_6x_2, x_7] = e$ to get:

$$x_2^{-1}x_6^{-1}x_2x_6x_2x_2^{-1}x_6^{-1}x_2x_7x_6x_2x_7 = x_2^{-1}x_6^{-1}x_2x_7x_6x_2x_2^{-1}x_6^{-1}x_2x_7x_6x_2,$$

namely $x_2x_7x_6x_2x_7 = x_7x_2x_7x_6x_2$. We add the expression $x_2x_2^{-1} = e$ to get $x_2x_7(x_2x_2^{-1})x_6x_2x_7 = x_7x_2x_7x_6x_2$, and using $[x_2^{-1}x_6x_2, x_7] = e$, we get $(x_2x_7)^2 = (x_7x_2)^2$.

Now we proceed with $i = 6, j = 8$ and then $i = 7, j = 8$ in relations (10) to get $[x_2^{-1}x_6x_2, x_8] = e$ and $[x_2^{-1}x_7x_2, x_8] = e$ respectively. These relations enable us to simplify the relation $(x_2x_6^{-1}x_7^{-1}x_8x_7x_6)^2 = (x_6^{-1}x_7^{-1}x_8x_7x_6x_2)^2$ (which we get from relations (6) for $i = 8$) to $(x_2x_8)^2 = (x_8x_2)^2$.

In the same manner, we conclude that relations (6) and (10) can be simplified to $(x_2x_i)^2 = (x_ix_2)^2$ for $6 \le i \le n+4$ and $[x_2^{-1}x_ix_2, x_j] = e$ for $6 \le i < j \le n+4$ respectively.

Relation (12) gets the form $[x_2, x_{n+5}] = e$ by substituting relation (1) in it and by using relations (9).

Relations (14) can be rewritten by substituting relation (1) in them (for $n+6 \le i \le n+m+4$):

$$[x_5x_{n+4}x_{n+3}\cdots x_6x_2x_6^{-1}\cdots x_{n+3}^{-1}x_{n+4}^{-1}x_5^{-1}, x_{n+6}^{-1}\cdots x_{i-2}^{-1}x_{i-1}^{-1}x_ix_{i-1}x_{i-2}\cdots x_{n+6}] = e.$$

Now, by relations (9), we get:

$$[x_5x_2x_5^{-1}, x_{n+6}^{-1}\cdots x_{i-2}^{-1}x_{i-1}^{-1}x_ix_{i-1}x_{i-2}\cdots x_{n+6}] = e.$$

For $i = n+6$, we get $[x_5x_2x_5^{-1}, x_{n+6}] = e$. Now, for $i = n+7$, we get $[x_5x_2x_5^{-1}, x_{n+6}^{-1}x_{n+7}x_{n+6}] = e$. By $[x_5x_2x_5^{-1}, x_{n+6}] = e$, it is simplified to $[x_5x_2x_5^{-1}, x_{n+7}] = e$. In the same way, we get that relations (14) are equivalent to $[x_5x_2x_5^{-1}, x_i] = e$, for $n+6 \le i \le n+m+4$.

Therefore, we have the following list of relations:

1. $x_1^{-1}x_2x_1 = x_{n+4}x_{n+3}\cdots x_6x_2x_6^{-1}\cdots x_{n+3}^{-1}x_{n+4}^{-1}$
2. $x_4 = x_2x_{n+5}x_5x_{n+5}^{-1}x_2^{-1}$

3. $x_1^{-1}x_2^{-1}x_1x_4x_1^{-1}x_2x_1 = x_5^{-1} \cdot x_{n+6}^{-1} \cdots x_{n+m+4}^{-1} \cdot x_5 \cdot x_{n+m+4} \cdots x_{n+6} \cdot x_5$
4. $(x_2x_i)^2 = (x_ix_2)^2$, where $i = 1, 4$
5. $(x_1^{-1}x_2x_1x_4)^2 = (x_4x_1^{-1}x_2x_1)^2$
6. $(x_2x_i)^2 = (x_ix_2)^2$, where $6 \leq i \leq n+4$
7. $(x_5x_i)^2 = (x_ix_5)^2$, where $n+5 \leq i \leq n+m+4$
8. $[x_2x_1x_2^{-1}, x_i] = e$, where $5 \leq i \leq n+m+4$
9. $[x_i, x_j] = e$, where $6 \leq i \leq n+4$ and $j = 4, 5, n+5, \ldots, n+m+4$
10. $[x_2^{-1}x_ix_2, x_j] = e$, where $6 \leq i < j \leq n+4$
11. $[x_4x_2x_4^{-1}, x_{n+5}] = e$
12. $[x_2, x_{n+5}] = e$
13. $[x_5x_4x_2x_4^{-1}x_5^{-1}, x_{n+6}^{-1} \cdots x_{i-2}^{-1}x_{i-1}^{-1}x_ix_{i-1}x_{i-2} \cdots x_{n+6}] = e$, where $n+6 \leq i \leq n+m+4$
14. $[x_5x_2x_5^{-1}, x_i] = e$, where $n+6 \leq i \leq n+m+4$
15. $[x_{n+5}, x_i] = e$, where $n+6 \leq i \leq n+m+4$
16. $[x_5^{-1}x_ix_5, x_j] = e$, where $n+6 \leq i < j \leq n+m+4$
17. $x_{n+m+4}x_{n+m+3} \cdots x_5x_4x_2^2x_1 = e.$

By relation (2), we can omit x_4 and replace it everywhere by

$$x_2x_{n+5}x_5x_{n+5}^{-1}x_2^{-1}.$$

We start with relations (4). Take $(x_2x_4)^2 = (x_4x_2)^2$. This relation is rewritten as:

$$\begin{aligned} &x_2 \cdot x_2x_{n+5}x_5x_{n+5}^{-1}x_2^{-1} \cdot x_2 \cdot x_2x_{n+5}x_5x_{n+5}^{-1}x_2^{-1} = \\ &= x_2x_{n+5}x_5x_{n+5}^{-1}x_2^{-1} \cdot x_2 \cdot x_2x_{n+5}x_5x_{n+5}^{-1}x_2^{-1} \cdot x_2. \end{aligned}$$

By relation (12) we easily get: $(x_2x_5)^2 = (x_5x_2)^2$.

Now, we consider relation (11):

$$\begin{aligned} e &= [x_4x_2x_4^{-1}, x_{n+5}] = \\ &= [x_2x_{n+5}x_5x_{n+5}^{-1}x_2^{-1} \cdot x_2 \cdot x_2x_{n+5}x_5^{-1}x_{n+5}x_2, x_{n+5}] = \\ &= [x_5x_2x_5^{-1}, x_{n+5}]. \end{aligned}$$

By the above resulting relations, we can simplify also relations (13):

$$\begin{aligned} e &= [x_5 \cdot x_2x_{n+5}x_5x_{n+5}^{-1}x_2^{-1} \cdot x_2 \cdot x_2x_{n+5}x_5^{-1}x_{n+5}^{-1}x_2^{-1} \cdot x_5^{-1}, x_{n+6}^{-1} \cdots x_{i-1}^{-1}x_ix_{i-1} \cdots x_{n+6}] = \\ &= [x_2, x_{n+6}^{-1} \cdots x_{i-1}^{-1}x_ix_{i-1} \cdots x_{n+6}], \end{aligned}$$

for $n+6 \leq i \leq n+m+4$. For $i = n+6$, we get $[x_2, x_{n+6}] = e$. For $i = n+7$, we get $[x_2, x_{n+6}^{-1}x_{n+7}x_{n+6}] = e$, and by the previous relation, we get $[x_2, x_{n+7}] = e$. In the same manner, we get $[x_2, x_i] = e$, for $n+6 \leq i \leq n+m+4$.

From relations (9), we have $[x_i, x_4] = e$, where $6 \leq i \leq n+4$. By relation (2), $e = [x_i, x_4] = [x_i, x_2x_{n+5}x_5x_{n+5}^{-1}x_2^{-1}]$. By relations (9) and (12), we get for $6 \leq i \leq n+4$: $[x_i, x_2x_5x_2^{-1}] = e$.

We rewrite now relation (5), by substituting relations (1) and (2) in it:

$$x_{n+4} \cdots x_6 x_2 x_6^{-1} \cdots x_{n+4}^{-1} \cdot x_2 x_{n+5} x_5 x_{n+5}^{-1} x_2^{-1} \cdot x_{n+4} \cdots x_6 x_2 x_6^{-1} \cdots x_{n+4}^{-1} \cdot x_2 x_{n+5} x_5 x_{n+5}^{-1} x_2^{-1}$$
$$= x_2 x_{n+5} x_5 x_{n+5}^{-1} x_2^{-1} \cdot x_{n+4} \cdots x_6 x_2 x_6^{-1} \cdots x_{n+4}^{-1} \cdot x_2 x_{n+5} x_5 x_{n+5}^{-1} x_2^{-1} \cdot x_{n+4} \cdots x_6 x_2 x_6^{-1} \cdots x_{n+4}^{-1}.$$

Since $[x_{n+5}, x_i] = e$, for $i = 2, 6, \ldots, n+4$ (by relations (9) and (12)),

$$x_{n+4} \cdots x_6 x_2 x_6^{-1} \cdots x_{n+4}^{-1} x_2 x_5 x_2^{-1} \cdot x_{n+4} \cdots x_6 x_2 x_6^{-1} \cdots x_{n+4}^{-1} x_2 x_5 x_2^{-1}$$
$$= x_2 x_5 x_2^{-1} x_{n+4} \cdots x_6 x_2 x_6^{-1} \cdots x_{n+4}^{-1} \cdot x_2 x_5 x_2^{-1} x_{n+4} \cdots x_6 x_2 x_6^{-1} \cdots x_{n+4}^{-1}.$$

Now, since we proved above that $[x_i, x_2 x_5 x_2^{-1}] = e$, for $6 \le i \le n+4$, we get $x_2 x_5 x_2 x_5 = x_5 x_2 x_5 x_2$, which is a consequence from relations (4) (see above).

Therefore, we have:
Generators: $\{x_1, x_2, x_5, x_6, x_7, \ldots, x_{n+4}, x_{n+5}, \ldots, x_{n+m+4}\}$.
Relations:

1. $x_1^{-1} x_2 x_1 = x_{n+4} x_{n+3} \cdots x_6 x_2 x_6^{-1} \cdots x_{n+3}^{-1} x_{n+4}^{-1}$
2. $x_1^{-1} x_2^{-1} x_1 \cdot x_2 x_{n+5} x_5 x_{n+5}^{-1} x_2^{-1} \cdot x_1^{-1} x_2 x_1 = x_5^{-1} \cdot x_{n+6}^{-1} \cdots x_{n+m+4}^{-1} \cdot x_5 \cdot x_{n+m+4} \cdots x_{n+6} \cdot x_5$
3. $(x_2 x_i)^2 = (x_i x_2)^2$, where $i = 1, 5, 6, \ldots, n+4$
4. $(x_5 x_i)^2 = (x_i x_5)^2$, where $n+5 \le i \le n+m+4$
5. $[x_2 x_1 x_2^{-1}, x_i] = e$, where $5 \le i \le n+m+4$
6. $[x_i, x_j] = e$, where $6 \le i \le n+4$ and $j = 5, n+5, \ldots, n+m+4$
7. $[x_i, x_2 x_5 x_2^{-1}]$, where $6 \le i \le n+4$
8. $[x_2^{-1} x_i x_2, x_j] = e$, where $6 \le i < j \le n+4$
9. $[x_2, x_i] = e$, where $n+6 \le i \le n+m+4$
10. $[x_5 x_2 x_5^{-1}, x_i] = e$, where $n+5 \le i \le n+m+4$
11. $[x_{n+5}, x_i] = e$, where $i = 2, n+6, \ldots, n+m+4$
12. $[x_5^{-1} x_i x_5, x_j] = e$, where $n+6 \le i < j \le n+m+4$
13. $x_{n+m+4} x_{n+m+3} \cdots x_5 x_2 x_{n+5} x_5 x_{n+5}^{-1} x_2 x_1 = e$.

Now we use the projective relation (relation (13)) in order to omit the generator x_1.

We rewrite relation (1), by substituting x_1:

$$x_{n+m+4} \cdots x_5 x_2 x_{n+5} x_5 x_{n+5}^{-1} x_2 x_2 x_2^{-1} x_{n+5} x_5^{-1} x_{n+5}^{-1} x_2^{-1} x_5^{-1} \cdots x_{n+m+4}^{-1}$$
$$= x_{n+4} \cdots x_6 x_2 x_6^{-1} \cdots x_{n+4}^{-1}.$$

By relations (6) and (11), we get:

$$x_{n+m+4} \cdots x_{n+5} x_5 x_2 x_{n+5} x_5 x_2 x_5^{-1} x_{n+5}^{-1} x_2^{-1} x_5^{-1} x_{n+5}^{-1} \cdots x_{n+m+4}^{-1} = x_2.$$

By relations (9) and (10), we have:

$$x_{n+5} x_5 x_2 x_5 x_2 x_5^{-1} x_2^{-1} x_5^{-1} x_{n+5}^{-1} = x_2.$$

Now, by relations (3),

$$x_{n+5}x_2x_{n+5}^{-1} = x_2,$$

which is known already by relation (11).

Now we simplify $(x_1x_2)^2 = (x_2x_1)^2$, which appears in relations (3). By substituting x_1 and some immediate cancellations,

$$x_{n+4}\cdots x_6x_5x_2x_{n+5}x_5x_{n+m+4}\cdots x_{n+6}x_{n+4}\cdots x_6$$
$$= x_2^{-1}x_{n+4}\cdots x_6x_5x_2x_{n+5}x_5x_{n+m+4}\cdots x_{n+6}x_{n+4}\cdots x_6x_2.$$

By relations (6) and (9),

$$x_{n+4}\cdots x_6x_5x_2x_{n+5}x_5x_{n+4}\cdots x_6 = x_2^{-1}x_{n+4}\cdots x_6x_5x_2x_{n+5}x_5x_{n+4}\cdots x_6x_2.$$

Since $[x_i, x_5] = e$ for $6 \leq i \leq n+4$ (by relations (6)), we get

$$x_{n+4}\cdots x_6x_5x_2x_{n+5}x_{n+4}\cdots x_6x_5 = x_2^{-1}x_{n+4}\cdots x_6x_5x_2x_{n+5}x_{n+4}\cdots x_6x_5x_2,$$

which is:

$$x_{n+4}\cdots x_6x_5x_2x_{n+5}x_{n+4}\cdots x_6x_5x_2^{-1}x_5^{-1}$$
$$= x_2^{-1}x_{n+4}\cdots x_6x_5x_2x_{n+5}x_{n+4}\cdots x_6.$$

By relations (6) and (10),

$$x_{n+4}\cdots x_6x_5x_2x_{n+4}\cdots x_6x_5x_2^{-1}x_5^{-1} = x_2^{-1}x_{n+4}\cdots x_6x_5x_2x_{n+4}\cdots x_6.$$

Now we add $x_2x_2^{-1} = e$ in two locations:

$$x_2x_{n+4}\cdots x_6(x_2x_2^{-1})x_5x_2x_{n+4}\cdots x_6x_5x_2^{-1}x_5^{-1}$$
$$= x_{n+4}\cdots x_6(x_2x_2^{-1})x_5x_2x_{n+4}\cdots x_6.$$

By relations (7),

$$x_2x_{n+4}\cdots x_6x_2x_{n+4}\cdots x_6x_2^{-1}x_5x_2x_5x_2^{-1}x_5^{-1}$$
$$= x_{n+4}\cdots x_6x_2x_{n+4}\cdots x_6x_2^{-1}x_5x_2.$$

Since $(x_2x_5)^2 = (x_5x_2)^2$ (see relations (3)), we get:

$$x_2x_{n+4}\cdots x_6x_2x_{n+4}\cdots x_6 = x_{n+4}\cdots x_6x_2x_{n+4}\cdots x_6x_2.$$

Now we add again $x_2x_2^{-1} = e$ as follows:

$$x_2(x_2x_2^{-1})x_{n+4}(x_2x_2^{-1})\cdots(x_2x_2^{-1})x_7(x_2x_2^{-1})x_6x_2x_{n+4}\cdots x_6$$
$$= x_{n+4}(x_2x_2^{-1})\cdots(x_2x_2^{-1})x_6x_2x_{n+4}\cdots x_6x_2.$$

We can rewrite the relation as follows:

$$x_2(x_2^{-1}x_{n+4}x_2)x_2^{-1}\cdots x_2(x_2^{-1}x_7x_2)(x_2^{-1}x_6x_2)x_{n+4}\cdots x_6$$
$$= (x_2^{-1}x_{n+4}x_2)x_2^{-1}\cdots x_2(x_2^{-1}x_7x_2)(x_2^{-1}x_6x_2)x_{n+4}\cdots x_6x_2,$$

and using relations (8), we get:

$$x_2(x_2^{-1}x_{n+4}x_2)x_{n+4}\cdots(x_2^{-1}x_7x_2)x_7(x_2^{-1}x_6x_2)x_6$$
$$= (x_2^{-1}x_{n+4}x_2)x_{n+4}\cdots(x_2^{-1}x_7x_2)x_7(x_2^{-1}x_6x_2)x_6x_2.$$

Since by relations (3) we have $(x_2x_i)^2 = (x_ix_2)^2$ for $6 \leq i \leq n+4$, this relation is redundant.

Relation (2) can be simplified by substituting relation (1) and using relations (6) and (11):

$$x_{n+5}\cdot x_{n+4}\cdots x_6\cdot x_2^{-1}\cdot x_6^{-1}\cdots x_{n+4}^{-1}\cdot x_2x_5x_2^{-1}\cdot x_{n+4}\cdots x_6\cdot x_2\cdot x_6^{-1}\cdots x_{n+4}^{-1}\cdot x_{n+5}^{-1}$$
$$= x_5^{-1}\cdot x_{n+6}^{-1}\cdots x_{n+m+4}^{-1}\cdot x_5\cdot x_{n+m+4}\cdots x_{n+6}\cdot x_5.$$

By relations (6) and (7),

$$x_5x_{n+5}x_5x_{n+5}^{-1}x_5^{-1} = x_{n+6}^{-1}\cdots x_{n+m+4}^{-1}\cdot x_5\cdot x_{n+m+4}\cdots x_{n+6}.$$

Using again $(x_5x_{n+5})^2 = (x_{n+5}x_5)^2$, we get:

$$x_{n+5}^{-1}x_5x_{n+5} = x_{n+6}^{-1}\cdots x_{n+m+4}^{-1}\cdot x_5\cdot x_{n+m+4}\cdots x_{n+6},$$

which can be rewritten as:

$$x_{n+m+4}\cdots x_{n+6}\cdot x_5\cdot x_{n+5} = x_{n+5}\cdot x_5\cdot x_{n+m+4}\cdots x_{n+6}. \qquad (*)$$

We will see now that relations (5) are all redundant. We split it into four subcases, according to the value of i:

- If we substitute $i = 5$ in relations (5), we have $[x_1, x_2^{-1}x_5x_2] = e$, which can be simplified to:

$$x_{n+m+4}\cdots x_5x_2x_{n+5}x_5x_{n+5}^{-1}x_5 = x_2^{-1}x_5x_2x_{n+m+4}\cdots x_5x_2x_{n+5}x_5x_{n+5}^{-1}.$$

  By adding $x_5^{-1}x_5 = e$ and using $(x_5x_{n+5})^2 = (x_{n+5}x_5)^2$, we get:

$$x_{n+m+4}\cdots x_{n+6}x_{n+5}x_{n+4}\cdots x_6x_5x_2x_5^{-1}x_{n+5}^{-1}x_5x_{n+5}x_5$$
$$= x_2^{-1}x_5x_2x_{n+m+4}\cdots x_{n+6}x_{n+5}x_{n+4}\cdots x_6x_5x_2x_5^{-1}x_{n+5}^{-1}x_5x_{n+5}.$$

  By relations (6), (7) and (10),

$$x_{n+m+4}\cdots x_{n+6}x_5x_2x_{n+5}x_5x_{n+5}^{-1}x_5^{-1} = x_2^{-1}x_5x_2x_{n+m+4}\cdots x_{n+6}x_5x_2x_5^{-1}.$$

By relations (10), (11) and $(x_2x_5)^2 = (x_5x_2)^2$:

$$x_{n+m+4}\cdots x_{n+6}x_5x_{n+5}x_5x_{n+5}^{-1}x_5^{-1}x_2 = x_5x_2x_{n+m+4}\cdots x_{n+6}.$$

By relations (9) and $(x_5x_{n+5})^2 = (x_{n+5}x_5)^2$,

$$x_{n+5}^{-1}x_5x_{n+5} = x_{n+6}^{-1}\cdots x_{n+m+4}^{-1}x_5x_{n+m+4}\cdots x_{n+6},$$

which is known already (by relation $(*)$).

- For $6 \le i \le n+4$, we have $[x_1, x_2^{-1}x_ix_2] = e$. By some cancellations, we have:

$$x_{n+m+4}\cdots x_5x_2x_{n+5}x_5x_{n+5}^{-1}x_i = x_2^{-1}x_ix_2x_{n+m+4}\cdots x_5x_2x_{n+5}x_5x_{n+5}^{-1}.$$

By relations (6) and (9), $x_{n+4}\cdots x_6x_5x_2x_i = x_2^{-1}x_ix_2x_{n+4}\cdots x_6x_5x_2$. In this relation we substitute first $i = 6$ to get (by relations (8)):

$$x_6x_5x_2x_6 = x_2^{-1}x_6x_2x_6x_5x_2,$$

which is $[x_6, x_2^{-1}x_5x_2] = e$. This relation appears already in relations (7). Now we substitute $i = 7$ to get (again by relations (8)):

$$x_7x_6x_5x_2x_7 = x_2^{-1}x_7x_2x_7x_6x_5x_2,$$

which is $x_6x_5x_2x_7 = x_2x_7x_2^{-1}x_6x_5x_2$. The addition of $x_2x_2^{-1} = e$ in two locations, enables us to translate

$$x_2^{-1}x_6(x_2x_2^{-1})x_5x_2x_7 = x_7x_2^{-1}x_6(x_2x_2^{-1})x_5x_2$$

to $(x_2^{-1}x_6x_2)(x_2^{-1}x_5x_2)x_7 = x_7(x_2^{-1}x_6x_2)(x_2^{-1}x_5x_2)$, which is of course redundant by relations (8). As for the other indices, all relations are redundant in a similar way.

- If $i = n+5$, we have by relations (11): $[x_1, x_{n+5}] = e$. By some immediate cancellations, we get:

$$\begin{aligned} &x_{n+m+4}x_{n+m+3}\cdots x_5x_2 \\ &\quad = x_{n+5}x_{n+m+4}x_{n+m+3}\cdots x_5x_2x_{n+5}x_5x_{n+5}^{-1}x_5^{-1}x_{n+5}^{-1}. \end{aligned}$$

By relations (6) and (11), we have $x_5x_2 = x_{n+5}x_5x_2x_{n+5}x_5x_{n+5}^{-1}x_5^{-1}x_{n+5}^{-1}$. By relations (4) and (10), the relation is redundant.

- For $n+6 \leq i \leq n+m+4$, we have by relations (9): $[x_1, x_i] = e$. Substituting relation (13), we have:

$$x_{n+m+4} \cdots x_5 x_2 x_{n+5} x_5 x_{n+5}^{-1} x_2 x_i = x_i x_{n+m+4} \cdots x_5 x_2 x_{n+5} x_5 x_{n+5}^{-1} x_2,$$

which is (by relations (6), (9) and (11)):

$$x_{n+m+4} \cdots x_{n+6} x_5 x_2 x_{n+5} x_5 x_i = x_i x_{n+m+4} \cdots x_{n+6} x_5 x_2 x_{n+5} x_5.$$

We add $x_5^{-1} x_5 = e$ in two locations as follows:

$$x_{n+m+4} \cdots x_{n+6} x_5 x_2 (x_5^{-1} x_5) x_{n+5} x_5 x_i$$
$$= x_i x_{n+m+4} \cdots x_{n+6} x_5 x_2 (x_5^{-1} x_5) x_{n+5} x_5.$$

This enables us to use relations (10) and to cancel $x_5 x_2 x_5^{-1}$:

$$x_{n+m+4} \cdots x_{n+6} x_5 x_{n+5} x_5 x_i = x_i x_{n+m+4} \cdots x_{n+6} x_5 x_{n+5} x_5.$$

By relation $(*)$, we can rewrite it as:

$$x_{n+5} x_5 x_{n+m+4} \cdots x_{n+6} x_5 x_i = x_i x_{n+5} x_5 x_{n+m+4} \cdots x_{n+6} x_5.$$

By relations (11), we get $[x_5^{-1} x_i x_5, x_{n+m+4} \cdots x_{n+6} x_5^2] = e$.
Now we substitute $i = n+6$:

$$x_5^{-1} x_{n+6} x_5 x_{n+m+4} \cdots x_{n+6} x_5^2 = x_{n+m+4} \cdots x_{n+6} x_5^2 x_5^{-1} x_{n+6} x_5.$$

By relations (12), we get $x_{n+6} x_5 x_{n+6} x_5 = x_5 x_{n+6} x_5 x_{n+6}$, which is known already (by relations (4)).
Now we substitute $i = n+7$:

$$x_5^{-1} x_{n+7} x_5 x_{n+m+4} \cdots x_{n+6} x_5 = x_{n+m+4} \cdots x_{n+6} x_5 x_{n+7},$$

Again, by relations (12), we get:

$$x_5^{-1} x_{n+7} x_5 x_{n+7} x_{n+6} x_5 = x_{n+7} x_{n+6} x_5 x_{n+7},$$

and by $(x_5 x_{n+7})^2 = (x_{n+7} x_5)^2$ (relations (4)), we get:

$$x_5 x_{n+7} x_5^{-1} x_{n+6} x_5 = x_{n+6} x_5 x_{n+7}.$$

Now, since $[x_{n+7}, x_5^{-1} x_{n+6} x_5] = e$ (relations (12)), this relation is redundant too.
In a similar way, the relations $[x_1, x_i] = e$ for $n+6 \leq i \leq n+m+4$, are redundant.

Therefore, the resulting presentation is as follows:
Generators: $\{x_2, x_5, x_6, x_7, \dots, x_{n+4}, x_{n+5}, \dots, x_{n+m+4}\}$.
Relations:

1. $x_{n+5}^{-1} x_5 x_{n+5} = x_{n+6}^{-1} \cdots x_{n+m+4}^{-1} \cdot x_5 \cdot x_{n+m+4} \cdots x_{n+6}$
2. $(x_2 x_i)^2 = (x_i x_2)^2$, where $i = 5, 6, \dots, n+4$
3. $(x_5 x_i)^2 = (x_i x_5)^2$, where $n+5 \le i \le n+m+4$
4. $[x_i, x_j] = e$, where $6 \le i \le n+4$ and $j = 5, n+5, \dots, n+m+4$
5. $[x_i, x_2 x_5 x_2^{-1}]$, where $6 \le i \le n+4$
6. $[x_2^{-1} x_i x_2, x_j] = e$, where $6 \le i < j \le n+4$
7. $[x_2, x_i] = e$, where $n+6 \le i \le n+m+4$
8. $[x_5 x_2 x_5^{-1}, x_i] = e$, where $n+5 \le i \le n+m+4$
9. $[x_{n+5}, x_i] = e$, where $i = 2, n+6, \dots, n+m+4$
10. $[x_5^{-1} x_i x_5, x_j] = e$, where $n+6 \le i < j \le n+m+4$,

as stated in Theorem 1.1. □

Corollary 3.1. *The affine and projective fundamental groups of the real conic-line arrangement $T_{n,m}$ are big.*

Proof. First, observe that if a group has a big quotient, then the group itself is big.

We start by proving that $\langle a, b | (ab)^2 = (ba)^2 = e\rangle$ is big (see also [1]). Let us take new generators $x = ab$, $y = b$, then the relation $(ab)^2 = (ba)^2 = e$ becomes: $x^2 = (yxy^{-1})^2 = e$, which is equal to: $x^2 = yx^2y^{-1} = e$. Hence, we have:

$$\langle a, b \mid (ab)^2 = (ba)^2 = e\rangle \cong \langle x, y \mid x^2 = e\rangle \cong \mathbb{Z} * \mathbb{Z}/2,$$

where $*$ is the free product. Now, the quotient of $G/\langle x^2 = e\rangle$ by the subgroup generated by y^3 is $\mathbb{Z}/2 * \mathbb{Z}/3$, which is known to be big. By the observation, $\langle a, b | (ab)^2 = (ba)^2 = e\rangle$ is big too.

The projective fundamental group of the arrangement $T_{0,0}$ is

$$\langle a, b | (ab)^2 = (ba)^2 = e\rangle \cong \mathbb{Z} * \mathbb{Z}/2,$$

which is big, as we proved above. Since the projective fundamental group is a quotient of the affine fundamental group by the projective relation, the affine fundamental group is big too.

Since $T_{0,0}$ is a sub-arrangement of $T_{n,m}$, and the fundamental groups of the arrangement $T_{0,0}$ are big, we have that the fundamental groups of $T_{n,m}$ are big too. □

Remark 3.2. Note that if we substitute $m = 0$ in the presentation of Theorem 1.1, we get the following simplified presentation for the fundamental group of $T_{n,0}$ (which is the real conic-line arrangement in $\mathbb{CP}^2$ with two conics, which are tangent to each other at two points, and with an arbitrary number of tangent lines to one of the conics):

$$\pi_1(\mathbb{CP}^2 - T_{n,0}) \cong \left\langle x_1, \dots, x_n, x_{n+2} \;\middle|\; \begin{array}{ll} (x_n x_{n+2})^2 = (x_{n+2} x_n)^2 & \\ (x_i x_{n+2})^2 = (x_{n+2} x_i)^2, & 1 \le i \le n-1 \\ [x_i, x_n] = [x_i, x_{n+2} x_n x_{n+2}^{-1}] = e, & 1 \le i \le n-1 \\ [x_i, x_{n+2} x_j x_{n+2}^{-1}] = e, & 1 \le i < j \le n-1 \end{array} \right\rangle.$$

References

[1] M. AMRAM, D. GARBER and M. TEICHER, *Fundamental groups of tangent conic-line arrangements with singularities up to order 6*, Math. Z. **256** (2007), 837–870.

[2] M. AMRAM and S. OGATA, *Degenerations and fundamental groups related to some special toric varieties*, Michigan Math. J. **54** (2006), 587–610.

[3] M. AMRAM and M. TEICHER, *Braid monodromy of special curves*, J. Knot Theory & Ramifications **10** (2) (2001), 171–212.

[4] M. AMRAM and M. TEICHER, *Fundamental groups of some special quadric arrangements*, Rev. Mat. Complut. **19** (2) (2006), 259–276.

[5] M. AMRAM, M. TEICHER and M. ULUDAG, *Fundamental groups of some quadric-line arrangements*, Topology Appl. **130** (2003), 159–173.

[6] E. ARTAL-BARTOLO, *Sur les couples de Zariski*, J. Alg. Geom. **3** (2) (1994), 223–247.

[7] E. ARTAL-BARTOLO and J. CARMONA-RUBER, *Zariski pairs, fundamental groups and Alexander polynomials*, J. Math. Soc. Japan **50** (3) (1998), 521–543.

[8] E. ARTAL-BARTOLO, J.I. COGOLLUDO and H. TOKUNAGA, *A survey on Zariski pairs*, In: “Algebraic Geometry in East Asia, Hanoi 2005”, Adv. Stud. Pure Math. **50**, Math. Soc. Japan, Tokyo (2008), 1–100.

[9] O. CHISINI, *Sulla identità birazionale delle funzioni algebriche di due variabili dotate di una medesima curva di diramazione*, Rend. Ist. Lombardo **77** (1944), 339–356.

[10] A. I. DEGTYAREV, *Quintics in $\mathbb{CP}^2$ with non-abelian fundamental group*, Algebra i Analiz **11** (5) (1999), 130–151 [Russian]; *Translation in:* St. Petersburg Math. J. **11** (5) (2000), 809–826.

[11] M. FRIEDMAN and M. TEICHER, *On non fundamental group equivalent surfaces*, Alg. Geom. Topo. **8** (2008), 397–433.
[12] D. GARBER, *Plane curves and their fundamental groups: Generalizations of Uludag's Construction*, Alg. Geom. Topo. **3** (2003), 593–622.
[13] D. GARBER and M. TEICHER, *The fundamental group's structure of the complement of some configurations of real line arrangements*, In: "Complex Analysis and Algebraic Geometry", T. Peternell and F.-O. Schreyer (eds.), de Gruyter (2000), 173–223.
[14] D. GARBER, M. TEICHER and U. VISHNE, *Classes of wiring diagrams and their invariants*, J. Knot Theory Ramifications **11** (8) (2002), 1165–1191.
[15] F. HIRZEBRUCH, *Some examples of algebraic surfaces*, Contemp. Math. **9** (1982), 55–71.
[16] E. R. VAN KAMPEN, *On the fundamental group of an algebraic curve*, Amer. J. Math. **55** (1933), 255–260.
[17] V. S. KULIKOV, *On Chisini's conjecture*, Izv. Ross. Akad. Nauk Ser. Mat. **63** (6) (1999), 83–116 [Russian]; *Translation in*: Izv. Math. **63** (1999), 1139–1170.
[18] V. S. KULIKOV, *On Chisini's conjecture II*, Izv. Ross. Akad. Nauk Ser. Mat. **72** (5) (2008), 63–76 [Russian]; *English translation*: Izv. Math. **72** (5) (2008), 901–913.
[19] V. S. KULIKOV and M. TEICHER, *Braid monodromy factorizations and diffeomorphism types*, Izv. Ross. Akad. Nauk Ser. Mat. **64** (2) (2000), 89–120 [Russian]; Translation in Izv. Math. **64** (2) (2000), 311–341.
[20] J. MILNOR, *Morse Theory*, Ann. Math. Stud. **51**, Princeton University Press, Princeton, NJ, 1963.
[21] B. MOISHEZON, *Stable branch curves and braid monodromies*, Lect. Notes in Math. **862** (1981), 107–192.
[22] B. MOISHEZON and M. TEICHER, *Braid group techniques in complex geometry I, Line arrangements in* $\mathbb{CP}^2$, Contemp. Math. **78** (1988), 425–555.
[23] B. MOISHEZON and M. TEICHER, *Braid group techniques in complex geometry II, From arrangements of lines and conics to cuspidal curves*, Algebraic Geometry, Lect. Notes in Math. **1479** (1990), 131–180.
[24] B. MOISHEZON and M. TEICHER, *Braid group techniques in complex geometry III, Projective degeneration of* V_3, Contemp. Math. **162** (1994), 313–332.
[25] B. MOISHEZON and M. TEICHER, *Braid group techniques in complex geometry IV, Braid monodromy of the branch curve* S_3 *of*

$V_3 \to \mathbb{CP}^2$ *and application to* $\pi_1(\mathbb{CP}^2 - S_3, *)$, Contemp. Math. **162** (1994), 332–358.

[26] M. OKA, *Two transforms of plane curves and their fundamental groups*, J. Math. Sci. Univ. Tokyo **3** (2) (1996), 399–443.

[27] I. SHIMADA, *A note on Zariski pairs*, Compositio Math. **104** (2) (1996), 125–133.

[28] H. TOKUNAGA, *Some examples of Zariski pairs arising from certain elliptic $K3$ surfaces*, Math. Z. **227** (3) (1998), 465–477.

[29] A. M. ULUDAĞ, *More Zariski pairs and finite fundamental group of curve complements*, Manu. Math. **106** (3) (2001), 271–277.

[30] O. ZARISKI, *On the problem of existence of algebraic functions of two variables possessing a given branch curve*, Amer. J. Math. **51** (1929), 305–328.

[31] O. ZARISKI, *The topological discriminant group of a Riemann surface of genus p*, Amer. J. Math. **59** (1937), 335–358.

Intersection cohomology of a rank one local system on the complement of a hyperplane-like divisor

Dimitry Arinkin[†] and Alexander Varchenko[◇]

Abstract. Under a certain condition A we give a construction to calculate the intersection cohomology of a rank one local system on the complement to a hyperplane-like divisor.

Let X be a smooth connected complex manifold and D a divisor. The divisor D is *hyperplane-like* if X can be covered by coordinate charts such that in each chart D is the union of hyperplanes. Such charts are called *linearizing*.

Let D be a hyperplane-like divisor, V a linearizing chart. A *local edge* of D in V is any nonempty irreducible intersection in V of hyperplanes of D in V. A local edge is *dense* if the subarrangement of all hyperplanes in V containing the edge is irreducible: the hyperplanes cannot be partitioned into nonempty sets so that, after a change of coordinates, hyperplanes in different sets are in different coordinates. An *edge* of D is the maximal analytic continuation in X of a local edge. Any edge is an immersed submanifold in X. An edge is called dense if it is locally dense.

Let $U = X - D$ be the complement to D. Let $\mathcal{L}$ be a rank one local system on U with nontrivial monodromy around each irreducible component of D. We want to calculate the intersection cohomology $H^*(X; j_{!*}\mathcal{L})$ where $j : U \to X$ is the embedding. In this calculation a role will be played by the dual local system $\mathcal{L}^\vee$ on U with the inverse monodromy.

Let $\pi : \tilde{X} \to X$ be a resolution of singularities of D in X.

Remark. Note that among the resolutions of singularities there is the minimal one. The minimal resolution is constructed by first blowing-up

[†] Supported in part by the Sloan Fellowship.
[◇] Supported in part by NSF grant DMS-1101508

dense vertices of D, then by blowing-up the proper preimages of dense one-dimensional edges of D and so on, see [4,5].

The preimage $\pi^{-1}(D)$ is a hyperplane-like divisor in $\tilde{X}$. The inverse image $\pi^*\mathcal{L}$ is a rank one local system on the complement $\tilde{X} - \pi^{-1}(D)$. However, its monodromy around some components of $\pi^{-1}(D)$ may be trivial. Denote by $\tilde{D} \subset \pi^{-1}(D) \subset \tilde{X}$ the maximal divisor in $\tilde{X}$ such that $\pi^*\mathcal{L}$ has nontrivial monodromy around $\tilde{D}$. The local systems $\pi^*\mathcal{L}$ and $\pi^*\mathcal{L}^\vee$ extend to local systems on $\tilde{U} = \tilde{X} - \tilde{D}$ denoted by $\tilde{\mathcal{L}}$ and $\tilde{\mathcal{L}}^\vee$, respectively. For $x \in X$, denote $\tilde{U}_x = \pi^{-1}(x) \cap \tilde{U}$.

Definition 1. We say that the local system $\mathcal{L}$ on U satisfies *condition A* with respect to the resolution of singularities $\pi : \tilde{X} \to X$, if for any edge F of D and any point x of F that is not contained in any smaller edge, we have $H^\ell(\tilde{U}_x; \tilde{\mathcal{L}}|_{\tilde{U}_x}) = 0$ for $\ell > \operatorname{codim} F - 1$. Similarly, we say that the local system $\mathcal{L}^\vee$ on U satisfies *condition A* with respect to the resolution of singularities $\pi : \tilde{X} \to X$, if for any edge F of D and any point $x \in F$ that is not contained in any smaller edge, we have $H^\ell(\tilde{U}_x; \tilde{\mathcal{L}}^\vee|_{\tilde{U}_x}) = 0$ for $\ell > \operatorname{codim} F - 1$.

It is not hard to check that to verify condition A, it suffices to consider only dense edges F.

Notice that if the monodromy of $\mathcal{L}$ lies in $\{z \in \mathbb{C}^\times \mid |z| = 1\}$ (in other words, $\mathcal{L}$ is a unitary local system), then $\mathcal{L}$ satisfies condition A with respect to π if and only if $\mathcal{L}^\vee$ satisfies condition A with respect to π.

Theorem 1. *If both local systems $\mathcal{L}$ and $\mathcal{L}^\vee$ on the complement U to the hyperplane-like divisor D in X satisfy condition A with respect to a resolution of singularities $\pi : \tilde{X} \to X$, then the intersection cohomology $H^*(X; j_{!*}\mathcal{L})$ is naturally isomorphic to $H^*(\tilde{U}; \tilde{\mathcal{L}})$.*

Proof. Let $\tilde{j} : \tilde{U} \to \tilde{X}$ be the open embedding. Since $\tilde{\mathcal{L}}$ has non-trivial monodromy around every component of $\tilde{D}$ and $\tilde{D}$ has normal crossings, we have $\tilde{j}_*\tilde{\mathcal{L}} = \tilde{j}_!\tilde{\mathcal{L}} = \tilde{j}_{!*}\tilde{\mathcal{L}}$ as well as $\tilde{j}_*\tilde{\mathcal{L}}^\vee = \tilde{j}_!\tilde{\mathcal{L}}^\vee = \tilde{j}_{!*}\tilde{\mathcal{L}}^\vee$. We also have $H^*(X; \pi_*\tilde{j}_*\tilde{\mathcal{L}}) = H^*(\tilde{U}; \tilde{\mathcal{L}})$ and $H^*(X; \pi_*\tilde{j}_*\tilde{\mathcal{L}}^\vee) = H^*(\tilde{U}; \tilde{\mathcal{L}}^\vee)$. To simplify notation, we write π_* for the direct image in the derived category $R\pi_*$. Thus, $\pi_*\tilde{j}_*\tilde{\mathcal{L}}$ is a complex of sheaves (more precisely, an object of the corresponding derived category), and $H^*(X; \pi_*\tilde{j}_*\tilde{\mathcal{L}})$ is its hypercohomology. □

The restriction of $\pi_*\tilde{j}_*\tilde{\mathcal{L}}$ to U is $\mathcal{L}$. The theorem follows from the next lemma.

Lemma 1. *Let $\pi : \tilde{X} \to X$ be a proper holomorphic map of complex manifolds. For $x \in X$, denote $\tilde{i}_x : \pi^{-1}(x) \to \tilde{X}$ the embedding of* *the fiber. Let $\tilde{\mathcal{F}}$ be a complex of sheaves with constructible cohomology sheaves on $\tilde{X}$. (In most examples, $\tilde{\mathcal{F}}$ is a perverse sheaf, see [1] for the definition.) Set $\tilde{\mathcal{F}}^{\vee} = \mathbb{D}D\tilde{\mathcal{F}}$, where $\mathbb{D}D$ is the Verdier duality functor.*

Suppose $\tilde{\mathcal{F}}$ satisfies the following condition B: for every $\ell > 0$ there is an analytic subset $X_\ell \subset X$, $\operatorname{codim} X_\ell = \ell + 1$, such that for any $x \in X - X_\ell$ we have $H^i(\pi^{-1}(x); \tilde{i}_x^* \tilde{\mathcal{F}}) = 0$ and $H^i(\pi^{-1}(x); \tilde{i}_x^* \tilde{\mathcal{F}}^{\vee}) = 0$ for all $i \geq \ell$ or $i < 0$. Then $\pi_* \tilde{\mathcal{F}} = \pi_! \tilde{\mathcal{F}}$ is a perverse IC-sheaf on X.

Proof. This lemma is a slight generalization of the results of Goresky and MacPherson ([2, Section 6.2]) about small maps. The argument of [2] applies without change. Indeed, by [2, Second theorem of Section 6.1], we need to verify that the complex $\pi_* \tilde{\mathcal{F}}$ is such that $H^i(\pi_* \tilde{\mathcal{F}}) = 0$ for $i < 0$ and $\operatorname{codim} \operatorname{supp}(H^i(\pi_* \tilde{\mathcal{F}})) > i$ for $i > 0$, and that the same is true for the dual complex $\mathbb{D}D\pi_* \tilde{\mathcal{F}} = \pi_* \tilde{\mathcal{F}}^{\vee}$. But these conditions are clear by base change. (Note that unlike [2], we use the non-self-dual normalization: for instance, a local system on a smooth manifold is an IC-sheaf in our convention, but it requires cohomological shift in the self-dual normalization.)

In particular, in the settings of the theorem, the lemma applies to $\mathcal{F} = \tilde{j}_* \tilde{\mathcal{L}}$, because both $\mathcal{L}$ and $\mathcal{L}^{\vee}$ satisfy condition A. This concludes the proof of the theorem. □

Corollary. *If the local systems $\mathcal{L}$ and $\mathcal{L}^{\vee}$ satisfy condition A with respect to each of two resolutions of singularities $\pi : \tilde{X} \to X$ and $\pi' : \tilde{X}' \to X$, then $H^*(\tilde{U}; \tilde{\mathcal{L}})$ and $H^*(\tilde{U}'; \tilde{\mathcal{L}}')$ are canonically isomorphic.*

Example 1. Let X be the projective space of dimension k. Let $D \subset X$ be the union of hyperplanes and $\mathcal{L}$ a rank one local system on $U = X - D$ with nontrivial monodromy around each hyperplane. Assume that D has normal intersections at all edges except at vertices. Then both $\mathcal{L}$ and $\mathcal{L}^{\vee}$ satisfy condition A with respect to the minimal resolution of singularities. Indeed, to obtain $\tilde{X}$ one has to blow-up dense vertices of D. If $x \in D$ is a dense vertex, then $\tilde{U}_x$ is nonempty only if $\mathcal{L}$ has trivial monodromy across x. In that case $\tilde{U}_x$ is an affine variety of dimension $k - 1$ and $H^\ell(\tilde{U}_x; \tilde{\mathcal{L}}|_{\tilde{U}_x}) = 0$, $H^\ell(\tilde{U}_x; \tilde{\mathcal{L}}^{\vee}|_{\tilde{U}_x}) = 0$ for $\ell \geq k$.

Example 2. Let X be $\mathbb{C}^3$. Let C be the central arrangement of six planes

$$H_1 : x_1 - x_3 = 0, \qquad H_2 : x_1 + x_3 = 0,$$
$$H_3 : x_2 - x_3 = 0, \qquad H_4 : x_2 + x_3 = 0,$$
$$H_5 : x_1 - x_2 = 0, \qquad H_6 : x_1 + x_2 = 0$$

with weights $a_1 = a_2, \quad a_3 = a_4, \quad a_5 = a_6, \quad a_1 + a_3 + a_5 = 0$. Let $\mathcal{L}$ be the local system on U with the monodromy $e^{2\pi i a_j} \neq 1$ around H_j. Let $\pi : \tilde{X} \to X$ be the minimal resolution of singularities. For $x = (0, 0, 0)$, the space $\tilde{U}_x$ is the projective plane with four blown-up points and six lines removed. We have $\dim H^3(\tilde{U}_x, \tilde{\mathcal{L}}) = 1$. This weighted arrangement $\mathcal{C}$ does not satisfies condition A with respect to the minimal resolution of singularities.

Remark. If $D \subset \mathbb{C}^k$ is the union of hyperplanes of a central arrangement and the monodromy of $\mathcal{L}$ is close to 1, then the intersection cohomology $H^*(X; j_{!*}\mathcal{L})$ was computed in [3] as the cohomology of the complex of flag forms of the arrangement.

The following equivariant version of Theorem 1 holds. Let G be a finite group, ρ an irreducible representation of G. For a representation M denote by $M^\rho \subset M$ the ρ-isotypical component.

Let $\pi : \tilde{X} \to X$ be a resolution of singularities of D as before. Assume that G acts on X and $\tilde{X}$ so that the actions preserve $U, \mathcal{L}, \tilde{U}, \tilde{\mathcal{L}}$ and commute with the map π. Then G acts on $H^*(X; j_{!*}\mathcal{L})$ and $H^*(\tilde{U}; \tilde{\mathcal{L}})$.

For $x \in X$, we denote by O_x the G-orbit of x.

Definition 2. We say that the local system $\mathcal{L}$ satisfies *condition A* with respect to ρ and a resolution of singularities π if for any edge F of D and any point $x \in F$ that is not contained in any smaller edge, we have $H^\ell(\cup_{y \in O_x} \tilde{U}_y; \tilde{\mathcal{L}}|_{\cup_{y \in O_x} \tilde{U}_y})^\rho = 0$ for $\ell > \operatorname{codim} F - 1$. Similarly, we say that the local system $\mathcal{L}^\vee$ on U satisfies *condition A* with respect to ρ and a resolution of singularities π if for any edge F of D and any point $x \in F$ that is not contained in any smaller edge, we have $H^\ell(\cup_{y \in O_x} \tilde{U}_y; \tilde{\mathcal{L}}^\vee|_{\cup_{y \in O_x} \tilde{U}_y})^\rho = 0$ for $\ell > \operatorname{codim} F - 1$.

Generally speaking, we can no longer consider dense edges only in the equivariant version of condition A. In principle, in the equivariant case, for each edge F, it suffices to check only the generic points $x \in F$, even though the stabilizer might be different for other points.

Theorem 2. *If both local systems $\mathcal{L}$ and $\mathcal{L}^\vee$ on U satisfy condition A with respect ρ and a resolution of singularities π, then the intersection cohomology $H^*(X; j_{!*}\mathcal{L})^\rho$ is naturally isomorphic to $H^*(\tilde{U}; \tilde{\mathcal{L}})^\rho$* as G-modules.

Proof. Consider the quotient X/G, which may be singular. The quotient map $q : X \to X/G$ is finite; therefore, the derived direct image $q_*(j_{!*}\mathcal{L})$ is a perverse IC-sheaf on X/G. The direct image carries a fiber-wise action of G, so we can take the ρ-isotypical component, which is a direct summand $q_*(j_{!*}\mathcal{L})^\rho \subset q_*(j_{!*}\mathcal{L})$. Thus $q_*(j_{!*}\mathcal{L})^\rho$ is itself an IC-sheaf.

Similarly, there is a direct summand $(q_*\pi_*\tilde{j}_*\tilde{\mathcal{L}})^\rho \subset q_*\pi_*\tilde{j}_*\tilde{\mathcal{L}}$. By the argument used in the proof of Theorem 1, $(q_*\pi_*\tilde{j}_*\tilde{\mathcal{L}})^\rho$ is a perverse IC-sheaf. It is thus identified with $q_*(j_{!*}\mathcal{L})^\rho$. □

ACKNOWLEDGEMENTS. The authors thank A. Levin and M. Falk for helping develop Example 2.

References

[1] A. BEILINSON, J. BERNSTEIN and P. DELIGNE, *Faisceaux pervers*, Astérisque **100**, Paris, Soc. Math. Fr. 1982.
[2] M. GORESKY and R. MACPHERSON, *Intersection homology. II*, Invent. Math. **72** (1983), no. 1, 77–129.
[3] S. KHOROSHKIN and A. VARCHENKO, *Quiver D-modules and homology of local systems over an arrangement of hyperplanes*, IMRP, 2006, doi: 10.1155.
[4] V. SCHECHTMAN, H. TERAO and A. VARCHENKO, *Local systems over complements of hyperplanes and the Kac-Kazhdan conditions for singular vectors*, J. Pure Appl. Algebra **100** (1995), 93–102.
[5] A. VARCHENKO, "Multidimensional Hypergeometric Functions and Representation Theory of Lie Algebras and Quantum Groups", Advances in Math. Phys., Vol. 21, World Scientific, 1995.

A survey of some recent results concerning polyhedral products

Anthony P. Bahri, Martin Bendersky, Frederick R. Cohen
and Samuel Gitler

Abstract. The purpose of this article is an exposition of recent results and applications arising from decompositions of suspensions of generalized moment-angle complexes also known as polyhedral products. Special examples of these spaces are complements of arrangements of complex coordinate planes as described in detail below.

The main features of this expository article are (1) a description of these decompositions together with (2) applications to the structure of the cohomology rings of these spaces. Definitions, examples, as well as one proof for a classical decomposition are included in this report.

1 Introduction

The main goal of this paper is to give an exposition of decompositions of suspensions of moment-angle complexes together with their generalizations also known as the polyhedral product functor [3,4,12,13,21,22,30,39].

The topological spaces now known as moment-angle complexes have origins dating back to work of Poincaré [44] in which linear differential equations over the real numbers were considered. C. L. Siegel [47,48] addressed similar constructions. Camacho, Kuiper, and Palis [15] considered analogous differential equations over the complex numbers in which certain polyhedral products arise as invariants. This setting was explained in beautiful, deep work of Santiago Lôpez de Medrano [39] who gave methods to analyze topological invariants of these spaces. H. M. S. Coxeter also developed some basic examples in 1938 [16] as explained by A. Suciu.

The first author was partially supported by the award of a research leave from Rider University. The third author was partially supported by DARPA grant number 2006-06918-01. The fourth author was partially supported by the Department of Mathematics at Princeton University.

Davis-Januszkiewicz introduced manifolds which are now known as moment-angle manifolds over a polytope [18]. That work led to striking developments in toric topology. Buchstaber-Panov introduced and extensively studied moment-angle complexes for any abstract simplicial complex K [12]. They completely described the rational cohomology ring structure in terms of the Tor-algebra of the Stanley-Reisner algebra [12]. Based on the classical moment map within mathematical physics, Buchstaber-Panov introduced the terminology of "moment-angle complex". These two independent works led to many interesting consequences.

On the other-hand, these spaces were the subject of important, independent developments in different directions of combinatorics, geometry, and topology. A small sample includes work of Goresky-MacPherson [30], Ziegler-Zivaljević, [56], and M. Hochster [32].

The moment-angle complexes denoted $Z(K; (D^2, S^1))$ here are the subject of substantial, elegant work of Denham-Suciu [22] as well as Notbohm-Ray [41], Grbiĉ-Theriault [31], Strickland [51], Baskakov [6, 7], Buchstaber-Panov [12], Panov [42], Baskakov-Buchstaber-Panov [8], Buchstaber-Panov-Ray [14], Franz [26,27], Panov-Ray-Vogt [43], Swartz [52] as well as Wang-Zheng [54]. Kamiyama-Tsukuda [37] considered similar constructions in their work on "arachnoid mechanisms".

Fundamental work on algebraic/geometric aspects of this subject is given in Danilov [17], De Concini-Procesi [20], Deligne-Goresky-MacPherson [21], Jewell-Orlik-Shapiro [35, 36], and de Longueville-Schultz [38]. Work at the interface with toric topology was given in Bosio-Meersseman [11] and followed by developments in Lôpez de Medrano-Gitler [29].

An earlier example of related decompositions of suspensions is given by C. Shaper [46]; a much earlier case is given by the feature that ordered configuration spaces of points in any Euclidean space admit stable decompositions as wedges of spheres. Elegant later work via homotopy theoretic techniques was given in Dobrinskaya [23], Félix-Tanré [24], as well as Berglund [9]. Currently, polyhedral products are the subject of broad investigation by A. Suciu, T. Holm, B. Matschke, T. Panov-V. Buchstaber, and S. Choi as well as many others.

The point of view of this paper is that the suspension of a polyhedral product admits further properties in the sense that the polyhedral product usually decomposes into smaller, simpler spaces known as bouquets. These decompositions in turn impact the cohomology algebras of polyhedral products for any cohomology theory. For example, the real K-theory of the Davis-Januszkiewicz spaces was worked out in [2] using these decompositions.

One reason that these decompositions inform on the product structure in the cohomology algebra is that this ring structure is induced by the diagonal map $\Delta : Z \to Z \times Z$ while the diagonal map for polyhedral products admits a finer decomposition after suspension in terms of the summands by the results stated below. Namely, the diagonal after suspension can be factored in terms of the decompositions maps of [5]. These features then provide a setting for analysis of the cup-product structure from knowledge of the cohomology for the individual summands of these decompositions together with their cup-product structure.

The purpose of this paper is to describe some of these results, how they fit together as well as to record basic questions which interweave these topics.

ACKNOWLEDGEMENTS. The authors would like to thank the referee for numerous excellent suggestions devoted to improving this paper. Time constraints force the authors to not take advantage of some of the referee's insightful suggestions. The authors would like to thank the Centro di Ricerca Matematica Ennio De Giorgi for providing a wonderful environment to learn and to develop some mathematics. The first, third, and fourth author especially thank Mario Salvetti, Filippo Callegaro, and Corrado de Concini. The authors would also like to thank the Institute for Advanced Study for also providing a wonderful environment to develop this mathematics.

2 Definitions

This section is a catalogue of definitions for the main results stated below. The main constructions in this article are defined in this section. First recall the definitions of an abstract simplicial complex, and a polyhedral product.

Definition 2.1. 1. Let K denote an abstract simplicial complex with m vertices labeled by the set $[m] = \{1, 2, \cdots, m\}$. Thus, K is a subset of the power set of $[m]$ such that an element given by a $(k-1)$-simplex σ of K is given by an ordered sequence $\sigma = (i_1, \cdots, i_k)$ with $1 \leq i_1 < \cdots < i_k \leq m$ such that if $\tau \subset \sigma$, then τ is a simplex of K. In particular the empty set $\emptyset$ is a subset of σ and so it is in K. The set $[m]$ is minimal in the sense that every $i \in [m]$ belongs to at least one simplex of K. The length k of σ is denoted $|\sigma|$.

2. Let $\Delta[m-1]$ denote the abstract simplicial complex given by the power set of $[m] = \{1, 2, \ldots, m\}$. Let $\Delta[m-1]_q$ denote the subset of the power set of $[m]$ given by all subsets of cardinality at most $q+1$. Thus $\Delta[m-1]_q$ is the q-skeleton of $\Delta[m-1]$.

3. A polyhedron is defined to be the geometric realization of a simplicial complex.

Let $(\underline{X}, \underline{A})$ denote the collection of spaces $\{(X_i, A_i, x_i)\}_{i=1}^m$ of the homotopy type of CW-complexes.

Definition 2.2. 1. The *generalized moment-angle complex or polyhedral product* determined by $(\underline{X}, \underline{A})$ and K denoted

$$Z(K; (\underline{X}, \underline{A}))$$

is defined using the functor

$$D : K \to CW_*$$

as follows: For every σ in K, let

$$D(\sigma) = \prod_{i=1}^{m} Y_i, \quad \text{where} \quad Y_i = \begin{cases} X_i & \text{if } i \in \sigma \\ A_i & \text{if } i \in [m] - \sigma. \end{cases}$$

with $D(\emptyset) = A_1 \times \ldots \times A_m$.

2. The *polyhedral product or generalized moment-angle complex* is

$$Z(K; (\underline{X}, \underline{A})) = \bigcup_{\sigma \in K} D(\sigma) = \text{colim} D(\sigma)$$

where the colimit is defined by the inclusions, $d_{\sigma,\tau} : D(\sigma) \to D(\tau)$ with $\sigma \subset \tau$ and $D(\sigma)$ is topologized as a subspace of the product $X_1 \times \ldots \times X_m$. Thus the *polyhedral product* is the underlying space $Z(K; (\underline{X}, \underline{A}))$ with base-point $\underline{*} = (x_1, \ldots, x_m) \in Z(K; (\underline{X}, \underline{A}))$.

3. Note that the definition of $Z(K; (\underline{X}, \underline{A}))$ did not require spaces to be either based or CW-complexes. In the special case where $X_i = X$ and $A_i = A$ for all $1 \leq i \leq m$, it is convenient to denote the polyhedral product by $Z(K; (X, A))$ to coincide with the notation in [22].

A direct variation of the structure of the polyhedral product follows next. Products of pointed spaces admit natural quotients called smash products. Spaces analogous to polyhedral products are given next where products of spaces are replaced by smash products, a setting in which base-points are required.

Definition 2.3. The smash product

$$X_1 \wedge X_2 \wedge \ldots \wedge X_m$$

is given by the quotient space $(X_1 \times \ldots \times X_m)/S(X_1 \times \ldots \times X_m)$ where $S(X_1 \times \ldots \times X_m)$ is the subspace of the product with at least one coordinate given by the base-point $x_j \in X_j$.

The (reduced) suspension of a (pointed) space $(X, *)$

$$\Sigma(X)$$

is the smash product

$$S^1 \wedge X.$$

Definition 2.4. Given a polyhedral product $Z(K; (\underline{X}, \underline{A}))$ obtained from $(\underline{X}, \underline{A}, \underline{*})$, the *smash polyhedral product*

$$\widehat{Z}(K; (\underline{X}, \underline{A}))$$

is defined to be the image of $Z(K; (\underline{X}, \underline{A}))$ in the smash product $X_1 \wedge X_2 \wedge \ldots \wedge X_m$.

The image of $D(\sigma)$ in $\widehat{Z}(K; (\underline{X}, \underline{A}))$ is denoted by $\widehat{D}(\sigma)$ and is

$$Y_1 \wedge Y_2 \wedge \ldots \wedge Y_m$$

where

$$Y_i = \begin{cases} X_i & \text{if } i \in \sigma \\ A_i & \text{if } i \in [m] - \sigma. \end{cases}$$

As in the case of $Z(K; (\underline{X}, \underline{A}))$, note that $\widehat{Z}(K; (\underline{X}, \underline{A}))$ is the colimit obtained from the spaces $\widehat{D}(\sigma)$ with $\widehat{D}(\sigma) \cap \widehat{D}(\tau) = \widehat{D}(\sigma \cap \tau)$.

Next, make further assumptions to be used later.

Definition 2.5. The pair $(\underline{X}, \underline{A})$ is said to be **wedge decomposable** provided

$$(\underline{X}, \underline{A}) = (\underline{B \vee C}, \underline{B \vee E}) = \{B_i \vee C_i, B_i \vee E_i, x_i)\}_{i=1}^{m}$$

are pointed triples of CW-complexes for all i for which E_i a pointed subspace of C_i such that the natural inclusion $E_i \subset C_i$ is null-homotopic.

For fixed

$$I = (i_1, \ldots, i_k) = \sigma \in K, \ \sigma \neq \emptyset.$$

define

$$W(\sigma; \underline{B \vee C}, \underline{B \vee E})) = Y_1 \wedge \cdots \wedge Y_m, \tag{2.1}$$

where

$$Y_j = \begin{cases} C_j & \text{if } \ j \in \sigma, \\ B_j \vee E_j & \text{if } \ j \in [m] - \sigma. \end{cases} \tag{2.2}$$

The point of this last definition is to provide an easily analyzed setting for further decompositions of smash polyhedral products. The next example provides a classical, elementary case of a polyhedral product.

Example 2.6. Let K denote the simplicial complex given by $\{\{1\},\{2\},\emptyset\}$ with

$$(X_i, A_i) = (D^n, S^{n-1}) \quad \text{for} \quad i = 1, 2$$

where D^n is the n-disk and S^{n-1} is its boundary sphere. Then

$$Z(K; (D^n, S^{n-1})) = D^n \times S^{n-1} \cup S^{n-1} \times D^n$$

with the boundary of $D^n \times D^n$ given by $\partial(D^n \times D^n) = D^n \times S^{n-1} \cup S^{n-1} \times D^n$ thus $Z(K; (D^n, S^{n-1})) = S^{2n-1}$.

Definition 2.7. Consider an ordered sequence $I = (i_1, \cdots, i_k)$ with $1 \leq i_1 < cldots < i_k \leq m$ together with pointed spaces $Y_1, \cdots, Y_m$. Then

1. the length of I is $|I| = k$,
2. the notation $I \subseteq [m]$ means I is any increasing subsequence of $(1, \cdots, m)$,
3. $Y^{[m]} = Y_1 \times \cdots \times Y_m$,
4. $Y^I = Y_{i_1} \times Y_{i_2} \times \cdots \times Y_{i_k}$,
5. $\widehat{Y}^I = Y_{i_1} \wedge \cdots \wedge Y_{i_k}$,

Two more conventions are listed next.

1. The symbol $X * Y$ denotes the join of two topological spaces X and Y. If X and Y are pointed spaces of the homotopy type of a CW-complex, then $X * Y$ has the homotopy type of the suspension $\Sigma(X \wedge Y)$.
2. Given the family of pairs $(\underline{X}, \underline{A}) = \{(X_i, A_i)\}_{i=1}^m$ and $I \subseteq [m]$, define

$$(\underline{X_I}, \underline{A_I}) = \{(X_{i_j}, A_{i_j})\}_{j=1}^{j=|I|}$$

which is the subfamily of $(\underline{X}, \underline{A})$ determined by I.

Standard constructions for simplicial complexes and associated posets are recalled next.

Definition 2.8. Let K denote a simplicial complex with m vertices.

1. Recall that the empty simplex $\emptyset$ is required to be in K.
2. Given a sequence $I = (i_1, \cdots, i_k)$ with $1 \leq i_1 < \cdots < i_k \leq m$, define $K_I \subseteq K$ to be the *full sub-complex* of K consisting of all simplices of K which have all of their vertices in I, that is $K_I = \{\sigma \cap I \mid \sigma \in K\}$.
3. Let $|K|$ denote the geometric realization of the simplicial complex K.

4. Associated to a simplicial complex K, there is a partially ordered set (poset) $\overline{K}$ given as follows. A point σ in $\overline{K}$ corresponds to a simplex $\sigma \in K$ with order given by *reverse* inclusion of simplices. Thus $\sigma_1 \leq \sigma_2$ in $\overline{K}$ if and only if $\sigma_2 \subseteq \sigma_1$ in K. The empty simplex $\emptyset$ is the unique maximal element of $\overline{K}$.
   Let P be a poset with $p \in P$. There are further posets given by

$$P_{\leq p} = \{q \in P \mid q \leq p\}$$

   as well as

$$P_{<p} = \{q \in P \mid q < p\}.$$

   Thus

$$\overline{K}_{<\sigma} = \{\tau \in \overline{K} \mid \tau < \sigma\} = \{\tau \in K \mid \tau \supset \sigma,\ \tau \neq \sigma\}.$$

On the other hand, given a poset P, there is an associated simplicial complex $\Delta(P)$ called the order complex of P which is defined as follows.

Definition 2.9. Given a poset P, the *order complex* $\Delta(P)$ is the simplicial complex with vertices given by the set of points of P and k-simplices given by the ordered $(k+1)$-tuples $(p_1, p_2, \ldots, p_{k+1})$ in P with $p_1 < p_2 < \ldots < p_{k+1}$. It follows that $\Delta(\overline{K}) = \mathrm{cone}(K')$ where K' denotes the barycentric subdivision of K.

In what follows, it is useful to define variations for a fixed, ambient I where the analogue of $D(\sigma)$ is replaced as follows.

Definition 2.10. Let K be an abstract simplicial complex with m vertices, and $(\underline{X}, \underline{A})$ pairs of pointed topological spaces.

For fixed $I = (i_1, \ldots, i_k)$, and every $\sigma \in K$, define

$$Y^I(\sigma \cap I) = Y_{i_1} \times \cdots \times Y_{i_k},$$

and

$$\widehat{Y}^I(\sigma \cap I) = Y_{i_1} \wedge \cdots \wedge Y_{i_k},$$

where

$$Y_j = \begin{cases} X_j & \text{if } j \in \sigma \cap I, \text{ and} \\ A_j & \text{if } j \in I - \sigma \cap I. \end{cases}$$

Furthermore,

$$\begin{aligned} Y^I(\Phi) &= A^I = A_{i_1} \times \cdots \times A_{i_k} \\ \widehat{Y}^I(\Phi) &= \widehat{A}^I = A_{i_1} \wedge \cdots \wedge A_{i_k}. \end{aligned}$$

Then the generalized moment-angle complexes are

$$Z(K_I; (\underline{X}, \underline{A})_I) = \bigcup_{\sigma \in K} Y^I(\sigma \cap I)$$

and

$$\widehat{Z}(K_I; (\underline{X}, \underline{A})_I) = \bigcup_{\sigma \in K} \widehat{Y}^I(\sigma \cap I).$$

Note: The notation $Z(K_I; (\underline{X_I}, \underline{A_I}))$ was used in [3] for $Z(K_I; (\underline{X}, \underline{A})_I)$.
To simplify notation below, the following notation

$$Z(K),\ \widehat{Z}(K),\ Z(K_I) \text{ and } \widehat{Z}(K_I)$$

is used to denote $Z(K; (\underline{X}, \underline{A}))$, $\widehat{Z}(K; (\underline{X}, \underline{A}))$, $Z(K_I; (\underline{X}, \underline{A})_I)$, and $\widehat{Z}(K_I; (\underline{X}, \underline{A})_I)$ respectively.

3 Four examples

Four additional important examples of polyhedral products are listed next. The first arises in considering complements of coordinate planes, the second arises from related $K(\pi, 1)$'s while the third and fourth arise in connection with the variety of commuting q-tuples in a group G.

Example 3.1. The first example is the space $Z(K; (D^2, S^1))$ which has the homotopy type of the complement of unions of certain coordinate planes in $\mathbb{C}^m$ corresponding to 'coordinate subspace arrangements' as developed in work of Buchstaber-Panov [12, 13].

1. Given a simplicial complex K with m vertices, and a simplex $\omega \in \Delta[m-1]$, define

$$L_\omega = \{(z_1, \cdots, z_m) \in \mathbb{C}^m |\ z_{i_1} = \cdots = z_{i_k} = 0\}$$

   for $(i_1, \cdots, i_k) \in \omega$.
2. Let

$$U(K) = \mathbb{C}^m - \bigcup_{\omega \notin K} L_\omega$$

3. The following is proven in [12, 13]. The natural inclusion

$$Z(K; (D^2, S^1)) \to U(K)$$

   is a homotopy equivalence (more precisely, a strong deformation retract), and $(S^1)^m$-equivariant. As suggested by the referee, a proof also can be given using the standard strong deformation retraction of $\mathbb{C}$ to D^2 together with the features in [22] top of page 31.

Example 3.2. A second example is given by $Z(K; (\underline{X}, \underline{A}))$ where

$$(\underline{X}, \underline{A}) = (\underline{BG}, \underline{*})$$

for $X_i = BG_i$ with $A_i = \{*\}$ where each G_i is a finite, discrete group. In case K is the 0-skeleton of the $(m-1)$-simplex, denoted $K(0)$, the space $Z(K(0); (\underline{X}, \underline{A}))$ is a $K(\pi, 1)$ where π is the free product of the groups G_i.

Recall that $Z(K; (\underline{BG}, \underline{*}))$ is homotopy equivalent to the Borel construction

$$E(G_1 \times \cdots \times G_m) \times_G Z(K; (\underline{EG}, \underline{G}))$$

thus there is a fibration

$$Z(K; (\underline{BG}, \underline{*})) \to BG_1 \times \cdots \times BG_m$$

with homotopy fibre $Z(K; (\underline{EG}, \underline{G}))$ where

$$G = G_1 \times \cdots \times G_m.$$

The last statement is basically Lemma 2.3.2 from Denham-Suciu [22].

In case K is a flag complex, $Z(K; (\underline{BG}, \underline{*}))$ is a $K(\pi, 1)$ as proven in work of Davis-Okun [19]. The monodromy representation for the natural fibration

$$Z(K; (\underline{BG}, \underline{*})) \to BG_1 \times \cdots \times BG_m$$

with homotopy fibre $Z(K; (\underline{EG}, \underline{G}))$ appears in work of Ali Al-Raisi and Mentor Stafa (in preparation) and arises in the context described next. If K has a minimal non-face of dimension at least 2, it is an exercise to see that $Z(K; (\underline{BG}, \underline{*}))$ is not a $K(\pi, 1)$.

Example 3.3. If G_i is a subgroup of G, then the natural maps in example 3.2 induce maps

$$BG_i \to BG$$

which extend to give a map

$$Z(K(0); (\underline{BG}, \underline{*})) \to BG$$

where $K(0)$ denotes the zero skeleton, the set of vertices of K. Depending on whether elements in G_i commute with elements in G_j, the previous map extends for some choices of K to a map

$$Z(K; (\underline{BG}, \underline{*})) \to BG.$$

In the special case for which G is a transitively commutative finite group with trivial center, the map $Z(K(0); (\underline{BG}, \underline{*})) \to BG$ gives an example of the fibrations developed in [1]. This case gives a comparison to spaces arising from the variety of commuting q-tuples in G a space denoted by $B(2, G)$ in [1]. Namely, there is a choice $(\underline{BG}, \underline{*})$ of abelian p-groups G_i such that $B(2, G)$ is $Z(K(0); (\underline{BG}, \underline{*}))$. The monodromy representation associated to the fibration $Z(K(0); (\underline{BG}, \underline{*})) \to BG$ is a basic problem posed in [1] recasting the question of whether G is a solvable group.

Example 3.4. The spaces $Z(K; (D^2, S^1))$ also are fibres of natural bundles over flag varieties. Namely, the space D^2 has a natural action of S^1 via rotations. Thus if K has m vertices, and T^m denotes a maximal torus in $U(m)$, the space

$$U(m) \times_{T^m} (K; Z((D^2, S^1))$$

fibres over the flag variety

$$U(m)/T^m.$$

The real analogue is given by

$$O(m) \times_{(\mathbb{Z}/2\mathbb{Z})^m} Z(K; (D^1, S^0)) \to O(m)/(\mathbb{Z}/2\mathbb{Z})^m.$$

Properties of these spaces will be developed elsewhere.

4 Decompositions of suspensions

The purpose of this section is to give a partial list of decompositions of suspensions of the spaces $Z(K; (\underline{X}, \underline{A}))$, starting with the following classical, well-known decomposition [34, 40]. Here $\bigvee$ denotes the wedge, and $\Sigma(X)$ denotes the reduced suspension of X. A proof of the next theorem is included here in section 6 for the convenience of the reader.

Theorem 4.1. *Let (Y_i, y_i) be pointed, CW-complexes. There is a pointed, natural homotopy equivalence*

$$H : \Sigma(Y_1 \times \ldots \times Y_m) \to \Sigma\left(\bigvee_{I \subseteq [m]} \widehat{Y}^I\right)$$

where I runs over all the non–empty sub-sequences of $(1, 2, \ldots, m)$. *Furthermore, the map H commutes with colimits.*

Recall from Definition 2.7 that $(\underline{X_I}, \underline{A_I})$ denotes the sub-collection of $(\underline{X}, \underline{A})$ determined by I. An application of Theorem 4.1 together with naturality yields the following decomposition theorem [4].

Theorem 4.2. *Let K be an abstract simplicial complex with m vertices. Given $(\underline{X}, \underline{A}) = \{(X_i, A_i)\}_{i=1}^m$ where (X_i, A_i, x_i) are pointed triples of CW-complexes, the homotopy equivalence of Theorem* 4.1 *induces a natural pointed homotopy equivalence*

$$H : \Sigma(Z(K; (\underline{X}, \underline{A}))) \to \Sigma\left(\bigvee_{I\subseteq[m]} \widehat{Z}(K_I; (\underline{X_I}, \underline{A_I}))\right).$$

The summands $\widehat{Z}(K_I; (\underline{X_I}, \underline{A_I})))$ are frequently quite tractable as well as useful. The combinatorial features of the associated homology groups tend to be simpler than that of $Z(K; (\underline{X}, \underline{A}))$ itself. For example, the Poincaré series for each of the summands is easily described in many cases as given below.

The next result from [3, 4] provides a determination of the homotopy type of the spaces $\widehat{Z}(K; (\underline{X}, \underline{A}))$ in case the inclusions $A_i \hookrightarrow X_i$ are null-homotopic for every $i \in [m]$. Let $|lk_\sigma(K)|$ denote the geometric realization of the link of σ in K.

Theorem 4.3. *Let K be an abstract simplicial complex with m vertices and $\overline{K}$ its associated poset. Let $(\underline{X}, \underline{A})$ have the property that the inclusion $A_i \subset X_i$ is null-homotopic for all i. Then there is a homotopy equivalence*

$$\widehat{Z}(K; (\underline{X}, \underline{A})) \to \bigvee_{\sigma\in K} |\Delta(\overline{K}_{<\sigma})| * \widehat{D}(\sigma)$$

where

$$|\Delta(\overline{K}_{<\sigma})| = |lk_\sigma(K)|$$

the link of σ in K.

Two special cases of Theorem 4.3 are presented next.

Theorem 4.4. *Let K be an abstract simplicial complex with m vertices and $(\underline{X}, \underline{A})$ have the property that all the A_i are contractible. Then there is a homotopy equivalence*

$$\widehat{Z}(K; (\underline{X}, \underline{A})) = \begin{cases} * & \text{if } K \text{ is not the simplex } \Delta[m-1], \text{ and} \\ \widehat{X}^{[m]} & \text{if } K \text{ is the simplex } \Delta[m-1]. \end{cases}$$

Notice that the spaces $\widehat{Z}(K_I; (\underline{X_I}, \underline{A_I}))$ of Theorem 4.2 are all contractible unless K_I is a simplex of K by Theorem 4.4. The simplices of K have been identified with certain increasing sequences $I \subseteq [m]$. Thus, the next result follows immediately from Theorems 4.2 and 4.4.

Theorem 4.5. *Let K be an abstract simplicial complex with m vertices and $(\underline{X}, \underline{A})$ have the property that all the A_i are contractible. Then there is a homotopy equivalence*

$$\Sigma\big(Z\big(K; (\underline{X}, \underline{A})\big)\big) \to \Sigma\left(\bigvee_{I \in K} \widehat{X}^I\right).$$

The next result follows at once from Theorem 4.5 [4].

Corollary 4.6. *If (1) $X_i = X$ and $A_i = A$ for all i are CW-complexes with A contractible, and (2) K_1 and K_2 are simplicial complexes both having m vertices as well as the same number of simplices in every dimension, then there is a homotopy equivalence*

$$\Sigma(Z(K_1; (X, A))) \to \Sigma(Z(K_2; (X, A))).$$

Thus there is an isomorphism

$$E_*(Z(K_1; (X, A))) \to E_*(Z(K_2; (X, A)))$$

for any homology theory $E_*(-)$.

The next theorem addresses the case for which all of the X_i are contractible. Recall the join of two spaces $X * Y$ as well as the notation $\widehat{A}^{[m]} = A_1 \wedge \ldots \wedge A_m$ from Definition 2.7.

Theorem 4.7. *Let K be an abstract simplicial complex with m vertices. Let*

$$(\underline{X}, \underline{A}) = \{(X_i, A_i, x_i)\}_{i=1}^m$$

denote m choices of triples of pointed CW-complexes for which X_i, and A_i are connected for all i. If all of the X_i are contractible, then there are homotopy equivalences

$$\widehat{Z}(K; (\underline{X}, \underline{A})) \to |K| * \widehat{A}^{[m]} \to \Sigma(|K| \wedge \widehat{A}^{[m]}).$$

Theorem 4.8. *Let K be an abstract simplicial complex with m vertices and $(\underline{X}, \underline{A})$ have the property that the the X_i are contractible for all i. Then there is a homotopy equivalence*

$$\Sigma Z(K; (\underline{X}, \underline{A})) \to \Sigma\left(\bigvee_{I \notin K} |K_I| * \widehat{A}^I\right).$$

Corollary 4.9. *Let* (X_i, A_i, x_i) *denote the triple* $(D^2, S^1, *)$ *for all* i. *Then there are homotopy equivalences*

$$\Sigma(Z(K; (D^2, S^1))) \to \Sigma\left(\bigvee_{I \notin K} |K_I| * S^{|I|}\right) \to \bigvee_{I \notin K} \Sigma^{2+|I|}|K_I|.$$

An analogous consequence is given next.

Corollary 4.10. *Let* (X_i, A_i, x_i) *denote the triple* $(D^{n+1}, S^n, *)$ *for all* i, $n \geq 1$. *Then there are homotopy equivalences*

$$\Sigma(Z(K; (D^{n+1}, S^n))) \to \Sigma\left(\bigvee_{I \notin K} |K_I| * S^{n|I|}\right) \to \bigvee_{I \notin K} \Sigma^{2+n|I|}|K_I|.$$

Remark 4.11. The multiplicative structure of the cohomology ring for $Z(K; (D^2, S^1))$ does not follow directly from Corollary 4.9. A further analysis of the ring structure follows by using the interplay between the maps defining the stable decomposition maps and the diagonal map as given in [5]. The cohomology ring structure was studied rationally by Buchstaber and Panov in [12] and integrally by Franz in [25] and [26] as well as Baskakov-Buchstaber-Panov in [8] and Lôpez de Medrano [39].

The definitions of wedge-decomposable pairs $(\underline{X}, \underline{A})$ as well as spaces $W(\sigma; \underline{B \vee C}, \underline{B \vee E}))$ were given in 2.10. A direct application of this structure is stated next,

Theorem 4.12. *Let* K *be an abstract simplicial complex with* m *vertices. Assume that the pair* $(\underline{X}, \underline{A}, \underline{*})$ *is wedge decomposable and is given by pointed triples of CW-complexes*

$$(\underline{X}, \underline{A}, \underline{*}) = \{(B_i \vee C_i, B_i, x_i)\}_{i=1}^m.$$

Then there is a pointed homotopy equivalence

$$\Theta : (B_1 \wedge \cdots \wedge B_m) \vee (\vee_{\emptyset \neq \sigma \in K} W(\sigma; \underline{B \vee C}, \underline{B}))) \to \widehat{Z}(K; (\underline{B \vee C}, \underline{B})).$$

Theorem 4.12 admits applications to the additive structure of the homology of the polyhedral product $Z(K; (\underline{X}, \underline{A}))$ where the inclusion $A_i \to X_i$ admits certain homological features which are described next.

Theorem 4.13. *Let* K *be an abstract simplicial complex with* m *vertices. Assume that*

$$(\underline{X}, \underline{A})$$

are pointed triples of connected CW-complexes for all i where the inclusions $A_i \to X_i$ induce a split monomorphism on the level of homology with field coefficients $\mathbb{F}$ for all i. Then there is an isomorphism on the level of homology groups with coefficients $\mathbb{F}$ given by

$$H_*((A_1 \wedge \cdots \wedge A_m) \vee (\vee_{\emptyset \neq \sigma \in K} W(\sigma; \underline{A \vee B}, \underline{A}))) \to H_*(\widehat{Z}(K; (\underline{X}, \underline{A})))$$

where B_i is the mapping cone of $A_i \to X_i$ for all i.

With additional finite type hypotheses, this theorem also gives the ring structure for the cohomology of $Z(K; (\underline{X}, \underline{A}))$ as long as the maps $A_i \to X_i$ induce split monomorphisms in homology.

Theorem 4.14. *Let K be an abstract simplicial complex with m vertices. Assume that $(\underline{X}, \underline{A})$ are pointed triples of CW-complexes for all i. The homology of $Z(K; (\underline{X}, \underline{A}))$ with any field coefficients $\mathbb{F}$ is a functor of K and*

$$H_*(A_i; \mathbb{F}) \to H_*(X_i; \mathbb{F})$$

for all i.

5 Applications to cohomology rings

The decompositions given in Section 4 admit applications to the structure of the cohomology ring of $Z(K; (\underline{X}, \underline{A}))$ in many cases. There are two extreme cases addressed first where either (1) each A_i is contractible, and (2) each X_i is contractible.

Definition 5.1. Let R be a ring.

1. A finite sequence of spaces $X_1, \ldots, X_m$ is said to satisfy the *strong form* of the Künneth Theorem over R provided that the natural map

$$\bigotimes_{1 \leq j \leq k} H^*(X_{i_j}; R) \to H^*(X_{i_1} \times \cdots \times X_{i_k}; R)$$

is an isomorphism for every sequence of integers $1 \leq i_1 < i_2 < \cdots, i_k \leq m$.

2. A finite sequence of path-connected spaces $X_1, \cdots, X_m$ is said to satisfy the *strong smash form* of the Künneth Theorem for the cohomology theory E^* provided that the natural map

$$\bigotimes_{1 \leq j \leq k} \overline{E}^*(X_{i_j}; R) \to \overline{E}^*(X_{i_1} \wedge \cdots \wedge X_{i_k}; R)$$

is an isomorphism for every sequence of integers $1 \leq i_1 < i_2 < \cdots, i_k \leq m$.

Observe that if a finite sequence of path-connected spaces $X_1, \cdots, X_m$ satisfies the strong form of the Künneth Theorem over R, then they satisfy the strong smash form of the Künneth Theorem over R. A similar result holds for any cohomology theory. Examples in the case $R = \mathbb{Z}$ are given by spaces X_i of finite type which have torsion-free cohomology over $\mathbb{Z}$.

Notice that there is a natural inclusion $j : Z(K; (\underline{X}, \underline{A})) \subset \prod_{i=1}^{m} X_i$. Theorem 4.5 implies that

$$j^* : H^*\left(\prod X_i; R\right) \to H^*(Z(K; (\underline{X}, \underline{A})); R)$$

is onto and that the kernel of j^* is defined to be *the generalized Stanley-Reisner ideal* $I(K)$ which is generated by all elements $x_{j_1} \otimes x_{j_2} \otimes \cdots \otimes x_{j_l}$ for which $x_{j_t} \in \overline{H}^*(X_{j_t}; R)$ and the sequence $J = (j_1, \ldots, j_l)$ is not a simplex of K. This construction provides a useful extension of the Stanley-Reisner ring [18] as stated next.

Theorem 5.2. *Let K be an abstract simplicial complex with m vertices and let*

$$(\underline{X}, \underline{A}) = \{(X_i, A_i, x_i)\}_{i=1}^{m}$$

be m pointed, connected CW-pairs. If all of the A_i are contractible and coefficients are taken in a ring R for which the spaces $X_1, \ldots, X_m$ satisfy the strong form of the Künneth Theorem over R, then there is an isomorphism of algebras

$$\bigotimes_{i=1}^{m} H^*(X_i; R)/I(K) \to H^*(Z(K; (\underline{X}, \underline{A})); R).$$

Furthermore, there are isomorphisms of underlying abelian groups given by

$$E^*(Z(K; (\underline{X}, \underline{A}))) \to \bigoplus_{I \in K} E^*(\widehat{X}^I)$$

for any reduced cohomology theory E^*.

Remark 5.3. An analogous result for the cohomology ring structure holds for any cohomology theory $E^*(-)$ for which $X_1, \ldots, X_m$ satisfy the analog of the strong form of the Künneth Theorem with respect to $E^*(-)$ [2].

The above theorem handles the cohomology ring of $Z(K; (X, A))$ in case all of the A_i are contractible. The methods here apply to more general cases by comparing diagonals with all of the $(\underline{X}, \underline{A})$.

Cup-products in the cohomology of any space W are induced by the diagonal map

$$\Delta : W \to W \times W$$

in cohomology. It suffices to identify the behavior of the diagonal map in cohomology after suspending

$$\Sigma(\Delta) : \Sigma(W) \to \Sigma(W \times W).$$

Thus the main direction of this section is a description of the behavior of the diagonal map for the polyhedral product after suspending together with the properties of the diagonal map which are preserved by the stable decomposition of Theorem 4.2 above [3, 4]. That point of view is developed next by first recalling the Theorem 4.2 which gives a natural, pointed homotopy equivalence

$$H : \Sigma(Z(K; (\underline{X}, \underline{A}))) \to \Sigma\left(\bigvee_{I \subseteq [m]} \widehat{Z}(K_I; (\underline{X}, \underline{A})_I)\right)$$

for an abstract simplicial complex and pointed triples of CW-complexes

$$(\underline{X}, \underline{A}) = (X_i, A_i, x_i)\}_{i=1}^m.$$

Let

$$\Delta_I : Y^I \to Y^I \wedge Y^I$$

denote the reduced diagonal of Y^I and let

$$\widehat{\Delta}_I : \widehat{Y}^I \to \widehat{Y}^I \wedge \widehat{Y}^I$$

denote the reduced diagonal of $\widehat{Y}^I$. In the paper [5], natural *partial diagonals* are defined using diagonals and projection maps below

$$\widehat{\Delta}_I^{J,L} : \widehat{Y}^I \to \widehat{Y}^J \wedge \widehat{Y}^L,$$

and by restriction

$$\widehat{\Delta}_I^{J,L} : \widehat{Z}(K_I) \to \widehat{Z}(K_J) \wedge \widehat{Z}(K_L)$$

where $J \cup L = I$. If $I = J = L$, these maps coincide with the *reduced* diagonal $\widehat{\Delta}_I$. Furthermore, if $\widehat{\Pi}_I : Y^{[m]} \to \widehat{Y}^I$ is the projection, there are commutative diagrams of CW-complexes and based, continuous maps

$$\begin{array}{ccc} Y^{[m]} & \xrightarrow{\Delta_{[m]}} & Y^{[m]} \wedge Y^{[m]} \\ \widehat{\Pi}_I \downarrow & & \downarrow \widehat{\Pi}_J \wedge \widehat{\Pi}_L \\ \widehat{Y}^I & \xrightarrow{\widehat{\Delta}_I^{J,L}} & \widehat{Y}^J \wedge \widehat{Y}^L \end{array}$$

and by restriction to $Z(K) \subset X^{[m]}$

$$\begin{array}{ccc} Z(K) & \xrightarrow{\Delta_K} & Z(K) \wedge Z(K) \\ \widehat{\Pi}_I \downarrow & & \downarrow \widehat{\Pi}_J \wedge \widehat{\Pi}_L \\ \widehat{Z}(K_I) & \xrightarrow{\widehat{\Delta}_I^{J,L}} & \widehat{Z}(K_J) \wedge \widehat{Z}(K_L). \end{array}$$

A definition is given next.

Definition 5.4. Assume given a family of based CW-pairs $(\underline{X}, \underline{A}) = \{(X_i, A_i)\}_{i=1}^m$. Given cohomology classes $u \in H^p(Z(K_J)), v \in H^q(Z(K_L))$, define

$$u * v = (\Delta_I^{J,L})^*(u \otimes v) \in H^{p+q}(\widehat{Z}(K_I)),$$
$$u * v = (\widehat{\Delta}_I^{J,L})^*(u \otimes v) \quad \text{thus} \quad u * v \in H^{p+q}(\widehat{Z}(K_I)).$$

The element $u * v \in H^{p+q}(\widehat{Z}(K_I))$ is called the $*$-product.

Let

$$\mathcal{H}^q(K; (\underline{X}, \underline{A})) = \bigoplus_{I \subseteq m} H^q(\widehat{Z}(K_I))$$

with

$$\mathcal{H}^*(K; (\underline{X}, \underline{A})) = \bigoplus_{I \subseteq m} H^*(\widehat{Z}(K_I)).$$

Define a map

$$\eta : \mathcal{H}^*(K; (\underline{X}, \underline{A})) \to H^*(Z(K; (\underline{X}, \underline{A}))$$

where η restricted to $H^*(\widehat{Z}(K_I))$ is $\widehat{\Pi}_I^*$.

By the decomposition given in Theorem 2.8 of [3] stated as Theorem 4.2 here, $\eta = \bigoplus_{I \leq [m]} \Pi_I^*$ is an additive isomorphism. The $*$-product gives $\mathcal{H}^*(K; (\underline{X}, \underline{A}))$ the structure of an algebra, a fact which is checked in [5] where the next result is proven.

Theorem 5.5. *Let K be an abstract simplicial complex with m vertices. Assume that $(\underline{X}, \underline{A}) = \{(X_i, A_i, x_i)\}_{i=1}^m$ is a family of based CW-pairs. Then*

$$\eta : \mathcal{H}^*(K; (\underline{X}, \underline{A})) \to H^*(Z(K; (\underline{X}, \underline{A}))$$

is a ring isomorphism.

Definition 5.6. Assume that $(\underline{X}, \underline{A}) = \{(X_i, A_i, x_i)\}_{i=1}^m$ is a family of based CW-pairs. The pair $(\underline{X}, \underline{A})$ is a *suspension pair* if $(X_i, A_i) = (\Sigma(U_i), \Sigma(V_i))$ for each i where, in addition, each inclusion $A_i \subset X_i$ given as the suspension of a map $f_i : V_i \to U_i$.

If the pair $(\underline{X}, \underline{A})$ is a suspension pair, then the reduced diagonal

$$\Delta_i : Y_i \to Y_i \wedge Y_i$$

for $Y_i = X_i$ or A_i is null-homotopic. This fact was used to prove the next result [5].

Theorem 5.7. *Let K be an abstract simplicial complex with m vertices. Assume that $(\underline{X}, \underline{A}) = \{(X_i, A_i, x_i)\}_{i=1}^m$ is a family of based CW-pairs. If $(\underline{X}, \underline{A}) = \{(\Sigma(U_i), \Sigma(V_i))\}_{i\in[m]}$ is a suspension pair, and $J \cap L \neq \emptyset$, then*

$$\widehat{\Delta}_I^{J,L} : \widehat{Z}(K_I) \to \widehat{Z}(K_J) \wedge \widehat{Z}(K_L)$$

is null-homotopic, and thus $u * v = 0$ for classes $u \in H^p\widehat{Z}(K_J)$ and $v \in H^q\widehat{Z}(K_L)$.

The next result identifies further information about cup-products in special cases [5].

Theorem 5.8. *Let K be an abstract simplicial complex with m vertices. Assume that $(\underline{CX}, \underline{X}) = \{(CX_i, X_i, x_i)\}_{i=1}^m$ is a family of based CW-pairs such that the finite product*

$$(X_1 \times \cdots \times X_m) \times (Z(K_{I_1}; (D^1, S^0)) \times \cdots \times Z(K_{I_t}; (D^1, S^0)))$$

for all $I_j \subseteq [m]$ satisfies the strong form of the Künneth theorem. Then the cup-product structure for the cohomology algebra $H^*(Z(K;(\underline{CX},\underline{X})))$ is a functor of the cohomology algebras of X_i for all i, and $Z(K_I;(D^1,S^0))$ for all I.

6 Proof of a classical decomposition

One feature which is shared in common with many of the decompositions above is that the decompositions follow from (1) analogous decompositions of a product, and (2) naturality.

The purpose of this section is to provide a proof of the **classical, standard** decomposition given in Theorem 4.1 for the convenience of the reader. This result, frequently stated for path-connected CW-complexes, also holds in the case of spaces, not necessarily connected, but with non-degenerate base-points. The main point here is that these classical decompositions are satisfied for many spaces which are not necessarily connected.

Theorem 6.1. *Let* (Y_i, y_i) *be pointed spaces with non-degenerate base-points. There is a pointed, natural homotopy equivalence*

$$H : \Sigma(Y_1 \times \ldots \times Y_m) \to \Sigma\left(\bigvee_{I\subseteq[m]} \widehat{Y}^I\right)$$

where I *runs over all the non–empty sub-sequences of* $(1, 2, \ldots, m)$. *Furthermore, the map* H *commutes with colimits.*

Proof. It suffices to give the details for the case of $m = 2$. The method here is to construct natural maps

$$\Theta : \Sigma(Y_1 \times Y_2) \to \Sigma(Y_1 \wedge Y_2) \vee \Sigma(Y_1) \vee \Sigma(Y_2),$$

and

$$\Phi : \Sigma(Y_1 \wedge Y_2) \vee \Sigma(Y_1) \vee \Sigma(Y_2) \to \Sigma(Y_1 \times Y_2)$$

such that both $\Theta \circ \Phi$, and $\Phi \circ \Theta$ are homotopic to the identity, thus give homotopy equivalences.

The map Θ is defined first. Consider the following three maps.

1. $\Sigma(p_1) : \Sigma(Y_1 \times Y_2) \to \Sigma(Y_1)$ given by suspending the projection map $p_1 : Y_1 \times Y_2 \to Y_1$,
2. $\Sigma(p_2) : \Sigma(Y_1 \times Y_2) \to \Sigma(Y_2)$ given by suspending the projection map $p_2 : Y_1 \times Y_2 \to Y_2$, and
3. $\Sigma(q) : \Sigma(Y_1 \times Y_2) \to \Sigma(Y_1 \wedge Y_2)$ given by suspending the collapse map $q : Y_1 \times Y_2 \to Y_1 \wedge Y_2$.

The map

$$\Theta : \Sigma(Y_1 \times Y_2) \to \Sigma(Y_1 \wedge Y_2) \vee \Sigma(Y_1) \vee \Sigma(Y_2)$$

is the sum

$$\Theta = \Sigma(q) + (\Sigma(p_1) + \Sigma(p_2)).$$

Next consider the maps

1. the natural inclusion $\iota_1 : Y_1 \to Y_1 \times Y_2$,
2. the natural inclusion $\iota_2 : Y_2 \to Y_1 \times Y_2$, and
3. the identity map $\iota : Y_1 \times Y_2 \to Y_1 \times Y_2$.

Observe that the two maps

- $\iota_1 \vee \iota_2 : Y_1 \vee Y_2 \to Y_1 \times Y_2$, and

- the restriction of ι to $Y_1 \vee Y_2$,

$$\iota|_{Y_1 \vee Y_2} : Y_1 \vee Y_2 \to Y_1 \times Y_2$$

are equal.

Thus

$$\Sigma(\iota) - (\Sigma(\iota_1 \circ p_1) + \Sigma(\iota_2 \circ p_2))$$

when restricted to $\Sigma(Y_1 \vee Y_2)$ is null. Since all base-points are assumed to be non-degenerate, there is a cofibration [50]

$$Y_1 \vee Y_2 \xrightarrow{\iota} Y_1 \times Y_2 \xrightarrow{q} Y_1 \wedge Y_2,$$

a feature basic in what follows. Since the composite

$$\Sigma(Y_1 \vee Y_2) \xrightarrow{\Sigma(\iota_1 \vee \iota_2)} \Sigma(Y_1 \times Y_2) \xrightarrow{\Sigma(\iota)-(\Sigma(\iota_1 \circ p_1)+\Sigma(\iota_2 \circ p_2))} \Sigma(Y_1 \times Y_2)$$

is null-homotopic as noted above, it follows from the previous cofibration sequence that there is an induced map

$$\gamma : \Sigma(Y_1 \wedge Y_2) \to \Sigma(Y_1 \times Y_2)$$

which gives a homotopy commutative diagram:

$$\begin{array}{ccc} \Sigma(Y_1 \times Y_2) & \xrightarrow{\Sigma(\iota)-(\Sigma(\iota_1 \circ p_1)+\Sigma(\iota_2 \circ p_2)))} & \Sigma(Y_1 \times Y_2) \\ {\scriptstyle \Sigma(q)}\downarrow & & \downarrow{\scriptstyle 1} \\ \Sigma(Y_1 \wedge Y_2) & \xrightarrow{\gamma} & \Sigma(Y_1 \times Y_2). \end{array}$$

The map γ is not unique up to homotopy; a choice of map

$$\Sigma(Y_1 \wedge Y_2) \longrightarrow \Sigma^2(Y_1 \vee Y_2) \longrightarrow \Sigma(Y_1 \times Y_2)$$

can be added to γ to give an analogous co-extension of

$$\Sigma(\iota) - (\Sigma(\iota_1 \circ p_1) + \Sigma(\iota_2 \circ p_2)) : \Sigma(Y_1 \times Y_2) \to \Sigma(Y_1 \times Y_2)$$

up to homotopy.

Define

$$\Phi : \Sigma(Y_1 \wedge Y_2) \vee \Sigma(Y_1) \vee \Sigma(Y_2) \to \Sigma(Y_1 \times Y_2)$$

as the sum

$$\Phi = \gamma + (\Sigma(\iota_1) + \Sigma(\iota_2)).$$

Consider the composite

$$\Phi \circ \Theta = \Phi \circ (\Sigma(q)) + \Phi \circ (\Sigma(p_1)) + \Phi \circ (\Sigma(p_2))$$

in $[\Sigma(Y_1 \times Y_2), \Sigma(Y_1 \times Y_2)]$ the group of homotopy classes of pointed maps.

1. The composite $\Phi \circ (\Sigma(q))$ is homotopic to $\Sigma(\iota) - (\Sigma(\iota_1 \circ p_1) + \Sigma(\iota_2 \circ p_2))$.
2. The composite $\Phi \circ \Sigma(p_1)$ is homotopic to $(\Sigma(\iota_1 \circ p_1)$.
3. The composite $\Phi \circ \Sigma(p_2)$ is homotopic to $(\Sigma(\iota_2 \circ p_2)$.
4. Thus the composite $\Phi \circ \Theta$ is homotopic to the identity of $\Sigma(Y_1 \times Y_2)$.

Next consider the composite

$$\Theta \circ \Phi = \Theta \circ (\gamma + (\Sigma(\iota_1) + \Sigma(\iota_2)))$$

in the group $[\Sigma(Y_1 \wedge Y_2) \vee \Sigma(Y_1) \vee \Sigma(Y_2), \Sigma(Y_1 \wedge Y_2) \vee \Sigma(Y_1) \vee \Sigma(Y_2)]$ in the group of homotopy classes of pointed maps. Thus $\Theta \circ \Phi$ is equal to the sum

$$\Theta \circ (\gamma) + \Theta \circ (\Sigma(\iota_1)) + \Theta \circ (\Sigma(\iota_2))$$

by definition. Further notice that

1. $\Theta \circ (\Sigma(\iota_1))$ is homotopic to the natural inclusion of $\Sigma(Y_1)$ in $\Sigma(Y_1) \vee \Sigma(Y_2) \vee \Sigma(Y_1 \wedge Y_2)$, and
2. $\Theta \circ (\Sigma(\iota_2))$ is homotopic to the natural inclusion of $\Sigma(Y_2)$ in $\Sigma(Y_1) \vee \Sigma(Y_2) \vee \Sigma(Y_1 \wedge Y_2)$.
3. Since $\Theta \circ \gamma \circ \Sigma(q)$ is homotopic to $\Sigma(\iota) - (\Sigma(\iota_1 \circ p_1) + \Sigma(\iota_2 \circ p_2))$, $\Theta \circ \Phi$ is homotopic to the identity of $\Sigma(Y_1 \wedge Y_2) \vee \Sigma(Y_1) \vee \Sigma(Y_2)$.

□

7 Problems

1. Find stable decompositions for natural choices of bundles

$$E_K \to Z(K; (\underline{X}, \underline{A}))$$

which specialize to the stable decompositions of $Z(K; (\underline{X}, \underline{A}))$. These bundles may be different than those in Lemma 2.3.3 of [22]. Example 3.4 is one such choice with

$$E_K = U(m) \times_{T^m} Z(K; (\underline{D^2}, \underline{S^1}))$$

where K has m vertices, and T^m is a maximal torus in $U(m)$. The motivation is that these choices of spaces E_K arise with the variety of commuting m tuples in $U(m)$.

2. The types of decompositions of suspensions described here apply to other spaces such as many configuration spaces, $Z(K; (\underline{X}, \underline{A}))$ as well as some semi-algebraic sets. For example, spaces of "sphere packings" defined below by $X(q, r, n)$ can regarded as the ways of packing q distinct spheres of fixed radius r in a disk D^n of radius 1.

A natural extension of the stable decompositions above is to find conditions which guarantee analogous decompositions for certain semi-algebraic sets such as $X(q,n,r)$ defined for fixed $0 < r < 1$, given by

$$X(q,r,n) = \{(z_1,\ldots,z_q) \in \mathbb{R}^n \mid \text{which satisfy conditions } (a)\ (b)\}$$

where

(a) $2r \leq |z_i - z_j|$ if $i \neq j$, and

(b) $|z_i| \leq 1 - r$ for all i.

The path-components of $X(q,n,r)$ frequently admit non-trivial stable decompositions. It is natural to conjecture that the path-components of $X(q,r,2)$ admit stable decompositions as bouquets of spheres. Furthermore, it is natural to conjecture that if $n \geq 2$, then the path-components of $X(q,n,r)$ admit stable decompositions as wedges of spheres, and smash products of real projective spaces $\mathbb{RP}^{n-1}$.
The motivation for this question is naive: do the above stable decompositions impact the subspace of packings which are "tight" ?

References

[1] A. ADEM, F. R. COHEN, and E. TORRES-GIESE, *Commuting elements, simplicial spaces and filtrations of classifying spaces*, Math. Proc. Camb. Phil. Soc. **152** (2012), 91–114.

[2] L. ASTEY, A. BAHRI, M. BENDERSKY, F. R. COHEN, D. M. DAVIS, S. GITLER, M. MAHOWALD, N. RAY and R. WOOD, *The KO-rings of BT^m, the Davis-Januszkiewicz Spaces and certain toric manifolds,* arXiv:1007.0069.

[3] A. BAHRI, M. BENDERSKY, F. R. COHEN and S. GITLER, *The polyhedral product functor: a method of decomposition for moment-angle complexes, arrangements and related spaces*, Advances in Math. **225** (2010), 1634–1668.

[4] A. BAHRI, M. BENDERSKY, F. R. COHEN and S. GITLER, *Decompositions of the polyhedral product functor with applications to moment-angle complexes and related spaces*, PNAS **106** (2009), 12241–12244.

[5] A. BAHRI, M. BENDERSKY, F. R. COHEN and S. GITLER, *Cup-products in generalized moment-angle complexes*, to appear in Math Proc. Camb. Phil. Soc. **152** (2012), 9–34.

[6] I. BASKAKOV, *Cohomology of K-powers of spaces and the combinatorics of simplicial divisions*, Russian Math. Surveys **57** (2002), 989–990.

[7] I. BASKAKOV, *Triple Massey products in the cohomology of moment-angle complexes*, Russian Math. Surveys **58** (2003), 1039–1041.

[8] I. BASKAKOV, V. BUCHSTABER and T. PANOV, *Algebras of cellular cochains, and torus actions*, Russian Math. Surveys **59** (2004), 562–563.

[9] A. BERGLUND, *Poincarśeries of monomial rings*, J. Algebra **295** (2006), 211–230.

[10] A. BJÖRNER and K. S. SARKARIA, *The zeta function of a simplicial complex*, Israel Journal of Math. **103** (1998) 29–40.

[11] F. BOSIO and L. MEERSSEMAN, *Real quadrics in Cn, complex manifolds and convex polytopes*, Acta Math. **197** (2006), 53–127.

[12] V. BUCHSTABER and T. PANOV, *Actions of tori, combinatorial topology and homological algebra*, Russian Math. Surveys, **55** (2000), 825–921.

[13] V. BUCHSTABER and T. PANOV, *Torus actions and their applications in topology and combinatorics*, AMS University Lecture Series **24** (2002).

[14] V. BUCHSTABER, T. PANOV and N. RAY, *Spaces of polytopes and cobordism of quasitoric manifolds*, Moscow Math. J. **7** (2007), 219–242.

[15] C. CAMACHO, N. KUIPER and J. PALIS, *The topology of holomorphic flows with singularity*, Pub. I. H.E.S. **48** (1978), 5–38.

[16] H. M. S. COXETER, *Regular Skew Polyhedra in Three and Four Dimension, and their Topological Analogues*, Proc. London Math. Soc. (1938) s2-43 (1): 33–62.

[17] V. I. DANILOV, *The geometry of toric varieties* (Russian), Uspekhi Mat. Nauk, **33** (1978), 85-134, translated in Russian Math. Surveys, **33** (1978), 97–154.

[18] M. DAVIS and T. JANUSZKIEWICZ, *Convex polytopes, Coxeter orbifolds and torus actions*, Duke Math. J. **62** (1991), 417–451.

[19] M. DAVIS and B. OKUN, *Cohomology computations for Artin groups, Bestvina.Brady groups, and graph products*, arxiv.

[20] C. DE CONCINI and C. PROCESI, *Wonderful models of subspace arrangements*, Selecta Math. (N.S.) **1** (1995), 459–494.

[21] P. DELIGNE, M. GORESKY and R. MACPHERSON, *L´algèbre de cohomologie du complément, dans un espace affine, d´une famille finie de sous-espaces affines*, Michigan Math. J. **48** (2000), 121–136.

[22] G. DENHAM and A. SUCIU, *Moment-angle complexes, monomial ideals and Massey products*, Pure Appl. Math. Q. **3** (2007), 25–60.

[23] N. DOBRINSKAYA, *Loops on polyhedral products and diagonal arrangements*, arXiv:0901.2871v1.

[24] Y. FÉLIX and D. TANRÉ, *Rational homotopy of the polyhedral product functor*, P.A.M.S. **137**, (2009), 891–898.

[25] M. FRANZ, *The integral cohomology of smooth toric varieties*, arXiv:math/0308253v1.

[26] M. FRANZ, *The integral cohomology of toric manifolds*, Proc. Steklov Inst. Math. **252** (2006), 53–62 [Proceedings of the Keldysh Conference, Moscow 2004].

[27] M. FRANZ, *Koszul Duality for Tori*, Thesis, Universität Konstanz, 2001.

[28] T. GANEA, *A generalization of the homology and homotopy suspension*, Comment. Math. Helv. **39** (1965), 295–322.

[29] S. GITLER and S. LÔPEZ DE MEDRANO, *Intersections of Quadrics, Moment-angle Manifolds and Connected Sums*, arXiv:0901.2580.

[30] M. GORESKY and R. MACPHERSON, "Stratified Morse Theory", Ergebnisse Der Mathematik Und Ihrer Grenzgebiete 3rd series, Vol. 14, Springer-Verlag, Berlin, 1988.

[31] J. GRBIĈ and S. THERIAULT, *Homotopy type of the complement of a coordinate subspace arrangement of codimension two*, Russian Math. Surveys **59** (2004), 1207–1209.

[32] M. HOCHSTER, *Cohen-Macaulay rings, combinatorics, and simplicial complexes*, In: "Ring Theory, II" (Proc. Second Conf., Univ. Oklahoma, Norman, Okla., 1975), 171–223, Lecture Notes in Pure and Appl. Math., Vol. 26, Dekker, New York, 1977.

[33] Y. HU, *On the homology of complements of arrangements of subspaces and spheres*, P.A.M.S. **122** (1994), 285–290.

[34] I. JAMES, *Reduced product spaces*, Ann. of Math. **62** (1955), 170–197.

[35] K. JEWELL, *Complements of sphere and sub-space arrangements*, Topology and its Applications **56** (1994) 199–214.

[36] K. JEWELL, P. ORLIK and B. Z. SHAPIRO, *On the complements of affine sub-space arrangements*, Topology Appl. **56** (1994), 215–233.

[37] Y. KAMIYAMA and S. TSUKUDA, *The configuration space of the n-arms machine in the Euclidean space*, Topology and its Applications **154** (2007), 1447–1464.

[38] M. DE LONGUEVILLE and C. SCHULTZ, *The cohomology rings of complements of subspace arrangements*, Math. Ann. **319** (2001), 625–646.

[39] S. LÔPEZ DE MEDRANO, *Topology of the Intersection of Quadrics in* $\mathbb{R}^n$, In: "Algebraic Topology" (Arcata Ca), LNM, Vol. 1370 (1989), Springer Verlag, 280-292.

[40] J. MILNOR, *On the construction F[K]*, In: "A Student's Guide to Algebraic Topology", J. F. Adams, editor, Lecture Notes of the London Mathematical Society, Vol. 4, 1972, 119–136.

[41] D. NOTBOHM and N. RAY, *On Davis-Januszkiewicz homotopy types. I. Formality and rationalization*, AGT **5** (2005), 31–51.

[42] T. PANOV, *Cohomology of face rings, and torus actions*, London Math. Soc. Lecture Notes, **347** (2008), Surveys in Contemporary Mathematics, Cambridge, 165–201, arXiv:math/0506526.

[43] T. PANOV, N. RAY and R. VOGT, *Colimits, Stanley-Reisner algebras, and loop spaces*, In: "Categorical Decomposition Techniques in Algebraic Topology", Progress in Mathematics, Vol. 215, 2004, 261–291, Birkhäuser, Basel.

[44] H. POINCARÉ, *Sur les properieties des fonctions definies par les equations aux differences partiells*, thése, Paris, 1879.

[45] G. PORTER, *The homotopy groups of wedges of suspensions*, Amer. J. Math. **88** (1966), 655–663.

[46] C. SHAPER, *Suspensions of affine arrangements*, Math. Ann. **209**, (1997), 463–473,

[47] C. L. SIEGAL, *Über die Normalform analytischer Differentialgleichungen in der Nähe einer Gleichgewischtslösung* Göttingen, Nachr. Akad.Wiss. Phys. Kl, (1952) 21–30.

[48] C. L. SIEGAL and J. MOSER, "Celestial Mechanics", Springer-Verlag, 1954.

[49] R. P. STANLEY, *Combinatorics and Commutative Algebra*, Second edition, Progress in Mathematics, Vol. 41, 1996, Birkhäuser, Boston, MA.

[50] N. E. STEENROD, *A convenient category of topological spaces*, Mich. J. Math. **34** (1968), 105–112

[51] N. STRICKLAND, *Notes on toric spaces*, preprint 1999, available at Strickland's webpage.

[52] E. SWARTZ, *Topological representations of matroids* , JAMS **16** (2003), 427–442.

[53] E. B. VINBERG, *Discrete linear groups generated by reflections* (Russian), Izv. Akad. Nauk SSSR Ser. Mat. **35** (1971), 1072–1112. English translation in Math. USSR Izvestija **5** (1971), 1083–1119.

[54] X. WANG and Q. ZHENG, *The homology of simplicial complement and the cohomology of the moment-angle-complexes*, arXiv 1006.3904.

[55] V. WELKER, G. ZIEGLER and R. ŽIVALJEVIĆ, *Homotopy colimits-comparison lemmas for combinatorial applications*, J. Reine Angew. Math. **509** (1999), 117–149.
[56] G. ZIEGLER and R. ŽIVALJEVIĆ, *Homotopy types of sub-space arrangements via diagrams of spaces*, Math. Ann. **295** (1993), 527–548.

Characters of fundamental groups of curve complements and orbifold pencils

Enrique Artal Bartolo, Jose Ignacio Cogolludo-Agustín
and Anatoly Libgober

Abstract. The present work is a user's guide to the results of [7], where a description of the space of characters of a quasi-projective variety was given in terms of global quotient orbifold pencils.

Below we consider the case of plane curve complements. In particular, an infinite family of curves exhibiting characters of any torsion and depth 3 will be discussed. Also, in the context of line arrangements, it will be shown how geometric tools, such as the existence of orbifold pencils, can replace the group theoretical computations via fundamental groups when studying characters of finite order, specially order two. Finally, we revisit an Alexander-equivalent Zariski pair considered in the literature and show how the existence of such pencils distinguishes both curves.

1 Introduction

Let $\mathcal{X}$ be the complement of a reduced (possibly reducible) projective curve $\mathcal{D}$ in the complex projective plane $\mathbb{P}^2$. The space of characters of the fundamental group $\operatorname{Char}(\mathcal{X}) := \operatorname{Hom}(\pi_1(\mathcal{X}), \mathbb{C}^*)$ has an interesting stratification by subspaces, given by the cohomology of the rank one local system associated with the character:

$$\mathring{V}_k(\mathcal{X}) := \{\chi \in \operatorname{Char}(\mathcal{X}) \mid \dim H^1(\mathcal{X}, \chi) = k\}. \tag{1.1}$$

The closures $V_k(\mathcal{X})$ of these jumping loci in $\operatorname{Char}(\mathcal{X})$ were called in [23] the characteristic varieties of $\mathcal{X}$. More precisely, the characteristic varieties associated to $\mathcal{X}$ were defined in [23] as the zero sets of Fitting ideals of the $\mathbb{C}[\pi_1/\pi_1']$-module which is the complexification $\pi_1'/\pi_1'' \otimes \mathbb{C}$ of the abelianized commutator of the fundamental group $\pi_1(\mathcal{X})$ (*cf.* Section 2 for more details). In particular the characteristic varieties (un-

Partially supported by the Spanish Ministry of Education MTM2010-21740-C02-02. The third named author was also partially supported by NSF grant.

like the jumping sets of the cohomology dimension greater than one) depend only on the fundamental group. Fox calculus provides an effective method for calculating the characteristic varieties if a presentation of the fundamental group by generators and relators is known.

For each character $\chi \in \mathrm{Char}(\mathcal{X})$ the *depth* was defined in [23] as

$$d(\chi) := \dim H^1(\mathcal{X}, \chi) \tag{1.2}$$

so that the strata (1.1) are the sets on which $d(\chi)$ is constant.

In [7], we describe a geometric interpretation of the depth by relating it to the pencils on $\mathcal{X}$ *i.e.* holomorphic maps $\mathcal{X} \to C$, $\dim C = 1$ having multiple fibers. In fact the discussion in [7] is in a more general context in which $\mathcal{X}$ is a smooth quasi-projective variety. [1] The viewpoint of [7] (and [8]) is that such a pencil can be considered as a map in the category of orbifolds. The orbifold structure of the curve C is matched by the structure of multiple fibers of the pencil. The main result of [7] can be stated as follows:

Theorem 1.1. *Let $\mathcal{X}$ be a quasi-projective manifold and let χ be a character of $\pi_1(\mathcal{X})$.*

(1) *Assume that there are n marked orbifold pencils* i.e. *maps $f_i : \mathcal{X} \to \mathcal{C}$ $(i = 1, ...n)$ where $\mathcal{C}$ is a fixed orbicurve, $\rho \in \mathrm{Char}^{\mathrm{orb}}(\mathcal{C})$ and $\chi = f_i^*(\rho)$. If these pencils are strongly independent, then $d(\chi) \geq nd(\rho)$.*
(2) *If χ is a character of order two and weight two, then there are exactly $d(\chi)$ strongly independent orbifold pencils on $\mathcal{X}$ whose target is the global $\mathbb{Z}_2$-orbifold $\mathcal{C} = \mathbb{C}_{2,2}$. These pencils are marked with the character ρ of $\pi_1^{\mathrm{orb}}(\mathbb{C}_{2,2})$ characterized by the condition that ρ is non-trivial on both standard generators of the latter orbifold fundamental group.*

We refer to Section 2 for all the required definitions, and in particular, the definition of strongly independent pencils.

According to this result, the orbifold pencils on $\mathcal{X}$ whose targets have an orbifold fundamental group with characters of positive depth, induce characters in $\mathrm{Char}(\mathcal{X})$ whose depth have the lower bound given in 1.1. One can compare this statement with previous results on pencils on quasi-projective manifolds. For example, consider a character χ which belongs

[1] Much of the discussion in the first two sections below applies to general quasi-projective varieties (*cf.* [7]), but in the present paper we will stay in the hypersurface complement context. As was noted, the characteristic varieties only depend on the fundamental group, hence, by the Lefschetz-type theorems it is enough to consider the curve complement class.

to a positive dimensional component of the characteristic variety. Then the results in [2] can be applied to such a component to obtain a pencil $f : \mathcal{X} \to \mathcal{C}$ and a character $\rho \in \mathrm{Char}(\mathcal{C})$ such that $\chi = f^*(\rho)$. Here $\mathcal{C}$ is the complement in $\mathbb{P}^1$ to a finite set containing say $d > 2$ points. Moreover, the number of independent pencils in the sense of Section 2 is equal to one (*cf.* [7, Lemma 4.15]; note that the depth of $\rho \in \mathrm{Char}(\mathcal{C})$ is equal to $d-2$). Hence in this case, the inequality in Theorem 1.1 (1) is equivalent to the shown in [2, Prop.1.7] inequality $\dim H^1(\mathcal{X}, \chi) \geq \dim H^1(\mathcal{C}, \rho)$.

The orbifold structure involved in Theorem 1.1 is essential since the orbifold pencils described there and considered *without the orbifold structure*, are just rational pencils whose target might have trivial fundamental group and thus the connection with the jumping loci disappears.

The part (2) asserts a partial converse for characters of order two *i.e.* the characters of order two having positive depth are pull-backs of orbifold characters on $\mathbb{C}_{2,2}$ by orbifold pencils. Note that, as shown in [6] for characters of order 5 on an affine quintic, not all characters on complements of plane curves can be described as pull-backs of orbifold pencils.

The goal of this paper is to illustrate in detail both parts (1) and (2) of Theorem 1.1 with examples in which orbifolds are unavoidable. We start with a section reviewing mainly known results on the cohomology of local systems, characteristic varieties, orbifolds, and Zariski pairs making possible to read the rest of the paper unless one is interested in the proofs of mentioned results. Then in Section 3, a family of curves is considered for which the characteristic variety contains isolated characters having torsion of arbitrary finite order and whose depth is 3. The calculations illustrate the use of Fox calculus for finding an explicit description of the characteristic varieties. Next, in the context of line arrangements, examples of Ceva and augmented Ceva arrangements are considered in Section 4. Their characteristic varieties have been studied in the literature via computer aided calculations based on fundamental group presentations and Fox calculus. Here we present an alternative way to study such varieties independent of the fundamental group illustrating the geometric approach of Theorem 1.1. Finally, in Section 5 we discuss a Zariski pair of sextic curves whose Alexander polynomials coincide. We determine this Zariski pair by the existence of orbifold pencils.

ACKNOWLEDGEMENTS. The authors want to express thanks to the organizers of the *Intensive research period on Configuration Spaces: Geometry, Combinatorics and Topology,* at the *Centro di Ricerca Matematica Ennio De Giorgi*, Pisa, in May 2010 for their hospitality and excellent working atmosphere. The second and third named authors want to thank the University of Illinois and Zaragoza respectively, where part of the work on the material of this paper was done during their visit in Spring and Summer 2011.

2 Preliminaries

In this section the necessary definitions used in Theorem 1.1 will be reviewed together with material on the characteristic varieties and Zariski pairs with the aim to keep the discussion of the upcoming sections in a reasonably self-contained manner.

2.1 Characteristic varieties

Characteristic varieties appeared first in the literature in the context of algebraic curves in [22]. They can be defined as follows.

Let $\mathcal{D} := \mathcal{D}_1 \cup \cdots \cup \mathcal{D}_r$ be the decomposition of a reduced curve $\mathcal{D}$ into irreducible components and let $d_i := \deg \mathcal{D}_i$ denote the degrees of the components $\mathcal{D}_i$. Let $\tau := \gcd(d_1, \ldots, d_r)$ and $\mathcal{X} = \mathbb{P}^2 \setminus \mathcal{D}$. Then (*cf.* [23])

$$H_1(\mathcal{X};\mathbb{Z}) = \left\langle \bigoplus_{i=1}^{r} \gamma_i \mathbb{Z} \right\rangle / \langle d_1\gamma_1 + \cdots + d_r\gamma_r \rangle \approx \mathbb{Z}^{r-1} \oplus \mathbb{Z}/\tau\mathbb{Z}, \quad (2.1)$$

where γ_i is the homology class of a meridian of $\mathcal{D}_i$ (*i.e.* the boundary of small disk transversal to $\mathcal{D}_i$ at a smooth point).

Let $\mathbf{ab} : G := \pi_1(\mathcal{X}) \to H_1(\mathcal{X};\mathbb{Z})$ be epimorphism of abelianization. The kernel G' of $\mathbf{ab}$, *i.e.* the commutator of G, defines the universal Abelian covering of $\mathcal{X}$, say $\mathcal{X}_{\mathbf{ab}} \xrightarrow{\pi} \mathcal{X}$, whose group of deck transformations is $H_1(\mathcal{X};\mathbb{Z}) = G/G'$. This group of deck transformations, since it acts on $\mathcal{X}_{\mathbf{ab}}$, also acts on $H_1(\mathcal{X}_{\mathbf{ab}};\mathbb{Z}) = G'/G''$. [2] This allows to endow $M_{\mathcal{D},\mathbf{ab}} := H_1(\mathcal{X}_{\mathbf{ab}};\mathbb{Z}) \otimes \mathbb{C}$ (as well as $\tilde{M}_{\mathcal{D},\mathbf{ab}} := H_1(\mathcal{X}_{\mathbf{ab}}, \pi^{-1}(*);\mathbb{Z}) \otimes \mathbb{C}$) with a structure of $\Lambda_{\mathcal{D}}$-module where

$$\Lambda_{\mathcal{D}} := \mathbb{C}[G/G'] \approx \mathbb{C}[t_1^{\pm 1}, \ldots, t_r^{\pm 1}]/(t_1^{d_1} \cdot \ldots \cdot t_r^{d_r} - 1). \quad (2.2)$$

Note that $\operatorname{Spec}\Lambda_{\mathcal{D}}$ can be identified with the commutative affine algebraic group $\operatorname{Char}\pi_1(X)$ having τ tori $(\mathbb{C}^*)^{r-1}$ as connected components. Indeed, the elements of $\Lambda_{\mathcal{D}}$ can be viewed as the functions on the group of characters of G.

Since G is a finitely generated group, the module $M_{\mathcal{D},\mathbf{ab}}$ (respectively $\tilde{M}_{\mathcal{D},\mathbf{ab}}$) is a finitely generated $\Lambda_{\mathcal{D}}$-module: [3] in fact one can construct

[2] This action corresponds to the action of G/G' on G'/G'' by conjugation.

[3] In most interesting examples with non-cyclic G/G' the group G'/G'' is infinitely generated.

a presentation of $M_{\mathcal{D},\mathbf{ab}}$ (respectively $\tilde{M}_{\mathcal{D},\mathbf{ab}}$) with the number of $\Lambda_{\mathcal{D}}$-generators at most $\binom{n}{2}$ (respectively n), where n is the number of generators of G. If G/G' is not cyclic (*i.e.* $r > 2$ or $r \geq 2$ and $\tau > 1$) then $\Lambda_{\mathcal{D}}$ is not a Principal Ideal Domain. One way to approach the $\Lambda_{\mathcal{D}}$-module structure of both $M_{\mathcal{D},\mathbf{ab}}$ and $\tilde{M}_{\mathcal{D},\mathbf{ab}}$ is to study their Fitting ideals (*cf.* [17]).

Let us briefly recall the relevant definitions. Let R be a commutative Noetherian ring with unity and M a finitely generated R-module. Choose a finite free presentation for M, say $\phi : R^m \to R^n$, where $M = \operatorname{coker} \phi$. The homomorphism ϕ has an associated $(n \times m)$ matrix A_ϕ with coefficients in R such that $\phi(x) = A_\phi x$ (the vectors below are represented as the column matrices).

Definition 2.1. The k-th *Fitting ideal* $F_k(M)$ of M is defined as the ideal generated by

$$\begin{cases} 0 & \text{if } k \leq \max\{0, n-m\} \\ 1 & \text{if } k > n \\ \text{minors of } A_\phi \text{ of order } (n-k+1) & \text{otherwise.} \end{cases}$$

It will be denoted F_k if no ambiguity seems likely to arise.

Definition 2.2. [22] With the above notations the *k-th characteristic variety* ($k > 0$) of $\mathcal{X} = \mathbb{P}^2 \setminus \mathcal{D}$ can be defined as the zero-set of the ideal $F_k(M_{\mathcal{D},\mathbf{ab}})$

$$V_k(\mathcal{X}) := Z(F_k(M_{\mathcal{D},\mathbf{ab}})) \subset \operatorname{Spec} \Lambda_{\mathcal{D}} = \operatorname{Char}(\mathbb{P}^2 \setminus \mathcal{D}).$$

Then $\mathring{V}_k(\mathcal{X})$ is the set of characters in $V_k(\mathcal{X})$ which do not belong to $V_j(\mathcal{X})$ for $j > k$. If a character χ belongs to $\mathring{V}_k(\mathcal{X})$ then k is called the *depth* of χ and denoted by $d(\chi)$ (*cf.* [23]).

An alternative notation for $\mathring{V}_k(\mathbb{P}^2 \setminus \mathcal{D})$ (respectively $V_k(\mathbb{P}^2 \setminus \mathcal{D})$) is $\mathring{V}_{k,\mathbb{P}}(\mathcal{D})$ (respectively $V_{k,\mathbb{P}}(\mathcal{D})$).

Remark 2.3. Essentially without loss of generality one can consider only the cases when the quotient by an ideal in the definition of the ring $\Lambda_{\mathcal{D}}$ in (2.2) is absent *i.e.* consider only the modules of the ring of Laurent polynomials. Indeed, consider a line L not contained in $\mathcal{D}$ and in general position (*i.e.* which does not contain singularities of $\mathcal{D}$ and is transversal to it). Then $\Lambda_{L\cup\mathcal{C}}$ is isomorphic to $\mathbb{C}[t_1^{\pm 1}, \ldots, t_r^{\pm 1}]$. Moreover, since we assume transversality $L \pitchfork \mathcal{D}$, then the $\Lambda_{L\cup\mathcal{D}}$-module $M_{L\cup\mathcal{D},\mathbf{ab}}$ does not depend on L (see for instance [9, Proposition 1.16]). The characteristic variety $V_{k,\mathbb{P}}(L \cup \mathcal{D})$ determines $V_{k,\mathbb{P}}(\mathcal{D})$ (*cf.* [9, 23]). By abuse of language it is called the *k-th affine characteristic variety* and denoted simply by $V_k(\mathcal{D})$.

One can also use the module $\tilde{M}_{\mathcal{D},\mathbf{ab}}$ to obtain the characteristic varieties of $\mathcal{D}$. One has the following connection

$$V_k(\mathcal{X}) \setminus \overline{1} = Z(F_{k+1}(\tilde{M}_{\mathcal{D},\mathbf{ab}})) \setminus \overline{1},$$

where $\overline{1}$ denotes the trivial character.

Remark 2.4. The depth of a character appears in explicit formulas for the first Betti number of cyclic and abelian unbranched and branched covering spaces (*cf.* [20, 22, 27])

Remark 2.5. One can also define the k-th characteristic variety $V_k(G)$ of any finitely generated group G (such that the abelianization $G/G' \neq 0$ or, more generally, for a surjection $G \to A$ where A is an abelian group) as the k-th characteristic variety of the $\Lambda_G = \mathbb{C}[G/G']$-module $M_G = H_1(\mathcal{X}_{G,\mathbf{ab}})$ obtained by considering the CW-complex $\mathcal{X}_G$ associated with a presentation of G and its universal abelian covering space $\mathcal{X}_{G,\mathbf{ab}}$ (respectively considering the covering space of $\mathcal{X}_G$ associated with the kernel of the map to A). Such invariant is independent of the finite presentation of G (respectively depends only on $G \to A$). This construction will be applied below to the orbifold fundamental groups of one dimensional orbifolds.

Remark 2.6. Note that one has:

- $V_k(\mathcal{D}) = \operatorname{Supp}_{\Lambda_{\mathcal{D}}} \wedge^i (H_1(\mathcal{X}_{\mathbf{ab}}; \mathbb{C}))$,
- $\operatorname{Spec} \Lambda_{L \cup \mathcal{D}} = \mathbb{T}^r = (\mathbb{C}^*)^r$, for the affine case, and
- $\operatorname{Spec} \Lambda_{\mathcal{D}} = \mathbb{T}_{\mathcal{D}} = \{\omega^i\}_{i=0}^{\tau-1} \times (\mathbb{C}^*)^{r-1} = V(t_1^{d_1} \cdot \ldots \cdot t_r^{d_r} - 1) \subset \mathbb{T}^r$, where ω is a τ-th primitive root of unity for the curves in projective plane.

Note also that in the case of a finitely presented group G such that $G/G' = \mathbb{Z}^r \oplus \mathbb{Z}/\tau_1\mathbb{Z} \oplus \cdots \oplus \mathbb{Z}/\tau_s\mathbb{Z}$ one has

$$\operatorname{Spec} \Lambda_G = \mathbb{T}_G = \{(\omega_1^{i_1}, \ldots, \omega_s^{i_s}) | i_k = 0, \ldots, \tau_k - 1, k = 1, \ldots, s\} \times (\mathbb{C}^*)^r, \tag{2.3}$$

where as above $\Lambda_G = \mathbb{C}[G/G']$ and ω_i is a τ_i-th primitive root of unity.

Let $\mathcal{X}$ be a smooth quasi-projective variety such that for its smooth compactification $\overline{\mathcal{X}}$ one has $H^1(\overline{\mathcal{X}}, \mathbb{C}) = 0$. This of course includes the cases $\mathcal{X} = \mathbb{P}^2 \setminus \mathcal{D}$. The structure of $V_k(\mathcal{X})$ is given by the following fundamental result.

Theorem 2.7 ([2]). *Each $V_k(\mathcal{X})$ is a finite union of cosets of subgroups of* Char($\mathcal{X}$). *Moreover, for each irreducible component W of $V_k(\mathcal{X})$ having a positive dimension there is a pencil $f : \mathcal{X} \to C$, where C is a $\mathbb{P}^1$ with deleted points, and a torsion character $\chi \in V_k(\mathcal{X})$ such that $W = \chi f^* H^1(C, \mathbb{C}^*)$.*

2.2 Essential coordinate components

Let $\mathcal{D}' \subsetneq \mathcal{D}$ be curve whose components form a subset of the set of components of $\mathcal{D}$. There is a natural epimorphism $\pi_1(\mathbb{P}^2 \setminus \mathcal{D}) \twoheadrightarrow \pi_1(\mathbb{P}^2 \setminus \mathcal{D}')$ induced by the inclusion. This surjection induces a natural inclusion $\operatorname{Spec} \Lambda_{\mathcal{D}'} \subset \operatorname{Spec} \Lambda_{\mathcal{D}}$. With identification of the generators of $\Lambda_{\mathcal{D}}$ with components of $\mathcal{D}$ as above, this embedding is obtained by assigning 1 to the coordinates corresponding to those irreducible components of $\mathcal{D}$ which are not in $\mathcal{D}'$ (*cf.* [23]).

The embedding $\operatorname{Spec} \Lambda_{\mathcal{D}'} \subset \operatorname{Spec} \Lambda_{\mathcal{D}}$ induces the inclusion $\operatorname{Char}(\mathcal{D}') \subset \operatorname{Char}(\mathcal{D})$ (*cf.* [23]); any irreducible component of $V_k(\mathcal{D}')$ is the intersection of an irreducible component of $V_k(\mathcal{D})$ with $\Lambda_{\mathcal{D}'}$.

Definition 2.8. Irreducible components of $V_k(\mathcal{D})$ contained in $\Lambda_{\mathcal{D}'}$ for some curve $\mathcal{D}' \subset \mathcal{D}$ are called *coordinate components* of $V_k(\mathcal{D})$. If an irreducible coordinate component V of $V_k(\mathcal{D}')$ is also an irreducible component of $V_k(\mathcal{D})$, then V is called a *non-essential coordinate component*, otherwise it is called an *essential coordinate component*.

See [4] for examples. A detailed discussion of more examples is done in Sections 3, 4, and 5.

As shown in [23, Lemma 1.4.3] (see also [15, Proposition 3.12]), essential coordinate components must be zero dimensional.

2.3 Alexander invariant

In Section 2.1 the characteristic varieties of a finitely presented group G are defined as the zeroes of the Fitting ideals of the module $M := G'/G''$ over G/G'. This module is referred to in the literature as the *Alexander invariant* of G. Note, however, that this is not the module represented by the matrix of Fox derivatives called the *Alexander module* of G.

Our purpose in this section is to briefly describe the Alexander invariant for fundamental groups of complements of plane curves and give a method to obtain a presentation of such a module from a presentation of G. In order to do so, consider $G := \pi_1(\mathbb{P}^2 \setminus \mathcal{D})$ the fundamental group of the curve $\mathcal{D}$. Without loss of generality one might assume that

(Z1) G/G' is a free group of rank r generated by meridians $\gamma_1, \gamma_2, \ldots, \gamma_r$,

then one has the following

Lemma 2.9 ([5, Proposition 2.3]). *Any group G as above satisfying (Z1) admits a presentation*

$$\langle x_1, \ldots, x_r, y_1, \ldots, y_s : R_1(\overline{x}, \overline{y}) = \ldots = R_m(\overline{x}, \overline{y}) = 1 \rangle, \tag{2.4}$$

where $\overline{x} := \{x_1, ..., x_r\}$ *and* $\overline{y} := \{y_1, ..., y_s\}$ *satisfying:*

(Z2) $\mathbf{ab}(x_i) = \gamma_i$, $\mathbf{ab}(y_j) = 0$, *and* R_k *can be written in terms of* $\overline{y}$ *and* $x_k[x_i, x_j]x_k^{-1}$, *where* $[x_i, x_j]$ *is the commutator of* x_i *and* x_j.

A presentation satisfying (Z2) is called a *Zariski presentation* of G.

From now on we will assume G admits a Zariski presentation as in (2.4). In order to describe elements of the module M it is sometimes convenient to see $\mathbb{Z}[G/G']$ as the ring of Laurent polynomials in r variables $\mathbb{Z}[t_1^{\pm 1}, ..., t_r^{\pm 1}]$, where t_i represents the action induced by γ_i on M as a multiplicative action, that is,

$$t_i g \overset{M}{=} x_i g x_i^{-1} \tag{2.5}$$

for any $g \in G'$.

Remark 2.10.

1. One of course needs to convince oneself that action (2.5) is independent, up to an element of G'', of the representative x_i as long as $\mathbf{ab}(x_i) = \gamma_i$. This is an easy exercise.
2. We denote by "$\overset{M}{=}$" equalities that are valid in M.

Example 2.11. Note that

$$[xy, z] \overset{M}{=} [x, z] + t_x[y, z], \tag{2.6}$$

where x, y, and z are elements of G and t_x denotes $\mathbf{ab}(x)$ in the multiplicative group. This is a consequence of the following

$$[xy, z] = xyzy^{-1}x^{-1}z^{-1} = x(yzy^{-1}z^{-1})x^{-1}xzx^{-1}z^{-1} \overset{M}{=} t_x[y, z] + [x, z].$$

As a useful application of (2.6) one can check that

$$[x^y, z] \overset{M}{=} [x, z] + (t_z - 1)[y, x], \tag{2.7}$$

where $x^y := yxy^{-1}$.

Note that $x_{ij} := [x_i, x_j]$, $1 \leq i < j \leq r$ and y_k, $k = 1, ..., s$ are elements in G', since $\mathbf{ab}(x_{ij}) = \mathbf{ab}(y_k) = 0$. Therefore

$$x_k[x_i, x_j]x_k^{-1} \overset{M}{=} t_k x_{i,j} \tag{2.8}$$

(see (2.5) and (Z2)). Moreover,

Proposition 2.12. *For a group* G *as above, the module* M *is generated by* $\overline{x}_{i,j} := \{x_{ij}\}_{1 \leq i < j \leq r}$ *and* $\overline{y} := \{y_k\}_{k=1,...,s}$.

Example 2.13. The module M is not freely generated by the set mentioned above, for instance, note that according to $(Z2)$ and (2.8) any relation in G, say $R_i(\overline{x},\overline{y})=1$ (as in (2.4)) can be written (in M) in terms of $\overline{\{x_{ij}\}}$ and $\overline{y}$ as $\mathcal{R}_i(\overline{x}_{ij},\overline{y})$. In other words, $\mathcal{R}_i(\overline{x}_{ij},\overline{y})=0$ is a relation in M.

Example 2.14. Even if G were to be the free group $\mathcal{F}_r$, M would not be freely generated by $\overline{\{x_{ij}\}}$ and $\overline{y}$. In fact,

$$J(x,y,z):=(t_x-1)[y,z]+(t_y-1)[z,x]+(t_z-1)[x,y]\overset{M}{=}0 \quad (2.9)$$

for any x, y, z in G. Using Example 2.11 repeatedly, one can check the following

$$[xy,z]=\begin{cases}\overset{M}{=}[x,z]+t_x[y,z]\\ =[y^{x^{-1}}x,z]\overset{M}{=}[y^{x^{-1}},z]+t_y[x,z]\\ \quad\overset{M}{=}[y,z]-(t_z-1)[x,y]+t_y[x,z],\end{cases} \quad (2.10)$$

where $a^b=bab^{-1}$. The difference between both equalities results in $J(x,y,z)=0$. Such relations will be referred to as *Jacobian relations* of M.

A combination of Examples 2.13 and 2.14 gives in fact a presentation of M.

Proposition 2.15 ([9, Proposition 2.39]). *The set of relations* $\mathcal{R}_1,\ldots,\mathcal{R}_m$ *as described in Example 2.13 and* $J(i,j,k)=J(x_i,x_j,x_k)$ *as described in Example 2.14 is a complete system of relations for* M.

Example 2.16. Let $G=\mathcal{F}_r$ be the free group in r generators, for instance, the fundamental group of the complement to the union of $r+1$ concurrent lines. According to Propositions 2.12 and 2.15, M has a presentation matrix A_r of size $\binom{r}{3}\times\binom{r}{2}$ whose columns correspond to the generators $x_{ij}=[x_i,x_j]$ and whose rows correspond to the coefficients of the Jacobian relations $J(i,j,k)$, $1\le i<j<k\le r$. For instance, if $r=4$

$$A_4:=\begin{bmatrix}(t_3-1) & -(t_2-1) & 0 & (t_1-1) & 0 & 0\\ (t_4-1) & 0 & -(t_2-1) & 0 & (t_1-1) & 0\\ 0 & (t_4-1) & -(t_3-1) & 0 & 0 & (t_1-1)\\ 0 & 0 & 0 & (t_4-1) & -(t_3-1) & (t_2-1)\end{bmatrix}.$$

Such matrices have rank $\binom{r-1}{2}$ if $t_i\neq 1$ for all $i=1,\ldots,r$, and hence the depth of a non-coordinate character is $r-1$. On the other hand, for the trivial character $\overline{1}$, the matrix A_n has rank 0 and hence $\overline{1}$ has depth $\binom{r}{2}$ (see Definitions 2.1 and 2.2 for details on the connection between the rank of A_n and the depth of a character).

2.4 Orbicurves

As a general reference for orbifolds and orbifold fundamental groups one can use [1], see also [19, 28]. A brief description of what will be used here follows.

Definition 2.17. An *orbicurve* is a complex orbifold of dimension equal to one. An orbicurve $\mathcal{C}$ is called a *global quotient* if there exists a finite group G acting effectively on a Riemann surface C such that $\mathcal{C}$ is the quotient of C by G with the orbifold structure given by the stabilizers of the G-action on C.

We may think of $\mathcal{C}$ as a Riemann surface with a finite number of points $R := \{P_1, \ldots, P_s\} \subset \mathcal{C}$ labeled with positive integers $\{m(P_1), \ldots, m(P_s)\}$ (for global quotients those are the orders of stabilizers of action of G on C). A neighborhood of a point $P \in \mathcal{C}$ with $m(P) > 0$ is the quotient of a disk (centered at P) by an action of the cyclic group of order $m(P)$ (a rotation).

A small loop around P is considered to be trivial in $\mathcal{C}$ if its lifting in the above quotient map bounds a disk. Following this idea, orbifold fundamental groups can be defined as follows.

Definition 2.18. (*cf.* [1,19,28]) Consider an orbifold $\mathcal{C}$ as above, then the *orbifold fundamental group* of $\mathcal{C}$ is

$$\pi_1^{\text{orb}}(\mathcal{C}) := \pi_1(\mathcal{C} \setminus \{P_1, \ldots, P_s\}) / \langle \mu_j^{m_j} = 1 \rangle$$

where μ_j is a meridian of P_j and $m_j := m(P_j)$.

According to Remark 2.5 the Definition 2.2 can be applied to the case of finitely generated groups. In particular one defines the k-th characteristic variety $V_k^{\text{orb}}(\mathcal{C})$ of an orbicurve $\mathcal{C}$ as $V_k(\pi_1^{\text{orb}}(\mathcal{C}))$. Therefore also the concepts of a character χ on $\mathcal{C}$ and its depth are well defined.

Example 2.19. Let us denote by $\mathbb{P}^1_{m_1,\ldots,m_s,k\infty}$ an orbicurve for which the underlying Riemann surface is $\mathbb{P}^1$ with k points removed and s labeled points with labels $m_1, \ldots, m_s$. If $k \geq 1$ (respectively $k \geq 2$) we also use the notation $\mathbb{C}_{m_1,\ldots,m_s,(k-1)\infty}$ (respectively $\mathbb{C}^*_{m_1,\ldots,m_s,(k-2)\infty}$) for $\mathbb{P}^1_{m_1,\ldots,m_s,k\infty}$. We suppress specification of actual points on $\mathbb{P}^1$. Note that

$$\pi_1^{\text{orb}}(\mathbb{P}^1_{m_1,\ldots,m_s,k\infty}) = \begin{cases} \mathbb{Z}_{m_1}(\mu_1) * \ldots * \mathbb{Z}_{m_s}(\mu_s) * \mathbb{Z} * \overset{k-1}{\ldots} * \mathbb{Z} & \text{if } k > 0 \\ \mathbb{Z}_{m_1}(\mu_1) * \ldots * \mathbb{Z}_{m_s}(\mu_s) / \prod \mu_i & \text{if } k = 0 \end{cases}$$

(here $\mathbb{Z}_m(\mu)$ denotes a cyclic group of order m with a generator μ). Note that a global quotient orbifold of $\mathbb{P}^1 \setminus \{nk \text{ points}\}$ by the cyclic action of order n on $\mathbb{P}^1$ that fixes two points, that is, $[x : y] \mapsto [\xi_n x : y]$ (which fixes $[0 : 1]$ and $[1 : 0]$) is $\mathbb{P}^1_{n,n,k\infty}$.

Interesting examples of elliptic global quotients occur for $\mathbb{P}^1_{2,3,6,k\infty}$, $\mathbb{P}^1_{3,3,3,k\infty}$, and $\mathbb{P}^1_{2,4,4,k\infty}$, which are global orbifolds of elliptic curves $E \setminus \{6k \text{ points}\}$, $E \setminus \{3k \text{ points}\}$, and $E \setminus \{4k \text{ points}\}$ respectively, see [11] for a study of the relationship between these orbifolds ($k = 0$) and the depth of characters of fundamental groups of the complements to plane singular curves.

Definition 2.20. A *marking* on an orbicurve $\mathcal{C}$ (respectively a quasiprojective variety $\mathcal{X}$) is a non-trivial character of its orbifold fundamental group (respectively its fundamental group) of positive depth k, that is, an element of $\operatorname{Char}^{\text{orb}}(\mathcal{C}) := \operatorname{Hom}(\pi_1^{\text{orb}}(\mathcal{C}), \mathbb{C}^*)$ (respectively $\operatorname{Char}(\mathcal{X}) := \operatorname{Hom}(\pi_1(\mathcal{X}), \mathbb{C}^*)$) which is in $V_k^{\text{orb}}(\mathcal{C})$ (respectively $V_k(\mathcal{X})$).

A *marked orbicurve* is a pair $(\mathcal{C}, \rho)$, where $\mathcal{C}$ is an orbicurve and ρ is a marking on $\mathcal{C}$. Analogously, one defines a *marked quasi-projective manifold* as a pair $(\mathcal{X}, \chi)$ consisting of a quasi-projective manifold $\mathcal{X}$ and a marking on it.

A marked orbicurve $(\mathcal{C}, \rho)$ is *a global quotient* if $\mathcal{C}$ is a global quotient of C, where C is a branched cover of $\mathcal{C}$ associated with the unbranched cover of $\mathcal{C} \setminus \{P_1, ..., P_s\}$ corresponding to the kernel of $\pi_1(\mathcal{C} \setminus \{P_1, ..., P_s\}) \to \pi_1^{\text{orb}}(\mathcal{C}) \stackrel{\rho}{\to} \mathbb{C}^*$. In other words, the covering space in Definition 2.17 corresponds to the kernel of ρ.

2.5 Orbifold pencils on quasi-projective manifolds

Definition 2.21. Let $\mathcal{X}$ be a quasi-projective variety, C be a quasi-projective curve, and $\mathcal{C}$ an orbicurve which is a global quotient of C. A *global quotient orbifold pencil* is a map $\phi : \mathcal{X} \to \mathcal{C}$ such that there exists $\Phi : X_G \to C$ where X_G is a quasi-projective manifold endowed with an action of the group G making the following diagram commute:

$$\begin{array}{ccc} X_G & \stackrel{\Phi}{\to} & C \\ \downarrow & & \downarrow \\ \mathcal{X} & \stackrel{\phi}{\to} & \mathcal{C} \end{array} \tag{2.11}$$

The vertical arrows in (2.11) are the quotients by the action of G.

If, in addition, $(\mathcal{X}, \chi)$ and $(\mathcal{C}, \rho)$ are marked, then the global quotient orbifold pencil $\phi : \mathcal{X} \to \mathcal{C}$ called *marked* if $\chi = \phi^*(\rho)$. We will refer to the map of pairs $\phi : (\mathcal{X}, \chi) \to (\mathcal{C}, \rho)$ as a *marked global quotient orbifold pencil* on $(\mathcal{X}, \chi)$ with target $(\mathcal{C}, \rho)$.

Definition 2.22. Global quotient orbifold pencils $\phi_i : (\mathcal{X}, \chi) \to (\mathcal{C}, \rho)$, $i = 1, ..., n$ are called *independent* if the induced maps $\Phi_i : X_G \to C$

define $\mathbb{Z}[G]$-independent morphisms of modules

$$\Phi_{i*} : H_1(X_G, \mathbb{Z}) \to H_1(C, \mathbb{Z}), \tag{2.12}$$

that is, independent elements of the $\mathbb{Z}[G]$-module $\mathrm{Hom}_{\mathbb{Z}[G]}(H_1(X_G, \mathbb{Z}), H_1(C, \mathbb{Z}))$.

In addition, if $\bigoplus \Phi_{i*} : H_1(X_G, \mathbb{Z}) \to H_1(C, \mathbb{Z})^n$ is surjective we say that the pencils ϕ_i are *strongly independent*.

Remark 2.23. Note that if either $n = 1$ or $H_1(C, \mathbb{Z}) = \mathbb{Z}[G]$, then independence is equivalent to strong independence (this is the case for Corollary 2.26(2) and Theorem 1.1(2)).

2.6 Structure of characteristic varieties (revisited)

The following are relevant improvements or additions to Theorem 2.7:

Theorem 2.24 ([8,24]). *The isolated zero-dimensional characters of $V_k(\mathcal{D})$ are torsion characters of* $\mathrm{Char}(\mathcal{D})$.

In [15, Theorem 3.9] (see also [16]) there is a description of one-dimensional components $\chi f^* H^1(C, \mathbb{C}^*) \subset V_k(\mathcal{X})$ mentioned in Theorem 2.7 and most importantly, of the order of χ in terms of multiple fibers of the rational pencil f.

In [23], an algebraic method is described to detect the irregularity of abelian covers of $\mathbb{P}^2$ ramified along $\mathcal{D}$. This method is very useful to compute *non-coordinate components* of $V_k(\mathcal{D})$ independently of a presentation of the fundamental group of the complement $\mathcal{X}$ of $\mathcal{D}$.

Theorem 1.1 (see [7]) has [15, Theorem 3.9] as a consequence, but uses the point of view of orbifold pencils. Using this result also the zero-dimensional components can be detected (in particular essential coordinate components) and in some cases characterized (see Section 4).

Another improvement of Theorem 2.7 was given in [8] were the point of view of orbifolds was first introduced as follows:

Theorem 2.25 ([8]). *Let $\mathcal{X}$ be a smooth quasi-projective variety. Let V be an irreducible component of $V_k(\mathcal{X})$. Then one of the two following statements holds:*

(1) *There exists an orbicurve $\mathcal{C}$, a surjective orbifold morphism $\rho : X \to \mathcal{C}$ and an irreducible component W of $V_k^{\mathrm{orb}}(\mathcal{C})$ such that $V = \rho^*(W)$.*
(2) *V is an isolated torsion point not of type* (1).

One has the following consequences from 1.1 (2) that allows us to characterize certain elements of $V_k(\mathcal{D})$:

Corollary 2.26. *Let $(\mathcal{X}, \chi)$ be a marked complement of $\mathcal{D}$. Then possible targets for marked orbifold pencils are $(\mathcal{C}, \rho)$ with $\mathcal{C} = \mathbb{P}^1_{m_1,\dots,m_s,k\infty}$ (see Example 2.19). Assume that there are n strongly independent marked orbifold pencils with such a fixed target $(\mathcal{C}, \rho)$. Then,*

(1) *In case $\mathcal{C}$ has no orbifold points, that is $s = 0$, the character χ belongs to a positive dimensional component V of* $\operatorname{Char}(\mathcal{X})$ *containing the trivial character. In this case, $d(\chi) = \dim V - 1 = n - 2$.*
(2) *In case χ is a character of order two, there is a unique marking on $\mathcal{C} = \mathbb{C}_{2,2}$ and $d(\chi)$ is the maximal number of strongly independent orbifold pencils with target $\mathcal{C}$.*
(3) *In case χ has torsion 3,4, or 6, there is a unique marking on $\mathcal{C} = \mathbb{P}^1_{3,3,3}$, $\mathcal{C} = \mathbb{P}^1_{2,4,4}$, or $\mathcal{C} = \mathbb{P}^1_{2,3,6}$ respectively and $d(\chi)$ is the maximal number of strongly independent orbifold pencils with target $\mathcal{C}$.*

Part (1) is a direct consequence of Theorem 2.7 and part (3) had already appeared in the context of Alexander polynomials in [11].

In Section 4 we will describe in detail examples of Corollary 2.26 (2) for line arrangements.

2.7 Zariski pairs

We will give a very brief introduction to Zariski pairs. For more details we refer to [9] and the bibliography therein.

Definition 2.27 ([3]). Two plane algebraic curves $\mathcal{D}$ and $\mathcal{D}'$ form a *Zariski pair* if there are homeomorphic tubular neighborhoods of $\mathcal{D}$ and $\mathcal{D}'$, but the pairs $(\mathbb{P}^2, \mathcal{D})$ and $(\mathbb{P}^2, \mathcal{D})$ are not homeomorphic.

The first example of a Zariski pair was given by Zariski [33], who showed that the fundamental group of the complement to an irreducible sextic (a curve of degree six) with six cusps on a conic is isomorphic to $\mathbb{Z}_2 * \mathbb{Z}_3$ whereas the fundamental group of any other sextic with six cusps is $\mathbb{Z}_6$. This paved the way for intensive research aimed to understand the connection between the topology of $(\mathbb{P}^2, \mathcal{D})$ and the position of the singularities of $\mathcal{D}$ (whether algebraically, geometrically, combinatorially...). This research has been often in the direction of a search for finer invariants of $(\mathbb{P}^2, \mathcal{D})$.

Characteristic varieties (described above) and the Alexander polynomials (*i.e.* the one variable version of the characteristic varieties), twisted polynomials [13], generalized Alexander polynomials [11, 26], dihedral covers of $\mathcal{D}$ ([32]) among many others are examples of such invariants.

Definition 2.28. If the Alexander polynomials $\Delta_{\mathcal{D}}(t)$ and $\Delta_{\mathcal{D}'}(t)$ coincide, then we say $\mathcal{D}$ and $\mathcal{D}'$ form an *Alexander-equivalent Zariski pair*.

In Section 5 we will use Theorem 1.1 to give an alternative proof that the curves in [4] Alexander-equivalent Zariski pair, without computing the fundamental group.

3 Examples of characters of depth 3: Fermat Curves

Consider the following family of plane curves:

$$\begin{aligned}
\mathcal{F}_n &:= \{f_n := x_1^n + x_2^n - x_0^n = 0\},\\
\mathcal{L}_1 &:= \{\ell_1 := x_0^n - x_2^n = 0\},\\
\mathcal{L}_2 &:= \{\ell_2 := x_0^n - x_2^n = 0\}.
\end{aligned}$$

We will study the characteristic varieties of the quasi-projective manifolds $\mathcal{X}_n := \mathbb{P}^2 \setminus \mathcal{D}_n$, where $\mathcal{D}_n := \mathcal{F}_n \cup \mathcal{L}_1 \cup \mathcal{L}_2$, in light of the results given in the previous sections, in particular the essential torsion characters will be considered and their depth will be exhibited as the number of strictly independent orbifold pencils.

3.1 Fundamental group

Note that $\mathcal{D}_n$ is nothing but the preimage by the Kummer cover $[x_0 : x_1 : x_2] \overset{\kappa_n}{\mapsto} [x_0^n : x_1^n : x_2^n]$ of the following arrangement of three lines in general position given by the equation

$$(x_0 - x_1)(x_0 - x_2)(x_0 - x_1 - x_2) = 0.$$

Such a map ramifies along $\mathcal{B} := \{x_0x_1x_2 = 0\}$. We will compute the fundamental group of $\mathcal{X}_n$ as a quotient of the subgroup K_n of $\pi_1(\mathbb{P}^2 \setminus \mathcal{L})$ associated with the Kummer cover, where

$$\mathcal{L} := \{x_0x_1x_2(x_0 - x_2)(x_0 - x_1)(x_0 - x_1 - x_2) = 0\} \tag{3.1}$$

is a Ceva arrangement. More precisely, the quotient is obtained as a factor of K_n by the normal subgroup generated by the meridians of the ramification locus $\kappa_n^{-1}(\mathcal{B})$ in $\mathcal{X}_n$.

The fundamental group of the complement to the Ceva arrangement $\mathcal{L}$ is given by the following presentation of G.

$$\langle e_0,\dots, e_5 : [e_1,e_2]=[e_3,e_5,e_1]=[e_3,e_4]=[e_5,e_2,e_4]=e_4e_3e_5e_2e_1e_0=1\rangle \tag{3.2}$$

where e_i is a meridian of the component appearing in the $(i+1)$-th place in (3.1), $[\alpha, \beta]$ denotes the commutator $\alpha\beta\alpha^{-1}\beta^{-1}$, and $[\alpha, \beta, \delta]$ denotes the triple of commutators $[\alpha\beta\delta, \alpha]$, $[\alpha\beta\delta, \beta]$, and $[\alpha\beta\delta, \delta]$ leading to a triple of relations in (3.2).

In other to obtain (3.2) one can use the *non-generic* Zariski-Van Kampen method on Figure 3.1 (see [9, Section 1.4]). The dotted line ℓ represents a generic line where the meridians $e_0, \dots, e_5$ are placed (note that the last relation on (3.2) is the relation in the fundamental group of $\ell \setminus (\mathcal{L} \cap \ell) \approx \mathbb{P}^1_{6\infty}$). The first two relations on (3.2) appear when moving the generic line around ℓ_1. The third and fourth relations come from moving the generic line around ℓ_4.

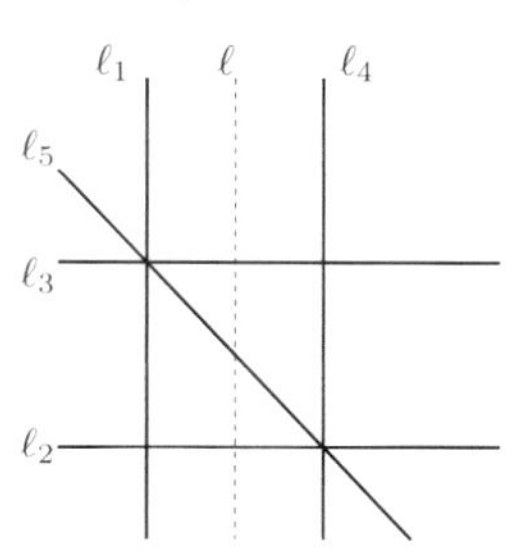

Figure 3.1. Ceva arrangement.

The fundamental group of the complement to $\mathcal{D}_n \cup \mathcal{B}$ is equal to the kernel K_n of the epimorphism

$$
\begin{array}{ccc}
G & \stackrel{\alpha}{\to} & \mathbb{Z}_n \times \mathbb{Z}_n \\
e_0 & \mapsto & (1,1) \\
e_1 & \mapsto & (1,0) \\
e_2 & \mapsto & (0,1) \\
e_3 & \mapsto & (0,0) \\
e_4 & \mapsto & (0,0) \\
e_5 & \mapsto & (0,0)
\end{array}
\tag{3.3}
$$

since it is the fundamental group of the abelian cover with covering transformations $\mathbb{Z}_n \times \mathbb{Z}_n$. Therefore a presentation of the fundamental group of the complement to $\mathcal{D}_n$ can be obtained by taking a factor of K_n by the normal subgroup generated by e_0^n, e_1^n, and e_2^n (which are the meridians to the preimages of the lines x_0, x_1, and x_2 respectively). Using the Reidemeister-Schreier method (*cf.* [21]) combined with the triviality of e_0^n, e_1^n, and e_2^n one obtains the following presentation for $G_n := \pi_1(\mathcal{X}_n)$:

$$
G_n = \left\langle\, e_{3,i,j}, e_{4,i,j}, e_{5,i,j} : \begin{array}{l}
(R1)\ e_{3,i+1,j} = e_{5,i,j}^{-1} e_{3,i,j} e_{5,i,j}, \\
(R2)\ e_{4,i,j+1} = e_{5,i,j}^{-1} e_{4,i,j} e_{5,i,j}, \\
(R3)\ e_{5,i+1,j} = e_{5,i,j}^{-1} e_{3,i,j}^{-1} e_{5,i,j} e_{3,i,j} e_{5,i,j}, \\
(R4)\ e_{5,i,j+1} = e_{5,i,j}^{-1} e_{4,i,j}^{-1} e_{5,i,j} e_{4,i,j} e_{5,i,j}, \\
(R5)\ [e_{3,i,j}, e_{4,i,j}] = 1, \\
(R6)\ \prod_{k=0}^{n-1} e_{4,k,k} e_{3,k,k} e_{5,k,k} = 1
\end{array} \right\rangle
\tag{3.4}
$$

where $i, j \in \mathbb{Z}_n$ and

$$e_{k,i,j} := e_1^i e_2^j e_k e_2^{-j} e_1^{-i}, \quad k = 3, 4, 5.$$

As a brief description of the Reidemeister-Schreier method, we recall that the generators of G_n are obtained from a set-theoretical section of α in (3.3) (in our case $s : \mathbb{Z}_n \times \mathbb{Z}_n \to G$ is given by $(i, j) \mapsto e_1^i e_2^j$) as follows

$$s(i, j)\, e_k\, (\alpha(e_k)s(i, j))^{-1}.$$

Thus the set $\{e_{k,i,j}\}$ above forms a set of generators of G_n. Finally a complete set of relations can be obtained by rewriting the relations of G in (3.2) (and their conjugates by $s(i, j)$) in terms of the generators of the subgroup G_n.

Example 3.1. In order to illustrate the rewriting method we will proceed with the second relation of G in (3.2).

$$\begin{aligned} s(i, j)[e_1, e_2]s(i, j)^{-1} &= e_1^i e_2^j (e_3 e_4 e_3^{-1} e_4^{-1}) e_2^{-j} e_1^{-i} \\ &= (e_1^i e_2^j e_3 e_2^{-j} e_1^{-i})\, (e_1^i e_2^j e_4 e_2^{-j} e_1^{-i})\, (e_1^i e_2^j e_3^{-1} e_2^{-j} e_1^{-i})\, (e_1^i e_2^j e_4^{-1} e_2^{-j} e_1^{-i}) \\ &= [e_{3,i,j}, e_{4,i,j}] \end{aligned}$$

3.2 Essential coordinate characteristic varieties

Now we will discuss a presentation of G'_n/G''_n as a module over G_n/G'_n, which will be referred to as $M_{\mathcal{D}_n,\mathbf{ab}}$. For details we refer to Section 2.3. Note that G_n/G'_n is isomorphic to $\mathbb{Z}^{2n}$ and is generated by the cycles γ_5, $\gamma_{3,j}$, $\gamma_{4,i}$, $(i, j \in \mathbb{Z}_n)$ where $\gamma_5 = \mathbf{ab}(e_{5,i,j})$, $\gamma_{3,j} = \mathbf{ab}(e_{3,i,j})$, and $\gamma_{4,i} = \mathbf{ab}(e_{4,i,j})$ satisfying $n\gamma_5 + \sum_j \gamma_{3,j} + \sum_i \gamma_{4,i} = 0$[4]. Let t_5 (respectively $t_{3,j}$, $t_{4,i}$) be the generators of G_n/G'_n viewed as a multiplicative group corresponding to the additive generators γ_5 (respectively $\gamma_{3,j}$, $\gamma_{4,i}$). The characteristic varieties of G_n are contained in

$$(\mathbb{C}^*)^{2n} = \operatorname{Spec} \mathbb{C}[t_5^{\pm 1}, t_{3,i}^{\pm 1}, t_{4,j}^{\pm 1}] \Big/ \Big(t_5^n \prod_j t_{3,j} \prod_i t_{4,i} - 1 \Big).$$

As generators of $M_{\mathcal{D}_n,\mathbf{ab}}$ we select commutators of the generators of G_n as given in (3.2). In order to do so, note that using relations $(R1) - (R4)$ in (3.4), a presentation of G_n can be given in terms the $2n + 1$ generators $e_5 := e_{5,0,0}$, $e_{3,j} := e_{3,0,j}$, and $e_{4,i} := e_{4,i,0}$. Hence, by Proposition 2.12, $M_{\mathcal{D}_n,\mathbf{ab}}$ is generated by the $\binom{2n+1}{2}$ commutators

$$\{[e_5, e_{3,j}], [e_5, e_{4,i}], [e_{4,i}, e_{3,j}], [e_{4,i_1}, e_{4,i_2}], [e_{3,j_1}, e_{3,j_2}]\}_{i_*, j_* \in \mathbb{Z}_n}, \tag{3.5}$$

[4] Recall that **ab** is the morphism of abelianization.

as a $\mathbb{C}[\mathbb{Z}[t_1^{\pm 1}, \ldots, t_4^{\pm 1}, t_5^{\pm 1}]]$-module. Also, according to Proposition 2.15, a complete set of relations of $M_{\mathcal{D}_n,\mathbf{ab}}$ is given by rewriting the following relations

$$\begin{array}{lll}
(M1) & [\prod_{i=0}^{n-1} e_{5,i,j}, e_{3,j}] & = 0 \\
(M2) & [\prod_{i=0}^{n-1} e_{5,i,j}, e_{4,i}] & = 0 \\
(M3) & [e_{5,i,j+1}, e_{3,i,j+1} e_{5,i,j+1}] e_{5,i,j+1}^{-1} & = [e_{5,i+1,j}, e_{4,i+1,j} e_{5,i+1,j}] e_{5,i+1,j}^{-1} \\
(M4) & \prod_{i=0}^{n-1} e_{4,i,i} e_{3,i,i} e_{5,i,i} & = 0
\end{array} \tag{3.6}$$

in terms of commutators (3.5) and by the Jacobian relations:

$$\begin{array}{ll}
(t_{3,j} - 1)[e_5, e_{4,i}] + (t_{4,i} - 1)[e_{3,j}, e_5] + (t_5 - 1)[e_{4,i}, e_{3,j}] & = 0, \\
(t_{3,j_1} - 1)[e_5, e_{3,j_2}] + (t_{3,j_2} - 1)[e_{3,j_1}, e_5] + (t_5 - 1)[e_{3,j_2}, e_{3,j_1}] & = 0, \\
(t_{4,i_1} - 1)[e_5, e_{4,i_2}] + (t_{4,i_2} - 1)[e_{4,i_1}, e_5] + (t_5 - 1)[e_{4,i_2}, e_{4,i_1}] & = 0, \\
\ldots &
\end{array} \tag{3.7}$$

In order to rewrite relations $(M1) - (M4)$ one needs to use (3.5) repeatedly. In what follows, we will concentrate on the characters of $\mathrm{Char}(\mathcal{D}_n)$ contained in the coordinate axes $t_{3,j} = t_{4,i} = 1$. Computations for the general case can also be performed, but are more technical and tedious.

Since we are assuming $t_{3,j} = t_{4,i} = 1$, and $t_5 \neq 1$, relations in (3.7) become $[e_{4,i}, e_{3,j}] = [e_{3,j_2}, e_{3,j_1}] = [e_{4,i_2}, e_{4,i_1}] = 0$ and hence $(R5)$ in (3.4) become redundant. A straightforward computation gives the following matrix where each line is a relation from (3.6) written in terms of the commutators $\{[e_5, e_{3,i}], [e_5, e_{4,i}]\}_{i \in \mathbb{Z}_n}$.

$$A_n := \left[\begin{array}{cccccc|cccccc}
\phi_n & 0 & 0 & \ldots & 0 & 0 & 0 & 0 & 0 & \ldots & 0 & 0 \\
0 & \phi_n & 0 & \ldots & 0 & 0 & 0 & 0 & 0 & \ldots & 0 & 0 \\
 & & & \vdots & & & & & & \vdots & & \\
0 & 0 & 0 & \ldots & 0 & \phi_n & 0 & 0 & 0 & \ldots & 0 & 0 \\
\hline\hline
0 & 0 & 0 & \ldots & 0 & 0 & \phi_n & 0 & 0 & \ldots & 0 & 0 \\
0 & 0 & 0 & \ldots & 0 & 0 & 0 & \phi_n & 0 & \ldots & 0 & 0 \\
 & & & \vdots & & & & & & \vdots & & \\
0 & 0 & 0 & \ldots & 0 & 0 & 0 & 0 & 0 & \ldots & 0 & \phi_n \\
\hline\hline
1 & -1 & 0 & \ldots & 0 & 0 & 1 & -1 & 0 & \ldots & 0 & 0 \\
1 & -1 & 0 & \ldots & 0 & 0 & 0 & t & -t & \ldots & 0 & 0 \\
 & & & \vdots & & & & & & \vdots & & \\
1 & -1 & 0 & \ldots & 0 & 0 & 0 & 0 & 0 & \ldots & t^{n-2} & -t^{n-2} \\
1 & -1 & 0 & \ldots & 0 & 0 & -t^{n-1} & 0 & 0 & \ldots & 0 & t^{n-1} \\
\hline
 & & & \vdots & & & & & & \vdots & & \\
\hline
0 & 0 & 0 & \ldots & t^{n-2} & -t^{n-2} & 1 & -1 & 0 & \ldots & 0 & 0 \\
0 & 0 & 0 & \ldots & t^{n-2} & -t^{n-2} & 0 & t & -t & \ldots & 0 & 0 \\
 & & & \vdots & & & & & & \vdots & & \\
0 & 0 & 0 & \ldots & t^{n-2} & -t^{n-2} & 0 & 0 & 0 & \ldots & t^{n-2} & -t^{n-2} \\
0 & 0 & 0 & \ldots & t^{n-2} & -t^{n-2} & -t^{n-1} & 0 & 0 & \ldots & 0 & t^{n-1} \\
\hline\hline
1 & t & t^2 & \ldots & t^{n-2} & t^{n-1} & 1 & t & t^2 & \ldots & t^{n-2} & t^{n-1}
\end{array}\right]$$

More precisely, the first (respectively second) block of A_n corresponds to the n relations given in $(M1)$ (respectively $(M2)$) of (3.6), $\phi_n := \frac{t^n-1}{t-1}$, and $t = t_5$. The following n blocks of A_n (between double horizontal lines) correspond to the n^2 relations given in $(M3)$ of (3.6). Note that the last row of each of these blocks is a consequence of the remaining $n-1$ rows. The last block corresponds to the relation given in $(M4)$ of (3.6).

Example 3.2. In order to illustrate A_n we will show how to rewrite the first relation for $n = 3$, that is,

$$[e_{5,0,j}e_{5,1,j}e_{5,2,j}, e_{3,j}] \overset{M}{=} \phi_n[e_5, e_{3,j}].$$

Using (2.6) one has

$$[e_{5,0,j}e_{5,1,j}e_{5,2,j}, e_{3,j}] \overset{M}{=} [e_{5,0,j}, e_{3,j}] + t[e_{5,1,j}, e_{3,j}] + t^2[e_{5,0,j}, e_{3,j}].$$

Therefore, it is enough to show that $[e_{5,i,j}, e_{3,j}] = [e_5, e_{3,j}]$. Note that $e_{5,i,j}$ is a conjugate of e_5 (using $(R3)$ and $(R4)$), hence, by (2.7) one obtains $[e_{5,i,j}, e_{3,j}] = [e_5, e_{3,j}]$ (since we are assuming $t_{3,j} = 1$).

Also note that, performing row operations, one can obtain the following equivalent matrix

$$B_n := \left[\begin{array}{cccccc|cccccc}
\phi_n & 0 & 0 & \dots & 0 & 0 & 0 & 0 & 0 & \dots & 0 & 0 \\
0 & \phi_n & 0 & \dots & 0 & 0 & 0 & 0 & 0 & \dots & 0 & 0 \\
 & & & \vdots & & & & & & \vdots & & \\
0 & 0 & 0 & \dots & 0 & \phi_n & 0 & 0 & 0 & \dots & 0 & 0 \\
\hline
0 & 0 & 0 & \dots & 0 & 0 & \phi_n & 0 & 0 & \dots & 0 & 0 \\
0 & 0 & 0 & \dots & 0 & 0 & 0 & \phi_n & 0 & \dots & 0 & 0 \\
 & & & \vdots & & & & & & \vdots & & \\
0 & 0 & 0 & \dots & 0 & 0 & 0 & 0 & 0 & \dots & 0 & \phi_n \\
\hline
1 & -1 & 0 & \dots & 0 & 0 & 1 & -1 & 0 & \dots & 0 & 0 \\
1 & -1 & 0 & \dots & 0 & 0 & 0 & t & -t & \dots & 0 & 0 \\
 & & & \vdots & & & & & & \vdots & & \\
1 & -1 & 0 & \dots & 0 & 0 & 0 & 0 & 0 & \dots & t^{n-2} & -t^{n-2} \\
0 & t & -t & \dots & 0 & 0 & 0 & 0 & 0 & \dots & t^{n-2} & -t^{n-2} \\
 & & & \vdots & & & & & & \vdots & & \\
0 & 0 & 0 & \dots & t^{n-2} & -t^{n-2} & 0 & 0 & 0 & \dots & t^{n-2} & -t^{n-2} \\
\hline
1 & t & t^2 & \dots & t^{n-2} & t^{n-1} & 1 & t & t^2 & \dots & t^{n-2} & t^{n-1}
\end{array}\right]$$

Finally, one can write the presentation matrix B_n in terms of the basis

$$\Big\{[e_5,e_{3,i}] - [e_5,e_{3,i+1}], [e_5,e_{3,n-1}], [e_5,e_{4,i}] \\ - [e_5,e_{4,i+1}], [e_5,e_{4,n-1}]\Big\}_{i=0,\dots,n-2}$$

resulting in

$$\left[\begin{array}{cccccc|cccccc}
\phi_n & \phi_n & 0 & \dots & 0 & 0 & 0 & 0 & 0 & \dots & 0 & 0\\
0 & \phi_n & \phi_n & \dots & 0 & 0 & 0 & 0 & 0 & \dots & 0 & 0\\
 & & & \vdots & & & & & & \vdots & & \\
0 & 0 & 0 & \dots & \phi_n & \phi_n & 0 & 0 & 0 & \dots & 0 & 0\\
0 & 0 & 0 & \dots & 0 & \phi_n & 0 & 0 & 0 & \dots & 0 & 0\\
\hline
0 & 0 & 0 & \dots & 0 & 0 & \phi_n & \phi_n & 0 & \dots & 0 & 0\\
0 & 0 & 0 & \dots & 0 & 0 & 0 & \phi_n & \phi_n & \dots & 0 & 0\\
 & & & \vdots & & & & & & \vdots & & \\
0 & 0 & 0 & \dots & 0 & 0 & 0 & 0 & 0 & \dots & \phi_n & \phi_n\\
0 & 0 & 0 & \dots & 0 & 0 & 0 & 0 & 0 & \dots & 0 & \phi_n\\
\hline
1 & 0 & 0 & \dots & 0 & 0 & 1 & 0 & 0 & \dots & 0 & 0\\
1 & 0 & 0 & \dots & 0 & 0 & 0 & t & 0 & \dots & 0 & 0\\
1 & 0 & 0 & \dots & 0 & 0 & 0 & 0 & t^2 & \dots & 0 & 0\\
 & & & \vdots & & & & & & \vdots & & \\
1 & 0 & 0 & \dots & 0 & 0 & 0 & 0 & 0 & \dots & t^{n-2} & 0\\
0 & t & 0 & \dots & 0 & 0 & 0 & 0 & 0 & \dots & t^{n-2} & 0\\
0 & 0 & t^2 & \dots & 0 & 0 & 0 & 0 & 0 & \dots & t^{n-2} & 0\\
 & & & \vdots & & & & & & \vdots & & \\
0 & 0 & 0 & \dots & t^{n-2} & 0 & 0 & 0 & 0 & \dots & t^{n-2} & 0\\
\hline
\phi_1 & \phi_2 & \phi_3 & \dots & \phi_{n-1} & \phi_n & \phi_1 & \phi_2 & \phi_3 & \dots & \phi_{n-1} & \phi_n
\end{array}\right]$$

One can use the units in the third block to eliminate columns, leaving the equivalent matrix

$$\left[\begin{array}{cccccc|cccccc}
\phi_n & 0 & 0 & \dots & 0 & 0 & 0 & 0 & 0 & \dots & 0 & 0\\
0 & 0 & 0 & \dots & 0 & 0 & 0 & 0 & 0 & \dots & -\phi_n & 0\\
0 & 0 & 0 & \dots & 0 & \phi_n & 0 & 0 & 0 & \dots & 0 & 0\\
\hline
0 & 0 & 0 & \dots & 0 & 0 & 0 & 0 & 0 & \dots & 0 & \phi_n\\
\hline
1 & 0 & 0 & \dots & 0 & 0 & 1 & 0 & 0 & \dots & 0 & 0\\
1 & 0 & 0 & \dots & 0 & 0 & 0 & 0 & 0 & \dots & t^{n-2} & 0
\end{array}\right] \cong \left[\begin{array}{c|ccc}
0 & -\phi_n & 0 & 0\\
0 & 0 & -\phi_n & 0\\
\phi_n & 0 & 0 & 0\\
\hline
0 & 0 & 0 & \phi_n\\
\hline
0 & -1 & t^{n-2} & 0
\end{array}\right].$$

Finally, a last combination of row operations using the units to eliminate columns results in

$$\left[\begin{array}{c|ccc}
0 & 0 & -\phi_n t^{n-2} & 0\\
0 & 0 & -\phi_n & 0\\
\phi_n & 0 & 0 & 0\\
\hline
0 & 0 & 0 & \phi_n\\
\hline
0 & -1 & t^{n-2} & 0
\end{array}\right] \cong \left[\begin{array}{c|cc}
0 & -\phi_n t^{n-2} & 0\\
0 & -\phi_n & 0\\
\phi_n & 0 & 0\\
\hline
0 & 0 & \phi_n
\end{array}\right] \cong \left[\begin{array}{c|cc}
0 & \phi_n & 0\\
\phi_n & 0 & 0\\
\hline
0 & 0 & \phi_n
\end{array}\right].$$

Hence the $n-1$ non-trivial torsion characters $\chi_n^i := (\xi_n^i, 1, \dots, 1)$, $i = 1, \dots, n$ belong to $\mathrm{Char}(\mathcal{D}_n)$ and have depth 3, that is, $\chi_n^i \in V_3(\mathcal{D}_n)$.

3.3 Marked orbifold pencils

By Theorem 1.1 (1) we know there are at most three strongly independent marked orbifold pencils from the marked variety $(\mathcal{X}_n, \chi_n)$. Our purpose is to explicitly show such three strongly independent pencils. Note that

$$\begin{aligned} j_k : \mathbb{P}^2 \setminus (\mathcal{F}_n \cup \mathcal{L}_k) &\to \mathbb{C}_n^* = \mathbb{P}^1_{(n,[1:0]),(\infty,[0:1]),(\infty,[1:1])} \\ [x:y:z] &\mapsto [f_n : x_k^n], \end{aligned} \tag{3.8}$$

for $j = 1, 2$ are two natural orbifold pencils coming from the n-ordinary points of $\mathcal{F}_n$ coming form the triple points of the Ceva arrangement $\mathcal{L}$ which are in $\mathcal{B}$. Consider the marked orbicurve $(\mathbb{C}_{n,n}, \rho_n)$, where $\rho_n = (\xi_n, 1)$, the first coordinate corresponds to the image of a meridian μ_1 around $[0:1] \in \mathbb{P}^1_{(n,[0:1]),(n,[1:0]),(\infty,[1:1])}$ and the second coordinate corresponds to the image of a meridian μ_2 around $[1:0]$ (note that $\pi_1^{\text{orb}}(\mathbb{C}_{n,n}) = \mathbb{Z}_n(\mu_1) * \mathbb{Z}_n(\mu_2)$).

In order to obtain marked orbifold pencils with target $(\mathbb{C}_{n,n}, \rho_n)$ one simply considers the following composition, where i_k and j_k are inclusions

$$\begin{aligned}\psi_k : \mathcal{X}_n \overset{i_k}{\hookrightarrow} \mathbb{P}^2 \setminus (\mathcal{F}_n \cup \mathcal{L}_k) &\overset{j_k}{\to} \mathbb{P}^1_{(n,[1:0]),(\infty,[0:1]),(\infty,[1:1])} \\ &\overset{i}{\hookrightarrow} \mathbb{P}^1_{(n,[0:1]),(n,[1:0]),(\infty,[1:1])}.\end{aligned}$$

Such pencils are clearly marked global quotient orbifold pencils from $(\mathcal{X}_n, \chi_n)$ to $(\mathbb{C}_{n,n}, \rho_n)$, where $(\mathbb{C}_{n,n}, \rho_n)$ is the marked quotient of $C_n := \mathbb{P}^1 \setminus \{[\xi_n^j : 1]\}_{j \in \mathbb{Z}_n}$ by the cyclic action $[x : y] \mapsto [\xi_n x : y]$. The resulting commutative diagrams are given by

$$\begin{array}{ccc} X_n & \overset{\Psi_k}{\to} & C_n \\ [x_0 : x_1 : x_2 : w] & \mapsto & [w : x_k] \\ \downarrow \pi & & \downarrow \\ \mathcal{X}_n & \overset{\psi_k}{\to} & \mathbb{C}_{n,n} \\ [x_0 : x_1 : x_2] & \mapsto & [f_n : x_k^n], \end{array} \tag{3.9}$$

$k = 1, 2$, where X_n is the smooth open surface given by $\{[x_0 : x_1 : x_2 : w] \in \mathbb{P}^3 \mid w^n = f_n\} \setminus \{f_n \ell_1 \ell_2 = 0\}$.

Note that there is a third quasitoric relation involving all components of $\mathcal{D}_n$, namely,

$$f_n x_0^n + \ell_1 \ell_2 = x_1^n x_2^n \tag{3.10}$$

and hence a global quotient marked orbifold map

$$\begin{array}{rl} \psi_3 : \quad \mathcal{X}_n & \to \mathbb{C}_{n,n} = \mathbb{P}^1_{(n,[0:1]),(n,[1:0]),(\infty,[1:1])} \\ [x : y : z] & \mapsto [-f_n x_0^n : x_1^n x_2^n], \end{array} \tag{3.11}$$

which gives rise to the following diagram

$$\begin{array}{ccc} X_n & \overset{\Psi_3}{\to} & C_n \\ [x_0 : x_1 : x_2 : w] & \mapsto & [-w x_0 : x_1 x_2] \\ \downarrow \pi_n & & \downarrow \\ \mathcal{X}_n & \overset{\psi_k}{\to} & \mathbb{C}_{n,n} \\ [x_0 : x_1 : x_2] & \mapsto & [-f_n x_0^n : x_1^n x_2^n]. \end{array} \tag{3.12}$$

Note that, when extending π_n to a branched covering, the preimage of each line $\{\ell_{k,i} = 0\} \subset \mathcal{L}_k$ $(k = 1, 2)$ in $\mathcal{D}_n$ $(\ell_{k,i} := x_0 - \xi_n^i x_k)$ decomposes into n irreducible components $\bigcup_{j\in\mathbb{Z}_n} \ell_{k,i,j}$ and thus allows to consider $\gamma_{k,i,j}$ $(k = 1, 2, i, j \in \mathbb{Z}_n)$ meridians around each component of $\{\ell_{k,i,j} = 0\}$. Also consider a meridian γ_0 around the preimage of $\mathcal{F}_n$.

Theorem 3.3. *The marked orbifold pencils ψ_1, ψ_2, and ψ_3 described above are strongly independent and hence they form a maximal set of strongly independent pencils.*

Proof. Consider $\Psi_{\varepsilon,*} : H_1(X_n; \mathbb{Z}) \to H_1(C_n; \mathbb{Z}) = \mathbb{Z}[\xi_n]$, $\varepsilon = 1, 2, 3$ the three equivariant morphisms described above. Using the commutative diagrams (3.9) and (3.12) one can easily see that

$$\Psi_{\varepsilon,*}(\gamma_{k,i,j}) = \begin{cases} \xi_n^j & \text{if } \varepsilon = k \in \{1, 2\} \\ \xi_n^{i+j} & \text{if } k = 3 \\ 0 & \text{otherwise} \end{cases} \tag{3.13}$$

and

$$\Psi_{\varepsilon,*}(\gamma_0) = 0$$

and therefore $\Psi_{\varepsilon,*}$ are surjective $\mathbb{Z}[\xi_n]$-module morphisms. Also note that $[\gamma_{k,i,j}] = \mu_n^j[\gamma_{k,i,0}] \in H_1(X_n; \mathbb{Z})$. Consequently according to (3.13) one has

$$\left(\Psi_{1,*} \oplus \Psi_{2,*} \oplus \Psi_{3,*}\right)(\gamma_{k,i,0}) = \begin{cases} (1, 0, \xi_n^i) & \text{if } k = 1 \\ (0, 1, \xi_n^i) & \text{if } k = 2 \end{cases}$$

which implies that $\Psi_{1,*} \oplus \Psi_{2,*} \oplus \Psi_{3,*}$ is surjective. After the discussion of Section 3.2, since the depth of ξ_n^i is three, the set of strongly independent pencils is indeed maximal. □

4 Order two characters: augmented Ceva

From Theorem 1.1(2), for any order two character χ of depth k in the characteristic variety of the complement of a curve there exist k independent pencils associated with χ whose target is a global quotient orbifold of type $\mathbb{C}_{2,2}$.

Interesting examples for $k > 1$ of this scenario are the augmented Ceva arrangements CEVA$(2, s)$, $s = 1, 2, 3$ (or *erweiterte Ceva cf.* [10, Section 2.3.J, page 81]). Consider the following set of lines:

$$\begin{array}{lll} \ell_1 := x & \ell_4 := (y - z) & \ell_7 := (x - y - z) \\ \ell_2 := y & \ell_5 := (x - z) & \ell_8 := (y - z - x) \\ \ell_3 := z & \ell_6 := (x - y) & \ell_9 := (z - x - y). \end{array} \tag{4.1}$$

The curve $\mathcal{C}_6 := \left\{\prod_{i=1}^6 \ell_i = 0\right\}$ is a realization of the Ceva arrangement CEVA(2) (a.k.a. braid arrangement or B_3-reflection arrangement). Note that this realization is different from the one used in Section 3. The curve $\mathcal{C}_7 := \left\{\prod_{i=1}^7 \ell_i = 0\right\}$ is the augmented Ceva arrangement CEVA(2, 1) (a.k.a. a realization of the non-Fano plane). The curve $\mathcal{C}_8 := \left\{\prod_{i=1}^8 \ell_i = 0\right\}$ is the augmented Ceva arrangement CEVA(2, 2) (a.k.a. a deleted B_3-arrangement). Finally, $\mathcal{C}_9 := \left\{\prod_{i=1}^9 \ell_i = 0\right\}$ is the augmented Ceva arrangement CEVA(2, 3).

The characteristic varieties of such arrangements of lines are well known (*cf.* [15, 30, 31]). Such computations are done via a presentation of the fundamental group and using Fox derivatives. In most cases (except for the simplest ones) the need of computer support is basically unavoidable. In [15, Example 3.11] there is an alternative calculation of the positive dimensional components of depth 1 via pencils.

Here we will give an interpretation via orbifold pencils of the characters of depth 2, which will account for the appearance of these components of the characteristic varieties independently of computation of the fundamental group.

4.1 Ceva pencils and augmented Ceva pencils

Note that $x(y-z) - y(x-z) + z(x-y) = 0$ and hence

$$\begin{array}{rccc} f_C: & \mathbb{P}^2 & \to & \mathbb{P}^1 \\ & [x:y:z] & \mapsto & [\ell_1\ell_4 : \ell_2\ell_5] \end{array}$$

is a pencil of conics such that ($f_C^*([0:1]) = \ell_1\ell_4$, $f_C^*([1:0]) = \ell_2\ell_5$, $f_C^{-1}([1:1]) = \ell_3\ell_6$) (we will refer to it as the *Ceva pencil*). Analogously

$$x(y-z)(x-y-z)^2 - y(x-z)(y-z-x)^2 + z(x-y)(z-x-y)^2 = 0$$

and hence

$$\begin{array}{rccc} f_{SC}: & \mathbb{P}^2 & \to & \mathbb{P}^1 \\ & [x:y:z] & \mapsto & [\ell_1\ell_4\ell_7^2 : \ell_2\ell_5\ell_8^2] \end{array}$$

is a pencil of quartics such that ($f_{SC}^*([0:1]) = \ell_1\ell_4\ell_7^2$, $f_{SC}^*([1:0]) = \ell_2\ell_5\ell_8^2$, $f_{SC}^*([1:1]) = \ell_3\ell_6\ell_9^2$) (we will refer to it as the *augmented Ceva pencil*).

4.2 Characteristic varieties of $\mathcal{C}_i$, $i = 6, 7, 8, 9$

We include the structure of the characteristic varieties of these curves for the reader's convenience. As reference for such computations see [14, 18, 23, 25, 30, 31].

We will denote by $\mathcal{X}_*$ the complement of the curve $\mathcal{C}_*$ in $\mathbb{P}^2$, for $* = 6, 7, 8, 9$.

4.2.1 Arrangement $\mathcal{C}_6$. The characteristic variety $\operatorname{Char}(\mathcal{C}_6)$ consists of four non-essential coordinate components associated with the four triple points of $\mathcal{C}_6$ (see Remark 2.26 (1))[5] and one essential component of dimension 2 and depth 1 given by the Ceva pencil

$$\psi_6 := f_C|_{\mathcal{X}_6} : \mathcal{X}_6 \to \mathbb{P}^1 \setminus \{[0:1], [1:0], [1:1]\}.$$

4.2.2 Arrangement $\mathcal{C}_7$. The characteristic variety $\operatorname{Char}(\mathcal{C}_7)$ consists of six (respectively four) non-essential coordinate components associated with the six triple points of $\mathcal{C}_7$ (respectively four $\mathcal{C}_6$-subarrangements) of dimension 2 and depth 1. In addition, there is one extra character of order two, namely,

$$\chi_7 := (1, -1, -1, 1, -1, -1, 1)$$

of depth 2.[6] In order to check the value of the depth, one needs to find all marked orbifold pencils in $(\mathcal{X}_7, \chi_7)$ of target $(\mathbb{C}_{2,2}, \rho)$ where $\rho := (-1, -1)$ is the only possible non-trivial character of $\mathbb{C}_{2,2}$. Two such independent pencils are the following,

$$\psi_{7,1} := f_C|_{\mathcal{X}_7} : \mathcal{X}_7 \to \mathbb{P}^1 \setminus \{[0:1], [1:0], [1:1]\} \to \mathbb{P}^1_{(2,[1:0]),(2,[1:1]),(\infty[0:1])}$$

and

$$\psi_{7,2} := f_{SC}|_{\mathcal{X}_7} : \mathcal{X}_7 \to \mathbb{P}^1_{(2,[1:0]),(2,[1:1]),(\infty[0:1])}.$$

This is the maximal number of independent pencils by Theorem 1.1.

4.2.3 Arrangement $\mathcal{C}_8$. The characteristic variety $\operatorname{Char}(\mathcal{C}_8)$ consists of six (respectively five) non-essential coordinate components associated with the six triple points of $\mathcal{C}_8$ (respectively four $\mathcal{C}_6$-subarrangements) of dimension 2 and depth 1. In addition, there is one 3-dimensional non-essential coordinate component of depth 2 associated with its quadruple point (see Corollary 2.26(2)).

Consider the following augmented Ceva pencil

$$\psi_{8,1} := f_{SC}|_{\mathcal{X}_8} : \mathcal{X}_8 \to \mathbb{P}^1_{(2,[1:1]),(\infty[0:1]),(\infty[1:0])}.$$

---

[5] a.k.a. local components.

[6] The subscript 7 refers to the arrangement $\mathcal{C}_7$. Similar notation will be used in the examples that follow. A second subscript (when necessary) will be used to index the characters considered.

Computation of the induced map on the variety of characters shows that this map yields the only non-coordinate translated component of dimension 1 and depth 1 observed in the references above. Finally, there are two characters of order two, namely,

$$\begin{aligned}\chi_{8,1} &:= (1, -1, -1, 1, -1, -1, 1, 1) \text{ and}\\ \chi_{8,2} &:= (-1, 1, -1, -1, 1, -1, 1, 1)\end{aligned}$$

of depth 2. In order to check the value of the depth, one needs to find two marked orbifold pencils on $(\mathcal{X}_8, \chi_{8,1})$ with target $(\mathbb{C}_{2,2}, \rho)$, where

$$\mathbb{C}_{2,2} := \mathbb{P}^1_{(2,[1:0]),(2,[1:1]),(\infty[0:1])}$$

and $\rho := (-1, -1, 1)$ is the only non-trivial character of $\mathbb{C}_{2,2}$. Two such independent pencils can, for example, be given as follows

$$\psi_{8,2} := f_C|_{\mathcal{X}_8} : \mathcal{X}_8 \to \mathbb{P}^1 \setminus \{[0:1], [1:0], [1:1]\} \to \mathbb{P}^1_{(2,[1:0]),(2,[1:1]),(\infty[0:1])}$$

and

$$\psi_{8,3} := f_{SC}|_{\mathcal{X}_8} : \mathcal{X}_8 \to \mathbb{P}^1_{(2,[1:1])} \setminus \{[1:0], [0:1]\} \to \mathbb{P}^1_{(2,[1:0]),(2,[1:1]),(\infty[0:1])}.$$

4.2.4 Arrangement $\mathcal{C}_9$. The characteristic variety $\operatorname{Char}(\mathcal{C}_9)$ consists of four (respectively eleven) non-essential coordinate components associated with the four triple points of $\mathcal{C}_9$ (respectively eleven $\mathcal{C}_6$-subarrangements), which have dimension 2 and depth 1. In addition, there are three 3-dimensional non-essential coordinate components of depth 2 associated with the quadruple points of $\mathcal{C}_9$. Consider the following augmented Ceva pencil

$$\psi_{9,1} := f_{SC}|_{\mathcal{X}_9} : \mathcal{X}_9 \to \mathbb{P}^1 \setminus \{[1:0], [0:1], [1:1]\}.$$

Computations of the induced map on the variety of characters show that this pencil yields the only non-coordinate translated component of dimension 2 and depth 1 observed in the references above.

Finally, there are also three characters of order two

$$\begin{aligned}\chi_{9,1} &:= (-1, -1, 1, -1, -1, 1, 1, 1, 1),\\ \chi_{9,2} &:= (-1, 1, -1, -1, 1, -1, 1, 1, 1), \text{ and}\\ \chi_{9,3} &:= (1, -1, -1, 1, -1, -1, 1, 1, 1)\end{aligned}$$

of depth 2. In order to check the value of the depth, one needs to find two independent marked orbifold pencils on $(\mathcal{X}_9, \chi_{9,1})$ with target $(\mathbb{C}_{2,2}, \rho)$ where $\mathbb{C}_{2,2} := \mathbb{P}^1_{(2,[0:1]),(2,[1:0]),(\infty[1:1])}$ and $\rho := (-1, -1, 1)$ is the only non-trivial character on $\mathbb{C}_{2,2}$. Two such independent pencils can be given, for example, as follows

$$\psi_{9,2} := f_C|_{\mathcal{X}_9} : \mathcal{X}_9 \to \mathbb{P}^1 \setminus \{[0:1],[1:0],[1:1]\} \to \mathbb{P}^1_{(2,[0:1]),(2,[1:0]),(\infty[1:1])}$$

and

$$\psi_{9,3} := f_{SC}|_{\mathcal{X}_9} : \mathcal{X}_9 \to \mathbb{P}^1 \setminus \{[0:1],[1:0],[1:1]\} \to \mathbb{P}^1_{(2,[0:1]),(2,[1:0]),(\infty[1:1])}.$$

Remark 4.1. Note that the depth 2 characters in $\mathrm{Char}(\mathcal{C}_8)$ and $\mathrm{Char}(\mathcal{C}_9)$ lie in the intersection of positive dimensional components and this fact forces them to have depth greater than 1, see [8, Proposition 5.9].

4.3 Comments on independence of pencils

- **Depth conditions on the target:** First of all note that the condition on the target $(\mathcal{C}, \rho)$ to have $d(\rho) > 0$ is essential in the discussion above, *i.e.* pencils with target satisfying $d(\rho) = 0$ may not contribute to the characteristic varieties. For instance, the space $\mathcal{X}_6$ also admits several global quotient pencils coming from the augmented Ceva pencil, namely

$$\psi'_6 := f_{SC}|_{\mathcal{X}_6} : \mathcal{X}_6 \to \mathbb{P}^1_{(2,[0:1]),(2,[1:0]),(2,[1:1])} \to \mathbb{P}^1_{(2,[0:1]),(2,[1:0])}.$$

  However, the orbifold $\mathbb{P}_{2,2}$ is a global quotient orbifold whose orbifold fundamental group is abelian, so no non-trivial characters belong to its characteristic variety.
- **Independence of pencils.** Here is an explicit argument for independence of pencils for one of the cases discussed in last section. Consider the pencils $\psi_{9,2}$ and $\psi_{9,3}$ described above as marked pencils from $(\mathcal{X}_9, \chi_{9,1})$ having $(\mathbb{C}_{2,2}, \rho)$ as target. The marking produces the following commutative diagrams:

$$\begin{array}{ccc} X_{9,2} & \overset{\Psi_{9,2}}{\to} & C_2 \\ [x:y:z:w] & \mapsto & [\ell_1\ell_4 : w] \\ \downarrow \pi & & \downarrow \tilde{\pi} \\ \mathcal{X}_9 & \overset{\psi_{9,2}}{\to} & \mathbb{C}_{2,2} \\ [x:y:z] & \mapsto & [\ell_1\ell_4 : \ell_2\ell_5], \end{array}$$

  and

$$\begin{array}{ccc} X_{9,2} & \overset{\Psi_{9,3}}{\to} & C_2 \\ [x:y:z:w] & \mapsto & [\ell_1\ell_4\ell_7 : w\ell_8] \\ \downarrow \pi & & \downarrow \tilde{\pi} \\ \mathcal{X}_9 & \overset{\psi_{9,3}}{\to} & \mathbb{C}_{2,2} \\ [x:y:z] & \mapsto & [\ell_1\ell_4\ell_7^2 : \ell_2\ell_5\ell_8^2], \end{array}$$

where $X_{9,2}$ is contained in $\{[x : y : z : w] \mid w^2 = \ell_1\ell_4\ell_2\ell_5\}$, $C_2 := \mathbb{P}^1 \setminus \{[1:1],[1:-1]\}$ and $\tilde{\pi}$ is given by $[u:v] \mapsto [u^2 : v^2]$. Consider $\gamma_{i,k}$, $i = 3,6,7,8,9$, $k = 1,2$ the lifting of meridians around ℓ_i in $X_{9,2}$. Also denote by $\mathbb{Z}[\mathbb{Z}_2]$ the ring of deck transformations of $\tilde{\pi}$ as before, where $\mathbb{Z}_2$ acts by multiplication by $\xi_2 = (-1)$. Note that, as before $\Psi_{9,2}(\gamma_{3,k}) = \Psi_{9,2}(\gamma_{3,k}) = (-1)^k$ and $\Psi_{9,3}(\gamma_{4,k}) = \Psi_{9,3}(\gamma_{4,k}) = (-1)^{k+1}$. However, $\Psi_{9,2}(\gamma_{9,k}) = 0$ and $\Psi_{9,3}(\gamma_{9,k}) = (-1)^k$. Therefore $\psi_{9,2}$ and $\psi_{9,3}$ are independent pencils of $(\mathcal{X}_9, \chi_{9,1})$ with target $(\mathbb{C}_{2,2}, \rho)$.

5 Curve arrangements

Consider the space $\mathcal{M}$ of sextics with the following combinatorics:

1. $\mathcal{C}$ is a union of a smooth conic $\mathcal{C}_2$ and a quartic $\mathcal{C}_4$;
2. $\operatorname{Sing}(\mathcal{C}_4) = \{P, S\}$ where S is a cusp of type $\mathbb{A}_4$ and P is a node of type $\mathbb{A}_1$;
3. $\mathcal{C}_2 \cap \mathcal{C}_4 = \{S, R\}$ where S is a $\mathbb{D}_7$ on $\mathcal{C}$ and R is a $\mathbb{A}_{11}$ on $\mathcal{C}$.

In [4] it is shown that $\mathcal{M}$ has two connected components, say $\mathcal{M}^{(1)}$ and $\mathcal{M}^{(2)}$. The following are equations for curves in each connected component:

$$f_6^{(1)} = f_2^{(1)} f_4^{(1)} := \left((y+3x)\,z + \tfrac{3y^2}{2}\right) \left(x^2z^2 - \left(xy^2 + \tfrac{15}{2}\,x^2y + \tfrac{9}{2}x^3\right) z - 3x\,y^3 - \tfrac{9x^2y^2}{4} + \tfrac{y^4}{4}\right)$$

for $\mathcal{C}_6^{(1)} \in \mathcal{M}^{(1)}$ and

$$f_6^{(2)} = f_2^{(2)} f_4^{(2)} := \left(\left(y + \tfrac{x}{3}\right) z - \tfrac{y^2}{6}\right) \left(xz^2 - \left(xy^2 + \tfrac{9x^2y}{2} + \tfrac{3x^3}{2}\right) z + \tfrac{y^4}{4} + \tfrac{3x^2y^2}{4}\right)$$

for $\mathcal{C}_6^{(2)} \in \mathcal{M}^{(2)}$.

The curves $\mathcal{C}_6^{(1)}$ and $\mathcal{C}_6^{(2)}$ form a Zariski pair since their fundamental groups are not isomorphic. This cannot be detected by Alexander polynomials since both are trivial. In [4] the existence of an essential coordinate character of order two in the characteristic variety of $\mathcal{C}_6^{(2)}$ was shown enough to distinguish both fundamental groups, since the characteristic variety of $\mathcal{C}_6^{(1)}$ is trivial.

By Theorem 1.1 (2) this fact can also be obtained by looking at possible orbifold pencils. Note that there exists a conic $\mathcal{Q} := \{q = 0\}$ passing through S and R such that $(\mathcal{Q}, \mathcal{C}_4^{(1)})_S = 4$, $(\mathcal{Q}, \mathcal{C}_4^{(2)})_S = 5$, and

$(\mathcal{Q}, \mathcal{C}_2^{(2)})_R = 3$, $(\mathcal{Q}, \mathcal{C}_2^{(2)})_R = 3$. Consider $L := \{\ell = 0\}$ the tangent line to $\mathcal{Q}$ at S. One has the following list of multiplicities of intersection:

$$\begin{array}{ll}
(\mathcal{Q}, \mathcal{C}_2^{(2)} + 2L)_S = (\mathcal{Q}, \mathcal{C}_4^{(2)})_S = 5 & (\mathcal{Q}, \mathcal{C}_2^{(2)} + 2L)_R = (\mathcal{Q}, \mathcal{C}_4^{(2)})_R = 3 \\
(\mathcal{C}_4^{(2)}, 2\mathcal{Q})_S = (\mathcal{C}_4^{(2)}, \mathcal{C}_2^{(2)} + 2L)_S = 10 & (\mathcal{C}_4^{(2)}, 2\mathcal{Q})_R = (\mathcal{C}_4^{(2)}, \mathcal{C}_2^{(2)} + 2L)_R = 6 \\
(\mathcal{C}_2^{(2)}, \mathcal{C}_4^{(2)})_S = (\mathcal{C}_2^{(2)}, 2\mathcal{Q})_S = 2 & (\mathcal{C}_2^{(2)}, \mathcal{C}_4^{(2)})_R = (\mathcal{C}_2^{(2)}, 2\mathcal{Q})_R = 6 \\
(L, \mathcal{C}_4^{(2)})_S = (L, 2\mathcal{Q})_S = 4 & (L, \mathcal{C}_4^{(2)})_R = (L, 2\mathcal{Q})_R = 0.
\end{array}$$

By [12], this implies that $(\mathcal{C}_2^{(2)} + 2L, \mathcal{C}_2^{(2)}, 2\mathcal{Q})$ are members of a pencil of quartics. In other words, there is a marked orbifold pencil from $\mathcal{C} := \mathbb{P}^2 \setminus \mathcal{C}_6^{(2)}$ marked with $\chi := (-1, 1)$ to $\mathbb{P}^1_{(2,[0,1]),(2,[1:0]),(\infty[1:1])}$ given by $[x : y : z] \mapsto [f_2^{(2)}\ell^2 : q^2]$ whose target mark is the character $\rho := (-1, -1, 1)$.

References

[1] A. ADEM, J. LEIDA and Y. RUAN, "Orbifolds and Stringy Topology", Cambridge University Press. 2007.

[2] D. ARAPURA, *Geometry of cohomology support loci for local systems I*, J. of Alg. Geom. **6** (1997), 563–597.

[3] E. ARTAL, *Sur les couples de Zariski*, J. Algebraic Geom. **4** (1994), 223–247.

[4] E. ARTAL, J. CARMONA and J. I. COGOLLUDO, *Essential coordinate components of characteristic varieties*, Math. Proc. Cambridge Philos. Soc. **136** (2004), 287–299.

[5] E. ARTAL, J. CARMONA, J. I. COGOLLUDO and M.Á. MARCO, *Invariants of combinatorial line arrangements and Rybnikov's example*, Singularity theory and its applications, Adv. Stud. Pure Math., vol. 43, Math. Soc. Japan, Tokyo, 2006, pp. 1–34.

[6] E. ARTAL and J. I. COGOLLUDO, *On the connection between fundamental groups and pencils with multiple fibers*, J. Singul. **2** (2010), 1–18.

[7] E. ARTAL, J. I. COGOLLUDO-AGUSTÍN and A. LIBGOBER, *Depth of cohomology support loci for quasi-projective varieties via orbifold pencils*, J. Reine Angew. Math., to appear, also available at arXiv:1008.2018 [math.AG].

[8] E. ARTAL, J. I. COGOLLUDO-AGUSTÍN and D. MATEI, *Characteristic varieties of quasi-projective manifolds and orbifolds*, Preprint available at `arXiv:1005.4761v2 [math.AG]`, 2010.

[9] E. ARTAL, J. I. COGOLLUDO and H.O TOKUNAGA, *A survey on Zariski pairs*, Algebraic geometry in East Asia—Hanoi 2005, Adv. Stud. Pure Math., vol. 50, Math. Soc. Japan, Tokyo, 2008, pp. 1–100.

[10] G. BARTHEL, F. HIRZEBRUCH and T. HÖFER, "Geradenkonfigurationen und Algebraische Flächen", Friedr. Vieweg & Sohn, Braunschweig, 1987.

[11] J. I. COGOLLUDO-AGUSTÍN and A. LIBGOBER, *Mordell-Weil groups of elliptic threefolds and the Alexander module of plane curves*, Preprint available at `arXiv:1008.2018v2 [math.AG]`, 2010.

[12] J. I. COGOLLUDO-AGUSTÍN and M.Á. MARCO BUZUNÁRIZ, *The Max Noether fundamental theorem is combinatorial*, Preprint available at `arXiv:1002.2325v1 [math.AG]`, 2009.

[13] J. I. COGOLLUDO and V. FLORENS, *Twisted Alexander polynomials of plane algebraic curves*, J. Lond. Math. Soc. (2) **76** (2007), 105–121.

[14] D. C. COHEN and A. I. SUCIU, *Characteristic varieties of arrangements*, Math. Proc. Cambridge Philos. Soc. **127** (1999), 33–53.

[15] A. DIMCA, *Pencils of plane curves and characteristic varieties* Preprint available at `math.AG/0606442`, 2006.

[16] A. DIMCA, S. PAPADIMA and A.I. SUCIU, *Formality, Alexander invariants, and a question of Serre*, Preprint available at `arXiv:math/0512480v3 [math.AT]`, 2005.

[17] D. EISENBUD, "Commutative Algebra. With a View Toward Algebraic Geometry", Graduate Texts in Mathematics, Vol. 150. Springer-Verlag, New York, 1995.

[18] M. FALK, *Arrangements and cohomology*, Ann. Comb. **1** (1997), 135–157.

[19] R. FRIEDMAN and J. W. MORGAN, "Smooth Four-manifolds and Complex Surfaces", Ergebnisse der Mathematik und ihrer Grenzgebiete (3) [Results in Mathematics and Related Areas (3)], vol. 27, Springer-Verlag, Berlin, 1994.

[20] E. HIRONAKA, *Abelian coverings of the complex projective plane branched along configurations of real lines*, Mem. Amer. Math. Soc. **105** (1993), vi+85.

[21] W. MAGNUS, A. KARRASS and D. SOLITAR, *Combinatorial group theory*, second ed., Dover Publications Inc., Mineola, NY, 2004, Presentations of groups in terms of generators and relations.

[22] A. LIBGOBER, *On the homology of finite abelian coverings*, Topology Appl. **43** (1992), 157–166.

[23] A. LIBGOBER *Characteristic varieties of algebraic curves*, Applications of algebraic geometry to coding theory, physics and computation (Eilat, 2001), Kluwer Acad. Publ., Dordrecht, 2001, pp. 215–254.

[24] A. LIBGOBER, *Non vanishing loci of Hodge numbers of local systems*, Manuscripta Math. **128** (2009), 1–31.

[25] A. LIBGOBER and S. YUZVINSKY, *Cohomology of local systems*, Arrangements—Tokyo 1998, Kinokuniya, Tokyo, 2000, pp. 169–184.

[26] M. OKA, *Alexander polynomial of sextics*, J. Knot Theory Ramifications **12** (2003), 619–636.

[27] M. SAKUMA, *Homology of abelian coverings of links and spatial graphs*, Canad. J. Math. **47** (1995), 201–224.

[28] P. SCOTT, *The geometries of 3-manifolds*, Bull. London Math. Soc. **15** (1983), 401–487.

[29] C. SIMPSON, *Subspaces of moduli spaces of rank one local systems*, Ann. Sci. Scuola Norm. Sup. (4) **26** (1993), 361–401.

[30] A.I. SUCIU, *Fundamental groups of line arrangements: enumerative aspects*, Advances in algebraic geometry motivated by physics (Lowell, MA, 2000), Contemp. Math., vol. 276, Amer. Math. Soc., 2001, pp. 43–79.

[31] A.I. SUCIU, *Translated tori in the characteristic varieties of complex hyperplane arrangements*, Topology Appl. **118** (2002), no. 1-2, 209–223, Arrangements in Boston: a Conference on Hyperplane Arrangements (1999).

[32] H. O. TOKUNAGA, *Some examples of Zariski pairs arising from certain $K3$ surfaces*, Math. Z. **227** (1998), no. 3, 465–477.

[33] O. ZARISKI, *O*n the problem of existence of algebraic functions of two variables possessing a given branch curve, Amer. J. of Math. **51** (1929).

Analytic continuation of a parametric polytope and wall-crossing

Nicole Berline and Michèle Vergne

Abstract. We define a set theoretic "analytic continuation" of a polytope defined by inequalities. For the regular values of the parameter, our construction coincides with the parallel transport of polytopes in a mirage introduced by Varchenko. We determine the set-theoretic variation when crossing a wall in the parameter space, and we relate this variation to Paradan's wall-crossing formulas for integrals and discrete sums. As another application, we refine the theorem of Brion on generating functions of polytopes and their cones at vertices. We describe the relation of this work with the equivariant index of a line bundle over a toric variety and Morelli constructible support function.

Contents **111**

1 Introduction

Consider a polytope $\mathfrak{q}(b)$ in $\mathbb{R}^d$ defined by a system of N linear inequalities:

$$\mathfrak{q}(b) := \{y \in \mathbb{R}^d;\ \langle \mu_i, y\rangle \leq b_i,\quad 1 \leq i \leq N.\} \tag{1.1}$$

In this article, we study the variation of the polytope $\mathfrak{q}(b)$ when the parameter $b = (b_i)$ varies in $\mathbb{R}^N$, but the linear forms μ_i are fixed (the parametric arrangement of hyperplanes $\langle \mu_i, y\rangle = b_i$ so obtained is called a mirage in [20]).

Our main construction is the following. Starting with a parameter b^0 which is regular (this is defined below), we construct a function $\mathcal{X}(x_1, x_2, \dots, x_N)$ on $\mathbb{R}^N$ which is a linear combination of characteristic functions of various semi-open coordinate quadrants in $\mathbb{R}^N$. Define

$$A(b)(y) = \mathcal{X}(b_1 - \langle \mu_1, y\rangle, \dots, b_N - \langle \mu_N, y\rangle).$$

The crucial feature of the function $\mathcal{X}$ is that, for b near b^0, $A(b)(y)$ is the characteristic function of the polytope $\mathfrak{q}(b)$, but $A(b)$ enjoys analyticity properties with respect to the parameter b when b moves in $\mathbb{R}^N$, that we will explain below. So we say that $A(b)$ is the "analytical continuation "of the polytope $\mathfrak{q}(b)$ (with initial value b^0).

Before stating these properties, let us give two examples. We denote by p_i the characteristic function of the closed coordinate half-space, $p_i = [x_i \geq 0]$, and we set $q_i = 1 - p_i = [x_i < 0]$. First, let $\mathfrak{q}$ be the d-dimensional simplex defined by the $d+1$ inequalities $y_i \geq 0$, $\sum_{i=1}^d y_i \leq 1$. In this case we have, (see Example 3.1),

$$\mathcal{X}(x) = p_1 \cdots p_{d+1} + (-1)^d q_1 \cdots q_{d+1}.$$

Thus $\mathcal{X}(x)$ is the sum of the [characteristic function of the] closed positive coordinate quadrant in $\mathbb{R}^{d+1}$ and of $(-1)^d$ times the open negative one. Let $b = (b_1, \dots, b_{d+1})$. If $b_1 + \cdots + b_{d+1} \geq 0$, then $A(b)(y) =$

$\mathcal{X}(b_1+y_1,\cdots,b_d+y_d,b_{d+1}-(y_1+\cdots+y_d))$ is the characteristic function of the simplex $\{y_i \geq -b_i, \sum_{i=1}^d y_i \leq b_{d+1}\}$, while if $b_1+\cdots+b_{d+1} < 0$, then $A(b)(y)$ is equal to $(-1)^d$ times the characteristic function of the symmetric open simplex $\{y_i < -b_i, \sum_{i=1}^d y_i > b_{d+1}\}$. In particular, in dimension $d = 1$, starting with the closed interval $[0,1]$, the analytic continuation $A(b)$ is the closed interval $\{-b_1 \leq y \leq b_2\}$ when $b_1+b_2 \geq 0$, while $A(b)$ is (-1) times the open interval $\{b_2 < y < -b_1\}$ when $b_1+b_2 < 0$ (Figure 1.1)

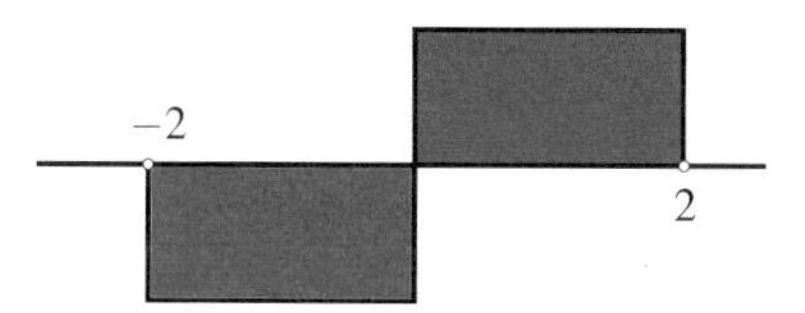

Figure 1.1. In blue for $b = (0,2)$, $\mathfrak{q}(b) = [0,2]$, in red for $b = (0,-2)$, $A(b) = (-1)$ times $]-2,0[$.

For the second example, we start with the tetragon illustrated in Figure 1.2 defined by the 4 inequalities $y_2+2 \geq 0$, $y_1+1 \geq 0$, $y_1+y_2 \leq 0$, $y_1-y_2 \geq 0$. In this case we have (see Example 4.2 and Subsection 4.5)

$$\mathcal{X}(x) = p_1p_2p_3p_4 - p_1q_2q_3p_4 - q_1p_2p_3q_4 + q_1q_2q_3q_4,$$

a signed sum of characteristic functions of 4 semi-open quadrants.

Some values of the analytic continuation $A(b)$ are illustrated in Figures 1.2 and 1.3. For each value of b, it is a signed sum of semi-open polygons. Components with a $+$ sign are colored in blue, components with a -1 sign are colored in red. Semi-openness is indicated by dotted lines.

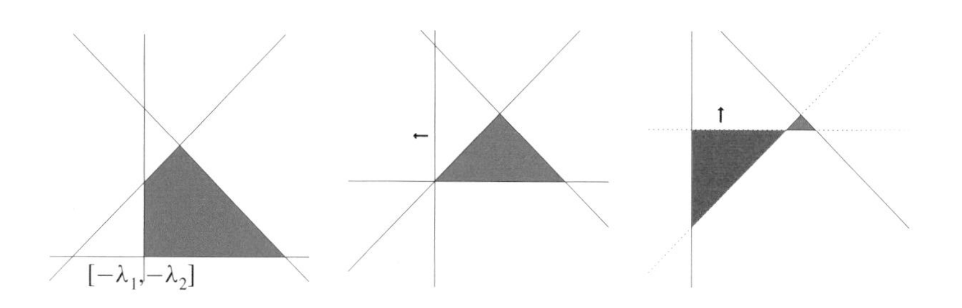

Figure 1.2. Analytic continuation of a tetragon.

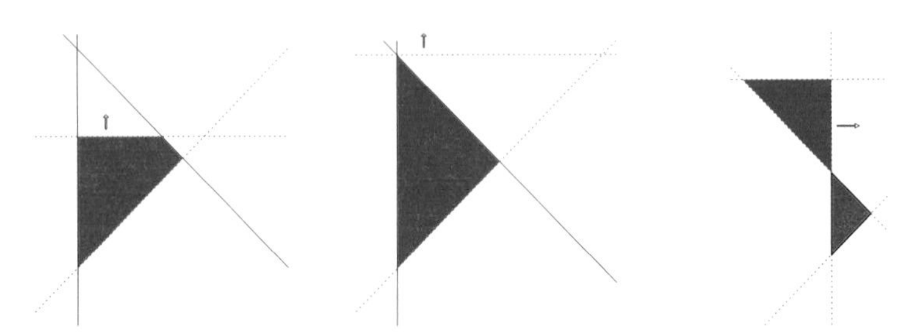

Figure 1.3. More analytic continuation of a tetragon.

Let us describe now some of the properties of $A(b)$.

A point $b = (b_i) \in \mathbb{R}^N$ is called regular (with respect to the sequence of linear forms μ_i) if a subset of k equations among the equations $\{\mu_i = b_i\}$ do not have a common solution if $k > d$. We define a tope τ to be a connected component of the open set of regular points b in $\mathbb{R}^N$. Topes are separated by hyperplanes which we call walls.

Let $b^0 \in \mathbb{R}^N$ be regular. Recall that we assume that $\mathfrak{q}(b^0)$ is compact. In this case, each vertex of the polytope $\mathfrak{q}(b^0)$ belongs to exactly d facets, in other words the polytope $\mathfrak{q}(b^0)$ is simple. Loosely speaking, the shape of the polytope $\mathfrak{q}(b)$ does not change when b remains close to b^0. The facets of $\mathfrak{q}(b)$ remain parallel to those of $\mathfrak{q}(b^0)$, while its vertices depend linearly on b. When b crosses a wall, the shape of $\mathfrak{q}(b)$ changes.

Let $h(y)$ be a polynomial function on $\mathbb{R}^d$. The integral

$$\int_{\mathfrak{q}(b)} h(y)dy,$$

and the discrete sum

$$\sum_{y \in \mathfrak{q}(b) \cap \mathbb{Z}^d} h(y)$$

are classical topics. In particular, if h is the constant function 1, these quantities are respectively the volume of the polytope $\mathfrak{q}(b)$ and the number of integral points in the polytope $\mathfrak{q}(b)$. It is well-known that the function $b \to \int_{\mathfrak{q}(b)} h(y)dy$ is given on each tope by a polynomial function of b. Moreover, if we assume that the linear forms μ_i are rational, the discrete sum $b \to \sum_{y \in \mathfrak{q}(b) \cap \mathbb{Z}^d} h(y)$ is given on each tope by a quasi-polynomial function of b. These results follow for instance from Brion's theorem of decomposing a polytope as a sum of its tangent cones at vertices, [6, 9]. When the parameter b crosses a wall of the tope τ, the integral $b \to \int_{\mathfrak{q}(b)} h(y)dy$ is given by a different polynomial, the discrete sum by a different quasi-polynomial. Their wall-crossing variations have been computed by Paradan, in a more general context of Hamiltonian geometry, using transversally elliptic operators, [18].

The function $\mathcal{X}$ which we construct in this article depends on the tope τ which contains the starting value b^0, and we will study its dependence with respect to τ. Therefore, we write $\mathcal{X}(\tau)(x)$ and $A(\tau, b)(y) = \mathcal{X}(\tau)(b_1 - \langle\mu_1, y\rangle, \dots, b_N - \langle\mu_N, y\rangle)$ instead of $\mathcal{X}(x)$ and $A(b)(y)$ from now on. The function $y \mapsto A(\tau, b)(y)$ enjoys the following properties.

- When b is in the closure $\overline{\tau}$ of the tope τ, $A(\tau, b)$ coincides with the characteristic function $[\mathfrak{q}(b)]$ of $\mathfrak{q}(b)$.
- The function $A(\tau, b)(y)$ is a linear combination with integral coefficients of characteristic functions of bounded faces of various dimensions of the arrangement of hyperplanes $\langle\mu_i, y\rangle = b_i, \quad 1 \le i \le N$.
- The integral

$$\int_{\mathbb{R}^d} A(\tau, b)(y) e^{\langle\xi, y\rangle} dy$$

  is an analytic function of $(\xi, b) \in (\mathbb{R}^d)^* \times \mathbb{R}^N$. For $b \in \overline{\tau}$, it coincides with $\int_{\mathfrak{q}(b)} e^{\langle\xi, y\rangle} dy$. If $h(y)$ is a polynomial function, then $b \mapsto \int_{\mathbb{R}^d} A(\tau, b)(y) h(y) dy$ is a polynomial function of $b \in \mathbb{R}^N$ which coincides with $\int_{\mathfrak{q}(b)} h(y) dy$ when $b \in \overline{\tau}$.
- Moreover, if we assume that the μ_i are rational, the discrete sum

$$\sum_{y \in \mathbb{Z}^d} A(\tau, b)(y) h(y)$$

  is a quasi-polynomial function of b, (see Definition 5.1 of quasi-polynomial functions). It coincides with $\sum_{y \in \mathfrak{q}(b) \cap \mathbb{Z}^d} h(y)$ for b in the initial tope and even in a neighborhood of its closure (see the precise statement in Corollary 3.1).

For instance, let us look again at the closed interval $[0, b]$. For $b \in \mathbb{N}$, the number of integral points in $[0, b]$ is given by the polynomial function $b + 1$. For a negative integer $b < 0$, the value $b + 1$ is indeed equal to (-1) times the number of integral points in the open interval $b < y < 0$.

The key idea is to define $A(\tau, b)$ as a signed sum of closed affine cones, shifted when b varies, so that their vertices depend linearly on the parameter b. We use decompositions of a polytope $\mathfrak{p}$ as a signed sum of cones, such as the Brianchon-Gram decomposition, (see for instance [8]).

Theorem 1.1 (Brianchon-Gram decomposition). *Let $\mathfrak{p} \subset \mathbb{R}^d$ be a polytope. For each face $\mathfrak{f}$ of $\mathfrak{p}$, let $\mathfrak{t}_{\mathrm{aff}}(\mathfrak{p}, \mathfrak{f}) \subseteq \mathbb{R}^d$ be the affine tangent cone to $\mathfrak{p}$ at the face $\mathfrak{f}$. Then*

$$[\mathfrak{p}] = \sum_{\mathfrak{f} \in \mathcal{F}(\mathfrak{p})} (-1)^{\dim \mathfrak{f}} [\mathfrak{t}_{\mathrm{aff}}(\mathfrak{p}, \mathfrak{f})], \tag{1.2}$$

where $\mathcal{F}(\mathfrak{p})$ is the set of faces of $\mathfrak{p}$.

Here, for a set $E \subset \mathbb{R}^d$, we denote by $[E]$ the function on $\mathbb{R}^d$ which is the characteristic function of the set E.

For regular values of b, our construction of $A(\tau, b)$ coincides with the parallel transport of Varchenko [20], the idea of which is quite simple. For instance, write the Brianchon-Gram formula for the closed interval $0 \leq y \leq b$,

$$[0 \leq y \leq b] = [y \leq b] + [y \geq 0] - [\mathbb{R}].$$

If the vertex b moves to the left, crosses the origin and becomes negative, the right hand side of the Brianchon-Gram formula becomes first, for $b = 0$, the characteristic function of the point 0, then for $b < 0$, the characteristic function of the open interval $b < y < 0$ with a minus sign.

Actually, instead of the Brianchon-Gram decomposition, Varchenko uses the polarized decomposition into semi-closed cones at vertices which he obtains in [20]. However, we go beyond [20] in several ways. First, as we already mentioned, we introduce (and compute) the "precursor" function $\mathcal{X}(\tau)$, a sum of characteristic functions of semi-open quadrants, which gives rise to $A(\tau, b)$ for all b. Moreover, we compute explicitly the wall-crossing variation

$$[\mathfrak{q}(b)] - A(\tau, b)$$

when b belongs to a tope adjacent to the starting tope τ. Actually, we compute the wall crossing variation at the level of the "precursor" function $\mathcal{X}(\tau)$ itself.

Finally, we show that "analytic continuation" of the faces of the polytope $\mathfrak{q}(b^0)$ occurs naturally, when one wants to compute $\sum_{y\in\mathfrak{q}(b_0)\cap\mathbb{Z}^d} e^{\langle\xi, y\rangle}$ for a degenerate value of ξ.

Let us now summarize the results of this article. We need some notations. It is more convenient to work in the framework of partition polytopes. So, let us first recall how one goes from the framework of linear inequalities $\langle\mu_i, y\rangle \leq b_i$ to that of partition polytopes. A partition polytope $\mathfrak{p}(\Phi, \lambda)$ is determined by a sequence $\Phi = (\phi_j)_{1\leq j\leq N}$ of elements of a vector space F (of dimension r) and an element $\lambda \in F$, as follows:

Definition 1.1.

$$\mathfrak{p}(\Phi, \lambda) = \left\{x \in \mathbb{R}^N;\ \sum_{j=1}^{N} x_j\phi_j = \lambda,\quad x_j \geq 0.\right\}$$

We assume that the cone $\mathfrak{c}(\Phi)$ generated by the ϕ_j's, is salient and that Φ generates F. Thus the set $\mathfrak{p}(\Phi, \lambda)$ is compact whenever $\lambda \in \mathfrak{c}(\Phi)$ (if λ is not in $\mathfrak{c}(\Phi)$, then $\mathfrak{p}(\Phi, \lambda)$ is empty.) The polytope $\mathfrak{p}(\Phi, \lambda)$ is, by definition, the intersection of the affine subspace

$$V(\Phi, \lambda) = \left\{ x \in \mathbb{R}^N ; \sum_{j=1}^{N} x_j \phi_j = \lambda \right\}$$

with the standard quadrant

$$Q := \left\{ x \in \mathbb{R}^N ; x_j \geq 0 \right\}.$$

A wall in F is a hyperplane generated by $r - 1$ linearly independent elements of Φ. An element $\lambda \in F$ is called Φ-regular, if λ does not lie on any wall. If $\lambda \in \mathfrak{c}(\Phi)$ is regular, the polytope $\mathfrak{p}(\Phi, \lambda)$ is a simple polytope of dimension $d = N - r$ contained in the affine space $V(\Phi, \lambda)$.

Consider the map $M : \mathbb{R}^N \to F$ given by $M(x) = \sum_i x_i \phi_i$. Let $V \subset \mathbb{R}^N$ be the kernel of M.

$$V = \left\{ x \in \mathbb{R}^N ; \sum_{j=1}^{N} x_j \phi_j = 0 \right\}.$$

So V has dimension $d = N - r$.

If E is a subset of $\mathbb{R}^N$, we denote now by $[E]$ the function on $\mathbb{R}^N$ which is the characteristic function of E. Thus, for any subset E of $\mathbb{R}^N$, $[E \cap V] = [E][V]$.

If $\lambda = M(b) = \sum_i b_i \phi_i$, the map

$$x \to x + b \tag{1.3}$$

is an isomorphism between V and the affine space $V(\Phi, \lambda)$.

Let μ_i be the linear form $-x_i$ restricted to V. The bijection $V \to V(\Phi, \lambda)$ maps the polytope $\mathfrak{q}(b) = \{y \in V;\ \langle \mu_i, y \rangle \leq b_i\}$ onto $\mathfrak{p}(\Phi, \lambda)$. Indeed, the point $(y_1 + b_1, \ldots, y_N + b_N)$ is in $\mathfrak{p}(\Phi, \lambda)$ if and only if $-y_i \leq b_i$.

Moreover, b is regular with respect to the sequence of linear forms μ_i on V if and only if $\lambda = M(b)$ is Φ-regular in F. A connected component of the set of Φ-regular elements of F will be called a Φ-tope. Thus a subset $\tau \subset F$ is a Φ-tope if and only if $M^{-1}(\tau) \subset \mathbb{R}^N$ is a connected component of the set of regular parameters, *i.e.* a tope with respect to (μ_i).

It is clearly equivalent to study the variation of the polytope $\mathfrak{q}(b)$ when b varies, or the variation of the partition polytope $\mathfrak{p}(\Phi, \lambda)$, when λ varies.

In this framework, the inequations $x_j \geq 0$ are fixed, while the affine space $V(\Phi, \lambda)$ varies. For example, Figure 1.4 shows the interval $[0, b]$, in blue, now realized as $\{x_1 \geq 0, x_2 \geq 0, x_1 + x_2 = b\}$. The analytic continuation $A(\tau, b)$ for $b < 0$ is colored in red on this figure, where a minus sign is assigned to red.

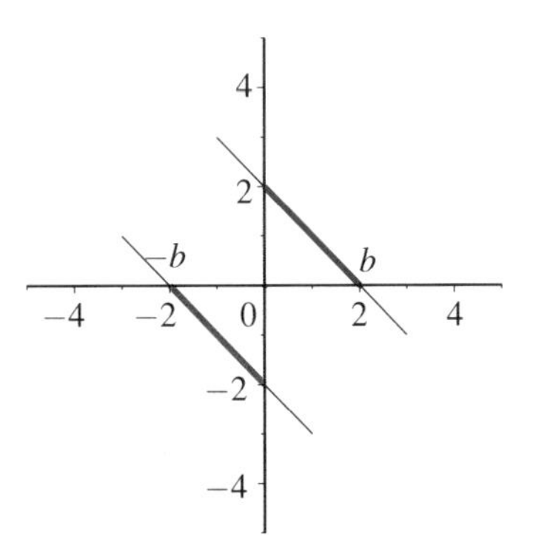

Figure 1.4. $F = \mathbb{R}$, $\Phi = (1, 1)$.

We fix a Φ-tope $\tau \subset F$, and consider $\lambda \in \tau$. Recall the combinatorial description of the faces of the partition polytope $\mathfrak{p}(\Phi, \lambda)$. We denote by $\mathcal{G}(\Phi, \tau)$ (respectively $\mathcal{B}(\Phi, \tau)$) the set of $I \subseteq \{1, \ldots, N\}$ such that $\{\phi_i, i \in I\}$ generates F (respectively is a basis of F) and such that τ is contained in the cone generated by $\{\phi_i, i \in I\}$. The set of faces (respectively vertices) of $\mathfrak{p}(\Phi, \lambda)$ is in one-to-one correspondence with $\mathcal{G}(\Phi, \tau)$ (respectively $\mathcal{B}(\Phi, \tau)$). The face which corresponds to I is

$$\mathfrak{f}_I(\Phi, \lambda) = \left\{ x \in \mathbb{R}^N_{\geq 0}, \sum_{j=1}^{N} x_j \phi_j = \lambda, \quad x_j = 0 \text{ for } j \in I^c \right\}. \tag{1.4}$$

The affine tangent cone to $\mathfrak{p}(\Phi, \lambda)$ at the face $\mathfrak{f}_I(\Phi, \lambda)$ is

$$\mathfrak{t}_I(\Phi, \lambda) = \left\{ x \in \mathbb{R}^N, \sum_{j=1}^{N} x_j \phi_j = \lambda, x_j \geq 0 \text{ for } j \in I^c \right\}. \tag{1.5}$$

If λ is in $\mathfrak{c}(\Phi)$, but is not in the tope τ, then the partition polytope $\mathfrak{p}(\Phi, \lambda)$ is not empty, but its faces are no longer in one-to-one correspondence with $\mathcal{G}(\Phi, \tau)$, (see Figure 1.2). Nevertheless, the cone in (1.5) makes sense **for every** $\lambda \in F$: it remains "the same cone" $\{x \in V; x_j \geq 0, j \in I^c\}$ up to a shift, under the map $V(\Phi, \lambda) \to V$ (see Formula (2.2)).

We introduce now the main character of this story, the function on $\mathbb{R}^N$ previously denoted by $\mathcal{X}(\tau)$.

Definition 1.2. The Geometric Brianchon-Gram function is

$$\mathcal{X}(\Phi, \tau) = \sum_{I \in \mathcal{G}(\Phi,\tau)} (-1)^{|I|-\dim F} \prod_{j \in I^c} [x_j \geq 0].$$

Let us compute this function for the case of $\Phi = (1, 1)$ in $F = \mathbb{R}$. Then

$$\mathcal{X}(\Phi, \tau) = [x_1 \geq 0] + [x_2 \geq 0] - [\mathbb{R}^2]$$

is equal to

$$[x_1 \geq 0, x_2 \geq 0] - [x_1 < 0, x_2 < 0],$$

the characteristic function of the closed positive quadrant minus the characteristic function of the open negative quadrant, (Figure 1.5).

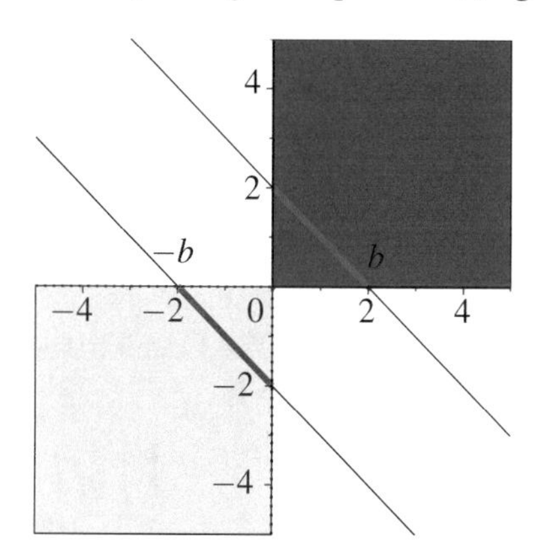

Figure 1.5. The function $\mathcal{X}(\Phi, \tau)$ for $\Phi = (1, 1)$.

If $\lambda \in \tau$, the Brianchon-Gram theorem implies

$$\mathcal{X}(\Phi, \tau)[V(\Phi, \lambda)] = [\mathfrak{p}(\Phi, \lambda)], \tag{1.6}$$

the characteristic function of the partition polytope $\mathfrak{p}(\Phi, \lambda)$. However, the function $\mathcal{X}(\Phi, \tau)[V(\Phi, \lambda)]$ is defined for any $\lambda \in F$. It is a signed sum of characteristic functions of closed cones intersected with the affine space $V(\Phi, \lambda)$.

For instance, in the case of $\Phi = (1, 1)$, taking the product of $\mathcal{X}(\Phi, \tau)$ with the characteristic function of the affine line $x_1 + x_2 = b$, we clearly recover the analytic continuation pictured in Figure 1.4.

One of our first results (and our main technical tool) (Theorem 3.3) is the fact that the Brianchon-Gram combinatorial function $X(\Phi, \tau)(p, q)$ coincides with the analogous function associated with any Lawrence-Varchenko polarized decomposition of a polytope into semi-closed cones at vertices [15, 20].

From this result, we deduce that the function $\mathcal{X}(\Phi, \tau)[V(\Phi, \lambda)]$ is the signed sum of characteristic functions of semi-open polytopes, in particular the support of this function is bounded for any $\lambda \in F$ (Corollary 3.2).

Reverting to the framework of linear inequalities, we define now $A(\tau, b)$ to be the inverse image of $\mathcal{X}(\Phi, \tau)$ under the map $v \to v + b$ from V to $\mathbb{R}^N$. For $b \in \tau$, $A(\tau, b)$ is the characteristic function of the polytope $\mathfrak{q}(b)$. For any value of b, it follows from the definition that $A(\tau, b)$ is the signed sum of the characteristic functions of the tangent cones to the faces of the initial polytope $\mathfrak{q}(b_0)$, with $b_0 \in \tau$, followed"by continuity ". The above qualitative result implies that $A(\tau, b)$ is a signed sum of bounded faces of various dimensions of the mirage $\mu_i = b_i$. It is easy to see that $A(\tau, b)$ enjoys the analyticity properties stated above.

Our main result is a wall crossing formula which we prove in a purely combinatorial context.

As the space $\mathbb{R}^N$ is the disjoint union of the semi-closed quadrants $Q_{\text{neg}}^B := \{x = (x_i); x_i < 0 \text{ for } i \in B,\ x_i \geq 0 \text{ for } i \in B^c\}$, we write $\mathcal{X}(\Phi, \tau)$ in terms of the characteristic functions of these quadrants.

We introduce the following polynomial in the variables p_i and q_i.

Definition 1.3. Let τ be a Φ-tope. The Combinatorial Brianchon-Gram function associated to the pair (Φ, τ) is

$$X(\Phi, \tau)(p, q) = \sum_{I \in \mathcal{G}(\Phi, \tau)} (-1)^{|I| - \dim F} \prod_{i \in I^c} p_i \prod_{i \in I} (p_i + q_i). \tag{1.7}$$

We recover $\mathcal{X}(\Phi, \tau)$ when we substitute $p_i = [x_i \geq 0]$ and $q_i = [x_i < 0]$ in $X(\Phi, \tau)(p, q)$ (so that $p_i + q_i = 1$).

For example, when $\Phi = (1, 1)$, we have

$$X(\Phi, \tau) = p_1(p_2+q_2)+p_2(p_1+q_1)-(p_1+q_1)(p_2+q_2) = p_1p_2-q_1q_2.$$

The polynomial $X(\Phi, \tau)$ enjoys remarkable properties. Let us say that the quadrant Q_{neg}^B is Φ-bounded, if the intersection of its closure $\overline{Q_{\text{neg}}^B}$ with V is reduced to 0. Equivalently, the intersection of Q_{neg}^B with the affine space $V(\Phi, \lambda)$ is bounded for any $\lambda \in F$.

We have

$$X(\Phi, \tau)(p, q) = \sum_B z_B \prod_{i \in B} p_i \prod_{i \in B^c} q_i.$$

where, for any subset $B \subseteq \{1, \ldots, N\}$ such that $z_B \neq 0$, the associated quadrant Q_{neg}^B is Φ-bounded. The coefficients z_B are in $\mathbb{Z}$ and we give an algorithmic formula for them.

As we will observe in the last section, the decomposition in Φ-bounded quadrants of $X(\Phi, \tau)$ is an analogue of the fact that the $\overline{\partial}$ cohomology spaces of a compact complex manifold are finite dimensional.

Our main result is Theorem 4.1, where we compute the function $X(\Phi, \tau_1) - X(\Phi, \tau_2)$, when τ_1 and τ_2 be two adjacent topes (meaning that the intersection of their closures is contained in a wall H and spans this wall).

We will not state the formula for $X(\Phi, \tau_1) - X(\Phi, \tau_2)$ in this introduction, but let us just mention a significant corollary, the wall-crossing formula for the polytope $\mathfrak{p}(\Phi, \lambda)$. Let A be the set of $i \in \{1, \ldots, N\}$ such that ϕ_i belongs to the open side of H which contains τ_1 (hence $-\phi_i$ belongs to the side of τ_2). Let

$$\mathfrak{p}_{\text{flip}}(\Phi, A, \lambda) = \{x \in V(\Phi, \lambda);\, x_i < 0 \text{ if } i \in A, x_i \geq 0 \text{ if } i \notin A\}.$$

Thus $\mathfrak{p}_{\text{flip}}(\Phi, A, \lambda)$ is a semi-closed bounded polytope in $V(\Phi, \lambda)$.

Theorem 1.2. *Let τ_1 and τ_2 be adjacent topes. If $\lambda \in \tau_2$, we have*

$$\mathcal{X}(\Phi, \tau_1)[V(\Phi, \lambda)] = [\mathfrak{p}(\Phi, \lambda)] - (-1)^{|A|}[\mathfrak{p}_{\text{flip}}(\Phi, A, \lambda)]. \qquad (1.8)$$

This formula is clearly inspired by the results of Paradan [18]. In turn, we show that it implies the convolution formula of Paradan which involves the number of lattice points of some lower dimensional polytopes associated to $\Phi \cap H$."

The above formula implies that, after crossing a wall, the analytic continuation of the original polytope $\mathfrak{p}(\Phi, \lambda)$ is the signed sum of two polytopes, among which one, but no more than one, may be empty. As illustrated in Section 4.5, we see the new polytope $\mathfrak{p}_{\text{flip}}(\Phi, A, \lambda)$ starting to show his nose when λ crosses the wall. To be precise, the wall H must separate two chambers, not just two topes, (as explained in Remark 4.2) in order for the new polytope $\mathfrak{p}_{\text{flip}}(\Phi, A, \lambda)$ to be not empty.

When F is provided with a lattice Λ, and the ϕ_i's are in Λ, the data (Φ, λ) parameterize a toric variety together with a line bundle. The zonotope

$$\mathfrak{b}(\Phi) := \left\{\sum_{i=1}^{N} t_i\phi_i;\, 0 \leq t_i \leq 1\right\}$$

plays an important role in the "continuity properties" of our formulae in the discrete case, where, for a tope τ, the "neighborhood" of $\tau \cap \Lambda$ is the fattened tope $(\tau - \mathfrak{b}(\Phi)) \cap \Lambda$. Remark indeed that the semi-closed flipped polytope $\mathfrak{p}_{\text{flip}}(\Phi, A, \lambda)$ of Formula (1.8) may not contain any integral point when λ stays near the wall between τ_1 and τ_2.

In Section 5, where we study discrete sums over partition polytopes, we recover the quasi-polynomiality over fattened topes which was previously obtained in [12, 13, 19], as well as wall crossing formulae. Remarkably, the proofs which we give in the present article are based only on the Brianchon-Gram decomposition of a polytope and some set theoretic computations.

Our original motivation for the present work was to understand Brion's formula when specialized at a degenerate point. Let $\mathfrak{p} \subset V$ be a full-dimensional polytope in a vector space V equipped with a lattice $V_{\mathbb{Z}}$. Consider the discrete sum

$$S(\mathfrak{p})(\xi) = \sum_{x \in V_{\mathbb{Z}} \cap \mathfrak{p}} e^{\langle \xi, x \rangle}.$$

Brion's theorem expresses the analytic function $S(\mathfrak{p})(\xi)$ as the sum

$$S(\mathfrak{p})(\xi) = \sum_{s \in \mathcal{V}(\mathfrak{p})} S(s + \mathfrak{c}_s)(\xi). \tag{1.9}$$

Here s runs over the set of vertices of $\mathfrak{p}$, and $s + \mathfrak{c}_s$ is the tangent cone at $\mathfrak{p}$ at the vertex s. Now the function $S(s + \mathfrak{c}_s)(\xi)$ is a meromorphic function of ξ. Its poles are the points $\xi \in V^*$ such that ξ vanishes on some edge generator of the cone $\mathfrak{c}_s$ (or equivalently, such that ξ takes the same value at the vertex s and some adjacent vertex s' of $\mathfrak{p}$).

It is well known that if ξ is regular with respect to $\mathfrak{p}$, (*i.e.* $\langle \xi, s \rangle \neq \langle \xi, s' \rangle$ for adjacent vertices), Brion's formula is the combinatorial translation of the localization formula in equivariant cohomology [4], in a case where the fixed points are isolated. The case where ξ is not regular corresponds to the case where the variety of fixed points has components of positive dimension. We obtain indeed the combinatorial translation of the localization formula in this degenerate case. The vertices must be replaced by the faces on which ξ is constant which are maximal with respect to this property. For such a face $\mathfrak{f}$, the tangent cone must be replaced by the transverse cone to $\mathfrak{p}$ along $\mathfrak{f}$. However, the formula is "nice" only under some conditions (satisfied for example when the polytope $\mathfrak{p}$ is simple). The formula involves the "analytic continuation" of the face $\mathfrak{f}$ obtained by slicing the polytope $\mathfrak{p}$ by affine subspaces parallel to $\mathfrak{f}$, Figure 6.2

Finally, in the last section, we sketch the relation of this work with the cohomology of line bundles over a toric variety. In the case where the ϕ_i's generate a lattice in F, a Φ-tope τ gives rise to a toric variety M_τ. Then, the value of the function $\mathcal{X}(\Phi, \tau)$ computed at a point $m \in$

$\mathbb{Z}^N \subset \mathbb{R}^N$ is the multiplicity of the character m in the alternate sum of the cohomology groups of the line bundle L_λ on M_τ which corresponds to $\lambda = \sum_i m_i \phi_i$. In other words, the function $\mathcal{X}(\Phi, \tau)$ induces on each affine space $V(\Phi, \lambda)$ the constructible function associated by Morelli [16] to the line bundle L_λ on M_τ.

The continuity result (Corollary 3.1) implies that the function $\lambda \to \dim H^0(M_\tau, \lambda)$ is a quasi-polynomial on the fattened tope $(\tau - \mathfrak{b}(\Phi)) \cap \Lambda$. We give some examples of computations in the last section.

These results have been presented by the second author M.V. in the Workshop: Arrangements of Hyperplanes held in Pisa in June 2010. M.V. thanks C. De Concini, H. Schenck and M. Wachs for numerous discussions on posets, cohomology of line bundles on toric varieties, during this special period, and thanks the Centro de Giorgi for providing such a stimulating atmosphere. The idea of this article arose while both authors were enjoying a Research in Pairs stay at Mathematisches Forschungsinstitut Oberwolfach in March/April 2010. The support of MFO is gratefully acknowledged.

ACKNOWLEDGEMENTS. We thank P. Johnson for drawing our attention to Varchenko's work and to the paper [10] where applications of Varchenko's work to wall crossing formulae for Hurwicz numbers are obtained.

List of notations

$A(b)$, $A(\tau, b)$	a function on $\mathbb{R}^d$, the analytic continuation of a polytope
$[E]$	characteristic function of a set $E \subseteq R^d$ or $E \subseteq \mathbb{R}^N$
F	r-dimensional real vector space; $\lambda \in F$
Φ	a sequence of N non zero vectors ϕ_i in F
e_i	canonical basis of $\mathbb{R}^N$
x_i	coordinates functions on $\mathbb{R}^N$
M	the map $\mathbb{R}^N \to F$; $M(e_i) = \phi_i$
$V(\Phi)$, V	$\{x \in \mathbb{R}^N; \sum_i x_i\phi_i = 0\}$
d	$N - r$; the dimension of V
$V(\Phi, \lambda)$	$\{x \in \mathbb{R}^N; \sum_i x_i\phi_i = \lambda\}$
$\mathfrak{p}(\Phi, \lambda)$	$\{x \in \mathbb{R}^N; x_i \geq 0; \sum_i x_i\phi_i = \lambda\}$
Partition polytope	a polytope $\mathfrak{p}(\Phi, \lambda)$
Λ	lattice in F; $\lambda \in \Lambda$
$k(\Phi)$	the function $\lambda \to$ cardinal $\mathfrak{p}(\Phi, \lambda)$
Partition function	the function $k(\Phi)$
Q	the standard quadrant $\{x \in \mathbb{R}^N; x_i \geq 0\}$
I, J, K, A, B	subsets of $\{1, 2, \ldots, N\}$
I^c	complementary subsets to I in $\{1,2,\ldots,N\}$
Φ_I	$(\phi_i, i \in I)$
$\mathfrak{c}(\Phi)$, $\mathfrak{c}(\Phi_I)$	cone generated by Φ, Φ_I
$\mathfrak{a}(K)$	the cone in $\mathbb{R}^N$ defined as $\{x \in \mathbb{R}^N; x_j \geq 0$ for $j \in K^c\}$
$\mathfrak{a}_0(K)$	the cone $V \cap \mathfrak{a}(K)$
$\mathfrak{t}_K(\Phi, \lambda)$	$\mathfrak{a}(K) \cap V(\Phi, \lambda)$
Φ-basic subset I	a subset I such that ϕ_i, $i \in I$, is a basis of F
Φ-generating subset I	a subset I such that ϕ_i, $i \in I$, generates F
$\mathcal{B}(\Phi)$	the set of Φ-basic subsets
$\mathcal{G}(\Phi)$	the set of Φ-generating subsets
$\rho_{\Phi,I}$	$\rho_{\Phi,I} : \mathbb{R}^N \to V(\Phi)$ with kernel $\oplus_{i\in I}\mathbb{R}e_i$
g_j^I	$\rho_{\Phi,I}(e_j)$, $j \in I^c$
wall H	hyperplane in F generated by $r - 1$ vectors in Φ
regular λ	λ does not belong to any wall H
tope τ	$\tau \subset F$, a connected component of the set of regular elements
$\mathcal{B}(\Phi, \tau)$	the set of Φ-basic subsets I such that $\tau \subset \mathfrak{c}(\Phi_I)$

$\mathcal{G}(\Phi, \tau)$ the set of Φ-generating subsets I such that $\tau \subset \mathfrak{c}(\Phi_I)$

arrangement$\mathcal{H}(\lambda)$ the collection of the hyperplanes $x_i = 0$ in $V(\Phi, \lambda)$

vertex s of the arrangement$\mathcal{H}(\lambda)$; s belongs to d hyperplanes of $\mathcal{H}(\lambda)$

$s_I(\Phi, \lambda)$ the vertex of $\mathcal{H}(\lambda)$ such that $s_j = 0$ for $j \in I^c$

$\mathfrak{f}_I(\Phi, \lambda)$, $\mathfrak{f}_I$ the face of $\mathfrak{p}(\Phi, \lambda)$ indexed by I; defined by $\mathfrak{p}(\Phi, \lambda) \cap \{x_j = 0,\ j \in I^c\}$

$\mathfrak{t}_{\mathrm{aff}}(\mathfrak{p}, \mathfrak{f})$ tangent affine cone to a polytope $\mathfrak{p}$ at the face $\mathfrak{f}$

$\mathcal{X}(\Phi, \tau)$ $\sum_{I \in \mathcal{G}(\Phi,\tau)} (-1)^{|I| - \dim F} \prod_{j \in I^c} [x_j \geq 0]$

$X(\Phi, \tau)(p, q)$ $\sum_{I \in \mathcal{G}(\Phi,\tau)} (-1)^{|I| - \dim F} \prod_{j \in I^c} p_j \prod_{i \in I} (p_i + q_i)$

w_B $\prod_{j \in B^c} p_j \prod_{i \in B} q_i$

W space of polynomials with basis w_B

Geom substituting $p_i = [x_i \geq 0]$, $q_j = [x_j < 0]$ in w_B

$\mathfrak{b}(\Phi)$ the zonotope generated by Φ; $\left\{ \sum_{i=1}^{N} t_i \phi_i ; 0 \leq t_i \leq 1 \right\}$

Q^B_{neg} $\{x = (x_i),\ x_i < 0 \text{ for } i \in B\ ;\ x_i \geq 0 \text{ for } i \in B^c\}$

Φ^B_{flip} the sequence $(\sigma_i \phi_i,\ 1 \leq i \leq N)$, where $\sigma_i = -1$ if $i \in B$; $\sigma_i = 1$ if $i \notin B$

$\widetilde{\mathfrak{c}}(\Phi^B_{\mathrm{flip}})$ $\left\{ \sum_i x_i \phi_i,\ x \in Q^B_{\mathrm{neg}} \right\}$

$\widetilde{\mathfrak{c}}_{\mathbb{Z}}(\Phi^B_{\mathrm{flip}})$ $\left\{ \sum_i x_i \phi_i,\ x \in Q^B_{\mathrm{neg}} \cap \mathbb{Z}^N \right\}$

β linear form on $\mathbb{R}^N$

$K^{c,+}_\beta$ $\{j \in K^c;\ \langle \beta, g^K_j \rangle > 0\}$

$K^{c,-}_\beta$ $\{j \in K^c;\ \langle \beta, g^K_j \rangle < 0\}$

$\mathfrak{a}(K, \beta)$ $\{x \in \mathbb{R}^N;\ x_i \geq 0\ ,\ i \in K^{c+}_\beta;\ x_i < 0\ ,\ i \in K^{c-}_\beta\}$

$\mathfrak{a}_0(K, \beta)$ the cone $V \cap \mathfrak{a}(K, \beta)$

$Y(\Phi, \tau, \beta)$ $\sum_{K \in \mathcal{B}(\Phi,\tau)} (-1)^{|K^{c-}_\beta|} \prod_{i \in K^{c+}_\beta} p_i \prod_{i \in K^{c-}_\beta} q_i \prod_{i \in K} (p_i + q_i)$

$\mathfrak{p}(\Phi, A, \lambda)$ $\{x \in V(\Phi, \lambda),\ \ x_i > 0 \text{ for } i \in A, x_i \geq 0 \text{ for } i \in A^c\}$

$\mathfrak{p}_{\mathrm{flip}}(\Phi, A, \lambda)$ $\{x \in V(\Phi, \lambda)\ \ x_i < 0 \text{ for } i \in A, x_i \geq 0 \text{ for } i \in A^c\}$

2 Definition of the analytic continuation

2.1 Some cones related to a partition polytope

In this article, there will be plenty of cones. A cone will always be an affine polyhedral convex cone. A cone will be called flat if it contains an affine line, otherwise, it will be called salient.

Let F be a real vector space of dimension r, and let $\Phi = (\phi_1, \ldots, \phi_N)$ be a sequence of N non zero elements of F. We assume that Φ generates F as a vector space.

The standard basis of $\mathbb{R}^N$ is denoted by e_i with dual basis the linear forms x_i. We denote by $M : \mathbb{R}^N \to F$ the surjective map which sends the vector e_i to the vector ϕ_i. The kernel of M is a subspace of dimension $d = N - r$ which will be denoted by $V(\Phi)$ or simply V when Φ is understood.

$$V(\Phi) := \{x \in \mathbb{R}^N; \sum_i x_i \phi_i = 0\}.$$

We denote by Q the standard quadrant

$$Q := \{x \in \mathbb{R}^N; x_i \geq 0\}.$$

The cone $\mathfrak{c}(\Phi)$ generated by Φ is the image of Q by M. Assume that the cone $\mathfrak{c}(\Phi)$ is salient. In other words, there exists a linear form $a \in F^*$ such that $\langle a, \phi_i \rangle > 0$ for all $1 \leq i \leq N$. This is also equivalent to the fact that $V \cap Q = 0$.

If I is a subset of $\{1, 2, \ldots, N\}$, we denote by I^c the complementary subset to I in $\{1, 2, \ldots, N\}$.

Definition 2.1. If I is a subset of $\{1, 2, \ldots, N\}$, let

$$\mathfrak{a}(I) = \{x \in \mathbb{R}^N; x_j \geq 0 \text{ for } j \in I^c\}$$

and let

$$\mathfrak{a}_0(I) = V \cap \mathfrak{a}(I) = \{x \in V; \ x_j \geq 0 \text{ for } j \in I^c\}$$

be the intersection of V with the cone $\mathfrak{a}(I)$.

Thus $\mathfrak{a}(I)$ is the product of the positive quadrant in the variables I^c, with a vector space of dimension $|I|$. The cone $\mathfrak{a}(I)$ is called an angle by Varchenko. It is never salient, except if $I = \emptyset$. With this notation, the positive quadrant Q is $\mathfrak{a}(\emptyset)$.

We now analyze the cone $\mathfrak{a}_0(I) \subseteq V$. A subset $I \subseteq \{1, 2, \ldots, N\}$ such that $\{\phi_i, i \in I\}$ is a basis of F will be called Φ-basic. We denote by $\mathcal{B}(\Phi)$ the set of Φ-basic subsets. A subset $I \subseteq \{1, 2, \ldots, N\}$ such that $\{\phi_i, i \in I\}$ generates F will be called Φ-generating. We denote by $\mathcal{G}(\Phi)$ the set of Φ-generating subsets.

Let I be Φ-basic. Then the cardinal of I^c is $d = N - r$ and the restrictions to V of the linear forms x_j, with $j \in I^c$, form a basis of V^*. Hence $\mathfrak{a}_0(I)$ is a cone of dimension d in V with d generators, in other words a simplicial cone of full dimension in the vector space V. Let us describe the edges of the simplicial cone $\mathfrak{a}_0(I)$. We have

$$\mathbb{R}^N = V(\Phi) \oplus (\oplus_{i\in I}\mathbb{R}e_i),$$

and we denote by $\rho_{\Phi,I}$ the corresponding linear projection $\mathbb{R}^N \to V$. For $j \in I^c$, we write $\phi_j = \sum_{i\in I} u_{i,j}\phi_i$.

Lemma 2.1. *Let I be Φ-basic. For $j \in I^c$, let*

$$g_j^I = \rho_{\Phi,I}(e_j) = e_j - \sum_{i\in I} u_{i,j}e_i.$$

Then the d vectors g_j^I are the generators of the edges of the simplicial cone $\mathfrak{a}_0(I)$.

Now, let I be a generating subset. Then the restrictions to V of the linear forms x_j, $j \in I^c$, are linearly independent elements of V^*. The cone $\mathfrak{a}_0(I)$ is again the product of a simplicial cone of dimension $|I^c|$ by a vector space of dimension $|I| - r$ More precisely, if K is any Φ-basic subset contained in I, the cone $\mathfrak{a}_0(I)$ is the product of the cone generated by $\rho_{\Phi,K}(e_j)$, $j \in I^c$, by the vector space generated by $\rho_{\Phi,K}(e_i)$ with $i \in I \setminus K$.

2.2 Vertices and faces of a partition polytope

Recall that, for $\lambda \in F$, we denote by $V(\Phi, \lambda) \subset \mathbb{R}^N$ the affine subspace

$$\{x \in \mathbb{R}^N; \sum_i x_i\phi_i = \lambda\}.$$

The intersections of the coordinates hyperplanes $\{x_i = 0\}$ with $V(\Phi, \lambda)$ form an arrangement $\mathcal{H}(\lambda)$ of N affine hyperplanes of $V(\Phi, \lambda)$.

By definition, a vertex of this arrangement is a point $s \in V(\Phi, \lambda)$ such that s belongs to at least d independent hyperplanes. The arrangement $\mathcal{H}(\lambda)$ is called regular if no vertex belongs to more than d hyperplanes. A Φ-wall H is a hyperplane of F spanned by $r - 1$ linearly independent elements of Φ. Thus $\mathcal{H}(\lambda)$ is regular if and only if λ does not belong to any Φ-wall, that is, if λ is regular.

By definition, a face of the arrangement $\mathcal{H}(\lambda)$ is the set of elements $x \in V(\Phi, \lambda)$ which satisfy a subset of the set of relations $\{x_i \geq 0, x_j \leq 0, x_k = 0\}$.

Recall that the partition polytope $\mathfrak{p}(\Phi, \lambda)$ is the intersection of the affine space $V(\Phi, \lambda)$ with the positive quadrant Q. Thus it is a bounded face of the arrangement of hyperplanes $\mathcal{H}(\lambda)$.

If $I \subset \{1, \ldots, N\}$ is Φ-basic, then λ has a unique decomposition $\lambda = \sum_{i \in I} x_i \phi_i$. If λ is regular, $x_i \neq 0$ for all i.

Definition 2.2. Let I be a Φ-basic subset, let $\lambda = \sum_{i \in I} x_i \phi_i$. Then $s_I(\Phi, \lambda)$ is the vertex of the arrangement $\mathcal{H}(\lambda)$ defined by $s_I(\Phi, \lambda) = (s_i)$ where $s_i = x_i$ if $i \in I$, and $s_j = 0$ if $j \in I^c$.

Observe that $s_I(\Phi, \lambda)$ depends linearly on λ.

If λ is regular, the vertices of the arrangement $\mathcal{H}(\lambda)$ are in one to one correspondence $I \mapsto s_I(\Phi, \lambda)$ with the set $\mathcal{B}(\Phi)$ of Φ-basic subsets of $\{1, \ldots, N\}$.

Definition 2.3. For I a subset of $\{1, 2, \ldots, N\}$, define

$$\mathfrak{t}_I(\Phi, \lambda) = \mathfrak{a}(I) \cap V(\Phi, \lambda).$$

If I is a Φ-basic subset, then the cone $\mathfrak{t}_I(\Phi, \lambda)$ is the shift $s_I(\Phi, \lambda) + \mathfrak{a}_0(I)$ of the **fixed simplicial cone** $\mathfrak{a}_0(I)$ by the vertex $s_I(\Phi, \lambda)$ which depends linearly of λ. We will use the formula

$$\mathfrak{t}_I(\Phi, \lambda) = s_I(\Phi, \lambda) + \mathfrak{a}_0(I). \tag{2.1}$$

Similarly, if I is a Φ-generating subset, choose a Φ-basic subset K contained in I, then the cone $\mathfrak{t}_I(\Phi, \lambda)$ is the shift of the **fixed cone** $\mathfrak{a}_0(I)$ by the vertex $s_K(\Phi, \lambda)$ which depends linearly of λ.

$$\mathfrak{t}_I(\Phi, \lambda) = \mathfrak{a}(I) \cap V(\Phi, \lambda) = s_K(\Phi, \lambda) + \mathfrak{a}_0(I). \tag{2.2}$$

So, one can say that the set $\mathfrak{t}_I(\Phi, \lambda)$ varies analytically with λ, whenever I is a generating subset. At least it "keeps the same shape". This is not the case when I is not generating, for example when $I = \emptyset$. Indeed $\mathfrak{t}_\emptyset(\Phi, \lambda)$ is the partition polytope $\mathfrak{p}(\Phi, \lambda)$, and it certainly does not vary "analytically".

We now analyze the faces of the partition polytope $\mathfrak{p}(\Phi, \lambda)$ and the corresponding tangent cones.

If τ is a Φ-tope, we denote by $\mathcal{B}(\Phi, \tau) \subseteq \mathcal{B}(\Phi)$ the set of basic subsets I such that τ is contained in the cone $\mathfrak{c}(\phi_I)$ generated by the ϕ_i, $i \in I$. In other words, the equation $\lambda = \sum_{i \in I} x_i \phi_i$ can be solved with positive x_i. Equivalently, the corresponding vertex $s_I(\Phi, \lambda)$ belongs to the polytope $\mathfrak{p}(\Phi, \lambda)$. Thus when λ is regular, there is a one-to-one correspondence between the elements $I \in \mathcal{B}(\Phi, \tau)$ and the vertices of the polytope $\mathfrak{p}(\Phi, \lambda)$.

When λ belongs to the closure of a tope τ, every vertex of $\mathfrak{p}(\Phi, \lambda)$ is still of the form $s_I(\Phi, \lambda)$ with $I \in \mathcal{B}(\Phi, \tau)$, but two Φ-basic subsets can give rise to the same vertex.

Let $I \in \mathcal{B}(\Phi, \tau)$. Assume that λ is regular, so that all coordinates s_i of $s_I(\Phi, \lambda)$ with $i \in I$ are positive. Then it is clear that the tangent cone to $\mathfrak{p}(\Phi, \lambda)$ at the vertex $s_I(\Phi, \lambda)$ is the cone determined by the inequations $x_i \geq 0$ for $i \in I^c$, while the sign of the coordinates x_i with $i \in I$ are arbitrary, In other words, it is the simplicial affine cone $\mathfrak{t}_I(\Phi, \lambda)$

We denote by $\mathcal{G}(\Phi, \tau) \subseteq \mathcal{G}(\Phi)$ the set of generating subsets I such that τ is contained in the cone $\mathfrak{c}(\phi_I)$ generated by the ϕ_i, $i \in I$.

If $I \in \mathcal{G}(\Phi, \tau)$, the intersection of $\mathfrak{p}(\Phi, \lambda)$ with $\{x_j = 0, j \in I^c\}$ is a face $\mathfrak{f}_I(\Phi, \lambda)$ of dimension $|I| - r$ of the polytope $\mathfrak{p}(\Phi, \lambda)$. The vertices of this face are the points $s_K(\Phi, \lambda)$ corresponding to all the Φ-basic subsets K contained in I. The affine tangent cone $\mathfrak{t}_{\mathrm{aff}}(\mathfrak{p}(\Phi, \lambda), f_K(\Phi, \lambda))$ to the polytope $\mathfrak{p}(\Phi, \lambda)$ along the face $f_K(\Phi, \lambda)$ is

$$\mathfrak{t}_{\mathrm{aff}}(\mathfrak{p}(\Phi, \lambda), f_K(\Phi, \lambda)) = \mathfrak{t}_I(\Phi, \lambda) = \mathfrak{a}(I) \cap V(\Phi, \lambda).$$

2.3 The Brianchon-Gram function

Summarizing, for $\lambda \in \tau$, there is a one-to one correspondence between the set of faces of the polytope $\mathfrak{p}(\Phi, \lambda)$ and the set $\mathcal{G}(\Phi, \tau)$. The Brianchon-Gram theorem implies, for $\lambda \in \tau$,

$$[\mathfrak{p}(\Phi, \lambda)] = \left(\sum_{I \in \mathcal{G}(\Phi, \tau)} (-1)^{|I| - \dim F} [\mathfrak{a}(I)] \right) [V(\Phi, \lambda)].$$

When λ varies, the right hand side is obtained by intersecting a number of fixed cones in $\mathbb{R}^N$ with the varying affine space $V(\Phi, \lambda)$. It is natural to introduce the function on $\mathbb{R}^N$

$$\begin{aligned} \mathcal{X}(\Phi, \tau) &= \sum_{I \in \mathcal{G}(\Phi, \tau)} (-1)^{|I| - \dim F} [\mathfrak{a}(I)] \\ &= \sum_{I \in \mathcal{G}(\Phi, \tau)} (-1)^{|I| - \dim F} \prod_{j \in I^c} [x_j \geq 0]. \end{aligned} \tag{2.3}$$

that is, the Geometric Brianchon-Gram function which we mentioned in the introduction.

For $\lambda \in \tau$, we have

$$\mathcal{X}(\Phi, \tau)[V(\Phi, \lambda)] = [\mathfrak{p}(\Phi, \lambda)], \tag{2.4}$$

the characteristic function of the partition polytope $\mathfrak{p}(\Phi, \lambda) \subset \mathbb{R}^N$.

Let us now consider the function $\mathcal{X}(\Phi, \tau)[V(\Phi, \lambda)]$ for any $\lambda \in F$. By Equations (2.1) and (2.2), we have

$$\mathcal{X}(\Phi, \tau)[V(\Phi, \lambda)] = \sum_{I \in \mathcal{G}(\Phi,\tau)} (-1)^{|I|-\dim F}[s_K(\Phi, \lambda) + \mathfrak{a}_0(I)].$$

Here, for each $I \in \mathcal{G}(\Phi, \tau)$, we choose $K \subset I$, a basic subset contained in I.

We thus see that $\mathcal{X}(\Phi, \tau)[V(\Phi, \lambda)]$ is constructed as follows. Start from the polytope $\mathfrak{p}(\Phi, \lambda_0)$ with $\lambda_0 \in \tau$, write the characteristic function of $\mathfrak{p}(\Phi, \lambda_0)$ as the alternating sum of its tangent cones at faces, and when moving λ in the whole space F, follow these cones by moving their vertex linearly in function of λ. As all the sets I entering in the formula for $\mathcal{X}(\Phi, \tau)$ are generating, the individual pieces $\mathfrak{a}(I) \cap V(\Phi, \lambda) = s_K(\Phi, \lambda) + \mathfrak{a}_0(I)$ keep the same shape.

It is clear that the support of the function $\mathcal{X}(\Phi, \tau)[V(\Phi, \lambda)]$ is a union of faces of various dimensions of the arrangement $\mathcal{H}(\lambda)$. We will show that it is is a union of bounded faces of this arrangement, for any $\lambda \in F$ (Corollary 3.2).

Remark 2.1. Chambers rather than topes are relevant to wall crossing. However, we preferred to use topes, because topes are naturally related to the whole set of vertices of the arrangement $\mathcal{H}(\lambda)$. A chamber is a connected component of the complement in F of the union of all the **cones** spanned by $(r-1)$-elements of $\Phi \cup (-\Phi)$. Chambers are bigger than topes, the closure of a chamber is a union of closures of topes. See Figure 2.1. But if τ_1 and τ_2 are contained in the same chamber, we have $\mathcal{G}(\Phi, \tau_1) = \mathcal{G}(\Phi, \tau_2)$, hence $X(\Phi, \tau_1) = X(\Phi, \tau_2)$.

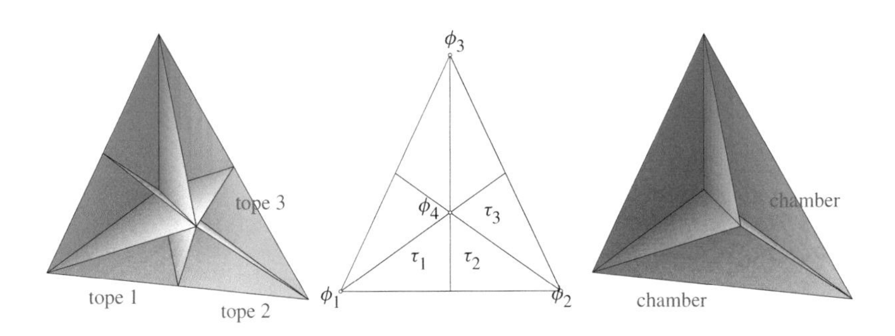

Figure 2.1. Left, topes for $\Phi = (\phi_1, \phi_2, \phi_3, \phi_1 + \phi_2 + \phi_3)$. Right, chambers.

3 Signed sums of quadrants

3.1 Continuity properties of the Brianchon-Gram function

Recall that we defined in the introduction the following polynomial in the variables p_i and q_i.

Definition 3.1. Let τ be a Φ-tope. The Combinatorial Brianchon-Gram function associated to the pair (Φ, τ) is

$$X(\Phi, \tau)(p, q) = \sum_{I \in \mathcal{G}(\Phi,\tau)} (-1)^{|I|-\dim F} \prod_{j \in I^c} p_j \prod_{i \in I} (p_i + q_i). \tag{3.1}$$

If the tope τ is not contained in $\mathfrak{c}(\Phi)$, the set $\mathcal{G}(\Phi, \tau)$ is empty and $X(\Phi, \tau) = 0$. Otherwise, if $\tau \subset \mathfrak{c}(\Phi)$, the sum defining $X(\Phi, \tau)$ is indexed by all the faces of the polytope $\mathfrak{p}(\Phi, \lambda_0)$ (for any choice of $\lambda_0 \in \tau$).

We recover $\mathcal{X}(\Phi, \tau)$ when we substitute $[x_i \geq 0]$ for p_i and $[x_i < 0]$ for q_i in $X(\Phi, \tau)(p, q)$ (so that $p_i + q_i = 1$).

The Combinatorial Brianchon-Gram function is a particular element of the space W below.

Definition 3.2. Let W be the subspace of $\mathbb{Q}[p_1, ..., p_N, q_1, ..., q_N]$ which consists of linear combinations of the monomials

$$w_B = \prod_{j \in B^c} p_j \prod_{i \in B} q_i$$

where B runs over the subsets of $\{1, \ldots, N\}$.

Thus we have

$$X(\Phi, \tau) = \sum_B z(\Phi, \tau, B) w_B, \tag{3.2}$$

with coefficients $z(\Phi, \tau, B) \in \mathbb{Z}$.

Remark: For the subset $B = \{1, 2, \ldots, N\}$, the coefficient $z(\Phi, \tau, B)$ is $(-1)^d = (-1)^{N-r}$.

Example 3.1 (The standard knapsack). Let $F = \mathbb{R}$, $\phi_i = 1$ for $i = 1, \ldots, N$, and $\tau = \mathbb{R}_{>0}$. From the usual inclusion-exclusion relations, we get

$$X(\Phi, \tau) = p_1 \cdots p_N - (-1)^N q_1 \cdots q_N. \tag{3.3}$$

An element in W gives a function on $\mathbb{R}^N$ by the following substitution.

Definition 3.3. We denote by Geom the map from W to the space of functions on $\mathbb{R}^N$ defined by substituting $[x_i \geq 0]$ for p_i and $[x_j < 0]$ for q_j.

Thus if w is an element of W, we may denote by w_{geom} the corresponding function on $\mathbb{R}^N$ so obtained. In particular, we may denote $\mathcal{X}(\Phi, \tau)$ by $X_{\text{geom}}(\Phi, \tau)$.

Later, we will use other substitutions.

We prove now some "continuity" properties of the Combinatorial Brianchon-Gram function when λ reaches the closure of the tope τ. Actually, these properties are shared by any element of the space W which satisfies the hypothesis of Proposition 3.1 below. We first introduce some definitions and prove an easy lemma.

Definition 3.4. The zonotope $\mathfrak{b}(\Phi)$ is the subset of F defined by

$$\mathfrak{b}(\Phi) = \left\{ \sum_{i=1}^{N} t_i \phi_i ; 0 \leq t_i \leq 1 \right\}.$$

When τ is a tope contained in $\mathfrak{c}(\Phi)$, the domain $\tau - \mathfrak{b}(\Phi) := \{x - y, x \in \tau, y \in \mathfrak{b}(\Phi)\}$ will play a crucial role in "continuity properties" of our functions. Remark that $\tau - \mathfrak{b}(\Phi)$ is a fattening of τ which contains the closure of the tope τ. Usually the set of integral points in $\tau - \mathfrak{b}(\Phi)$ is larger than the set of integral points in $\overline{\tau}$.

Definition 3.5. For $B \subseteq \{1, \ldots, N\}$,

- $Q^B_{\text{neg}} \subset \mathbb{R}^N$ is the semi-closed quadrant

$$Q^B_{\text{neg}} = \{x = (x_i), x_i < 0 \text{ for } i \in B, x_i \geq 0 \text{ for } i \in B^c\},$$

- Φ^B_{flip} is the sequence $[\sigma_i \phi_i, 1 \leq i \leq N]$, where $\sigma_i = -1$ if $i \in B$ and $\sigma_i = 1$ if $i \notin B$.
- $\widetilde{\mathfrak{c}}(\Phi^B_{\text{flip}}) \subset F$ is the semi-closed cone

$$\widetilde{\mathfrak{c}}(\Phi^B_{\text{flip}}) = \left\{ \sum_i x_i \phi_i, x \in Q^B_{\text{neg}} \right\}$$

and

$$\widetilde{\mathfrak{c}}_{\mathbb{Z}}(\Phi^B_{\text{flip}}) = \left\{ \sum_i x_i \phi_i, x \in Q^B_{\text{neg}} \cap \mathbb{Z}^N \right\}.$$

With this notation, the standard quadrant is

$$Q = Q^{\emptyset}_{\text{neg}}.$$

Remark that the closure of the semi-closed cone $\widetilde{\mathfrak{c}}(\Phi^B_{\text{flip}})$ is the closed cone $\mathfrak{c}(\Phi^B_{\text{flip}})$.

We recall the following lemma.

Lemma 3.1. *The following conditions are equivalent:*

(i) *The cone* $\mathfrak{c}(\Phi^B_{\text{flip}})$ *is salient*
(ii) $\overline{Q}^B_{\text{neg}} \cap V = \{0\}$
(iii) *For any* $\lambda \in F$, $V(\Phi, \lambda) \cap \overline{Q}^B_{\text{neg}}$ *is bounded.*

Lemma 3.2. *Let* $\tau \subset \mathfrak{c}(\Phi)$ *be a tope and* $\overline{\tau}$ *its closure. Let B be a subset of* $\{1, 2, \ldots, N\}$. *Assume that the semi-open cone* $\widetilde{\mathfrak{c}}(\Phi^B_{\text{flip}})$ *and the tope* τ *are disjoint. Then*

(i) τ *is disjoint from the closed cone* $\mathfrak{c}(\Phi^B_{\text{flip}})$.
(ii) *The closure* $\overline{\tau}$ *of* τ *is disjoint from the semi-open cone* $\widetilde{\mathfrak{c}}(\Phi^B_{\text{flip}})$.
(iii) $\tau - \mathfrak{b}(\Phi)$ *is disjoint from* $\widetilde{\mathfrak{c}}_{\mathbb{Z}}(\Phi^B_{\text{flip}})$.

Proof. (i) Assume that the semi-open cone $\widetilde{\mathfrak{c}}(\Phi^B_{\text{flip}})$ and the tope τ are disjoint. As τ is open, it is disjoint from the closure $\mathfrak{c}(\Phi^B_{\text{flip}})$ of $\widetilde{\mathfrak{c}}(\Phi^B_{\text{flip}})$.

(ii) Choose z small in τ. As $\tau \subset \mathfrak{c}(\Phi)$, we can write $z = \sum_{a\in A} \epsilon_a \phi_a$ with A a subset of $\{1, 2, \ldots, N\}$ and $\epsilon_a > 0$. As τ is a cone, we may assume the ϵ_a very small. Let $\lambda \in \overline{\tau}$. Then $\lambda + z \in \tau$. Now, if λ belongs also to $\widetilde{\mathfrak{c}}(\Phi^B_{\text{flip}})$, we may write $\lambda = \sum_{i=1}^N x_i \phi_i$ with $x_i < 0$ if $i \in B$ and $x_i \geq 0$ if $i \in B^c$ and we see that $\lambda + z$ is still in $\widetilde{\mathfrak{c}}(\Phi^B_{\text{flip}})$ if ϵ_a are sufficiently small. This contradicts the fact that $\widetilde{\mathfrak{c}}(\Phi^B_{\text{flip}}) \cap \tau$ is empty. So (ii) is proven.

Let us prove (iii). Assume that there exist $(n_i) \in \mathbb{Z}^N$, with $n_i < 0$ for $i \in B$ and $n_i \geq 0$ for $i \notin B$, such that $\sum_i n_i \phi_i \in \tau - \mathfrak{b}(\Phi)$. Thus there exist (t_i) with $0 \leq t_i \leq 1$, for $i = 1, \ldots, N$, and $\lambda \in \tau$, such that $\sum_i (n_i + t_i)\phi_i = \lambda$. As n_i are integers, we have $n_i \leq -1$ hence $n_i + t_i \leq 0$ for $i \in B$. We have also $n_i + t_i \geq 0$ for $i \notin B$. It follows that $\lambda \in \tau \cap \mathfrak{c}(\Phi^B_{\text{flip}})$. This contradicts (ii). □

The following proposition states continuity properties on closures and beyond.

Proposition 3.1. *Let* $\tau \subset \mathfrak{c}(\Phi)$ *be a tope and let* $Z = \sum_B z_B w_B \in W$ *be such that*

$$\sum_B z_B[Q^B_{\text{neg}}]\,[V(\Phi, \lambda)] = [\mathfrak{p}(\Phi, \lambda)] \text{ for every } \lambda \in \tau. \qquad (3.4)$$

Then

(i) $z_\emptyset = 1$.
(ii) *The equation*

$$\sum_B z_B[Q^B_{\text{neg}}]\,[V(\Phi,\lambda)] = [\mathfrak{p}(\Phi,\lambda)]$$

still holds for every $\lambda \in \overline{\tau}$.
(iii) *For* $\lambda \in \tau - \mathfrak{b}(\Phi)$, *we have*

$$\sum_B z_B[Q^B_{\text{neg}}][V(\Phi,\lambda)\cap\mathbb{Z}^N] = [\mathfrak{p}(\Phi,\lambda)\cap\mathbb{Z}^N].$$

Proof. Let $\lambda \in \tau$ and $x \in \mathfrak{p}(\Phi,\lambda)$. Then $x \in Q = Q^{\emptyset}_{\text{neg}}$. As the quadrants Q^B_{neg} are pairwise disjoint, $x \notin Q^B_{\text{neg}}$ for $B \neq \emptyset$ hence (3.4) implies (i).

Next, let $B \neq \emptyset$. Let $\lambda \in \tau$. Assume there is an $x \in Q^B_{\text{neg}} \cap V(\Phi,\lambda)$. Then $x \notin \mathfrak{p}(\Phi,\lambda)$, thus (3.4) implies that $z_B = 0$. Hence, if $z_B \neq 0$, the semi-open cone $\widetilde{\mathfrak{c}}(\Phi^B_{\text{flip}})$ and the tope τ are disjoint. We can then apply Lemma 3.2. As $\overline{\tau}$ is disjoint from $\widetilde{\mathfrak{c}}(\Phi^B_{\text{flip}})$, we see that $V(\Phi,\lambda)$ does not intersect any of the Q^B_{neg} with $z_B \neq 0$ and Q^B_{neg} different of Q. This implies (ii). In the same way, we obtain (iii). □

Example 3.2 (See Figure 3.1). Let $N=3$, $\dim F=2$, $\Phi=(\phi_1,\phi_2,\phi_3=\phi_1+\phi_2)$. If τ_1 is the open cone generated by (ϕ_1,ϕ_3), we have $X(\Phi,\tau_1) = (p_1+q_1)(p_2+q_2)p_3 + (p_1+q_1)p_2 - (p_1+q_1)(p_2+q_2)(p_3+q_3) = p_1p_2p_3 - p_1q_2q_3 + q_1p_2p_3 - q_1q_2q_3$. We can check that $X(\phi,\tau_1)$ satisfies the properties (ii) and (iii) of Proposition 3.1 on Figure 3.1.

Corollary 3.1 (Continuity on the closure of a tope τ). *Let* $\Phi=(\phi_j)_{1\le j\le N}$ *be a sequence of non zero elements of a vector space* F, *generating* F, *and spanning a salient cone and let* $\tau \subset \mathfrak{c}(\Phi)$ *be a tope relative to* Φ .

(i) *If* λ *belongs to the closure* $\overline{\tau}$ *of the tope* τ, *then we have the equality of characteristic functions of sets*

$$\mathcal{X}(\Phi,\tau)[V(\Phi,\lambda)] = [\mathfrak{p}(\Phi,\lambda)].$$

(ii) *If* $\lambda \in \tau - \mathfrak{b}(\Phi)$, *we still have the equality of characteristic functions of sets of lattice points*

$$\mathcal{X}(\Phi,\tau)[V(\Phi,\lambda)]\,[\mathbb{Z}^N] = [\mathfrak{p}(\Phi,\lambda)\cap\mathbb{Z}^N].$$

Remark 3.1. There are other elements $Z=\sum z_B w_B$ which satisfy (3.4). The simplest one is $Z = w_\emptyset = p_1\cdots p_N$. However it does not enjoy the analytic properties of $\mathcal{X}(\Phi,\tau)$, see Theorem 5.1.

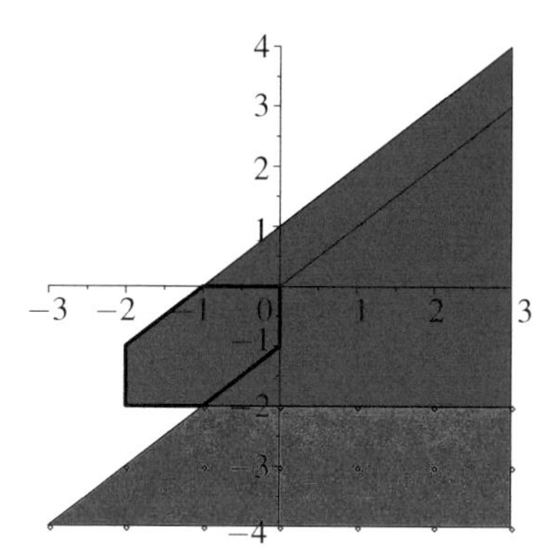

Figure 3.1. (Example 3.2). In blue, $-\mathfrak{b}(\Phi)$ is the closed hexagon, τ and $\tau - \mathfrak{b}(\Phi)$ are open sets. For $B = (2, 3)$, $\tilde{\mathfrak{c}}_Z(\Phi^B_{\text{neg}})$ is the set of lattice points in the red zone, $m\phi_1 - n\phi_3$ with $m \geq 1$ and $n \geq 2$.

3.2 Polarized sums

We now introduce another function on $\mathbb{R}^N$ related to the polarized decomposition of a polytope as a signed sums of polarized semi-closed cones at the vertices.

In addition to the data of the previous section, we use here a linear form β on $\mathbb{R}^N$, regular with respect to Φ in the following sense.

Let I be a basic subset for Φ, and recall the description of the cone $\mathfrak{a}_0(I)$ given in Lemma 2.1 with generators $g_j^I = \rho_{\Phi,I}(e_j)$:

$$\mathfrak{a}_0(I) = \sum_{j \in I^c} \mathbb{R}_{\geq 0} g_j^I.$$

We assume that β is such that its restriction to V does not vanish on any edge g_j^I of the simplicial cones $\mathfrak{a}_0(I)$ when I varies in $\mathcal{B}(\Phi)$. That is $\langle \beta, \rho_{\Phi,I}(e_j) \rangle \neq 0$ for all $I \in \mathcal{B}(\Phi)$ and $j \in I^c$.

We associate to β the "polarized" cone

$$\mathfrak{a}_0(I, \beta) = \sum_{j; \langle \beta, g_j^I \rangle > 0} \mathbb{R}_{\geq 0} g_j^I + \sum_{j; \langle \beta, g_j^I \rangle < 0} \mathbb{R}_{< 0} g_j^I.$$

This is the cone obtained by reversing the direction of some of the generators of the simplicial cone $\mathfrak{a}_0(I)$, in order that β takes positive value on all of them. Note however the delicate condition on signs.

Now all the cones $\mathfrak{a}_0(I, \beta)$ are contained in the half space of V determined by $\beta \geq 0$.

For each $K \in \mathcal{B}(\Phi)$, we denote by K_β^{c+} (respectively K_β^{c-}) the set of $j \in K^c$ such that $\langle \beta, \rho_{\Phi,K} e_j \rangle > 0$ (respectively $\langle \beta, \rho_{\Phi,K} e_j \rangle < 0$).

Definition 3.6. If K is a subset of $\{1, 2, \dots, N\}$, we denote by

$$\mathfrak{a}(K, \beta) = \{x \in \mathbb{R}^N; x_i \geq 0 \text{ for } i \in K_\beta^{c+}, x_i < 0 \text{ for } i \in K_\beta^{c-}\}.$$

Thus the set $\mathfrak{a}(K, \beta)$ is the product of three terms: the closed quadrant in the variables in K_β^{c+}, the opposite of the open quadrant in the variables in K_β^{c-}, and a vector space in the variable in K.

If $K \in \mathcal{B}(\Phi)$, the cone $\mathfrak{a}(K, \beta) \cap V(\Phi, \lambda)$ is the translate by the vertex $s_K(\Phi, \lambda)$ of the semi-open cone $\mathfrak{a}_0(K, \beta)$ of dimension d. In particular $\mathfrak{a}(K, \beta) \cap V(\Phi, \lambda)$ is contained in the half space $s_K(\Phi, \lambda) + \beta \geq 0 \cap V(\Phi, \lambda)$ of $V(\Phi, \lambda)$.

If $\lambda \in \tau$ and $K \in \mathcal{B}(\Phi, \tau)$, the cone $\mathfrak{a}(K, \beta) \cap V(\Phi, \lambda)$ is obtained by reversing some of the generators of the tangent cone to the polytope $\mathfrak{p}(\Phi, \lambda)$ at the vertex $s_K(\Phi, \lambda)$ so that β takes positive values on them. We say that it is the polarized tangent cone.

Recall the Lawrence-Varchenko polarized decomposition of $\mathfrak{p}(\Phi, \lambda)$, (actually, we will give a proof below).

$$[\mathfrak{p}(\Phi, \lambda)] = \sum_{K \in \mathcal{B}(\Phi, \tau)} (-1)^{|K_\beta^{c-}|} [\mathfrak{a}(K, \beta) \cap V(\Phi, \lambda)].$$

Again this equality is obtained by intersecting a number of fixed cones in $\mathbb{R}^N$ with the varying affine subspace $V(\Phi, \lambda)$. Therefore it is natural to define the following function.

Definition 3.7. The Combinatorial Lawrence-Varchenko function is the following element of W:

$$Y(\Phi, \tau, \beta) = \sum_{K \in \mathcal{B}(\Phi, \tau)} (-1)^{|K_\beta^{c-}|} \prod_{i \in K_\beta^{c+}} p_i \prod_{i \in K_\beta^{c-}} q_i \prod_{i \in K} (p_i + q_i). \quad (3.5)$$

If the tope τ is not contained in $\mathfrak{c}(\Phi)$, then $Y(\Phi, \tau, \beta) = 0$. Otherwise, if $\tau \subset \mathfrak{c}(\Phi)$, the sum defining $Y(\Phi, \tau, \beta)$ is indexed by all the vertices of the polytope $\mathfrak{p}(\Phi, \lambda_0)$ (for any choice of $\lambda_0 \in \tau$).

If we replace p_i by the characteristic function of $x_i \geq 0$ and q_i by the characteristic function of $x_i < 0$ ($p_i + q_i = 1$), we obtain a function $Y_{\text{geom}}(\Phi, \tau, \beta)$ on $\mathbb{R}^N$.

By construction, if $\lambda \in \tau$, the product $[V(\Phi, \lambda)] Y_{\text{geom}}(\Phi, \tau, \beta)$ is the signed sum of polarized semi-closed cones at the vertices of the (simple) partition polytope $\mathfrak{p}(\Phi, \lambda)$. Lawrence-Varchenko's theorem can be restated as $[V(\Phi, \lambda)] Y_{\text{geom}}(\Phi, \tau, \beta) = [\mathfrak{p}(\Phi, \lambda)]$ while Brianchon-Gram's theorem is $[V(\Phi, \lambda)] X_{\text{geom}}(\Phi, \tau) = [\mathfrak{p}(\Phi, \lambda)]$ for any $\lambda \in \tau$. Indeed by construction, $X_{\text{geom}}(\Phi, \tau)$ is our function $\mathcal{X}(\Phi, \tau)$, our main object of study.

The Lawrence-Varchenko decomposition of a simple polytope can be derived from the Brianchon-Gram one, by grouping some faces with a common vertex, [15]. It is remarkable that their combinatorial precursors actually coincide as elements of the space W, as we show in the next theorem.

Theorem 3.3. *Let $\Phi = (\phi_j)_{1\le j\le N}$ be a sequence of non zero elements of a vector space F, generating F, and spanning a salient cone, and let $\tau \subset \mathfrak{c}(\Phi)$ be a Φ-tope. Let $X(\Phi,\tau)$ be the Combinatorial Brianchon-Gram function. For any linear form β which is regular with respect to Φ, let $Y(\Phi,\tau,\beta)$ be the Combinatorial Lawrence-Varchenko function. Then*

$$Y(\Phi,\tau,\beta) = X(\Phi,\tau).$$

Proof. In the sum $X(\Phi,\tau)$, for a given $K \in \mathcal{B}(\Phi,\tau)$, we group together the $I \in \mathcal{G}(\Phi,\tau)$ such that $K \subseteq I$ and $\langle \beta, \rho_{\Phi,K} e_i\rangle < 0$, for every $i \in I \setminus K$. We denote the set of these I by $\mathcal{G}(\Phi,\tau)^K_\beta$. □

Lemma 3.3. *Let $I \in \mathcal{G}(\Phi,\tau)$ and $K \in \mathcal{B}(\Phi,\tau)$ such that $K \subseteq I$. Then $I \in \mathcal{G}(\Phi,\tau)^K_\beta$ if and only if for any $\lambda \in \tau$, on the face $\mathfrak{f}_I(\Phi,\lambda)$ of $\mathfrak{p}(\Phi,\lambda)$ which is indexed by I, the linear form β reaches its maximum at the vertex s_K indexed by K.*

Proof. Let $\lambda \in \tau$. Let $x = \sum_{i\in I} x_i e_i \in \mathfrak{f}_I(\Phi,\lambda)$, with $x \neq s_K$. Then $x - s_K$ is the projection $\rho_{\Phi,K}(x) = \sum_{i\in I\setminus K} x_i \rho_{\Phi,K}(e_i)$. Hence

$$\langle \beta, x - s_K\rangle = \sum_{i\in I\setminus K} x_i \langle \beta, \rho_{\Phi,K}(e_i)\rangle.$$

Assume that $\langle \beta, \rho_{\Phi,K} e_i\rangle < 0$, for every $i \in I \setminus K$. As $x_i \ge 0$ and $x_i > 0$ for at least one index $i \in I \setminus K$, we have $\langle \beta, x - s_K\rangle < 0$.

Conversely, let $i \in I \setminus K$. Take an $x = \sum_{k\in K} x_k e_k + x_i e_i \in \mathfrak{f}_I(\Phi,\lambda)$ with $x_i > 0$. Then by assumption we have $\langle \beta, x - s_K\rangle < 0$, hence $x_i\langle \beta, \rho_{\Phi,K} e_i\rangle < 0$.

It follows that, when K runs over $\mathcal{B}(\Phi,\tau)$ (the set of vertices of $\mathfrak{p}(\Phi,\lambda)$), the subsets $\mathcal{G}(\Phi,\tau)^K_\beta$ form a partition of $\mathcal{G}(\Phi,\tau)$ (the set of faces). Therefore, in order to prove Theorem 3.3, there remains to show that for every $K \in \mathcal{B}(\Phi,\tau)$, we have

$$\sum_{I\in\mathcal{G}(\Phi,\tau)^K_\beta} (-1)^{|I|-\dim F} \prod_{i\in I^c} p_i \prod_{i\in I}(p_i+q_i)$$
$$= (-1)^{|K^{c-}_\beta|} \prod_{i\in K^{c+}_\beta} p_i \prod_{i\in K^{c-}_\beta} q_i \prod_{i\in K}(p_i+q_i). \quad (3.6)$$

We factor out $\prod_{i\in K}(p_i+q_i)$. We need to prove

$$\sum_{I\in\mathcal{G}(\Phi,\tau)^K_\beta}(-1)^{|I|-\dim F}\prod_{i\in I^c}p_i\prod_{i\in I\setminus K}(p_i+q_i)$$
$$=(-1)^{|K^{c-}_\beta|}\prod_{i\in K^{c+}_\beta}p_i\prod_{i\in K^{c-}_\beta}q_i. \quad (3.7)$$

We make several observations. First, $\mathcal{G}(\Phi,\tau)^K_\beta$ is precisely the set of $I\subseteq\{1,\dots,N\}$ such that $K\subseteq I$ and $I\setminus K\subseteq K^{c-}_\beta$. Moreover, for each $I\in\mathcal{G}(\Phi,\tau)^K_\beta$, the set of indices $I^c\bigsqcup(I\setminus K)$ is exactly the complement K^c and $|I|-\dim F=|I\setminus K|$, as $\dim F=|K|$. Let $B\subseteq K^c$. A given monomial $\prod_{i\in K^c\setminus B}p_i\prod_{i\in B}q_i$ appears on the left hand side of (3.7) with coefficient

$$\sum_{\{I\in\mathcal{G}(\Phi,\tau)^K_\beta,\ B\subseteq I\setminus K\}}(-1)^{|I\setminus K|}.$$

By the usual inclusion-exclusion relations applied to the subsets $I\setminus K$ of K^c, this sum is equal to $(-1)^{|K^{c-}_\beta|}$ if $B=K^{c-}_\beta$ and to 0 otherwise. □

Example 3.4. In Example 3.1 of the standard knapsack with $N=3$, we take $\langle\beta,x\rangle=x_1+\frac{x_2}{2}+\frac{x_3}{3}$. We obtain $Y(\Phi,\tau,\beta)=(p_1+q_1)q_2q_3+p_1(p_2+q_2)p_3-p_1q_2(p_3+q_3)$. It is indeed equal to $X(\Phi,\tau)=p_1p_2p_3+q_1q_2q_3$.

Example 3.5. (continues Example 3.2). The subspace V is generated by the vector $e_1+e_2-e_3$. Hence, the projections $\rho_{(\Phi,K)}(e_i)$ are $\rho_{\Phi,(1,2)}(e_3)=e_3-e_1-e_2$, $\rho_{\Phi,(1,3)}(e_2)=\rho_{\Phi,(2,3)}(e_1)=e_1+e_2-e_3$. We can take $\langle\beta,x\rangle=x_1+x_2+x_3$. Let τ_1 be the cone generated by (ϕ_1,ϕ_3). We obtain

$$Y(\Phi,\tau_1,\beta)=-(p_1+q_1)(p_2+q_2)q_3+(p_1+q_1)p_2(p_3+q_3) \quad (3.8)$$
$$=p_1p_2p_3+p_1q_2p_3-q_1p_2q_3-q_1q_2q_3. \quad (3.9)$$

Comparing with Example 3.2, we check that $Y(\Phi,\tau_1,\beta)=X(\Phi,\tau_1)$.

Corollary 3.2. *For any $\lambda\in F$, $\mathcal{X}(\Phi,\tau)[V(\Phi,\lambda)]$ is a signed sum of bounded polytopes.*

Proof. Choose any β regular, then the function $\mathcal{X}(\Phi,\tau)[V(\Phi,\lambda)]$ is equal to $Y_{\mathrm{geom}}(\Phi,\tau,\beta)[V(\Phi,\lambda)]$. Taking $m(\beta)$ to be the minimum of the values $\langle\beta,s_K(\Phi,\lambda)\rangle$ over the $K\in\mathcal{B}(\Phi,\lambda)$, we see that the support of the function $\mathcal{X}(\Phi,\tau)[V(\Phi,\lambda)]$ is contained in the half space $\{\langle\beta,x\rangle\geq m(\beta)\}$ of $V(\Phi,\lambda)$. As this equality holds for any regular linear form β, this implies that the support of $\mathcal{X}(\Phi,\tau)[V(\Phi,\lambda)]$ is bounded. □

4 Wall-crossing

4.1 Combinatorial wall-crossing

In this section we prove the main theorem of this article: a formula for $X(\Phi, \tau_2) - X(\Phi, \tau_1)$, when τ_1 and τ_2 are adjacent topes.

The computation comes out nicely when we use the polarized expression $Y(\Phi, \tau, \beta)$ as a sum over $\mathcal{B}(\Phi, \tau)$ (Theorem 3.3), because it is easy to analyze how $\mathcal{B}(\Phi, \tau)$ changes as we cross the wall H between the two topes.

We recall that two topes τ_1 and τ_2 are called adjacent if the intersection of their closures spans a wall H. We denote by $\Phi \cap H$ the subsequence of Φ formed by the elements ϕ_i belonging to H.

Lemma 4.1. *Let τ_1 and τ_2 be adjacent Φ-topes such that $\tau_1 \subset \mathfrak{c}(\Phi)$. Let $K \in \mathcal{B}(\Phi, \tau_1)$ such that $K \notin \mathcal{B}(\Phi, \tau_2)$.*

(i) *For all $k \in K$ but one, say k_1, we have $\phi_k \in H$. The vector ϕ_{k_1} is in the open side of H which contains τ_1.*
(ii) *Let τ_{12} be the unique tope of $\Phi \cap H$ such that $\overline{\tau_1} \cap \overline{\tau_2} \subset \overline{\tau_{12}}$. Then τ_{12} is contained in the cone generated by the vectors ϕ_k for $k \in K, k \neq k_1$.*

Proof. Up to renumbering, we can assume that $K = \{1, \ldots, r\}$. Let $x_1, \ldots, x_r$ be the coordinates on F relative to this basic subset. If $K \notin \mathcal{B}(\Phi, \tau_2)$, at least one of these coordinates, say x_1, is < 0 on τ_2. Then the wall H must be the hyperplane $\{x_1 = 0\}$. (i) follows immediately.

The proof of (ii) is also easy. □

Recall Definition 3.5 of flipped systems Φ^A_{flip}.

Lemma 4.2. *Let τ_1 and τ_2 be adjacent Φ-topes such that $\tau_1 \subset \mathfrak{c}(\Phi)$. Let H be their common wall. Let A be the set of $i \in \{1, \ldots, N\}$ such that ϕ_i belongs to the open side of H which contains τ_1, (hence $-\phi_i$ belongs to the side of τ_2). Then $\mathcal{B}(\Phi^A_{\text{flip}}, \tau_2)$ is equal to the symmetric difference $\mathcal{B}(\Phi, \tau_1) \triangle \mathcal{B}(\Phi, \tau_2)$. More precisely*

$$K \in \mathcal{B}(\Phi, \tau_1), K \notin \mathcal{B}(\Phi, \tau_2) \Leftrightarrow K \in \mathcal{B}(\Phi^A_{\text{flip}}, \tau_2), K \cap A \neq \emptyset,$$

$$K \in \mathcal{B}(\Phi, \tau_2), K \notin \mathcal{B}(\Phi, \tau_1) \Leftrightarrow K \in \mathcal{B}(\Phi^A_{\text{flip}}, \tau_2), K \cap A = \emptyset.$$

Moreover, the cone $\mathfrak{c}(\Phi^A_{\text{flip}})$ is salient , and τ_2 is contained in at least one of the cones $\mathfrak{c}(\Phi)$ or $\mathfrak{c}(\Phi^A_{\text{flip}})$.

Proof. It follows easily from Lemma 4.1 (i) and the definition of Φ^A_{flip}. □

It will be convenient to have a notation.

Definition 4.1. Let τ_1, τ_2 be adjacent Φ-topes. We denote by $A(\Phi, \tau_1, \tau_2)$ the set of $i \in \{1, \dots, N\}$ such that ϕ_i belongs to the open side of the common wall which contains τ_1 .

Theorem 4.1. *Let $\Phi = (\phi_j)_{1 \le j \le N}$ be a sequence of non zero elements of a vector space F, generating F, and spanning a salient cone. Let τ_1 and τ_2 be adjacent Φ-topes such that $\tau_1 \subset \mathfrak{c}(\Phi)$. Let H be their common wall. Let A be the set of $i \in \{1, \dots, N\}$ such that ϕ_i is in the open side of H which contains τ_1. Let Φ^A_{flip} be the sequence $\sigma_i^A \phi_i$, where $\sigma_i^A = -1$ if $i \in A$ and $\sigma_i^A = 1$ if $i \notin A$. Let $X(\Phi, \tau_1)$, $X(\Phi, \tau_2)$ and $X(\Phi^A_{\text{flip}}, \tau_2)$ be the corresponding Combinatorial Brianchon-Gram polynomials $\in \mathbb{Z}[p_i, q_i]$. Let Flip_A be the ring homomorphism from $\mathbb{Z}[p_i, q_i]$ to itself defined by exchanging p_i and q_i for $i \in A$. Then we have the wall-crossing formula*

$$X(\Phi, \tau_1) = X(\Phi, \tau_2) - (-1)^{|A|} \, \text{Flip}_A \, X(\Phi^A_{\text{flip}}, \tau_2). \tag{4.1}$$

Remark 4.1. If the tope τ_2 is not contained in $\mathfrak{c}(\Phi)$, (res $\mathfrak{c}(\Phi^A_{\text{flip}})$), then $\mathcal{G}(\Phi, \tau_2)$, (respectively $\mathcal{G}(\Phi^A_{\text{flip}}, \tau_2)$), is empty, hence $X(\Phi, \tau_2) = 0$, (resp $X(\Phi^A_{\text{flip}}, \tau_2) = 0$).

Remark 4.2. By Remark 2.1, the actual wall crossing variations occur only on walls between chambers. This in agreement with this formula: indeed if H is not a wall between chambers, the tope τ_{12} is not contained in $\mathfrak{c}(\Phi \cap H)$ and τ_2 is not contained in $\mathfrak{c}(\Phi^A_{\text{flip}})$.

Remark 4.3. The theorem is trivially true if $\tau_1 \not\subseteq \mathfrak{c}(\Phi)$. Indeed, in this case, we have $X(\phi, \tau_1) = 0$, and $A(\Phi, \tau_1, \tau_2) = \emptyset$, so that the right hand side of (4.1) is $X(\phi, \tau_2) - X(\phi, \tau_2) = 0$.

Proof of Theorem 4.1. Let β be a linear form on $\mathbb{R}^N$ which is regular for Φ. Let β^A be the linear form defined by

$$\langle \beta^A, e_i \rangle = \sigma_i^A \langle \beta, e_i \rangle$$

where $\sigma_i^A = -1$ if $i \in A$ and $\sigma_i^A = 1$ if $i \in A^c$. We have, for every i,

$$\langle \beta^A, \rho_{\Phi^A_{\text{flip}}, K} e_i \rangle = \sigma_i^A \langle \beta, \rho_{\Phi, K} e_i \rangle. \tag{4.2}$$

It follows, in particular, that β^A is regular for Φ^A_{flip}.

First we will prove the following relation between the polarized sums

$$Y(\Phi, \tau_2, \beta) - Y(\Phi, \tau_1, \beta) = (-1)^{|A|} \, \text{Flip}_A \, Y(\Phi^A_{\text{flip}}, \tau_2, \beta^A). \tag{4.3}$$

Then we obtain (4.1) by applying Theorem 3.2. We write

$$\begin{aligned} Y(\Phi,\tau_2,\beta) - Y(\Phi,\tau_1,\beta) \\ = \sum_{K\in\mathcal{B}(\Phi,\tau_2)} (-1)^{|K_\beta^{c-}|} \prod_{i\in K_\beta^{c+}} p_i \prod_{i\in K_\beta^{c-}} q_i \prod_{i\in K}(p_i+q_i) \\ - \sum_{K\in\mathcal{B}(\Phi,\tau_1)} (-1)^{|K_\beta^{c-}|} \prod_{i\in K_\beta^{c+}} p_i \prod_{i\in K_\beta^{c-}} q_i \prod_{i\in K}(p_i+q_i) \end{aligned} \quad (4.4)$$

The terms for which $K \in \mathcal{B}(\Phi,\tau_1) \cap \mathcal{B}(\Phi,\tau_2)$ cancel out. For the other terms, we apply Lemma 4.2. Take a K in $\mathcal{B}(\Phi,\tau_1) \,\triangle\, \mathcal{B}(\Phi,\tau_2) = \mathcal{B}(\Phi^A_{\text{flip}},\tau_2)$. Using (4.2), we check easily that the unique K-term in $Y(\Phi,\tau_2,\beta) - Y(\Phi,\tau_1,\beta)$ is equal to the K-term in $(-1)^{|A|}\,\mathrm{Flip}_A\, Y(\Phi^A_{\text{flip}},\tau_2,\beta^A)$. □

Example 4.2. $N = 4$ and $\Phi = (\phi_1,\phi_2,\phi_3 = \frac{1}{2}(\phi_2-\phi_1),\phi_4 = \frac{1}{2}(\phi_1+\phi_2))$. The tope τ_1 is the open cone generated by ϕ_2 and ϕ_4. The adjacent tope τ_2 is the open cone generated by ϕ_4 and ϕ_1, see Figure 4.1. Then ϕ_2 and ϕ_3 lie on the τ_1-side of the common wall, so that $A = \{2,3\}$ and $\Phi^A_{\text{flip}} = (\phi_1,-\phi_2,-\phi_3,\phi_4)$. We obtain

$$\begin{aligned} X(\Phi,\tau_1) &= p_1p_2p_3p_4 - p_1q_2q_3p_4 - q_1p_2p_3q_4 + q_1q_2q_3q_4, \\ X(\Phi,\tau_2) &= p_1p_2p_3p_4 + p_1q_2q_3q_4 + q_1p_2p_3p_4 + q_1q_2q_3q_4, \\ X(\Phi^A_{\text{flip}},\tau_2) &= p_1p_2p_3p_4 + p_1p_2p_3q_4 + q_1q_2q_3p_4 + q_1q_2q_3q_4, \\ \mathrm{Flip}_A\, X(\Phi^A_{\text{flip}},\tau_2) &= p_1q_2q_3p_4 + p_1q_2q_3q_4 + q_1p_2p_3p_4 + q_1p_2p_3q_4, \\ &= X(\Phi,\tau_2) - X(\Phi,\tau_1). \end{aligned}$$

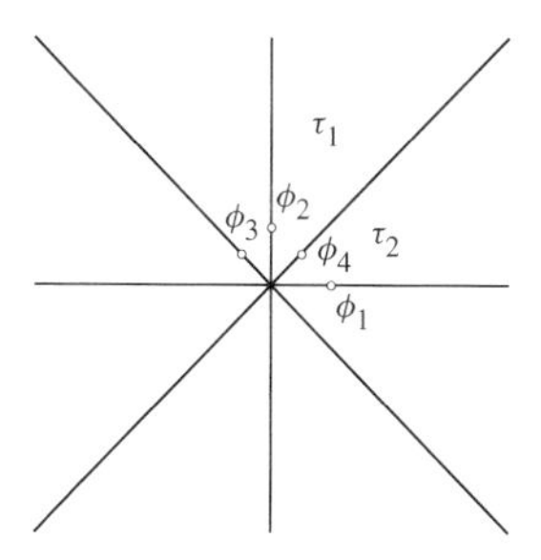

Figure 4.1. Topes τ_1,τ_2 for $\Phi = (\phi_1,\phi_2,\phi_3 = \frac{1}{2}(\phi_2-\phi_1),\phi_4 = \frac{1}{2}(\phi_1+\phi_2))$.

Later, in order to give a formula for the decomposition of $\mathcal{X}(\Phi,\tau)$ in quadrants, we will need to iterate the wall-crossing through a sequence of consecutively adjacent topes. To help understand what we obtain, let us first cross two consecutive walls.

Corollary 4.1. *Let τ_1, τ_2, τ_3 be pairwise different Φ-topes such that τ_2 is adjacent to τ_1 and to τ_3. Let $A_1 = A(\Phi, \tau_1, \tau_2)$, $A_2 = A(\Phi, \tau_2, \tau_3)$ and let $A_{1,2}$ be the symmetric difference $A_1 \bigtriangleup A(\Phi_{\text{flip}}^{A_1}, \tau_2, \tau_3)$. Then the three cones $\mathfrak{c}(\Phi_{\text{flip}}^{A_1})$, $\mathfrak{c}(\Phi_{\text{flip}}^{A_2})$ and $\mathfrak{c}(\Phi_{\text{flip}}^{A_{1,2}})$ are salient.*

We have

$$\begin{aligned} X(\Phi, \tau_1) = X(\Phi, \tau_3) &- (-1)^{|A_1|} \operatorname{Flip}_{A_1} X(\Phi_{\text{flip}}^{A_1}, \tau_3) \\ &- (-1)^{|A_2|} \operatorname{Flip}_{A_2} X(\Phi_{\text{flip}}^{A_2}, \tau_3) \\ &+ (-1)^{|A_{1,2}|} \operatorname{Flip}_{A_{1,2}} X(\Phi_{\text{flip}}^{A_{1,2}}, \tau_3). \end{aligned} \tag{4.5}$$

Proof. We apply the wall-crossing theorem to (Φ, τ_1, τ_2). We obtain

$$X(\Phi, \tau_1) = X(\Phi, \tau_2) - (-1)^{|A_1|} \operatorname{Flip}_{A_1} X(\Phi_{\text{flip}}^{A_1}, \tau_2).$$

In the right hand side, we transform each term by crossing the wall from τ_2 into τ_3. First,

$$X(\Phi, \tau_2) = X(\Phi, \tau_3) - (-1)^{|A_2|} \operatorname{Flip}_{A_2} X(\Phi_{\text{flip}}^{A_2}, \tau_3).$$

In order to apply the wall-crossing to $(\Phi_{\text{flip}}^{A_1}, \tau_2, \tau_3)$, we observe that the sign rule implies

$$\begin{aligned} (\Phi_{\text{flip}}^{A_1})^{A(\Phi_{\text{flip}}^{A_1}, \tau_2, \tau_3)} &= \Phi_{\text{flip}}^{A_1 \Delta A(\Phi_{\text{flip}}^{A_1}, \tau_2, \tau_3)}, \\ \operatorname{Flip}_{A_1} \circ \operatorname{Flip}_{A(\Phi_{\text{flip}}^{A_1}, \tau_2, \tau_3)} &= \operatorname{Flip}_{A_1 \Delta A(\Phi_{\text{flip}}^{A_1}, \tau_2, \tau_3)}. \end{aligned}$$

Moreover, $(-1)^{|A_1|}(-1)^{A(\Phi_{\text{flip}}^{A_1}, \tau_2, \tau_3)} = (-1)^{A_1 \Delta A(\Phi_{\text{flip}}^{A_1}, \tau_2, \tau_3)}$. Hence we obtain

$$\begin{aligned} -(-1)^{|A_1|} \operatorname{Flip}_{A_1} X(\Phi_{\text{flip}}^{A_1}, \tau_2) = &-(-1)^{|A_1|} \operatorname{Flip}_{A_1} X(\Phi_{\text{flip}}^{A_1}, \tau_3) \\ &+ (-1)^{|A_{1,2}|} \operatorname{Flip}_{A_{1,2}} X(\Phi_{\text{flip}}^{A_{1,2}}, \tau_3). \end{aligned}$$

This proves (4.5). The cones are salient by the very definition of the flipped systems $\Phi_{\text{flip}}^{A_i}$ and $(\Phi_{\text{flip}}^{A_1})^{A(\Phi_{\text{flip}}^{A_1}, \tau_2, \tau_3)}$. □

We now cross a number of walls to go from a tope τ to another tope ν. A signed subset of $\{1, 2, \ldots, N\}$ is a list $[\epsilon, I]$ where I is a subset of $\{1, 2, \ldots, N\}$ and $\epsilon = \pm 1$ a sign.

Definition 4.2. Let τ and ν be two topes, and let us choose a sequence τ_k, $k = 1, \ldots, \ell$ of topes such that τ_{k+1} is adjacent to τ_k for every $1 \leq k \leq \ell - 1$, and $\tau_1 = \tau$, $\tau_\ell = \nu$.

For every sequence $K = (1 \leq k_1 < \cdots < k_s \leq \ell - 1)$, let $A_K \subseteq \{1, \ldots, N\}$ be the subset defined recursively as follows. If $s = 0$, that is $K = \emptyset$, then $A_\emptyset = \emptyset$. If $K = (k_1, k_2, \ldots, k_s)$ with $s \geq 1$, let $A = A_{(k_1, \ldots, k_{s-1})}$, then

$$A_K = A \bigtriangleup A(\Phi_{\text{flip}}^A, \tau_{k_s}, \tau_{k_s+1}).$$

We define the list $\mathcal{A}(\nu, \tau)$ to be the list of signed subsets $[(-1)^{|K|}, A_K]$ of $\{1, 2, \ldots, N\}$ so obtained.

Remark 4.4. The list $\mathcal{A}(\nu, \tau)$ depends of the choice of path of adjacent topes from τ to ν , but we do not indicate this in the notation.

We obtain the following result, if there are $\ell - 1$ wall crossings to go from τ to ν.

Corollary 4.2. *Let ν be a tope. Then for every $[\epsilon, A] \in \mathcal{A}(\nu, \tau)$, the cone $\mathfrak{c}(\Phi_{\text{flip}}^A)$ is salient. Furthermore, we have*

$$X(\Phi, \tau) = \sum_{[\epsilon, A] \in \mathcal{A}(\nu, \tau)} \epsilon \operatorname{Flip}_A X(\Phi_{\text{flip}}^A, \nu).$$

Proof. The recursion rule means that, when we travel through the sequence of topes $\tau_i, i = 1, \ldots, \ell$, and apply Formula (4.1), we choose the flipped term when we cross the wall between τ_{k_i} and τ_{k_i+1}, and the unflipped term for the other walls. For instance, $A_{\{1\}} = A(\Phi, \tau_1, \tau_2)$, and $A_{\{1,2\}} = A_{\{1\}} \bigtriangleup A(\Phi_{\text{flip}}^{A_{\{1\}}}, \tau_2, \tau_3)$, in agreement with the two-step wall-crossing formula. The general case is immediate by induction. □

4.2 Semi-closed partition polytopes

In order to state the geometric consequences of the above combinatorial wall-crossing formulas, we introduce some semi-closed partition polytopes, to which the Brianchon-Gram theorem extends naturally.

Definition 4.3.

$$\mathfrak{p}(\Phi, A, \lambda) = \{x \in V(\Phi, \lambda), \quad x_i > 0 \text{ for } i \in A, x_i \geq 0 \text{ for } i \in A^c.\}$$

When λ is regular, the closure of $\mathfrak{p}(\Phi, A, \lambda)$ is the partition polytope $\mathfrak{p}(\Phi, \lambda)$.

Definition 4.4. For $A \subseteq \{1, \ldots, N\}$ we denote by Geom_A the map from W to the space of functions on $\mathbb{R}^N$ defined by substituting $1 - p_i$ for q_i, then $[x_i \geq 0]$ for p_i if $i \notin A$ and $[x_i > 0]$ for p_i if $i \in A$.

When A is the empty set, the substitution $\mathrm{Geom}_\emptyset$ coincide with the usual substitution Geom defined before. When we consider non empty subsets $A \subseteq \{1, \dots, N\}$, we obtain an extension of the Brianchon-Gram theorem to these semi-closed polytopes.

Proposition 4.1. *Let $A \subseteq \{1, \dots, N\}$. For $\lambda \in \tau$, we have*

$$\mathrm{Geom}_A\, X(\Phi, \tau)\, [V(\Phi, \lambda)] = [\mathfrak{p}(\Phi, A, \lambda)]. \tag{4.6}$$

Proof. When $A = \emptyset$, it is exactly the Brianchon-Gram theorem. We proceed by induction on the cardinality of A. If $A \neq \emptyset$, we can assume that $N \in A$, up to renumbering. Let $A' = A \setminus \{N\}$.

We write $[\mathfrak{p}(\Phi, A, \lambda)] = [\mathfrak{p}(\Phi, A', \lambda)] + \big([\mathfrak{p}(\Phi, A, \lambda)] - [\mathfrak{p}(\Phi, A', \lambda)]\big)$. We have

$$[\mathfrak{p}(\Phi, A', \lambda)] - [\mathfrak{p}(\Phi, A, \lambda)] = [\mathfrak{p}(\Phi, A', \lambda)][x_N = 0].$$

Let us show that we have

$$\begin{aligned}(\mathrm{Geom}_{A'}\, X(\Phi, \tau) - \mathrm{Geom}_A\, X(\Phi, \tau))[V(\Phi, \lambda)]\\ = [\mathfrak{p}(\Phi, A', \lambda)][x_N = 0].\end{aligned} \tag{4.7}$$

We first prove (4.7) in the case where $A = \{N\}$, hence $A' = \emptyset$.

We observe that the right hand side of (4.7) is the characteristic function of the face of $\mathfrak{p}(\Phi, \lambda)$ defined by $x_N = 0$. If we identify the hyperplane $\{x_N = 0\}$ with $\mathbb{R}^{N-1}$, this face is the partition polytope $\mathfrak{p}(\Phi', \lambda)$ corresponding to $\Phi' = (\phi_i)$, $1 \leq i \leq N - 1$ and $\lambda \in F$.

We now look at the left hand side of (4.7). We see that

$$\begin{aligned}&\mathrm{Geom}_\emptyset\, X(\Phi, \tau) - \mathrm{Geom}_{\{N\}}\, X(\Phi, \tau)\\ &= \left(\sum_{\substack{I \in \mathcal{G}(\Phi,\tau),\\ I^c \ni N}} (-1)^{|I| - \dim F} \prod_{i \in I^c, i \neq N} [x_i \geq 0] \right) [x_N = 0],\end{aligned} \tag{4.8}$$

because the terms indexed by the subsets I such that $N \notin I^c$ cancel out in the difference.

If Φ' does not generate Φ, we see that both sides of the equation (4.7) are equal to 0. Indeed as λ is regular, it cannot be contained in the smaller dimensional space generated by Φ', and every generating subset in $\mathcal{G}(\Phi, \tau)$ contains the index N.

Now assume that Φ' generates F. The Φ-tope τ is contained in a unique Φ'-tope τ'. The set $\mathcal{G}(\Phi', \tau')$ consists precisely of the subsets $I' \subseteq \{1, \dots, N - 1\}$ such that $I' \in \mathcal{G}(\Phi, \tau)$.

Therefore the right hand side of (4.8) is the Brianchon-Gram decomposition of the facet $\mathfrak{p}(\Phi', \lambda)$. Thus we have proved (4.7) in the case where $A = \{N\}$.

The general case when $A' \neq \emptyset$ is similar. We have now

$$\begin{aligned}&\operatorname{Geom}_{A'} X(\Phi, \tau) - \operatorname{Geom}_A X(\Phi, \tau)\\ &= \left(\sum_{\substack{I \in \mathcal{G}(\Phi,\tau),\\ I^c \ni N}} (-1)^{|I| - \dim F} \prod_{i \in I^c \cap A'} [x_i > 0] \prod_{\substack{i \in I^c \cap A'^c,\\ i \neq N}} [x_i \geq 0] \right) [x_N = 0].\end{aligned}$$

By the induction hypothesis, the right hand side of this equality is the Brianchon-Gram decomposition of the semi-closed polytope $\mathfrak{p}(\Phi', A', \lambda) = \mathfrak{p}(\Phi, A', \lambda) \cap \{x_N = 0\}$. □

Remark 4.5. The formula is not necessarily true on the boundary of τ, as shown by the trivial example $\Phi = (\phi_1)$, $A = \{1\}$, $\lambda = 0$.

It will be useful to rephrase Proposition 4.1 in the terms which arise in the combinatorial wall-crossing Theorem 4.1.

Definition 4.5. For $A \subseteq \{1, \ldots, N\}$ such that $\mathfrak{c}(\Phi^A_{\text{flip}})$ is salient and $\lambda \in F$, let

$$\begin{aligned}&\mathfrak{p}_{\text{flip}}(\Phi, A, \lambda)\\ &= \{x \in \mathbb{R}^N \,;\, \sum_i x_i \phi_i = \lambda, x_i < 0 \text{ for } i \in A, x_i \geq 0 \text{ for } i \notin A\}. \quad (4.9)\end{aligned}$$

Proposition 4.2. *Let $A \subseteq \{1, \ldots, N\}$ be such that the cone $\mathfrak{c}(\Phi^A_{\text{flip}})$ is salient. Let τ be a Φ-tope. Then, for $\lambda \in \tau$ we have*

$$\operatorname{Geom}_\emptyset \operatorname{Flip}_A X(\Phi^A_{\text{flip}}, \tau)[V(\Phi, \lambda)] = [\mathfrak{p}_{\text{flip}}(\Phi, A, \lambda)]. \quad (4.10)$$

Proof. For any polynomial $Z \in \mathbb{C}[p_i, q_i]$, we have

$$\operatorname{Geom}_\emptyset \operatorname{Flip}_A(Z) = \operatorname{Geom}_A(Z) \circ \sigma^A \quad (4.11)$$

where $\sigma^A x = (\sigma_i^A x_i)$ with $\sigma_i^A = -1$ if $i \in A$ and $\sigma_i^A = 1$ if $i \notin A$. Moreover we have

$$[V(\Phi, \lambda)] = [V(\Phi^A_{\text{flip}}, \lambda)] \circ \sigma^A.$$

Thus

$$\begin{aligned}&\operatorname{Geom}_\emptyset \operatorname{Flip}_A X(\Phi^A_{\text{flip}}, \tau)[V(\Phi, \lambda]\\ &\qquad = \left(\operatorname{Geom}_A X(\Phi^A_{\text{flip}}, \tau)[V(\Phi^A_{\text{flip}}, \lambda)]\right) \circ \sigma^A. \quad (4.12)\end{aligned}$$

We apply Proposition 4.1 to the sequence Φ^A_{flip} and the Φ^A_{flip}-tope τ. We obtain that the right hand side of (4.12) is equal to

$$[\mathfrak{p}(\Phi^A_{\text{flip}}, A, \lambda] \circ \sigma^A.$$

By definition of $\mathfrak{p}(\Phi^A_{\text{flip}}, A, \lambda)$, this is precisely the characteristic function of the set of x such that $\sigma_i^A x_i = -x_i > 0$ for $i \in A$ and $\sigma_i^A x_i = x_i \geq 0$ for $i \notin A$, and $\sigma^A x \in V(\Phi^A_{\text{flip}}, \lambda)$, *i.e.* $x \in \mathfrak{p}_{\text{flip}}(\Phi, A, \lambda)$. □

4.3 Decomposition in quadrants

Recall that when τ and ν are two topes, we have defined a list $\mathcal{A}(\nu, \tau)$ of signed subsets $[\epsilon, A]$ of $\{1, 2, \ldots, N\}$.

Theorem 4.3. *Let $z(\Phi, \tau, B)$, $B \subset \{1, \ldots, N\}$ be the collection of coefficients of the Combinatorial Brianchon-Gram polynomial associated to the Φ-tope τ.*

$$X(\Phi, \tau) = \sum_B z(\Phi, \tau, B) \prod_{i \notin B} p_i \prod_{i \in B} q_i.$$

(i) *If $B = \emptyset$, then $z(\Phi, \tau, B) = 1$ while if $B = \{1, 2, \ldots, N\}$, then $z(\Phi, \tau, B) = (-1)^d$.*
(ii) *If $z(\Phi, \tau, B) \neq 0$, then the cone $\mathfrak{c}(\Phi^B_{\text{flip}})$ is salient .*
(iii) *More precisely, if $z(\Phi, \tau, B)$ is not equal to 0, choose $x = (x_i) \in Q^B_{\text{neg}}$ such that $\lambda = \sum_i x_i \phi_i$ is a regular element in F and let ν be the tope containing λ. Then B occurs in the list in $\mathcal{A}(\nu, \tau)$ and we have*

$$z(\Phi, \tau, B) = (-1)^{|B|} \sum_{\{[\epsilon, B] \in \mathcal{A}(\nu, \tau)\}} \epsilon. \tag{4.13}$$

Proof. We already remarked (i).

For the choice of x as in (iii), the coefficient $z_B = z(\Phi, \tau, B)$ is the value of Geom $X(\Phi, \tau)[V(\Phi, \lambda)]$ at such a point x. We now apply Corollary 4.2 and Proposition 4.2 using the tope ν, where λ belongs.

We obtain that the value z_B is the signed sum of the values of $\mathfrak{p}_{\text{flip}}(\Phi, A, \lambda) = [Q^A_{\text{neg}}] \cap [V(\Phi, \lambda)]$ at x. By definition, this value is zero if the set of i with $x_i < 0$ is different from B. Otherwise, is equal to 1. □

Example 4.4. Let τ_1 and τ_2 be the adjacent topes and $A = A(\Phi, \tau_1, \tau_2)$. If $\tau_2 \subset \mathfrak{c}(\Phi^A_{\text{flip}})$, then $z(\Phi, \tau, A) = -(-1)^{|A|}$. An example of this situation is the standard knapsack, (Example 3.1), where there are exactly two topes, $\mathbb{R}_{>0}$ and $\mathbb{R}_{<0}$. The theorem implies that the Brianchon-Gram polynomial is $p_1 \cdots p_N - (-1)^N q_1 \cdots q_N$, as we found directly.

Example 4.5. If there is no subset K in the sum (4.13), or if there are more than one, then the coefficient $z(\Phi, \tau, B)$ may be 0 although the cone $\mathfrak{c}(\Phi^B_{\text{flip}})$ is salient . Let us take Example 3.2. The direct computation gave $X(\Phi, \tau_1) = p_1p_2p_3 - p_1q_2q_3 + q_1p_2p_3 - q_1q_2q_3$. We see that there are no terms corresponding to $B = \{2\}$ and $B = \{1, 3\}$. For $B = \{2\}$, the tope $\tau_2 = \overset{\circ}{\mathfrak{c}}\ (\phi_1, -\phi_2)$ is contained in $\mathfrak{c}(\phi_1, -\phi_2, \phi_3)$, hence we take the sequence (τ_1, τ_2). For $K = (1)$ we have $A_{(1)} = \{2, 3\} \neq B$, so there is no K such that $A_K = B$. For $B = \{1, 3\}$, we need a sequence of three topes, (τ_1, τ_2, τ_3) where now $\tau_2 = \overset{\circ}{\mathfrak{c}}\ (\phi_2, \phi_3)$ and $\tau_3 = \overset{\circ}{\mathfrak{c}}\ (-\phi_1, \phi_2) \subset \mathfrak{c}(\Phi^B_{\text{flip}})$. The K such that $A_K = \{1, 3\}$ are $K = (2)$ and $K = (1, 2)$, which indeed lead to opposite signs in the sum (4.13).

4.4 Geometric wall-crossing

By "intersecting" Geom $X(\Phi, \tau)$ with $[V(\Phi, \lambda)]$, we translate the results on $X(\Phi, \tau)$ in geometric terms.

Corollary 4.3 (Notations of Theorem 4.1 and Definition 4.5). *Let τ_1 be a Φ-tope. For λ in the adjacent tope τ_2, we have the geometric wall-crossing for.mula*

$$\mathcal{X}(\Phi, \tau_1)[V(\Phi, \lambda)] = [\mathfrak{p}(\Phi, \lambda)] - (-1)^{|A|}[\mathfrak{p}_{\text{flip}}(\Phi, A, \lambda)]. \tag{4.14}$$

Proof. We apply the map Geom on both sides of the combinatorial wall-crossing formula (4.1), then we multiply by the characteristic function $[V(\Phi, \lambda)]$. We obtain, by definition,

$$\mathcal{X}(\Phi,\tau_1)[V(\Phi,\lambda)] = [\mathfrak{p}(\Phi,\lambda] - (-1)^{|A|} \operatorname{Geom} \operatorname{Flip}_A X(\Phi^A_{\text{flip}},\tau_2)[V(\Phi,\lambda)],$$

hence (4.14) by applying the semi-closed Brianchon-Gram formula, as stated in Proposition 4.2, to the tope τ_2. □

Corollary 4.4. *For any $\lambda \in F$, the function*

$$\mathcal{X}(\Phi, \tau)[V(\Phi, \lambda)] = \sum_{\{B, \lambda \in \mathfrak{c}(\Phi^B_{\text{flip}})\}} z(\Phi, \tau, B)[\mathfrak{p}_{\text{flip}}(\Phi, B, \lambda)] \tag{4.15}$$

is a linear combination with integral coefficients of semi-closed partition polytopes.

4.5 An example

We return to Example 4.2, see Figure 4.1, with $\Phi = (\phi_1, \phi_2, \phi_3 = \frac{1}{2}(\phi_2 - \phi_1), \phi_4 = \frac{1}{2}(\phi_1 + \phi_2))$.

For $\lambda = \lambda_1\phi_1 + \lambda_2\phi_2$, we parametrize the 2-dim subspace $V(\Phi, \lambda) \subset \mathbb{R}^4$ by $(y_1, y_2) \mapsto (y_1+\lambda_1, y_2+\lambda_2, y_1-y_2, -(y_1+y_2))$. We start with λ in the tope τ_1 generated by ϕ_2 and ϕ_4, *i.e.* $\lambda_2 > \lambda_1 > 0$. Then $\mathfrak{p}(\lambda)$ corresponds under the parameterization to the tetragon in Figure 4.2, defined by the inequations

$$\begin{aligned} y_1 + \lambda_1 &\geq 0 \\ y_2 + \lambda_2 &\geq 0 \\ y_1 - y_2 &\geq 0 \\ y_1 + y_2 &\leq 0 \end{aligned}$$

We describe its analytic continuation, as the parameter λ visits the topes, one after the other. In the figures, the polytopes which are counted positively are coloured in blue, those wich are counted negatively are coloured in red. Semi-openness is indicated with dashed lines. When λ moves to the right and reaches the wall generated by ϕ_4, the tetragon $\mathfrak{p}(\lambda)$ transforms into a triangle (Figure 4.2). When λ enters the adjacent tope τ_2 generated by (ϕ_1, ϕ_4), *i.e.* $\lambda_1 > \lambda_2 > 0$, the wall-crossing polytope appears (Figure 4.3). It is (with sign -1) the semi-closed triangle

$$\begin{aligned} y_1 + \lambda_1 &\geq 0 \\ y_2 + \lambda_2 &< 0 \\ y_1 - y_2 &< 0 \\ y_1 + y_2 &\leq 0, \text{ (this condition is redundant).} \end{aligned}$$

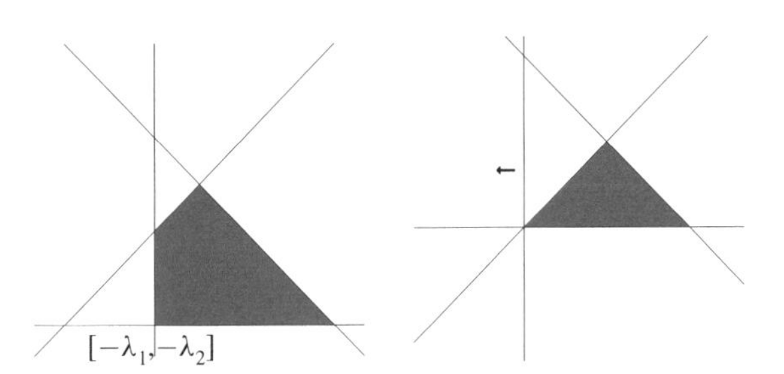

Figure 4.2. $\lambda_2 > \lambda_1 > 0$, (tope (ϕ_2, ϕ_4)), then $\lambda_2 = \lambda_1 > 0$, (wall (ϕ_4)).

Then λ moves downwards towards the wall generated by ϕ_1. The positive closed triangle shrinks while the negative semi-closed one increases. When λ reaches the wall, the closed triangle is reduced to a point (Figure 4.3). When λ enters the tope $(\phi_1, -\phi_3)$, the negative semi-closed triangle deforms into a negative semi-closed quadrileral (Figure 4.4).

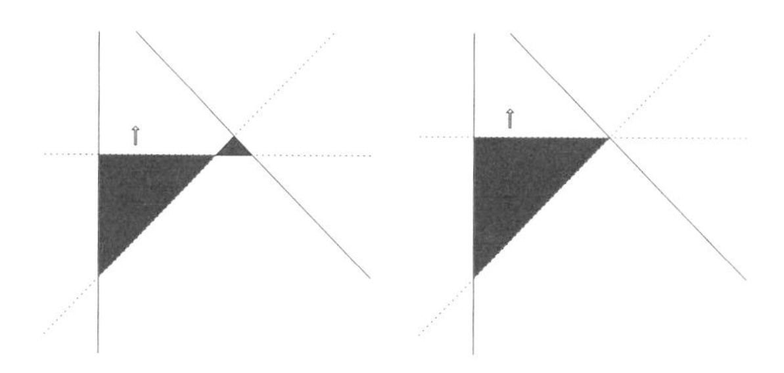

Figure 4.3. $\lambda_1 > \lambda_2 > 0$, (tope (ϕ_4, ϕ_1)), then $\lambda_1 > 0 = \lambda_2$, (wall (ϕ_1)).

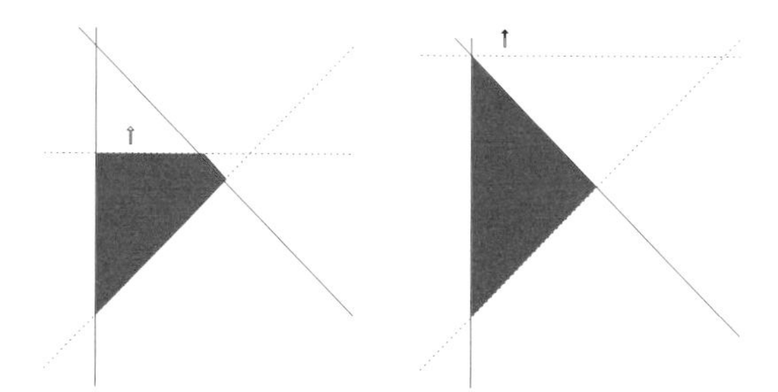

Figure 4.4. $\lambda_1 > -\lambda_2 > 0$, (tope $(\phi_1, -\phi_3)$), then $\lambda_1 = -\lambda_2 > 0$, (wall $(-\phi_3)$).

When λ reaches the wall generated by $-\phi_3$ (Figure 4.5), then enters the tope $(-\phi_3, -\phi_2)$, a new positive open triangle appears. Then λ reaches the wall generated by $-\phi_2$ (Figure 4.5) and enters the tope $(-\phi_2, -\phi_4)$ (Figure 4.6). The analytic continuation is now a positive open tetragon opposite to the (closed) initial one. (This case is pointed out in [20]).

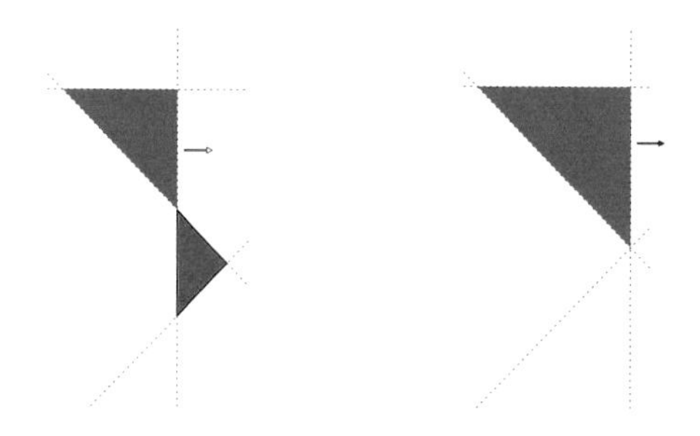

Figure 4.5. $-\lambda_2 > \lambda_1 > 0$, (tope $(-\phi_3, -\phi_2)$), then $-\lambda_2 > 0 = \lambda_1$, (wall $-\phi_2$).

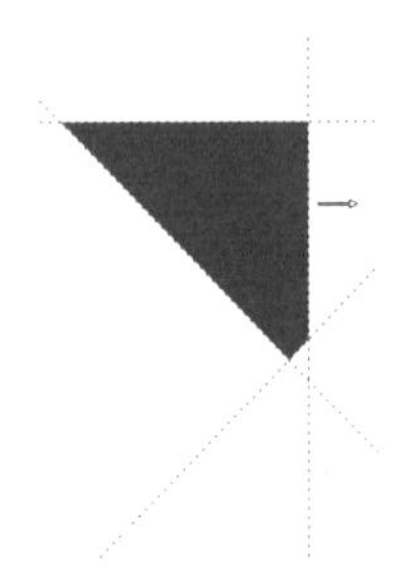

Figure 4.6. $\lambda_1 < \lambda_2 < 0$, (tope $(-\phi_2, -\phi_4)$). The polygon is now the interior of the opposite of the initial tetragon of Figure 4.2, cf. [20].

5 Integrals and discrete sums over a partition polytope

As a consequence of the compactness result of Corollary 4.4, together with the set theoretic relations of Corollary 3.1, we recover properties of sums and integrals over partition polytopes which were previously obtained in [9] and [12, 13, 19]. Moreover, the set-theoretic wall-crossing formula has obvious implications for sums and integrals. In particular, when applied to the number of points of a partition polytope, it implies the wall-crossing formula of [18, Theorem 5.2]. We will explain in more details this last point in Subsection 5.3.

5.1 Generating functions of polyhedra and Brion's theorem

Let V be a real dimensional vector space. We choose a Lebesgue measure dv on V. Let us recall the notion of valuations and of generating functions of cones (see the survey [3]).

Recall that a *valuation* F is a map from a set of polyhedra $\mathfrak{p} \subset V$ to a vector space $\mathcal{M}$ such that whenever the characteristic functions $[\mathfrak{p}_i]$ of a family of polyhedra $\mathfrak{p}_i$ satisfy a linear relation $\sum_i r_i[\mathfrak{p}_i] = 0$, then the elements $F(\mathfrak{p}_i)$ satisfy the same relation $\sum_i r_i F(\mathfrak{p}_i) = 0$. Thus any valuation defined on the set of all polyhedra can be extended to the "analytic continuation", which is a signed sum of polytopes. In particular, the valuation defined on the set of polyhedra by the Euler characteristic (see [3]) is identically equal to 1 on the "analytic continuation" as follows from Brianchon-Gram decomposition and Euler relations.

We now study two other classical instances of valuations.

There exists a unique valuation $\mathfrak{p} \mapsto I(\mathfrak{p})$ which associates to every polyhedron $\mathfrak{p} \subseteq V$ a meromorphic function $I(\mathfrak{p})(\xi)$ on V^*, so that the following properties hold:

(i) If $\mathfrak{p}$ contains a straight line, then $I(\mathfrak{p}) = 0$.

(ii) If $\xi \in V^*$ is such that $e^{\langle \xi, x\rangle}$ is integrable over $\mathfrak{p}$ for the measure dv, then

$$I(\mathfrak{p})(\xi) = \int_{\mathfrak{p}} e^{\langle \xi, x\rangle}\, dv.$$

Moreover, for every point $s \in V$, one has

$$I(s + \mathfrak{p})(\xi) = e^{\langle \xi, s\rangle} I(\mathfrak{p})(\xi).$$

$I(\mathfrak{p})(\xi)$ is called the *continuous generating function* of $\mathfrak{p}$.

Assume that V is equipped with a lattice $V_{\mathbb{Z}}$.

There exists a unique valuation $\mathfrak{p} \mapsto S(\mathfrak{p})$ which associates to every rational polyhedron $\mathfrak{p} \subseteq V$ a meromorphic function $S(\mathfrak{p})(\xi)$ on V^*, so that

(i) if $\mathfrak{p}$ contains a straight line, then $S(\mathfrak{p}) = 0$;
(ii) if $\xi \in V^*$ is such that $e^{\langle \xi, x\rangle}$ is summable over the set of lattice points of $\mathfrak{p}$, then

$$S(\mathfrak{p})(\xi) = \sum_{x \in \mathfrak{p} \cap V_{\mathbb{Z}}} e^{\langle \xi, x\rangle}.$$

Moreover, for every point $s \in V_{\mathbb{Z}}$, one has

$$S(s + \mathfrak{p})(\xi) = e^{\langle \xi, s\rangle} S(\mathfrak{p})(\xi).$$

$S(\mathfrak{p})(\xi)$ is called the *(discrete) generating function* of $\mathfrak{p}$.

These valuations are easily constructed, either by algebraic methods (see [3]), or by introducing the Fourier transforms of discrete or continuous measures associated to the polyhedron $\mathfrak{p}$ (see Section 6). Furthermore, there is an important property of the generating functions $S(\mathfrak{p})(\xi)$ and $I(\mathfrak{p})(\xi)$. Introduce the space $\mathcal{M}_\ell(V^*)$ of meromorphic functions on V^* which can be written as the quotient of a function which is holomorphic near $\xi = 0$ by a product of linear forms. The functions $I(\mathfrak{p})(\xi)$ and $S(\mathfrak{p})(\xi)$ belong to the space $\mathcal{M}_\ell(V^*)$. Then a function $f(\xi) \in \mathcal{M}_\ell(V^*)$ has a unique expansion into homogeneous rational functions

$$f(\xi) = \sum_{m \geq m_0} f_{[m]}(\xi),$$

where the summands $f_{[m]}(\xi)$ have degree m as we define now: if P is a homogeneous polynomial on V^* of degree p, and D a product of r linear forms, then $\frac{P}{D}$ is an element in $\mathcal{M}_\ell(V^*)$ homogeneous of degree $m = p - r$.

Let $\mathfrak{p}$ be a polytope with set of faces $\mathcal{F}(\mathfrak{p})$, and affine tangent cones $\mathfrak{t}_{\text{aff}}(\mathfrak{p}, \mathfrak{f})$ at $\mathfrak{f}$. We obtain from the Brianchon-Gram theorem:

$$\int_{\mathfrak{p}} e^{\langle \xi, v \rangle} dv = \sum_{\mathfrak{f}} (-1)^{\dim \mathfrak{f}} I(\mathfrak{t}_{\text{aff}}(\mathfrak{p}, \mathfrak{f}))(\xi).$$

Furthermore, as the cone $\mathfrak{t}_{\text{aff}}(\mathfrak{p}, \mathfrak{f})$ contains a straight line, when the dimension of $\mathfrak{f}$ is strictly greater than 0, this gives the well-known Brion's formula:

$$\int_{\mathfrak{p}} e^{\langle \xi, v \rangle} dv = \sum_{s} I(s + \mathfrak{c}_s)(\xi).$$

Here s runs through the vertices of $\mathfrak{p}$ and $\mathfrak{c}_s$ is the tangent cone at s.

Similarly, when V is a rational vector space with lattice $V_{\mathbb{Z}}$, and $\mathfrak{p}$ a rational polytope, we have

$$\sum_{x \in \mathfrak{p} \cap V_{\mathbb{Z}}} e^{\langle \xi, x \rangle} = \sum_{s} S(s + \mathfrak{c}_s)(\xi). \tag{5.1}$$

These formulae are at the heart of Varchenko's "analytic continuation procedure": we see intuitively that if the vertices of a polytope $\mathfrak{q}(b)$ vary "analytically" with a parameter b, the integrals and discrete sums will also vary "analytically". We will state precise results in the next section.

5.2 Polynomiality and wall-crossing for integrals and sums

Recall the following definition

Definition 5.1. If F is equipped with a lattice Λ, a quasi-polynomial function f on Λ is a function such that there exists a sublattice $\Lambda' \subset \Lambda$ so that, for any $\lambda_0 \in \Lambda$, the function $\lambda' \to f(\lambda_0 + \lambda')$ is given by the restriction to Λ' of a polynomial function f_{λ_0} on F.

Theorem 5.1. *Let $\Phi = (\phi_j)_{1 \leq j \leq N}$ be a sequence of non zero elements of a vector space F, generating F, and spanning a salient cone. Let $\tau \subset F$ be a Φ-tope such that τ is contained in the cone $\mathfrak{c}(\Phi)$ generated by Φ. For $\lambda \in F$, let $V(\Phi, \lambda)$ be the affine subspace of $\mathbb{R}^N$ defined by $\sum_{i=1}^N x_i \phi_i = \lambda$. Let*

$$\mathcal{X}(\Phi, \tau) = \sum_{I \in \mathcal{G}(\Phi, \tau)} (-1)^{|I| - \dim F} \prod_{i \in I^c} [x_i \geq 0],$$

where $\mathcal{G}(\Phi, \tau)$ is the set of $I \subseteq \{1, \ldots, N\}$ such that $\{\phi_i, i \in I\}$ generates F and such that τ is contained in the cone generated by $\{\phi_i, i \in I\}$.

Let $h(x)$ be a polynomial function on $\mathbb{R}^N$. Fix a Lebesgue measure on the subspace V and let $dm_\Phi(x)$ be the corresponding Lebesgue measure on $V(\Phi, \lambda)$. Define

$$I(\Phi, \tau, h)(\lambda) = \int_{V(\Phi,\lambda)} \mathcal{X}(\Phi, \tau)(x)h(x)dm_\Phi(x). \tag{5.2}$$

In the case where F is a rational space with lattice Λ and that the ϕ_i are lattice vectors, define

$$S(\Phi, \tau, h)(\lambda) = \sum_{x \in V(\Phi,\lambda)\cap\mathbb{Z}^N} \mathcal{X}(\Phi, \tau)(x)h(x). \tag{5.3}$$

Then

(i) $\lambda \mapsto I(\Phi, \tau, h)(\lambda)$ *is a polynomial function on F.*
(ii) $\lambda \mapsto S(\Phi, \tau, h)(\lambda)$ *is a quasi-polynomial function on the lattice $\Lambda \subset F$.*
(iii) *If λ belongs to the closure of the tope τ, we have*

$$I(\Phi, \tau, h)(\lambda) = \int_{\mathfrak{p}(\Phi,\lambda)} h(x)dm_\Phi(x). \tag{5.4}$$

Let $\mathfrak{b}(\Phi)$ be the zonotope generated by Φ. If $\lambda \in (\tau - \mathfrak{b}(\Phi)) \cap \Lambda$, we have

$$S(\Phi, \tau, h)(\lambda) = \sum_{x \in \mathfrak{p}(\Phi,\lambda)\cap\mathbb{Z}^N} h(x). \tag{5.5}$$

(iv) *Furthermore, we have the following wall-crossing formulas (with the notations of Theorem 4.1). For $\lambda \in \tau_2$, we have*

$$I(\Phi, \tau_1, h)(\lambda) = \int_{\mathfrak{p}(\Phi,\lambda)} h(x)dm_\Phi(x) - (-1)^{|A|} \int_{\mathfrak{p}_{\mathrm{flip}}(\Phi,A,\lambda)} h(x)dm_\Phi(x). \tag{5.6}$$

$$S(\Phi, \tau_1, h)(\lambda) = \sum_{x \in \mathfrak{p}(\Phi,\lambda)\cap\mathbb{Z}^N} h(x) - (-1)^{|A|} \sum_{x \in \mathfrak{p}_{\mathrm{flip}}(\Phi,A,\lambda)\cap\mathbb{Z}^N} h(x). \tag{5.7}$$

Proof. (iii) follows immediately from Corollary 3.1 and (iv) from the wall crossing formulas of Corollary 4.3, together with Corollary 3.1.

The proof of the polynomiality in (i) and (ii) relies on the properties of generating functions, as we explain in [2] for the weighted Ehrhart theory.

To begin with, observe that it is enough to prove the theorem in the case where the weight $h(x)$ is a power of a linear form

$$h(x) = \frac{\langle \xi, x \rangle^M}{M!},$$

for $\xi \in (\mathbb{R}^N)^*$. This is the term of ξ-degree M of the exponential $e^{\langle \xi, x \rangle}$. Thus, we consider the functions of $\xi \in (\mathbb{R}^N)^*$

$$I(\Phi, \tau)(\xi, \lambda) = \int_{V(\Phi,\lambda)} \mathcal{X}(\Phi, \tau)(x) e^{\langle \xi, x \rangle} dm_\Phi(x). \tag{5.8}$$

$$S(\Phi, \tau)(\xi, \lambda) = \sum_{x \in V(\Phi,\lambda) \cap \mathbb{Z}^N} \mathcal{X}(\Phi, \tau)(x) e^{\langle \xi, x \rangle}. \tag{5.9}$$

As $\mathcal{X}(\Phi, \tau)(x)[V(\Phi, \lambda)](x)$ has bounded support by Corollary 3.2, (5.8) and (5.9) are holomorphic functions of ξ. We recover $I(\Phi, \tau, h)(\lambda)$ and $S(\Phi, \tau, h)(\lambda)$ by taking their term of ξ-degree M.

However the dependence on λ can be analyzed by looking at each summand in

$$\mathcal{X}(\Phi, \tau) \cap [V(\Phi, \lambda)] = \sum_{K \in \mathcal{G}(\Phi,\tau)} (-1)^{|K| - \dim F} [\mathfrak{t}_K(\Phi, \lambda)]. \tag{5.10}$$

Indeed, it is immediate to extend the valuation $I(\mathfrak{p})$ defined in the preceding section to a valuation $I(\mathfrak{p}, \lambda)$ defined on polyhedrons contained in the affine space $V(\Phi, \lambda)$.

Namely, there exists a unique valuation $I(\mathfrak{p}, \lambda)$ associating to every polyhedron $\mathfrak{p} \subseteq V(\Phi, \lambda)$ a meromorphic function $I(\mathfrak{p}, \lambda)(\xi)$ on $\mathbb{C}^N$, so that the following properties hold:

(i) If $\mathfrak{p}$ contains a straight line, then $I(\mathfrak{p}, \lambda) = 0$.
(ii) If $\xi \in \mathbb{C}^N$ is such that $\mathrm{e}^{\langle \xi, x \rangle}$ is integrable over $\mathfrak{p}$ for the measure dm_Φ, then

$$I(\mathfrak{p}, \lambda)(\xi) = \int_{\mathfrak{p}} \mathrm{e}^{\langle \xi, x \rangle} dm_\Phi(x).$$

Moreover, for every point $s \in \mathbb{R}^N$, one has

$$I(s + \mathfrak{p}, \lambda + \sum_{i=1}^N s_i \phi_i)(\xi) = \mathrm{e}^{\langle \xi, s \rangle} I(\mathfrak{p}, \lambda)(\xi).$$

Similarly, if F is a space with a lattice Λ, and the elements ϕ_i belongs to Λ, then, for $\lambda \in \Lambda$, there exists a unique valuation $\mathfrak{p} \mapsto S(\mathfrak{p}, \lambda)$ associating to any $\lambda \in \Lambda$ and every rational polyhedron $\mathfrak{p} \subseteq V(\Phi, \lambda)$ a meromorphic function $S(\mathfrak{p}, \lambda)(\xi)$ on $\mathbb{C}^N$, so that
(i) if $\mathfrak{p}$ contains a straight line, then $S(\mathfrak{p}, \lambda) = 0$;
(ii) if $\xi \in \mathbb{C}^N$ is such that $\mathrm{e}^{\langle \xi, x\rangle}$ is summable over the set $\mathfrak{p} \cap \mathbb{Z}^N$, then

$$S(\mathfrak{p}, \lambda)(\xi) = \sum_{x \in \mathfrak{p} \cap \mathbb{Z}^N} \mathrm{e}^{\langle \xi, x\rangle}.$$

Moreover, for every point $s \in \mathbb{Z}^N$, one has

$$S(s + \mathfrak{p}, \lambda + \sum_i s_i \phi_i)(\xi) = \mathrm{e}^{\langle \xi, s\rangle} S(\mathfrak{p}, \lambda)(\xi).$$

Look at Equation (5.10). The polyhedron $\mathfrak{t}_K(\Phi, \lambda)$ contains a straight line as soon as if $K \in \mathcal{G}(\Phi, \tau)$ is not a basic subset. Thus, in terms of the valuations $I(\mathfrak{p}, \lambda)$, $S(\mathfrak{p}, \lambda)$, we have

$$I(\Phi, \tau)(\xi, \lambda) = \sum_{K \in \mathcal{B}(\Phi, \tau)} I(\mathfrak{t}_K(\Phi, \lambda), \lambda)(\xi),$$

and

$$S(\Phi, \tau)(\xi, \lambda) = \sum_{K \in \mathcal{B}(\Phi, \tau)} S(\mathfrak{t}_K(\Phi, \lambda), \lambda)(\xi).$$

Each of these equations expresses a holomorphic function as a sum of meromorphic ones whose poles cancel out. Furthermore, we can recover the term of ξ-degree M by taking the homogeneous de degree in ξ in each of these functions of ξ.

Regarding the dependence on λ, we have already observed the following crucial fact: the cone $\mathfrak{t}_K(\Phi, \lambda)$ is the shift $s_K(\Phi, \lambda) + \mathfrak{a}_0(K)$ of the **fixed cone** $\mathfrak{a}_0(K)$ by the vertex $s_K(\Phi, \lambda)$ depending linearly of λ.

Let us first study the integral. Using the translation property of the valuation $I(\mathfrak{p}, \lambda)$, we can express $I(\mathfrak{p}, \lambda)$ in function of the valuation $I(\mathfrak{p})$ defined on polyhedrons contained in the fixed space V. We then have

$$I(\mathfrak{t}_K(\Phi, \lambda), \lambda)(\xi) = e^{\langle \xi, s_K(\Phi, \lambda)\rangle} I(\mathfrak{a}_0(K))(\xi).$$

Only the first factor $e^{\langle \xi, s_K(\Phi, \lambda)\rangle}$ depends on λ. Actually, it is easy to see that $I(\mathfrak{a}_0(K))(\xi)$ is homogeneous of degree $-d$. Hence, the term of ξ-degree M of $I(\mathfrak{t}_K(\Phi, \lambda), \lambda)(\xi)$ is given by

$$I(\mathfrak{t}_K(\Phi, \lambda), \lambda)(\xi)_{[M]} = \frac{\langle \xi, s_K(\Phi, \lambda)\rangle^{M+d}}{(M+d)!} I(\mathfrak{a}_0(K))(\xi)_{[-d]}.$$

Thus we have, for $h(x) = \frac{\langle \xi, x\rangle^M}{M!}$,

$$I(\Phi, \tau, h)(\lambda) = \sum_{K \in \mathcal{B}(\Phi,\tau)} \frac{\langle \xi, s_K(\Phi, \lambda)\rangle^{M+d}}{(M+d)!} I(\mathfrak{a}_0(K))(\xi)_{[-d]}.$$

The right hand side of this formula is a polynomial function of λ of degree $M+d$, with coefficients which are polynomial functions of ξ of degree M, (although each K summand has poles). Thus we have proved (i).

Let us now study the discrete sum

$$S(\Phi, \tau)(\xi, \lambda) = \sum_{K \in \mathcal{B}(\Phi,\tau)} S(\mathfrak{t}_K(\Phi, \lambda), \lambda)(\xi).$$

Consider a sublattice Λ' of Λ such that all elements $s_K(\Phi, \lambda')$ have integral coefficients for $\lambda' \in \Lambda'$ and all $K \in \mathcal{B}(\Phi, \tau)$. If D is the least common multiple of all determinants of the Φ-basic subsets I in $\mathcal{B}(\Phi)$, we can choose $\Lambda' = D\Lambda$.

Thus, if $\lambda = \lambda_0 + \lambda'$, with $\lambda_0 \in \Lambda$,$\lambda' \in \Lambda'$, we obtain

$$S(\mathfrak{t}_K(\Phi, \lambda_0 + \lambda'), \lambda_0 + \lambda')(\xi) = \mathrm{e}^{\langle \xi, s_K(\Phi,\lambda')\rangle} S(\mathfrak{t}_K(\Phi, \lambda_0), \lambda_0)(\xi).$$

Indeed $\mathfrak{t}_K(\Phi, \lambda_0 + \lambda') = s_K(\Phi, \lambda') + \mathfrak{t}_K(\Phi, \lambda_0)$.

Here again the dependence in λ' is only through the factor $\mathrm{e}^{\langle \xi, s_K(\Phi,\lambda')\rangle}$ and $s_K(\Phi, \lambda')$ depends linearly on λ'.

If $f_{\lambda_0}(K, \xi) = S(\mathfrak{t}_K(\Phi, \lambda_0), \lambda_0)(\xi)$, a meromorphic function of ξ of degree greater or equal to $-d$, we obtain:

$$S(\mathfrak{t}_K(\Phi, \lambda_0 + \lambda'), \lambda_0 + \lambda')(\xi)_{[M]} = \sum_{k=0}^{M+d} \frac{\langle \xi, s_K(\Phi, \lambda')\rangle^k}{k!} f_{\lambda_0}(K, \xi)_{[M-k]}.$$

This is a polynomial function of λ' of degree $M+d$.

Adding up the contributions, we see that we obtain that

$$\lambda' \to S(\Phi, \tau)(\xi, \lambda_0 + \lambda')$$

is a polynomial function of λ' and ξ. □

Remark 5.1. Consider Equation (4.15):

$$\mathcal{X}(\Phi, \tau)[V(\Phi, \lambda)] = \sum_{\{B, \lambda \in \mathfrak{c}(\Phi^B_{\text{flip}})\}} z(\Phi, \tau, B)[Q^B_{\text{neg}}][V(\Phi, \lambda)]. \quad (5.11)$$

Consider the case where the elements ϕ_i are in a lattice Λ of F. Summing up the function $h = 1$ over $V(\Phi, \lambda) \cap \mathbb{Z}^N$ on both sides, we obtain an expression for the quasi polynomial function $S(\Phi, \tau, h)(\lambda)$ in function of the partition functions associated to the flipped systems Φ^B_{flip}.

The functions $S(\Phi, \tau, h)(\lambda)$ are elements of the Dahmen-Micchelli space associated to Φ and Λ. It was proved in [13] that any Dahmen-Micchelli quasi polynomial can be expressed as a linear combination of partition functions associated to flipped systems Φ^B_{flip}. The equation (5.11) can be considered as a "set-theoretic" generalization of this theorem.

5.3 Paradan's convolution wall-crossing formulas

We assume that F is equipped with a lattice Λ.

The convolution of two functions f_1, f_2 (satisfying adequate support conditions) on Λ is defined by

$$(f_1 * f_2)(\mu) = \sum_{\lambda_1+\lambda_2=\mu} f_1(\lambda_1) f_2(\lambda_2).$$

If $\mu \in \Lambda$, we write δ_μ for the function on Λ such that $f(\lambda) = \delta^\lambda_\mu$.

Let h be a polynomial function on $\mathbb{R}^N$, and consider

$$E(\Phi, h)(\lambda) = \sum_{x \in \mathfrak{p}(\Phi,\lambda)\cap\mathbb{Z}^N} h(x).$$

When h is the constant function 1, then

$$k(\Phi)(\lambda) = E(\Phi, 1)(\lambda) = \text{Card}(\mathfrak{p}(\Phi, \lambda) \cap \mathbb{Z}^N)$$

is the partition function associated to the sequence Φ. The function $k(\Phi)(\lambda)$ is the convolution product $f_{\phi_1} * \cdots * f_{\phi_N}$, where, for $\phi \in F$,

$$f_\phi := \sum_{n=0}^{\infty} \delta_{n\phi}.$$

Indeed, by definition $k(\Phi)(\lambda)$ is the number of solutions in integers $n_i \geq 0$ of the equation $\sum_i n_i\phi_i = \lambda$.

The case of a polynomial function h can be treated similarly. Assume h is a product $h(x_1, x_2, \ldots, x_N) = \prod_{i=1}^N h_i(x_i)$ where h_i are polynomial functions on $\mathbb{R}$. For h a polynomial function on $\mathbb{R}$, and ϕ a non zero element in Λ, introduce

$$f^h_\phi = \sum_{n=0}^{\infty} h(n)\delta_{n\phi}.$$

Then we see that
$$E(\Phi, h) = f_{\phi_1}^{h_1} * \cdots * f_{\phi_N}^{h_N}.$$

With the notations of Theorem 5.1, for each tope τ, the function $S(\Phi, \tau, h)(\lambda)$ is a quasi-polynomial function on the lattice Λ such that $E(\Phi, h)(\lambda) = S(\Phi, h, \tau)(\lambda)$ for $\lambda \in (\tau - \mathfrak{b}(\Phi)) \cap \Lambda$.

Let τ_1, τ_2 be two adjacent topes separated by a wall H. Let τ_{12} be the unique tope of $\Phi \cap H$ such that $\overline{\tau_1} \cap \overline{\tau_2} \subset \overline{\tau_{12}}$. Paradan's formula is a formula for $S(\Phi, h, \tau_1) - S(\Phi, h, \tau_2)$ when τ_1, τ_2 are adjacent topes in terms of the convolution of the quasi-polynomial function $S(\Phi \cap H, h, \tau_{1,2})$ on $\Lambda \cap H$ with some functions f_ϕ^h.

Before stating the formula, we note one property of the function $S(\Phi, h, \tau)$.

Assume that H is a face of the cone $\mathfrak{c}(\Phi)$ and that τ is a tope with one of its wall equal to H. Let τ_H be the unique tope of $\Phi \cap H$ so that $\overline{\tau} \cap H$ is contained in τ_H. Let $S(\Phi \cap H, h, \tau_H)$ be the quasi-polynomial function on $\Lambda \cap H$ associated to this data.

Let us denote the subsequence of elements ϕ_i not in H by $\Phi \setminus H = (\phi_1, \ldots, \phi_M)$. Then if $n_1, n_2, \ldots, n_M$ are non negative integers, and $\lambda \in \Lambda$, there are only a finite number of n_i such that $\lambda - \sum_{i=1}^M n_i \phi_i$ belongs to H, as the elements ϕ_i are all on one side of H.

Let $H^{\geq 0}$ be the closed half space delimited by H and containing τ.

Proposition 5.1. *For* $\lambda \in H^{\geq 0}$, $S(\Phi, h, \tau)(\lambda)$ *is equal to*

$$\sum_{n_i \geq 0, \lambda - \sum_i n_i \phi_i \in H} (h_1(n_1) \cdots h_M(n_M)) S(\Phi \cap H, h, \tau_H) \left(\lambda - \sum_{i=1}^M n_i \phi_i \right).$$

In other words, on $H^{\geq 0} \cap \Lambda$,

$$S(\Phi, h, \tau) = S(\Phi \cap H, h, \tau_H) * f_{\phi_1}^{h_1} * \cdots * f_{\phi_M}^{h_M}.$$

Proof. As follows from [5], the right hand side, being the convolution of a quasi-polynomial function on the lattice $\Lambda \cap H$ with products $f_{\phi_1}^{h_1} * \cdots * f_{\phi_M}^{h_M}$, coincides with a quasi-polynomial function on the domain $H^{\geq 0} \cap \Lambda$.

Now, to prove that the left hand side coincide with the right hand side, we will use the fact that two quasi-polynomial functions agreeing on $\mathfrak{c} \cap \Lambda$, where $\mathfrak{c}$ is a cone with non empty interior, coincide on Λ.

If $\lambda \in \tau$ is sufficiently near a point of $\overline{\tau} \cap H$, then the set $(\lambda - \sum_{i=1}^M \mathbb{R}_{\geq 0} \phi_i) \cap H$ is contained in τ_H. We see that the set $\mathfrak{c} := \{\lambda \in \tau; (\lambda - \sum_{i=1}^M \mathbb{R}_{\geq 0} \phi_i) \cap H \subset \tau_H\}$ is an open cone in τ. On $\mathfrak{c} \cap \Lambda$, the function $S(\Phi, h, \tau)$ coincide with $E(\Phi, h)$. On the other hand, if we compute $E(\Phi, h)$ and $E(\Phi \cap H, h)$ by their respective convolution formulae, we obtain that the right hand side coincide also with $E(\Phi, h)$ for $\lambda \in \mathfrak{c} \cap \Lambda$. This proves the proposition. □

Let A be the set of indices k such that ϕ_k belongs to the τ_2 open side of H. Let $\mathcal{X}(\Phi, A, \tau_2) = \text{Geom}(\text{Flip}_A\, X(\Phi^A_{\text{flip}}, \tau_2))$ and let

$$S(\Phi, A, h, \tau_2)(\lambda) = \sum_{x \in V(\Phi,\lambda)\cap\mathbb{Z}^N} h(x)\mathcal{X}(\Phi, A, \tau_2)(x).$$

We then obtain

$$S(\Phi, h, \tau_1) - S(\Phi, h, \tau_2) = S(\Phi, A, h, \tau_2).$$

Observe that $\overline{\tau}_2 \cap H$ is contained in the boundary of the cone $\mathfrak{c}(\Phi^A_{\text{flip}})$. Using a slight modification of Proposition 5.1 above, we then can give the following "convolution description" of the function $S(\Phi, A, h, \tau_2)$.

Let B be the set of indices k such that ϕ_k belongs to the τ_1 open side of H. Thus the vectors ϕ_a for $a \in A$ and $-\phi_b$ for $b \in B$ all belong to the τ_2 open side of H.

Define

$$\text{Flip}\, f^h_\phi = -\sum_{n=1}^{\infty} h(-n)\delta_{-n\phi}.$$

Then we have

Proposition 5.2. *The quasi-polynomial function $S(\Phi, A, h, \tau_2)$ is given by the convolution formula:*

$$S(\Phi, A, h, \tau_2) =$$

$$S(\Phi \cap H, h, \tau_{1,2}) * \left(\prod_{a\in A} \text{Flip}\, f^{h_a}_{\phi_a} * \prod_{b\in B} f^{h_b}_{\phi_b} - \prod_{b\in B} \text{Flip}\, f^{h_b}_{\phi_b} * \prod_{a\in A} f^{h_a}_{\phi_a}\right).$$

This is the wall crossing convolution formula given by Paradan in [18, Theorem 5.2]. It expresses the wall crossing variation in terms of the function $S(\Phi \cap H, h, \tau_{1,2})$ associated to a lower dimensional system (see also [5]).

6 A refinement of Brion's theorem

Let $\mathfrak{p} \subset V$ be a full-dimensional polytope in a vector space V provided with a lattice $V_{\mathbb{Z}}$. Recall Brion's Formula (1.9) for the generating function of a polytope.

$$S(\mathfrak{p})(\xi) = \sum_{s\in\mathcal{V}(\mathfrak{p})} S(s + \mathfrak{c}_s)(\xi).$$

As $\mathfrak{p}$ is compact, the function $\xi \in V^* \mapsto S(\mathfrak{p})(\xi)$ is holomorphic, but the contribution of each cone is a meromorphic function with singularities along hyperplanes. More precisely, an element $\xi \in V^*$ is singular for $S(s + \mathfrak{c}_s)$ if and only if ξ is constant on some face $\mathfrak{f}$ of $\mathfrak{p}$ such that s is a vertex of $\mathfrak{f}$ and $\dim \mathfrak{f} > 0$.

It is well known that Brion's formula is the combinatorial translation of the localization formula in equivariant cohomology, in the case of isolated fixed points. In this section, we generalize (1.9) to the combinatorial case which corresponds to non isolated fixed points [4]. In this degenerate case, the connected components of the set of fixed points correspond to the faces of $\mathfrak{p}$ on which ξ is constant which are maximal with respect to this property. The contribution of such a face to the sum $S(\mathfrak{p})(\xi)$ is

$$\sum_{s \in \mathcal{V}(\mathfrak{f})} S(s + \mathfrak{c}_s)(\xi).$$

We will study this sum by relating it to a Brianchon-Gram continuation of the face $\mathfrak{f}$. We will assume that the polytope $\mathfrak{p}$ is simple. The general case needs more efforts.

We need to introduce some meromorphic functions similar to the function $S(s + \mathfrak{c})(\xi)$. Let $\mathfrak{q} = s + \mathfrak{c}$ be a polyhedral cone in V, where $\mathfrak{c}$ is a cone generated by elements $g_j \in V_{\mathbb{Z}}$. Let P be a quasi polynomial function on $V_{\mathbb{Z}}$. The following sum $\sum_{x \in V_{\mathbb{Z}} \cap \mathfrak{q}} P(\xi) e^{\langle \xi, x \rangle}$ defines a generalized function F of the variable $\xi \in i V^*$: the above sum converges when ξ is imaginary in the distribution sense.

It is easy to see that $\prod_i (1 - e^{\langle \xi, g_j \rangle}) F(\xi)$ is an analytic function of ξ. Thus, outside the affine hyperplanes in $i V^*$ defined by $\langle \xi, g_j \rangle \in 2i\pi\mathbb{Z}$, the generalized function $F(\xi)$ is equal to $S(\mathfrak{q}, P)(\xi)$, where $S(\mathfrak{q}, P)(\xi)$ is a meromorphic function of ξ with poles on $\langle g_j, \xi \rangle \in 2i\pi\mathbb{Z}$. In particular this function belongs to the space $\mathcal{M}_\ell(V^*)$ introduced before. We write

$$S(\mathfrak{q}, P)(\xi) = \sum_{x \in V_{\mathbb{Z}} \cap \mathfrak{q}} P(\xi) e^{\langle \xi, x \rangle}$$

and depending on the context, we consider $S(\mathfrak{q}, P)$ either as a generalized function of $\xi \in i V^*$ or as a meromorphic function of $\xi \in V_{\mathbb{C}}$. If $\mathfrak{q}$ is a cone invariant by translation by a vector $v \in V_{\mathbb{Z}}$, it is easy that the generalized function $S(\mathfrak{q}, P)(\xi)$ is annihilated by a power of $(1 - e^{\langle v, \xi \rangle})$. The simplest case is when $P = 1$, $V = \mathfrak{c} = \mathbb{R}$, $V_{\mathbb{Z}} = \mathbb{Z}$, so that the equality is simply $(1 - e^{i\theta}) \sum_{n \in \mathbb{Z}} e^{in\theta} = 0$. In particular, if $\mathfrak{q}$ is a flat cone, the meromorphic function $S(\mathfrak{q}, P)(\xi)$ is equal to 0.

If $\mathfrak{f}$ is a face of $\mathfrak{p}$, we denote by $\operatorname{aff}(\mathfrak{f})$ the affine space generated by $\mathfrak{f}$ and by $\operatorname{lin} \mathfrak{f}$ the linear space parallel to $\operatorname{aff}(\mathfrak{f})$, that is the space spanned by elements $x - y$ with $x, y \in \mathfrak{f}$. The projection $\mathfrak{t}_{\text{trans}}(\mathfrak{p}, \mathfrak{f})$ of $\mathfrak{t}_{\text{aff}}(\mathfrak{p}, \mathfrak{f})$ in $V/\operatorname{lin}\mathfrak{f}$ is called the transverse cone. Note that this transverse cone is a salient cone in $V/\operatorname{lin}\mathfrak{f}$ with vertex y_0 the projection of any $y \in \mathfrak{f}$.

Theorem 6.1. *Let V be a rational vector space with lattice $V_{\mathbb{Z}}$. Let $\mathfrak{p} \subset V$ be a simple rational polytope and let $\mathfrak{f}$ be a face of $\mathfrak{p}$. Let $\mathfrak{t}_{\text{trans}}(\mathfrak{p}, f) \subset V/\operatorname{lin}\mathfrak{f}$ be the transverse cone. The tangent cone to $\mathfrak{p}$ at the vertex s is denoted by $s + \mathfrak{c}_s$. For $\xi \in V^*$, let*

$$S(s + \mathfrak{c}_s)(\xi) = \sum_{x \in (s+\mathfrak{c}_s) \cap V_{\mathbb{Z}}} e^{\langle \xi, x \rangle}.$$

The set of vertices of $\mathfrak{f}$ is denoted by $\mathcal{V}(\mathfrak{f})$.

(i) *The sum $\sum_{s \in \mathcal{V}(\mathfrak{f})} S(s + \mathfrak{c}_s)(\xi)$ restricts to a meromorphic function on $\operatorname{lin}\mathfrak{f}^{\perp} \subset V^*$, which is given by*

$$\sum_{s \in \mathcal{V}(\mathfrak{f})} S(s + \mathfrak{c}_s)(\xi) = \sum_{y \in \mathfrak{t}_{\text{trans}}(\mathfrak{p}, \mathfrak{f}) \cap (V/\operatorname{lin}\mathfrak{f})_{\mathbb{Z}}} e^{\langle \xi, y \rangle} P(y),$$

where $P(y)$ is a quasi-polynomial function on the projected lattice $(V/\operatorname{lin}\mathfrak{f})_{\mathbb{Z}} \subset V/\operatorname{lin}\mathfrak{f}$. Moreover if ξ is regular with respect to the cone $\mathfrak{t}_{\text{trans}}(\mathfrak{p}, \mathfrak{f})$, that is if ξ is not constant on a face strictly containing $\mathfrak{f}$, then $\sum_{s \in \mathcal{V}(\mathfrak{f})} S(s + \mathfrak{c}_s)(\xi)$ is holomorphic at ξ.

(ii) *For y close enough to the vertex y_0 of the transverse cone, $P(y)$ is the number of lattice points of the slice $\mathfrak{p} \cap (\operatorname{lin}\mathfrak{f} + y)$.*

Proof. We compute the signed sum of the generating functions of the tangent cones $\mathfrak{t}_{\text{aff}}(\mathfrak{p}, \mathfrak{g})$ where $\mathfrak{g}$ runs over the set $\mathcal{F}(\mathfrak{f}) \subset \mathcal{F}(\mathfrak{p})$ of faces of $\mathfrak{f}$. Since $\mathfrak{t}_{\text{aff}}(\mathfrak{p}, \mathfrak{g})$ contains lines if $\mathfrak{g}$ is not a vertex, we have

$$\sum_{s \in \mathcal{V}(\mathfrak{f})} S(s + \mathfrak{c}_s)(\xi) = \sum_{\mathfrak{g} \in \mathcal{F}(\mathfrak{f})} (-1)^{\dim \mathfrak{g}} S(\mathfrak{t}_{\text{aff}}(\mathfrak{p}, \mathfrak{g}))(\xi). \tag{6.1}$$

We will relate the right hand side to sums over slices of $\mathfrak{p}$ by affine subspaces parallel to $\mathfrak{f}$.

We define

$$\mathcal{T}(y)(x) = \sum_{\mathfrak{g} \in \mathcal{F}(\mathfrak{f})} (-1)^{\dim \mathfrak{g}} [\mathfrak{t}_{\text{aff}}(\mathfrak{p}, \mathfrak{g}) \cap (\operatorname{aff}(\mathfrak{f}) + y)](x). \tag{6.2}$$

The support of $\mathcal{T}(y)$ is illustrated in Figure 6.2.

Let us only observe that, as $\mathfrak{t}_{\text{aff}}(\mathfrak{p}, \mathfrak{g}) \cap \operatorname{aff}(\mathfrak{f})$ is the tangent cone of the polytope $\mathfrak{f} \subset \operatorname{aff}(\mathfrak{f})$ along its face $\mathfrak{g}$, we have, by Brianchon-Gram theorem,

$$\mathcal{T}(0) = [\mathfrak{f}].$$

Moreover, if y is small enough, then $\mathcal{T}(y)$ is the characteristic function of the intersection $\mathfrak{p} \cap (\mathrm{aff}(\mathfrak{f}) + y)$. This result can be deduced from the Euler relations. In the next section, in the case where $\mathfrak{p}$ is simple, we will obtain it as a consequence of Corollary 3.1 which is of course itself based on the Euler relations via the Brianchon-Gram theorem.

Let us compute the right hand side of Equation (6.1). If $\xi \in \mathrm{lin}\, \mathfrak{f}^{\perp}$ then $e^{\langle \xi, x\rangle}$ is constant on $\mathrm{lin}\, \mathfrak{f} + y$. Identifying $\mathrm{lin}\, \mathfrak{f}^{\perp}$ with $(V/\mathrm{lin}\, \mathfrak{f})^*$, we denote this constant value by $e^{\langle \xi, y\rangle}$.

Thus, we slice the lattice $V_{\mathbb{Z}}$ in slices parallel to the subspace $\mathrm{lin}\, \mathfrak{f}$. The slices are indexed by the projected lattice $(V/\mathrm{lin}\, \mathfrak{f})_{\mathbb{Z}}$. We write

$$\sum_{\mathfrak{g}\in\mathcal{F}(\mathfrak{f})} (-1)^{\dim \mathfrak{g}} S(\mathfrak{t}_{\mathrm{aff}}(\mathfrak{p}, \mathfrak{g}))(\xi) = \sum_{\mathfrak{g}\in\mathcal{F}(\mathfrak{f})} (-1)^{\dim \mathfrak{g}} \sum_{x\in \mathfrak{t}_{\mathrm{aff}}(\mathfrak{p},\mathfrak{g})} e^{\langle \xi, x\rangle}$$
$$= \sum_{y\in(V/\mathrm{lin}\,\mathfrak{f})_{\mathbb{Z}}} \sum_{\mathfrak{g}\in\mathcal{F}(\mathfrak{f})} (-1)^{\dim \mathfrak{g}} \sum_{x\in \mathfrak{t}_{\mathrm{aff}}(\mathfrak{p},\mathfrak{g})\cap(\mathrm{lin}\,\mathfrak{f}+y)\cap V_{\mathbb{Z}}} e^{\langle \xi, x\rangle}. \quad (6.3)$$

Let $y_0 \in V/\mathrm{lin}\, \mathfrak{f}$ be the projection of the face $\mathfrak{f}$. From (6.3), we obtain

$$\sum_{s\in\mathcal{V}(\mathfrak{f})} S(s + \mathfrak{c}_s)(\xi) = \sum_{y\in(V/\mathrm{lin}\,\mathfrak{f})_{\mathbb{Z}}} e^{\langle \xi, y\rangle} \sum_{x\in V_{\mathbb{Z}}} \mathcal{T}(y - y_0)(x). \quad (6.4)$$

The shift by y_0 is there to make further notations simpler. At this point, we postpone the proof of Theorem 6.1 until the next section, where we will relate $\mathcal{T}(y)$ to the Brianchon-Gram continuation of the face $\mathfrak{f}$, under the assumption that $\mathfrak{p}$ is simple. □

6.1 Brianchon-Gram continuation of a face of a partition polytope

Let $\mathfrak{p} = \mathfrak{p}(\Phi, \lambda) \subset \mathbb{R}^N$. Let $\mathfrak{f}$ be a face of $\mathfrak{p}$. If λ is regular and belongs to a tope τ, then there is a unique $I \in \mathcal{G}(\Phi, \tau)$ such that $\mathfrak{f} = \mathfrak{f}(\Phi, \lambda, I)$ is the corresponding face. We have $\dim \mathfrak{f} = |I| - \dim F$. If λ is on a wall, there may be several such pairs (τ, I).

Definition 6.1. Let $I \subseteq \{1, \ldots, N\}$ be such that the sequence Φ_I generates F. Let $\widetilde{\Phi}_I = (\widetilde{\phi}_i)$, $1 \le i \le N$, be the sequence of elements in $F \oplus \mathbb{R}^{I^c}$ defined by $\widetilde{\phi}_i = \phi_i$ if $i \in I$ and $\widetilde{\phi}_i = \phi_i \oplus e_i$, if $i \in I^c$.

Lemma 6.1.

(i) *The sequence $\widetilde{\Phi}_I$ generates a salient cone of full dimension in $F \oplus \mathbb{R}^{I^c}$.*
(ii) $V(\widetilde{\Phi}_I, (\lambda, y)) = \{x \in \mathbb{R}^N;\ \sum_{i=1}^N x_i\phi_i = \lambda,\ x_i = y_i$ *for* $i \in I^c\}$.
(iii) *Let τ be a Φ_I-tope. Let R be an open quadrant in $\mathbb{R}^{I^c}$. Then $\{(\lambda, y) \in F \oplus \mathbb{R}^{I^c};\ y \in R, \lambda - \sum_{i\in I^c} y_i\phi_i \in \tau\}$ is a $\widetilde{\Phi}_I$-tope and all $\widetilde{\Phi}_I$-topes are of this form.*

(iv) *Let τ be a Φ_I-tope, let τ_I be the $\widetilde{\Phi}_I$-tope which consists of (λ, y) such that $y_i > 0$ for $i \in I^c$ and $\lambda - \sum_{i\in I^c} y_i\phi_i \in \tau$. Then $\mathcal{G}(\widetilde{\Phi}_I, \tau_I)$ is the set of $K \cup I^c$, where $K \subseteq I$ and $K \in \mathcal{G}(\Phi_I, \tau)$. Hence*

$$X(\widetilde{\Phi}_I, \tau_I) = \sum_{K\in\mathcal{G}(\Phi_I,\tau)} (-1)^{|K|-\dim F} \prod_{i\in I\setminus K} p_i \prod_{i\in K\cup I^c} (p_i + q_i).$$

Proof. (i) follows from the fact that Φ_I generates F. (ii) is immediate. Consider the linear bijection from $F \oplus \mathbb{R}^{I^c}$ to itself defined by $(\lambda, y) \mapsto (\lambda - \sum_{i\in I^c} y_i\phi_i, y)$. The image of $\widetilde{\phi}_i$ is ϕ_i if $i \in I$, and e_i if $i \in I^c$. Therefore the $\widetilde{\Phi}_I$-topes are the pull-backs of the topes relative to the sequence $\psi_i = \phi_i$ if $i \in I$ and $\psi_i = e_i$ if $i \in I^c$. The latter are the products of Φ_I-topes in F with the quadrants in $\mathbb{R}^{I^c}$. This proves (iii).

Let $\widetilde{K} \subseteq \{1, \dots, N\}$. Then $\widetilde{\Phi}_{\widetilde{K}}$ generates $F \oplus \mathbb{R}^{I^c}$ if and only if $\widetilde{K} = K \cup I^c$, where $K \subseteq I$ is such that Φ_K generates F. Moreover $\tau_I \subset c((\widetilde{\Phi}_I)_{\widetilde{K}})$ if and only if $\tau \subset \mathfrak{c}(\Phi_K)$, whence (iv). □

Proposition 6.1. *Let τ be a Φ-tope and let $\lambda \in \overline{\tau}$. Let $\mathfrak{p} = \mathfrak{p}(\Phi, \lambda)$ and $\mathfrak{f} = \mathfrak{f}_I(\Phi, \lambda)$ be a face of $\mathfrak{p}$. Assume that* $\dim \mathfrak{f} = |I| - \dim F$. *We identify the quotient space $V/\operatorname{lin}\mathfrak{f}$ with $\mathbb{R}^{I^c}$ by the projection parallel to $\mathbb{R}^I$. Let τ_I be the Φ_I-tope which contains τ. If $y_i \geq 0$ for $i \in I^c$ and $\lambda - \sum_{i\in I^c} y_i\phi_i \in \overline{\tau_I}$, then*

$$\begin{aligned}\mathcal{X}(\widetilde{\Phi}_I, \tau_I)[V(\widetilde{\Phi}_I, (\lambda, y))] &= [\mathfrak{p}(\widetilde{\Phi}_I, (\lambda, y))]\\ &= [\mathfrak{p}(\Phi, \lambda) \cap (\operatorname{aff}(\mathfrak{f}) + y)].\end{aligned} \tag{6.5}$$

In particular, if λ is regular, the conditions $y_i \geq 0$ for $i \in I^c$ and $\lambda - \sum_{i\in I^c} y_i\phi_i \in \overline{\tau_I}$ define a neighborhood of $y = 0$ in $\mathbb{R}^{I^c}_{\geq 0}$ on which (6.5) holds.

Proof. The conditions $y_i \geq 0$ for $i \in I^c$ and $\lambda - \sum_{i\in I^c} y_i\phi_i \in \overline{\tau_I}$ mean that (λ, y) belongs to the closure of the $\widetilde{\Phi}_I$-tope τ_I associated to τ_I. Therefore by Corollary 3.1, we have

$$\mathcal{X}(\widetilde{\Phi}_I, \tau_I)[V(\widetilde{\Phi}_I, (\lambda, y))] = [\mathfrak{p}(\widetilde{\Phi}_I, (\lambda, y))].$$

Moreover, as $\dim \mathfrak{f} = |I| - \dim F$, the affine span $\operatorname{aff}(\mathfrak{f})$ is given by $\operatorname{aff}(\mathfrak{f}) = \{x \in V(\Phi, \lambda); x_i = 0 \text{ for } i \in I^c\}$. It follows that $V(\widetilde{\Phi}_I, (\lambda, y)) = \operatorname{aff}(\mathfrak{f}) + y$, hence $\mathfrak{p}(\widetilde{\Phi}_I, (\lambda, y)) = \mathfrak{p}(\Phi, \lambda) \cap (\operatorname{aff}(\mathfrak{f}) + y)$. □

Remark 6.1. Define $\mathfrak{q}_0(\mathfrak{p}, \mathfrak{f}, \tau) \subseteq \mathbb{R}^{I^c}_{\geq 0}$ by

$$\mathfrak{q}_0(\mathfrak{p}, \mathfrak{f}, \tau) = \{y = (y_i) \in \mathbb{R}^{I^c}\ ;\ y_i \geq 0 \text{ for } i \in I^c, \lambda - \sum_{i\in I^c} y_i\phi_i \in \overline{\tau_I}\}.$$

The set $\mathfrak{q}_0(\mathfrak{p}, \mathfrak{f}, \tau)$ is a polytope in $V/\operatorname{lin}\mathfrak{f} \simeq \mathbb{R}^{I^c}$. Let us denote its cone at vertex 0 by $\mathfrak{t}_0(\mathfrak{p}, \mathfrak{f}, \tau)$.

$$\mathfrak{t}_0(\mathfrak{p}, \mathfrak{f}, \tau) = \{y = (y_i) \in \mathbb{R}^{I^c} \ ; \ y_i \geq 0 \text{ for } i \in I^c, \\ \lambda - \epsilon \sum_{i \in I^c} y_i \phi_i \in \overline{\tau_I} \text{ for } \epsilon > 0 \text{ small enough } \}.$$

Then $\mathfrak{t}_0(\mathfrak{p}, \mathfrak{f}, \tau)$ is a subcone of the transverse cone $\mathfrak{t}_0(\mathfrak{p}, \mathfrak{f})$. If $\lambda \in \tau$ is regular, then $\mathfrak{t}_0(\mathfrak{p}, \mathfrak{f}, \tau) = \mathfrak{t}_0(\mathfrak{p}, f) = \mathbb{R}^{I^c}_{\geq 0}$.

If λ lies on a wall of a tope τ, then $\mathfrak{t}_0(\mathfrak{p}, \mathfrak{f}, \tau)$ may be strictly contained in the transverse cone $\mathfrak{t}_0(\mathfrak{p}, \mathfrak{f})$. When we consider all the topes τ such that $\lambda \in \overline{\tau}$, the cones $\mathfrak{t}_0(\mathfrak{p}, \mathfrak{f}, \tau')$ form a subdivision of $\mathfrak{t}_0(\mathfrak{p}, f)$. An example is illustrated in Figure 6.1. The polytope $\mathfrak{p} \subset \mathbb{R}^3$ is a tipi with four poles, with vertices $(0,0,0)$, $(1,0,0)$, $(0,1,0)$, $(0,0,1)$, $(1,1,0)$ and $\mathfrak{f}$ is the vertical edge with vertices $(0,0,0)$, $(0,0,1)$. The picture shows also the corresponding system Φ such that $\mathfrak{p}$ corresponds to a partition polytope $\mathfrak{p}(\Phi, \lambda)$. In this case, λ belongs to the wall generated by ϕ_3, thus λ belongs to two tope closures τ_1 and τ_2. We identify the quotient $V/\operatorname{lin}(\mathfrak{f})$ with the ground. Then the sets $\mathfrak{q}_0(\mathfrak{p}, \mathfrak{f}, \tau_i)$ are the two triangles which subdivide the ground face of the tipi.

This remark suggests how to modify Proposition 6.2 in the case of a non simple polytope.

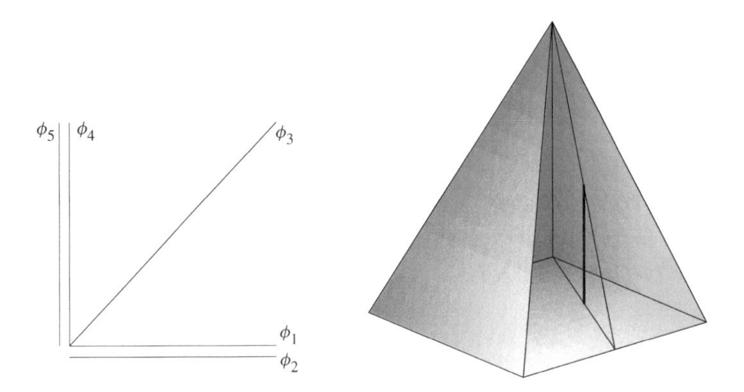

Figure 6.1.

Proposition 6.2. *Let $\Phi = (\phi_j)_{1 \leq j \leq N}$ be a sequence of non zero elements of a vector space F, generating F, and spanning a salient cone. Let τ be a Φ-tope, $\lambda \in \tau$ a regular element and $I \in \mathcal{G}(\Phi, \tau)$. Let $\mathfrak{p} = \mathfrak{p}(\Phi, \lambda)$ and $\mathfrak{f} = \mathfrak{f}(\Phi, \lambda, I)$.*

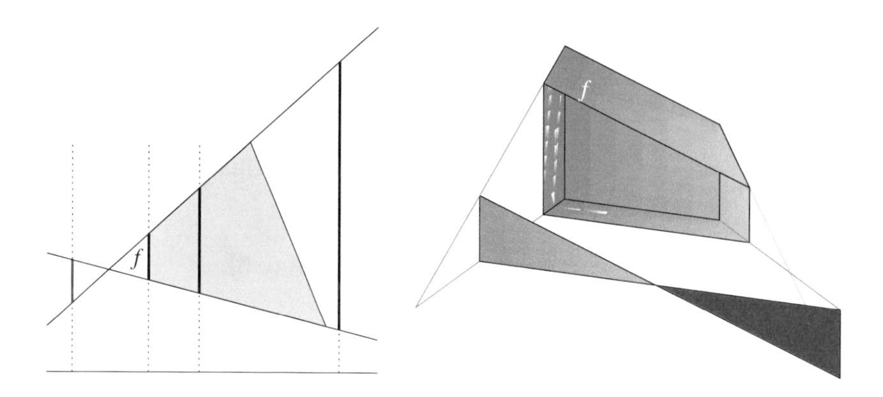

Figure 6.2. Brianchon-Gram continuation of a face. The segment and the triangle in red come with a minus sign. The end-points of the segment have to be deleted and two edges of the triangle also.

We identify the quotient space $V/\operatorname{lin}\mathfrak{f}$ with $\mathbb{R}^{I^c}$ by the projection parallel to $\mathbb{R}^I$. For $y \in \mathbb{R}^{I^c}$, let

$$\mathcal{T}(y) = \sum_{\mathfrak{g}\in\mathcal{F}(\mathfrak{f})} (-1)^{\dim \mathfrak{g}}[\mathfrak{t}_{\mathrm{aff}}(\mathfrak{p}, \mathfrak{g}) \cap (\mathrm{aff}(\mathfrak{f}) + y)], \tag{6.6}$$

where the set of faces of $\mathfrak{f}$ is denoted by $\mathcal{F}(\mathfrak{f})$.

Let $\widetilde{\Phi}_I = (\widetilde{\phi}_i)$, $1 \leq i \leq N$, be the sequence of elements in $F \oplus \mathbb{R}^{I^c}$ defined by $\widetilde{\phi}_i = \phi_i$ if $i \in I$ and $\widetilde{\phi}_i = \phi_i \oplus e_i$, if $i \in I^c$. Let τ_I be the $\widetilde{\Phi}_I$-tope which consists of elements $(\lambda, y) \in F \oplus \mathbb{R}^{I^c}$ such that $y_i > 0$ for $i \in I^c$ and $\lambda - \sum_{i\in I^c} y_i\phi_i \in \tau_I$, where τ_I is the unique Φ_I-tope which contains τ. Then

$$\mathcal{T}(y)(x) = \mathcal{X}(\widetilde{\Phi}_I, \tau_I)(x)[V(\widetilde{\Phi}_I, (\lambda, y))](x) \prod_{i\in I^c} [y_i \geq 0], \tag{6.7}$$

Proof. The faces $\mathfrak{g}$ of $\mathfrak{f} = \mathfrak{f}(\Phi, \lambda, I)$ are indexed by the subsets $K \in \mathcal{G}(\Phi, \tau)$ which are contained in I. For $\mathfrak{g} = \mathfrak{f}(\Phi, \lambda, K)$, we have

$$\mathfrak{t}_{\mathrm{aff}}(\mathfrak{p}, \mathfrak{g}) = \{x \in V(\Phi, \lambda); x_i \geq 0 \text{ for } i \in K^c\}.$$

We write (6.6) as

$$\mathcal{T}(y) = \sum_{K\in\mathcal{G}(\Phi,\tau), K\subseteq I} (-1)^{|K|-\dim F} \prod_{i\in K^c} [x_i \geq 0]\, [\mathrm{aff}(\mathfrak{f}) + y]. \tag{6.8}$$

We observe that $\mathcal{G}(\Phi_I, \tau_I) = \{K \in \mathcal{G}(\Phi, \tau), K \subseteq I\}$. Therefore, by Lemma 6.1, we have

$$\mathcal{X}(\widetilde{\Phi}_I, \tau_I)[V(\widetilde{\Phi}_I, (\lambda, y))] = \sum_{K \in \mathcal{G}(\Phi,\tau), K \subseteq I} (-1)^{|K|-\dim F} \prod_{i \in I \setminus K} [x_i \geq 0][V(\widetilde{\Phi}_I, (\lambda, y))].$$

We factor out $\prod_{i \in I^c} [x_i \geq 0]$ in each summand of (6.8).
As $V(\widetilde{\Phi}_I, (\lambda, y)) = \mathrm{aff}(\mathfrak{f}) + y$, we obtain

$$\mathcal{T}(y)(x) = \mathcal{X}(\widetilde{\Phi}_I, \tau_I)(x)[V(\widetilde{\Phi}_I, (\lambda, y))](x) \prod_{i \in I^c} [x_i \geq 0].$$

As $x_i = y_i$ for $i \in I^c$ if $x \in V(\widetilde{\Phi}_I, (\lambda, y))$, we obtain (6.7) □

We resume the proof of Theorem 6.1.

Proof of Theorem 6.1. We identify $\mathfrak{p}$ with a partition polytope $\mathfrak{p}(\Phi, \lambda)$ by an affine map $V \simeq V(\Phi, \lambda)$. We can assume that λ is regular. Some care is needed with respect to the lattice $V_{\mathbb{Z}}$. In general, its image $V(\Phi, \lambda)_{\mathbb{Z}}$ in $V(\Phi, \lambda)$ is not $\mathbb{Z}^N \cap V(\Phi, \lambda)$. However we can always write $V(\Phi, \lambda)_{\mathbb{Z}} = (b + \Gamma) \cap V(\Phi, \lambda)$, where $b \in \mathbb{Q}^N$ and Γ is a lattice in $\mathbb{R}^N$, (Γ is a fixed lattice and b projects on λ). Let τ be the Φ-tope which contains λ and let $I \in \mathcal{G}(\Phi, \tau)$ such that $\mathfrak{f}$ is identified with the face $\mathfrak{f}(\Phi, \lambda, I)$. Then $V/\operatorname{lin}\mathfrak{f}$ is identified with $V/\operatorname{lin}\mathfrak{f} \simeq \mathbb{R}^{I^c}$ and the projected lattice $(V/\operatorname{lin}\mathfrak{f})_{\mathbb{Z}}$ is identified with a lattice in $\mathbb{R}^{I^c}$. By Proposition 6.2, we have, for every $x \in \mathbb{R}^N$,

$$\mathcal{T}(y)(x) = \mathcal{X}(\widetilde{\Phi}_I, \tau_I)[V(\widetilde{\Phi}_I, (\lambda, y))](x) \prod_{i \in I^c} [y_i \geq 0].$$

So we define

$$P(y) = \sum_{x \in b + \Gamma} \mathcal{X}(\widetilde{\Phi}_I, \tau_I)[V(\widetilde{\Phi}_I, (\lambda, y - y_0))](x).$$

Then $P(y)$ is a quasi-polynomial function of $y \in (V/\operatorname{lin}\mathfrak{f})_{\mathbb{Z}}$. This fact follows from a minor generalization of Theorem 5.1 (ii). We only have to take care of the shifts: the summation is over $x \in b + \Gamma$ and the parameter $y - y_0$ in the Brianchon-Gram function runs over the shifted lattice $(V/\operatorname{lin}\mathfrak{f})_{\mathbb{Z}} - y_0$.

The equalities (6.3) and (6.4) of generalized functions imply equalities of holomorphic functions of ξ in an open subset of $(\operatorname{lin}\mathfrak{f})^{\perp}$, hence

$\sum_{s\in\mathcal{V}(\mathfrak{f})} S(s+\mathfrak{c}_s)(\xi)$ restricts to a meromorphic function on $(\mathrm{lin}\,\mathfrak{f})^\perp$, given by

$$\sum_{s\in\mathcal{V}(\mathfrak{f})} S(s+\mathfrak{c}_s)(\xi) = \sum_{y\in\mathfrak{t}_{\mathrm{trans}}(\mathfrak{p},\mathfrak{f})\cap(V/\mathrm{lin}\,\mathfrak{f})_{\mathbb{Z}}} e^{\langle\xi,y\rangle}P(y). \tag{6.9}$$

So we have proved (i).

By Proposition 6.1, for $y\in\mathfrak{t}_{\mathrm{trans}}(\mathfrak{p},\mathfrak{f})$ close to the vertex, $\mathcal{T}(y-y_0)$ is the characteristic function of the slice $\mathfrak{p}\cap(\mathrm{aff}(\mathfrak{f})+y)$, hence (ii). □

7 Cohomology of line bundles over a toric variety

Let us indicate the relation of our work with toric varieties. Let $\Phi = (\phi_j)_{1\le j\le N}$ be a sequence of non zero elements of a vector space F, generating F, and spanning a salient cone. Assume that the ϕ_i's belong to a lattice Λ and let T be the torus with character group Λ embedded in $T^N = S_1^N$ by the characters of T associated to (ϕ_i). This determines an action of T in the complex space $\mathbb{C}^N$. Each tope τ determines a toric variety M_τ (with orbifold singularities) for the quotient torus T^N/T, in the following way. If $\lambda\in\tau$, M_τ is the reduced manifold $\mathbb{C}^N//_\lambda T$ at $\lambda\in\mathfrak{t}^*$. Then the vectors ϕ_i parameterize the boundary divisors D_i in M_τ and each element $\lambda\in\Lambda$ determines a T^N-equivariant sheaf $\mathcal{O}(\lambda)$ on M_τ.

The lattice of characters of the d-dimensional torus T^N/T is identified with $V\cap\mathbb{Z}^N$.

The torus T^N acts on the cohomology groups $H^i(M_\tau,\mathcal{O}(\lambda))$. When $\lambda\in\tau$, then all the cohomology groups H^i for $i>0$ vanish, and a weight $m\in\mathbb{Z}^N$ of T^N occurs in $H^0(M_\tau,\mathcal{O}(\lambda))$ if and only if $m\in\mathfrak{p}(\Phi,\lambda)\cap\mathbb{Z}^N$. Thus the dimension of the space $H^0(M_\tau,\mathcal{O}(\lambda))$ is just the number of integral points in $\mathfrak{p}(\Phi,\lambda)$.

If $\lambda\in\Lambda$ does not belong to the tope τ, and $i>0$, the cohomology space $H^i(M_\tau,\mathcal{O}(\lambda))$ is in general not zero. It is natural to introduce the virtual space

$$\mathcal{H}(\tau,\lambda) := \sum_{i=0}^{d}(-1)^i H^i(M_\tau,\mathcal{O}(\lambda)).$$

It follows from the Kawasaki-Riemann-Roch theorem that the virtual dimension of $\mathcal{H}(\tau,\lambda)$ is a quasi polynomial function of λ.

More precisely, we can use the fixed point theorem to compute the character of T^N in $\mathcal{H}(\tau,\lambda)$ (see [7]). As the construction of the present article reproduces this fixed point theorem at the level of sets, we obtain the weight decomposition of the T^N-module

$$\mathcal{H}(\tau,\lambda) = \sum_{m\in\mathbb{Z}^N\cap V(\Phi,\lambda)} \mathcal{X}(\Phi,\tau)(m)e^m.$$

In other words, the function $\mathcal{X}(\Phi, \tau)$ on $\mathbb{Z}^N$ computes simultaneously (for all sheaves $\mathcal{O}(\lambda)$) the multiplicity of a weight m in the alternating sum of cohomology spaces. In particular, the function $\mathcal{X}(\Phi, \tau) \cap [V(\Phi, \lambda)]$ is the constructible function on $V(\Phi, \lambda)$ associated by Morelli [16] to the sheaf $\mathcal{O}(\lambda)$.

Recall the formula

$$\mathcal{X}(\Phi, \tau) = \sum_B z(\Phi, \tau, B)[\mathcal{Q}^B_{\text{neg}}].$$

Let us comment on the explicit computation of the coefficients $z(\Phi, \tau, B)$ of $\mathcal{X}(\Phi, \tau)$. We wrote a brute force Maple program to compute $X(\Phi, \tau)$, out of its definition (Equation (2.3)), by enumerating the generating subsets of the system Φ and checking which ones are in $\mathcal{G}(\Phi, \tau)$. It would be certainly more efficient to use Theorem 3.3, and then determine $\mathcal{B}(\Phi, \tau)$ using the reverse-search algorithm of Avis-Fukuda [1]. Anyway, we obtain the decomposition as a sum of monomials

$$X(\Phi, \tau) = \sum_B z(\Phi, \tau, B) \prod_{i \in B^c} p_i \prod_{j \subset B} q_j.$$

If $m \in \mathbb{Z}^N$, we denote by B_m the set of indices i such that $m_i < 0$. Then the multiplicity of m in the T^N module $\mathcal{H}(\tau, \lambda)$ is obtained by computing the coefficient $z(\Phi, \tau, B_m)$ of the monomial $\prod_{i \in B_m^c} p_i \prod_{i \in B_m} q_i$ in $X(\Phi, \tau)$.

Y. Karshon and S. Tolman [14] have studied the representation space $\mathcal{H}(\tau, \lambda)$ associated to a non-ample line bundle on the manifold M_τ, and they have given an algorithm to compute a weight in this representation space by wall crossing. Our algorithm (Theorem 4.3) to determine $z(\Phi, \tau, B)$ is probably very similar. However, as we deal with arbitrary "weights" ϕ_i (not assumed rational), our methods use "only linear algebra", not geometry.

By summing up the multiplicities of the weights in $\mathcal{H}(\tau, \lambda)$, we obtain the expression of the function

$$\lambda \mapsto \dim \mathcal{H}(\tau, \lambda) = \sum_B z(\Phi, \tau, B) \text{cardinal}(\mathfrak{p}_{\text{flip}}(\Phi, B, \lambda) \cap \mathbb{Z}^N)$$

as a sum of partition functions with respect to particular flipped systems.

Remark that if $\lambda \in (\tau - \mathfrak{b}(\Phi)) \cap \Lambda$, the continuity property asserts that the dimension of $\mathcal{H}(\tau, \lambda)$ is still equal to the dimension of H^0, that is the cardinal of $\mathfrak{p}(\Phi, \lambda) \cap \mathbb{Z}^N$. This is in accordance with the following vanishing theorem [17].

Theorem 7.1. *If $\lambda \in (\tau - \mathfrak{b}(\Phi)) \cap \Lambda$ then $H^i(M_\tau, \mathcal{O}(\lambda)) = 0$ for $i > 0$.*

It would be interesting to study the locally quasi polynomial function $h_i(\Phi, \tau)(\lambda) = \dim H^i(M_\tau, \mathcal{O}(\lambda))$ for each i. From Demazure's description of the individual cohomology groups $H^i(M_\tau, \mathcal{O}(\lambda))$ (see for example the forthcoming book [11, Chapter 9]), we see that it is a locally quasi-polynomial function, sum of partition functions of flipped systems. Thus each locally quasi polynomial function $h_i(\Phi, \tau)$ is a particular element of the generalized Dahmen-Micchelli space $\mathcal{F}(\Phi)$ introduced in [13]. It would be interesting to study the relations between these different locally quasi polynomial functions on Λ.

Let us give a last example to illustrate the method. We consider the hexagon defined by the following inequalities in $\mathbb{R}^2$. $x_1 \geq 0$, $x_1 \leq 2$, $x_2 \geq 0$, $x_1 + x_2 \geq 1$, $x_1 + x_2 \leq 4$, $x_1 - x_2 \geq -2$. The corresponding toric variety M_{hex} of dimension 2 is defined by the fan with edges $(1,0)$, $(1,1)$, $(0,1)$, $(-1,0)$, $(-1,-1)$, $(1,-1)$.

We can also describe M_{hex} as a reduced Hamiltonian manifold, with the help of an ample line bundle. We consider the standard torus of dimension 4 acting in $\mathbb{C}^6$ with the following list Φ of weights

$$((1,0,0,0), (0,1,0,0), (0,0,1,0), (0,0,0,1), (-1,-1,1,1), (1,-1,0,1)).$$

If τ is the tope which contains the vector $[2,-1,2,4]$, then the reduced manifold M_τ is the manifold M_{hex}.

We compute $X(\Phi, \tau)$ (by brute force) and obtain:

$$\begin{aligned}
X(\Phi,\tau) = \\
& p_1p_2p_3p_4p_5p_6 - p_1p_2p_3p_4q_5q_6 - p_1p_2p_3p_5q_4q_6 - p_1p_2p_3p_6q_4q_5 \\
& - 2\, p_1p_2p_3q_4q_5q_6 \\
& - p_1p_2p_4p_6q_3q_5 - p_1p_2p_4q_3q_5q_6 - p_1p_2p_6q_3q_4q_5 - p_1p_2q_3q_4q_5q_6 - p_1p_3p_5p_6q_2q_4 \\
& - p_1p_3p_5q_2q_4q_6 - p_1p_3p_6q_2q_4q_5 - p_1p_3q_2q_4q_5q_6 - p_1p_4p_5p_6q_2q_3 - p_1p_4p_6q_2q_3q_5 \\
& - p_1p_5p_6q_2q_3q_4 - p_1p_6q_2q_3q_4q_5 - p_2p_3p_4p_5q_1q_6 - p_2p_3p_4q_1q_5q_6 - p_2p_3p_5q_1q_4q_6 \\
& - p_2p_3q_1q_4q_5q_6 - p_2p_4p_5p_6q_1q_3 - p_2p_4p_5q_1q_3q_6 - p_2p_4p_6q_1q_3q_5 - p_2p_4q_1q_3q_5q_6 \\
& - p_3p_4p_5p_6q_1q_2 - p_3p_4p_5q_1q_2q_6 - p_3p_5p_6q_1q_2q_4 - p_3p_5q_1q_2q_4q_6 \\
& - 2\, p_4p_5p_6q_1q_2q_3 \\
& - p_4p_5q_1q_2q_3q_6 - p_4p_6q_1q_2q_3q_5 - p_5p_6q_1q_2q_3q_4 + q_1q_2q_3q_4q_5q_6.
\end{aligned}$$

We can immediately read on this expression the multiplicity of a weight $m = (m_1, m_2, m_3, m_4, m_5, m_6)$ in the space $\mathcal{H}(\tau, \lambda)$ for any m and any λ. We see that the multiplicities of m can be 0,1,-1,-2 depending on the quadrant in which m lies.

For example, for $\lambda = (200, 434, 378, -400)$, the weight

$$m = (200, 234, 478, -200, -100, -100)$$

has multiplicity -2 in the space $\mathcal{H}(\tau, \lambda)$. Indeed the coefficient of $p_1p_2p_3q_4q_5q_6$ in $X(\Phi, \tau)$ is -2.

Given $\lambda \in \mathbb{Z}^4$, we parameterize the integral points in $V(\Phi, \lambda)$ by $(x_1, x_2) \in \mathbb{Z}^2$, with corresponding $m \in \mathbb{Z}^6$ given by

$$m = (\lambda_1 + x_1 - x_2, \lambda_2 + x_1 + x_2, \lambda_3 - x_1, \lambda_4 - x_1 - x_2, x_1, x_2).$$

With this parametrization, the Figures 7.1, 7.2 and 7.3 describe the support of the module $\mathcal{H}(\tau, \lambda)$ as λ moves along the line joining $\lambda_0 =$ (200, −100, 200, 400 (in the ample cone) to $\lambda_1 =$ (200, 434, 378, −400). The line crosses six walls.

We assign colors to the multiplicities: $blue = 1$, $yellow := -1$, $red := -1$, $green := -1$, $magenta := -2$, $black := -1$, $khaki = -1$. In the first figure 7.1, λ is in the starting tope (the ample cone). In the last three steps, a polygon with multiplicity −2 (colored in magenta) has appeared in the middle of the picture.

Figure 7.1. At the beginning $\lambda = (200, -100, 200, 400)$ is in the ample cone. The partition polytope is an hexagon.

Figure 7.2. From left to right, λ crosses three walls, one at a time. The new triangles have multiplicity -1.

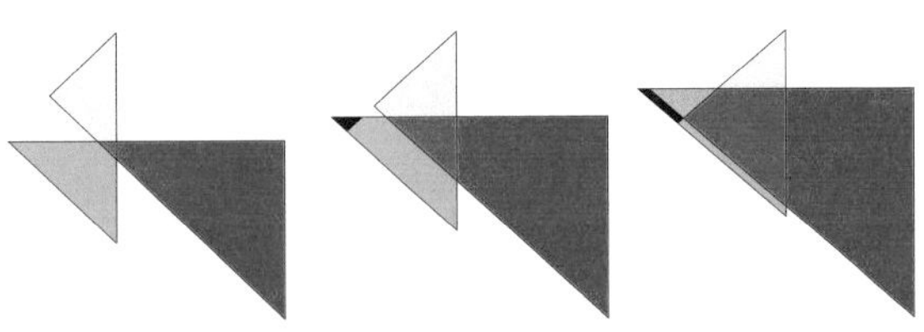

Figure 7.3. λ crosses three more walls. The polytope colored in magenta has multiplicity -2.

References

[1] D. AVIS and K. FUKUDA, *A pivoting algorithm for convex hulls and vertex enumeration of arrangements and polyhedra*, Discrete Comput. Geom. **8** (1992), no. 3, 295–313, ACM Symposium on Computational Geometry (North Conway, NH, 1991).

[2] V. BALDONI, N. BERLINE, J. D. LOERA, M. KÖPPE and M. VERGNE, *Computation of the highest coefficients of weighted Ehrhart quasi-polynomials for a rational polytope*, arXiv:1011.1602 [math.CO], 2010.

[3] A. I. BARVINOK and J. E. POMMERSHEIM, *An algorithmic theory of lattice points in polyhedra*, In: "New Perspectives in Algebraic Combinatorics", L. J. Billera, A. Björner, C. Greene, R. E. Simion, and R. P. Stanley (eds.), Math. Sci. Res. Inst. Publ., vol. 38, Cambridge Univ. Press, Cambridge, 1999, pp. 91–147.

[4] N. BERLINE and M. VERGNE, *Classes caractéristiques équivariantes. Formule de localisation en cohomologie équivariante*, C. R. Acad. Sci. Paris Sér. I Math. **295** (1982), no. 9, 539–541.

[5] A. BOYSAL and M. VERGNE, *Paradan's wall crossing formula for partition functions and Khovanski-Pukhlikov differential operator*, Ann. Inst. Fourier (Grenoble) **59** (2009), no. 5, 1715–1752.

[6] M. BRION, *Points entiers dans les polyèdres convexes*, Ann. Sci. École Norm. Sup. **21** (1988), no. 4, 653–663.

[7] M. BRION and M. VERGNE, *An equivariant Riemann-Roch theorem for complete, simplicial toric varieties*, J. Reine Angew. Math. **482** (1997), 67–92.

[8] M. BRION and M. VERGNE, *Lattice points in simple polytopes*, J.A.M.S. **10** (1997).

[9] M. BRION and M. VERGNE, *Residue formulae, vector partition functions and lattice points in rational polytopes*, J. Amer. Math. Soc. **10** (1997), 797–833.

[10] R. CAVALIERI, P. JOHNSON and H. MARKWIG, *Chamber structure of double Hurwitz numbers*, 2010, arXiv:1003.1805v1.

[11] D. COX, J. LITTLE and H. SCHENCK, *Toric varieties*, 2010, Available at http://www.cs.amherst.edu/ dac/toric.html.

[12] W. DAHMEN and C. A. MICCHELLI, *The number of solutions to linear Diophantine equations and multivariate splines*, Trans. Amer. Math. Soc. **308** (1988), no. 2, 509–532.

[13] C. DE CONCINI, C. PROCESI and M. VERGNE, *Vector partition function and generalized Dahmen-Micchelli spaces*, 2008, arXiv:0805.2907v2.

[14] Y. KARSHON and S. TOLMAN, *The moment map and line bundles over presymplectic toric manifolds*, J. Differential Geom. **38** (1993), no. 3, 465–484.
[15] J. LAWRENCE, *Polytope volume computation*, Math. Comp. **57** (1991), no. 195, 259–271.
[16] R. MORELLI, *The K-theory of a toric variety*, Adv. Math. **100** (1993), no. 2, 154–182.
[17] M. MUSTATA, *Vanishing theorems on toric varieties*, Tohoku Math. J. (2) **54** (2002), no. 3, 451–470.
[18] P.-E. PARADAN, *Jump formulas in Hamiltonian geometry*, 2004, arXiv math 0411306, to appear in Geometric Aspects of Analysis and Mechanics, Proceedings of the conference in honor of H. Duistermaat, Utrecht 2007.
[19] A. SZENES and M. VERGNE, *Residue formulae for vector partitions and Euler-Maclaurin sums*, Formal power series and algebraic combinatorics - Scottsdale, AZ, 2001, Adv. in Appl. Math., vol. 30, 2003, arXiv math 0202253, pp. 295–342.
[20] A. N. VARCHENKO, *Combinatorics and topology of the arrangement of affine hyperplanes in the real space*, Functional Anal. Appl. **21, no. 1** (1987), 9–19, Russian original publ. Funktsional. Anal. i Prilozhen. 21 (1987), no. 1, p.11–22.

Embeddings of braid groups into mapping class groups and their homology

Carl-Friedrich Bödigheimer and Ulrike Tillmann

Abstract. We construct several families of embeddings of braid groups into mapping class groups of orientable and non-orientable surfaces and prove that they induce the trivial map in stable homology in the orientable case, but not so in the non-orientable case. We show that these embeddings are non-geometric in the sense that the standard generators of the braid group are not mapped to Dehn twists.

1 Introduction

Let $\Gamma_{g,n}$ denote the mapping class group of an oriented surface $\Sigma_{g,n}$ of genus g with n parametrized boundary components, *i.e.*, $\Gamma_{g,n}$ is the group of connected components of the group of orientation preserving diffeomorphisms of $\Sigma_{g,n}$ that fix the boundary point-wise. For a simple closed curve a on the surface, let D_a denote the Dehn twist around a. When two simple closed curves a and b intersect in one point the associated Dehn twists satisfy the braid relation $D_a D_b D_a = D_b D_a D_b$, and if they do not intersect, the corresponding Dehn twists commute $D_a D_b = D_b D_a$.

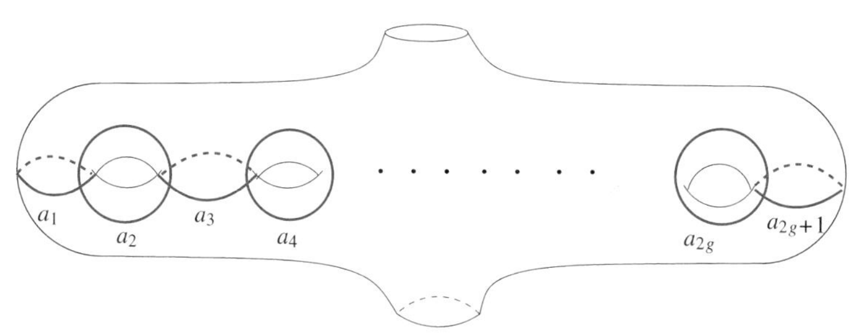

Figure 1.1. *The standard geometric embedding* $\phi : B_{2g+2} \to \Gamma_{g,2}$.

Thus a chain of n interlocking simple closed curves $a_1, \dots, a_n$ on some surface defines a map from the braid group B_{n+1} on $n+1$ strands into the mapping class group of a subsurface Σ containing the union of theses curves; these mapping classes fix the boundary of Σ point-wise. The smallest such subsurface Σ is a neighbourhood of the union of the curves. When $n = 2g + 1$ is odd, this is $\Sigma_{g,2}$, and when $n = 2g$ is even, this is $\Sigma_{g,1}$. Thus we have homomorphisms of groups

$$\phi : B_{2g+2} \longrightarrow \Gamma_{g,2} \quad \text{and } \phi : B_{2g+1} \longrightarrow \Gamma_{g,1}. \tag{1.1}$$

These are injections by a theorem of Birman and Hilden [4, 5].

Wajnryb [23] calls such embeddings that send the standard generators of the braid group to Dehn twists *geometric*. He asks in [24] whether there are non-geometric embeddings. The first example of such a non-geometric embedding was given by Szepietowski [18]. Our first goal in this paper is to produce many more such non-geometric embeddings and show that they are ubiquitous. We construct these in Section 2 and prove in Section 3 that they are non-geometric.

Motivated by a conjecture of Harer and following some ideas of F.R. Cohen [8], in [17] and [16] it was shown that the geometric embedding ϕ induces the trivial map in stable homology, that is the map in homology is zero in positive degrees as long as the genus of the underlying surface is large enough relative to the degree. Our second goal here is to show that this is also the case for the non-geometric embeddings constructed in Section 2. (For one of these maps, this answers a question left open in [17].) To this purpose we show in Section 4 how our embeddings from Section 2 induce maps of algebras over an E_2-operad, and deduce in Section 5 that all of them induce the trivial map in stable homology. Furthermore, while it may be expected that all maps from a braid group to the mapping class group of an orientable surface will induce the trivial map on stable homology, we show that this is not true for embeddings of a braid group into the mapping class group of a non-orientable surface by explicitly computing the image of one such embedding in stable homology. Finally, in Section 6, we analyse the induced maps in unstable homology. Only partial results are obtained here. In particular, it remains an open question whether ϕ induces the trivial map in unstable homology for field coefficients.

ACKNOWLEDGEMENTS. We would like to thank Blazej Szepietowski for sending his paper and Mustafa Korkmaz for e-mail correspondence.

2 Non-geometric embeddings

We will construct various injections of braid groups into mapping class groups. All embeddings that we know of, geometric or not, initially start with the standard identification of the braid groups with mapping class groups. The pure braid group on g strands is the mapping class group $\Gamma_{0,1}^{g}$ of a disk with g marked points, and the pure ribbon braid group is the mapping class group $\Gamma_{0,g+1} \simeq \mathbb{Z}^g \times \Gamma_{0,1}^{g}$ of a disk with g holes the boundary of which are parametrised. The factors of $\mathbb{Z}$ correspond to the Dehn twists around the boundary circles of the holes. Similarly, the braid group B_g can be identified with the mapping class group $\Gamma_{0,1}^{(g)}$ of a disk with g punctures (or g unordered marked points), and the ribbon braid group $\mathbb{Z} \wr B_g$ with the mapping class group $\Gamma_{0,(g),1}$ of the disk with g unordered holes for which the underlying diffeomorphisms may in a parametrisation preserving way interchange the boundaries of the holes and fix the outer boundary curve point-wise. Note that the resulting inclusion

$$\gamma : B_g \hookrightarrow \mathbb{Z} \wr B_g \simeq \Gamma_{0,(g),1} \tag{2.1}$$

maps the standard generator that interchanges the i-th and $(i+1)$-st strand in the braid group to half of the Dehn twists around a simple closed curve enclosing the i-th and $(i+1)$-st holes followed by a half Dehn twist around each of these holes in the opposite direction; see Figure 2.1.

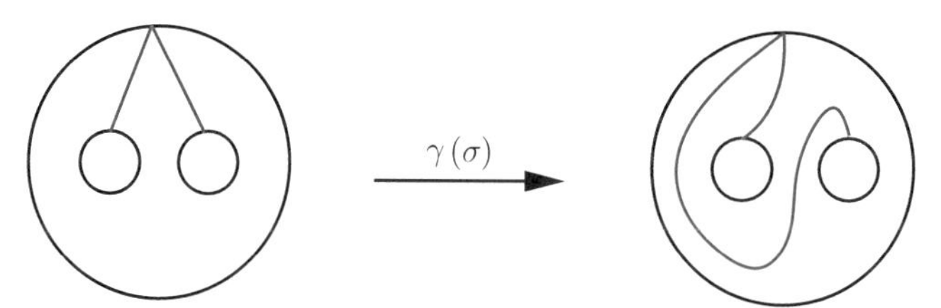

Figure 2.1. *Image of a generator $\sigma \in B_g$ under γ.*

Clearly, any genus zero subsurface $\Sigma_{0,g+1}$ of a surface Σ defines for us a map from the pure ribbon braid group $\Gamma_{0,g+1}$ into the mapping class group of Σ. Not all such maps, however, can be extended to the ribbon braid group $\Gamma_{0,(g),1}$. Below we explore various constructions of surfaces from $\Sigma_{0,g+1}$ that allow such an extension.

2.1 Mirror construction

Our first example of a non-geometric embedding was also considered in [17].

We double the disk with g holes by reflecting it in a plane containing the boundary circles of the holes to obtain an oriented surface $\Sigma_{g-1,2}$ as indicated in Figure 2.2, and extend diffeomorphisms of the disk with holes to $\Sigma_{g-1,2}$ by reflection in the plane. This defines a map on mapping class groups

$$R : \Gamma_{0,(g),1} \longrightarrow \Gamma_{g-1,2}.$$

Precomposing R with the natural inclusion $B_g \subset \Gamma_{0,(g),1}$ defines the map

$$B_g \xrightarrow{\gamma} \Gamma_{0,(g),1} \xrightarrow{R} \Gamma_{g-1,2}. \tag{2.2}$$

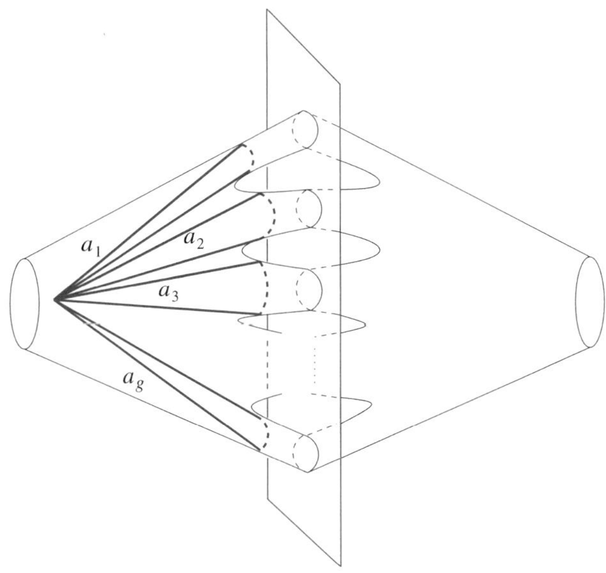

Figure 2.2. *Mirror construction R and subgroup* $\langle a_1, \ldots, a_g \rangle \subset \pi_1(\Sigma_{g-1,2})$.

Lemma 2.1. *The composition* $R \circ \gamma$ *is an injection.*

Proof. We are going to detect the elements in the image by their action on the fundamental group of the underlying surface. Recall, the action of the braid group on the fundamental group of the disk with g holes defines Artin's inclusion [1] of the braid group into the automorphism group of $\pi_1(\Sigma_{0,g+1}) = F_g = \langle a_1, \ldots, a_g \rangle$, a free group of rank g:

$$A : B_g \hookrightarrow \mathrm{Aut}(F_g).$$

The fundamental group of $\Sigma_{0,g+1}$ in turn injects into the fundamental group of $\Sigma_{g-1,2}$. Indeed, each standard generator of $F_g = \langle a_1, \ldots, a_g \rangle$ is mapped to the corresponding generator of $\pi_1(\Sigma_{g-1,2})$ which is the free group $F_{2g-1} = \langle a_1, b_1, \ldots, b_{g-1}, a_g \rangle$ of rank $2g - 1$; see Figure 2.2.

The action of diffeomorphisms on the fundamental group of the surface induces a group homomorphisms $\Gamma_{g-1,2} \to \mathrm{Aut}(F_{2g-1})$, and by restricting to the subgroup $F_g = \langle a_1, \ldots, a_g \rangle$ a map to the set $\mathrm{Hom}(F_g, F_{2g-1})$. We follow this map by the projection map to the monoid $\mathrm{Hom}(F_g, F_g)$, induced from the group homomorphisms $F_{2g-1} \to F_g$ that maps each a_i to itself and each b_i to the identity element. We have the following commutative diagram; note that the right vertical arrow is only a map of sets.

$$\begin{array}{ccc} B_g & \xrightarrow{R\circ\gamma} & \Gamma_{g-1,2} \\ {\scriptstyle A}\downarrow & & \downarrow \\ \mathrm{Aut}(F_g) & \longrightarrow & \mathrm{Hom}(F_g, F_g). \end{array}$$

As A and the bottom horizontal map are injective this shows that $R \circ \gamma$ is injective. □

2.2 Szepietowski's construction

We next recall the construction from [18]. Starting with $\Sigma_{0,g+1}$, the disk with g holes, we glue to the boundary of each hole a Möbius band $N_{1,1}$ so that the resulting surface is a non-orientable surface $N_{g,1}$ of genus g with one parametrised boundary component. Any diffeomorphisms $f : \Sigma_{0,g+1} \to \Sigma_{0,g+1}$ which fixes the boundary pointwise can be extended across the boundary by the identity of $N_{1,1} \sqcup \ldots \sqcup N_{1,1}$ and gives thus a diffeomorphism of $N_{g,1}$. This defines (via γ) a homomorphism from the pure braid group to the mapping class group of $N_{g,1}$. This latter homomorphism can be extended to a map from the whole braid group as follows: extend a diffeomorphism f which permutes the g inner boundaries by the corresponding permutation diffeomorphism of $N_{1,1} \sqcup \ldots \sqcup N_{1,1}$. This defines

$$\varphi : B_g \longrightarrow \mathcal{N}_{g,1}, \tag{2.3}$$

where $\mathcal{N}_{g,1}$ denotes the mapping class group of $N_{g,1}$. By a result of Birman and Chillingworth [2,3], extended to surfaces with boundary in [18], the lift of diffeomorphisms of a non-orientable surface to its double cover induces an injection of mapping class groups:

$$L : \mathcal{N}_{g,1} \hookrightarrow \Gamma_{g-1,2}.$$

Lemma 2.2. *The composition $L \circ \varphi$ is an injection.*

Proof. We give an alternative proof to the one in [18]. As in the proof of Lemma 2.1 we consider the action of the mapping class group on the fundamental group of the underlying surfaces. In this case the fundamental group of the disk with g holes $\pi_1(\Sigma_{0,g+1}) = F_g = \langle a_1, \ldots, a_g \rangle$ injects into the fundamental group $\pi_1(N_{g,1}) = F_g = \langle c_1, \ldots, c_g \rangle$ by sending a_i to $2c_i$. As by [1] the action of the braid group on the subgroup $\langle a_1, \ldots, a_g \rangle$ is faithful it will be so on the whole group. Hence φ is injective, and so is $L \circ \varphi$. □

2.3 Geometric embedding and orientation cover

The closed non-orientable surface N_{2g+1} can be obtained by sewing a Möbius band to the orientable surface $\Sigma_{g,1}$.

Lemma 2.3. *The inclusion of surfaces $\Sigma_{g,1} \subset N_{2g+1}$ induces an inclusion of mapping class groups*

$$\Gamma_{g,1}/\langle D_\partial \rangle \simeq \Gamma_g^1 \hookrightarrow \mathcal{N}_{2g+1}.$$

Here D_∂ denotes the Dehn twist around the boundary curve in $\Sigma_{g,1}$. This is a central element in $\Gamma_{g,1}$. We identify the quotient group with the mapping class group of a surface with a marked point (or a surface with one non-parametrized boundary component).

Remark. This subgroup of $\mathcal{N}_{2g+1}$ can be identified with the point-wise stabiliser of the core of the Möbius band that has been sewn on to $\Sigma_{g,1}$ to form N_{2g+1}. For $g > 1$ this subgroup has infinite index but has the same virtual cohomological dimension as the whole group [10],

$$\text{vcd}(\Gamma_g^1) = 4g - 3 = \text{vcd}(\mathcal{N}_{2g+1}).$$

Proof. It is an elementary fact that the Dehn twist around the boundary of a Möbius strip is isotopic to the identity. This implies that D_∂ is in the kernel of $\Gamma_{g,1} \to \mathcal{N}_{2g+1}$. To show that D_∂ generates the kernel, consider the composition

$$\Gamma_{g,1} \longrightarrow \mathcal{N}_{2g+1} \stackrel{L}{\hookrightarrow} \Gamma_{2g}.$$

Let J denote the fix-point free orientation reversing involution of Σ_{2g} with quotient N_{2g+1}. As in [2], we embed Σ_{2g} in $\mathbb{R}^3$ symmetrically around the origin and take $J = -\text{Id}$ to be the reflection through the origin, see Figure 2.3. The image of any element $x \in \Gamma_{g,1}$ is the product of $\overline{x}$ with $J\overline{x}J$, where $\overline{x}$ acts on the left side of the surface via x and via the identity on the right side. So if x is in the kernel then $\overline{x}^{-1} = J\overline{x}J$ in Γ_{2g}. But $\overline{x}^{-1}$ can be represented by a diffeomorphism with support entirely in the left half of the surface and $J\overline{x}J$ by a diffeomorphism with support entirely in the right half of the surface. As the diffeomorphisms are isotopic, so must be their supports. Hence, $\overline{x}$ has a representing diffeomorphism supported in a tubular neighbourhood of the boundary, *i.e.* $x \in \langle D_\partial \rangle$. □

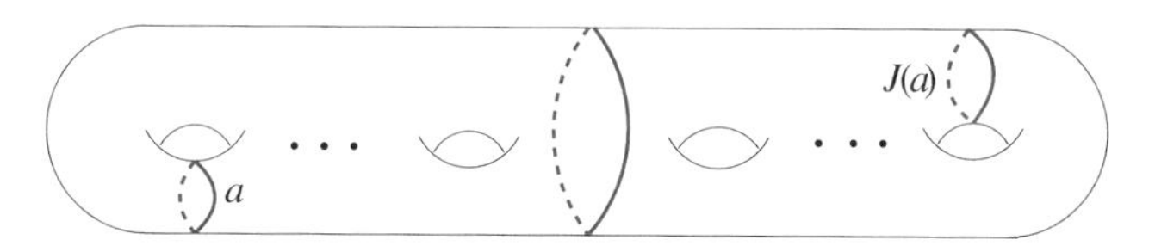

Figure 2.3. *The involution* $J = -\mathrm{Id}$ *in* $\mathbb{R}^3$.

Our third construction is the composition

$$L' \circ \phi : B_{2g} \overset{\phi}{\hookrightarrow} \Gamma_{g-1,2} \hookrightarrow \Gamma_g^1 \longrightarrow \mathcal{N}_{2g+1} \overset{L}{\longrightarrow} \Gamma_{2g}. \tag{2.4}$$

The unlabelled map $\Gamma_{g-1,2} \to \Gamma_g^1$, which is induced by gluing a pair of pants to the two boundary circles of $\Sigma_{g-1,2}$, is an inclusion, see [13]. Thus, again we have constructed an embedding.

2.4 Geometric embedding and mirror construction

Similarly to our third example, we may combine the geometric embedding $\phi : B_{2g} \to \Gamma_{g-1,2}$ with a mirror construction. For this first glue a torus $\Sigma_{1,2}$ along one of its boundary circles to $\Sigma_{g-1,2}$ and embed the other boundary circle in the plane. Now double the resulting surface $\Sigma_{g,2}$ by reflection in the plane to yield a surface $\Sigma_{2g,2}$. We leave it as an exercise to prove the following result.

Lemma 2.4. *The composition* $B_{2g} \overset{\phi}{\longrightarrow} \Gamma_{g-1,2} \longrightarrow \Gamma_{g,2} \overset{R'}{\longrightarrow} \Gamma_{2g,2}$ *is an injection.*

2.5 Operadic embedding

The following embedding is well-known as part of an E_2-operad action, compare Section 4.

Starting with a disk $\Sigma_{0,g+1}$ with g holes we glue a torus $\Sigma_{1,1}$ with one disk removed to each of boundaries of the g holes of the disk. The result is a surface $\Sigma_{g,1}$. As in the construction of φ in (2.3) we may extend diffeomorphisms of the disk via the identity to the glued on tori to define a map

$$\varphi^+ : B_g \longrightarrow \Gamma_{g,1}. \tag{2.5}$$

Lemma 2.5. *The map φ^+ is an injection.*

Proof. The fundamental group of $\Sigma_{0,g+1}$ is freely generated by the g curves $c_1, \dots, c_g$ that start at a point on the outside boundary and wind around one of the holes: $\pi_1(\Sigma_{0,g+1}) = \langle c_1, \dots, c_g \rangle$. The fundamental group of $\Sigma_{g,1}$ is freely generated by $2g$ curves: $\pi_1(\Sigma_{g,1}) = \langle a_1, b_1, \dots, a_g, b_g \rangle$ with a_i and b_i the standard generators in the fundamental group of the i-th copy of the torus. Then under the inclusion of the disk into the genus g surface the generator c_i maps to $a_i b_i a_i^{-1} b_i^{-1}$. So the c_i are mapped to words on subsets of the alphabet that are disjoint. Hence their images generate a free group on g generators and the induced map on fundamental groups is therefore an injection. We argue as before, that therefore the action of B_g via φ^+ on $\pi_1(\Sigma_{g,1})$ remains faithful and hence φ^+ must be an injection. □

2.6 More constructions

Other inclusions of the braid group into mapping class groups can be constructed from the above ones by precomposing with an automorphism of the braid group or composition with an automorphism of the mapping class groups. We note here that conjugation of a non-geometric (or geometric) embedding by a mapping class yields again a non-geometric (or geometric) embedding as Dehn twists are conjugated to Dehn twists. Thus these will indeed produce new examples of non-geometric embeddings (or geometric ones). We also note that conjugation by a fixed element induces the identity in homology. The results of Section 5 and Section 6 are therefore also valid for these variations. We will not mention these additional embeddings any further.

3 Proving non-geometricity

All of our constructions in Section 2, with the exception of φ^+ in (2.5), use an orientation reversing diffeomorphism of the oriented surface associated to the target. This is key for proving that these embeddings are not geometric.

Lemma 3.1. *Let J be an orientation reversing involution of $\Sigma_{g,n}$ and $x \in \Gamma_{g,n}$ commute with J. Then x is not a power of a Dehn twist unless it is trivial.*

Proof. We borrow an argument from [18]. Assume x is the k-th power of a Dehn twist D_c around a simple closed curve c in $\Sigma_{g,n}$. As x commutes with J and J is orientation reversing,

$$x = D_c^k = J D_c^k J = D_{J(c)}^{-k}.$$

But this identity can only hold if c is isotopic to $J(c)$ and $k = -k$. Therefore $x = D_c^k$ is trivial. □

Theorem 3.2. *The embeddings $R \circ \gamma$, $L \circ \varphi$, $L' \circ \phi$ and $R' \circ \phi$ are not geometric.*

An embedding that sends the standard generators of the braid group to some powers of Dehn twists are also called *pseudo-geometric* [24]. The proof of the theorem will show that these maps are not even pseudo-geometric.

Proof. Consider $R \circ \gamma$ and let σ be the image under γ of one of the standard generators of the braid group. The image $R(\sigma)$ is by definition invariant under the reflection in the plane (see Figure 2.2) which is orientation reversing. Lemma 3.1 implies that $R(\sigma)$ cannot be a power of a Dehn twist. The arguments for $L \circ \varphi$, $L' \circ \phi$ and $R' \circ \phi$ are similar. □

Remark. The geometric embedding $\phi : B_{2g+2} \to \Gamma_{g,2}$ can also be constructed by a 'doubling' procedure as the maps above. For this, identify first the braid group B_{2g+2} as the mapping class group of a disk with $2g + 2$ unordered marked points which in turn we identify as the orbit space of a genus g surface $\Sigma_{g,2}$ under the hyper-elliptic involution, see for example [16]. However, in this case the involution is orientation preserving and the construction leads to a geometric inclusion.

We now turn to the standard embedding γ from (2.1) and the operadic embedding φ^+ constructed in (2.5).

Theorem 3.3. *The embeddings γ and φ^+ are neither geometric nor pseudo-geometric.*

Proof. Let σ be one of the standard generators for the braid group B_g and consider its image under γ. This is a mapping class supported on a disk $\Sigma_{0,3}$ with two holes; compare Figure 2.1. An application of the Jordan Curve Theorem shows that there are only three non-contractible non-isotopic simple closed curves on $\Sigma_{0,3}$, each isotopic to one of the boundary circles. It is straight forward to check that $\gamma(\sigma)$ is not isotopic to (a power of) a Dehn twist around any of these three curves. Hence, γ is not geometric (or pseudo-geometric).

We now turn to φ^+. By definition (2.5), it is the composition of γ and the map induced by the inclusion of surfaces $\Sigma_{0,g+1} \subset \Sigma_{g,1}$ achieved by sewing a torus with one boundary component to each of the interior boundaries of $\Sigma_{0,g+1}$. Thus $\varphi(\sigma)$ is still defined as pictured in Figure 2.1 and supported by the same $\Sigma_{0,3}$, now a subsurface of $\Sigma_{g,1}$.

The support of any Dehn twist D_a around a simple closed curve a is a neighbourhood of a. Thus, if $\varphi(\sigma) = D_a^k$ for some a and $k \in \mathbb{N}$, we must be able to isotope the curve a into $\Sigma_{0,3}$. But the argument above still applies and shows that $\varphi^+(\sigma)$ cannot be (a power of) a Dehn twist of any curve in $\Sigma_{0,3}$. Hence φ^+ is not geometric and not pseudo-geometric. $\square$

4 Action of the braid group operad

Consider the following group level version of the well-known E_2-operad. (See, *e.g.* [19] for details.)

As in the introduction to Section 2, identify the pure braid group on k strands with a subgroup $\mathcal{D}_k \subset \Gamma_{0,k+1}$ of the pure ribbon braid group, *i.e.* the mapping class group of a disk with k holes whose boundaries are parametrised. The collection $\mathcal{D} = \{\mathcal{D}_k\}_{k\geq 0}$ forms an operad with structure maps

$$\theta : \mathcal{D}_k \times (\mathcal{D}_{m_1} \times \cdots \times \mathcal{D}_{m_k}) \longrightarrow \mathcal{D}_{m_1+\cdots+m_k}$$

induced by sewing of the underlying surfaces. To be more precise, for each i, the boundary of the i-th hole in $\Sigma_{0,k+1}$ is sewn to the (outer) boundary of the i-th disk Σ_{0,m_i+1}.

The operad $\mathcal{D}$ acts naturally on $\mathcal{B} = \coprod_{m\geq 1} B_m$ where each braid group B_m is identified via γ as a subgroup of $\Gamma_{0,(m),1}$. The action is again induced by gluing of the underlying surfaces. Indeed

$$\theta_{\mathcal{B}} : \mathcal{D}_k \times (B_{m_1} \times \cdots \times B_{m_k}) \longrightarrow B_{m_1+\cdots+m_k}$$

agrees with the structure map θ on the pure braid subgroups.

This action of $\mathcal{D}$ can further be extended to an action on

$$\Gamma_R = \coprod_{m>1} \Gamma_{m-1,2}$$

via the mirror construction R from (2.2). To define

$$\theta_R : \mathcal{D}_k \times (\Gamma_{m_1-1,2} \times \cdots \times \Gamma_{m_k-1,2}) \longrightarrow \Gamma_{m_1+\cdots+m_k-1,2}$$

place each of the underlying surfaces across a plane, so that one half is reflected by the plane onto the other, as in Figure 2.2. Then sewing k-legged trousers to the k boundary components on the left halves and another one to the right halves gives a surfaces of type $\Sigma_{m_1+\cdots+m_k-1,2}$. An element in $\mathcal{D}_k$ defines a mapping class on the left k-legged trousers and by mirroring a class on the right k-legged trousers. The following result holds by construction.

Lemma 4.1. *The map $R \circ \gamma$ induces a map of $\mathcal{D}$-algebras $\mathcal{B} \to \Gamma_R$.*

Similarly, $\mathcal{N} := \coprod_{m>1} \mathcal{N}_{m,1}$ is a $\mathcal{D}$-algebra. Again, the action

$$\theta_{\mathcal{N}} : \mathcal{D}_k \times (\mathcal{N}_{m_1,1} \times \cdots \times \mathcal{N}_{m_k,1}) \longrightarrow \mathcal{N}_{m_1+\cdots+m_k,1}$$

is induced by sewing the legs of k-legged trousers to the boundary components of the k non-orientable surfaces. And again by construction, we obtain the following result.

Lemma 4.2. *The map φ induces a map of $\mathcal{D}$-algebras $\mathcal{B} \to \mathcal{N}$.*

To see that the lift L to the orientation cover is a map of $\mathcal{D}$-algebras we need to consider a variant Γ_L of the $\mathcal{D}$-algebra Γ_R. The underlying groups are the same but the action θ_L is such that it commutes with L. To achieve this the mapping class defined on one k-legged trousers is paired with that on the other via the lift L so that the following holds.

Lemma 4.3. *The map L induces a map of $\mathcal{D}$-algebras $\mathcal{N} \to \Gamma_L$.*

We recall from [16] that also the standard embedding ϕ from (1.1) induces a map of $\mathcal{D}$-algebras $\mathcal{B}^{ev} \to \Gamma_\phi$. Here $\mathcal{B}^{ev} = \coprod_{m>1} B_{2m}$ is a $\mathcal{D}$-subalgebra of $\mathcal{B}$, and Γ_ϕ is the same collection of groups as Γ_R and Γ_L but has a slightly different action. We think of ϕ as explained in the remark following Theorem 3.2 above as lifting mapping classes of the $2m$-punctured disk to the ramified double cover Σ_{m-1} associated to the hyper-elliptic involution. Thus, in this case θ_ϕ is defined so as to commute with the hyper-elliptic involution.

Lemma 4.4. *The map ϕ induces a map of $\mathcal{D}$-algebras $\mathcal{B}^{ev} \to \Gamma_\phi$.*

And we also recall the best-known $\mathcal{D}$-algebra structure on

$$\Gamma := \coprod_{m \geq 0} \Gamma_{m,1};$$

see [12] and also [7]. In this case the action θ_Γ is defined just as for $\theta_\mathcal{N}$ but with $\mathcal{N}_{m_i,1}$ replaced by $\Gamma_{m_i,1}$. As φ^+ is essentially part of the $\mathcal{D}$-algebra structure the following is immediate.

Lemma 4.5. *The map φ^+ induces a map of $\mathcal{D}$-algebras $\mathcal{B} \to \Gamma$.*

5 The maps in stable homology

We will determine the induced maps in stable homology of all the embeddings of braid groups into mapping class groups of orientable or non-orientable surfaces constructed earlier.

The Harer-Ivanov homology stability theorem states that the embedding $\Sigma_{g,1} \to \Sigma_{g+1,1}$ induces an isomorphism

$$H_*(\Gamma_{g,1}) \longrightarrow H_*(\Gamma_{g+1,1})$$

in a range of degrees, called the *stable range*. This range has recently been improved to $* \leq 2(g-1)/3$ by Boldsen [6] and Randal-Williams [15].

Theorem 5.1. *In the stable range, the maps $R \circ \gamma$, $L \circ \varphi$, ϕ, φ^+, $L' \circ \phi$, and $R' \circ \phi$ induce the zero map in any reduced, generalised homology theory.*

Proof. We sketch the argument here and refer for more details to [17], [16]. In Section 4 we showed that the maps $R \circ \gamma$, $L \circ \varphi$, ϕ and φ^+ are maps of $\mathcal{D}$-algebras. After taking classifying spaces and group completion they induce therefore maps of double loop spaces [1]

$$\Omega B\left(\coprod_{g>0} B(B_g)\right) \simeq \mathbb{Z} \times B(B_\infty)^+ \simeq \Omega^2 S^2 \longrightarrow \Omega B\left(\coprod_{g\geq 0} B\Gamma_{g,1}\right) \simeq \mathbb{Z} \times B\Gamma_\infty^+.$$

[1] The group completion of $\coprod_{m\geq 1} B(B_{2m})$ consists of all the even components in $\Omega^2 S^2$, and the argument goes through.

Here $B_\infty = \lim_{g\to\infty} B_g$ and $\Gamma_\infty = \lim_{g\to\infty} \Gamma_{g,1}$ are the infinite braid and mapping class groups and X^+ denotes the Quillenization of the space X. As $B(B_\infty)^+ \simeq \Omega_0^2 S^2 \simeq \Omega^2 S^3$ is the free object on the circle in the category of double loop spaces, on a connected component these maps are determined by their restriction to the circle. But these restrictions have to be homotopic to the constant map as $B(\Gamma_\infty)^+$ is simply connected. Hence maps $B(B_\infty)^+ \to B(\Gamma_\infty)^+$ that are maps of double loop spaces are null-homotopic. In particular, they induce the zero map in any reduced, generalised homology theory. Finally, the mapping class groups satisfy (ordinary) homology stability. By an application of the Atiyah-Hirzebruch spectral sequence, the statement of the theorem follows for the first four maps, including ϕ, and hence for $L' \circ \phi$ and $R' \circ \phi$. □

We now turn our attention to the mapping class group of non-orientable surfaces and the embedding $\varphi : B_g \to \mathcal{N}_{g,1}$ defined in (2.3). The commutator subgroup of $\mathcal{N}_g$ is generated by Dehn twists around two-sided curves. For $g \geq 7$ it has index two and thus $H_1(\mathcal{N}_g) = \mathbb{F}_2$ and is in particular not trivial; see [11].

The mapping class groups $\mathcal{N}_g$ also satisfy homology stability:

$$H_*(\mathcal{N}_g) = H_*(\mathcal{N}_{g,1}) = H_*(\mathcal{N}_{g+1,1}).$$

Here $* \leq (g-3)/3$ for the first equality and $* \leq g/3$ for the second, see [22] and [15].

If σ is a standard generator of the braid group, its image under φ interchanges two cross caps in $N_{g,1}$. Therefore, it is not in the index two subgroup of $\mathcal{N}_{g,1}$ generated by Dehn twists around two-sided curves. Indeed, the product of $\varphi(\sigma)$ with the Dehn twist around the two-sided curve that goes once through each cross cap is a cross-cap slide, see [18]. Hence, the map induced by φ on the first homology groups is surjective and not trivial. More generally we have the following result.

Theorem 5.2. *Let $g \geq 7$ and $0 < * \leq g/3$. When $\mathbb{F} = \mathbb{Q}$ or $\mathbb{F} = \mathbb{F}_p$ for an odd prime p, the map*

$$\varphi_* : H_*(B_g; \mathbb{F}) \longrightarrow H_*(\mathcal{N}_{g,1}; \mathbb{F})$$

is zero, while for $\mathbb{F} = \mathbb{F}_2$ it is an injection.

Proof. The basic idea of the proof is similar to that used in Theorem 5.1 but we need to use also some quite technical results from [21], [19] and [20]. We sketch the argument.

It is well-known that the map from the braid to the symmetric group induces after taking classifying spaces, stabilisation and Quillenization the canonical map

$$\mathbb{Z} \times BB_\infty^+ \simeq \Omega^2 S^2 \longrightarrow \Omega^\infty S^\infty \simeq \mathbb{Z} \times B\Sigma_\infty^+$$

from the free object generated by S^0 in the category of double loop spaces to the corresponding one in the category of infinite loop spaces. In homology with $\mathbb{F}_2$ coefficients it induces an inclusion and it is zero in reduced homology with field coefficients of characteristic other than 2.

By the main theorem of [19], the double loop space structure on $\mathbb{Z} \times B\mathcal{N}_\infty^+$ defined by the $\mathcal{D}$-algebra structure on $\mathcal{N}$ extends to an infinite loop space structure. This implies that the map $\Omega^2 S^2 \to \mathbb{Z} \times B\mathcal{N}_\infty^+$ induced by φ factors through $\Omega^\infty S^\infty$ via the above map.

Using cobordism categories of non-orientable surfaces one can show that there is another infinite loop space structure on $\mathbb{Z} \times B\mathcal{N}_\infty^+$. By a theorem of Wahl these two infinite loop space structures are the same up to homotopy. To be more precise, Wahl shows in [21] that the two constructions lead to the same infinite loop space structure up to homotopy in the orientable case, i.e., for $\mathbb{Z} \times B\Gamma_\infty^+$. Her argument goes through verbatim to prove the same result for non-orientable surfaces. Thus, the map of infinite loop spaces $\Omega^\infty S^\infty \to \mathbb{Z} \times B\mathcal{N}_\infty^+$ here is up to homotopy the same as the one used in [20]. In [20] we showed however that this map has a splitting up to homotopy and, in particular, induces an injection in homology.

Combining all this we have proved that the composition

$$\mathbb{Z} \times BB_\infty^+ \simeq \Omega^2 S^2 \longrightarrow \Omega^\infty S^\infty \longrightarrow \mathbb{Z} \times B\mathcal{N}_\infty^+$$

induces an injection on $\mathbb{F}_2$-homology and is trivial in reduced homology for $\mathbb{F} = \mathbb{Q}$ or $\mathbb{F}_p$, p odd. As $B_g \to B_\infty$ induces an injection in homology with any field coefficients, see [9], and by homology stability of the non-orientable mapping class group the theorem follows. □

6 Calculations in unstable homology

In this section we examine the homomorphisms induced by our embeddings in unstable homology with field coefficients.

We restrict our discussion to the orientable case though a similar analysis goes through also in the non-orientable case.

Consider any map

$$\alpha_* : H_*(B_m; \mathbb{F}) \longrightarrow H_*(\Gamma_{g,b}; \mathbb{F}) \tag{6.1}$$

induced by a homomorphism $\alpha : B_m \to \Gamma_{g,b}$ that is part of a $\mathcal{D}$-algebra map; here $b = 1, 2$ and $\mathbb{F}$ is any field. The main fact we will be using is that the homology of the braid group is generated by classes of degree one when taking the $\mathcal{D}$-algebra structure into account.

6.1. The rational case: Recall from [9] that for $m > 1$ the $\mathbb{Q}$-homology of B_m is of rank one in degrees 0 and 1, and zero otherwise. Thus rationally, the braid groups have the homology of a circle. We recall

$$H_1(\Gamma_{2,2}) = H_1(\Gamma_{2,1}) = \mathbb{Z}/10\mathbb{Z}, \quad \text{and} \quad H_1(\Gamma_{g,1}) = 0 \text{ when } g \geq 3. \tag{6.2}$$

The first identity follows from the stability results [6] and [15]; and the two computations are well-known. Thus α_* is trivial in rational homology.

We now turn to fields of finite characteristic. Recall from [9] that for $m > 1$ the $\mathbb{F}_p$-homology is generated by $H_1(B_m; \mathbb{F}_p) = \mathbb{F}_p$ and the homology operations induced from the action of $\mathcal{D}$. These operations are the product, the first Dyer-Lashof operation Q, and in the case of odd primes, the combination with the Bockstein operator βQ. More precisely, for $x \in H_*(B_m; \mathbb{F}_p)$, the operation

$$Q : H_r(B_m; \mathbb{F}_p) \longrightarrow H_{p(r+1)-1}(B_{pm}; \mathbb{F}_p)$$

is defined by the formula

$$Q(x) = \theta_*(e_1 \otimes x \otimes x) \quad \text{for} \quad p = 2 \tag{6.3}$$

$$Q(x) = \theta_*(e_{p-1} \otimes x^p) \quad \text{for} \quad p > 2 \tag{6.4}$$

where e_1 and e_{p-1} are chains in $B\mathcal{D}_2$ respectively $B\mathcal{D}_p$ of degree 1 respectively $p - 1$, and θ_* is induced by the action $\theta = \theta_{\mathcal{B}}$ of $\mathcal{D}$ on $\mathcal{B}$ as defined in Section 4.

6.2. The case p even: Quoting [9, page 347], the $\mathbb{F}_2$-homology of the braid group B_m can be described as

$$H_*(B_m; \mathbb{F}_2) = \mathbb{F}_2[x_i]/I$$

where $x_i = Q(x_{i-1})$ is of degree $2^i - 1$ and I is the ideal generated by all monomials $x_{i_1}^{k_1} \dots x_{i_t}^{k_t}$ such that $\sum_{j=1}^{t} k_j 2^{i_j} > m$. In particular,

$$x_i = 0 \quad \text{if} \quad 2^i > m.$$

6.3. The case p odd: Similarly, by [9, page 347], for p odd the F_p-homology of the braid group B_m is additively the quotient of an algebra on exterior generorotors λ and y_i, and on polynomial generators βy_i, of respective degrees $1, 2p^i - 1$ and $2p^i - 2$; the quotient is by an ideal J_m which includes y_i whenever $2p^i > m$. Furthermore, $y_1 = Q(\lambda)$ and $y_{i+1} = Q(y_i)$ for generators in the appropriate homology groups.

We consider now the image of the generators x_i and y_i under the map α_*. As an immediate consequence of Theorem 5.1, and using the best homology stability range available, we have

$$\alpha_*(x_i) = 0 \quad \text{for} \quad (3 \cdot 2^i - 1)/2 \leq g \quad \text{and} \quad p = 2 \tag{6.5}$$

$$\alpha_*(y_i) = 0 \quad \text{for} \quad (6\, p^i - 1)/2 \leq g \quad \text{and} \quad p > 2. \tag{6.6}$$

We now explain in two examples how the Dyer-Lashof algebra structure can be used to deduce similar results independent of Theorem 5.1, which in some cases lead to stronger vanishing results.

6.4. Example: Consider the operadic embedding (2.5) when $m = g$, $b = 1$ and α is

$$\varphi^+ : B_g \longrightarrow \Gamma_{g,1}.$$

The $\mathcal{D}$-algebra structure in this case is given by gluing the boundary of the surfaces $\Sigma_{g_i,1}$ to the boundaries of the holes in the disk $\Sigma_{0,k+1}$. The genus of the resulting surface is simply the sum of the genera g_i.

Observe that to use the $\mathcal{D}$-algebra structure, the genus of the surface corresponding to the target group has to be large enough to be able to decompose it. The equation $\varphi^+_*(Q(z)) = Q(\varphi^+_*(z))$ gives rise to the following inductive formula. Let d_i be the least genus of the target surface such that $\varphi^+_*(x_{i+1}) = 0$, respectively $\varphi^+_*(y_i) = 0$. Then using (6.3) and (6.4) we can conclude that

$$d_i \leq p\, d_{i-1} \quad \text{and thus} \quad d_i \leq p^i\, d_0,$$

where, by (6.2), $d_0 = 3$ when $p = 2, 5$, and $d_0 = 2$ when $p \neq 2, 5$. Thus we have that

$$\varphi^+_*(x_i) = 0 \quad \text{if} \quad 3 \cdot 2^{i-1} \leq g \quad \text{and} \quad p = 2, \tag{6.7}$$

$$\varphi^+_*(y_i) = 0 \quad \text{if} \quad 3\, p^i \leq g \quad \text{and} \quad p = 5, \tag{6.8}$$

$$\varphi^+_*(y_i) = 0 \quad \text{if} \quad 2\, p^i \leq g \quad \text{and} \quad p \neq 5. \tag{6.9}$$

This gives an improvement on (6.5) and (6.6) for primes other than 2 and 5.

As $x_i = 0$ for $2^i > m$ and $y_i = 0$ for $2\,p^i > m$, our computations show that the map in homology induced by φ^+ is zero in unstable homology of positive degree for characteristics $p \neq 2, 5$. When $p = 2$ or $p = 5$, we cannot always decide with the above methods whether the top dimensional x_i and y_i classes are mapped to zero or not.

6.5. Example: We now consider the geometric embedding (1.1) when $m = 2g + 2$, $b = 2$ and α is

$$\phi : B_{2g+2} \longrightarrow \Gamma_{g,2}.$$

The $\mathcal{D}$-algebra structure in this case glues the surfaces $\Sigma_{g_i,2}$ to two disks $\Sigma_{0,k+1}$ with k holes. The resulting surface has genus the sum of the genera g_i plus $k - 1$. Thus, the inductive formula for d_i is given by

$$d_i \leq p\,d_{i-1} + p - 1 \leq p^i\,(d_0 + 1) - 1.$$

By (6.2) we have $d_0 = 3$ and $d_i = 4\,p^i - 1$ for $p = 2, 5$, and we have $d_0 = 2$ and $d_i = 3\,p^i - 1$ for $p \neq 2, 5$.
The d_i are growing faster here then in Example 6.4. Indeed, as is easily checked, any class x_i and y_i that in this way can be shown to vanish under ϕ_* is already in the stable range. Thus no extra information can be gained in addition to what is known by Theorem 5.1; compare (6.5) and (6.6).

In particular, we cannot determine with our methods here whether (a) the top three x_i classes and whether (b) the top two y_i classes for $p = 3$ resp. the top y_i class for $p \neq 3$ vanish under ϕ_*. More precisely, we have $x_i \neq 0$ when $2^{i-1} - 1 \leq g$, while $\phi_*(x_i) = 0$ when $2^{i+2} - 1 \leq g$ and similarly $y_i \neq 0$ when $p^{i-1} - 1 \leq g$, while $\phi_*(y_i) = 0$ when $4p^i - 1 \leq g$ in case $p = 5$, or when $3p^i - 1 \leq g$ in case $p \neq 5$.

Therefore the conclusions drawn in [16, Corollary 4.1] (and [8, Corollary 2.7]) are too strong: *It remains an open question whether for $g \geq 3$ the homomorphism $\phi : B_{2g+2} \to \Gamma_{g,2}$ induces the zero map in reduced $\mathbb{F}_p$-homology for all primes p.*

In our explicit calculations above we have concentrated on the generators x_i and y_i. But a similar analysis can be given to determine when the image of products is trivial.

Finally, an analogous study can be given for $\mathcal{D}$-algebra maps from braid groups to mapping class groups of non-orientable surfaces in the case when $p \neq 2$.

References

[1] E. ARTIN, *Theorie der Zöpfe*, Abh. Math. Semin. Univ. Hamburg **4** (1925), 47–72.

[2] J. S. BIRMAN and D. R. J. CHILLINGWORTH, *On the homeotopy group of a non-orientable surface*, In: Proc. Cambridge Philos. Soc. **71** (1972), 437–448.

[3] J. S. BIRMAN and D. R. J. CHILLINGWORTH, *Erratum: "On the homeotopy group of a non-orientable surface"* [Proc. Cambridge Philos. Soc. 71 (1972), 437–448; MR0300288], Math. Proc. Cambridge Philos. Soc. **136** (2004), no. 2, 441.

[4] J. S. BIRMAN and H. M. HILDEN, *Lifting and projecting homeomorphisms*, Arch. Math. (Basel) **23** (1972), 428–434.

[5] J. S. BIRMAN and H. M. HILDEN, *On isotopies of homeomorphisms of Riemann surfaces*, Ann. of Math. (2) **97** (1973), 424–439.

[6] S. K. BOLDSEN, *Improved homological stability for the mapping class group with integral or twisted coefficients*, arXiv:0904.3269

[7] C.-F. BÖDIGHEIMER, *On the topology of moduli spaces of Riemann surfaces, Part II: homology operations*, Math. Gottingensis, Heft **23** (1990).

[8] F. R. COHEN, *Homology of mapping class groups for surfaces of low genus*, In: The Lefschetz Centennial Conference", Part II (Mexico City, 1984), Contemp. Math., 58, II, Amer. Math. Soc., Providence, RI, 1987, 21–30.

[9] F. R. COHEN, T. J. LADA and J. P. MAY, "The Homology of Iterated Loop Spaces", Lecture Notes in Mathematics, Vol. 533. Springer-Verlag, Berlin-New York, 1976.

[10] N. V. IVANOV, *Complexes of curves and the Teichmüller modular group*, Uspekhi Mat. Nauk **42** (1987), 49–91.

[11] M. KORKMAZ, *First homology group of mapping class groups of nonorientable surfaces*. Math. Proc. Cambridge Philos. Soc. **123** (1998), no. 3, 487–499.

[12] E. Y. MILLER, *The homology of the mapping class group*, J. Differential Geom. **24** (1986), 1–14.

[13] L. PARIS and D. ROLFSEN, *Geometric subgroups of mapping class groups*, J. Reine Angew. Math. **521** (2000), 47–83.

[14] J. POWELL, *Two theorems on the mapping class group of a surface*, Proc. Amer. Math. Soc. **68** (1978), no. 3, 347–350.

[15] O. RANDAL-WILLIAMS, *Resolutions of moduli spaces and homological stability*. arXiv:0909.4278

[16] G. Segal and U. Tillmann, *Mapping configuration spaces to moduli spaces*, In: "Groups of Diffeomorphisms", Adv. Stud. Pure Math., Vol. 52, Math. Soc. Japan, Tokyo, 2008, 469–477.
[17] Y. Song and U. Tillmann, *Braids, mapping class groups, and categorical delooping*, Math. Ann. **339** (2007), no. 2, 377–393.
[18] B. Szepietowski, *Embedding the braid group in mapping class groups*, Publ. Mat. **54** (2010), no. 2, 359–368.
[19] U. Tillmann, *Higher genus surface operad detects infinite loop spaces*, Math. Ann. 317 (2000), no. 3, 613–628.
[20] U. Tillmann, *The representation of the mapping class group of a surface on its fundamental group in stable homology*, Q. J. Math. **61** (2010), no. 3, 373–380.
[21] N. Wahl, *Infinite loop space structure(s) on the stable mapping class group*, Topology **43** (2004), no. 2, 343–368.
[22] N. Wahl, *Homological stability for the mapping class groups of non-orientable surfaces*, Invent. Math. **171** (2008), no. 2, 389–424.
[23] B. Wajnryb, *Artin groups and geometric monodromy*, Invent. Math. 138 (1999), no. 3, 563–571.
[24] B. Wajnryb, *Relations in the mapping class group*. In: "Problems on Mapping Class Groups and Related Topics", Proc. Sympos. Pure Math., Vol. 74, Amer. Math. Soc., Providence, RI, 2006, 115–120.

The cohomology of the braid group B_3 and of $SL_2(\mathbb{Z})$ with coefficients in a geometric representation

Filippo Callegaro, Frederick R. Cohen* and Mario Salvetti

Abstract. This article is a short version of a paper which addresses the cohomology of the third braid group and of $SL_2(\mathbb{Z})$ with coefficients in geometric representations. We give precise statements of the results, some tools and some proofs, avoiding very technical computations here.

1 Introduction

This article is a summary of a paper which addresses the cohomology of the third braid group or $SL_2(\mathbb{Z})$ with local coefficients in a classical geometric representation described below. These structures are at the intersection of several basic topics during the special period on "Configuration Spaces", which took place in Pisa.

The authors started to work together to develop the problem considered here precisely during the special period on "Configuration Spaces" in Pisa. So we believe that the work in this paper is particularly suitable for the Conference Proceedings.

This article contains precise statements of the results, some of the "main tools" and "general results", avoiding some long and delicate details. These details, to appear elsewhere, would have made the paper too long and technical, probably losing the spirit of these Proceedings. The purpose of this paper is to provide precise knowledge of the results as well as general indications of the proofs.

The precise problem considered here is the computation of the cohomology of the braid group B_3 and of the special linear group $SL_2(\mathbb{Z})$ with certain non-trivial local coefficients. First consider a particularly natural geometric, classical representation. In general, let $M_{g,n}$ be an orientable

*Partially supported by DARPA.

surface of genus g with n connected components in its boundary. Isotopy classes of Dehn twists around simple loops $c_1, \ldots, c_{2g}$ such that

$$|c_i \cap c_{i+1}| = 1 \ (i = 1, \ldots, 2g-1), \ \ c_i \cap c_j = \emptyset \text{ otherwise}$$

represents the braid group B_{2g+1} in the symplectic group $\mathrm{Aut}(H_1(M_{g,n}); <>)$ of all automorphisms preserving the intersection form. Such representations arise classically, for example in singularity theory, as it is related to some monodromy operators (See [22] for example) as well as in number theory in the guise of modular forms (See [20] for example.), and geometry (See [14] for example.).

In this paper we restrict to the case $g = 1$, $n = 1$, where the symplectic group equals $SL_2(\mathbb{Z})$. We extend the above representation to the symmetric power $M = \mathbb{Z}[x, y]$. Notice that this representation splits into irreducibles $M = \oplus_{n \geq 0} M_n$, according to polynomial-degree.

Our result is the complete integral computation of

$$H^*(B_3; M) \ = \ \oplus_n H^*(B_3; M_n)$$

and

$$H^*(SL_2(\mathbb{Z}); M) \ = \ \oplus_n H^*(SL_2(\mathbb{Z}); M_n).$$

The free part of the cohomology of $SL_2(\mathbb{Z})$ is a classical computation due to G. Shimura ([20], [11]) and is related to certain spaces of modular forms. A version for the "stabilized" mapping class group in characteristic zero was developed by Looijenga [14]. Here we give the complete calculation of the cohomology including the p-torsion groups.

These groups have an interesting description in terms of a variation of the so called "divided polynomial algebra". Unexpectedly, we found a strong relation between the cohomology in our case and the cohomology of some interesting spaces which were constructed in a completely different framework and which are basic to the homotopy groups of spheres.

An outline of the paper is as follows. In Section 2, we give the general results and the main tools which are used here. Section 3 is devoted to giving all precise statements together with some additional details concerning a mod p variation of the divided polynomial algebra. Section 4 provides sketches of required side results which may be of independent interest. Section 5 is a remark about potential extensions of the work here to principal congruence subgroups. Section 6 compares the cohomology determined here to the cohomology of some spaces which seem quite different. We are led to ask questions about possible connections as well as questions about analogous results concerning congruence subgroups.

ACKNOWLEDGEMENTS. The authors are grateful to the Ennio de Giorgi Mathematical Research Institute, the Scuola Normale Superiore as well as the University of Pisa for the opportunity to start this work together and the wonderful setting to do some mathematics. The second author would like to thank many friends for an interesting, wonderful time in Pisa.

2 General results and main tools

Let B_3 denote the braid group with 3 strands. We consider the geometric representation of B_3 into the group

$$\mathrm{Aut}(H_1(M_{1,1};\mathbb{Z}), <>) \cong SL_2(\mathbb{Z})$$

of the automorphisms which preserve the intersection form where $M_{1,1}$ is a genus-1 oriented surface with 1 boundary component. Explicitly, if the standard presentation of B_3 is

$$B_3 = <\sigma_1,\ \sigma_2 \ :\ \sigma_1\sigma_2\sigma_1 = \sigma_2\sigma_1\sigma_2>,$$

then one sends $\sigma_1,\ \sigma_2$ into the Dehn twists around one parallel and one meridian of $M_{1,1}$ respectively. Taking a natural basis for the H_1, the previous map is explicitly given by

$$\lambda : B_3 \to SL_2(\mathbb{Z}) : \quad \sigma_1 \to \begin{bmatrix} 1 & 0 \\ -1 & 1 \end{bmatrix}, \quad \sigma_2 \to \begin{bmatrix} 1 & 1 \\ 0 & 1 \end{bmatrix}.$$

Of course, any representation of $SL_2(\mathbb{Z})$ will induce a representation of B_3 by composition with λ. We will ambiguously identify σ_i with $\lambda(\sigma_i)$ when the meaning is clear from the context.

It is well known that λ is surjective and the kernel is infinite cyclic generated by the element $c := (\sigma_1\sigma_2)^6$ ([16, Theorem 10.5]) so that we have a sequence of groups

$$1 \longrightarrow \mathbb{Z} \xrightarrow{j} B_3 \xrightarrow{\lambda} SL_2(\mathbb{Z}) \longrightarrow 1. \tag{2.1}$$

Let V be a rank-two free $\mathbb{Z}$-module and $x,\ y$ a basis of $V^* := \mathrm{Hom}_{\mathbb{Z}}(V,\mathbb{Z})$. The natural action of $SL_2(\mathbb{Z})$ over V^*

$$\sigma_1 : \begin{cases} x \to x - y \\ y \to y \end{cases}, \quad \sigma_2 : \begin{cases} x \to x \\ y \to x + y \end{cases}$$

extends to the symmetric algebra $S[V^*]$, which we identify with the polynomial algebra $M := \mathbb{Z}[x, y]$. The homogeneous component M_n of degree n is an irreducible invariant free submodule of rank $n+1$ so we have a decomposition $M = \bigoplus_n M_n$.

In this paper we describe the cohomology groups

$$H^*(B_3; M) = \bigoplus_{n\geq 0} H^*(B_3; M_n)$$

and

$$H^*(SL_2(\mathbb{Z}); M) = \bigoplus_{n\geq 0} H^*(SL_2(\mathbb{Z}); M_n)$$

where the action of B_3 over M is induced by the above map λ.

The free part of the cohomology of $SL_2(\mathbb{Z})$ tensored with the real numbers is well-known to be isomorphic to the ring of modular forms based on the $SL_2(\mathbb{Z})$-action on the upper $1/2$ plane by fractional linear tranformations ([11]):

$$\begin{array}{lr} H^1(SL_2(\mathbb{Z});\ M_{2n}\otimes\mathbb{R}) \cong \mathcal{M}od^0_{2n+2}\oplus\mathbb{R} & n\geq 1 \\ H^i(SL_2(\mathbb{Z});\ M_n\otimes\mathbb{R}) \ = 0 & i>1 \text{ or } i=1, \text{ odd } n \end{array}$$

where $\mathcal{M}od^0_n$ is the vector space of cuspidal modular forms of weight n. In our computations, a basic first step is to rediscover the Poincaré series for the cohomology of $SL_2(\mathbb{Z})$ with coefficients in M.

Essentially, we use two "general" tools for the computations. The first one is the spectral sequence associated to (2.1), with

$$E_2^{s,t} = H^s(SL_2(\mathbb{Z}); H^t(\mathbb{Z}, M)) \Rightarrow H^{s+t}(B_3; M) \tag{2.2}$$

Notice that the element c above acts trivially on M, so we have

$$H^i(\mathbb{Z}; M) = H^i(\mathbb{Z}; \mathbb{Z})\otimes M = \begin{cases} M & \text{for } i=0,\ 1 \\ 0 & \text{for } i>1 \end{cases}$$

and the spectral sequence has a two-row E_2-page

$$E_2^{s,t} = \begin{cases} 0 & \text{for } t>1 \\ & \\ H^s(SL_2(\mathbb{Z}), M) & \text{for } t=0,\ 1. \end{cases} \tag{2.3}$$

The second main tool which we use comes from the following well-known presentation of $SL_2(\mathbb{Z})$ as an amalgamated product of torsion groups. We set for brevity $\mathbb{Z}_n := \mathbb{Z}/n\mathbb{Z}$.

Proposition 2.1 ([17,19]). *The group $SL_2(\mathbb{Z})$ is an amalgamated free product*

$$SL_2(\mathbb{Z}) = \mathbb{Z}_4 *_{\mathbb{Z}_2} \mathbb{Z}_6$$

where

$$\mathbb{Z}_4 \text{ is generated by the 4-torsion element } w_4 := \begin{bmatrix} 0 & 1 \\ -1 & 0 \end{bmatrix}$$

$$\mathbb{Z}_6 \text{ is generated by the 6-torsion element } w_6 := \begin{bmatrix} 1 & 1 \\ -1 & 0 \end{bmatrix}$$

$$\mathbb{Z}_2 \text{ is generated by the 2-torsion element } w_2 := \begin{bmatrix} -1 & 0 \\ 0 & -1 \end{bmatrix}.$$

So, for every $SL_2(\mathbb{Z})$-module N, we can use the associated a Mayer-Vietoris sequence

$$\to H^i(\mathbb{Z}_4; N) \oplus H^i(\mathbb{Z}_6; N) \to H^i(\mathbb{Z}_2; N) \to H^{i+1}(SL_2(\mathbb{Z}); N) \to \quad (2.4)$$

We start to deduce some "general" result on the cohomology of $SL_2(\mathbb{Z})$. From (2.4) and very well-known properties of cohomology groups, it immediately follows:

Proposition 2.2. *Assume that* 2 *and* 3 *are invertible in the module* N *(equivalent,* $1/6 \in N$*). Then*

$$H^i(SL_2(\mathbb{Z}); N) = 0 \quad \text{for } i > 1.$$

Corollary 2.3. *If* $1/6 \in N$ *and* N *has no* $SL_2(\mathbb{Z})$ *invariants (i.e.* $H^0(SL_2(\mathbb{Z}); N) = 0$*) then*

$$H^1(B_3; N) = H^1(SL_2(\mathbb{Z}); N) = H^2(B_3; N).$$

Proof. The corollary follows immediately from the above spectral sequence which degenerates at E_2. □

Remark 2.4. The group B_3 has cohomological dimension 2 (see for example [7, 18]) so we only need to determine cohomology up to H^2. It also follows that the differential

$$d_2^{s,1} : E_2^{s,1} \to E_2^{s+2,0}$$

is an isomorphism for all $s \geq 2$, and is surjective for $s = 1$.

For a given finitely generated $\mathbb{Z}$-module L and for a prime p define the p-torsion component of L as follows:

$$L_{(p)} = \{x \in L \mid \exists k \in \mathbb{N} \text{ such that } p^k x = 0\}.$$

Moreover write FL for the free part of L, that is FL defines the isomorphism class of a maximal free $\mathbb{Z}$-submodule of L.

Return to the module M defined before. The next result follows directly from the previous discussion (including Remark 2.4).

Corollary 2.5.

(1) *For all primes p we have*

$$H^1(B_3; M)_{(p)} = H^1(SL_2(\mathbb{Z}); M)_{(p)}$$

(where $H^i_{(p)}$ denotes the component of p-torsion of $H^i(-)$).

(2) *For all primes $p > 3$,*

$$H^1(B_3; M)_{(p)} = H^1(SL_2(\mathbb{Z}); M)_{(p)} = H^2(B_3; M)_{(p)}.$$

(3) *For the free parts for $n > 0$,*

$$FH^1(B_3; M_n) = FH^1(SL_2(\mathbb{Z}); M_n) = FH^2(B_3; M_n).$$

Notice that statement (1) of corollary 2.5 fails for the torsion-free summand: see theorem 3.7 below for the precise statement including it.

3 Main results

In this section we state the main results, namely the complete description of the cohomology of the braid group B_3 and of $SL_2(\mathbb{Z})$ with coefficients in the module M. First, we need some definitions.

Let $\mathbb{Q}[x]$ be the ring of rational polynomials in one variable. We define the subring

$$\Gamma[x] := \Gamma_{\mathbb{Z}}[x] \subset \mathbb{Q}[x]$$

as the subset generated, as a $\mathbb{Z}$-submodule, by the elements $x_n := \frac{x^n}{n!}$ for $n \in \mathbb{N}$. It follows from the next formula that $\Gamma[x]$ is a subring with

$$x_i x_j = \binom{i+j}{i} x_{i+j}. \tag{3.1}$$

The algebra $\Gamma[x]$ is usually known as the *divided polynomial algebra* over $\mathbb{Z}$ (see for example chapter 3C of [13] or the Cartan seminars as an early reference [4]). In the same way, by using (3.1), one can define $\Gamma_R[x] := \Gamma[x] \otimes R$ over any ring R.

Let p be a prime number. Consider the p-adic valuation $v := v_p : \mathbb{N} \setminus \{0\} \to \mathbb{N}$ such that $p^{v(n)}$ is the maximum power of p dividing n. Define the ideal I_p of $\Gamma[x]$ as

$$I_p := (p^{v(i)+1} x_i, \quad i \geq 1)$$

and call the quotient

$$\Gamma_p[x] := \Gamma[x]/I_p$$

the *p-local divided polynomial algebra* (mod p).

Proposition 3.1. *The ring $\Gamma_p[x]$ is naturally isomorphic to the quotient*

$$\mathbb{Z}[\xi_1, \xi_p, \xi_{p^2}, \xi_{p^3}, \dots]/J_p$$

where J_p is the ideal generated by the polynomials

$$p\xi_1 , \quad \xi_{p^i}^p - p\xi_{p^{i+1}} \; (i \geq 1).$$

The element ξ_{p^i} corresponds to the generator $x_{p^i} \in \Gamma_p[x]$, $i \geq 0$.

Proof. The ideal I_p is graded and there is a direct sum decomposition of $\mathbb{Z}$-modules with respect to the degree

$$\Gamma_p[x] \; = \; \oplus_{i \geq 0} \left[\mathbb{Z}/p^{v(i)+1}\mathbb{Z}\right](x_i).$$

We fix in this proof $\deg(x) = 1$, so the i-th degree component of I_p is given by the ideal $p\tilde{I}_i \subset \mathbb{Z}$, where

$$\tilde{I}_i := \left(p^{v(1)}\binom{i}{1}, \dots, p^{v(i-1)}\binom{i}{i-1}, \; p^{v(i)}\right).$$

It is easy to see that actually $\tilde{I}_i = (p^{v(i)})$.

Now we show that the x_{p^i}'s are generators of $\Gamma_p[x]$ as a ring. We will make use of the following well known (and easy) lemma.

Lemma 3.2. *Let $n = \sum_j n_j \, p^j$ be the p-adic expansions of n. Then*

$$v(n!) \; = \; \frac{n - \sum_j n_j}{p-1} \; = \; \sum_j n_j \, \frac{p^j - 1}{p-1}.$$

Let $n := \sum_{j=0}^{k} n_j \, p^j$ be the p-adic expansion of n. It follows from (3.1) that

$$\prod_{j=0}^{k} (x_{p^j})^{n_j} \; = \; \frac{n!}{(p!)^{n_1}(p^2!)^{n_2} \dots (p^k!)^{n_k}} \, x_n. \tag{3.2}$$

From Lemma 3.2 the coefficient of x_n in the latter expression is not divisible by p. This clearly shows that the x_{p^i}'s are generators.

Again by (3.1) we have

$$(x_{p^i})^p = \frac{p^{i+1}!}{(p^i!)^p} x_{p^{i+1}}.$$

By Lemma 3.2 the coefficient of $x_{p^{i+1}}$ in the latter expression is divisible by p to the first power. Therefore, up to multiplying each x_{p^i} by a number which is a unit in $\mathbb{Z}_{p^{v(i)+1}}$, we get the relations

$$(x_{p^i})^p = p\, x_{p^{i+1}}, \quad i \geq 1.$$

Notice that these relations, together with $px_1 = 0$, imply by induction that $p^{v(i)+1} x_{p^i} = 0$. Therefore one has a well defined surjection between $\tilde{\Gamma}_p[x] := \mathbb{Z}[\xi_1, \xi_p, \xi_{p^2}, \xi_{p^3}, \ldots]/J_p$ and $\Gamma_p[x]$.

In $\tilde{\Gamma}_p[x]$ there exist a unique normal form, which derives from the unique p-adic expansion of a natural number n: each class is uniquely represented as a combination of monomials $\prod_{j=0}^{k} (\xi_{p^j})^{n_j}$. Such monomials map to (3.2), an invertible multiple of x_n, as we have seen before.

This concludes the proof of the Proposition 3.1. □

Remark 3.3. Note that if $\Gamma[x]$ is graded with $\deg x = k$, then also $\mathbb{Z}[\xi_1, \xi_p, \xi_{p^2}, \ldots]/J_p$ is graded, with $\deg \xi_{p^i} = kp^i$.

Proposition 3.4. *Let $\mathbb{Z}_{(p)}$ be the local ring obtained by inverting numbers prime to p and let $\Gamma_{\mathbb{Z}_{(p)}}[x]$ be the divided polynomial algebra over $\mathbb{Z}_{(p)}$. One has an isomorphism:*

$$\Gamma_p[x] \cong \Gamma_{\mathbb{Z}_{(p)}}[x]/(px).$$

Proof. The ideal (px) is graded (here we also set $\deg(x) = 1$) with degree i component given by the ideal

$$(pi) = (p^{v(i)+1}) \subset \mathbb{Z}_{(p)}.$$

This statement follows directly from the natural isomorphism

$$\mathbb{Z}_{(p)}/(p^k) \cong \mathbb{Z}_{p^k}.$$

□

We can naturally define a divided polynomial ring in several variables as follows:

$$\Gamma[x, x', x'', \ldots] := \Gamma[x] \otimes_{\mathbb{Z}} \Gamma[x'] \otimes_{\mathbb{Z}} \Gamma[x''] \otimes_{\mathbb{Z}} \cdots$$

with the ring structure induced as subring of $\mathbb{Q}[x, x', x'', \ldots]$. In a similar way we have

$$\Gamma_p[x, x', x'', \ldots] := \Gamma_p[x] \otimes_{\mathbb{Z}} \Gamma_p[x'] \otimes_{\mathbb{Z}} \Gamma_p[x''] \otimes_{\mathbb{Z}} \cdots.$$

In the following we need only to consider the torsion part of Γ_p, which is that generated in degree greater than 0. So, we define the submodule

$$\Gamma_p^+[x, x', x'', \ldots] := \Gamma_p[x, x', x'', \ldots]_{\deg>0}.$$

For every prime p and for $k \geq 1$ define polynomials

$$\mathcal{P}_{p^k} := xy(x^{p^k-1} - y^{p^k-1}).$$

Set

$$\mathcal{P}_p := \mathcal{P}_{p^1}, \ \mathcal{Q}_p := \mathcal{P}_{p^2}/\mathcal{P}_p = \sum_{h=0}^{p} (x^{p-1})^{p-h}(y^{p-1})^h \ \in \mathbb{Z}[x, y].$$

Remark 3.5. There is a natural action of $SL_2(\mathbb{Z})$ on $(\mathbb{CP}^\infty)^2$ which induces on the cohomology ring

$$H^*((\mathbb{CP}^\infty)^2; \mathbb{Z}) = \mathbb{Z}[x, y]$$

the same action defined in Section 2 (see [11]). For coherence with this geometrical action, from now on we fix on $\mathbb{Z}[x,y]$ the grading $\deg x = \deg y = 2$. Then we have

$$\deg \mathcal{P}_p = 2(p+1), \deg \mathcal{Q}_p = 2p(p-1).$$

Remark 3.6. Let $\mathrm{Diff}_+ T^2$ the group of all orientation preserving diffeomorphisms of the 2-dimensional torus T^2 and let $\mathrm{Diff}_0 T^2$ be the connected component of the identity. As showed in [9] the inclusion $T^2 \subset \mathrm{Diff}_0 T^2$ is an homotopy equivalence. We recall from [11] the exact sequence of groups

$$1 \to \mathrm{Diff}_0 T^2 \to \mathrm{Diff}_+ T^2 \to SL_2(\mathbb{Z}) \to 1$$

that induces a Serre spectral sequence

$$E_2^{i,j} = H^i(SL_2(\mathbb{Z}); H^j((\mathbb{CP}^\infty)^2; \mathbb{Z}) \Rightarrow H^{i+j}(B\mathrm{Diff}_+ T^2; \mathbb{Z}).$$

The spectral sequence above collapses with coefficients $\mathbb{Z}[1/6]$. The total space $B\mathrm{Diff}_+ T^2$ is homotopy equivalent to the Borel construction

$$E \times_{SL_2(\mathbb{Z})} (\mathbb{CP}^\infty)^2$$

where E is a contractible space with free $SL_2(\mathbb{Z})$-action.

Let G be a subgroup of $SL_2(\mathbb{Z})$. For any prime p, we define the Poincaré series

$$P^i_{G,p}(t) := \sum_{n=0}^{\infty} \dim_{\mathbb{F}_p}(H^i(G; M_n) \otimes \mathbb{F}_p)t^n \tag{3.3}$$

and for rational coefficients

$$P^i_{G,0}(t) := \sum_{n=0}^{\infty} \dim_{\mathbb{Q}} H^i(G; M_n \otimes \mathbb{Q})t^n. \tag{3.4}$$

We obtain:

Theorem 3.7. *Let $M = \oplus_{n\geq 0} M_n$ be the $SL_2(\mathbb{Z})$-module defined in the previous section. Then*

1. $H^0(B_3; M) = H^0(SL_2(\mathbb{Z}), M) = \mathbb{Z}$ *concentrated in degree* $n = 0$;
2. $H^1(B_3; M)_{(p)} = H^1(SL_2(\mathbb{Z}); M)_{(p)} = \Gamma^+_p[\mathcal{P}_p, \mathcal{Q}_p]$;
3. $H^1(SL_2(\mathbb{Z}); M_0) = 0$; $H^1(B_3; M_0) = \mathbb{Z}$;
4. $FH^1(B_3; M_n) = FH^1(SL_2(\mathbb{Z}); M_n) = \mathbb{Z}^{f_n}$ *for* $n > 0$

where the rank f_n is given by the Poincaré series

$$P^1_{B_3,0}(t) = \sum_{n=0}^{\infty} f_n t^n = \frac{t^4(1 + t^4 - t^{12} + t^{16})}{(1 - t^8)(1 - t^{12})}. \tag{3.5}$$

Theorem 3.8. *For $i > 1$ the cohomology $H^i(SL_2(\mathbb{Z}), M)$ is 2-periodic. The free part is zero and there is no p-torsion for $p \neq 2, 3$. There is no 2^i-torsion for $i > 2$ and no 3^i-torsion for $i > 1$.*

There is a rank-1 module of 4-torsion in $H^{2n}(SL_2(\mathbb{Z}), M_{8m})$, all the others modules have no 4-torsion.

Finally the 2 and 3-torsion components are determined by the following Poincaré polynomials:

$$P^2_{SL_2(\mathbb{Z}),2}(t) = \frac{1 - t^4 + 2t^6 - t^8 + t^{12}}{(1 - t^2)(1 + t^6)(1 - t^8)} \tag{3.6}$$

$$P^3_{SL_2(\mathbb{Z}),2}(t) = \frac{t^4(2 - t^2 + t^4 + t^6 - t^8)}{(1 - t^2)(1 + t^6)(1 - t^8)} \tag{3.7}$$

$$P^2_{SL_2(\mathbb{Z}),3}(t) = \frac{1}{1 - t^{12}} \tag{3.8}$$

$$P^3_{SL_2(\mathbb{Z}),3}(t) = \frac{t^8}{1 - t^{12}}. \tag{3.9}$$

We recall (Corollary 2.3) that for any $SL_2(\mathbb{Z})$-module N the free part and part of p-torsion in $H^1(SL_2(\mathbb{Z}); N)$ and $H^2(B_3; N)$ are the same, with $p \neq 2$ or 3. We obtain the following theorems.

Theorem 3.9. *The following equalities hold:*

1. $H^2(B_3; M)_{(p)} = H^1(B_3; M)_{(p)} = \Gamma_p^+[\mathcal{P}_p, \mathcal{Q}_p]$ *for* $p \neq 2, 3;$
2. $FH^2(B_3; M_n) = FH^1(B_3; M_n) = \mathbb{Z}^{f_n}$

where the ranks f_n have been defined in Theorem 3.7.

The last isomorphism of part 1 follows from Theorem 4.4.

For the 2 and 3-torsion components, we find that the second cohomology group of B_3 differs from the first cohomology group as follows. We consider the $\mathbb{Z}$-module structure, and notice that $\Gamma_p^+[\mathcal{Q}_p]$ and $\mathcal{P}_p \cdot \Gamma_p^+[\mathcal{Q}_p]$ are direct summand submodules of $\Gamma_p^+[\mathcal{P}_p, \mathcal{Q}_p]$, for all p. In the second cohomology group we have modifications of these submodules.

Theorem 3.10. *For the 2 and 3-torsion components of the module $H^2(B_3; M)$ one has the following expressions:*

a) $H^2(B_3; M)_{(2)} = (\Gamma_2^+[\mathcal{P}_2, \mathcal{Q}_2] \oplus \mathbb{Z}[\overline{\mathcal{Q}}_2^2])/\sim$
Here $\overline{\mathcal{Q}}_2$ is a new variable of the same degree (= 4) as $\mathcal{Q}_2$; the quotient module is defined by the relations $\frac{\mathcal{Q}_2^n}{n!} \sim 2\overline{\mathcal{Q}}_2^n$ for n even and $\frac{\mathcal{Q}_2^n}{n!} \sim 0$ for n odd.

b) $H^2(B_3; M)_{(3)} = (\Gamma_3^+[\mathcal{P}_3, \mathcal{Q}_3] \oplus \mathbb{Z}[\overline{\mathcal{Q}}_3])/\sim$
Here $\overline{\mathcal{Q}}_3$ is a new variable of the same degree (= 12) as $\mathcal{Q}_3$; the quotient module is defined by the relations $\frac{\mathcal{Q}_3^n}{n!} \sim 3\overline{\mathcal{Q}}_3^n$ and $\mathcal{P}_3\frac{\mathcal{Q}_3^n}{n!} \sim 0$.

Remark 3.11. For $p = 2$ one has to kill all the submodules generated by elements of the form $\frac{\mathcal{Q}_2^n}{n!}$ for odd n. All these submodules are isomorphic to $\mathbb{Z}/2$. For n even the submodule generated by $\frac{\mathcal{Q}_2^n}{n!}$, that is isomorphic to $\mathbb{Z}/2^{m+1}$ where m is the greatest power of 2 that divides n, must be replaced by a submodule isomorphic to $\mathbb{Z}/2^{m+2}$.

For $p = 3$ one has to kill all the submodules generated by elements of the form $\mathcal{P}_3\frac{\mathcal{Q}_3^n}{n!}$, that are isomorphic to $\mathbb{Z}/3$. Moreover for all n the submodule generated by $\frac{\mathcal{Q}_3^n}{n!}$, that is isomorphic to $\mathbb{Z}/3^{m+1}$ where m is the greatest power of 3 that divides n, must be replaced by a submodule isomorphic to $\mathbb{Z}/3^{m+2}$.

4 Methods

A classical result ([8]) characterizes the polynomials in $\mathbb{F}_p[x, y]$ which are invariant under the action of $GL_n(\mathbb{F}_p)$ (or of $SL_n(\mathbb{F}_p)$). We state only the case $n = 2$, which is what we need here.

Theorem 4.1 ([8], [21]). *For p a prime number, the algebra of $SL_2(\mathbb{F}_p)$ -invariants in $\mathbb{F}_p[x, y]$ is the polynomial algebra*

$$\mathbb{F}_p[\mathcal{P}_p, \mathcal{Q}_p],$$

where $\mathcal{P}_p$, $\mathcal{Q}_p$ are the polynomials defined in part 3.

Of course, the algebra $\mathbb{F}_p[\mathcal{P}_p, \mathcal{Q}_p]$ is graded. We keep here $\deg x = \deg y = 2$, so the degree is that induced by $\deg(\mathcal{P}_p) = 2(p+1)$, and $\deg(\mathcal{Q}_p) = 2p(p-1)$.

As a consequence of Theorem 4.1 and other results we have:

Corollary 4.2. *A polynomial $P \in \mathbb{Z}_{p^r}[x, y]$ is invariant under the action of $SL_2(\mathbb{Z})$ on $\mathbb{Z}_{p^r}[x, y]$ induced by the action on M_n, if and only if*

$$P = \sum_{i=0}^{r-1} p^i \; F_i$$

where F_i is a polynomial in the p^{r-i-1}-th powers of the elements $\mathcal{P}_p$, $\mathcal{Q}_p$.

Hence it is easy to see that

Proposition 4.3. *If $n > 0$, there are no non-trivial invariants in M_n under the action of $SL_2(\mathbb{Z})$.*

Thus, by the Universal Coefficient Theorem we can prove

Theorem 4.4. *For all prime numbers p and all $n > 0$, the p-torsion component*

$$H^1(SL_2(\mathbb{Z}); M_n)_{(p)} \cong H^1(B_3; M_n)_{(p)}$$

is described as follows: each monomial

$$\mathcal{P}_p^k \mathcal{Q}_p^h, \quad 2k(p+1) + 2hp(p-1) = n$$

generates a $\mathbb{Z}_{p^{m+1}}$-summand, where p^m is the largest power of p which divides $gcd(k, h)$.

We then need some computations about the invariant of the subgroups $\mathbb{Z}_2, \mathbb{Z}_4, \mathbb{Z}_6 \subset SL_2(\mathbb{Z})$, acting on $\mathbb{Z}[x, y]$ and their Poincaré series. Almost all the computations are straightforward.

Proposition 4.5. *The module of invariants $H^0(\mathbb{Z}_2; M)$ is isomorphic to the polynomial ring*

$$\mathbb{Z}[a_2, b_2, c_2]/(a_2^2 b_2^2 = c_2^2)$$

under the correspondence $a_2 = x^2$, $b_2 = y^2$, $c_2 = xy$.

The module $M \otimes \mathbb{Z}_2$ is $\mathbb{Z}_2$-invariant.

Proposition 4.6. *The Poincaré series for $H^i(\mathbb{Z}_2, M)$ are given by*

$$P^0_{\mathbb{Z}_2,p}(t) = \frac{(1+t^4)}{(1-t^4)^2}$$

$$P^{2i}_{\mathbb{Z}_2,2}(t) = \frac{(1+t^4)}{(1-t^4)^2}$$

$$P^{2i+1}_{\mathbb{Z}_2,2}(t) = \frac{2t^2}{(1-t^4)^2}$$

Proposition 4.7. *The module of invariants $H^0(\mathbb{Z}_4, M)$ is isomorphic to the polynomial ring*

$$\mathbb{Z}[d_2, e_4, f_4]/(f_4^2 = (d_2^2 - 4e_4)e_4) \tag{4.1}$$

under the correspondence $d_2 = x^2 + y^2$, $e_4 = x^2y^2$, $f_4 = x^3y - xy^3$.

The module of mod 2-*invariants $H^0(\mathbb{Z}_4, M \otimes \mathbb{Z}_2)$ is isomorphic to the polynomial ring*

$$\mathbb{Z}[\sigma_1, \sigma_2] \tag{4.2}$$

under the correspondence $\sigma_1 = x + y$, $\sigma_2 = xy$.

Proposition 4.8. *With respect to the grading of M the Poincaré series for $H^i(\mathbb{Z}_4, M)$ are given by*

$$P^{2i}_{\mathbb{Z}_4,p}(t) = \frac{1+t^8}{(1-t^4)(1-t^8)} \text{ for } p = 2 \text{ or } i = 0$$

$$P^{2i+1}_{\mathbb{Z}_4,2}(t) = \frac{t^2+t^4+t^6-t^8}{(1-t^4)(1-t^8)}$$

Proposition 4.9. *The module of invariants $H^0(\mathbb{Z}_6, M)$ is isomorphic to the polynomial ring*

$$\mathbb{Z}[p_2, q_6, r_6]/(r_6^2 = q_6(p_2^3 - 13q_6 - 5r_6)) \tag{4.3}$$

via the correspondence $p_2 = x^2 + xy + y^2$, $q_6 = x^2y^2(x+y)^2$, $r_6 = x^5y - 5x^3y^3 - 5x^2y^4 - xy^5$.

The module of mod 2*-invariants* $H^0(\mathbb{Z}_6, M \otimes \mathbb{Z}_2)$ *is isomorphic to the polynomial ring*

$$\mathbb{Z}_2[s_2, t_3, u_3]/(u_3^2 = s_2^3 + t_3^2 + t_3 u_3) \tag{4.4}$$

with the correspondence $s_2 = x^2 + xy + y^2$, $t_3 = xy(x+y)$, $u_3 = x^3 + x^2 y + y^3$.

The module of mod 3*-invariants* $H^0(\mathbb{Z}_6, M \otimes \mathbb{Z}_3)$ *is isomorphic to the polynomial ring*

$$\mathbb{Z}_3[v_2, w_4, z_6]/(w_4^2 = v_2 z_6) \tag{4.5}$$

with the correspondence $v_2 = (x-y)^2$, $w_4 = xy(x+y)(x-y)$, $z_6 = x^2 y^2 (x+y)^2$.

Proposition 4.10. *With respect to the grading of M the Poincaré series for $H^i(\mathbb{Z}_6, M)$ are given by*

$$\begin{aligned}
P^{2i}_{\mathbb{Z}_6,p}(t) &= \frac{1+t^{12}}{(1-t^4)(1-t^{12})} \quad \text{for } p = 2 \text{ or } i = 0, \text{ any } p \\
P^{2i}_{\mathbb{Z}_6,3}(t) &= \frac{1}{1-t^{12}} \quad \text{for } i > 0 \\
P^{2i+1}_{\mathbb{Z}_6,2}(t) &= \frac{2t^6}{(1-t^4)(1-t^{12})} \\
P^{2i+1}_{\mathbb{Z}_6,3}(t) &= \frac{t^8}{1-t^{12}}
\end{aligned}$$

We use the previous results to compute the modules $H^i(SL_2(\mathbb{Z}); M \otimes \mathbb{Z}_p)$ and the module of p-torsion $H^i(SL_2(\mathbb{Z}); M)_{(p)}$ for any prime p.

In dimension 0 we have that $H^0(SL_2(\mathbb{Z}); M)$ is isomorphic to $\mathbb{Z}$ concentrated in degree 0. The module $H^0(SL_2(\mathbb{Z}); M \otimes \mathbb{Z}_p)$ has already been described (Theorem 4.1), as well as $H^1(SL_2(\mathbb{Z}); M)_{(p)}$ (Theorem 4.4).

We can understand the p-torsion in $H^1(SL_2(\mathbb{Z}); M)$ by means of the results cited in Section 4.1 for Dickson invariants. In fact the Universal Coefficients Theorem gives us the following isomorphism of $\mathbb{Z}$ modules:

$$H^1(SL_2(\mathbb{Z}); M)_{(p)} = \Gamma_p^+[\mathcal{P}_p, \mathcal{Q}_p]. \tag{4.6}$$

The Poincaré series for the previous module is:

$$P^1_{SL_2(\mathbb{Z}),p}(t) = \frac{t^{2(p+1)} + t^{2p(p-1)} - t^{2(p^2+1)}}{(1-t^{2(p+1)})(1-t^{2p(p-1)})}.$$

Since $H^i(\mathbb{Z}_4; M)_{(p)} = H^i(\mathbb{Z}_2; M)_{(p)} = 0$ for a prime $p \neq 2$ and $i \geq 1$ we have

Lemma 4.11. *For $i \geq 2$, $H^i(SL_2(\mathbb{Z}; M \otimes \mathbb{Z}_3) = H^i(\mathbb{Z}_6; M \otimes \mathbb{Z}_3)$ and we have the isomorphism*

$$H^i(SL_2(\mathbb{Z}); M)_{(3)} = H^i(\mathbb{Z}_6; M)_{(3)} \qquad i \geq 2. \tag{4.7}$$

For any prime $p \neq 2, 3$ and for $i \geq 2$ we have $H^i(SL_2(\mathbb{Z}; M \otimes \mathbb{Z}_p) = 0$ and $H^i(SL_2(\mathbb{Z}); M)_{(p)} = 0$.

For $p = 2$ we need to compute in detail the maps appearing in the Mayer-Vietoris long exact sequence and we get a detailed desctiption of the module $H^j(SL_2(\mathbb{Z}); M_i)_{(2)}$ (that we omit here).

Finally, in order to compute the cohomology of B_3 we use the Serre spectral sequence already described in Section 2 for the extension

$$0 \to \mathbb{Z} \to B_3 \to SL_2(\mathbb{Z}) \to 1.$$

We recall (equation (2.3)) that the E_2-page is represented in the following diagram:

$$\begin{array}{|llll}
H^0(SL_2(\mathbb{Z}), M) & H^1(SL_2(\mathbb{Z}), M) & H^2(SL_2(\mathbb{Z}), M) & \cdots \\
& \searrow^{d_2} & \searrow^{d_2} & \\
H^0(SL_2(\mathbb{Z}), M) & H^1(SL_2(\mathbb{Z}), M) & H^2(SL_2(\mathbb{Z}), M) & \cdots \\
\hline
\end{array}$$

where all the other rows are zero. Recall that $H^i(SL_2(\mathbb{Z}), M)$ is 2-periodic and with no free part for $i \geq 2$.

A detailed analysis of this spectral sequence and of the analogous with coefficients in the module $M \otimes \mathbb{Z}_{p^k}$ for $p = 2$ and $p = 3$ gives the result stated in Theorem 3.10.

Since the cohomology group $H^i(SL_2(\mathbb{Z}), M)$ has only 2 and 3-torsion for $i \geq 2$, the same argument implies that for the p-torsion with $p > 3$ the spectral sequence collapses at E_2. Hence for $p > 3$ we have the isomorphism

$$H^1(B_3; M)_{(p)} = H^2(B_3; M)_{(p)} = H^1(SL_2(\mathbb{Z}); M)_{(p)} = \Gamma_p^+[\mathcal{P}_p, \mathcal{Q}_p]$$

where $\deg \mathcal{P}_p = 2(p+1)$ and $\deg \mathcal{Q}_p = 2p(p-1)$ as in Section 3.

5 Principal congruence subgroups

Recall that the principal congruence subgroups of level N in $SL_2(\mathbb{Z})$ is given by the kernel of the mod-N reduction map

$$SL_2(\mathbb{Z}) \to SL_2(\mathbb{Z}/NZ)$$

denoted $\Gamma_2(N)$. These kernels are either finitely generated free groups or a product of a finitely generated free group with $\mathbb{Z}/2\mathbb{Z}$. Analogous computations apply in many of these cases, and will be addressed in the more detailed sequel to this paper.

6 Topological comparisons and speculations

The purpose of this section is to describe a topological space with the features that the reduced integral cohomology is precisely the torsion in $H^*(SL_2(\mathbb{Z}); M \otimes \mathbb{Z}[1/6])$.

In particular, there are spaces $T_p(2n+1)$ which are total spaces of p-local fibrations, $p > 3$, of the form

$$S^{2n-1} \to T_p(2n+1) \to \Omega(S^{2n+1})$$

for which the integral cohomology is recalled next where it is assumed that $p > 3$ throughout this section ([1], [12]).

The integral cohomology ring of $\Omega(S^{2n+1})$ is the divided power algebra $\Gamma[x(2n)]$ where the degree of $x(2n)$ is $2n$. The mod-p homology ring (as a coalgebra) of $T_p(2n+1)$ is

$$E[v] \otimes F_p[w]$$

where the exterior algebra $E[v]$ is generated by an element v of degree $2n-1$, and the mod-p polynomial ring $F_p[w]$ is generated by an element w of degree $2n$. Furthermore, the r-th higher order Bockstein defined by b_r is given as follows:

$$b_r(w^{p^s}) = \begin{cases} 0 & \text{if} \quad r < s \\ vw^{-1+p^s} & \text{if} \quad r = s. \end{cases} \tag{6.1}$$

This information specifies the integral homology groups of $T_p(2n+1)$ (and thus the integral cohomology groups) listed in the following theorem. Recall the algebra $\Gamma_p[x]$ of Proposition 3.1 In addition fix a positive even $2n$, and consider the graded ring $\Gamma_p[x(2n)]$ where $x(2n)$ has degree $2n > 0$.

Theorem 6.1. *Assume that p is prime with $p > 3$.*

1. *The reduced integral homology groups of $T_p(2n+1)$ are given as follows:*

$$\overline{H}_i(T_p(2n+1)) = \begin{cases} \mathbb{Z}/p^r\mathbb{Z} & \text{if } i = 2np^{r-1}k - 1 \\ & \text{where } k \text{ is not divisible by } p, \\ \{0\} & \text{otherwise} \end{cases} \tag{6.2}$$

2. *The additive structure for the cohomology ring of* $T_p(2n+1)$ *with* $\mathbb{Z}_{(p)}$ *coefficients is given as follows:*

$$\overline{H}^i(T_p(2n+1)) = \begin{cases} \mathbb{Z}/p^r\mathbb{Z} & \text{if } i = 2np^{r-1}k \\ & \text{where } k \text{ is not divisible by } p, \\ \{0\} & \text{otherwise} \end{cases} \tag{6.3}$$

3. *The cohomology ring of* $T_p(2n+1)$ *with* $\mathbb{Z}_{(p)}$ *coefficients is isomorphic to* $\Gamma_p[x(2n)]$ *in degrees strictly greater than* 0.

The next step is to compare the cohomology of the spaces $T_p(2n+1)$ to the earlier computations.

Theorem 6.2. *Assume that* p *is prime with* $p > 3$. *The* p*-torsion in the cohomology of*

$$EB_3 \times_{B_3} (\mathbb{CP}^\infty)^2$$

is the p*-torsion of the reduced cohomology of the following space*

$$\Sigma^2(T_p(2p+3) \times T_p(2p^2-2p+1)) \vee \Sigma(T_p(2p+3) \vee T_p(2p^2-2p+1)).$$

References

[1] D. ANICK, *Differential algebras in topology*, Research Notes in Mathematics, Vol. 3, A K Peters Ltd., Wellesley, MA, 1993. MR1213682 (94h:55020)

[2] E. BRIESKORN, *Sur les groupes de tresses [d'après V. I. Arnol'd]*, Séminaire Bourbaki, 24ème année (1971/1972), Exp. No. 401, Springer, Berlin, 1973, pp. 21–44. Lecture Notes in Math., Vol. 317. MR0422674 (54 #10660)

[3] K. S. BROWN, *Cohomology of groups*, Graduate Texts in Mathematics, vol. 87, Springer-Verlag, New York, 1994, Corrected reprint of the 1982 original. MR1324339 (96a:20072)

[4] H. CARTAN, *Puissances dividées*, Séminaire Henri Cartan; 7e année: 1954/55. Algébre d'Eilenberg-Maclane et homotopie, Exposé no. 7, Secrétariat mathématique, Paris, 1956, p. 11.

[5] F. R. COHEN, *On genus one mapping class groups, function spaces, and modular forms*, Topology, geometry, and algebra: interactions and new directions (Stanford, CA, 1999), Contemp. Math., vol. 279, Amer. Math. Soc., Providence, RI, 2001, pp. 103–128. MR1850743 (2002f:55037)

[6] C. DE CONCINI and C. PROCESI, *Wonderful models of subspace arrangements*, Selecta Math. (N.S.) **1** (1995), no. 3, 459–494. MR1366622 (97k:14013)

[7] C. DE CONCINI and M. SALVETTI, *Cohomology of Artin groups: Addendum: "The homotopy type of Artin groups" [Math. Res. Lett.* **1** *(1994), no. 5, 565–577; MR1295551 (95j:52026)] by Salvetti*, Math. Res. Lett. **3** (1996), no. 2, 293–297. MR1386847 (97b:52015)

[8] L. EU. DICKSON, *A fundamental system of invariants of the general modular linear group with a solution of the form problem*, Trans. Amer. Math. Soc. **12** (1911), no. 1, 75–98. MR1500882

[9] C. J. EARLE and J. EELLS, *A fibre bundle description of Teichmüller theory*, J. Differential Geometry **3** (1969), 19–43. MR0276999 (43 #2737a)

[10] M. EICHLER, *Eine Verallgemeinerung der Abelschen Integrale*, Math. Z. **67** (1957), 267–298. MR0089928 (19,740a)

[11] M. FURUSAWA, M. TEZUKA and N. YAGITA, *On the cohomology of classifying spaces of torus bundles and automorphic forms*, J. London Math. Soc. (2) **37** (1988), no. 3, 520–534. MR939127 (89f:57060)

[12] B. GRAY and S. THERIAULT, *An elementary construction of Anick's fibration*, Geom. Topol. **14** (2010), no. 1, 243–275. MR2578305 (2011a:55013)

[13] A. HATCHER, *Algebraic topology*, Cambridge University Press, Cambridge, 2002. MR1867354 (2002k:55001)

[14] E. LOOIJENGA, *The complement of the bifurcation variety of a simple singularity*, Invent. Math. **23** (1974), 105–116. MR0422675 (54 #10661)

[15] E. LOOIJENGA, *Stable cohomology of the mapping class group with symplectic coefficients and the universal Abel-Jacobi map*, J. Algebraic Geom. **5** (1996), no. 1, 135–150. MR1358038 (97g:14026)

[16] J. MILNOR, *Introduction to algebraic K-theory*, Princeton University Press, Princeton, N.J., 1971, Annals of Mathematics Studies, No. 72. MR0349811 (50 #2304)

[17] W. MAGNUS, A. KARRASS and D. SOLITAR, *Combinatorial group theory: Presentations of groups in terms of generators and relations*, Interscience Publishers [John Wiley & Sons, Inc.], New York-London-Sydney, 1966. MR0207802 (34 #7617)

[18] M. SALVETTI, *The homotopy type of Artin groups*, Math. Res. Lett. **1** (1994), no. 5, 565–577. MR1295551 (95j:52026)

[19] J.-P. SERRE, *Trees*, Springer Monographs in Mathematics, Springer-Verlag, Berlin, 2003, Translated from the French original

by John Stillwell, Corrected 2nd printing of the 1980 English translation. MR1954121 (2003m:20032)

[20] G. SHIMURA, *Introduction to the arithmetic theory of automorphic functions*, Publications of the Mathematical Society of Japan, No. 11. Iwanami Shoten, Publishers, Tokyo, 1971, Kanô Memorial Lectures, No. 1. MR0314766 (47 #3318)

[21] R. STEINBERG, *On Dickson's theorem on invariants*, J. Fac. Sci. Univ. Tokyo Sect. IA Math. **34** (1987), no. 3, 699–707. MR927606 (89c:11177)

[22] B. WAJNRYB, *On the monodromy group of plane curve singularities*, Math. Ann. **246** (1979/80), no. 2, 141–154. MR564684 (81d:14016)

Pure braid groups are not residually free

Daniel C. Cohen[†], Michael Falk and Richard Randell

Abstract. We show that the Artin pure braid group P_n is not residually free for $n \geq 4$. Our results also show that the corank of P_n is equal to 2 for $n \geq 3$.

1 Introduction

A group G is residually free if for every $x \neq 1$ in G, there is a homomorphism f from G to a free group F so that $f(x) \neq 1$ in F. Equivalently, G embeds in a product of free groups (of finite rank). Examples of residually free groups include the fundamental groups of orientable surfaces. In this note, we show that the Artin pure braid group, the kernel $P_n = \ker(B_n \to \Sigma_n)$ of the natural map from the braid group to the symmetric group, is not residually free for $n \geq 4$. (It is easy to see that the pure braid groups P_2 and P_3 are residually free.) We also classify all epimorphisms from the pure braid group to free groups, and determine the corank of the pure braid group. For $n \geq 5$, the fact that P_n is not residually free was established independently by L. Paris (unpublished), see Remark 5.4.

For $n \geq 3$, the braid groups themselves are not residually free. Indeed, the only nontrivial two-generator residually free groups are $\mathbb{Z}$, $\mathbb{Z}^2$, and F_2, the nonabelian free group of rank two, see Wilton [26]. Since B_n can be generated by two elements for $n \geq 3$, it is not residually free. (For $n = 2$, $B_2 = \mathbb{Z}$ is residually free.)

[†]Partially supported by Louisiana Board of Regents grant NSF(2010)-PFUND-171

Portions of this project were completed during during the intensive research period "Configuration Spaces: Geometry, Combinatorics and Topology," May-June, 2010, at the Centro di Ricerca Matematica Ennio De Giorgi in Pisa. The authors thank the institute and the organizers of the session for financial support and hospitality, and for the excellent working environment.

If G is a group which is not residually free, then any group $\tilde{G}$ with a subgroup isomorphic to G cannot be residually free. Consequently, the (pure) braid groups are "poison" groups for residual freeness. In particular, a group with a subgroup isomorphic to the 4-strand pure braid group P_4 or the 3-strand braid group B_3 is not residually free. Since $P_4 < P_n$ for every $n \geq 4$, our main result follows from the special case $n = 4$. Moreover, the same observation enables us to show that a number of other groups are not residually free. These include (pure) braid groups of orientable surfaces, the (pure) braid groups associated to the full monomial groups, and a number of irreducible (pure) Artin groups of finite type.

This research was motivated by our work in [8, Section 3], which implies the residual freeness of fundamental groups of the complements of certain complex hyperplane arrangements. In particular, the proof of the assertion in the last sentence of [8, Example 3.25] gives the last step in the proof of Theorem 5.2 below.

ACKNOWLEDGEMENTS. The authors are grateful to Luis Paris and Ivan Marin for helpful conversations, and to the referee for pertinent observations.

2 Automorphisms of the pure braid group

Let B_n be the Artin braid group, with generators $\sigma_1, \ldots, \sigma_{n-1}$ and relations $\sigma_i\sigma_{i+1}\sigma_i = \sigma_{i+1}\sigma_i\sigma_{i+1}$ for $1 \leq i \leq n-2$, and $\sigma_j\sigma_i = \sigma_i\sigma_j$ for $|j-i| \geq 2$. The Artin pure braid group P_n has generators

$$\omega_{i,j} = \sigma_{j-1} \cdots \sigma_{i+1}\sigma_i^2\sigma_{i+1}^{-1} \cdots \sigma_{j-1}^{-1} = \sigma_i^{-1} \cdots \sigma_{j-2}^{-1}\sigma_{j-1}^2\sigma_{j-2} \cdots \sigma_i,$$

and relations

$$A_{r,s}^{-1}\omega_{i,j}A_{r,s} = \begin{cases} \omega_{i,j} & \text{if } i < r < s < j, \\ \omega_{i,j} & \text{if } r < s < i < j, \\ A_{r,j}\omega_{i,j}A_{r,j}^{-1} & \text{if } r < s = i < j, \\ A_{r,j}A_{s,j}\omega_{i,j}A_{s,j}^{-1}A_{r,j}^{-1} & \text{if } r = i < s < j, \\ [A_{r,j}, A_{s,j}]\omega_{i,j}[A_{r,j}, A_{s,j}]^{-1} & \text{if } r < i < s < j, \end{cases} \tag{2.1}$$

where $[u, v] = uvu^{-1}v^{-1}$ denotes the commutator. See, for instance, Birman [4] as a general reference on braid groups. It is well known that the pure braid group admits a direct product decomposition

$$P_n = Z \times P_n/Z, \tag{2.2}$$

where $Z = Z(P_n) \cong \mathbb{Z}$ is the center of P_n, generated by

$$Z_n = (A_{1,2})(A_{1,3}A_{2,3}) \cdots (A_{1,n} \cdots A_{n-1,n}). \tag{2.3}$$

Note that $P_3 = Z(P_3) \times P_3/Z(P_3) \cong \mathbb{Z} \times F_2$, where F_2 is the free group on two generators. For any $n \geq 3$, by (2.2), there is a split, short exact sequence

$$1 \to \mathrm{tv}(P_n) \to \mathrm{Aut}(P_n) \leftrightarrows \mathrm{Aut}(P_n/Z) \to 1, \tag{2.4}$$

where the subgroup $\mathrm{tv}(P_n)$ of $\mathrm{Aut}(P_n)$ consists of those automorphisms which become trivial upon passing to the quotient P_n/Z.

For a group G with infinite cyclic center $Z = Z(G) = \langle z \rangle$, a transvection is an endomorphism of G of the form $x \mapsto xz^{t(x)}$, where $t\colon G \to \mathbb{Z}$ is a homomorphism, see Charney and Crisp [5]. Such a map is an automorphism if and only if its restriction to Z is surjective, which is the case if and only if $z \mapsto z$ or $z \mapsto z^{-1}$, that is, $t(z) = 0$ or $t(z) = -2$. For the pure braid group P_n, the transvection subgroup $\mathrm{tv}(P_n)$ of $\mathrm{Aut}(P_n)$ consists of automorphisms of the form $\omega_{i,j} \mapsto \omega_{i,j} Z_n^{t_{i,j}}$, where $t_{i,j} \in \mathbb{Z}$ and $\sum t_{i,j}$ is either equal to 0 or -2. In the former case, $Z_n \mapsto Z_n$, while $Z_n \mapsto Z_n^{-1}$ in the latter. This yields a surjection $\mathrm{tv}(P_n) \to \mathbb{Z}_2$, with kernel consisting of transvections for which $\sum t_{i,j} = 0$. Since P_n has $\binom{n}{2} = N + 1$ generators, this kernel is free abelian of rank N. The choice $t_{1,2} = -2$ and all other $t_{i,j} = 0$ gives a splitting $\mathbb{Z}_2 \to \mathrm{tv}(P_n)$. Thus, $\mathrm{tv}(P_n) \cong \mathbb{Z}^N \rtimes \mathbb{Z}_2$. Explicit generators of $\mathrm{tv}(P_n)$ are given below.

For $n \geq 4$, Bell and Margalit [3] show that the automorphism group of the pure braid group admits a semidirect product decomposition

$$\mathrm{Aut}(P_n) \cong (\mathbb{Z}^N \rtimes \mathbb{Z}_2) \rtimes \mathrm{Mod}(\mathbb{S}_{n+1}). \tag{2.5}$$

Here, $\mathbb{Z}^N \rtimes \mathbb{Z}_2 = \mathrm{tv}(P_n)$ is the transvection subgroup of $\mathrm{Aut}(P_n)$ described above, $\mathbb{S}_{n+1}$ denotes the sphere S^2 with $n+1$ punctures, and $\mathrm{Mod}(\mathbb{S}_{n+1})$ is the extended mapping class group of $\mathbb{S}_{n+1}$, the group of isotopy classes of all self-diffeomorphisms of $\mathbb{S}_{n+1}$. The semidirect product decomposition (2.5) is used in [7] to determine a finite presentation for $\mathrm{Aut}(P_n)$. From this work, it follows that $\mathrm{Aut}(P_n)$ is generated by automorphisms

$$\xi,\ \beta_k\ (1 \leq k \leq n),\ \psi,\ \phi_{p,q}\ (1 \leq p < q \leq n,\ \{p,q\} \neq \{1,2\}), \tag{2.6}$$

given explicitly by

$$\xi\colon \omega_{i,j} \mapsto (A_{i+1,j}\cdots A_{j-1,j})^{-1}\omega_{i,j}^{-1}(A_{i+1,j}\cdots A_{j-1,j}),$$

$$\beta_k\colon \omega_{i,j} \mapsto \begin{cases} A_{i-1,j} & \text{if } k=i-1,\\ A_{i,i+1}^{-1}A_{i+1,j}A_{i,i+1} & \text{if } k=i<j-1,\\ A_{i,j-1} & \text{if } k=j-1>i,\\ A_{j,j+1}^{-1}A_{i,j+1}A_{j,j+1} & \text{if } k=j,\\ \omega_{i,j} & \text{otherwise,}\end{cases} \quad \text{for } 1<k\leq n-1,$$

$$\beta_n\colon \omega_{i,j} \mapsto \begin{cases} \omega_{i,j} & \text{if } j\neq n,\\ (A_{i,n}A_{1,i}\cdots A_{i-1,i}A_{i,i+1}\cdots A_{i,n-1})^{-1} & \text{if } j=n,\end{cases}$$

$$\psi\colon \omega_{i,j} \mapsto \begin{cases} A_{1,2}Z_n^{-2} & \text{if } i=1 \text{ and } j=2,\\ \omega_{i,j} & \text{otherwise,}\end{cases}$$

$$\phi_{p,q}\colon \omega_{i,j} \mapsto \begin{cases} A_{1,2}Z_n & \text{if } i=1 \text{ and } j=2,\\ A_{p,q}Z_n^{-1} & \text{if } i=p \text{ and } j=q,\\ \omega_{i,j} & \text{otherwise.}\end{cases} \tag{2.7}$$

It is readily checked that these are all automorphisms of P_n. The automorphisms ψ and $\phi_{p,q}$ are transvections. For $k \leq n-1$, $\beta_k \in \mathrm{Aut}(P_n)$ arises from the conjugation action of B_n on P_n, $\beta_k(\omega_{i,j}) = \sigma_k^{-1}\omega_{i,j}\sigma_k$, see Dyer and Grossman [14].

Remark 2.1. The presentation of $\mathrm{Aut}(P_n)$ found in [7] is given in terms of the generating set ϵ, ω_k $(1 \leq k \leq n)$, ψ, $\phi_{p,q}$ $(1 \leq p < q \leq n,\ \{p,q\} \neq \{1,2\})$, where $\xi = \epsilon \circ \psi$, $\beta_2 = \omega_2 \circ \phi_{1,3}^{-1}$, $\beta_n = \omega_n \circ \psi \circ \phi_{1,n} \circ \phi_{2,n}$, and $\beta_k = \omega_k$ for $k \neq 2, n$. This presentation exhibits the semidirect product structure (2.5) of $\mathrm{Aut}(P_n)$.

3 Epimorphisms to free groups

We study surjective homomorphisms from the pure braid group P_n to the free group F_k on $k \geq 2$ generators. Since $P_2 = \mathbb{Z}$ is infinite cyclic, we assume that $n \geq 3$. We begin by exhibiting a number of specific such homomorphisms.

Let $F_2 = \langle x, y\rangle$ be the free group on two generators, and write $[n] = \{1, 2, \ldots, n\}$. If $I = \{i, j, k\} \subset [n]$ with $i < j < k$, define $f_I\colon P_n \to F_2$

by

$$f_I(A_{r,s}) = \begin{cases} x & \text{if } r = i \text{ and } s = j, \\ y & \text{if } r = i \text{ and } s = k, \\ y^{-1}x^{-1} & \text{if } r = j \text{ and } s = k, \\ 1 & \text{otherwise.} \end{cases} \tag{3.1}$$

If $I = \{i, j, k, l\} \subset [n]$ with $i < j < k < l$, define $f_I \colon P_n \to F_2$ by

$$f_I(A_{r,s}) = \begin{cases} x & \text{if } r = i \text{ and } s = j, \\ y & \text{if } r = i \text{ and } s = k, \\ y^{-1}x^{-1} & \text{if } r = j \text{ and } s = k, \\ y^{-1}x^{-1} & \text{if } r = i \text{ and } s = l, \\ xyx^{-1} & \text{if } r = j \text{ and } s = l, \\ x & \text{if } r = k \text{ and } s = l, \\ 1 & \text{otherwise.} \end{cases} \tag{3.2}$$

In either case (3.1) or (3.2), note that f_I is surjective by construction. It is readily checked that f_I is a homomorphism. We will show that these are, in an appropriate sense, the only epimorphisms from the pure braid group to a nonabelian free group.

Remark 3.1. The epimorphisms $f_I \colon P_n \to F_2$ are induced by maps of topological spaces. Let

$$F(\mathbb{C}, n) = \{(z_1, \dots, z_n) \in \mathbb{C}^n \mid z_i \neq z_j \text{ if } i \neq j\}$$

be the configuration space of n distinct ordered points in $\mathbb{C}$. It is well known that $P_n = \pi_1(F(\mathbb{C}, n))$ and that $F(\mathbb{C}, n)$ is a $K(P_n, 1)$-space.

For a subset I of $[n]$ of cardinality k, let $p_I \colon F(\mathbb{C}, n) \to F(\mathbb{C}, k)$ denote the projection which forgets all coordinates not indexed by I. The induced map on pure braid groups forgets the corresponding strands. Additionally, let $q_n \colon F(\mathbb{C}, n) \to F(\mathbb{C}, n)/\mathbb{C}^*$ denote the natural projection, where $\mathbb{C}^*$ acts by scalar multiplication. In particular, $q_3 \colon F(\mathbb{C}, 3) \to F(\mathbb{C}, 3)/\mathbb{C}^* \cong \mathbb{C} \times (\mathbb{C} \setminus \{\text{two points}\})$. Finally, define $g \colon F(\mathbb{C}, 4) \to F(\mathbb{C}, 3)$ by

$$g(z_1,z_2,z_3,z_4) = \big((z_1+z_2-z_3-z_4)^2, (z_1+z_3-z_2-z_4)^2, (z_1+z_4-z_2-z_3)^2\big).$$

One can check that if $|I| = 3$, then $f_I = (q_3 \circ p_I)_*$, while if $|I| = 4$, $f_I = (q_3 \circ g \circ p_I)_*$.

In the case $n = 4$, $F(\mathbb{C}, n)$ is diffeomorphic to $\mathbb{C} \times M$, where M is the complement of the Coxeter arrangement $\mathcal{A}$ of type D_3. With this identification, the mappings p_I and g correspond to the components of the mapping constructed in [8, Example 3.25]. The map g is the pencil associated with the non-local component of the first resonance variety of $H^*(M;\mathbb{C})$ as in [16, 19], see below.

One can also check that the homomorphism $g_*\colon P_4 \to P_3$ is the restriction to pure braid groups of the famous homomorphism $B_4 \to B_3$ of full braid groups, given by $\sigma_1 \mapsto \sigma_1, \sigma_2 \mapsto \sigma_2, \sigma_3 \mapsto \sigma_1$.

Let G and H be groups, and let f and g be (surjective) homomorphisms from G to H. Call f and g equivalent if there are automorphisms $\phi \in \mathrm{Aut}(G)$ and $\psi \in \mathrm{Aut}(H)$ so that $g \circ \phi = \psi \circ f$. If f and g are equivalent, we write $f \sim g$.

Proposition 3.2. *If I and J are subsets of $[n]$ of cardinalities 3 or 4, then the epimorphisms f_I and f_J from P_n to F_2 are equivalent.*

Proof. If $n = 3$, there is nothing to prove. So assume that $n \geq 4$.

Let $I = \{i_1, \dots, i_q\} \subset [n]$ with $q \geq 2$ and $i_1 < i_2 < \cdots < i_q$. Define $\alpha_I \in B_n$ by

$$\alpha_I = (\sigma_{i_1-1}\cdots\sigma_1)(\sigma_{i_2-1}\cdots\sigma_2)\cdots(\sigma_{i_q-1}\cdots\sigma_q),$$

where $\sigma_{i_k-1}\cdots\sigma_k = 1$ if $i_k = k$. Then $\alpha_I^{-1}A_{i_r,i_s}\alpha_I = A_{r,s}$ for $1 \leq r < s \leq q$. This can be seen by checking that, for instance, the geometric braids $\alpha_I A_{r,s}\alpha_I^{-1}$ and A_{i_r,i_s} are equivalent. Denote the automorphism $\omega_{i,j} \mapsto \alpha_I^{-1}\omega_{i,j}\alpha_I$ of P_n by the same symbol,

$$\alpha_I = (\beta_{i_1-1}\cdots\beta_1)(\beta_{i_2-1}\cdots\beta_2)\cdots(\beta_{i_q-1}\cdots\beta_q) \in \mathrm{Aut}(P_n).$$

Then, $\alpha_I(A_{i_r,i_s}) = A_{r,s}$ for $1 \leq r < s \leq q$.

If I has cardinality 3, then, by the above, we have $f_I = f_{[3]} \circ \alpha_I$, so $f_I \sim f_{[3]}$. Similarly, if $|I| = 4$, then $f_I = f_{[4]} \circ \alpha_I$ and $f_I \sim f_{[4]}$. Thus it suffices to show that $f_I \sim f_{[3]}$ for some I with $|I| = 4$. This can be established by checking that $f_I = f_{[3]} \circ \beta_n$ for $I = \{1, 2, 3, n\}$. □

Remark 3.3. The homomorphisms f_I also have a natural interpretation in terms of the moduli space $\mathfrak{M}_{0,n}$ of genus-zero curves with n marked points. By definition, $\mathfrak{M}_{0,n}$ is the quotient of the configuration space $F(S^2,n)$ of the Riemann sphere $S^2 = \mathbb{C}\cup\{\infty\}$ by the action of $\mathrm{PSL}(2,\mathbb{C})$. The map $h_n\colon F(\mathbb{C},n) \to \mathfrak{M}_{0,n+1}$ given by $h_n(z_1, \dots, z_n) = [(z_1, \dots, z_n, \infty)]$ induces a homeomorphism

$$F(\mathbb{C},n)/\,\mathrm{Aff}(\mathbb{C}) \to \mathfrak{M}_{0,n+1},$$

where $\mathrm{Aff}(\mathbb{C}) \cong \mathbb{C} \rtimes \mathbb{C}^*$ is the affine group, and a homotopy equivalence

$$\overline{h}_n \colon F(\mathbb{C}, n)/\mathbb{C}^* \to \mathfrak{M}_{0,n+1}.$$

For $1 \le i \le 5$, let $\delta_i : \mathfrak{M}_{0,5} \to \mathfrak{M}_{0,4}$ be defined by forgetting the i-th point. Then, in the notation of Remark 3.1, for $1 \le i \le 4$, $\delta_i \circ h_4 = \overline{h}_3 \circ q_3 \circ p_I$, where $I = [4] \setminus \{i\}$. Up to a linear change of coordinates in $F(\mathbb{C}, 3)$, $\delta_5 \circ h_4 = \overline{h}_3 \circ q_3 \circ g$. (See also Pereira [24, Example 3.1].) The maps $\delta_i : \mathfrak{M}_{0,5} \to \mathfrak{M}_{0,4}$, $1 \le i \le 5$, are clearly equivalent up to diffeomorphism of the source. Applying Remark 3.1, this gives an alternate proof of Proposition 3.2.

For our next result, we require some properties of the cohomology ring of the pure braid group. Let $A = \bigoplus_{k \ge 0} A^k$ be a graded algebra over a field $\Bbbk$ that is connected ($A^0 \cong \Bbbk$), graded-commutative ($b \cdot a = (-1)^{pq} a \cdot b$ for $a \in A^p$ and $b \in A^q$), and satisfies $\dim A^1 < \infty$. Since $a \cdot a = 0$ for each $a \in A^1$, multiplication by a defines a cochain complex (A, δ_a):

$$A^0 \xrightarrow{\delta_a} A^1 \xrightarrow{\delta_a} A^2 \xrightarrow{\delta_a} \cdots\cdots \xrightarrow{\delta_a} A^\ell,$$

where $\delta_a(x) = ax$. The resonance varieties $\mathcal{R}^d(A)$ of A are defined by

$$\mathcal{R}^d(A) = \{a \in A^1 \mid H^d(A, \delta_a) \ne 0\}.$$

If $\dim A^d < \infty$, then $\mathcal{R}^d(A)$ is an algebraic set in A^1.

In the case where $A = H^*(M(\mathcal{A}); \Bbbk)$ is the cohomology ring of the complement of a complex hyperplane arrangement, and $\Bbbk$ has characteristic zero, work of Libgober and Yuzvinsky [19] (see also [16]) shows that $\mathcal{R}^1(A)$ is the union of the maximal isotropic subspaces of A^1 for the quadratic form

$$\mu \colon A^1 \otimes A^1 \to A^2, \ \mu(a \otimes b) = ab \tag{3.3}$$

having dimension at least two. Note that, for any field $\Bbbk$, any isotropic subspace of A^1 of dimension at least two is contained in $\mathcal{R}^1(A)$.

For our purposes, it will suffice to take $\Bbbk = \mathbb{Q}$. Let $A = H^*(P_n; \mathbb{Q})$ be the rational cohomology ring of the pure braid group, that is, the cohomology of the complement of the braid arrangement in $\mathbb{C}^n$. By work of Arnold [1] and F. Cohen [10], A is generated by degree one elements $\omega_{i,j}$, $1 \le i < j \le n$, which satisfy (only) the relations

$$\omega_{i,j}\omega_{i,k} - \omega_{i,j}\omega_{j,k} + \omega_{i,k}\omega_{j,k} \quad \text{for} \quad 1 \le i < j < k \le n,$$

and their consequences. The irreducible components of the first resonance variety $\mathcal{R}^1(A)$ may be obtained from work of Cohen and Suciu [9] (see also [24]), and may be described as follows. The rational vector space $A^1 = H^1(P_n; \mathbb{Q})$ is of dimension $\binom{n}{2}$, and has basis $\{\omega_{i,j} \mid 1 \leq i < j \leq n\}$. If $I = \{i, j, k\} \subset [n]$ with $i < j < k$, let V_I be the subspace of A^1 defined by

$$V_I = \operatorname{span}\{\omega_{i,j} - \omega_{j,k}, \omega_{i,k} - \omega_{j,k}\}.$$

If $I = \{i, j, k, l\} \subset [n]$ with $i < j < k < l$, let V_I be the subspace of A^1 defined by

$$V_I = \operatorname{span}\{\omega_{i,j} + a_{k,l} - \omega_{j,k} - a_{i,l}, \omega_{i,k} + a_{j,l} - \omega_{j,k} - a_{i,l}\}.$$

The 2-dimensional subspaces V_I of A^1, where $|I| = 3$ or $|I| = 4$, are the irreducible components of $\mathcal{R}^1(A) = \bigcup_{k=3}^4 \bigcup_{|I|=k} V_I$.

Remark 3.4. One can check that $V_I = f_I^*(H^1(F_2; \mathbb{Q}))$, where $f_I\colon P_n \to F_2$ is defined by (3.1) if $|I| = 3$ and by (3.2) if $|I| = 4$. Work of Schenck and Suciu [25, Lemma 5.3] implies that for any two components V_I and V_J of $\mathcal{R}^1(A)$, there is an isomorphism of $A^1 = H^1(P_n; \mathbb{Q})$ taking V_I to V_J. This provides an analog, on the level of (degree-one) cohomology, of Proposition 3.2.

Theorem 3.5. *Let $n \geq 3$ and $k \geq 2$, and consider the pure braid group P_n and the free group F_k.*

(a) *If $k \geq 3$, there are no epimorphisms from P_n to F_k.*
(b) *If $k = 2$, there is a single equivalence class of epimorphisms from P_n to F_2.*

Proof. For part (a), if $f: P_n \to F_k$ is an epimorphism, then f splits, so

$$f^*\colon H^1(F_k; \mathbb{Q}) \to H^1(P_n; \mathbb{Q})$$

is injective. Consequently, $f^*(H^1(F_k; \mathbb{Q}))$ is a k-dimensional isotropic subspace of $A^1 = H^1(P_n; \mathbb{Q})$ for the form (3.3). Since this subspace is isotropic, it must be contained in an irreducible component of $\mathcal{R}^1(A)$. Since these components are all of dimension 2, we cannot have $k \geq 3$.

For part (b), by Proposition 3.2, it suffices to show that an arbitrary epimorphism $f: P_n \to F_2$ is equivalent to f_I for some I of cardinality 3 or 4. We will extensively use that fact that if $[a, b] = 1$ in F_2, then $\langle a, b\rangle < F_2$ is free and abelian, so $a = z^m$ and $b = z^n$ for some $z \in F_2$ and $m, n \in \mathbb{Z}$. Additionally, if the homology class $[a]$ of a in $H_1(F_2; \mathbb{Z})$

is part of a basis and $[a, b] = 1$, then b is a power of a. Indeed, suppose $\{[a], [a']\}$ is a basis for $H_1(F_2; \mathbb{Z})$. Write $[z] = c_1[a] + c_2[a']$ with $c_1, c_2 \in \mathbb{Z}$. Then we have $[a] = m[z] = mc_1[a] + mc_2[a']$, which implies $c_2 = 0$ and $m = c_1 = \pm 1$, yielding the assertion.

Let $f\colon P_n \to F_2$ be an epimorphism. Then $f^*(H^1(F_2; \mathbb{Q}))$ is a 2-dimensional isotropic subspace of $H^1(P_n; \mathbb{Q})$ for the form (3.3). Since the irreducible components of $\mathcal{R}^1(A)$ are all of dimension 2, we must have $f^*(H^1(F_2; \mathbb{Q})) = V_I$ for some $I \subset [n]$ of cardinality 3 or 4. Since the cohomology rings of P_n and F_2 are torsion-free, passing to integer coefficients, we have $f^*(H^1(F_2; \mathbb{Z})) = V_I \cap \mathbb{Z}^{\binom{n}{2}} \subset H^1(P_n; \mathbb{Z})$. Consequently, there is an automorphism φ^* of $H^1(F_2; \mathbb{Z})$ so that $f^* = f_I^* \circ \varphi^*$. Passing to homology (again using torsion-freeness), we have $f_* = \varphi_* \circ (f_I)_*$, where $\varphi_* \in \mathrm{Aut}(H_1(F_2; \mathbb{Z}))$ is dual to φ^*. Let $\varphi \in \mathrm{Aut}(F_2)$ be an automorphism which induces φ_*.

From the definitions (3.1) and (3.2) of the epimorphisms f_I, there exists $\{i, j, k\}$ with $1 \le i < j < k \le n$, $f_I(\omega_{i,j}) = x$, and $f_I(\omega_{i,k}) = y$, where $F_2 = \langle x, y\rangle$. Let $u = f(\omega_{i,j})$ and $v = f(\omega_{i,k})$. Using the equation $f_* = \varphi_* \circ (f_I)_*$, we have

$$[u] = [f(\omega_{i,j})] = f_*([\omega_{i,j}]) = \varphi_*([f_I(\omega_{i,j})] = \varphi_*([x]) = [\varphi(x)],$$

and similarly $[v] = [\varphi(y)]$. Thus $\{[u], [v]\}$ is a basis for $H_1(F_2; \mathbb{Z})$.

Let $w = f(\omega_{j,k})$. Using the pure braid relations (2.1), we have $\omega_{i,j}\omega_{i,k}\omega_{j,k} = \omega_{i,k}\omega_{j,k}\omega_{i,j} = \omega_{j,k}\omega_{i,j}\omega_{i,k}$. Applying f, these imply that $[uv, w] = 1$ and $[u, vw] = 1$ in F_2. Since $\{[u], [u] + [v]\}$ is a basis for $H_1(F_2; \mathbb{Z})$, these imply that $w = (uv)^m$ and $vw = u^n$ for some $m, n \in \mathbb{Z}$. A calculation with homology classes reveals that $m = n = -1$. Hence $w = v^{-1}u^{-1}$, i.e., $uvw = 1$.

Suppose that $f(A_{r,s}) = 1$ for all $\{r, s\} \not\subset \{i, j, k\}$. Then the image of f is contained in the subgroup $\langle u, v\rangle$ of F_2. Since f is by hypothesis an epimorphism, we have $\langle u, v\rangle = F_2$. Letting λ be the automorphism of F_2 taking u to x and v to y, we have $\lambda \circ f = f_{\{i,j,k\}}$.

Now suppose that $f(A_{r,s}) \neq 1$ for some $\{r, s\} \not\subset \{i, j, k\}$. First assume that $\{r, s\} \cap \{i, j, k\} = \emptyset$. We claim that $f(A_{r,s}) = 1$. There are various cases depending on the relative positions of $r < s$ and $i < j < k$. We consider the case $i < r < j < k < s$ and leave the remaining analogous cases to the reader. In this instance, we have relations $[\omega_{j,k}, A_{r,s}] = 1$ and $\omega_{i,j}^{-1} A_{r,s}\omega_{i,j} = [A_{i,s}, A_{j,s}]A_{r,s}[A_{i,s}, A_{j,s}]^{-1}$. The second of these, together with the pure braid relations (2.1) may be used to show that $[\omega_{i,j}, A_{j,s}^{-1}A_{r,s}A_{j,s}] = 1$. Applying f, we have $[w, f(A_{r,s})] = 1$ and $[u, z^{-1}f(A_{r,s})z] = 1$ in F_2, where $z = f(A_{j,s})$. Since any two element subset of $\{[u], [v], [w]\}$ forms a basis for $H_1(F_2; \mathbb{Z})$, these relations imply that

$f(A_{r,s}) = w^m$ and $z^{-1}f(A_{r,s})z = u^n$ for some $m, n \in \mathbb{Z}$. Consequently, $m[w]=n[u]$ in $H_1(F_2;\mathbb{Z})$, which forces $m = n = 0$ and $f(A_{r,s}) = 1$.

Thus, we must have $f(A_{r,s}) \neq 1$ for some $\{r, s\}$ with $|\{r, s\} \cap \{i, j, k\}| = 1$. As above, there are several cases, and we consider a representative one, leaving the other, similar, cases to the reader.

Assume that $r = k$, so that $i < j < k < s$, and that $f(A_{k,s}) \neq 1$. Applying f to the pure braid relations $[\omega_{i,j}, A_{k,s}] = 1$, $[\omega_{j,k}, A_{i,s}] = 1$, and $[\omega_{i,k}, A_{k,s}^{-1}A_{j,s}A_{k,s}] = 1$ yields $[u, f(A_{k,s})] = 1$, $[w, f(A_{i,s})] = 1$, and $[v, f(A_{k,s}^{-1}A_{j,s}A_{k,s})] = 1$ in F_2. It follows that $f(A_{k,s}) = u^m$, $f(A_{i,s}) = w^n$, and $f(A_{j,s}) = u^m v^l u^{-m}$ for some $m, n, l \in \mathbb{Z}$. Since $f(A_{k,s}) \neq 1$, we have $m \neq 0$. Then, applying f to the pure braid relations $[\omega_{i,k}, A_{i,s}A_{k,s}] = 1$ and $[\omega_{j,k}, A_{j,s}A_{k,s}] = 1$, we obtain $[v, w^n u^m] = 1$ and $[w, u^m v^l] = 1$ in F_2. Thus, $w^n u^m = v^p$ and $u^m v^l = w^q$ for some $p, q \in \mathbb{Z}$. Passing to homology, using the fact that $[u]+[v]+[w] = 0$ in $H_1(F_2;\mathbb{Z})$ since $uvw = 1$ in F_2, reveals that $m = n = l$.

In P_n, we also have the relation $[A_{j,s}, A_{k,s}\omega_{j,k}] = 1$. Applying f we obtain the relation $[u^m v^m u^{-m}, u^m w] = 1$ in F_2. Hence, $u^m v^m u^{-m} = z^p$ and $u^m w = z^q$ for some $z \in F_2$ and $p, q \in \mathbb{Z}$. Writing $[z] = c_1[u] + c_2[v]$, we have

$$m[v] = pc_1[u] + pc_2[v] \text{ and } (m-1)[u] - [v] = qc_1[u] + qc_2[v]$$

in $H_1(F_2;\mathbb{Z})$. It follows that $m = n = l = 1$, and therefore $f(A_{i,s}) = w = v^{-1}u^{-1}$, $f(A_{j,s}) = uvu^{-1}$, and $f(A_{k,s}) = u$. Thus the image of f is contained in the subgroup $\langle u, v\rangle$ of F_2. As before, $\langle u, v\rangle = F_2$ since f is an epimorphism. Recalling that λ is the automorphism of F_2 taking u to x and v to y, we have $\lambda \circ f = f_{\{i,j,k,s\}}$. □

The proof of Theorem 3.5(a) actually yields the following general result.

Theorem 3.6. *Let G be a finitely generated group, and $\mathbb{k}$ an algebraically closed field. Then there are no epimorphisms from G to F_k for $k > \dim \mathcal{R}^1(H^*(G, \mathbb{k}))$.*

Recall that the corank of a group G is the largest natural number k for which the free group F_k is an epimorphic image of G. The corank of $P_2 \cong \mathbb{Z}$ is 1. For larger n, as an immediate consequence of Theorem 3.5, we obtain the following.

Corollary 3.7. *For $n \geq 3$, the corank of the pure braid group P_n is equal to 2.*

Theorem 3.6 yields a more general result.

Corollary 3.8. *Let G be a finitely generated group and $\Bbbk$ an algebraically closed field. Then the corank of G is bounded above by* $\dim \mathcal{R}^1(H^*(G, \Bbbk))$.

In fact, results of Dimca, Papadima, and Suciu [12] imply that the corank of G is equal to $\dim \mathcal{R}^1(H^*(G, \mathbb{C}))$ for a wide class of quasi-Kahler groups, including fundamental groups of complex projective hypersurface complements.

4 Epimorphisms of the lower central series Lie algebra

An analogue of Theorem 3.5 holds on the level of lower central series Lie algebras. For a group G, let G_k be the k-th lower central series subgroup, defined inductively by $G_1 = G$ and $G_{k+1} = [G_k, G]$ for $k \geq 1$. Let $\mathfrak{g}(G) = \bigoplus_{k\geq 1} G_k/G_{k+1}$. The map $G\times G \to G$ given by the commutator, $(x, y) \mapsto [x, y]$, induces a bilinear map $\mathfrak{g}(G) \times \mathfrak{g}(G) \to \mathfrak{g}(G)$ which defines a Lie algebra structure on $\mathfrak{g}(G)$.

The structure of the lower central series Lie algebra $\mathfrak{g}(P_n)$ of the pure braid group was first determined rationally by Kohno [17]. The following description holds over the integers as well, see Papadima [21]. For each $j \geq 1$, let $L[j]$ be the free Lie algebra generated by elements $a_{1,j+1}, \dots, a_{j,j+1}$. Then $\mathfrak{g}(P_n)$ is additively isomorphic to $\bigoplus_{j=1}^{n-1} L[j]$, and the Lie bracket relations in $\mathfrak{g}(P_n)$ are the infinitesimal pure braid relations, given by

$$\begin{aligned} &[\omega_{i,j} + \omega_{i,k} + \omega_{j,k}, a_{m,k}] = 0, \quad \text{for } m \in \{i, j\},\\ &[\omega_{i,j}, a_{k,l}] = 0, \quad \text{for } \{i, j\} \cap \{k, l\} = \emptyset. \end{aligned} \tag{4.1}$$

Let $\mathfrak{f}_k$ be the free Lie algebra (over $\mathbb{Z}$) generated by $x_1, \dots, x_k$. The homomorphisms $f_I \colon P_n \to F_2$, $|I| = 3, 4$, induce surjective Lie algebra homomorphisms $\mathfrak{g}(P_n) \to \mathfrak{f}_2$. Calculations with (3.1) and (3.2) reveal that these are given by

$$(f_{\{i,j,k\}})_*(a_{r,s}) = \begin{cases} x_1 & \text{if } \{r, s\} = \{i, j\},\\ x_2 & \text{if } \{r, s\} = \{i, k\},\\ -x_1 - x_2 & \text{if } \{r, s\} = \{j, k\},\\ 0 & \text{otherwise,} \end{cases}$$

$$(f_{\{i,j,k,l\}})_*(a_{r,s}) = \begin{cases} x_1 & \text{if } \{r, s\} = \{i, j\} \text{ or } \{r, s\} = \{k, l\},\\ x_2 & \text{if } \{r, s\} = \{i, k\} \text{ or } \{r, s\} = \{j, l\},\\ -x_1 - x_2 & \text{if } \{r, s\} = \{j, k\} \text{ or } \{r, s\} = \{i, l\},\\ 0 & \text{otherwise.} \end{cases}$$

Theorem 4.1. *Up to isomorphism, the maps* $(f_I)_*\colon \mathfrak{g}(P_n) \to \mathfrak{f}_2$, $|I| = 3, 4$, *are the only epimorphisms from the lower central series Lie algebra* $\mathfrak{g}(P_n)$ *to a free Lie algebra of rank at least* 2.

Sketch of Proof. If $\varphi\colon \mathfrak{g}(P_n) \to \mathfrak{f}_k$ is an epimorphism, there is an induced epimorphism $\mathfrak{g}(P_n) \otimes \mathbb{Q} \to \mathfrak{f}_k \otimes \mathbb{Q}$ which we denote by the same symbol. Let $b_{i,j} = \varphi(\omega_{i,j})$. Then the relations (4.1) imply that

$$\begin{aligned} &[b_{r,s}, b_{i,j} + b_{i,k} + b_{j,k}] = 0 \quad \text{if } \{r, s\} \subset \{i, j, k\}, \text{ and} \\ &[b_{i,j}, b_{k,l}] = 0 \quad \text{if } \{i, j\} \cap \{k, l\} = \emptyset. \end{aligned} \tag{4.2}$$

Since $\mathfrak{f}_k \otimes \mathbb{Q}$ is free, we conclude that $b_{r,s}$ is a scalar multiple of $b_{i,j} + b_{i,k} + b_{j,k}$ for each $\{r, s\} \subset \{i, j, k\}$ and that $b_{k,l}$ is a scalar multiple of $b_{i,j}$ if $\{i, j\} \cap \{k, l\} = \emptyset$. Write $b_{r,s} = \sum_{j=1}^{k} c_{r,s;j} x_j$ as a linear combination of the generators $x_1, \ldots, x_k$ of $\mathfrak{f}_k \otimes \mathbb{Q}$. For $1 \leq p \leq k$, let $\eta_p = \sum_{1\leq r<s\leq n} c_{r,s;p}\omega_{r,s} \in H^1(P_n; \mathbb{Q})$, where $\omega_{r,s} = a^*_{r,s}$ is dual to $a_{r,s} = [A_{r,s}] \in H_1(P_n; \mathbb{Q})$.

If $1 \leq p < q \leq k$, it follows from (4.2) that the 2×2 determinants

$$\begin{vmatrix} c_{r,s;p} & c_{r,s;q} \\ c_{i,j;p} + c_{i,k;p} + c_{j,k;p} & c_{i,j;q} + c_{i,k;q} + c_{j,k;q} \end{vmatrix}$$

and

$$\begin{vmatrix} c_{u,v;p} & c_{u,v;q} \\ c_{i,j;p} + c_{l,k;p} & c_{i,j;q} + c_{k,l;q} \end{vmatrix}$$

vanish for $\{r, s\} \subset \{i, j, k\}$, and $\{u, v\} = \{i, j\}$ or $\{k, l\}$ where $\{i, j\} \cap \{k, l\} = \emptyset$. By [15, 19], this implies that $\eta_p \wedge \eta_q = 0$. Consequently, $\{\eta_1, \ldots, \eta_k\}$ spans an isotropic subspace of A^1, where $A = H^*(P_n; \mathbb{Q})$. Then $k \leq \dim \mathcal{R}^1(A, \mathbb{Q})$ as in the proof of Theorem 3.5. One concludes that if $\varphi\colon \mathfrak{g}(P_n) \twoheadrightarrow \mathfrak{f}_k$, then k is at most 2.

Finally, one shows that every surjection $\mathfrak{g}(P_n) \to \mathfrak{f}_2$ is induced by f_I for some I with $|I| = 3$ or 4, using a linear version of the argument in the proof of Theorem 3.5 and the infinitesimal pure braid relations (4.1). □

5 Pure braid groups are not residually free

Recall that a group G is residually free if for every $x \neq 1$ in G, there is a homomorphism f from G to a free group F so that $f(x) \neq 1$ in F. In this section, we show that P_n is not residually free for $n \geq 4$, and derive some consequences. Since $P_2 \cong \mathbb{Z}$ and $P_3 \cong \mathbb{Z} \times F_2$, these groups are residually free.

For each $I \subset [n]$ of cardinality 3 or 4, let $K_I = \ker(f_I \colon P_n \to F_2)$. Define

$$K_n = \bigcap_{k=3}^{4} \bigcap_{|I|=k} K_I.$$

Proposition 5.1. *The subgroup K_n of P_n is characteristic.*

Proof. It suffices to show that if β is one of the generators of $\mathrm{Aut}(P_n)$ listed in (2.6), then $\beta(K_n) = K_n$.

If $\beta = \psi$ or $\beta = \phi_{p,q}$ is a transvection, then for each I, $f_I \circ \beta = f_I$ since $f_I(Z_n) = 1$, where Z_n is the generator of the center $Z(P_n)$ of P_n recorded in (2.3). It follows that $\beta(K_I) = K_I$ for each I, which implies that $\beta(K_n) = K_n$.

If $\beta = \xi$, then for each I, it is readily checked that $f_I \circ \xi = \lambda \circ f_I$, where $\lambda \in \mathrm{Aut}(F_2)$ is defined by $\lambda(x) = x^{-1}$ and $\lambda(y) = xy^{-1}x^{-1}$. Thus, $\xi(K_I) = K_I$, and $\xi(K_n) = K_n$.

If $\beta = \beta_k$ for $1 \le k \le n-1$, let τ_k denote the permutation induced by β_k. Let $I = \{i_1, \ldots, i_l\}$ where $l = 3$ or 4, and let $\tau_k(I)$ denote the set $\{\tau_k(i_1), \ldots, \tau_k(i_l)\}$ with the elements in increasing order. Define automorphisms $\lambda_1, \lambda_2 \in \mathrm{Aut}(F_2)$ by

$$\lambda_1 \colon \begin{cases} x \mapsto x, \\ y \mapsto x^{-1}y^{-1}, \end{cases} \quad \text{and} \quad \lambda_2 \colon \begin{cases} x \mapsto xyx^{-1}, \\ y \mapsto x, \end{cases}$$

and set $\lambda_3 = \lambda_1$. Then, calculations with the definitions of the automorphism β_k and the epimorphisms $f_I \colon P_n \to F_2$ (see (2.7), (3.1), and (3.2)) reveal that

$$f_I \circ \beta_k = \begin{cases} \lambda_j \circ f_{\tau_k(I)} & \text{if } k = i_j = i_{j+1} - 1, \\ f_{\tau_k(I)} & \text{otherwise.} \end{cases}$$

Note that, in the first case above, $\tau_k(I) = I$ and $j < l$. Thus, $K_I = \beta_k(K_{\tau_k(I)})$ for each I, and β_k permutes the subgroups K_I of P_n (for $|I| = 3$ and $|I| = 4$ respectively). It follows that $\beta_k(K_n) = K_n$.

Finally, if $\beta = \beta_n$, calculations with (2.7), (3.1), and (3.2)) reveal that

$$f_I \circ \beta_n = \begin{cases} f_{I \cup \{n\}} & \text{if } |I| = 3 \text{ and } n \notin I, \\ \lambda_1 \circ f_I & \text{if } |I| = 3 \text{ and } n \in I, \\ f_I & \text{if } |I| = 4 \text{ and } n \notin I, \\ f_{I \setminus \{n\}} & \text{if } |I| = 4 \text{ and } n \in I. \end{cases}$$

Thus, $\beta_n(K_I) = K_I$ if either $|I| = 3$ and $n \in I$ or $|I| = 4$ and $n \notin I$, while $\beta_n(K_I) = K_{I \cup \{n\}}$ and $\beta_n(K_{I \cup \{n\}}) = K_I$ if $|I| = 3$ and $n \notin I$. It follows that $\beta_n(K_n) = K_n$, and K_n char P_n. □

Theorem 5.2. *For $n \geq 4$, the pure braid group P_n is not residually free.*

Proof. Let $f\colon P_n \to F$ be a homomorphism from the pure braid group to a nonabelian free group. Since P_n is finitely generated and subgroups of free groups are free, we may assume without loss that $f\colon P_n \to F_k$ is a surjection onto a finitely generated nonabelian free group. We claim that K_n is contained in the kernel of f. By Theorem 3.5, $k = 2$ and $f \sim f_{[3]}$. Thus there are automorphisms $\alpha \in \mathrm{Aut}(P_n)$ and $\lambda \in \mathrm{Aut}(F_2)$ so that $\lambda \circ f = f_{[3]} \circ \alpha$. Let $x \in K_n$. Then $\alpha(x) \in K_n$ since K_n is characteristic in P_n by Proposition 5.1. Since $K_n \subset K_{[3]} = \ker(f_{[3]})$ by definition, we have $\alpha(x) \in \ker(f_{[3]})$. Hence, $\lambda \circ f(x) = f_{[3]} \circ \alpha(x) = 1$, and $x \in \ker(f)$.

To complete the proof, it suffices to exhibit a nontrivial element of K_n that is in the kernel of every homomorphism $g\colon P_n \to \mathbb{Z}$ from the pure braid group to an abelian free group. This is straightforward since is it easy to see that the intersection $K_n \cap [P_n, P_n]$ of K_n with the commutator subgroup of P_n is nontrivial. For instance, a calculation reveals that $x = [[A_{1,2}, A_{2,3}], [A_{2,3}, A_{3,4}]] \in K_n \cap [P_n, P_n]$. The pure braid x is nontrivial (one can check that the braids $[A_{1,2}, A_{2,3}][A_{2,3}, A_{3,4}]$ and $[A_{2,3}, A_{3,4}][A_{1,2}, A_{2,3}]$ are distinguished by the Artin representation). Thus $x \neq 1$ is in the kernel of every homomorphism from P_n to a free group, and P_n is not residually free. □

Remark 5.3. When viewed as an element of the 4-strand pure braid group, the braid $[[A_{1,2}, A_{2,3}], [A_{2,3}, A_{3,4}]] \in P_4$ is an example of a Brunnian braid. The deletion of any strand trivializes the braid, see Figure 5.1. (Compare [8, Section 3].)

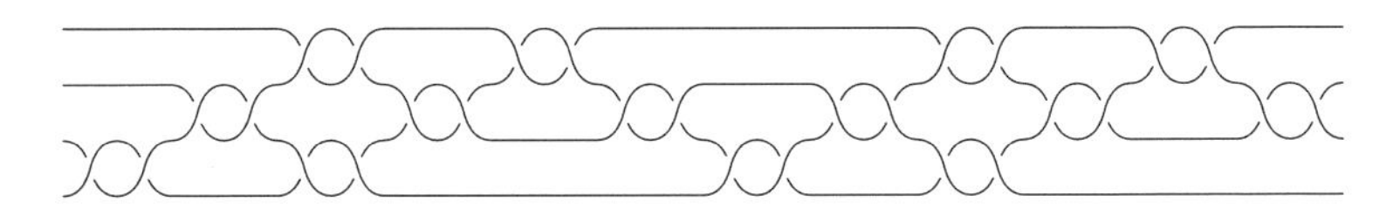

Figure 5.1. The braid $[[A_{1,2}, A_{2,3}], [A_{2,3}, A_{3,4}]]$ in P_4.

Remark 5.4. I. Marin showed us an argument he credited to L. Paris, showing the P_5 is not residually free, implying that P_n is not residually free for $n \geq 5$. Paris' argument uses the solution of the Tits conjecture for B_5 due to Droms, Lewin, and Servatius [13] (see also Collins [11]) to produce a subgroup of P_5 isomorphic to the free product $\mathbb{Z} * (\mathbb{Z} \times F_2)$, as explained in [20, Proposition 1.1]. This latter group is not residually free, see [2, Theorems 6 and 3].

Let Σ be an orientable surface, possibly with punctures. Let $\Sigma^{\times n} = \Sigma \times \cdots \times \Sigma$ denote the n-fold Cartesian product. The pure braid group

$P_n(\Sigma)$ of the surface Σ is the fundamental group of the configuration space

$$F(\Sigma, n) = \{(x_1, \ldots, x_n) \in \Sigma^{\times n} \mid x_i \neq x_j \text{ if } i \neq j\}.$$

of n distinct ordered points in Σ.

Corollary 5.5. *For $n \geq 4$, the pure braid group $P_n(\Sigma)$ is not residually free.*

Proof. If $\Sigma \neq S^2$, the pure braid group P_4 embeds in $P_n(\Sigma)$, see Paris and Rolfsen [23]. If $\Sigma = S^2$, then $P_4 < P_n(S^2)$ for $n \geq 5$, and $P_4(S^2) \cong P_4/Z(P_4)$ (see, for instance, [4]). So the result follows from Theorem 5.2. □

Remark 5.6. The fundamental group of the orbit space $F(\Sigma, n)/\Sigma_n$, where Σ_n denotes the symmetric group, is the (full) braid group $B_n(\Sigma)$ of the surface Σ. Recall from the Introduction that B_3 is not residually free, and that the only two-generator residually free groups are $\mathbb{Z}$, $\mathbb{Z}^2$, and F_2. If $\Sigma \neq S^2$ and $n \geq 3$, then $B_n(\Sigma)$ has a B_3 subgroup, so is not residually free. If $\Sigma = S^2$, then $B_3 < B_n(S^2)$ for $n \geq 4$, and $B_3(S^2) \cong B_3/Z(B_3)$. So $B_n(S^2)$ is not residually free for $n \geq 3$.

A complex hyperplane arrangement $\mathcal{A} = \{H_1, \ldots, H_m\}$ is a finite collection of codimension one subspaces of $\mathbb{C}^n$. Fix coordinates $(z_1, \ldots, z_n)$ on $\mathbb{C}^n$, and for $1 \leq i \leq m$, let $\ell_i(z_1, \ldots, z_n)$ be a linear form with $\ker(\ell_i) = H_i$. The product $Q = Q(\mathcal{A}) = \prod_{i=1}^m \ell_i$ is a defining polynomial for $\mathcal{A}$. The group $G(\mathcal{A})$ of the arrangement is the fundamental group of the complement $M(\mathcal{A}) = \mathbb{C}^n \setminus \bigcup_{i=1}^m H_i = \mathbb{C}^n \setminus Q^{-1}(0)$.

The arrangement $\mathcal{A}_{r,1,n}$ with defining polynomial

$$Q = Q(\mathcal{A}_{r,1,n}) = z_1 \cdots z_n \prod_{1 \leq i < j \leq n} (z_i^r - z_j^r)$$

is known as the full monomial arrangement (it is the reflection arrangement corresponding to the full monomial group $G(r, 1, n)$). Note that the arrangement $\mathcal{A}_{2,1,n}$ is the Coxeter arrangement of type B_n. The complement $M(\mathcal{A}_{r,1,n})$ of the full monomial arrangement may be realized as the orbit configuration space

$$F_\Gamma(\mathbb{C}^*, n) = \{(x_1, \ldots, x_n) \in (\mathbb{C}^*)^{\times n} \mid \Gamma \cdot x_i \cap \Gamma \cdot x_j = \emptyset \text{ if } i \neq j\}$$

of ordered n-tuples of points in $\mathbb{C}^*$ which lie in distinct orbits of the free action of $\Gamma = \mathbb{Z}_r$ on $\mathbb{C}^*$ by multiplication by the primitive r-th root of unity $\exp(2\pi\sqrt{-1}/r)$.

Call the fundamental group $P(r,1,n) = G(\mathcal{A}_{r,1,n})$ the pure monomial braid group. For $n = 1$, $P(r,1,1) \cong \mathbb{Z}$, and for $n = 2$, it is well known that $P(r,1,2) \cong \mathbb{Z} \times F_{r+1}$. Hence, $P(r,1,n)$ is residually free for $n \leq 2$.

Corollary 5.7. *For $n \geq 3$, the pure monomial braid group $P(r,1,n)$ is not residually free.*

Proof. For $n \geq 3$, it follows from [6] that the pure braid group P_4 embeds in $P(r,1,n)$. So the result follows from Theorem 5.2. □

Remark 5.8. The fundamental group of the orbit space $M(\mathcal{A}_{r,1,n})/G(r,1,n)$ is the (full) monomial braid group $B(r,1,n)$. This group admits a presentation with generators $\rho_0, \rho_1, \dots, \rho_{n-1}$ and relations

$$\begin{aligned}(\rho_0\rho_1)^2 &= (\rho_1\rho_0)^2, \ \rho_i\rho_{i+1}\rho_i \\ &= \rho_{i+1}\rho_i\rho_{i+1} \ (1 \leq i < n), \\ \rho_i\rho_j &= \rho_j\rho_i \ (|j-i| > 1).\end{aligned}$$

Observe that $B(r,1,n)$ is independent of r, and is the Artin group of type B_n. For $n \geq 3$, the group $B(r,1,n)$ has a B_3 subgroup, so is not residually free. The group $B(r,1,2)$ is not residually free, since it is a two-generator group which is not free or free abelian.

Let Γ be a Coxeter graph, with associated Artin group A_Γ and pure Artin group P_Γ. We say that Γ contains an A_k subgraph if it contains a path of length k with unlabelled edges as a vertex-induced subgraph.

Corollary 5.9. *If Γ contains an A_3 subgraph, then the associated pure Artin group P_Γ is not residually free.*

Proof. If Γ contains an A_3 subgraph, it follows from van der Lek [18] (see also [22]) that P_Γ has a P_4 subgroup. So the result follows from Theorem 5.2. □

If Γ is a connected Coxeter graph of finite type different from B_3, H_3, or F_4, then Γ contains an A_3 subgraph, hence P_Γ is not residually free by Corollary 5.9. If Γ is of type B_3, then P_Γ is the pure monomial braid group $P(2,1,3)$, so is not residually free by Corollary 5.7. If Γ is of type F_4, then Γ contains a B_3 subgraph, hence P_Γ is not residually free by the same argument as in the proof of Corollary 5.9.

Remark 5.10. If Γ contains an A_2 subgraph, then the (full) Artin group A has a B_3 subgroup, so is not residually free. This includes all irreducible Artin groups of finite type and rank at least 2, except type $I_2(m)$.

References

[1] V. ARNOLD, *The cohomology ring of the group of dyed braids*, Mat. Zametki **5** (1969), 227–231. MR0242196

[2] B. BAUMSLAG, *Residually free groups*, Proc. London Math. Soc. (3) **17** (1967), 402–418. MR0215903

[3] R. BELL and D. MARGALIT, *Injections of Artin groups*, Comment. Math. Helv. **82** (2007), 725–751. MR2341838

[4] J. BIRMAN, *Braids, Links and Mapping Class Groups*, Ann. of Math. Studies, Vol. 82, Princeton Univ. Press, Princeton, NJ, 1975. MR375281

[5] R. CHARNEY and J. CRISP, *Automorphism groups of some affine and finite type Artin groups*, Math. Res. Lett. **12** (2005), 321–333. MR2150887

[6] D. COHEN, *Monodromy of fiber-type arrangements and orbit configuration spaces*, Forum Math. **13**, 2001, 505–530. MR1830245

[7] D. COHEN, *Automorphism groups of some pure braid groups*, preprint 2011. `arxiv:1106.4316`

[8] D. COHEN, M. FALK and R. RANDELL, *Discriminantal bundles and representations of arrangement groups*, preprint 2010. `arxiv:1008.0417`

[9] D. COHEN and A. SUCIU, *Characteristic varieties of arrangements*, Math. Proc. Cambridge Phil. Soc. **127** (1999), 33–53. MR1692519

[10] F. COHEN, *The homology of $\mathcal{C}_{n+1}$-spaces, $n \geq 0$*, in: The homology of iterated loop spaces, pp. 207–352, Lect. Notes in Math., Vol. 533, Springer-Verlag, Berlin-New York, 1976. MR0436146

[11] D. COLLINS, *Relations among the squares of the generators of the braid group*, Invent. Math. **117** (1994), 525–529. MR1283728

[12] A. DIMCA, Ş. PAPADIMA and A. SUCIU, *Topology and geometry of cohomology jump loci*, Duke Math. J. **148** (2009), 405–457. MR2527322

[13] C. DROMS, J. LEWIN and H. SERVATIUS, *The Tits conjecture and the five string braid group*, In: "Topology and Combinatorial Group Theory" (Hanover, NH, 1986/1987; Enfield, NH, 1988), Lect. Notes in Math., Vol. 1440, Springer, Berlin, 1990, 48–51. MR1082979

[14] J. DYER and E. GROSSMAN, *The automorphism groups of the braid groups*, Amer. J. Math. **103** (1981), 1151–1169. MR0636956

[15] M. FALK, *Arrangements and cohomology*, Ann. Comb. **1** (1997), 135–157. MR1629681

[16] M. FALK and S. YUZVINSKY, *Multinets, resonance varieties, and pencils of plane curves*, Compositio Math. **143** (2007), 1069–1088. MR2339840
[17] T. KOHNO, *Series de Poincaré-Koszul associée aux groupes de tresses pures*, Invent. Math. **82** (1985), 57–75. MR0808109
[18] H. VAN DER LEK, *The homotopy type of complex hyperplane complements* Ph. D. Thesis, University of Nijmegen, Nijmegen, Netherlands, 1983.
[19] A. LIBGOBER and S. YUZVINSKY, *Cohomology of the Orlik-Solomon algebras and local systems*, Compositio Math. **121** (2000), 337–361. MR1761630
[20] I. MARIN, *Residual nilpotence for generalizations of pure braid groups*, In: "Configuration Spaces", Proceedings, A. Bjorner, F. Cohen, C. De Concini, C. Procesi and M. Salvetti (eds.), Edizioni della Normale, Pisa, 2012, 390–401.
[21] Ş. PAPADIMA, *The universal finite-type invariant for braids, with integer coefficients*, in: Arrangements in Boston: a Conference on Hyperplane Arrangements (1999), Topology Appl. **118** (2002), 169–185. MR1877723
[22] L. PARIS, *Parabolic subgroups of Artin groups*, J. Algebra **196** (1997), 369–399. MR1475116
[23] L. PARIS and D. ROLFSEN, *Geometric subgroups of surface braid groups*, Ann. Inst. Fourier (Grenoble) **49** (1999), 417–472. MR1697370
[24] J. PEREIRA, *Resonance webs of hyperplane arrangements*, In: "Arrangements of Hyperplanes (Sapporo 2009)", Adv. Stud. Pure Math., Vol. 62, Math. Soc. Japan, 2012, 261–291.
[25] H. SCHENCK and A. SUCIU, *Resonance, linear syzygies, Chen groups, and the Bernstein-Gelfand-Gelfand correspondence*, Trans. Amer. Math. Soc. **358** (2006), 2269–2289. MR2197444
[26] H. WILTON, *Residually free 3-manifolds*, Algebr. Geom. Topol. **8** (2008), 2031–2047. MR2449007

Hodge-Deligne equivariant polynomials and monodromy of hyperplane arrangements

Alexandru Dimca[1,2] and Gus Lehrer[2]

Abstract. We investigate the interplay between the monodromy and the Deligne mixed Hodge structure on the Milnor fiber of a homogeneous polynomial. In the case of hyperplane arrangement Milnor fibers, we obtain a new result on the possible weights. For line arrangements, we prove in a new way the fact due to Budur and Saito that the spectrum is determined by the weak combinatorial data, and show that such a result fails for the Hodge-Deligne polynomials. In an appendix, we also establish a connection between the Hodge-Deligne polynomials and rational points over finite fields.

1 Introduction

Let $E = \mathbb{C}^{n+1}$, with $\mathcal{A}$ a finite set of hyperplanes through 0 in E, $Z = Z_{\mathcal{A}} = \cup_{H \in \mathcal{A}} H$ and $N = N_{\mathcal{A}} = E \setminus Z$ the corresponding complement. To keep notation simple, we will also denote by $\mathcal{A}$ the associated hyperplane arrangement in the complex projective space $\mathbb{P}^n$ and use H for an affine hyperplane in E and also for the associated projective hyperplane in $\mathbb{P}^n$.

Let $Q = 0$ be a reduced equation for the union $V = \cup_{H \in \mathcal{A}} H \subset \mathbb{P}^n$ of (projective) hyperplanes in $\mathcal{A}$ and $F = \{x \in \mathbb{C}^{n+1} \mid Q(x) = 1\}$ the associated Milnor fiber. If $d = |\mathcal{A}|$ is the number of hyperplanes in $\mathcal{A}$, then $d = \deg Q$ and there is a monodromy isomorphism

$$h : F \to F, \quad h(x) = \lambda \cdot x, \tag{1.1}$$

with $\lambda = \exp(2\pi i/d)$. This may be regarded as an action on F of the cyclic group μ_d generated by λ. Any irreducible representation of μ_d is one-dimensional, and has the form $\mathbb{C}_\alpha$ ($\alpha \in \mu_d$), where for $1 \in \mathbb{C}_\alpha$,

[1] Partially supported by the ANR-08-BLAN-0317-02 (SEDIGA).

[2] Partially supported by Australian Research Council Grants DP0559325 and DP110103451.

$\lambda \cdot 1 = \alpha$. If V is any μ_d-module, we shall denote by V_β ($\beta \in \mu_d$) its $\mathbb{C}_\beta$-isotypic component.

It is an open question whether the Betti numbers $b_j(F)$ of F or, more generally, the dimension of the isotypic components $H^j(F, \mathbb{C})_\beta$ for $0 < j \leq n$ and $\beta \in \mu_d$, are determined by the combinatorics of $\mathcal{A}$. This is a natural question, since the cohomology algebra $H^*(M, \mathbb{Z})$ of the complement $M = \mathbb{P}^n \setminus V$ is known to be determined by the combinatorics of $\mathcal{A}$, see [16]. The same applies to $N = M \times \mathbb{C}^\times$. In particular, using the degree d covering projection $p : F \to M$, we see that $\chi(F) = d \cdot \chi(M)$ is determined by the combinatorics.

A recent result of Budur and Saito in [3] asserts that a related invariant, the spectrum of a hyperplane arrangement $\mathcal{A}$ in $\mathbb{P}^n$ defined by

$$Sp(\mathcal{A}) = \sum_{a \in \mathbb{Q}} m_a t^a, \tag{1.2}$$

with $m_a = \sum_j (-1)^{j-n} \dim Gr_{\mathbf{F}}^p \tilde{H}^j(F, \mathbb{C})_\beta$ where $p = [n + 1 - a]$ and $\beta = \exp(2\pi i a)$, is also determined by the combinatorics.

On the other hand, it was shown in [10] and in [4], that for $n = 2$ (*i.e.* for a line arrangement) the eigenspace decomposition

$$H^1(F, \mathbb{Q}) = H^1(F, \mathbb{Q})_1 \oplus H^1(F, \mathbb{Q})_{\neq 1}, \tag{1.3}$$

is actually a decomposition of mixed Hodge structures (denoted henceforth by MHS) such that $H^1(F, \mathbb{Q})_1 = p^*(H^1(M, \mathbb{Q}))$ is pure of type $(1, 1)$, and $H^1(F, \mathbb{Q})_{\neq 1}$ is pure of weight 1.

One may consider a more general setting where the union of hyperplanes V is replaced by a degree d hypersurface V in $\mathbb{P}^n$ given by a reduced equation $Q_V(x) = 0$, the associated Milnor fiber F_V being defined by $Q_V(x) = 1$, and ask which of the above results remain true. It is natural to replace the combinatorics of $\mathcal{A}$ by the local type of singularities of V and the topology of the dual complex associated to a normal crossing exceptional divisor arising in the resolution of singularities for V_{sing} as in [2]. Some questions have easy negative answers in this more general setting, for instance the classical example of Zariski of cuspidal sextics implies that $b_1(F)$ depends on the position of singularies in general.

In this paper we investigate the relationship between the monodromy action and the MHS on the Milnor fiber cohomology in this more general setting. To do this, we first refine the equivariant weight polynomials introduced in [9] to get the equivariant Hodge-Deligne polynomials $P^\Gamma(X)$ associated to a finite group Γ acting (algebraically) on a complex algebraic variety X.

More precisely, let X be a quasi-projective variety over $\mathbb{C}$ and consider the Deligne MHS on the rational cohomology groups $H^*(X, \mathbb{Q})$ of X. Since this MHS is functorial with respect to algebraic mappings, if Γ acts algebraically on X, each of the graded pieces

$$H^{p,q}(H^j(X, \mathbb{C})) := Gr_{\mathbf{F}}^p Gr_{p+q}^{\mathbf{W}} H^j(X, \mathbb{C}) \tag{1.4}$$

becomes a Γ-module, and these modules are the building blocks of the polynomial $P^{\Gamma}(X)$. In this situation we refer to $H^*(X, \mathbb{Q})$ as being a graded Γ-MHS.

This viewpoint is applied to the hypersurface X_V, the projective closure of F_V, given by the equation

$$Q_V(x) - t^d = 0 \tag{1.5}$$

in $\mathbb{P}^{n+1}$. The main results can be stated as follows.

Theorem 1.1. *Let $H_0^*(X_V, \mathbb{Q}) = \operatorname{coker}\{H^*(\mathbb{P}^{n+1}, \mathbb{Q}) \to H^*(X_V, \mathbb{Q})\}$ be the primitive cohomology of X_V, where the morphism is induced by the inclusion $i : X_V \to \mathbb{P}^{n+1}$. Then, for any hypersurface $V \subset \mathbb{P}^n$, there are natural isomorphisms of μ_d-MHS*

$$H^j(F_V, \mathbb{Q})_{\neq 1} = H_0^{2n-j}(X_V, \mathbb{Q})^{\vee}(-n)$$

for any $0 < j \leq n$.

Here, for a MHS H, we denote by $H^{\vee}$ the dual MHS, and $H(m)$ denotes the Tate twist, see [17] for details. We can restate this result more explicitly using the equivariant Hodge-Deligne numbers $h^{p,q}(H, \alpha)$, see Example 2.1 for the definition and the basic properties of these numbers.

Corollary 1.2. *For any hypersurface $V \subset \mathbb{P}^n$ of degree d, any $\alpha \in \mu_d$, $\alpha \neq 1$ and any $0 < j \leq n$,*

$$\begin{aligned} h^{p,q}(H^j(F_V, \mathbb{C}), \alpha) &= h^{n-p,n-q}(H_0^{2n-j}(X_V, \mathbb{C}), \overline{\alpha}) \\ &= h^{n-q,n-p}(H_0^{2n-j}(X_V, \mathbb{Q}), \alpha). \end{aligned}$$

Theorem 1.3. *Let F be the Milnor fiber of a hyperplane arrangement $\mathcal{A}$ in $\mathbb{P}^n$. Then*

$$Gr_{2j}^{\mathbf{W}} H^j(F, \mathbb{Q})_{\neq 1} = 0$$

for any $0 < j \leq n$.

Remark 1.4. (i) For any hyperplane arrangement $\mathcal{A}$ in $\mathbb{P}^n$ and any $j \geq 0$, it is known that $H^j(F,\mathbb{Q})_1 = p^*(H^j(M,\mathbb{Q}))$ is pure of type (j,j), see [13]. That is why we consider here only the summand $H^*(F,\mathbb{Q})_{\neq 1}$.

(ii) The result in Theorem 1.3 is optimal, since even for a line arrangement $H^2(F,\mathbb{Q})$ may have weights 2 and 3, see Example 5.3 below. Moreover, this result does not hold for a general hypersurface V, in fact not even for V an irreducible curve, see Example 4.3 below.

We have the following more precise result for some hypersurfaces V.

Theorem 1.5. *Let F_V be the Milnor fiber of a hypersurface V in $\mathbb{P}^n$ having only isolated singularities. Then the following hold.*
(i) $\tilde{H}^j(F_V,\mathbb{Q}) = 0$ *for* $0 \leq j \leq n-2$;
(ii) $H^{n-1}(F_V,\mathbb{Q})_{\neq 1}$ *is a pure Hodge structure of weight* $n-1$;
(iii) *If in addition the singularities of V are weighted homogeneous, then* $H^n(F_V,\mathbb{Q})_{\neq 1}$ *is a MHS with weights n and $n+1$*, i.e. $Gr_j^W H^n(F,\mathbb{Q})_{\neq 1} = 0$ *for* $j > n+1$.

The first part of the next Theorem gives a new proof of a result already present in [3]. For a line arrangement $\mathcal{A}$, let us refer to the following as the weak combinatorial data: d, the number of lines in $\mathcal{A}$ and m_k, the numbers of points of multiplicity k in V, for all $k \geq 2$.

Theorem 1.6. *For a line arrangement $\mathcal{A}$, one has the following.*
(i) *The spectrum $Sp(\mathcal{A})$ is determined by the weak combinatorial data. More generally, the spectrum $Sp(V)$ of a hypersurface V having only isolated singularities is determined by $d = deg(V)$ and by the local type of its singularities.*
(ii) *The Hodge-Deligne polynomial $P^{\mu_d}(F)$ corresponding to the monodromy action on F is not determined by the weak combinatorial data.*

In fact, a weaker invariant, the weight polynomial $W(F)$, as recalled in Remark 2.3 (take $\Gamma = 1$), is itself not determined by the weak combinatorial data (same example as in the proof of Theorem 1.6).

Explicit numerical formulas for the coefficients m_a in the spectrum of $\mathcal{A}$ were obtained in [3], Theorem 3. On the other hand, our proof gives a geometric description of these coefficients in terms of specific μ_d-actions on Milnor fibers and on a degree d smooth surface, see for instance (5.8).

We mention also Corollary 3.2, computing the Hodge-Deligne polynomial of the link (or deleted neighbourhood) of the singular locus Σ of a projective variety X in terms of the Hodge-Deligne polynomial of the exceptional divisor D of an embedded resolution of the pair (X,Σ).

In the Appendix, we use p-adic Hodge theory to prove that quite generally, whenever a Γ-variety X is defined over a number field, the number of rational points of its reductions modulo prime ideals can be used in certain cases to compute the Hodge-Deligne polynomial $P_c^{\Gamma}(X)(u, v)$. We thank Mark Kisin for discussions about this subject.

We would also like to thank Morihiko Saito for his help in proving Theorem 1.3.

2 Equivariant Hodge-Deligne polynomials

Let Γ be a finite group and denote by $R(\Gamma)$ the complex representation ring of Γ. If V is a (finite dimensional) Γ-module, we denote by the same symbol V the class of V in $R(\Gamma)$.

When E is a Γ-module, the dual module $E^{\vee}$ is defined in the usual way, that is

$$(g \cdot h)(v) = h(g^{-1} \cdot v), \tag{2.1}$$

for any $h \in E^{\vee}$, $v \in E$ and $g \in \Gamma$. This gives rise to an involution

$$\iota : R(\Gamma) \to R(\Gamma), \quad E \mapsto E^{\vee}$$

of the representation ring $R(\Gamma)$.

Example 2.1. If $\Gamma = \mu_d$, then as explained in the Introduction, any irreducible Γ-module is of the form $\mathbb{C}_{\alpha}$ for some $\alpha \in \mu_d$. One then has $\iota(\mathbb{C}_{\alpha}) = \mathbb{C}_{\overline{\alpha}} = \mathbb{C}_{\alpha^{-1}}$. Moreover, if a Γ-module E is defined over $\mathbb{Q}$, then $\dim E_{\alpha} = \dim E_{\overline{\alpha}}$ for any $\alpha \in \mu_d$. It follows that in this case $\iota(E) = E$.

If H is a μ_d-MHS and $\alpha \in \mu_d$, we write $h^{p,q}(H, \alpha)$ for the multiplicity of the irreducible representation $\mathbb{C}_{\alpha}$ in the representation $H^{p,q}(H)$. That is, $h^{p,q}(H, \alpha) = \dim H^{p,q}(H)_{\alpha}$. With this notation, we have

$$h^{p,q}(H^{\vee}, \alpha) = h^{-p,-q}(H, \overline{\alpha}). \tag{2.2}$$

To see this, note that $H^{p,q}(H^{\vee})$ is the dual of $H^{-p,-q}(H)$. For the Tate twist $H(m)$, one has

$$h^{p,q}(H(m), \alpha) = h^{p+m,q+m}(H, \alpha) \text{ and } h^{p,q}(H, \alpha) = h^{q,p}(H, \overline{\alpha}). \tag{2.3}$$

For the second equality, recall that complex conjugation establishes an $\mathbb{R}$-linear isomorphism $H^{p,q} \to H^{q,p}$.

Definition 2.2. The equivariant Hodge-Deligne polynomial of the Γ-variety X is the polynomial $P^{\Gamma}(X) \in R(\Gamma)[u, v]$ defined as the sum

$$P^{\Gamma}(X)(u, v) = \sum_{p,q} E^{\Gamma;p,q}(X) u^p v^q$$

where $E^{\Gamma;p,q}(X) = \sum_j (-1)^j H^{p,q}(H^j(X,\mathbb{C}))$. Similarly, the equivariant Hodge-Deligne polynomial with compact supports of the Γ-variety X is the polynomial $P_c^{\Gamma}(X) \in R(\Gamma)[u,v]$ defined by the sum

$$P_c^{\Gamma}(X)(u,v) = \sum_{p,q} E_c^{\Gamma;p,q}(X) u^p v^q$$

where $E_c^{\Gamma;p,q}(X) = \sum_j (-1)^j H^{p,q}(H_c^j(X,\mathbb{C}))$.

A similar notation $P^{\Gamma}(H^*)$ will be used when $H^*(X,\mathbb{Q})$ is replaced by any graded Γ-MHS H^*.

Remark 2.3. If we set $u = v$ in the above formulas, we get exactly the equivariant weight polynomials of the Γ-variety X introduced in [9], namely $P^{\Gamma}(X)(u,u) = W^{\Gamma}(X,u)$ and $P_c^{\Gamma}(X)(u,u) = W_c^{\Gamma}(X,u)$. So the equivariant Hodge-Deligne polynomials are refinements of the equivariant weight polynomials.

It follows from Poincaré Duality, see [17, Theorem 6.23, page 155], that when X is a smooth connected n-dimensional variety, one has the following relation

$$\iota(P^{\Gamma}(X)(u,v)) = u^n v^n P_c^{\Gamma}(X)(u^{-1},v^{-1}) \tag{2.4}$$

where the involution ι acts on the coefficients of these polynomials. In fact, the formula (1.6) in [9] should be modified to read

$$\iota(W^{\Gamma}(X,u)) = u^{2n} W_c^{\Gamma}(X,u^{-1}). \tag{2.5}$$

However, since the weight filtration is defined over $\mathbb{Q}$, in many cases, *e.g.* when $\Gamma = \mu_d$, the involution ι acts trivially on the coefficients of $W^{\Gamma}(X,u)$, recall Example 2.1 above.

Remark 2.4. Larger, hence more interesting, symmetry groups acting on the Milnor fiber of a hyperplane arrangement may occur as follows. Let $G \subseteq \mathrm{GL}(E)$ be a finite group which preserves $\mathcal{A}$, and leaves invariant a polynomial (not necessarily reduced) Q such that $\cup_{H\in\mathcal{A}} H = Q^{-1}(0)$; for example, G might be a unitary reflection group, which leaves invariant a suitable product of linear forms corresponding to $\mathcal{A}$ [14].

Write d for the degree of Q. Then $\Gamma := G \times \mu_d$ acts on E: $(g,\xi)\cdot v = \xi^{-1} g v$ for $v \in E$, and $\Gamma F \subseteq F$.

Define $\widetilde{N} := \{(v,\zeta) \in N \times \mathbb{C}^{\times} \mid Q(v) = \zeta^d\} = \{(v,\zeta) \mid \zeta^{-1} v \in F\}$. Then $\mathbb{C}^{\times}$ acts on $\widetilde{N}$ diagonally: $\alpha : (v,\zeta) \mapsto (\alpha v, \alpha\zeta)$, and we have the following commutative diagram

$$\begin{array}{ccc} \widetilde{N} & \xrightarrow{p_1} & N \\ {\scriptstyle \pi_1}\downarrow & & \downarrow{\scriptstyle \pi} \\ F & \xrightarrow{p} & M \end{array}$$

where p, p_1 are unramified μ_d-coverings, and the vertical arrows are 'division by $\mathbb{C}^\times$'. In fact, it is well known that $N = M \times \mathbb{C}^\times$, see for instance [8, Proposition 6.4.1, page 209]. It follows that $\widetilde{N} = F \times \mathbb{C}^\times$.

Remark 2.5. If the algebraic variety X is defined over an algebraic number field, it may be reduced modulo various primes, and properties of $P^\Gamma(X)$ may be deduced by considering rational points of twisted Frobenius maps in such reductions. Although we do not explore this theme extensively in this work, we provide some basic background results in the Appendix. In particular, we give a proof of the following result for a Γ-variety defined over a number field.

Theorem 2.6. *(See Theorem A.2 below)*
Suppose there are polynomials $P_X(t; w) = \sum_{i=0}^{2\dim X} a_{2i}(w)t^i \in \mathbb{C}[t]$ *such that for almost all* q*, and all* $w \in \Gamma$*, we have* $|X(\overline{\mathbb{F}_q})^{w\,\mathrm{Frob}_q}| = P_X(q; w)$*. Then* $E_c^{\Gamma;d,e} = 0$ *if* $d \neq e$*, and* $P_c^\Gamma(X)(x, y) = W_c^\Gamma(X)(\sqrt{xy})$*. Moreover* $P_X(t^2; w) = W_c^\Gamma(X)(t; w)$*.*

Here we write, for any polynomial $Q(x,y)=\sum_{i,j} Q_{i,j}x^i y^j \in R(\Gamma)[x,y]$ and $w \in \Gamma$, $Q(x, y; w) := \sum_{i,j} \mathrm{Trace}(w, Q_{i,j})x^i y^j$.

3 Localization at the singular locus

Let X be an n-dimensional projective algebraic variety, $n \geq 2$, with singular locus Σ. Let $X^* = X \setminus \Sigma$ be the regular part of X.

Assume that a finite group Γ acts algebraically on X; then Σ is Γ-invariant, *i.e.* for $g \in \Gamma$ and $a \in \Sigma$ one has $g \cdot a \in \Sigma$.

As in [7], we consider the following exact sequence of Γ-MHS, see for instance [17, page 136].

$$\ldots \to H^k_\Sigma(X) \to H^k(X) \to H^k(X^*) \to H^{k+1}_\Sigma(X) \to \ldots \tag{3.1}$$

For the equivariant Hodge-Deligne polynomials, this yields (with obvious notation)

$$P^\Gamma(X) = P^\Gamma(X^*) + P^\Gamma(H^*_\Sigma(X)). \tag{3.2}$$

On the other hand, let T be a Γ-stable algebraic neighbourhood of Σ in X, and $T^* = T \setminus \Sigma$ the corresponding deleted neighbourhood of Σ in X, which is homotopically equivalent to the link $K(\Sigma)$ of Σ in X. Then Remark 6.17 in [17, p.151], yields an exact sequence of Γ-MHS

$$\ldots \to H^k_\Sigma(X) \to H^k(\Sigma) \to H^k(T^*) \to H^{k+1}_\Sigma(X) \to \ldots \tag{3.3}$$

which, at the level of equivariant Hodge-Deligne polynomials, gives

$$P^\Gamma(H^*_\Sigma(X)) = P^\Gamma(\Sigma) - P^\Gamma(T^*). \tag{3.4}$$

Further, by the additivity of the equivariant Hodge-Deligne polynomials with compact supports, see for instance [9], one has

$$P_c^\Gamma(X^*) = P^\Gamma(X) - P^\Gamma(\Sigma).$$

Finally, the Poincaré Duality formula (2.4) for smooth varieties implies that

$$P^\Gamma(X^*)(u,v) = u^n v^n \iota(P_c^\Gamma(X^*))(u^{-1}, v^{-1}).$$

Assembling the above relations yields a proof of the following result, which is an equivariant form of Proposition 1.7 in [7].

Proposition 3.1. *Maintaining the above notation, we have*

$$\begin{aligned} &P^\Gamma(X)(u,v) - u^n v^n \iota(P^\Gamma(X))(u^{-1}, v^{-1}) \\ &\quad = P^\Gamma(\Sigma)(u,v) - u^n v^n \iota(P^\Gamma(\Sigma))(u^{-1}, v^{-1}) - P^\Gamma(T^*)(u,v). \end{aligned}$$

Now suppose we have a Γ-equivariant log resolution $\pi : (\tilde{X}, D) \to (X, \Sigma)$, see [1]. Then the deleted neighbourhoods T^* of Σ in X and T_1^* of D in $\tilde{X}$ clearly coincide. Applying Proposition 3.1 to $(\tilde{X}, D)$ and using (2.4), we obtain the following generalization of the known relation between the resolution and MHS on the link for an isolated surface singularity, see [6].

Corollary 3.2. *With the above notation, one has*

$$P^\Gamma(T^*)(u,v) = P^\Gamma(D)(u,v) - u^n v^n \iota(P^\Gamma(D))(u^{-1}, v^{-1}).$$

For the remainder of this section we confine attention to the following particular situation. Let X be a hypersurface in $\mathbb{P}^{n+1}$, with $n \geq 2$, having only isolated singularities

$$\Sigma = \{a_1, ..., a_m\}.$$

Assume that the finite group Γ acts algebraically on $\mathbb{P}^{n+1}$ in such a way that the hypersurface X is Γ-stable, *i.e.* for $g \in \Gamma$ one has $g \cdot X \subset X$, and Σ is point-wise invariant, *i.e.* for $g \in \Gamma$ and $a_j \in \Sigma$ one has $g \cdot a_j = a_j$.

Now the link K_s of each singular point a_s may be chosen to be Γ-invariant, and hence the cohomology groups $H^*(K_s)$ acquire a Γ-mixed Hodge structure. Moreover, for any k, one has the following isomorphism of Γ-MHS:

$$H^k(T^*, \mathbb{Q}) = \oplus_s H^k(K_s, \mathbb{Q}). \tag{3.5}$$

Let us look at the polynomial $P^{\Gamma}(X)$ in more detail. To do this, let $i : X \to \mathbb{P}^{n+1}$ be the inclusion and define the primitive cohomology of X to be

$$H_0^*(X,\mathbb{Q}) = \operatorname{coker}\{i^* : H^*(\mathbb{P}^{n+1},\mathbb{Q}) \to H^*(X,\mathbb{Q})\}.$$

This is clearly a graded Γ-MHS and one has

$$P^{\Gamma}(X) = P^{\Gamma}(H_0^*(X,\mathbb{Q})) + P_n \tag{3.6}$$

where $P_n(u,v) = 1 + uv + ... + u^n v^n$. To see this, notice that Γ acts trivially on $H^*(\mathbb{P}^{n+1})$ and recall the known facts on the cohomology of hypersurfaces with isolated singularities, see [6].

In particular, it is known that $H_0^j(X,\mathbb{Q}) = 0$ except for $j = n$ (here the weights are $\leq n$ since X is proper) and for $j = n+1$, when $H_0^{n+1}(X,\mathbb{Q})$ is pure of weight $n+1$ by Steenbrink's results, see [21].

Henceforth we assume in addition that each of the isolated singularities (X, a_s) is weighted homogeneous. Then the possible weights on $H_0^n(X,\mathbb{Q})$ are just $n-1$ and n, see [7, Example (B), formula (i), page 381].

Moreover, the only nonzero cohomology groups of a link K_s are the following: $H^0(K_s)$ (1-dimensional, of type (0,0)), $H^{n-1}(K_s)$ (of pure weight $n-1$), $H^n(K_s)$ (of pure weight $n+1$), and $H^{2n-1}(K_s)$ (1-dimensional, of type (n,n)).

As a consequence of this discussion, we see that knowledge of the Hodge-Deligne polynomials of the links K_s and of the primitive cohomology group $H_0^{n+1}(X,\mathbb{Q})$ determine the terms of weight $n-1$ in the Hodge-Deligne polynomial of X. More precisely, we have the following result extending Corollary 1.8 in [7].

Corollary 3.3. *For $p+q = n-1$, one has the following equality in $R(\Gamma)$:*

$$H^{p,q}(H_0^n(X)) = \sum_s H^{p+1,q+1}(H^n(F_s)) - H^{n-p,n-q}(\iota(H_0^{n+1}(X))).$$

Here F_s is the Milnor fibre corresponding to K_s, and to obtain the formula, we use the isomorphism of μ_d-representations

$$H^{p,q}(H^{n-1}(K_s)) = H^{p+1,q+1}(H^n(F_s))$$

valid for $p+q = n-1$ and any weighted homogeneous hypersurface singularity (X, a_s) of dimension n.

The Hodge-Deligne polynomials of the Milnor fibers F_s are local invariants easy to compute in general (since one has explicit bases for these cohomology groups in terms of algebraic differential forms), while the the Hodge-Deligne polynomial of $H_0^{n+1}(X, \mathbb{Q})$ may be computed in many cases using the results in [6, (6.3.15), page 202], see for instance [6, Theorem 6.4.15].

4 Monodromy of Milnor fibers of line arrangements

Recall the following notation from the Introduction: we consider a degree d reduced hypersurface V in $\mathbb{P}^n$, $n \geq 2$, given by $Q_V = 0$ and the associated Milnor fiber F_V defined in $\mathbb{C}^{n+1}$ by $Q_V(x) = 1$. We further consider X_V, the projective closure of F_V in $\mathbb{P}^{n+1}$, given by $Q_V(x) - t^d = 0$. When V is the union of the hyperplanes in $\mathcal{A}$, we may drop the subscript V from F_V and X_V.

We consider the (monodromy) μ_d-action on F_V given by

$$\alpha \cdot (x_0, ..., x_n) = (\alpha x_0, ..., \alpha x_n) \tag{4.1}$$

for any $\alpha \in \mu_d$ and $(x_0, ..., x_n) \in F_V$, and the related μ_d-action on $\mathbb{P}^{n+1}$ given by

$$\alpha \cdot (x_0 : ... : x_n : t) = (x_0 : ... : x_n : \alpha^{-1} t) \tag{4.2}$$

for any $\alpha \in \mu_d$ and $(x_0 : ... : x_n : t) \in \mathbb{P}^{n+1}$. Then the obvious isomorphism $F_V \to X_V \setminus V$ given by $(x_0, ..., x_n) \mapsto (x_0 : ... : x_n : 1)$ is μ_d-equivariant, in particular

$$P_c^{\mu_d}(F) = P^{\mu_d}(X) - P^{\mu_d}(V). \tag{4.3}$$

The last term $P^{\mu_d}(V)$ is easy to compute (and depends only on the combinatorics in the case of hyperplane arrangements), since the μ_d-action on V is trivial. In this way we arrive at the situation studied in the previous section.

4.1 Proof of Theorem 1.1

Proof. This proof is very simple and quite general; it does not require the results obtained above. We have the following exact sequence of Γ-MHS.

$$... \to H_0^{i-1}(V) \to H_c^i(F_V) \to H_0^i(X_V) \to H_0^i(V) \to ... \tag{4.4}$$

Recall that $H_c^i(F_V)$ is dual to $H^{2n-j}(F_V)$, and hence $\dim H_c^i(F_V)_1 = \dim H^{2n-i}(F_V)_1 = b_{2n-i}(M)$. On the other hand, Alexander Duality implies that

$$\dim H^{2n-i}(M) = \dim H^i(\mathbb{P}^n, V) = \dim H_0^{i-1}(V).$$

Moreover, since $X_V/\mu_d = \mathbb{P}^n$, it follows that $H_0^*(X)_1$, the fixed part under $\Gamma = \mu_d$, is trivial. As a result we get the following identification of μ_d-MHS

$$H_c^i(F)_{\neq 1} = H_0^i(X). \tag{4.5}$$

This identification yields Theorem 1.1 by Poincaré Duality. □

4.2 Proof of Theorem 1.3

Proof. This proof requires a number of results due to Budur and Saito in [3,19,20]. First, notice that by taking a generic hyperplane section and applying the affine version of the Lefschetz Theorem, see for instance [6, page 25], it is enough to prove Theorem 1.3 for $j = n$. To proceed, we need the following result, see Lemma (3.6) in [19].

Proposition 4.1. *If $Gr_{2n}^{\mathbf{W}} H^n(F, \mathbb{C})_\alpha \neq 0$, then $N^n \neq 0$ on $\psi_{Q,\alpha}\mathbb{C}_Z$, where $Z = \mathbb{C}^{n+1}$ and N is the logarithm of the unipotent part of the monodromy.*

For the general properties of the nearby cycles $\psi_{Q,\alpha}\mathbb{C}_Z$ we refer to [8], and for the weight filtration on them to [18]. Let $D \subset Z$ be the affine cone over V, *i.e.* $D = Q^{-1}(0)$. There is a canonical way to construct an embedded resolution of D in Z, see section (2.1) in [3] for a projective version and section (3.5) in [20] for a special affine case.

Let $Z_0 = Z$ and denote by $p_0 : Z_1 \to Z_0$ the blow-up of the origin in $Z_0 = \mathbb{C}^{n+1}$. Then for $1 \leq i \leq n-1$, let $p_i : Z_{i+1} \to Z_i$ be the blow-up with center C_i, the disjoint union of the proper transforms in Z_i of the linear spaces (edges) $V \in L(\mathcal{A})$ (regarded as subspaces in Z) with $\dim V = i$. Let $\tilde{Z} = Z_n$, $\tilde{p} : \tilde{Z} \to Z$ the composition of the p_i's and $\tilde{D} = \tilde{p}^{-1}(D)$. Then $\tilde{D}$ is a normal crossing divisor in $\tilde{Z}$, with irreducible components parametrized by all the edges $V \in L(\mathcal{A})$ with $\dim V \leq n-1$. Let $\tilde{D}_V$ denote the irreducible component of $\tilde{D}$ corresponding to the edge V. We need the following result, see Proposition (2.3) in [3] (our situation is slightly different, but the same proof applies).

Proposition 4.2. *The intersection of a family of irreducible components $(\tilde{D}_{V_k})_{k=1,r}$ is empty, unless $V_1 \subset V_2 \subset \ldots \subset V_r$ up to a permutation. In particular, the multiplicity of $\tilde{D}$ at any point $y \in \tilde{D}$ is bounded by n.*

Let $\tilde{Q} = Q \circ \tilde{p}$. Then one has an isomorphism

$$R\tilde{p}_* \psi_{\tilde{Q},\alpha}\mathbb{C}_{\tilde{Z}} = \psi_{Q,\alpha}\mathbb{C}_Z \tag{4.6}$$

compatible with the N-actions. The order of N, acting on the right hand side is bounded by n, by the results in the section (3.2) of [20] and Proposition 4.2. Hence $N^n = 0$ on both sides of (4.6). One may also use Theorem 2.14 in [18]. In view of Proposition 4.1 this completes the proof of Theorem 1.3. □

4.3 Proof of Theorem 1.5

Proof. The first claim is rather obvious in view of Kato-Matsumoto Theorem, see [6, Theorem (3.2.2)].

The second claim follows from Theorem 1.1: indeed, it follows from [21] that in this case $H_0^{n+1}(X_V)$ is a pure HS of weight $n+1$. This fact was also proved in [10].

For the last claim, using again Theorem 1.1, we have to show that $H_0^n(X_V, \mathbb{Q})$ has only weights $n-1$ and n. But this was already noticed in the final part of Section 3. □

The following example shows that Theorems 1.3 and 1.5 are quite sharp.

Example 4.3. We show that $Gr_4^{\mathbf{W}} H^2(F_V, \mathbb{Q}) \neq 0$ for some curves V in $\mathbb{P}^2$. As above, again using Theorem 1.1, we have to show that one may have $\mathbf{W}_0 H_0^2(X_V, \mathbb{Q}) \neq 0$. Let V be an irreducible plane curve having a singularity (V, a) whose local monodromy operator is not of finite order, *e.g.* suppose that a local equation for (V, a) is

$$(x^2 + y^3)(y^2 + x^3) = 0.$$

Then the resolution graph of the corresponding sigularity for the surface X_V has at least one cycle, see [15]. This implies that the cohomology $H^1(K)$ of the corresponding link has elements of weight 0 (dual to the elements of weight 4 in $H^2(K)$ described in [6, Example C29), page 245]. Then an application of Corollary 3.3 with $p = q = 0$ yields the claimed result.

5 Computation of Hodge-Deligne polynomials for line arrangements

Now we start the proof of Theorem 1.6. For this we use an idea already present in [7, page 380]. Let X be a hypersurface in $\mathbb{P}^{n+1}$, with $n \geq 2$, having only isolated singularities $\Sigma = \{a_1, ..., a_m\}$. Let F_s be the Milnor fiber of the singularity (X, a_s). Steenbrink [21] has constructed a MHS on $H^*(F_s)$ such that the following is a MHS exact sequence.

$$0 \to H^n(X) \to H^n(X_\infty) \to \oplus_s H^n(F_s) \to H^{n+1}(X) \to 0. \tag{5.1}$$

Here X_∞ is a smooth surface in $\mathbb{P}^{n+1}$, nearby X, regarded as a generic fiber in a 1-parameter smoothing X_w of X. Moreover, $H^n(X_\infty)$ is endowed with the Schmid-Steenbrink limit MHS, whose Hodge filtration will be denoted by $\mathbf{F}_{SS}$. The Hodge filtration $\mathbf{F}_{SS}$ on $H^n(X_\infty)$, being the limit of the Deligne Hodge filtration $\mathbf{F}$ on $H^n(X_w)$, yields for any p isomorphisms

$$Gr^p_{\mathbf{F}_{SS}} H^n(X_\infty) = Gr^p_{\mathbf{F}} H^n(X_w) \tag{5.2}$$

of $\mathbb{C}$-vector spaces (*i.e.* equality of dimensions). Note that our smoothing X_w can be constructed in a μ_d-equivariant way, *e.g.* just take X_w to be the zeroset in $\mathbb{P}^{n+1}$ of a polynomial of the form $Q_1(x) + wR_1(x)$ with R_1 a generic homogeneous polynomial of degree d in $\mathbb{C}[x]$. With such a choice, the isomorphisms (5.2) become equalities in the representation ring $R(\mu_d)$. Moreover, these representations can be explicitly determined, since they coincide with the representations computed as in the Example below (by a standard deformation argument).

Example 5.1. Let Y be the smooth surface in $\mathbb{P}^3$ defined by $x^d + y^d + z^d + t^d = 0$ with the μ_d-action induced by that on $\mathbb{P}^3$ described above. Using the description of the vector spaces $Gr^p_{\mathbf{F}} H^2_0(Y)$ in terms of rational differential forms à la Griffiths, see for instance [6], it follows that $H^{p,2-p}(d) = Gr^p_{\mathbf{F}} H^2_0(Y)$ is the following μ_d-representation:

(i) if $p = 2$, then the multiplicity of the representation $\mathbb{C}_{\lambda^k}$ is 0 for $k = 1, 2$ and $\binom{k-1}{2}$ for $k = 3, ..., d-1$.

(ii) if $p = 0$, since $\overline{H^{2,0}(d)} = H^{0,2}(d)$, it follows by conjugating (i) that the multiplicity of the representation $\mathbb{C}_{\lambda^k}$ is 0 for $k = d-1, d-2$ and $\binom{d-k-1}{2}$ for $k = 1, ..., d-3$.

(iii) If we denote by $h^{p,q}(\alpha)$ the multiplicity of the representation $\mathbb{C}_\alpha$ for $\alpha \in \mu_d$, $\alpha \neq 1$ in the representation $H^{p,q}(d)$ above, the multiplicities $h^{1,1}(\alpha)$ are determined using (i), (ii) and the formula

$$h^{2,0}(\alpha) + h^{1,1}(\alpha) + h^{0,2}(\alpha) = d^2 - 3d + 3.$$

For each $p = 0, 1, 2$, the exact sequence (5.1) yields an exact sequence of μ_d-modules

$$0 \to Gr^p_{\mathbf{F}} H^2_0(X) \to Gr^p_{\mathbf{F}_{SS}} H^2_0(X_\infty) \to \oplus_s Gr^p_{\mathbf{F}} H^2(F_s) \to Gr^p_{\mathbf{F}} H^3(X) \to 0. \tag{5.3}$$

If H is μ_d-MHS and $\alpha \in \mu_d$, we use the notation $h^{p,q}(H, \alpha)$ for the multiplicity of the representation $\mathbb{C}_\alpha$ in the representation $H^{p,q}(H)$. With this notation, the exact sequence (5.3) and Corollary 3.3 yield the following.

Proposition 5.2. *For $p+q=n$, one has*

$$h^{p,q}(H_0^n(X),\alpha) = h^{p,q}(\alpha) + h^{p,q+1}(H^{n+1}(X),\alpha) + h^{p+1,q}(H^{n+1}(X),\alpha) - \sum_s (h^{p,q}(H^n(F_s,\alpha) + h^{p,q+1}(H^n(F_s,\alpha) + h^{p+1,q}(H^n(F_s,\alpha)).$$

Example 5.3. The Ceva (or Fermat) arrangement $A(3,3,3)$ is defined by

$$Q = (x^3-y^3)(x^3-z^3)(y^3-z^3) = 0.$$

The monodromy action on $H^1(F,\mathbb{C})$ has only three eigenvalues: 1 (with multiplicity 8), $\alpha_1 = \exp(2\pi i/3) = \lambda^3$ (with multiplicity 2), and $\alpha_2 = \exp(4\pi i/3) = \lambda^6$ (with multiplicity 2), see [4, 3.2 (i)], and also [23] for more on this line arrangement.

This arrangement has nine lines, no ordinary double points, and twelve triple points. It follows that the surface X has in this case 12 singularities with local equation

$$a^3+b^3+c^9=0.$$

A local computation using [6, see particularly page 194], implies that each $H^2(F_s)$ has weights 2 and 3. The part of weight 3 has dimension 2 and the corresponding representations are

$$H^{2,1}(H^2(F_s)) = \mathbb{C}_{\lambda^6} \text{ and } H^{1,2}(H^2(F_s)) = \mathbb{C}_{\lambda^3} \tag{5.4}$$

with $\lambda = \exp(2\pi i/9)$. The part of weight 2 has dimension 30 and one has

$$H^{2,0}(H^2(F_s)) = \mathbb{C}_{\lambda^7} \oplus \mathbb{C}_{\lambda^8} \text{ and } H^{0,2}(H^2(F_s)) = \mathbb{C}_{\lambda} \oplus \mathbb{C}_{\lambda^2}. \tag{5.5}$$

The remaining representation $H^{1,1}(H^2(F_s))$ has dimension 26 and is determined by the equalities $h^{1,1}(H^2(F_s),\lambda^k) = 3$ for $0<k<4$ or $5<k<9$ and $h^{1,1}(H^2(F_s),\lambda^k) = 4$ for $k=4,5$. It follows that

$$H^{2,1}(H^3(X)) = 2\mathbb{C}_{\lambda^6} \text{ and } H^{1,2}(H^3(X)) = 2\mathbb{C}_{\lambda^3}. \tag{5.6}$$

Finally, using Corollary 3.3, we get

$$h^{1,2}(H^2(F),\alpha) = h^{1,0}(H_0^2(X),\overline{\alpha}) = 7$$

for $\alpha = \lambda^3$; in particular there are elements of weight 3 in $H^2(F)_{\neq 1}$. And using Proposition 5.2, we get also that

$$h^{0,2}(H^2(F),\alpha) = h^{2,0}(H_0^2(X),\overline{\alpha}) \neq 0$$

for $\alpha = \lambda^5$; in particular there are also elements of weight 2 in $H^2(F)_{\neq 1}$.

Finally, we prove Theorem 1.6. For the first claim we have to show that the multiplicities m_a can be expressed in terms of the listed invariants and $\beta = \exp(2\pi i a)$. The case when a is an integer is very easy, using the identification $H^*(F, \mathbb{Q})_1 = H^*(M, \mathbb{Q})$ and the well known fact that the Betti numbers of M can be expressed in terms of the listed invariants.

We treat the case $1 < a < 2$ and leave the other cases, which are entirely similar and easier, to the reader. In the case $1 < a < 2$, one has, with notation from the Introduction

$$m_a = -h^{1,0}(H^1(F), \beta) + h^{1,1}(H^2(F), \beta) + h^{1,2}(H^2(F), \beta).$$

Using Corollary 1.2, we have

$$m_a = -h^{1,2}(H^3(X), \overline{\beta}) + h^{1,1}(H_0^2(X), \overline{\beta}) + h^{1,0}(H_0^2(X), \overline{\beta}).$$

By Proposition 5.2 we get

$$h^{1,1}(H_0^2(X), \overline{\beta}) = h^{1,1}(\overline{\beta}) + h^{1,2}(H^3(X), \overline{\beta}) + h^{2,1}(H^3(X), \overline{\beta}) - LC1 \tag{5.7}$$

where $LC1$ is a number depending only on local invariants at the singularities, *i.e.* computable from the invariants d and m_k for $k \geq 2$ and β. Using Corollary 3.3 we also get

$$h^{1,0}(H_0^2(X), \overline{\beta}) = LC2 - h^{2,1}(H^3(X), \overline{\beta})$$

where $LC2$ is another local constant as above. It follows that in the last formula for m_a the subtle invariants related to X cancel out and the result involves only the local constants $LC1$ and $LC2$ in addition to the number $h^{1,1}(\overline{\beta})$ which depends only on d and β.

In fact, this computation yields the following formula

$$m_a = h^{1,1}(\gamma) - \sum_s (h^{1,1}(H^2(F_s), \gamma) + h^{1,2}(H^2(F_s), \gamma)) \tag{5.8}$$

with $1 < a < 2$ and $\gamma = \exp(-2\pi i a)$.

A similar argument applies to any any hypersurface having only isolated singularities, in view of our Theorem 1.5.

To show that the Hodge-Deligne polynomial $P^{\mu_d}(F)$ corresponding to the monodromy action is not determined by d and by the numbers m_k of points of multiplicity $k \geq 2$ in V, it is enough to find one coefficient which involves invariants associated to X. Indeed, it is known that there are line arrangements $\mathcal{A}$ and $\mathcal{A}'$, having the same list of invariants and with different values for some $h^{1,2}(H^3(X), \beta)$, see for instance [6, Theorem 6.4.15 and its proof, pages 212-213]. If $\beta \neq 1$, then the multiplicity of $\mathbb{C}_\beta$ in the virtual representation $E^{\mu_d;1,1}(F)$, which is the coefficient of uv in $P^{\mu_d}(F)$, is $h^{1,1}(H_0^2(X), \beta)$. Now the formula (5.7) with β replacing $\overline{\beta}$ completes the proof.

A Appendix

Rational points over finite fields and equivariant Hodge-Deligne polynomials

A.1 The setting

Let X be a variety over $\mathcal{O}[\frac{1}{n}]$, where $\mathcal{O}$ is the ring of integers of an algebraic number field F, and suppose that the finite group Γ acts as a group of scheme automorphisms on X. Then X has (compact support) equivariant Hodge-Deligne modules $H^{d,e}(H_c^j(X(\mathbb{C}), \mathbb{C}) \in R(\Gamma)$ defined as in 2.2 above, and correspondingly, the equivariant Hodge-Deligne polynomial $P_c^\Gamma(X)(x, y) \in R(\Gamma)[x, y]$, also defined in Definition 2.2. If $\Gamma = 1$, we have the usual Hodge-Deligne numbers [12] given by

$$h^{d,e}(j) := \dim_{\mathbb{C}} \mathrm{Gr}_{\mathbf{F}}^d \mathrm{Gr}_{\overline{\mathbf{F}}}^e H_c^j(X(\mathbb{C}), \mathbb{C}). \tag{A.1}$$

The Euler-Hodge numbers of X are given by

$$h^{d,e} := \sum_j (-1)^j h^{d,e}(j), \tag{A.2}$$

and the (non-equivariant, compact supports) Hodge-Deligne polynomial by

$$P_c(X)(x, y) := \sum_{d,e} h^{d,e} x^d y^e. \tag{A.3}$$

In this appendix, we amplify some of the results of [12] to make more explicit connections between the Hodge-Deligne polynomials of X and the eigenvalues of (possibly twisted) Frobenius endomorphisms on the p-adic étale cohomology of the reduction modulo various primes of X.

In particular, we show how to deduce a result of Katz [11] by this means, and give an equivariant analogue (Theorem A.2) of that result. Our argument uses the K group of representations of the Galois group, rather than the K group of schemes, as Katz did.

A.2 Background in p-adic Hodge theory

We recall the basic setup, and amplify some results of [12].

Notation. We shall use the notation of [12]. In particular, F is a number field, S a finite set of primes in F, $\overline{F}$ an algebraic closure of F, and $F_S \subset \overline{F}$ the maximal extension of F which is unramified outside S. Write $G_{F,S} = \mathrm{Gal}(F_S/F)$, and for a prime of F $v \notin S$, write Frob_v for the corresponding geometric Frobenius automorphism in $G_{F,S}$. Write q_v

for the cardinality of the residue field of v. We shall often write Frob_q for Frob_v if $q_v = q$.

For a rational prime p such that S contains all the prime divisors of p in F, denote by $\mathbb{Q}_p(i)$ the ith tensor power of the one dimensional cyclotomic representation of $G_{F,S}$ over $\mathbb{Q}_p$. For each prime $\mathfrak{p}$ dividing p, fix an algebraic closure $\overline{F}_{\mathfrak{p}}$ of the $\mathfrak{p}$-adic completion $F_{\mathfrak{p}}$ of F at $\mathfrak{p}$. Fix an embedding $\overline{F} \to \overline{F}_{\mathfrak{p}}$ and denote the corresponding decomposition group by $G_{F_{\mathfrak{p}}}$. There is then a canonical homomorphism $G_{F_{\mathfrak{p}}} \to G_{F,S}$, and representations of $G_{F,S}$ may therefore be restricted to $G_{F_{\mathfrak{p}}}$.

We refer to [12, Section 2] for properties of Fontaine's filtered field B_{dR}. The relevant notation we require is as follows. The field B_{dR} is discretely valued and contains $F_{\mathfrak{p}}$. Its residue field is denoted $\mathbb{C}_p$, and its decreasing filtration is denoted $\mathrm{Fil}^{\bullet} B_{dR}$. If V is a finite dimensional continuous $\mathbb{Q}_p G_{F_{\mathfrak{p}}}$-module, recall that V is said to be de Rham if $\dim_{F_{\mathfrak{p}}}(B_{dR} \otimes_{\mathbb{Q}_p} V)^{G_{F_{\mathfrak{p}}}} = \dim_{\mathbb{Q}_p}(V)$. We note that it is pointed out in [12] that it follows from arguments of Faltings, Tsuji and Kisin that any subquotient of $H_c^j(X, \mathbb{Q}_p)$ is de Rham.

A.3 Cohomology and eigenvalues of Frobenius

Recall that the de Rham cohomology $H^j_{c,dR}(X)$ is an F-vector space with a decreasing (Hodge) filtration $\mathbf{F}^{\bullet}$, whose complexification $H_c^j(X(\mathbb{C}),\mathbb{C}) := H^j_{c,dR}(X) \otimes_F \mathbb{C}$ correspondingly has two decreasing filtrations $\mathbf{F}^{\bullet}, \overline{\mathbf{F}}^{\bullet}$, as well as the increasing weight filtration $\mathbf{W}_{\bullet}$. The associated graded components of these filtrations are related by

$$\mathrm{Gr}^{\mathbf{W}}_m H^j_{c,dR}(X) \otimes_F \mathbb{C} = \oplus_{d+e=m} \mathrm{Gr}^d_{\mathbf{F}} \mathrm{Gr}^e_{\overline{\mathbf{F}}} H^j_c(X(\mathbb{C}), \mathbb{C}). \tag{A.4}$$

The p-adic cohomology $H_c^j(X, \mathbb{Q}_p)$ also has a weight filtration (*cf.* [12, (2.1.5)]) $\mathbf{W}_{\bullet} H_c^j(X, \mathbb{Q}_p)$, whose associated graded parts are denoted $\mathrm{Gr}^{\mathbf{W}}_m H_c^j(X,\mathbb{Q}_p)$. The eigenvalues of Frob_v (see above) on $\mathrm{Gr}^{\mathbf{W}}_m H_c^j(X,\mathbb{Q}_p)$ are known to be of the form $\zeta q_v^{\frac{m}{2}}$, where ζ is an algebraic number which has absolute value 1 in any embedding $\mathbb{Q}_p \to \mathbb{C}$. We fix such an embedding, and denote the eigenvalues of Frob_v on $\mathrm{Gr}^{\mathbf{W}}_m H_c^j(X, \mathbb{Q}_p)$ by $\zeta^j_{m,k} q^{\frac{m}{2}}$, $k = 1, 2, \ldots, d^j_m$, where $d^j_m = \dim_{\mathbb{Q}_p} \mathrm{Gr}^{\mathbf{W}}_m H_c^j(X, \mathbb{Q}_p)$.

A.4 Filtrations and comparison theorems

Recall the following facts from [12]. We have (*cf.* [12, (2.1.3)]) the following isomorphism of filtered $F_{\mathfrak{p}} G_{F_{\mathfrak{p}}}$-modules for each j:

$$H^j_{c,dR}(X) \otimes_F B_{dR} \xrightarrow{\sim} H_c^j(X, \mathbb{Q}_p) \otimes_{\mathbb{Q}_p} B_{dR}, \tag{A.5}$$

where on the left, the filtration is the tensor product of $\mathrm{Fil}^\bullet$ on B_{dR} and $\mathbf{F}^\bullet$ on $H^j_{c,dR}(X)$, while on the right side the filtration comes from just the filtration on B_{dR}. Moreover the isomorphism (A.5) respects the weight filtrations on the two cohomology theories (see [12, Lemma (2.1.4), Cor. (2.1.5)]).

Since $\mathrm{Gr}^k_{\mathrm{Fil}}(B_{dR}) \simeq \mathbb{C}_p(k)$ as $F_\mathfrak{p} G_{F_\mathfrak{p}}$-module, where $\mathbb{C}_p(k)$ denotes the kth Tate twist of the cyclotomic character, we obtain the following isomorphism of $F_\mathfrak{p} G_{F_\mathfrak{p}}$-modules by taking the weight m component of the degree d associated graded of the filtered spaces in (A.5).

$$\oplus_{i=0}^d \mathrm{Gr}^i_{\mathbf{F}} \mathrm{Gr}^{\mathbf{W}}_m H^j_{c,dR}(X) \otimes_F \mathbb{C}_p(d-i) \xrightarrow{\sim} \mathrm{Gr}^{\mathbf{W}}_m H^j_c(X, \mathbb{Q}_p) \otimes_{\mathbb{Q}_\mathfrak{p}} \mathbb{C}_p(d). \tag{A.6}$$

Now take $G_{F_\mathfrak{p}}$-fixed points of both sides of (A.6). Since $G_{F_\mathfrak{p}}$ has trivial action on $H^j_{c,dR}(X)$, and $\mathbb{C}_p(d)^{G_{F_\mathfrak{p}}} = 0$ if $d \neq 0$, while $\mathbb{C}_p(0)^{G_{F_\mathfrak{p}}} = F_\mathfrak{p}$ (see [22]) the left side becomes $\mathrm{Gr}^d_{\mathbf{F}} \mathrm{Gr}^{\mathbf{W}}_m H^j_{c,dR}(X) \otimes_F F_\mathfrak{p}$, which has $F_\mathfrak{p}$-dimension $h^{d,m-d}(j)$. We therefore have, for each j, m and d,

$$\mathrm{Gr}^d_{\mathbf{F}} \mathrm{Gr}^{\mathbf{W}}_m H^j_{c,dR}(X) \otimes_F F_\mathfrak{p} \xrightarrow{\sim} \left(\mathrm{Gr}^{\mathbf{W}}_m H^j_c(X, \mathbb{Q}_p) \otimes_{\mathbb{Q}_\mathfrak{p}} \mathbb{C}_p(d)\right)^{G_{F_\mathfrak{p}}}, \tag{A.7}$$

and taking dimensions over $F_\mathfrak{p}$ we obtain

$$h^{d,m-d}(j) = \dim_{F_\mathfrak{p}} \left(\mathrm{Gr}^{\mathbf{W}}_m H^j_c(X, \mathbb{Q}_p) \otimes_{\mathbb{Q}_\mathfrak{p}} \mathbb{C}_p(d)\right)^{G_{F_\mathfrak{p}}}. \tag{A.8}$$

We shall make use of the Grothendieck ring $R(\mathbb{Q}_p G_{F,S})$ of finite dimensional continuous $\mathbb{Q}_p G_{F,S}$ representations. If R is such a representation, we write $[R]$ for its class in $R(\mathbb{Q}_p G_{F,S})$. Every such element is equal to $\sum_i [S_i]$ where S_i is a simple $\mathbb{Q}_p G_{F,S}$-module, and if $[R] = \sum_i [S_i]$, we say that $\oplus_i S_i$ is the semisimplification of R, and write $R_{ss} = \oplus_i S_i$.

A.5 Katz's theorem

We shall show how the above considerations may be used to prove the following result of Katz.

Theorem A.1 (Katz, [11], Appendix). *Suppose there is a polynomial $P_X(t) \in \mathbb{C}[t]$ such that for almost all q, $|X(\mathbb{F}_q)| = P_X(q)$. Then $h^{d,e} = 0$ if $d \neq e$, and $P_c(X)(x, y) = P_X(xy)$.*

The term "almost all" here means that for all but finitely many rational primes ℓ, there is a power q_ℓ of ℓ such that the assertion holds for $q = q_\ell^r$, for any r.

Proof of Katz's theorem. We are given a polynomial $P_X(t) = \sum_{i=0}^{n} a_{2n} t^n$, such that for almost all q, $|X(\mathbb{F}_q)| = P_X(q)$. Define the constants c_i, $i = 0, 1, \ldots, n$ by

$$c_i = \begin{cases} a_i \text{ if } i \text{ is even} \\ 0 \text{ otherwise.} \end{cases} \tag{A.9}$$

By the Grothendieck fixed point theorem, we have

$$\begin{aligned} |X(\mathbb{F}_q)| &= \sum_{j=0}^{2\dim(X)} (-1)^j \operatorname{Trace}(\operatorname{Frob}_q, H_c^j(X, \mathbb{Q}_p)) \\ &= \sum_m \sum_{j=0}^{2\dim(X)} (-1)^j \operatorname{Trace}(\operatorname{Frob}_q, \operatorname{Gr}_m^{\mathbf{W}} H_c^j(X, \mathbb{Q}_p)). \end{aligned} \tag{A.10}$$

Now all modules $\operatorname{Gr}_m^{\mathbf{W}} H_c^j(X, \mathbb{Q}_p))$ are represented in the Grothendieck ring $R(\mathbb{Q}_p G_{F,S})$, and modules with equal trace functions on almost all Frob_v are equal in $R(\mathbb{Q}_p G_{F,S})$. Using the fact that the concept of weight as defined by the eigenvalues of Frobenius coincides with that arising from Hodge theory [5], it follows by taking the pieces of weight m in (A.10) that the following equation holds in $R(\mathbb{Q}_p G_{F,S})$.

$$\sum_j (-1)^j [\operatorname{Gr}_m^{\mathbf{W}} H_c^j(X, \mathbb{Q}_p)] = c_m \mathbb{Q}_p \left(-\frac{m}{2}\right). \tag{A.11}$$

Note that the right side of (A.11) is zero if $c_m = 0$, in particular if m is odd. Write

$$V_m^e := \oplus_{j \text{ even}} \operatorname{Gr}_m^{\mathbf{W}} H_c^j(X, \mathbb{Q}_p) \text{ and}$$

$$V_m^o := \oplus_{j \text{ odd}} \operatorname{Gr}_m^{\mathbf{W}} H_c^j(X, \mathbb{Q}_p).$$

It follows from (A.7) that (*cf.* (A.2))

$$h^{d,m-d} = \dim_{F_\mathfrak{p}} (V_m^e \otimes \mathbb{C}_p(d))^{G_{F_\mathfrak{p}}} - \dim_{F_\mathfrak{p}} (V_m^o \otimes \mathbb{C}_p(d))^{G_{F_\mathfrak{p}}}. \tag{A.12}$$

We observe next that in (A.12), we may replace V_m^e etc. by their semisimplifications. To see this, let $V = V_m^e$; then clearly $\left(V \otimes_{\mathbb{Q}_p} \mathbb{C}_p(d)\right)_{ss} = V_{ss} \otimes_{\mathbb{Q}_p} \mathbb{C}_p(d)$ as $G_{F_\mathfrak{p}}$-modules. But it follows from (A.6) that $V \otimes_{\mathbb{Q}_p} \mathbb{C}_p$, and hence also $V \otimes_{\mathbb{Q}_p} \mathbb{C}_p(d)$, is semisimple, and therefore equal to its semisimplification. Thus $V_{ss} \otimes_{\mathbb{Q}_p} \mathbb{C}_p(d) \simeq V \otimes_{\mathbb{Q}_p} \mathbb{C}_p(d)$.

It then follows from (A.12)and (A.11) that

$$h^{d,m-d} = c_m \dim_{F_\mathfrak{p}} \left(\mathbb{Q}_p \left(-\frac{m}{2}\right) \otimes \mathbb{C}_p(d)\right)^{G_{F_\mathfrak{p}}}. \tag{A.13}$$

Hence $h^{d,m-d} = 0$ unless m is even and $m = 2d$, and if this condition is satisfied, then $h^{d,d} = c_{2d}$. The result is now clear. $\square$

A.6 Equivariant theory

Let Γ be a finite group of automorphisms of X, where X is as in Section A.1. Then Γ preserves all the filtrations discussed above, and we define the equivariant Hodge numbers by

$$h^{d,e}(j, w) := \mathrm{Trace}\left(w, \mathrm{Gr}^d_{\mathbf{F}} \mathrm{Gr}^e_{\overline{\mathbf{F}}} H^j_c(X(\mathbb{C}), \mathbb{C})\right), \tag{A.14}$$

for $w \in \Gamma$.

Similarly we define

$$h^{d,e}(w) := \sum_j (-1)^j h^{d,e}(j, w), \tag{A.15}$$

and the equivariant Hodge polynomials by

$$P^{\Gamma}_c(X)(x, y; w) := \sum_{d,e} h^{d,e}(w) x^d y^e. \tag{A.16}$$

We shall prove the following equivariant generalization of Katz's theorem.

Theorem A.2. *Suppose there are polynomials* $P_X(t;w) = \sum_{i=0}^{2\dim X} a_{2i}(w)t^i \in \mathbb{C}[t]$ *such that for almost all* q*, and all* $w \in \Gamma$*, we have* $|X(\overline{\mathbb{F}}_q)^{w\,\mathrm{Frob}_q}| = P_X(q; w)$*. Then* $h^{d,e}(w) = 0$ *if* $d \neq e$*, and* $P^{\Gamma}_c(X)(x, y; w) = P_X(xy, w)$ *for each* $w \in \Gamma$*. Moreover the function* $w \mapsto a_{2j}(w)$ *is a virtual character of* Γ *for each* j*.*

Proof. This is similar to the proof of Katz's theorem above, and we maintain the above notation. Write $\Theta := R(\Gamma \times G_{F,S})$ be the Grothendieck group of finite dimensional continuous representations of $\Gamma \times G_{F,S}$ over $\mathbb{Q}_p$, with Γ having the discrete topology. Note that the set of elements (w, Frob_q) is dense in $\Gamma \times G_{F,S}$, so that two elements of Θ are equal if and only if each element (w, Frob_q) has equal traces on the two modules.

Now any element θ of Θ may be written uniquely in the form $\theta = \sum_\phi \chi_\phi \otimes \phi$, where the (finite) sum is over the simple representations ϕ of $G_{F,S}$, and for each ϕ, $\chi_\phi \in R(\Gamma)$ is a virtual representation of Γ. This applies in particular to the $\Gamma \times G_{F,S}$ modules $\mathrm{Gr}^{\mathbf{W}}_m H^j_c(X, \mathbb{Q}_p)$.

By the Grothendieck fixed point theorem, we have, for any $w \in \Gamma$,

$$\begin{aligned}
|X(\overline{\mathbb{F}_q})^{w\,\mathrm{Frob}_q}| &= \sum_{j=0}^{2\dim(X)} (-1)^j \operatorname{Trace}(w\,\mathrm{Frob}_q, H_c^j(X, \mathbb{Q}_p)) \\
&= \sum_m \sum_{j=0}^{2\dim(X)} (-1)^j \operatorname{Trace}(w\,\mathrm{Frob}_q, \operatorname{Gr}_m^{\mathbf{W}} H_c^j(X, \mathbb{Q}_p)) \\
&= P_X(q, w),
\end{aligned} \tag{A.17}$$

and taking the weight m piece of (A.17), it follows that we have the following equation in $\Theta = R(\Gamma \times G_{F,S})$.

$$\sum_j (-1)^j [\operatorname{Gr}_m^{\mathbf{W}} H_c^j(X, \mathbb{Q}_p)] = \chi_m \otimes \mathbb{Q}_p\left(-\frac{m}{2}\right), \tag{A.18}$$

where χ_m is a virtual representation of Γ whose character at $w \in \Gamma$ is $a_{2m}(w)$.

Next, for any element $\theta = \sum_\phi \chi_\phi \otimes \phi$ of Θ, define the $G_{F_\mathfrak{p}}$-invariant part $\theta^{G_{F_\mathfrak{p}}}$ by $\theta^{G_{F_\mathfrak{p}}} = \chi_1 \in R(\Gamma)$, the coefficient of the trivial representation of $G_{F,S}$. This coincides with the $1_{G_{F_\mathfrak{p}}}$-isotypic part of θ in the case of proper representations. It follows from (A.7) that as Γ-module,

$$\operatorname{Gr}_{\mathbf{F}}^d \operatorname{Gr}_m^{\mathbf{W}} H_{c,dR}^j(X) \otimes_F F_\mathfrak{p} = \left(\operatorname{Gr}_m^{\mathbf{W}} H_c^j(X, \mathbb{Q}_p) \otimes_{\mathbb{Q}_\mathfrak{p}} (1_\Gamma \otimes \mathbb{C}_p(d))\right)^{G_{F_\mathfrak{p}}}. \tag{A.19}$$

Hence by (A.18) we have the following equation in $R(\Gamma)$. Write $H^{d,m-d}$ for the element of $R(\Gamma)$ represented by $\sum_j (-1)^j \operatorname{Gr}_{\mathbf{F}}^d \operatorname{Gr}_m^{\mathbf{W}} H_{c,dR}^j(X) \otimes_F F_\mathfrak{p}$. Then

$$H^{d,m-d} = \left(\chi_m \otimes (\mathbb{Q}_p(-\frac{m}{2}) \otimes_{\mathbb{Q}_p} \mathbb{C}_p(d))\right)^{G_{F_\mathfrak{p}}}. \tag{A.20}$$

The right side of (A.20) is 0 unless $m = 2d$, and is equal to χ_m when $m = 2d$; the result is now clear. □

Remark A.3. In Remark 2.3 it was pointed out that for the equivariant weight polynomials $W_c^\Gamma(X)(x)$ of [9], we have the relation $P_c^\Gamma(X)(x,x) = W_c^\Gamma(X)(x)$. Hence given the conditions of Theorem A.2, the conclusion may be stated as $P_c^\Gamma(X)(x, y) = W_c^\Gamma(X)(\sqrt{xy})$.

A.7 Further remarks

We note that the following result is an easy consequence of [12].

Proposition A.4. *Let V be a continuous $\mathbb{Q}_p G_{F,S}$ module. Then*

(i) *For fixed integer i, let V_i be the subset of vectors $x \in V$ such that for almost all q,* $\mathrm{Frob}_q\, x = \zeta q^i x$, *for some root of unity ζ. Then V_i is a subspace of V.*

(ii) Frob_q *acts semisimply on the subspace $V_T := \sum_i V_i$ for almost all q.*

Proof. (i) If x and y are in V_i, then for almost all q, $(\mathrm{Frob}_q)^{n(x)} x = q^{n(x)i} x$ for some integer $n(x)$, and similarly for y; so $\mathrm{Frob}_q^{n(x)n(y)}(x+y) = q^{n(x)n(y)i}(x+y)$, whence $x+y \in V_i$.

(ii) The proof of [12, Proposition (1.2)] shows that Frob_v acts semisimply on V_i, and hence on $V_T := \sum_i V_i$. □

It follows from Proposition A.4 and (A.8) that $\dim\left(H_c^j(X, \mathbb{Q}_p)\right)_d \leq h^{d,d}(j)$.

When X is smooth and projective the space V_T (which in this case consists just of a single V_i) is the subject of the Tate conjecture, which asserts that it should be the subspace spanned by cycle classes. Thus equality above would mean that the cohomology is spanned by cycle classes. This is satisfied only in certain special cases - for example if X has a stratification by affine spaces.

References

[1] D. Abramovich and J. Wang, *Equivariant resolution of singularities in characteristic 0*, Math. Res. Lett. **4** (1997), 427–433.

[2] D. Arapura, P. Bakhtary and J. Wlodarczyk, *The combinatorial part of the cohomology of a singular variety*, arXiv:0902.4234.

[3] N. Budur and M. Saito, *Jumping coefficients and spectrum of a hyperplane arrangement*, Math. Ann. **347** (2010), 545–579.

[4] N. Budur, A. Dimca and M. Saito, *First Milnor cohomology of hyperplane arrangements*, Contemporary Mathematics **538** (2011), 279–292.

[5] P. Deligne, *Poids dans la cohomologie des variétés algébriques*, In: “Proceedings of the International Congress of Mathematicians”, (Vancouver, B. C., 1974), Vol. 1, Canad. Math. Congress, Montreal, Que., (1975), 79–85.

[6] A. Dimca, “Singularities and Topology of Hypersurfaces”, Universitext, Springer-Verlag, 1992.

[7] A. Dimca, *Hodge numbers of hypersurfaces*, Abh. Math. Sem. Hamburg **66** (1996), 377–386.

[8] A. DIMCA, "Sheaves in Topology", Universitext, Springer-Verlag, 2004.

[9] A. DIMCA and G. I. LEHRER, "Purity and Equivariant Weight Polynomials", In: *Algebraic Groups and Lie Groups*, editor G. I. Lehrer, Cambridge University Press, 1997.

[10] A. DIMCA and S. PAPADIMA, *Finite Galois covers, cohomology jump loci, formality properties, and multinets*, Ann. Scuola Norm. Sup. Pisa Cl. Sci (5), Vol. X (2011), 253–268.

[11] T. HAUSEL and F. RODRIGUEZ-VILLEGAS, *Mixed Hodge polynomials of character varieties. With an appendix by Nicholas M. Katz*, Invent. Math. **174** (2008), no. 3, 555–624.

[12] M. KISIN and G. I. LEHRER, *Eigenvalues of Frobenius and Hodge numbers*, Pure Appl. Math. Q. **2** (2006) 497–518.

[13] G. I. LEHRER, *The ℓ-adic cohomology of hyperplane complements*, Bull. London Math. Soc. **24** (1992), 76–82.

[14] G. I. LEHRER and D. E. TAYLOR, "Unitary Reflection Groups", Australian Mathematical Society Lecture Series, 20. Cambridge University Press, Cambridge, 2009.

[15] A. NÉMETHI, *The resolution of some surface singularities*, I. (cyclic coverings), Proceedings of the AMS Conference, San Antonio, 1999. Contemporary Mathematics 266, Singularities in Algebraic and Analytic Geometry, AMS 2000, 89–128.

[16] P. ORLIK and H. TERAO, "Arrangements of Hyperplanes", Springer-Verlag, Berlin Heidelberg New York, 1992.

[17] C. PETERS and J. STEENBRINK, "Mixed Hodge Structures", Ergeb. der Math. und ihrer Grenz. 3. Folge 52, Springer, 2008.

[18] M. SAITO, *Mixed Hodge modules*, Publ. RIMS, Kyoto Univ. 26 (1990), 221–333.

[19] M. SAITO, *Multiplier ideals, b-functions, and spectrum of a hypersurface singularity*, Compositio Math. **143** (2007), 1050–1068.

[20] M. SAITO, *Bernstein-Sato polynomials of hyperplane arrangements*, math.AG/0602527.

[21] J. STEENBRINK, *Mixed Hodge structures on the vanishing cohomology*, In: "Real and Complex Singularities", Oslo, 1976, 525–563.

[22] J. T. TATE, '*p-divisible groups*, In: "1967 Proc. Conf. Local Fields" (Driebergen, 1966) pp. 158–183 Springer, Berlin.

[23] H. ZUBER, *Non-formality of Milnor fibers of line arrangements*, Bull. London Math. Soc. **42** (2010), no. 5, 905–911.

The contravariant form on singular vectors of a projective arrangement

Michael J. Falk and Alexander N. Varchenko*

Abstract. We define the flag space and space of singular vectors for an arrangement $\mathcal{A}$ of hyperplanes in projective space equipped with a system of weights $a\colon \mathcal{A} \to \mathbb{C}$. We show that the contravariant bilinear form of the corresponding weighted central arrangement induces a well-defined form on the space of singular vectors of the projectivization. If $\sum_{H\in\mathcal{A}} a(H) = 0$, this form is naturally isomorphic to the restriction to the space of singular vectors of the contravariant form of any affine arrangement obtained from $\mathcal{A}$ by dehomogenizing with respect to one of its hyperplanes.

1 Introduction

Let $\mathcal{A} = \{H_1, \ldots, H_n\}$ be an arrangement of affine hyperplanes in $\mathbb{C}^\ell$. Let $f_i\colon \mathbb{C}^\ell \to \mathbb{C}$ be an affine linear functional with zero locus H_i, for $1 \leqslant i \leqslant n$. Let $M = M(\mathcal{A}) = \mathbb{C}^\ell - \bigcup_{i=1}^n H_i$ be the complement to the arrangement. If W is a $\mathbb{C}$-vector space, then W^* denotes its dual space. Let $\mathbb{C}^\times = \mathbb{C} - \{0\}$.

Let $\omega_i = d\log(f_i)$ for $1 \leqslant i \leqslant n$. Denote by A the $\mathbb{C}$-subalgebra of the holomorphic De Rham complex of M generated by the closed forms $1, \omega_1, \ldots, \omega_n$. The algebra A is graded, $A = \bigoplus_{p=0}^{\ell} A^p$, and called the Arnol'd-Brieskorn-Orlik-Solomon algebra or the OS algebra of $\mathcal{A}$. The dual space $\mathcal{F} = \mathcal{F}(\mathcal{A}) := \bigoplus_{p\geqslant 0} \mathcal{F}^p$ of A is called the *flag space* of $\mathcal{A}$, [6].

Let $a = (a_1, \ldots, a_n) \in \mathbb{C}^n$ be a vector of weights. The *contravariant form* of the weighted arrangement $(\mathcal{A}, a)$ is the symmetric bilinear form

*Partially supported by NSF grant DMS-1101508

This research was started during the intensive research period "Configuration Spaces: Geometry, Combinatorics and Topology," May-June, 2010, at the Centro di Ricerca Matematica Ennio De Giorgi in Pisa. The authors thank the institute, organizers, and staff for their hospitality and financial support.

$S = \oplus S_p \colon \mathcal{F} \otimes \mathcal{F} \to \mathbb{C}$, where $S_p \colon \mathcal{F}^p \otimes \mathcal{F}^p \to \mathbb{C}$ is defined by

$$S_p(F, F') = \sum_J a_J F(\omega_J) F'(\omega_J). \tag{1.1}$$

The sum is over all sequences $J = (j_1, \cdots, j_p)$ with $1 \leqslant j_1 < \cdots < j_p \leqslant n$, $a_J = \prod_{i=1}^p a_{j_i}$ and $\omega_J = \omega_{j_1} \wedge \cdots \wedge \omega_{j_p}$, [6].

In particular, if $\{F_1, \ldots, F_n\} \subseteq \mathcal{F}^1$ is the basis dual to the basis $\{\omega_1, \ldots, \omega_n\}$ of $A^1 \cong \mathbb{C}^n$, then

$$S_1(F_i, F_j) = a_i \delta_{ij}. \tag{1.2}$$

The contravariant form has many remarkable properties, see [6–9]. It is a generalization of the Shapovalov form associated to a tensor product of highest weight representations of a simple Lie algebra – for this application $\mathcal{A}$ is a discriminantal arrangement and a is determined by the representations.

The space $\mathcal{F}$ has a combinatorially defined differential $d \colon \mathcal{F}^p \to \mathcal{F}^{p+1}$. The space A has a differential $\delta_a \colon A^p \to A^{p+1}$ defined by multiplication by $\omega_a := \sum_{i=1}^n a_i \omega_i$. The contravariant form S induces a morphism of complexes $\psi \colon (\mathcal{F}, d) \to (A, \delta_a)$, see [6] and Section 2. The pair $(\mathcal{F}, d)$ is the *flag complex* of $\mathcal{A}$.

Let $\mathrm{Sing}(\mathcal{F}^\ell) = \mathrm{Sing}_a(\mathcal{F}^\ell) \subseteq \mathcal{F}^\ell$ be the annihilator of $\omega_a \wedge A^{\ell-1}$. It is called the subspace of *singular vectors* of $\mathcal{F}^\ell$, relative to a. This terminology is introduced in [8] and motivated by [6]. In [6] the subspace $\mathrm{Sing}(\mathcal{F}^\ell)$ for a discriminantal arrangement is interpreted as the subspace of singular vectors of a tensor product of Verma modules over a Kac-Moody algebra. The inclusion $\mathrm{Sing}(\mathcal{F}^\ell) \hookrightarrow (A^\ell)^*$ induces an isomorphism

$$\mathrm{Sing}(\mathcal{F}^\ell) \to \left(H^\ell(A, \delta_a)\right)^* = \left(A^\ell / (\omega_a \wedge A^{\ell-1})\right)^*.$$

Let $\Phi_a = \prod_i f_i^{-a_i}$ be the master function associated with $(\mathcal{A}, a)$, and let $\mathcal{L}_a$ be the rank-one local system on M whose local sections are the multiples of single-valued branches of Φ_a. The inclusion of (A, δ_{ca}) into the twisted algebraic de Rham complex of $\mathcal{L}_{ca}$ induces an isomorphism

$$H^*(A, \delta_{ca}) \cong H^*(M, \mathcal{L}_{ca})$$

for generic c [6]. Since $\mathrm{Sing}_{ca}(\mathcal{F}^\ell) = \mathrm{Sing}_a(\mathcal{F}^\ell)$ for any nonzero c, this implies that $\mathrm{Sing}_a(\mathcal{F}^\ell)$ is isomorphic to the local system homology $H_\ell(M, \mathcal{L}_{-ca})$ for generic c.

The restriction of the contravariant form S_ℓ to the subspace $\mathrm{Sing}(\mathcal{F}^\ell)$ is of special interest. It relates linear and nonlinear characteristics of the weighted arrangement $(\mathcal{A}, a)$. For example, the rank of the restriction of

S_ℓ to $\operatorname{Sing}(\mathcal{F}^\ell)$ bounds from above the number of critical points of the master function Φ_a, see [9]. For other properties see in [8,9].

The complement M to $\mathcal{A}$ in $\mathbb{C}^\ell$ can be identified with the complement to the arrangement $\mathcal{A}_\infty$ of projective hyperplanes in projective space $\mathbb{P}^\ell$ consisting of the closures of the hyperplanes of $\mathcal{A}$ together with the hyperplane H_∞ at infinity. The purpose of this note is to develop the notions of the flag space $\mathcal{F}$, the space $\operatorname{Sing}(\mathcal{F}^\ell)$ of singular vectors of $\mathcal{F}^\ell$, and the bilinear form $S_\ell|_{\operatorname{Sing}(\mathcal{F}^\ell)}$ on $\operatorname{Sing}(\mathcal{F}^\ell)$, starting with the projective arrangement $\mathcal{A}_\infty$ in $\mathbb{P}^\ell$. Namely, the purpose is to define these objects in such a way that they will not depend on the choice of the hyperplane at infinity of the projective arrangement.

The following example illuminates our constructions.

Example 1.1. Let $\mathcal{A}$ be the affine arrangement in $\mathbb{C}^1$ of n distinct points $z_1, \dots, z_n$. Then $\omega_i = d\log(x - z_i)$ for $1 \leqslant i \leqslant n$. We have $A^0 \cong \mathbb{C}$, $A^1 \cong \mathbb{C}^n$ and $A^p = 0$ for $p \geqslant 2$. Let $a \in \mathbb{C}^n$ be a vector of weights. The contravariant form of $(\mathcal{A}, a)$ on $\mathcal{F}^1$ is given by (1.2). The image $\omega_a \wedge A^0$ of δ_a in A^1 is spanned by ω_a, and the subspace $\operatorname{Sing}(\mathcal{F}^1) \subset \mathcal{F}^1$ of singular vectors is the subspace $\{\sum_{i=1}^n c_i F_i \in \mathcal{F}^1 \mid \sum_{i=1}^n c_i a_i = 0\}$.

Our construction identifies the pair $(\operatorname{Sing}(\mathcal{F}^1), S_1|_{\operatorname{Sing}(\mathcal{F}^1)})$ with the pair

$$\Big(\big(\operatorname{Ann}(\tilde\omega_a \wedge \tilde A^0) + \operatorname{Ann}(q^* A^1)\big)\Big/ \operatorname{Ann}(q^* A^1),\ \tilde S_1|_{(\operatorname{Ann}(\tilde\omega_a \wedge \tilde A^0) + \operatorname{Ann}(q^* A^1))/\operatorname{Ann}(q^* A^1)}\Big)$$

described below.

Let $[u : v]$ be homogeneous coordinates on $\mathbb{P}^1$, with $x = \frac{v}{u}$. The projectivization $\mathcal{A}_\infty$ of $\mathcal{A}$ is the arrangement in $\mathbb{P}^1$ of the points $p_1 = [1 : z_1], \dots, p_n = [1 : z_n]$ and the point $p_0 = [1 : 0]$ at infinity. The weight of p_i is a_i for $1 \leqslant i \leqslant n$ and the weight of p_0 is $a_0 = -\sum_{i=1}^n a_i$.

In our construction we use the associated central arrangement in $\mathbb{C}^2$, the cone $\tilde{\mathcal{A}}$ of $\mathcal{A}_\infty$, consisting of the lines $v - z_i u = 0$ for $1 \leqslant i \leqslant n$ and the line $u = 0$. Introduce the following one-forms on $\mathbb{C}^2$: $\tilde\omega_i = d\log(v - z_i u)$ for $1 \leqslant i \leqslant n$, and $\tilde\omega_0 = d\log(u)$. The arrangement $\tilde{\mathcal{A}}$ is weighted with the weights $\tilde a = (a_0, \dots, a_n)$. We will denote by $\tilde M, \tilde A, \tilde{\mathcal{F}}, \tilde\omega_a, \tilde S$ the complement, Orlik-Solomon algebra, flag space of $\tilde{\mathcal{A}}$, special element, and the contravariant form of $(\tilde{\mathcal{A}}, \tilde a)$, respectively.

The orbit map $q\colon \tilde M \to \tilde M/\mathbb{C}^\times = \mathbb{C} - \{z_1, \dots, z_n\}$ induces an injection $q^*\colon A \to \tilde A$ whose image is the subalgebra generated by $\{\sum_{i=0}^n \lambda_i \tilde\omega_i \in \tilde A^1 \mid \sum_{i=0}^n \lambda_i = 0\} \subset \tilde A^1$. One computes

$$q^*(\omega_i) = q^*(d\log(x - z_i)) = d\log(\frac{v}{u} - z_i) = \tilde\omega_i - \tilde\omega_0.$$

Then the special element ω_a of A^1 is mapped by q^* to the special element $\tilde{\omega}_a = \sum_{i=0}^n a_i \tilde{\omega}_i$ of $\tilde{A}^1$. Identifying A^1 with $q^* A^1$, the flag space $\mathcal{F}^1 = (A^1)^*$ is isomorphic to the quotient of $\tilde{\mathcal{F}}^1$ by the annihilator $\operatorname{Ann}(q^* A^1) \subset \tilde{F}^1$ of $q^* A^1 \subset \tilde{A}^1$. The subspace $\operatorname{Ann}(q^* A^1)$ is spanned by $\sum_{i=0}^n \tilde{F}_i$. (Notice that in this consideration the index 0 does not play any special role.) The subspace $\operatorname{Ann}(\tilde{\omega}_a \wedge \tilde{A}^0)$ of $\tilde{\mathcal{F}}^1$ consists of flags $\sum_{i=0}^n c_i \tilde{F}_i$ such that $\sum_{i=0}^n c_i a_i = 0$. This subspace is orthogonal to the subspace $\operatorname{Ann}(q^* A^1)$ relative to the contravariant form of $\tilde{\mathcal{A}}$. Indeed we have $\tilde{S}_1(\sum_{i=0}^n \tilde{F}_i, \sum_{i=0}^n c_i \tilde{F}_i) = \sum_{i=0}^n c_i a_i = 0$. Thus, the contravariant form $\tilde{S}_1$ induces a well-defined form on the image of $\operatorname{Ann}(\tilde{\omega}_a \wedge \tilde{A}^0)$ in $\tilde{\mathcal{F}}^1 / \operatorname{Ann}(q^* A^1)$, namely, a form on

$$\big(\operatorname{Ann}(\tilde{\omega}_a \wedge \tilde{A}^0) + \operatorname{Ann}(q^* A^1)\big) \Big/ \operatorname{Ann}(q^* A^1) \cong \operatorname{Ann}(\tilde{\omega}_a \wedge \tilde{A}^0) \Big/ \big(\operatorname{Ann}(\tilde{\omega}_a \wedge \tilde{A}^0) \cap \operatorname{Ann}(q^* A^1)\big).$$

The flags $\tilde{F}_1, ..., \tilde{F}_n$ induce a basis of $\tilde{\mathcal{F}}^1/\operatorname{Ann}(q^*A^1)$. Using this basis, we see that the form induced by $\tilde{S}_1$ on $\big(\operatorname{Ann}(\tilde{\omega}_a \wedge \tilde{A}^0) + \operatorname{Ann}(q^*A^1)\big)/\operatorname{Ann}(q^*A^1)$ corresponds to the restriction of the original form S_1 to the subspace $\operatorname{Sing}(\mathcal{F}^1)$ under the isomorphism of $\mathcal{F}^1$ with $\tilde{\mathcal{F}}^1 / \operatorname{Ann}(q^* A^1)$.

Notice that the form $\tilde{S}_1$ does not induce a well-defined form on $\mathcal{F}^1 = \tilde{\mathcal{F}} / \operatorname{Ann}(q^* A^1)$ – the extension of $S_1|_{\operatorname{Sing}(\mathcal{F}^1)}$ defined by (1.2) depends on the choice of hyperplane at infinity.

In general, for any weighted affine arrangement $(\mathcal{A}, a)$ in $\mathbb{C}^\ell$, we identify the pair $\big(\operatorname{Sing}(\mathcal{F}^\ell), S_\ell|_{\operatorname{Sing}(\mathcal{F}^\ell)}\big)$ with the pair

$$\Big(\big(\operatorname{Ann}(\tilde{\omega}_a \wedge \tilde{A}^{\ell-1}) + \operatorname{Ann}(q^* A^\ell)\big) \Big/ \operatorname{Ann}(q^* A^\ell),\ \tilde{S}_\ell|_{(\operatorname{Ann}(\tilde{\omega}_a \wedge \tilde{A}^{\ell-1}) + \operatorname{Ann}(q^* A^\ell)) / \operatorname{Ann}(q^* A^\ell)}\Big)$$

expressed in terms of the cone $\tilde{\mathcal{A}}$ of the projectivization $\mathcal{A}_\infty$ of $\mathcal{A}$.

Our statement that the pair $\big(\operatorname{Sing}(\mathcal{F}^\ell), S_\ell|_{\operatorname{Sing}(\mathcal{F}^\ell)}\big)$ can be constructed in terms of $\mathcal{A}_\infty$, without choosing a particular hyperplane at infinity, is analogous to the following fact from representation theory. Let V_{Λ_i}, $i = 0, \ldots, n$, be irreducible finite dimensional highest weight representations of a simple Lie algebra. Here Λ_i is the highest weight of V_{Λ_i}. Let $\Lambda_0^\vee$ be the highest weight of the representation dual to V_{Λ_0}. Let S_i be the Shapovalov form on V_{Λ_i}. Let $\operatorname{Sing}(\otimes_{i=0}^n V_{\Lambda_i})[0] \subset \otimes_{i=0}^n V_{\Lambda_i}$ be the subspace of singular vectors of weight zero and $\operatorname{Sing}(\otimes_{i=1}^n V_{\Lambda_i})[\Lambda_0^\vee] \subset \otimes_{i=1}^n V_{\Lambda_i}$ the subspace of singular vectors of weight $\Lambda_0^\vee$. Then the pair

$\left(\mathrm{Sing}(\otimes_{i=1}^n V_{\Lambda_i})[\Lambda_0^\vee], (\otimes_{i=1}^n S_i)|_{\mathrm{Sing}(\otimes_{i=1}^n V_{\Lambda_i})[\Lambda_0^\vee]}\right)$ is isomorphic to the pair $\left(\mathrm{Sing}(\otimes_{i=0}^n V_{\Lambda_i})[0], (\otimes_{i=0}^n S_i)|_{\mathrm{Sing}(\otimes_{i=1}^n V_{\Lambda_i})[0]}\right)$.

ACKNOWLEDGEMENTS. We are also grateful to Sergey Yuzvinsky, who took part in our initial discussions, for his helpful remarks. The second author thanks for hospitality IHES where this paper was finished.

2 Flag complex and contravariant form of a central arrangement

We recall in more detail some of the theory of flag complexes from [6]. The following notation, which differs from the notation of Section 1, will be used throughout the rest of the paper. For general background on arrangements see [5].

Suppose $\mathcal{A} = \{H_0, \dots, H_n\}$ is a central arrangement in $\mathbb{C}^{\ell+1}$. Let $f_0, \dots, f_n \in (\mathbb{C}^{\ell+1})^*$ with $H_i = \ker(f_i)$ for $0 \leqslant i \leqslant n$. Let $\omega_i = \frac{df_i}{f_i}$ for $0 \leqslant i \leqslant n$, and let A be the OS algebra of $\mathcal{A}$, as defined in Section 1. Let E be the graded exterior algebra over $\mathbb{C}$ with generators $e_0, \dots, e_n$ of degree one. Let $\partial\colon E^p \to E^{p-1}$ be defined by

$$\partial(e_{j_1} \wedge \cdots \wedge e_{j_p}) = \sum_{i=1}^{p} (-1)^{i-1} e_{j_1} \wedge \cdots \wedge \hat{e}_{j_i} \wedge \cdots \wedge e_{j_p},$$

where $\hat{}$ denotes deletion. If $J = (j_1, \dots, j_p)$, denote the product $e_{j_1} \wedge \cdots \wedge e_{j_p}$ by e_J. Say J is *dependent* if $\{f_i \mid i \in J\}$ is linearly dependent in $(\mathbb{C}^{\ell+1})^*$. Let I be the ideal of E generated by $\{\partial e_J \mid J \text{ is dependent}\}$. By [4], the surjection $E \to A$ sending e_i to ω_i has kernel I. We tacitly identify A with E/I. The map ∂ induces a well-defined map $\partial\colon A \to A$, a graded derivation of degree -1, and (A, ∂) is a chain complex.

Let $L = L(\mathcal{A})$ be the intersection lattice of $\mathcal{A}$, the set of intersections of subcollections of $\mathcal{A}$, partially-ordered by reverse inclusion. Let $\mathrm{Flag} = \oplus_{p=0}^{\ell+1} \mathrm{Flag}^p$ be the graded $\mathbb{C}$-vector space with Flag^p having basis consisting of chains $(X_0 < \cdots < X_p)$ of L satisfying $\mathrm{codim}(X_i) = i$ for $0 \leqslant i \leqslant p$. Such a chain will be called a *flag*. For each ordered subset $J = (j_1, \dots, j_p)$ of $\{0, \dots, n\}$, let $\xi(J)$ be the chain $(X_0 < \cdots < X_p)$ of L, where $X_0 = \mathbb{C}^{\ell+1}$ and $X_i = \bigcap_{k=1}^{i} H_{j_k}$ for $1 \leqslant i \leqslant p$. Note that $\xi(J)$ is a flag if and only if $\{H_i \mid i \in J\}$ is independent in $\mathcal{A}$. If π is a permutation of $\{1, \dots, p\}$, let $J^\pi = (j_{\pi(1)}, \dots, j_{\pi(p)})$. For any flag $F \in \mathrm{Flag}^p$ and any ordered p-subset J of $\{1, \dots, n\}$, there is at most one permutation π such that $F = \xi(J^\pi)$.

Define a bilinear pairing

$$\langle\ ,\ \rangle\colon \mathrm{Flag}^p \otimes E^p \to \mathbb{C}$$

by

$$\langle F, e_J\rangle = \begin{cases} \operatorname{sgn}(\pi) & \text{if } \xi(J^\pi) = F \\ 0 & \text{otherwise} \end{cases}, \tag{2.1}$$

for every flag F in Flag^p and ordered p-subset J of $\{0, \ldots, n\}$.

Proposition 2.1 ([6]). $\langle F, \partial e_J\rangle = 0$ *for every* $F \in \mathrm{Flag}^p$ *and dependent* $(p+1)$*-tuple* J. *Moreover, if* $(X_0 < \cdots < X_{i-1} < X_{i+1} < \cdots X_p)$ *is a chain in* L *with* $\operatorname{codim}(X_j) = j$ *then*

$$\left\langle \sum_{X_{i-1}<X<X_{i+1}} (X_0 < \cdots < X_{i-1} < X < X_{i+1} < \cdots X_p),\ e_J \right\rangle = 0, \tag{2.2}$$

for every ordered p*-subset* J *of* $\{0, \ldots, n\}$.

Let $\mathcal{F} = \bigoplus_{p=0}^{\ell+1} \mathcal{F}^p$ be the quotient of Flag by the homogeneous subspace spanned by the sums

$$\sum_{X_{i-1}<X<X_{i+1}} (X_0 < \cdots < X_{i-1} < X < X_{i+1} < \cdots X_p) \tag{2.3}$$

as $(X_0 < \cdots < X_{i-1} < X_{i+1} < \cdots X_p)$ ranges over all chains in L with $\operatorname{codim}(X_j) = j$. Denote the image of $(X_0 < \cdots < X_p)$ in $\mathcal{F}^p$ by $[X_0 < \cdots < X_p]$. By Proposition 2.1, $\langle\ ,\ \rangle$ induces a well-defined bilinear pairing $\langle\ ,\ \rangle\colon \mathcal{F}^p \otimes A^p \to \mathbb{C}$.

The pairing $\langle\ ,\ \rangle$ is a combinatorial model of the integration pairing of the ordinary homology and cohomology of the complement M with coefficients in $\mathbb{C}$ – see Theorem 4.3 and formula (4.4.3) in [6].

Theorem 2.2 ([6]). *The pairing* $\langle\ ,\ \rangle\colon \mathcal{F}^p \otimes A^p \to \mathbb{C}$ *is nondegenerate.*

Let $\varphi\colon A \to \mathcal{F}^*$ be defined by $\varphi(x) = \langle -, x\rangle\colon \mathcal{F} \to \mathbb{C}$. By Theorem 2.2, φ is an isomorphism. The value of $\varphi(\omega_J)$ in terms of the canonical basis of $\mathcal{F}$ is given in [6, (2.3.2)]. Similarly, $\varphi^*\colon \mathcal{F} \to \mathcal{A}^*$ is an isomorphism, with $\varphi^*(F) = \langle F, -\rangle\colon A \to \mathbb{C}$. $\mathcal{F}$ is called the *flag space* of $\mathcal{A}$.

Let $d\colon \mathcal{F}^p \to \mathcal{F}^{p+1}$ be the linear map defined by

$$d(X_0 < \cdots < X_p) = \sum_{\substack{X>X_p \\ \operatorname{codim}(X)=p+1}} (X_0 < \cdots < X_p < X). \tag{2.4}$$

Clearly d induces a linear map $d\colon \mathcal{F}^p \to \mathcal{F}^{p+1}$. Relations (2.3) imply $d \circ d = 0$. The pair $(\mathcal{F}, d)$ is called the *flag complex* of $\mathcal{A}$. The following result is a reformulation of Lemma 2.3.4 of [6].

Theorem 2.3. *For any $F \in \mathcal{F}^p$ and $x \in A^{p+1}$,*

$$\langle F, \partial x\rangle = \langle dF, x\rangle.$$

Let $d^*\colon (\mathcal{F}^p)^* \to (\mathcal{F}^{p-1})^*$ be the adjoint of $d\colon \mathcal{F}^{p-1} \to \mathcal{F}^p$.

Corollary 2.4 ([6, Lemma 2.3.4]). *The map $\varphi\colon (A, \partial) \to (\mathcal{F}^*, d^*)$ given by $\varphi(x) = \langle -, x\rangle$ is an isomorphism of chain complexes.*

Similarly $\varphi^*\colon (\mathcal{F}, d) \to (A^*, \partial^*)$ is an isomorphism of cochain complexes.

There is a decomposition of $\mathcal{F}$ dual to the Brieskorn decomposition [5, Lemma 5.91] of A. For $X \in L$ let $\mathcal{F}^p_X$ be the image in $\mathcal{F}^p$ of the subspace of Flag^p spanned by flags that terminate at X.

Then by [6, (2.12)],

$$\mathcal{F}^p = \bigoplus_{\mathrm{codim}(X)=p} \mathcal{F}^p_X. \tag{2.5}$$

Let $a = (a_0, \dots, a_n) \in \mathbb{C}^{n+1}$. Let $\omega_a = \sum_{j=0}^n a_j\omega_j$ and $\delta_a\colon A \to A$ with $\delta_a(x) = \omega_a \wedge x$. Let

$$S = \oplus S_p\colon \mathcal{F} \otimes \mathcal{F} \to \mathbb{C}$$

be the contravariant form of the weighted arrangement $(\mathcal{A}, a)$, defined by

$$S_p(F, F') = \sum_J a_J \langle F, e_J\rangle\langle F', e_J\rangle, \tag{2.6}$$

as in (1.1). S gives rise to the map $\mathcal{F} \to \mathcal{F}^*$ that sends F to $S(F, -)\colon \mathcal{F} \to \mathbb{C}$. By composing this map with the isomorphism $\varphi^{-1}\colon \mathcal{F}^* \to A$, one obtains a map $\psi\colon \mathcal{F} \to A$, characterized by the formula

$$S_p(F, F') = \langle F, \psi(F')\rangle, \tag{2.7}$$

for all $F, F' \in \mathcal{F}^p$, for each p. ψ is called the *contravariant map*.

Theorem 2.5 ([6, Lemma 3.2.5]). *The contravariant map $\psi\colon (\mathcal{F}, d) \to (A, \delta_a)$ is a morphism of cochain complexes.*

Corollary 2.6. *For every $F \in \mathcal{F}^p$ and $F' \in \mathcal{F}^{p-1}$,*

$$S_p(F, dF') = \langle F, \omega_a \wedge \psi(F')\rangle.$$

3 Projective OS algebra and flag space

Let $\mathcal{A}$ be a central arrangement as in Section 2. Let $\overline{\mathcal{A}}$ denote the projectivization of $\mathcal{A}$, consisting of the projective hyperplanes

$$\overline{H}_i := (H_i - \{0\})/\mathbb{C}^\times$$

in $\mathbb{P}^\ell = (\mathbb{C}^{\ell+1} - \{0\})/\mathbb{C}^\times$, for $0 \leqslant i \leqslant n$. Let $\overline{M} = \mathbb{P}^\ell - \cup_{i=0}^n \overline{H}_i$ be the complement to $\overline{\mathcal{A}}$ in $\mathbb{P}^\ell$.

Definition 3.1 ([2]). The OS algebra $\overline{A} = A(\overline{\mathcal{A}})$ of the projective arrangement $\overline{\mathcal{A}}$ is the kernel of $\partial : A \to A$.

Denote by $\iota : \overline{A} \to A$ the natural imbedding. Let $(\mathcal{F}, d)$ be the flag complex of $\mathcal{A}$.

Definition 3.2. The *flag space* $\overline{\mathcal{F}} = \mathcal{F}(\overline{\mathcal{A}})$ of the projective arrangement $\overline{\mathcal{A}}$ is the quotient $\mathcal{F}/\operatorname{im}(d)$.

Thus $\overline{\mathcal{F}}$ is obtained from $\mathcal{F}$ by introducing the additional relations

$$\sum_{\substack{X > X_p \\ \operatorname{codim}(X) = p+1}} (X_0 < \cdots < X_p < X) = 0, \tag{3.1}$$

where $(X_0 < \cdots < X_p)$ ranges over all flags of length p in L, for $0 \leqslant p \leqslant \ell$.

Let $\pi : \mathcal{F} \to \overline{\mathcal{F}}$ be the canonical projection. For $F \in \mathcal{F}$ we write $\overline{F} = \pi(F)$. Then, for instance, $\sum_{i=0}^n \overline{F}_i = 0$, where $\{F_0, \ldots, F_n\} \subseteq \overline{F}^1$ is the basis dual to $\{\omega_0, \ldots, \omega_n\} \subseteq A^1$.

Theorem 3.3. *Let $\iota^* : A^* \to \overline{A}^*$ be the adjoint of ι, given by restriction. Then the isomorphism $\varphi^* : \mathcal{F} \to A^*$ induces an isomorphism $\overline{\varphi}^* : \overline{\mathcal{F}} \to \overline{A}^*$, given by the commutative diagram*

$$\begin{array}{ccc} \mathcal{F} & \xrightarrow{\varphi^*} & A^* \\ {\scriptstyle \pi}\downarrow & & \downarrow{\scriptstyle \iota^*} \\ \overline{\mathcal{F}} & \xrightarrow{\overline{\varphi}^*} & \overline{A}^* \end{array}$$

Theorem 3.3 is proved below.

Lemma 3.4 ([5, Lemma 3.13]). *The complex*

$$0 \longrightarrow A^\ell \xrightarrow{\partial} A^{\ell-1} \xrightarrow{\partial} \cdots \xrightarrow{\partial} A^1 \xrightarrow{\partial} A^0 \longrightarrow 0$$

is exact.

Corollary 3.5. *The complex*

$$0 \longrightarrow \mathcal{F}^0 \xrightarrow{d} \mathcal{F}^1 \xrightarrow{d} \cdots \xrightarrow{d} \mathcal{F}^{\ell-1} \xrightarrow{d} \mathcal{F}^\ell \longrightarrow 0$$

is exact.

Corollary 3.5 also follows from [6, Cor. 2.8].

Proposition 3.6. $\mathrm{Ann}(\overline{A}) = \mathrm{im}(d\colon \mathcal{F} \to \mathcal{F})$.

Proof. By Lemma 3.4, $\overline{A} = \mathrm{im}(\partial)$. Then $F \in \mathrm{Ann}(\overline{A})$ if and only if $\langle F, \partial x\rangle = 0$ for all $x \in A$. By Theorem 2.3, this is equivalent to the statement $\langle dF, x\rangle = 0$ for every $x \in A$, or $dF = 0$ by Theorem 2.2. Then $\mathrm{Ann}(\overline{A}) = \ker(d)$, which equals $\mathrm{im}(d)$ by Corollary 3.5. □

Proof of Theorem 3.3. The assertion follows immediately from Proposition 3.6 and Definition 3.2. □

Lemma 3.4 also has the following consequence.

Corollary 3.7. $\overline{A}$ *is the subalgebra of* A *generated by* 1 *and* $\overline{A}^1 = \{\sum_{i=0}^n c_i\omega_i \mid \sum_{i=0}^n c_i = 0\}$.

Proof. By Lemma 3.4, we have $\overline{A} = \mathrm{im}(\partial)$. One can show by induction that

$$\partial\omega_J = (\omega_{j_2} - \omega_{j_1}) \wedge \cdots \wedge (\omega_{j_p} - \omega_{j_1}),$$

for any ordered subset $J = (j_1, \ldots, j_p)$ of $\{0, \ldots, n\}$ with $p \geqslant 2$. Each factor on the right-hand side lies in $(\overline{A})^1$. Since such ω_J (along with 1) span A, the result follows. □

Remark 3.8. The algebra $\overline{A}$ is naturally isomorphic to $H^*(\overline{M}, \mathbb{C})$ and $\iota : \overline{A} \to A$ is identified with the homomorphism $q^* : H^*(\overline{M}, \mathbb{C}) \to H^*(M, \mathbb{C})$ induced by the orbit map $q : M \to \overline{M}$. This follows from a general result about differential forms on complements of projective hypersurfaces, see [3, Section 6.1]. The space $\overline{\mathcal{F}}$ is naturally isomorphic to the homology space $H_*(\overline{M}, \mathbb{C})$ and the projection $\pi : \mathcal{F} \to \overline{\mathcal{F}}$ is identified with the homomorphism $q_* : H_*(M, \mathbb{C}) \to H_*(\overline{M}, \mathbb{C})$.

4 Singular subspace and contravariant form for projective arrangements

Let $\mathcal{A} = \{H_0, \ldots, H_n\}$ be a central arrangement in $\mathbb{C}^{\ell+1}$ as above. Let $a = (a_0, \ldots, a_n) \in \mathbb{C}^{n+1}$ and $\omega_a = \sum_{i=0}^n a_i\omega_i \in A^1$. We identify the flag space $\mathcal{F} = \mathcal{F}(\mathcal{A})$ with A^* via the map φ^* of Section 2.

Definition 4.1. The *singular subspace* $\mathrm{Sing}(\overline{\mathcal{F}}^{\ell}) \subset \overline{\mathcal{F}}^{\ell}$ is

$$\pi(\mathrm{Ann}(\omega_a \wedge A^{\ell-1})) = \big(\mathrm{Ann}(\omega_a \wedge A^{\ell-1}) + \mathrm{im}(d)\big)/\,\mathrm{im}(d) \subset \mathcal{F}^{\ell}/\,\mathrm{im}(d) = \overline{\mathcal{F}}^{\ell}.$$

Let $S\colon \mathcal{F} \otimes \mathcal{F} \to \mathbb{C}$ be the contravariant form of the central arrangement $\mathcal{A}$, as defined in (2.6).

Theorem 4.2. *The subspaces* $\mathrm{Ann}(\omega_a \wedge A)$ *and* $\mathrm{im}(d)$ *of* $\mathcal{F}$ *are orthogonal with respect to* S.

Proof. In Section 2 we constructed the contravariant map $\psi : \mathcal{F} \to A$ satisfying $S_p(F, F') = \langle F, \psi(F')\rangle$ for every $F, F' \in \mathcal{F}^p$. By Corollary 2.6 ψ satisfies $S_p(F, dF') = \langle F, \omega_a \wedge \psi(F')\rangle$ for all $F \in \mathcal{F}^p$ and $F' \in \mathcal{F}^{p-1}$. Suppose $F \in \mathrm{Ann}(\omega_a \wedge A^{p-1}) \subseteq \mathcal{F}^p$. Then, for every $F' \in \mathcal{F}^{p-1}$, $\langle F, \omega_a \wedge \psi(F')\rangle = 0$. Then $S_p(F, dF') = 0$ for every $F' \in \mathcal{F}^{p-1}$. Thus $\mathrm{Ann}(\omega_a \wedge A)$ is orthogonal to $\mathrm{im}(d)$. □

Define the bilinear form

$$\overline{S}_{\ell} : \mathrm{Sing}(\overline{\mathcal{F}}^{\ell}) \otimes \mathrm{Sing}(\overline{\mathcal{F}}^{\ell}) \to \mathbb{C}$$

by $\overline{S}_{\ell}(\overline{F}, \overline{F}') = S_{\ell}(F, F')$.

Corollary 4.3. *The form* $\overline{S}_{\ell}\colon \mathrm{Sing}(\overline{\mathcal{F}}^{\ell}) \otimes \mathrm{Sing}(\overline{\mathcal{F}}^{\ell}) \to \mathbb{C}$ *is well-defined.*

5 Dehomogenization

Throughout this section we assume $a \in \mathbb{C}^{n+1}$ satisfies $\sum_{i=0}^{n} a_i = 0$. Then $\omega_a \in \overline{A}^1$ and $\omega_a \wedge \overline{A} \subseteq \overline{A}$. Fix a hyperplane $H_j \in \mathcal{A}$. For simplicity of notation we assume $j = 0$, but the index 0 will play no special role. Choose coordinates $(x_0, \dots, x_{\ell})$ on $\mathbb{C}^{\ell+1}$ so that H_0 is defined by the equation $x_0 = 0$. The *decone* of $\mathcal{A}$ relative to H_0 is an affine arrangement $\mathrm{d}\mathcal{A} = \{\mathrm{d}H_1, \dots, \mathrm{d}H_n\}$ in $\mathbb{C}^{\ell}$. The affine hyperplane $\mathrm{d}H_i$ is defined by $\hat{f}_i(x_1, \dots, x_{\ell}) = 0$, where $\hat{f}_i(x_1, \dots, x_{\ell}) = f_i(1, x_1, \dots, x_{\ell})$ and $f_i\colon \mathbb{C}^{\ell+1} \to \mathbb{C}$ is a linear defining form for H_i. Let $\hat{M} = \mathbb{C}^{\ell} - \bigcup_{i=1}^{n} \mathrm{d}H_i$, $\hat{\omega}_i = d\log(\hat{f}_i)$, and let $\hat{A}$ be the algebra of differential forms on $\mathbb{C}^{\ell}$ generated by 1 and $\hat{\omega}_i$, $1 \leqslant i \leqslant n$. Let $\hat{a} = (a_1, \dots, a_n)$.

Lemma 5.1. *The map* $\epsilon\colon \hat{A} \to \overline{A}$, $\hat{\omega}_i \mapsto \omega_i - \omega_0$, *is a well-defined isomorphism. Moreover,* ϵ *sends* $\hat{\omega}_{\hat{a}} = \sum_{i=1}^{n} a_i \hat{\omega}_i$ *to* ω_a.

We note for future reference that

$$\epsilon(\hat{\omega}_{\hat{J}}) = (\omega_{j_1} - \omega_0) \wedge \cdots \wedge (\omega_{j_p} - \omega_0) = \partial\omega_{(0,\hat{J})}, \tag{5.1}$$

for any ordered p-subset $\hat{J} = (j_1, \ldots, j_p\}$ of $\{1, \ldots, n\}$, where $(0, \hat{J}) = (0, j_1, \ldots, j_p)$.

As in Section 2, the flag space $\hat{\mathcal{F}} = \mathcal{F}(\mathrm{d}\mathcal{A})$ of the affine arrangement $\mathrm{d}\mathcal{A}$ can be identified with $\hat{A}^*$, and the singular subspace $\mathrm{Sing}(\hat{\mathcal{F}}^\ell) \subset \hat{\mathcal{F}}^\ell$ relative to $\hat{a}$ is defined by $\mathrm{Sing}(\hat{\mathcal{F}}^\ell) = \mathrm{Ann}(\hat{\omega}_{\hat{a}} \wedge \hat{A}^{\ell-1})$. The contravariant form $\hat{S} = \oplus \hat{S}_p$ of $\mathrm{d}\mathcal{A}$ is given by

$$\hat{S}_p(\hat{F}, \hat{F}') = \sum_{\hat{J}} \hat{a}_{\hat{J}} \hat{F}(\hat{\omega}_{\hat{J}}) \hat{F}'(\hat{\omega}_{\hat{J}}),$$

summing over increasing p-tuples $\hat{J} = (j_1, \ldots, j_p)$ of elements of $\{1, \ldots, n\}$. We identify $\overline{\mathcal{F}}$ with $(\overline{A})^*$ via the isomorphism $\overline{\varphi}^*$ of Theorem 3.3.

In this section we prove the following theorem.

Theorem 5.2. *The map $\epsilon^*\colon \overline{\mathcal{F}} \to \hat{\mathcal{F}}$ restricts to an isomorphism of inner-product spaces*

$$\epsilon^*\colon \left(\mathrm{Sing}(\overline{\mathcal{F}}^\ell), \overline{S}_\ell|_{\mathrm{Sing}(\overline{\mathcal{F}}^\ell)}\right) \xrightarrow{\cong} \left(\mathrm{Sing}(\hat{\mathcal{F}}^\ell), \hat{S}_\ell|_{\mathrm{Sing}(\hat{\mathcal{F}}^\ell)}\right).$$

Recall that $\overline{A} = \ker(\partial\colon A \to A)$. Define $\sigma\colon \overline{A}^{p-1} \to A^p$ by $\sigma(x) = \omega_0 \wedge x$.

Lemma 5.3. *For each p, we have $A^p = (\omega_0 \wedge \overline{A}^{p-1}) \oplus \overline{A}^p$.*

Proof. By Lemma 3.4 the complex (A, ∂) is exact. Hence, for each p, there is a short exact sequence

$$0 \longrightarrow \overline{A}^p \xrightarrow{\iota} A^p \xrightarrow{\partial} \overline{A}^{p-1} \longrightarrow 0. \tag{5.2}$$

This sequence splits; the map $\sigma\colon \overline{A}^{p-1} \to A^p$ defined above is a section of $\partial\colon A^p \to \overline{A}^{p-1}$. Indeed, $\partial \circ \sigma(x) = \partial(\omega_0 \wedge x) = \partial\omega_0 \wedge x - \omega_0 \wedge \partial x = x$ for $x \in \overline{A}^{p-1}$. Then $A^p = \mathrm{im}(\sigma) \oplus \ker(\partial) = (\omega_0 \wedge \overline{A}^{p-1}) \oplus \overline{A}^p$ as claimed. □

Recall that $\ker(\pi) = \mathrm{Ann}(\overline{A})$. The map $\pi\colon \mathcal{F}^p \to \overline{\mathcal{F}}^p$ is the adjoint of the inclusion $\overline{A} \to A$. Let $\sigma^*\colon \mathcal{F}^p \to \overline{\mathcal{F}}^{p-1}$ be the adjoint of $\sigma\colon \overline{A}^{p-1} \to A^p$.

Lemma 5.4. *We have the following statements:*

(i) $\mathcal{F}^p = \ker(\sigma^*) \oplus \ker(\pi)$.
(ii) *The restriction* $\pi|_{\mathrm{Ann}(\omega_0 \wedge \overline{A}^{p-1})} : \mathrm{Ann}(\omega_0 \wedge \overline{A}^{p-1}) \to \overline{\mathcal{F}}^p$ *is an isomorphism.*
(iii) $\mathrm{Ann}(\omega_0 \wedge \overline{A}) = \mathrm{Ann}(\omega_0 \wedge A)$.

Proof. Taking duals in (5.2), we obtain the exact sequence

$$0 \longrightarrow \overline{\mathcal{F}}^{p-1} \xrightarrow{\partial^*} \mathcal{F}^p \xrightarrow{\pi} \overline{\mathcal{F}}^p \longrightarrow 0. \tag{5.3}$$

The map $\sigma^* \colon \mathcal{F}^p \to \overline{\mathcal{F}}^{p-1}$ satisfies $\sigma^* \circ \partial^* = (\partial \circ \sigma)^* = \mathrm{id}_{\overline{\mathcal{F}}^{p-1}}$. Then $\mathcal{F}^p = \ker(\sigma^*) \oplus \mathrm{im}(\partial^*)$. Statement (i) follows by exactness.

We have $\sigma^*(F)(x) = F\big(\sigma(x)\big) = F(\omega_0 \wedge x)$ for all $x \in \overline{A}$. Then $F \in \ker(\sigma^*)$ if and only if $F \in \mathrm{Ann}(\omega_0 \wedge \overline{A}^{p-1})$, i.e., $\ker(\sigma^*) = \mathrm{Ann}(\omega_0 \wedge \overline{A}^{p-1})$. Applying (i), we have $\mathrm{Ann}(\omega_0 \wedge \overline{A}^{p-1}) \cap \ker(\pi) = 0$ and $\overline{\mathcal{F}}^p = \pi(\mathcal{F}^p) = \pi\big(\ker(\sigma^*)\big) = \pi\big(\mathrm{Ann}(\omega_0 \wedge \overline{A}^{p-1})\big)$. This proves (ii).

For (iii), assume $F \in \mathrm{Ann}(\omega_0 \wedge \overline{A})$ and $x \in \omega_0 \wedge A$. Write $x = \omega_0 \wedge y$ for $y \in A$. By Lemma 5.3, we can write $y = y_1 + y_2$ with $y_1 \in \omega_0 \wedge \overline{A}$ and $y_2 \in \overline{A}$. Then $\omega_0 \wedge y_1 = 0$, so $x = \omega_0 \wedge y_2 \in \omega_0 \wedge \overline{A}$. Then $F(x) = 0$. Thus $\mathrm{Ann}(\omega_0 \wedge \overline{A}) \subseteq \mathrm{Ann}(\omega_0 \wedge A)$. The opposite inclusion holds because $\omega_0 \wedge \overline{A} \subseteq \omega_0 \wedge A$. □

Recall the decomposition (2.5) of $\mathcal{F}$. In this context Lemma 5.4 yields the following result, which we will consider in more detail in the next section.

Corollary 5.5. *For* $0 \leqslant p \leqslant \ell$,

$$\overline{\mathcal{F}}^p \cong \bigoplus_{\substack{\mathrm{codim}(X)=p \\ H_0 \nleqslant X}} \mathcal{F}_X^p$$

Proof. Using (2.1) one can check easily that $[X_0 < \cdots < X_p] \in \mathrm{Ann}(\omega_0 \wedge A^{p-1})$ if and only if $H_0 \nleqslant X_p$. Then, for each $X \in L$ of codimension p,

$$\mathrm{Ann}(\omega_0 \wedge A^{p-1}) \cap \mathcal{F}_X^p = \begin{cases} 0 & \text{if } H_0 \leqslant X \\ \mathcal{F}_X^p & \text{if } H_0 \nleqslant X \end{cases}.$$

The claim then follows from parts (ii) and (iii) of Lemma 5.4. □

Lemma 5.6. $\mathrm{Sing}(\overline{\mathcal{F}}^{\ell}) = \pi\big(\mathrm{Ann}(\omega_a \wedge \overline{A}^{\ell-1})\big)$.

Proof. First we claim that $\mathrm{Ann}(\omega_a \wedge \overline{A}) \cap \mathrm{Ann}(\omega_0 \wedge \overline{A}) \subseteq \mathrm{Ann}(\omega_a \wedge A)$. Let F be an element of $\mathrm{Ann}(\omega_a \wedge \overline{A}) \cap \mathrm{Ann}(\omega_0 \wedge \overline{A})$ and let $x \in \omega_a \wedge A$. Write $x = \omega_a \wedge y$ with $y \in A$. By Lemma 5.3 we can write $y = y_1 + y_2$ with $y_1 \in \omega_0 \wedge \overline{A}$ and $y_2 \in \overline{A}$. Write $y_1 = \omega_0 \wedge y_1'$ with $y_1' \in \overline{A}$. Since $\omega_a \wedge \overline{A} \subseteq \overline{A}$, $\omega_a \wedge y_1 = \omega_a \wedge (\omega_0 \wedge y_1') = \omega_0 \wedge \big(-(\omega_a \wedge y_1')\big) \in \omega_0 \wedge \overline{A}$. Then $x = \omega_a \wedge y = \omega_a \wedge y_1 + \omega_a \wedge y_2 \in \omega_0 \wedge \overline{A} + \omega_a \wedge \overline{A}$. Then $F(x) = 0$. This proves the claim.

Next we claim $\mathrm{Ann}(\omega_a \wedge \overline{A}) = \mathrm{Ann}(\omega_a \wedge A) + \mathrm{Ann}(\overline{A})$. Let $F \in \mathrm{Ann}(\omega_a \wedge \overline{A})$. By part (i) of Lemma 5.4, we can write $F = F_1 + F_2$ where $F_1 \in \ker(\sigma^*) = \mathrm{Ann}(\omega_0 \wedge \overline{A})$ and $F_2 \in \ker(\pi) = \mathrm{Ann}(\overline{A})$. Since $\omega_a \wedge \overline{A} \subseteq \overline{A}$, $F_2 \in \mathrm{Ann}(\omega_a \wedge \overline{A})$. Then $F_1 = F - F_2 \in \mathrm{Ann}(\omega_a \wedge \overline{A})$. Then $F_1 \in \mathrm{Ann}(\omega_a \wedge \overline{A}) \cap \mathrm{Ann}(\omega_0 \wedge \overline{A})$. Then $F_1 \in \mathrm{Ann}(\omega_a \wedge A)$ by the preceding claim. Thus $F = F_1 + F_2 \in \mathrm{Ann}(\omega_a \wedge A) + \mathrm{Ann}(\overline{A})$. This proves $\mathrm{Ann}(\omega_a \wedge \overline{A}) \subseteq \mathrm{Ann}(\omega_a \wedge A) + \mathrm{Ann}(\overline{A})$. The opposite inclusion follows easily from the fact that $\omega_a \wedge \overline{A} \subseteq (\omega_a \wedge A) \cap \overline{A}$. This proves our second claim. The assertion of the lemma follows immediately. □

We note the following consequence of Lemma 5.6 for later use. As observed earlier, $(\overline{A}, \delta_a)$ is a subcomplex of (A, δ_a).

Corollary 5.7. *The inclusion* $\mathrm{Sing}(\overline{\mathcal{F}}^{\ell}) \hookrightarrow \overline{\mathcal{F}}^{\ell} = (\overline{A}^{\ell})^*$ *induces an isomorphism*

$$\mathrm{Sing}(\overline{\mathcal{F}}^{\ell}) \xrightarrow{\cong} \big(H^{\ell}(\overline{A}, \delta_a)\big)^*.$$

Proof. Lemma 3.4 implies $\overline{A}^{\ell+1} = 0$, so $H^{\ell}(\overline{A}, \delta_a) = \overline{A}^{\ell}/(\omega_a \wedge \overline{A}^{\ell-1})$. Then $\big(H^{\ell}(\overline{A}, \delta_a)\big)^*$ is isomorphic to the annihilator of $\omega_a \wedge \overline{A}^{\ell-1}$ in $(\overline{A}^{\ell})^*$. This annihilator is equal to

$$\big(\mathrm{Ann}(\omega_a \wedge \overline{A}^{\ell-1}) + \mathrm{Ann}(\overline{A}^{\ell})\big) \big/ \mathrm{Ann}(\overline{A}^{\ell}).$$

By Definition 4.1, Proposition 3.6, and Lemma 5.6, this is equal to $\mathrm{Sing}(\overline{\mathcal{F}}^{\ell})$. □

Proof of Theorem 5.2. Let $F \in \mathrm{Ann}(\omega_a \wedge A^{\ell-1})$, and let $\hat{x} \in \hat{\omega}_a \wedge \hat{A}^{\ell-1}$. Then $\epsilon^*(\overline{F})(\hat{x}) = F\big(\epsilon(\hat{x})\big)$. Since $\epsilon(\hat{\omega}_{\hat{a}}) = \omega_a$, $\epsilon(\hat{x}) \in \omega_a \wedge A^{\ell-1}$, so $F\big(\epsilon(\hat{x})\big) = 0$. Then $\epsilon^*(\overline{F})(\hat{x}) = 0$. Thus $\epsilon^*(\mathrm{Sing}(\overline{\mathcal{F}}^{\ell})) \subseteq \mathrm{Sing}(\hat{\mathcal{F}}^{\ell})$.

Conversely, suppose $\hat{F} \in \mathrm{Sing}(\hat{\mathcal{F}}^{\ell})$. Write $\hat{F} = \epsilon^*(\overline{F})$ with $F \in \mathcal{F}^{\ell}$. Let $x \in \omega_a \wedge \overline{A}^{\ell-1}$. Then $x \in \overline{A}^{\ell}$, so $x = \epsilon(\hat{x})$ for some $\hat{x} \in \hat{A}^{\ell}$. Since

$x \in \omega_a \wedge \overline{A}^{\ell-1}$, $\hat{x} \in \hat{\omega}_a \wedge \hat{A}^{\ell-1}$. Then $\hat{F}(\hat{x}) = 0$ by definition of $\mathrm{Sing}(\hat{\mathcal{F}}^\ell)$. Then $F(x) = F\big(\epsilon(\hat{x})\big) = \epsilon^*(\overline{F})(\hat{x}) = \hat{F}(\hat{x}) = 0$. This shows that $F \in \mathrm{Ann}(\omega_a \wedge \overline{A}^{\ell-1})$. Then $F \in \mathrm{Ann}(\omega_a \wedge A^{\ell-1})$ by Lemma 5.6. Then $\overline{F} \in \mathrm{Sing}(\overline{\mathcal{F}}^\ell)$ by definition of $\mathrm{Sing}(\overline{\mathcal{F}}^\ell)$. Thus $\mathrm{Sing}(\hat{\mathcal{F}}^\ell) \subseteq \epsilon^*(\mathrm{Sing}(\overline{\mathcal{F}}^\ell))$, and ϵ^* restricts to an isomorphism $\mathrm{Sing}(\overline{\mathcal{F}}^\ell) \to \mathrm{Sing}(\hat{\mathcal{F}}^\ell)$.

It remains to prove that $\hat{S}_\ell(\epsilon^*(\overline{F}), \epsilon^*(\overline{F}')) = \overline{S}_\ell(\overline{F}, \overline{F}')$ for all $\overline{F}, \overline{F}' \in \mathrm{Sing}(\overline{\mathcal{F}}^\ell)$. By (5.1), we have

$$\begin{aligned}
\hat{S}_\ell\big(\epsilon^*(\overline{F}), \epsilon^*(\overline{F}')\big) &= \sum_{\hat{J}} \hat{a}_{\hat{J}} \epsilon^*(\overline{F})(\hat{\omega}_{\hat{J}}) \epsilon^*(\overline{F}')(\hat{\omega}_{\hat{J}}) \\
&= \sum_{\hat{J}} a_{\hat{J}} F\big(\epsilon(\hat{\omega}_{\hat{J}})\big) F'\big(\epsilon(\hat{\omega}_{\hat{J}})\big) \\
&= \sum_{\hat{J}} a_{\hat{J}} F(\partial\omega_{(0,\hat{J})}) F'(\partial\omega_{(0,\hat{J})}).
\end{aligned}$$

The sum is over increasing p-tuples $\hat{J}$ of elements of $\{1, \dots, n\}$. By parts (ii) and (iii) of Lemma 5.4, we may assume that $F, F' \in \mathrm{Ann}(\omega_0 \wedge A^{\ell-1})$. Since $\partial\omega_{(0,\hat{J})} = \omega_{\hat{J}} - \omega_0 \wedge \partial\omega_{\hat{J}}$, this implies $F(\partial\omega_{(0,\hat{J})}) = F(\omega_{\hat{J}})$ and similarly for F'. Then the last sum above is equal to $\sum_{\hat{J}} a_{\hat{J}} F(\omega_{\hat{J}}) F'(\omega_{\hat{J}})$. This sum is equal to $\sum_J a_J F(\omega_J) F'(\omega_J)$, summing now over all increasing p-tuples J of elements of $\{0, \dots, n\}$, again because $F, F' \in \mathrm{Ann}(\omega_0 \wedge A^{\ell-1})$. This equals $\overline{S}_\ell(\overline{F}, \overline{F'})$ by definition. $\square$

We close this section with a topological remark. Consider the (multi-valued) master function $\Phi_a = \prod_{i=0}^n f_i^{-a_i}$ on $\mathbb{C}^{\ell+1}$. Since $\sum_{i=0}^n a_i = 0$, Φ_a is invariant under the action of $\mathbb{C}^\times$, hence induces a (multi-valued) master function $\overline{\Phi}_a$ on $\overline{M}$. We have $\overline{\Phi}_a = \hat{\Phi}_{\hat{a}} \circ h$ where $\hat{\Phi}_{\hat{a}} = \prod_{i=1}^n \hat{f}_i^{-a_i}$ is the master function of $(\mathrm{d}\mathcal{A}, \hat{a})$ on $\hat{M}$, and $h\colon \overline{M} \to \hat{M}$ is the canonical diffeomorphism. The associated rank-one local systems $\hat{\mathcal{L}}_{\hat{a}}$ on $\hat{M}$ and $\overline{\mathcal{L}}_a$ on $\overline{M}$ then satisfy $h^*\hat{\mathcal{L}}_{\hat{a}} = \overline{\mathcal{L}}_a$. The inclusion of $(\overline{A}, \delta_{ca})$ in the twisted algebraic de Rham complex of $\overline{\mathcal{L}}_{ca}$ induces an isomorphism of $H^*(\overline{A}, \delta_{ca})$ with $H^*(\overline{M}, \overline{\mathcal{L}}_{ca})$ for generic c. As before, $\mathrm{Sing}_a(\overline{\mathcal{F}}^\ell)$ is equal to $\mathrm{Sing}_{ca}(\overline{\mathcal{F}}^\ell)$ for any nonzero scalar c. Then, by Corollary 5.7, we have the following corollary.

Corollary 5.8. *For generic c, the inclusion $\mathrm{Sing}_a(\overline{\mathcal{F}}^\ell) \hookrightarrow (\overline{A}^\ell)^*$ induces an isomorphism*

$$\mathrm{Sing}_a(\overline{\mathcal{F}}^\ell) \xrightarrow{\cong} H_\ell(\overline{M}, \overline{\mathcal{L}}_{ca}).$$

This isomorphism does not involve the choice of a hyperplane at infinity; the index 0 again plays no special role. We have the following commutative diagram of isomorphisms, for generic c:

$$\begin{array}{ccc} \mathrm{Sing}_a(\overline{\mathcal{F}}^{\ell}) & \xrightarrow{\epsilon^*} & \mathrm{Sing}_{\hat{a}}(\hat{\mathcal{F}}^{\ell}) \\ \downarrow & & \downarrow \\ H_\ell(\overline{M}, \overline{\mathcal{L}}_{ca}) & \xrightarrow{h_*} & H_\ell(\hat{M}, \hat{\mathcal{L}}_{c\hat{a}}) \end{array}$$

The effect on $\mathrm{Sing}_{\hat{a}}(\hat{\mathcal{F}}^{\ell})$ of different choices of H_0 is analyzed in the next section.

6 Transition functions

The right-hand side of the formula in Corollary 5.5 is the decomposition of the flag space $\hat{\mathcal{F}}^p$ of the decone d$\mathcal{A}$, see [6]. It can be considered to be the dehomogenization of the projective flag space $\overline{\mathcal{F}}$ relative to H_0. The dehomogenizations relative to different hyperplanes form a set of "affine charts" for $\overline{\mathcal{F}}$. We compute the transition functions.

For $0 \leqslant j \leqslant n$, let $\hat{A}_j, \hat{\mathcal{F}}_j$, and $\hat{S}^{(j)}$ denote the OS algebra, flag complex, and contravariant form of the affine arrangement obtained by deconing $\mathcal{A}$ with respect to H_j. Let $\epsilon_j \colon \hat{A}_j \to \overline{A}$ be the isomorphism determined by $\epsilon(\hat{\omega}_k) = \omega_k - \omega_j$, for $0 \leqslant k \leqslant n$ and $k \neq j$, as in Lemma 5.1. Let $\epsilon_j^* \colon \overline{\mathcal{F}} \to \hat{\mathcal{F}}_j$ be the adjoint of ϵ_j. For $0 \leqslant i < j \leqslant n$, set $\tau_{ij} = \epsilon_j^* \circ (\epsilon_i^*)^{-1}$. Then $\tau_{ij} \colon \hat{\mathcal{F}}_i \to \hat{\mathcal{F}}_j$ is an isomorphism. Theorem 5.2 has the following corollary.

Corollary 6.1. *The restriction of τ_{ij} is an isomorphism of inner product spaces*

$$\tau_{ij} \colon \left(\mathrm{Sing}(\hat{\mathcal{F}}_i^{\ell}), \hat{S}_\ell^{(i)}|_{\mathrm{Sing}(\hat{\mathcal{F}}_i^{\ell})}\right) \xrightarrow{\cong} \left(\mathrm{Sing}(\hat{\mathcal{F}}_j^{\ell}), \hat{S}_\ell^{(j)}|_{\mathrm{Sing}(\hat{\mathcal{F}}_j^{\ell})}\right).$$

According to Corollary 5.5, τ_{ij} can be considered to be an isomorphism

$$\tau_{ij} \colon \bigoplus_{\substack{\mathrm{codim}(X)=p \\ H_i \not\leqslant X}} \mathcal{F}_X^p \longrightarrow \bigoplus_{\substack{\mathrm{codim}(X)=p \\ H_j \not\leqslant X}} \mathcal{F}_X^p.$$

We describe this map explicitly.

In the special case $p = 1$ there is an easy formula for τ_{ij}. Let $\{F_0, ..., F_n\}$ be the canonical basis of $\mathcal{F}^1$, and suppose $k \neq i$. Then

$$\tau_{ij}(F_k) = \begin{cases} F_k \text{ if } k \neq j \\ -\sum_{r \neq j} F_r \text{ if } k = j. \end{cases}$$

To describe the general formula, we will use the following lemma.

Lemma 6.2. *Let* $X \in L$ *with* $\operatorname{codim}(X) = p$, *and let* $H \in \mathcal{A}$. *Then* $\mathcal{F}^p_X$ *is spanned by elements* $[X_0 < \cdots < X_{p-1} < X]$ *satisfying* $H \nleqslant X_{p-1}$.

Proof. We induct on p, the case $p = 0$ being trivial. Let $p > 0$ and $[X_0 < \cdots < X_{p-1} < X] \in \mathcal{F}^p$. By the inductive hypothesis, we may assume $H \nleqslant X_{p-2}$. (Here we rely on the fact that the assignment $[X_0 < \cdots < X_{p-1}] \mapsto [X_0 < \cdots < X_{p-1} < X_p]$ determines a well-defined linear map $\mathcal{F}^{p-1}_{X_{p-1}} \to \mathcal{F}^p_{X_p}$.) If $H \nleqslant X_{p-1}$ we are done. Otherwise, by (2.3), we have

$$[X_0 < \cdots < X_{p-1} < X] = \sum_{\substack{X_{p-2} < X' < X \\ X' \neq X_{p-1}}} -[X_0 < \cdots X_{p-2} < X' < X].$$

Since $H \nleqslant X_{p-2}$ and $H \leqslant X_{p-1}$, $H \nleqslant X'$ for any $X' \neq X$ satisfying $X_{p-2} < X'$ and $\operatorname{codim}(X') = p - 1$. Then every flag $(X_0 < \cdots X_{p-2} < X' < X)$ that appears on the right-hand side satisfies the required condition. This completes the inductive step. $\square$

Theorem 6.3. *Let* $[X_0 < \cdots < X_p] \in \mathcal{F}^p$ *with* $H_i \nleqslant X_p$. *If* $H_j \nleqslant X_p$, *then*

$$\tau_{ij}([X_0 < \cdots < X_p]) = [X_0 < \cdots < X_p].$$

If $H_j \leqslant X_p$ *and* $H_j \nleqslant X_{p-1}$, *then*

$$\tau_{ij}([X_0 < \cdots < X_p]) = \sum_{X_{p-1} < X', X' \neq X_p} -[X_0 < \cdots < X_{p-1} < X'].$$

Proof. By definition, $\tau_{ij}([X_0 < \cdots < X_p])$ is the unique element of $\bigoplus_{\substack{\operatorname{codim}(X)=p \\ H_j \nleqslant X}} \mathcal{F}^p_X$ that represents the same element of $\overline{\mathcal{F}}^p$ as $[X_0 < \cdots < X_p]$. By (3.1), the right-hand side represents the same element of $\overline{\mathcal{F}}^p$ as $[X_0 < \cdots < X_p]$, in either case. An argument similar to the one used in the preceding lemma shows that the right-hand side lies in

$$\bigoplus_{\substack{\operatorname{codim}(X)=p \\ H_j \nleqslant X}} \mathcal{F}^p_X$$

in either case. The claim follows. $\square$

By Lemma 6.2, this theorem is sufficient to determine τ_{ij} uniquely. By Corollary 6.1, τ_{ij} sends singular vectors of $\hat{\mathcal{F}}_i^\ell$ to singular vectors of $\hat{\mathcal{F}}_j^\ell$, and preserves the value of the contravariant form on such vectors.

Similarly, there is an algebra isomorphism $\tau_{ji}^*: \hat{A}_i \to \hat{A}_j$ determined by

$$\tau_{ji}^*(\hat{\omega}_k) = \begin{cases} \hat{\omega}_k - \hat{\omega}_i & \text{if } k \neq i \\ -\hat{\omega}_i & \text{if } k = i. \end{cases}$$

As in Section 2, there is an isomorphism $\hat{\mathcal{F}}_i^* \to \hat{A}_i$ defined by the affine version of (2.1), and the contravariant map $\psi_i: \hat{\mathcal{F}}_i \to \hat{A}_i$ characterized by the formula

$$S(\hat{F}, \hat{F}') = \langle \hat{F}, \psi_i(\hat{F}') \rangle.$$

The image of ψ_i is the *complex of flag forms* of $\hat{\mathcal{A}}_i$. It is a subcomplex of $(\hat{A}_i, \delta_{\hat{a}_i})$. Theorem 5.2 has the following consequence.

Corollary 6.4. *The following diagram commutes:*

$$\begin{array}{ccc} \mathrm{Sing}(\hat{\mathcal{F}}_i^\ell) & \xrightarrow{\psi_i} & \hat{A}_i^\ell \\ {\scriptstyle \tau_{ij}}\downarrow & & \downarrow{\scriptstyle \tau_{ji}^*} \\ \mathrm{Sing}(\hat{\mathcal{F}}_j^\ell) & \xrightarrow{\psi_j} & \hat{A}_j^\ell \end{array}$$

References

[1] V. I. ARNOL'D, *The cohomology ring of the colored braid group*, Mat. Zametki **5** (1969), 227–231.

[2] D. COHEN, G. DENHAM, M. FALK and A. VARCHENKO, *Vanishing products of one-forms and critical points of master functions*, In: "Arrangements of Hyperplanes (Sapporo 2009)", Adv. Stud. Pure Math., Math. Soc. Japan (2012), 75–107.

[3] A. DIMCA, "Singularities and Topology of Hypersurfaces", Springer-Verlag, New York, 1992.

[4] P. ORLIK and L. SOLOMON, *Topology and combinatorics of complements of hyperplanes*, Inv. Math. **56** (1980), 167–189.

[5] P. ORLIK and H. TERAO, "Arrangements of Hyperplanes", Springer-Verlag, 1992.

[6] V. V. SCHECHTMAN and A. N. VARCHENKO, *Arrangements of hyperplanes and Lie algebra homology*, Inv. Math. **106** (1991), 139–194.

[7] A. VARCHENKO, "Multidimensional Hypergeometric Functions and Representation Theory of Lie Algebras and Quantum Groups",

Advanced Series in Mathematical Physics, 21, World Scientific, 1995, x+371 pages.

[8] A. VARCHENKO, *Bethe Ansatz for arrangements of hyperplanes and the Gaudin model*, Mosc. Math. J. **6** (2006), 195–210, 223–224.

[9] A. VARCHENKO, *Quantum integrable model of an arrangement of hyperplanes*, SIGMA, 7, 032, 2011, 55 pages.

Fox-Neuwirth cell structures and the cohomology of symmetric groups

Chad Giusti and Dev Sinha

Abstract. We use the Fox-Neuwirth cell structure for one-point compactifications of configuration spaces as the starting point for understanding our recent calculation of the mod-two cohomology of symmetric groups. We then use that calculation to give short proofs of classical results on this cohomology due to Nakaoka and to Madsen.

1 Introduction

Group cohomology touches on a range of subjects within algebra and topology. It is thus amenable to study by a similar range of techniques, as we have found in our recent work on the cohomology of symmetric groups. The key organizational tool is the algebraic notion of a Hopf ring. But it is the geometry of bundles and that of the corresponding models for classifying spaces which have guided us, and in particular led to rediscovery of the Hopf ring structure.

We consider cohomology of symmetric groups as giving characteristic classes for finite-sheeted covering maps. Equivalently, we consider the configuration space models for their classifying spaces. There is a strong analogy with characteristic classes for vector bundles and the geometry of Grassmannians. In the vector bundle setting, we embed a bundle over a manifold base space in a trivial bundle and define characteristic classes which are Poincaré dual to the locus where the fiber intersects the standard flag in subspaces with a prescribed set of dimensions. These cohomology classes are pulled back from classes represented by Schubert cells. We may use the isomorphism $O(1) = S_2$, to translate to covering spaces. Embed a two-sheeted covering of a manifold once again in a trivial vector bundle. Then the characteristic classes, which are all powers of the first Stiefel-Whitney class, are Poincaré dual to the locus where the two points in a fiber share their first n coordinates. We will see that in general characteristic classes of finite sheeted covering spaces are defined by loci where the points in the fiber are (nested) partitioned into collections of points which share prescribed coordinates.

A critical component of the the theory of characteristic classes for vector bundles is the Schubert cell structure on Grassmannians. We begin this paper by developing an analogue for configuration spaces, namely the Fox-Neuwirth cell structure on their one-point compacitifications. While this cell structure in the two-dimensional setting dates back fifty years [14] and was developed in all dimensions mod-two by Vassiliev [30], we present what is to our knowledge the first treatment of the differential integrally. This cell structure is "Alexander dual" to one on configuration spaces themselves which has enjoyed a renaissance of interest lately, being studied in a number of contexts, namely by Tamaki [26, 27] related to work on the iterated cobar construction [28, 29], by Ayala and Hepworth [3] in the context of higher category theory and by Blagojević and Ziegler [8] in the context of partitioning convex bodies. After presenting the cell structure, we compute the mod-two cohomology of $B\mathcal{S}_4$, giving cochain representatives.

In order to understand arbitrary symmetric groups, we find it essential to consider them all together, much as in homology where their direct sum has an elegant description as a "Q-ring" [7]. In cohomology the relevant structure is that of a Hopf ring, due to Strickland and Turner [25]. This was the key organizing structure of our work with Salvatore in [15]. Our construction of a model for this Hopf ring structure at the level of Fox-Neuwirth cochains is new. After giving a Hopf ring presentation in terms of generators and relations, we give a graphical presentation of the resulting monomial basis which we call the skyline diagram basis, recapping the major results from our paper [15] and connecting them with Fox-Neuwirth cochain representatives first identified by Vassiliev [30]. We use the skyline diagram presentation to revisit the structure of the cohomology of S_∞ as an algebra over the Steenrod algebra, and in particular to see the Dickson algebras as an associated graded.

2 The Fox-Neuwirth cochain complexes

In this section we decompose configuration spaces into open cells, analogously to how one decomposes Grassmannians into Schubert cells. Consider $\mathbb{R}^m$ with its standard basis, and let $\mathrm{Conf}_n(\mathbb{R}^m)$ denote the space of configurations of n distinct, labelled points in $\mathbb{R}^m$ topologized as a subspace of $\mathbb{R}^{mn}$. The configuration space $\mathrm{Conf}_n(\mathbb{R}^m)$ admits a free, transitive action of $\mathcal{S}_n$, the symmetric group on n letters, by permutation of the labels. We denote by $\overline{\mathrm{Conf}_n}(\mathbb{R}^m)$ the quotient $\mathrm{Conf}_n(\mathbb{R}^m)/\mathcal{S}_n$, which is the space of unlabeled configurations of n points in $\mathbb{R}^m$.

The standard embeddings $\mathbb{R}^m \hookrightarrow \mathbb{R}^{m+1}$ give rise to a canonical directed system of configuration spaces $\mathrm{Conf}_n(\mathbb{R}^2) \hookrightarrow \mathrm{Conf}_n(\mathbb{R}^3) \hookrightarrow \cdots$ whose maps are all equivariant with respect to the $\mathcal{S}_n$ action. We denote the limit by $\mathrm{Conf}_n(\mathbb{R}^\infty)$. A map from a sphere of dimension less than $m-1$ to $\mathrm{Conf}_n(\mathbb{R}^m)$ can be first null-homotoped in $\mathbb{R}^{mn}$, but then by general position be null-homotoped in $\mathrm{Conf}_n(\mathbb{R}^m)$ itself. Thus $\mathrm{Conf}_n(\mathbb{R}^m)$ is $m-2$ connected, so $\mathrm{Conf}_n(\mathbb{R}^\infty)$ is weakly contractible. Moreover $\mathrm{Conf}_n(\mathbb{R}^\infty)$ inherits a free action of $\mathcal{S}_n$, so it is a model for $E\mathcal{S}_n$. Thus $\overline{\mathrm{Conf}}_n(\mathbb{R}^\infty) \simeq B\mathcal{S}_n$. Indeed, $B\mathcal{S}_n$ classifies finite-sheeted covering spaces, and by embedding such a bundle over some paracompact base in a trivial Euclidean bundle we can define the classifying map to $\overline{\mathrm{Conf}}_n(\mathbb{R}^\infty)$.

2.1 Fox-Neuwirth cells

We now describe "open" cellular decompositions of $\mathrm{Conf}_n(\mathbb{R}^m)$ and $\overline{\mathrm{Conf}}_n(\mathbb{R}^m)$ which then define CW-structures on their one-point compactifications. These cellular decompositions are due to Fox and Neuwirth [14] when $m = 2$, were considered by Vassiliev in higher dimensions [30], and have also been considered in the context of E_m-operads by Berger [5, 6]. They are in some sense Alexander dual to the Milgram decompositions of configuration spaces [20] (and more generally the Salvetti complex [23]) and the realizations of θ-categories of Joyal [18]. In the limit these cellular decompositions give rise to a cochain complex to compute the cohomology of symmetric groups - see Theorem 2.9.

The cell structure is based on the dictionary ordering of points in $\mathbb{R}^m$ using standard coordinates, which we denote by $<$. This ordering gives rise to an ordering of points in a configuration.

Definition 2.1. A depth-ordering is a series of labeled inequalities $i_1 <_{a_1} i_2 <_{a_2} i_3 <_{a_3} \cdots <_{a_{n-1}} i_n$, where the a_i are non-negative integers and the set of $\{i_k\}$ is exactly $\{1, \ldots, n\}$.

For any depth-ordering Γ define $\mathrm{Conf}_\Gamma(\mathbb{R}^m)$ to be the collection of all configurations $(x_1, x_2, \ldots, x_n) \in \mathrm{Conf}_n(\mathbb{R}^m)$ such that for any p, $x_{i_p} < x_{i_{p+1}}$ and their first a_p coordinates are equal while their $(a_p + 1)$st coordinate must differ.

The subspace $\mathrm{Conf}_\Gamma(\mathbb{R}^m)$ is empty if any a_i is greater than or equal to m.

Theorem 2.2 (after Fox-Neuwirth). *For any Γ the subspace $\mathrm{Conf}_\Gamma(\mathbb{R}^m)$ is homeomorphic to a Euclidean ball of dimension $mn - \sum a_i$. The images of the $\mathrm{Conf}_\Gamma(\mathbb{R}^m)$ are the interiors of cells in an equivariant CW structure on the one-point compactification $\mathrm{Conf}_n(\mathbb{R}^m)^+$.*

Because this cell structure is equivariant, it descends to one for $\overline{\mathrm{Conf}}_n(\mathbb{R}^m)^+$, where the cells are indexed only by the a_i. But to understand the boundary structure, we need to work with ordered configurations. While these cells are very easy to name, their boundary structure is intricate. Consider the cell $\mathrm{Conf}_\Gamma(\mathbb{R}^m)$ with $\Gamma = 1 <_1 2 <_0 3$. It is immediate to see the cell labeled by $1 <_2 2 <_0 3$ on the boundary of Γ, as the inequality between the second coordinates of x_1 and x_2 degenerates to an equality. But the cell labeled by $2 <_2 1 <_0 3$ is also on the boundary, since the third coordinates of x_1 and x_2 are unrestricted as the second coordinates become equal. There are even more possibilities when the inequality between the first coordinates of x_2 (and thus x_1 as well) and x_3 degenerates to an equality. The first inequality, based on the second coordiates of x_1 and x_2, will still hold. But the second coordinate of x_3 has been unrestricted by Γ, so the cells labeled by $1 <_1 2 <_1 3$, $1 <_1 3 <_1 2$ and $3 <_1 1 <_1 2$ all occur on the boundary of Γ. See Figure 2.1. The combinatorics of shuffles will play a central role in the boundary structure in general.

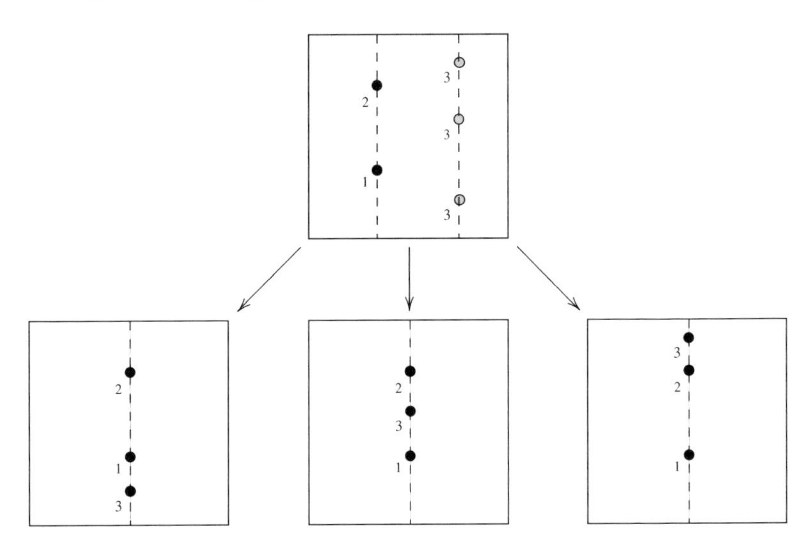

Figure 2.1. The degeneration of the second inequality in the cell $1 <_1 2 <_0 3$ to the bounding cells $3 <_1 < 1 <_1 2$, $1 <_1 3 <_1 2$ and $1 <_1 2 <_1 3$ respectively.

We introduce a data structure equivalent to depth orderings in terms of trees in order to more easily describe the boundary combinatorics. These trees were also used by Vassiliev [30] to label this cell structure, and have been of interest in higher category theory (see for example the Chapter 7 of the expository book [11]). Recall that in a rooted tree, one vertex is under another if it is in the unique path from the root to that vertex.

Definition 2.3. For a depth-ordering $\Gamma = i_1 <_{a_1} i_2 < \cdots <_{a_{n-1}} i_n$ define a planar level tree τ_Γ up to isotopy as follows.

- There is a root vertex at the origin, and all others vertices have positive integer-valued heights (that is, y-coordinates).

- There are n leaves which are of height m labelled $i_1, i_2, \cdots i_n$ in order from left to right (that is, in order of x-coordinates).
- Under each leaf there is a unique edge from a vertex of height i to one of height $i+1$ for each $0 \le i < m$. The leaves labelled i_p and i_{p+1} share the edges under them up to height a_p, but not above this height.

For example, the depth ordering $\Gamma = 3 <_2 5 <_0 1 <_3 2 <_0 4 <_1 6$ is represented by the following tree.

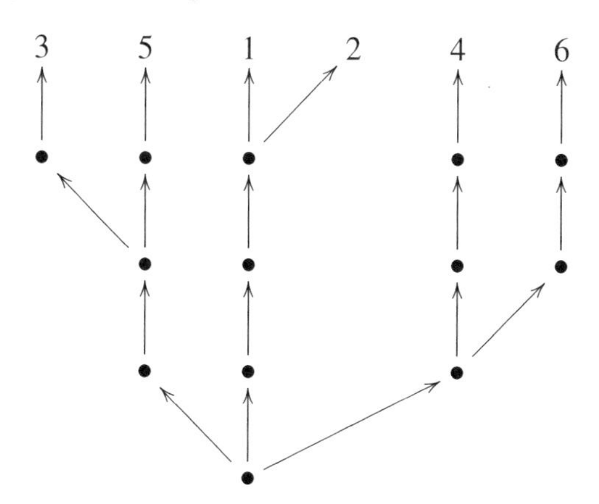

The tree τ_Γ uniquely determines Γ, so we use the two structures interchangeably. Recall as well that the planar isotopy class of a level tree is equivalent to an abstract level tree with an ordering (say from left to right) of the edges incident to each vertex.

Definition 2.4. Let τ be a planar level tree. If e and f are two edges which are incident to the internal vertex v, consecutive in the incident ordering with e before f, then the edge quotient $\tau_{/(e=f)}$ is the planar level tree described as an abstract level tree with edge orderings as follows.

- Its edge set is obtained from that of τ by identifying e and f. We call the resulting special edge the quotient edge.
- Its vertex set is obtained from that of τ by identifying the terminal vertices of e and f. We call the resulting vertex the quotient vertex.
- The edges with this terminal vertex as initial are ordered consistently with their previous ordering, with those incident to e before those incident before f.
- All other incidence relations and edge orderings are transported by the bijection away from e and f with the edges and vertices of τ.

Definition 2.5. A vertex permutation at v of a planar level tree τ is a tree which differs from τ only by changing the edge ordering of v. We denote such by $\sigma_v\tau$, where σ_v is the permutation of edges at v.

When σ_v is a shuffle, we call the resulting tree a vertex shuffle of τ. When the tree in question is an edge quotient $\tau_{/(e=f)}$, by convention we shuffle at the quotient vertex v using the initial partition into edges which were incident to e in τ followed by edges which were incident to f.

We can now describe the boundary structure of Fox-Neuwirth cells.

Proposition 2.6. *The cell* $\mathrm{Conf}_{\Gamma'}(\mathbb{R}^m)$ *is in the boundary of* $\mathrm{Conf}_{\Gamma}(\mathbb{R}^m)$ *if and only if* $\tau_{\Gamma'}$ *is isomorphic to* $\sigma_v \tau_{\Gamma/(e=f)}$ *for some consecutive edges* e *and* f *and some shuffle* σ_v *at the quotient vertex.*

For example, let $m = 3$ and $\Gamma = 1 <_1 2 <_0 3$. Then Γ is represented by the following tree.

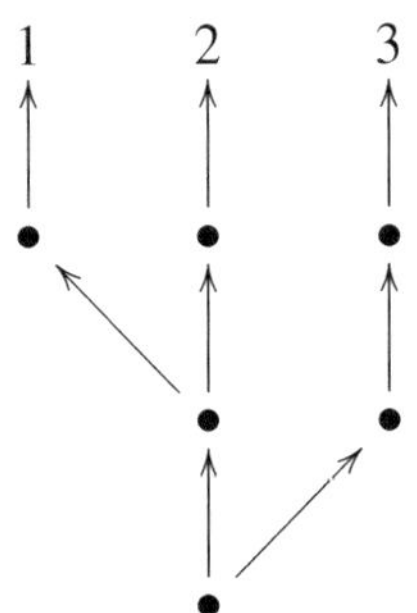

$\mathrm{Conf}_{\Gamma}(\mathbb{R}^3)$ has as its boundary precisely cells $\mathrm{Conf}_{\Gamma'}(\mathbb{R}^3)$ with Γ' represented by each of the following.

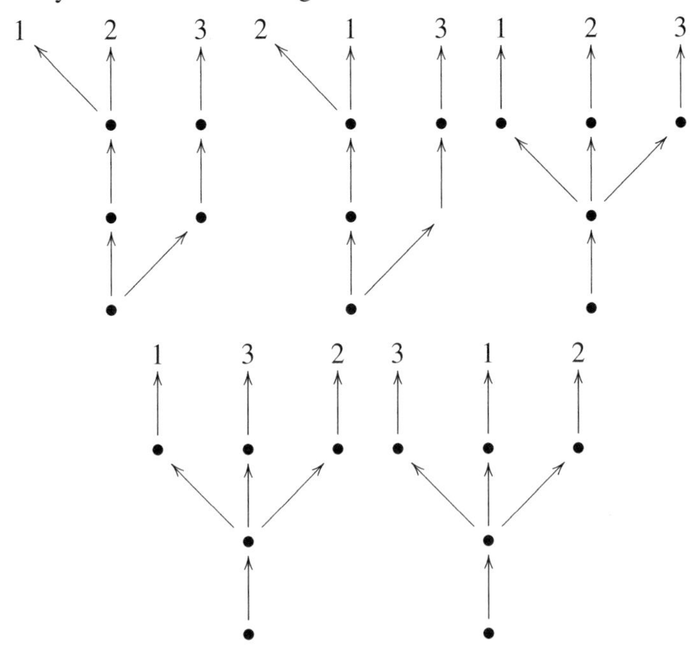

If we had instead considered $\mathrm{Conf}_\Gamma(\mathbb{R}^2)$, only the three cells corresponding to the last three in the above list would appear in the boundary since the other potential cells are empty.

There is one edge quotient for each consecutive pair of leaves in the tree τ_Γ, or equivalently for each inequality in Γ. For each edge quotient, the number of vertex shuffles varies and is always at least two.

2.2 The cochain complexes

We give what is to our knowledge the first explicit formula for the differential in the resulting integral cochain complexes. In light of Proposition 2.6, it remains to determine signs. Fox-Neuwrith cells are simple to endow with coordinates and thus orientations. Take the standard coordinates of $\mathrm{Conf}_n(\mathbb{R}^m)$ as a subspace of $\mathbb{R}^{mn}$, namely $(x_1)_1, (x_1)_2, \ldots, (x_1)_m, (x_2)_1, \ldots$, and omit any coordinate which from the definition of $\mathrm{Conf}_\Gamma(\mathbb{R}^m)$ must be equal to one which appears earlier in the list. Straightforward analysis suffices to understand the difference between the orientation of a cell and that induced by a cell which bounds it, leading to the following.

Definition 2.7. The sign of the quotient vertex of the pth edge quotient of $\Gamma = i_1 <_{a_1} i_2 < \cdots <_{a_{n-1}} i_n$, denoted $\mathrm{sgn}_m(p)$, $p \in \{1, \ldots, n-1\}$ is defined as $\mathrm{sgn}_m(p, \Gamma) = (-1)^\kappa$, where

$$\kappa = p + \sum_{k=1}^{p-1} \left(m - \min\{a_k, a_p + 1\}\right) + (m - (a_p + 1)) + \sum_{k=p+1}^{n-1} \left(m - \min\{a_k, a_p\}\right).$$

If we consider only configurations in $\mathbb{R}^m$ with m even, we can delete m from the definition of κ. We now only concern ourselves with the limit as m goes to infinity.

Definition 2.8. Define the cochain complex $\widetilde{\mathrm{FN}}_n^{\,*}$ as follows. As a free abelian group it is generated by depth-orderings, with $[\Gamma]$ in degree $\sum a_p$ when $\Gamma = i_1 <_{a_1} i_2 < \cdots <_{a_{n-1}} i_n$. The differential is

$$d[\Gamma] = \sum_p \mathrm{sgn}_0(p, \Gamma) a_p[\Gamma], \quad \text{where} \quad a_p[\Gamma] = \sum_{\sigma_v} [\Gamma'].$$

The first sum is over inequalities in Γ, indexed by p, which in turn determine an adjacent edge pair e, f in τ_Γ. For each such edge pair we sum over all possible vertex shuffles of the quotient tree at the quotient vertex, taking the depth-ordering Γ' associated to the resulting tree $\sigma_v \tau_{\Gamma/(e=f)}$.

Theorem 2.9. *$\widetilde{\mathrm{FN}_n}^*$ is a cochain model for $E\mathcal{S}_n$, so its quotient by $\mathcal{S}_n$ which we call $\mathrm{FN}_n{}^*$ has cohomology isomorphic to that of the nth symmetric group.*

In algebraic terms, $\widetilde{\mathrm{FN}_n}^*$ is a free $\mathbb{Z}[\mathcal{S}_n]$ resolution of the trivial module. It is a challenge to even show algebraically that $\widetilde{\mathrm{FN}_n}^*$ is a cochain complex, much less that it is acyclic.

Sketch of proof. By Theorem 2.2, the $[\Gamma]$ naturally span the corresponding cellular chain complex for the one-point compactification $\mathrm{Conf}_n(\mathbb{R}^m)^+$, with $[\Gamma]$ in $C^{CW}_{mn-\sum a_i}\left(\mathrm{Conf}_n(\mathbb{R}^m)^+\right)$. By Alexander and Spanier-Whitead duality for manifolds as explained by Atiyah [2], the resulting homology group is isomorphic to cohomology of $\mathrm{Conf}_n(\mathbb{R}^m)$ in degree $k = \sum a_i$. These cochain groups are isomorphic through the maps induced by the directed system in degrees k less than m, so the inverse limit is just $\widetilde{\mathrm{FN}_n}^k$.

The main work in proof is then to show that the boundary homomorphisms have the indicated signs. □

We now consider more closely the quotient complex $\mathrm{FN}_n{}^*$ whose cohomology is that of $B\mathcal{S}_n$. The equivalence class of some $\Gamma = i_1 <_{a_1} < i_2 < \cdots <_{a_{n-1}} i_n$ modulo $\mathcal{S}_n$, which acts by permuting the i_k, is given by the subscripts of the inequalities which we now denote $\mathbf{a} = [a_1, ..., a_{n-1}]$. These correspond to trees $\tau_{\mathbf{a}}$ as described above but with leaves now unlabelled. Thus $\mathrm{FN}_n{}^*$ is significantly smaller than the bar complex, having rank in degree k of the number of partitions of k into $n-1$ non-negative integers. We express the differential of $\mathrm{FN}_n{}^*$ in terms of $\mathbf{a}$.

Definition 2.10. An ℓ-block of $\mathbf{a}$ is a maximal (possibly empty) subsequence of consecutive a_i greater than ℓ in $\mathbf{a}$. Denote the ordered collection of all ℓ-blocks of $\mathbf{a}$ by $B_\ell(\mathbf{a})$.

For example, $B_0([0, 2, 6, 0, 1]) = (\emptyset, [2, 6], \emptyset, [1])$ and its B_1 is $(\emptyset, [2, 6], \emptyset, \emptyset)$. These collections $B_\ell(\mathbf{a})$ correspond to the forests of rooted trees obtained by deleting from $\tau_{\mathbf{a}}$ all vertices of height lower than ℓ and their incident edges.

Definition 2.11. Let $\mathbf{a} \in \mathrm{FN}_n{}^k$. Denote by $\mathbf{a}[i]$ the element $[a_1, a_2, \dots \dots, a_{i-1}, a_i + 1, a_{i+1}, \dots, a_{n-1}] \in \mathrm{FN}_n{}^{k+1}$.

Let $a_i \in \mathbf{a}$ and denote by $\mathbf{A}_i$ the a_i-block of $\mathbf{a}[i]$ containing a_i. Such a block $\mathbf{A}_i$ corresponds to a rooted subtree $\tau_{\mathbf{A}_i}$ of $\tau_{\mathbf{a}[i]}$. Let $N = \#B_{a_i+1}(\mathbf{A}_i)$, the number of trees in the forest obtained by removing the root from $\tau_{\mathbf{A}_i}$ along with its incident edges, and let k to be the number of trees in this forest whose roots were incident to the edge e in the tree $\tau_{\mathbf{a}}$ before the edge quotient $e = f$ which produced $\tau_{\mathbf{a}[i]}$. Define $\mathrm{Sh}(\mathbf{a}, i)$ to be the set of $(k, N-k)$-shuffles.

$\mathrm{Sh}(\mathbf{a}, i)$ acts on $\mathbf{a}[i]$ by shuffling the elements of $B_{a_i+1}(\mathbf{A_i})$. Equivalently, it acts via vertex shuffles on the tree $\tau_{\mathbf{a}[i]}$.

Proposition 2.12. *The differential in* $\mathrm{FN}_n{}^*$ *is given by*

$$\delta(\mathbf{a}) = \sum_{i=1}^{n-1} (-1)^{i+\alpha(i)} \delta_i(\mathbf{a}) \quad \textit{where} \quad \delta_i(\mathbf{a}) = \sum_{\sigma \in \mathrm{Sh}(\mathbf{a},i)} (-1)^{\kappa(\sigma,\mathbf{a})} \sigma \cdot \mathbf{a}[i].$$

If we define $h(\mathbf{v})$ to be the height in the tree $\tau_{\mathbf{a}}$ of the vertex $\mathbf{v}$ and $\#(\sigma, h)$ to be the total number of transpositions of vertices at height h in $\tau_{\mathbf{A}_i}$ which occur when σ acts on the tree, then we have

$$\alpha(i) = \sum_{j=1}^{i-1} \min\{a_j, a_i + 1\} + a_i + 1 + \sum_{j=i+1}^{n-1} \min\{a_j, a_i\}, \quad \textit{and} \quad \kappa(\sigma, \mathbf{a}) = \sum_{\mathbf{v}} (\#(\sigma, h) - \#(\sigma, h-1)) \cdot h(\mathbf{v}),$$

where the second sum is indexed over vertices $\mathbf{v}$ in the subtree $\tau_{\mathbf{A}_i}$.

The signs and the differential as a whole do not seem to simplify, as evidenced by the following examples of computations, which spite of the superficial similarity in cell names have substantially different boundaries.

$$\begin{aligned}
\delta([2,0,1,2]) &= -3[2,0,2,2] - 1[2,1,1,2] \\
\delta([1,0,2,2]) &= -1[1,1,2,2] + 1[1,2,2,1] + 2[2,0,2,2] - 1[2,2,1,1] \\
\delta([0,2,1,2]) &= -6[0,2,2,2] - 1[1,2,1,2] + 1[2,1,1,2] - 1[2,1,2,1] \\
\delta([0,1,2,2]) &= -4[0,2,2,2] - 1[1,2,2,1].
\end{aligned}$$

The first author has coded a program in Java to compute these differentials. It is appended at the end of the .tex file which was uploaded for the arXiv preprint of this paper, and is also available on his web site.

3 The cohomology of BS_4

In this section we establish the following.

Theorem 3.1. *There is an additive basis for $H^*(BS_4; \mathbb{F}_2)$ with representatives given by elements of* $\mathrm{FN}_4{}^*$ *of the following forms*

- $[a, 0, 0] + [0, a, 0] + [0, 0, a]$, *with* $a > 0$
- $[a, 0, b] + [b, 0, a]$, *with* $0 < a < b$

- $[b, a, b]$, *with* $0 \leq a \leq b$.

These Fox-Neuwirth representative cocycles along with the general representatives we find below were also found by Vassiliev [30], though he did not use them to determine multiplicative structure as we do and only sketched the proof that these cocycles form a basis. Because mod-two homology of symmetric groups is well-understood [12,22], we based our calculations in [15] on knowledge of homology, as did Vassiliev. We take the opportunity to give a complete proof of this result for $B\mathcal{S}_4$ to show that a self-contained treatment of cohomology is feasible.

Throughout this section, we label the entries of a basis element of $\mathrm{FN}_4{}^*$ by a, b, c with $a \leq b \leq c$.

Proof. That the cochains listed are cocycles is a straightforward calculation using the formula for the differential δ in the Fox-Neuwirth cochain complex given in Proposition 2.12.

Cocycles which are "disjoint from" (that is, project to zero in) the subspaces on which our generating cocycles are naturally defined are null-homologous, as we will establish through the construction of a chain homotopy operator. Consider the submodule S of FN_4^* spanned by cochains whose entries are all positive, and not of the form $[b, a, b]$, $[a, a, b]$ or $[b, a, a]$ with $a \leq b$. We define a chain operator $P : S \to \mathrm{FN}_4^*$ of degree -1 which lowers the smallest entry by one or maps to zero, depending on the order of the blocks with respect to the minimum entry. Explicitly, we have the following, where $0 < a < b < c$.

- $[a, b, c] \overset{P}{\mapsto} [a-1, b, c]$, and $[a, c, b] \overset{P}{\mapsto} [a-1, c, b]$, and $[a, b, b] \overset{P}{\mapsto} [a-1, b, b]$
- $[b, c, a]$, $[c, b, a]$ and $[b, b, a]$ all map to zero.
- $[b, a, c] \overset{P}{\mapsto} [b, a-1, c]$,
- $[c, a, b]$ maps to zero.

Let χ be the automorphism of S which exchanges a-blocks, which is a chain map. That is $\chi[a, b, c] = [b, c, a]$, while $\chi[b, a, c] = [c, a, b]$, and so forth. We claim that on S the map P is a chain null-homotopy of χ. For example if $b > a + 1$ and $c > b + 1$ then

$$\begin{aligned} P\delta[a,b,c] &= P\left([a+1,b,c] + [b,c,a+1] + [a,b+1,c] + [a,c,b+1]\right) \\ &= [a, b, c] + 0 + [a-1, b+1, c] + [a-1, c, b+1], \end{aligned}$$

while

$$\begin{aligned} \delta P[a, b, c] &= \delta[a-1, b, c] \\ &= [a, b, c] + [b, c, a] + [a-1, b+1, c] + [a-1, c, b+1], \end{aligned}$$

so their sum is $[b, c, a]$. Removing the restrictions $b > a+1$ and $c > b+1$ and treating other cases all follows from similar direct computation.

Thus for any cocycle γ in S,

$$\delta(P\chi(\gamma)) = P(\delta\chi(\gamma)) + \chi(\chi(\gamma)) = \gamma,$$

so any cocycle in S is trivial in cohomology.

We now show that a cycle is always homologous to one whose image in subspaces complementary to S is given by the cycles we have named, which seems to require ad-hoc analysis.

Consider chains whose first entry is zero, but whose second and third are not, which we call S_{f0}. When we compose the differential with projection onto S_{f0}, we obtain a complex which we claim is acyclic, as can be seen using a "decrease the smallest nonzero entry" nullhomtopy as above. Thus any chain which projects to S_{f0} non-trivially is homologous to one which projects to it trivially. Similarly, we can rule out cycles involving chains whose last entry is zero but first two are not, namely S_{l0}.

For chains whose middle entry alone is zero S_{m0}, we have $\delta[a, 0, b] = [a, 1, b]+[b, 1, a]$. Cochains with a middle entry of one do not otherwise appear in the image of δ, so the projection of a cycle onto S_{m0} must itself be a cycle, which direct calculation shows must be of the form $[a, 0, b]+[b, 0, a]$.

Similarly, if we consider chains with only one-nonzero entry and two zeros, which we call S_{00}, the boundary involves chains with a single zero and a single one. Only chains with two zeros can have such boundaries, so the projection of a cycle onto S_{00} must itself be a cycle, which direct calculation shows must be of the form $[a, 0, 0] + [0, a, 0] + [0, 0, a]$.

Finally, consider S_{2a}, which is the span of $[a, a, b]$, $[a, b, a]$, and $[a, a, b]$. Assume at first that $a + 1 < b$, in which case the image of $[a, a, b]$ under δ is $[a, a + 1, b] + [a, b, a + 1]$. The only other cochain for which either of these appears non-trivially in its coboundary is $[a, b, a]$, whose coboundary consists of those along with $[b, a + 1, a]$ and $[a + 1, b, a]$. These latter two terms are the coboundary of $[b, a, a]$, so the only cycle in which these occur nontrivially is $[a, a, b]+[a, b, a]+[a, a, b]$. But this is trivial in cohomology, equal to $\delta([a, a-1, b])$. Thus any cycle is homologous to one which projects trivially onto S_{2a}. The analysis with $a + 1 = b$ is similar.

Because $FN_4{}^*$ is spanned by S, S_{f0}, S_{l0}, S_{m0}, S_{00}, S_{2a}, and the cycles listed in the statement of the theorem, the result follows. □

3.1 Block symmetry and skyline diagrams

The cochain representatives for the cohomology of BS_4 all exhibit some symmetry. Define a block permutation to be a permutation of the entries of a Fox-Neuwirth basis cochain which does not change its (unordered) set of i-blocks for any i. The Fox-Neuwirth cochains which comprise each of the cocyles listed in Theorem 3.1 are invariant under block-permutations. In particular, if we let $\mathrm{Symm}(\mathbf{a})$ denote the sum of all distinct block-permutations of $\mathbf{a}$, then the first two sets of generators are $\mathrm{Symm}([a, 0, b])$ for $0 \leq a < b$ and the third is $\mathrm{Symm}([b, a, b])$, for $0 \leq a \leq b$. Symmetric representatives exist for the cohomology BS_n more generally, as we will see below. It might be enlightening to have a more direct proof of this block-symmetry property than through simply observing it holds after full computation.

The symmetry of these cochains allows the essential data defining them to be given in a graphical representation which we call *skyline diagrams*. The basis elements described in Theorem 3.1 correspond to skyline diagrams in Figure 3.1. These will be treated fully in Section 5.

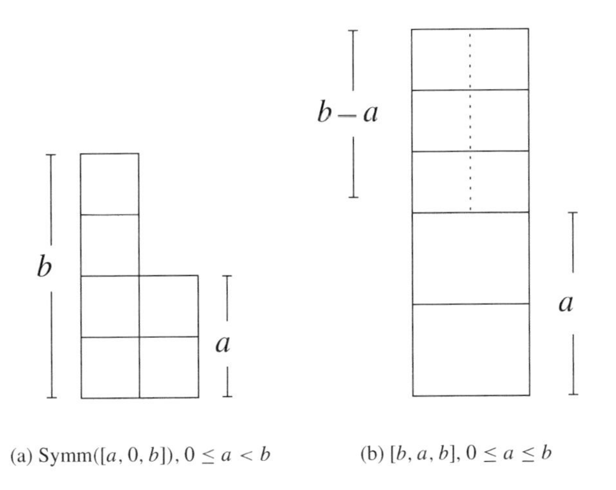

(a) $\mathrm{Symm}([a, 0, b]), 0 \leq a < b$ (b) $[b, a, b], 0 \leq a \leq b$

Figure 3.1. Skyline diagrams for the basis of $H^*(BS_4; \mathbb{F}_2)$ described in Theorem 3.1.

3.2 Geometry and characteristic classes

We can understand our cocycles geometrically through intersection theory, which is one of the standard ways to understand duality. If M is a manifold without boundary and M^+ is its one-point compactification, then at the cochain level Alexander-Spanier-Whitehead duality identifies a cellular codimension-d cell C in M^+ through a zig-zag of maps with a cochain on M whose value on chains $\sigma : \Delta^d \to M$ which are transversal to the interior of C is the mod-two count of $\sigma^{-1}(C)$. In our setting, these counts are of configurations whose points share prescribed coordinates.

For example, in $H_4(B\mathcal{S}_4)$ there is the fundamental class of a submanifold $\mathbb{R}P^2 \times \mathbb{R}P^2$ of $\mathrm{Conf}_4(\mathbb{R}^\infty)$. This submanifold parameterizes configurations of four points two of which are on the unit sphere say centered at the origin and two of which are on a unit sphere chosen to be disjoint from that one. The cocycle [2, 0, 2] evaluates non-trivially on this homology class, as can be seen by a count of configurations in this submanifold and in the cell, as we picture in Figure 3.2.

The corresponding characteristic class can be evaluated on a four-sheeted covering space by embedding it in a trivial Euclidean bundle and taking the Thom class of the locus where the fibers of the cover can be partitioned into two groups of two, each of which share their first two coordinates. For the cocyle [2, 1, 2], the corresponding characteristic class would be the Thom class of the locus where all four points in the fiber share their first coordinate, and they can be partitioned into two groups of two which share an additional coordinate.

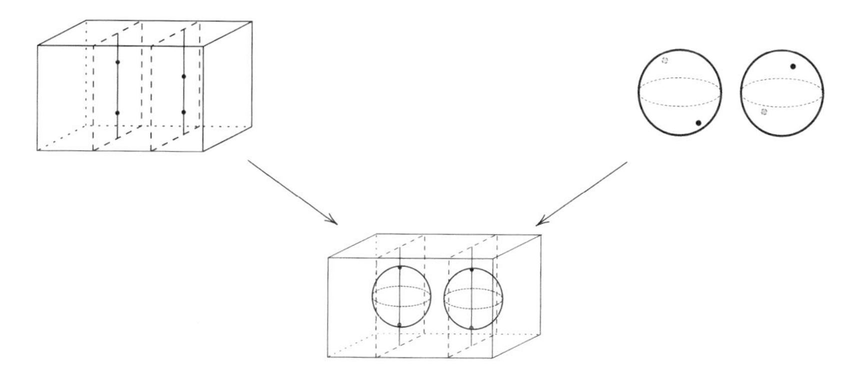

Figure 3.2. The cocycle $[2, 0, 2] \in H^4(B\mathcal{S}_4)$ pairs non-trivially with the fundamental class of $\mathbb{R}P^2 \times \mathbb{R}P^2$ considered as a submanifold of $\mathrm{Conf}_4(\mathbb{R}^\infty)$ as indicated.

4 Hopf ring structure and presentation

While cup product structure typically clarifies the calculation of cohomology, the cohomology rings of symmetric groups have been notoriously difficult to understand. Feshbach's description [13] is complicated, with recursively presented relations. We have found that the situation is clarified once one uses a second product structure on cohomology, along with a coproduct (and antipode which is trivial) in order to obtain what is known as a Hopf ring. This structure was first discovered by Strickland and Turner [25].

4.1 Cup product structure

Definition 4.1. Let $\mathbf{a} = [a_1, \ldots, a_{n-1}]$ and $\mathbf{b} = [b_1, \ldots, b_{n-1}]$ be basis elements in the mod-two Fox-Neuwirth cochain complex. Define their intersection product $\mathbf{a} \cdot \mathbf{b}$ to be

$$[a_1, \ldots, a_{n-1}] \cdot [b_1, \ldots, b_{n-1}] = [a_1 + b_1, \ldots, a_{n-1} + b_{n-1}],$$

Integrally, we would also have the usual sign accounting for the possible difference in orientations.

Lemma 4.2. *The intersection product makes* FN^* *a (commutative) differential graded algebra.*

Proposition 4.3. *Through the isomorphism of Theorem 2.9, the product on cohomology induced by the intersection product on* FN_n^* *agrees with the cup product on cohomology of* $B\mathcal{S}_n$.

We call this the intersection product in light of the fact mentioned above that the cohomology classes which arise from the Fox-Neuwirth cell structure on $\overline{\mathrm{Conf}}_n(\mathbb{R}^D)^+$ are Thom classes of the union of corresponding cells in $\overline{\mathrm{Conf}}_n(\mathbb{R}^D)$. These unions of cells correspond to images of manifolds whose fundamental classes in locally finite homology are dual to the Thom classes, as in [15, Definition], and thus are appropriate for elementary intersection theory. We can generalize the Fox-Neuwirth cell structure to a family of such in which specified sets of coordinates other than the first a_i coordinates must be equal, with the resulting cohomology classes being independent of which are chosen. We then use that the cup product of Thom classes of two varieties is the Thom class of their intersection (when transversal). For example the set of configurations in which the first two points have their first a_1 coordinates agree intersected with the set in which the the first two points have the next b_1 coordinates equal will of course be the set in which the first two points have their first $a_1 + b_1$ equal, and so forth, showing that the intersection product in the Fox-Neuwirth cell structure corresponds to the product structure of the associated Thom classes.

Using the intersection product, it is straightforward to compute the cup product structure appearing in Figure 4.1 on $H^*(B\mathcal{S}_4; \mathbb{F}_2)$, using the skyline basis for convenience. The computations use observations that were made in the proof of Theorem 3.1 of cocycles which are coboundaries. Using the graphical skyline presentation, multiplication consists simply of "stacking columns in all possible ways", with a result of zero if a vertical line does not continue for the entire height of the column. See the end

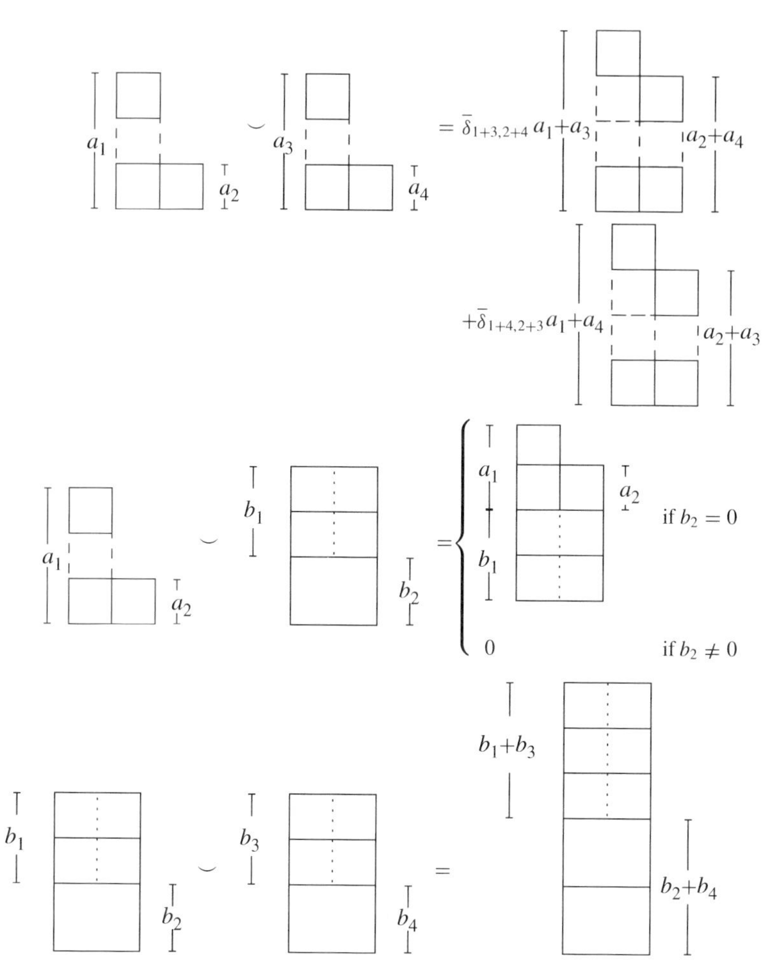

Figure 4.1. The cup product structure on $H(BS_4)$. Here, $\overline{\delta}_{i+j,k+\ell}$ is $1 + \delta(a_i + a_j, a_k + a_\ell) \in \mathbb{F}_2$ where δ is the usual Kronecker delta function.

of Section 5 for the general recipe, and Section 6 of [15] for a detailed account.

For a ring presentation, denote cohomology classes using only cochains, and let $\mathbf{x} = \mathrm{Symm}[1, 0, 0]$, $\mathbf{y} = [1, 0, 1]$ and $\mathbf{z} = [1, 1, 1]$. Then $H^*(BS_4; \mathbb{F}_2) \cong \mathbb{F}_2[\mathbf{x}, \mathbf{y}, \mathbf{z}]/(\mathbf{xz})$. This presentation is deceptively simple. At present there are only recursive presentations of the relations in the cohomology rings of general symmetric groups.

4.2 Hopf ring structure

We have found that the best way to organize the relations in the cohomology rings of symmetric groups, which seem inherently recursive, is through a Hopf ring structure

Definition 4.4. A Hopf ring is a ring object in the category of coalgebras. Explicitly, a Hopf ring is vector space V with two multiplications, one comultiplication, and an antipode $(\odot, \cdot, \Delta, S)$ such that the first multiplication forms a Hopf algebra with the comultiplication and antipode, the second multiplication forms a bialgebra with the comultiplication, and these structures satisfy the distributivity relation

$$\alpha \cdot (\beta \odot \gamma) = \sum_{\Delta\alpha = \sum a' \otimes a''} (a' \cdot \beta) \odot (a'' \cdot \gamma).$$

Hopf rings were introduced by Milgram [21], and arise in topology as one structure governing the homology of infinite loop spaces which represent ring spectra. We give examples arising from algebra in [15, Section 2] and will give some explicit calculations below.

Definition 4.5. Let $\mathbf{a} = [a_1, \ldots, a_{n-1}]$, and by convention set $a_0 = a_n = 0$. The coproduct of $\mathbf{a} \in \mathrm{FN}_n{}^*$, is the sum $\Delta\mathbf{a} = \sum_{a_i=0}[a_1, \ldots, a_{i-1}] \otimes [a_{i+1}, \ldots, a_n]$ in $\bigoplus_{i+j=n} \mathrm{FN}_i{}^* \otimes \mathrm{FN}_j{}^*$.

Thus if $a_i > 0$ for $1 \leq i \leq n-1$ then $\mathbf{a}$ will be primitive.

Definition 4.6. The transfer product of $\mathbf{a} \in \mathrm{FN}_i{}^*$ and $\mathbf{b} \in \mathrm{FN}_j{}^*$, denoted $\mathbf{a} \odot \mathbf{b}$, is the sum of cochains in $\mathrm{FN}_{i+j}{}^*$ whose zero blocks are shuffles of the zero blocks of $\mathbf{a}$ and $\mathbf{b}$.

Theorem 4.7. *The coproduct and transfer product induce a well-defined coproduct and product respectively on cohomology of $\bigoplus_n \mathrm{FN}_n{}^*$.*

Through the isomorphism of Theorem 2.9, the induced coproduct and product on cohomology agree with the direct sum over $i + j = n$ of natural maps and transfers respectively associated to the standard inclusions of symmetric groups $\mathcal{S}_i \times \mathcal{S}_j \hookrightarrow \mathcal{S}_n$.

Thus, the coproduct and transfer product on cohomology agree with the standard coproduct and the transfer product of Strickland and Turner [25]. They show that the cohomology of the disjoint union of the symmetric groups is a Hopf ring with these along with the cup product, so we have the following corollary, once we define the intersection product between cochains in different summands of $\bigoplus_n \mathrm{FN}_n{}^*$ to be zero.

Corollary 4.8. *Through the isomorphism of Theorem 2.9, the intersection product, transfer product, and coproduct induce a Hopf semiring* *structure on the cohomology of* $\bigoplus_n \mathrm{FN}_n{}^*$. *With mod two coefficients, the identity map defines an antipode which with these structures yields a Hopf ring.*

An algebraic proof of this corollary is also possible. At the chain level, $(\cdot, \Delta)$ and $(\odot, \Delta)$ both induce bialgebra structures on $\bigoplus_n \mathrm{FN}_n{}^*$, but the Hopf ring distributivity relation fails. However, there is a Hopf semiring structure on the subcomplex of symmetrized chains. This is analogous to rings of symmetric invariants, which are treated in Section 2 of [15]. For any algebra A which is flat over its ground ring R, the total symmetric invariants, $\bigoplus_n (A^{\otimes n})^{S_n}$, forms a Hopf semiring. The bialgebra structures hold before taking invariants, but Hopf ring distributivity requires taking invariants.

We can use Hopf ring distributivity of the cup product over the transfer product to understand the cup product. Recall that Symm($\mathbf{a}$) denote the sum of all cochains related to $\mathbf{a}$ by a block-permutation. To compute Symm([3, 0, 2, 2, 2]) $\cdot$ [Symm([4, 0, 2, 0, 2])], we can decompose Symm[4, 0, 2, 0, 2] as $[4] \odot [2, 0, 2]$. Now, $\Delta(\mathrm{Symm}([3, 0, 2, 2, 2])) = [3] \otimes [2, 2, 2]$ plus terms which must cup to zero with the chains in our decomposition of Symm[4, 0, 2, 0, 2]) because they lie in the cohomology of other symmetric groups. Thus we have

$$\begin{aligned}
&\mathrm{Symm}([3,0,2,2,2]) \cdot \mathrm{Symm}([4,0,2,0,2]) \\
&\quad = (([3]\cdot[4]) \odot ([2,2,2]\cdot[2,0,2])) \\
&\quad = [7] \odot [4,2,4] \\
&\quad = [7,0,4,2,4] + [4,2,4,0,7] \\
&\quad = \mathrm{Symm}([7,0,4,2,4]).
\end{aligned}$$

4.3 Hopf ring presentation

We now develop the most basic classes, whose cup and transfer products will yield all mod-two cohomology of symmetric groups.

Definition 4.9. Let $\mathbf{g}_{\ell,n}$ be the basis element of $(\mathrm{FN}_{n2^\ell})^{n(2^\ell-1)}$ with n 0-blocks, each of which consists of $(2^\ell - 1)$ consecutive entries of 1. Let $\gamma_{\ell,n} = [\mathbf{g}_{\ell,n}]$.

For example, $\mathbf{g}_{2,3} = [1,1,1,0,1,1,1,0,1,1,1] \in C^9(BS_{12})$. The following is an elaboration of the main result our paper with Salvatore [15], with Fox-Neuwirth cochain representatives now given.

Theorem 4.10. *As a Hopf ring, $H^*(\coprod_n BS_n; \mathbb{F}_2)$ is generated by the classes $\gamma_{\ell,n} \in H^{n(2^\ell-1)}(BS_{n2^\ell})$, along with unit classes on each component. The coproduct of $\gamma_{\ell,n}$ is given by*

$$\Delta\gamma_{\ell,n} = \sum_{i+j=n} \gamma_{\ell,i} \otimes \gamma_{\ell,j}.$$

Relations between transfer products of these generators are given by

$$\gamma_{\ell,n} \odot \gamma_{\ell,m} = \binom{n+m}{n} \gamma_{\ell,n+m}.$$

The antipode is the identity map. Cup products of generators on different components are zero, and there are no other relations between cup products of generators.

For example, the cocycle Symm([4, 3, 4, 1, 4, 3, 4, 0, 3, 3, 3, 0, 1, 0, 1, 0, 1, 0, 0]) is represented as the Hopf ring monomial $\gamma_{3,1}\gamma_{2,2}{}^2\gamma_{1,4} \odot \gamma_{2,1}{}^3 \odot \gamma_{1,2} \odot \gamma_{1,1} \odot 1_2$ and, as we will see in the next section, the skyline diagram

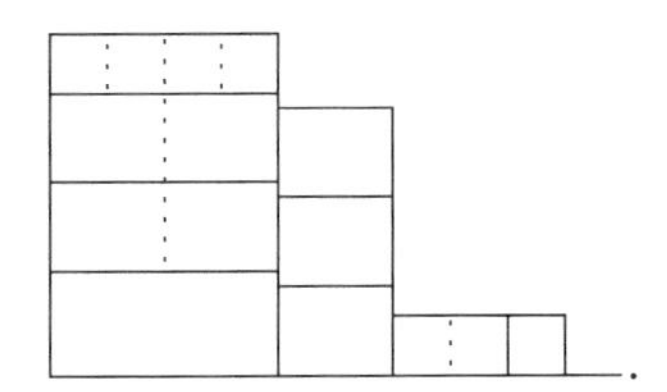

An immediate corollary of this theorem is that all cohomology classes have block-symmetric representatives, which was also found by Vassiliev [30].

5 The skyline basis

A Hopf ring monomial in classes x_i is one of the form $f_1 \odot f_2 \odot \cdots \odot f_k$, where each f_j is a monomial under the $\cdot$ product in the x_i. Because of Hopf ring distributivity, it is convenient to use Hopf ring monomials in chosen generators to span a Hopf ring. In this section, we give a graphical presentation of a Hopf ring monomial basis for the mod-two cohomology of symmetric groups, using both spatial dimensions to represent the two products.

Definition 5.1. A gathered monomial in the cohomology of symmetric groups is a Hopf ring monomial in the generators $\gamma_{\ell,n}$ where such n are maximal or equivalently the number of transfer products which appear is minimal.

For example, $\gamma_{1,4}\gamma_{2,2}{}^3 \odot \gamma_{1,2}\gamma_{2,1}{}^3 = \gamma_{1,6}\gamma_{2,3}{}^3$. Gathered monomials such as the latter in which no transfer products appear are building blocks for general gathered monomials.

Definition 5.2. A gathered block is a monomial of the form $\prod_i \gamma_{\ell_i,n_i}{}^{d_i}$, where the product is the cup product. Its profile is defined to be the collection of pairs (ℓ_i, d_i).

Non-trivial gathered blocks must have all of the numbers $2^{\ell_i} n_i$ equal, and we call this number divided by two the width. We assume that the factors are ordered from smallest to largest n_i (or largest to smallest ℓ_i), and then note that $n_i = 2^{\ell_1 - \ell_i} n_1$.

Proposition 5.3. *A gathered monomial can be written uniquely as the transfer product of gathered blocks with distinct profiles. Gathered monomials form a canonical additive basis for the cohomology of $\coprod_n BS_n$.*

Graphically, we represent $\gamma_{\ell,n}$ by a rectangle of width $n \cdot 2^\ell$ and height $1 - \frac{1}{2^\ell}$, so that its area corresponds to its degree. We represent 1_n by an edge of width n (a height-zero rectangle). A gathered block, which is a product of $\gamma_{\ell,n}$ for fixed $n \cdot 2^\ell$, is represented by a single column of such rectangles, stacked on top of each other, with order which does not matter. A gathered monomial is represented by placing such columns next to each other, which we call the skyline diagram of the monomial. We also refer to the gathered monomial basis as the skyline basis to emphasize this presentation.

In terms of skyline diagrams, the coproduct can be understood by introducing vertical dashed lines in the rectangles representing $\gamma_{\ell,n}$, dividing the rectangle into n equal pieces. The coproduct is then given by dividing along all existing columns and vertical dashed lines of full height and then partitioning them into two to make two new skyline diagrams.

The transfer product corresponds to placing two column Skyline diagrams next to each other and merging columns with the same constituent blocks, with a coefficient of zero if any of those column widths share a one in their dyadic expansion.

For cup product, we start with two column diagrams and consider all possible ways to split each into columns, along either original boundaries of columns or along the vertical lines of full height internal to the rectangles representing $\gamma_{\ell,n}$. We then match columns of each in all possible ways up to automorphism, and stack the resulting matched columns to get a new set of columns, as we saw for BS_4.

See Figure 5.1 for illustrations of cup product, which thus also involve the coproduct and transfer product. See Section 6 of [15] for thorough treatments of all of these structures.

$$\gamma_{1,1} \odot 1_4 \cdot \gamma_{1,2} \odot 1_2 = \gamma_{1,3} + \gamma_{1,1}^2 \odot \gamma_{1,1} \odot 1_2$$

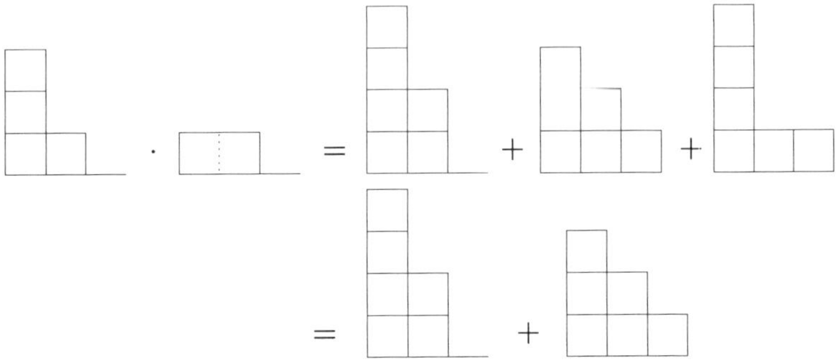

$$\gamma_{1,1}^3 \odot 1_2 \cdot \gamma_{1,2} \odot 1_2 = \gamma_{1,1}^4 \odot \gamma_{1,1}^2 \odot 1_2 + \gamma_{1,1}^3 \odot \gamma_{1,1}^2 \odot \gamma_{1,1}$$

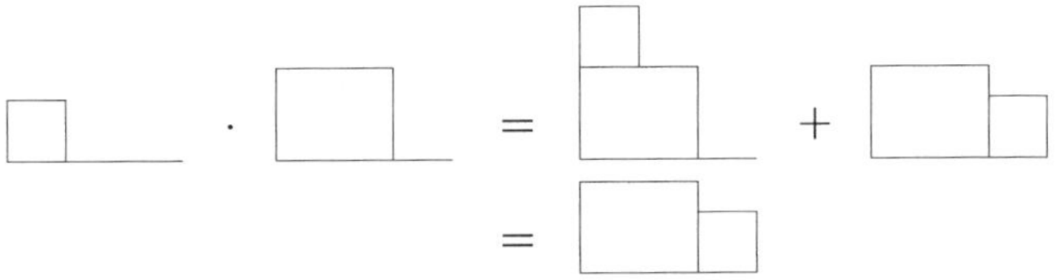

$$\gamma_{1,1} \odot 1_4 \cdot \gamma_{2,1} \odot 1_2 = \gamma_{2,1} \odot \gamma_{1,1}$$

Figure 5.1. Examples of product computations in $H^*(B\mathcal{S}_6; \mathbb{Z}/2)$ expressed in skyline and Hopf monomial notation

We can use our formula for multiplication to see for example that Vassiliev's conjecture that dth powers in the cohomology of $\overline{\mathrm{Conf}_n}(\mathbb{R}^d)$ are zero is not true. The cohomology of $\overline{\mathrm{Conf}_n}(\mathbb{R}^d)$ is the quotient of that of $\overline{\mathrm{Conf}_n}(\mathbb{R}^\infty)$ setting skyline diagrams with some block height greater than or equal to d to zero. Then for example in $\overline{\mathrm{Conf}_6}(\mathbb{R}^3)$ the cube of

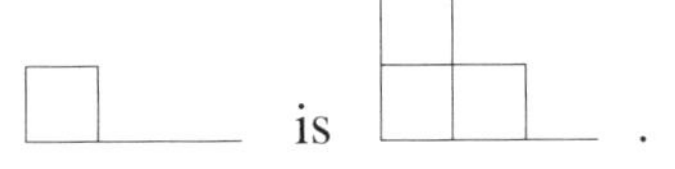

6 The cohomology of $B\mathcal{S}_\infty$ as an algebra over the Steenrod algebra

6.1 Nakaoka's theorem revisited

The infinite symmetric group plays a special role in algebraic topology. Let $\Omega^\infty S^\infty$ denote the direct limit, under suspension of maps, of the space of based maps from S^d to itself. The Barratt-Priddy-Quillen-Segal theorem [4] says that the cohomology of $\Omega^\infty S^\infty$ is isomorphic to that of $B\mathcal{S}_\infty$.

The map $B\mathcal{S}_n \to B\mathcal{S}_{n+1}$ induced by inclusion induces the map on cohomology which sends a skyline diagram with at least one empty column to that obtained by removing that column, and is zero on diagrams with no empty columns. The inverse limit is thus spanned by skyline diagrams with a finite number of non-empty columns, along with infinitely many empty columns, which we ignore. We let the width of such a diagram be the total width of the non-empty columns. Multiplication is through essentially the same algorithm as in for $B\mathcal{S}_n$, which generally increases width unless for example some diagram is raised to a power of two.

Definition 6.1. A column is even if every block type occurs an even number of times, and odd if at least one block type occurs an odd number of times. Define the two-root of a skyline diagram D consisting of a single column as the odd column R such that $R^{2^p} = D$.

Theorem 6.2 (after Nakaoka [22]). *The mod-two cohomology of $B\mathcal{S}_\infty$ is polynomial, generated by diagrams consisting of a single odd column.*

This theorem along with Theorem 4.10 extends Nakaoka's theorem by giving cochain representatives in $\mathrm{FN}_\infty{}^*$ for generators. This result should also be compared with the explicit calculation of homology primitives by Wellington [31].

Proof. We filter the cohomology of $B\mathcal{S}_\infty$, as represented by skyline diagrams, by width. If D is a diagram with columns $C_1, \ldots, C_n$, let $R_1, \ldots, R_n$ be their two-roots with ${R_i}^{2^{p_i}} = C_i$. By abuse let R_i denote the diagram which consists of R_i as its only non-empty column. Using the algorithm to multiply skyline diagrams by stacking columns, $D = \prod_i {R_i}^{2^{p_i}}$ modulo diagrams of lower filtration. Thus the associated graded to the width filtration is a polynomial algebra on diagrams consisting of a single odd column. So the cohomology of $B\mathcal{S}_\infty$ itself must be polynomial. □

In the course of proof, we see that the change of basis between the skyline basis and the monomial basis arising from the theorem is nontrivial but straightforward.

6.2 Steenrod structure

Next we focus on the Steenrod algebra structure on the cohomology of $B\mathcal{S}_\infty$ or equivalently $\Omega^\infty S^\infty$. To do so it is best to use the connection between the cohomology of groups and invariant theory which has been a fundamental tool in the subject. Recall for example from [1, Chapters 3 and 4] that if H is a subgroup of G then the Weyl group of H in G acts on the cohomology of H. Moreover, the restriction map from the cohomology of G to that of H has image in the invariants under this action, namely $(H^*(BH))^{W(H)}$.

In the study of the cohomology of symmetric groups, the invariant theory which arises is classical. Let the subgroup H in question be the subgroup $(\mathbb{Z}/2)^n \subset \mathcal{S}_{2^n}$ defined by having $(\mathbb{Z}/2)^n$ act on itself, which we call V_n. The cohomology of V_n is that of $\prod_n \mathbb{R}P^\infty$, namely $\mathbb{F}_2[x_1, \ldots, x_n]$. If we view the action of V_n on itself as given by linear translations on the $\mathbb{F}_2$-vector space $\oplus_n \mathbb{F}_2$, then we can see that the normalizer of this subgroup is isomorphic to all affine transformations of $(\mathbb{F}_2)^n$. The Weyl group is thus $GL_n(\mathbb{F}_2)$. Moreover, the action is by linear action on the variables $x_1, \ldots, x_n$.

The invariants $\mathbb{F}_2[x_1, \ldots, x_n]^{GL_n(\mathbb{F}_2)}$, studied a century ago, are known as Dickson algebras. Because permutation matrices are in $GL_n(\mathbb{F}_2)$ the invariants are in particular symmetric polynomials. But for example there is never a $GL_n(\mathbb{F}_2)$ invariant in degree one, since the lone symmetric invariant $x_1 + \cdots + x_n$ is not invariant under the linear substitution $x_1 \mapsto x_1 + x_2$ (and $x_i \mapsto x_i$ for $i > 1$). Dickson's theorem is that as rings these invariants are polynomial algebras on generators $d_{k,\ell}$ in dimensions $2^k(2^\ell - 1)$ where $k + \ell = n$. For example, $\mathbb{F}_2[x_1, x_2]^{GL_2(\mathbb{F}_2)}$ is generated by an invariant in degree two, namely ${x_1}^2 + {x_2}^2 + x_1 x_2$, along with ${x_1}^2 x_2 + x_1 {x_2}^2$ in degree three.

This connection to invariant theory allows us to determine the action of the Steenrod algebra. The standard starting point is that of the cohomology of $\mathbb{R}P^\infty$, which allows us to understand the Steenrod structure on that of $BV_n \simeq \prod_n \mathbb{R}P^\infty$ by the Cartan formula. Because the Steenrod action is defined by squaring individual variables, which is a linear operation over $\mathbb{F}_2$, the GL_n-invariants are preserved by the Steenrod action. For example

$$Sq^1({x_1}^2 + {x_2}^2 + x_1 x_2) = 0 \cdot {x_1}^3 + 0 \cdot {x_2}^2 + {x_1}^2 x_2 + x_1 {x_2}^2.$$

In [16] Hu'ng calculated the Steenrod squares on Dickson classes as given by

$$Sq^i d_{k,\ell} = \begin{cases} d_{k',\ell'} & i = 2^k - 2^{k'} \\ d_{k',\ell'} d_{k'',\ell''} & i = 2^n + 2^k - 2^{k'} - 2^{k''}, \quad k' \le k < k'' \\ {d_{k,\ell}}^2 & i = 2^k(2^\ell - 1) \\ 0 & \text{otherwise.} \end{cases} \tag{6.1}$$

Turning our attention back to symmetric groups, the transfer product in cohomology is induced by a stable map. Thus there is a Cartan formula for transfer product as well, and it suffices to understand Steenrod structure on Hopf ring generators. In [15] we prove the following.

Theorem 6.3. *The restriction of $\gamma_{\ell,2^k}$ with $k+\ell=n$ to the elementary abelian subgroup V_n is the Dickson invariant $d_{k,\ell}$.*

This theorem, Hu'ng's calculation above and the fact that the direct some of the restriction map to V_n and the coproduct Δ is injective allow us to understand the Steenrod on the $\gamma_{\ell,2^k}$.

Theorem 6.4. *A Steenrod square on $\gamma_{\ell,2^k}$ is represented in the skyline basis by the sum of all diagrams which are of full width, with at most two boxes stacked on top of each other, and with the width of columns delineated by any of the vertical lines (of full height) at least ℓ.*

See Figure 6.1 for some examples using the Cartan formula and this result.

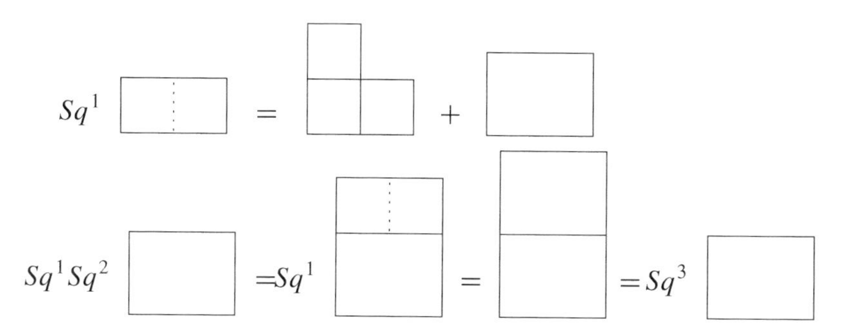

Figure 6.1. Steenrod algebra action on some elements of the skyline basis for BS_4. The second calculation uses the first through the Cartan formula, with the term $(\gamma_{1,1}^2 \odot \gamma_{1,1}) \cdot \gamma_{2,1}$ being zero.

6.3 Madsen's theorem revisited

One of the first questions to ask about an algebra R over the Steenrod algebra is its vector space of indecomposables, which we denote $\mathrm{Indec}_{\mathcal{A}-\mathrm{Alg}} R$. If R is the cohomology of a space, it is this vector space which can evaluate non-trivially on the Hurewicz homomorphism from homotopy to homology. In the case of the infinite symmetric group (and thus $\Omega^\infty S^\infty$), this question is reduced to one in invariant theory. This connection was first made by Madsen forty years ago [19], but there is still much to learn about the resulting algebraic question [17].

Our starting point is Theorem 6.2, which immediately determines the indecomposables of $H^*(B\mathcal{S}_\infty)$ as an algebra.

Definition 6.5. Define the width splitting of $\text{Indec}_{\text{Alg}} H^*(B\mathcal{S}_\infty)$, the algebra indecomposables of $H^*(B\mathcal{S}_\infty)$, by letting W_k denote the span of indecomposables which are represented by single columns of width 2^{k-1}. Define the width filtration through the increasing sums $\bigoplus_{k=1}^n W_k$.

We have the following remarkable connection between topology and invariant theory, which goes back at least to work of Selick (and Cohen and Peterson) [24].

Theorem 6.6. *The associated graded of the width filtration on* $\text{Indec}_{\text{Alg}} H^*(B\mathcal{S}_\infty)$ *is isomorphic as* $\mathcal{A}$*-modules to the direct sum of Dickson algebras* $\bigoplus_n \mathbb{F}_2[x_1, \ldots, x_n]^{GL_n(\mathbb{F}_2)}$.

Proof. In the description for Steenrod action on the Hopf ring generators $\gamma_{\ell,2^k}$ of Theorem 6.4 there is exactly one term in which there is just one column. This term corresponds to the formula for squares on Dickson generators as in Equation 6.1 (by replacing $d_{k,\ell}$ by $\gamma_{\ell,2^k}$). All other terms have more than one column, and thus modulo decomposables are lower in the width filtration. □

The fact that for an algebra R over the Steenrod algebra, the quotient map induces an isomorphism of graded vector spaces $\text{Indec}_{\mathcal{A}-\text{Alg}} R \cong \text{Indec}_{\mathcal{A}-\text{Mod}}(\text{Indec}_{\text{Alg}} R)$ implies that $\text{Indec}_{\mathcal{A}-\text{Alg}} H^*(\Omega^\infty S^\infty)$ is a quotient of $\bigoplus_n \text{Indec}_{\mathcal{A}-\text{Mod}} \mathbb{F}_2[x_1, \ldots, x_n]^{GL_n(\mathbb{F}_2)}$. To date, the most extensive calculations have been of the dual primitives in homology by Wellington [31] and recently by Zare [32]. The Steenrod structure on the algebra indecomposables was put to great use by Campbell, Cohen, Peterson and Selick [9,10]. The skyline basis in cohomology gives a distinct approach, which may be especially useful in tandem with homology.

References

[1] A. Adem and R. J. Milgram, "Cohomology of Finite Groups", Grundlehren der Mathematischen Wissenschaften [Fundamental Principles of Mathematical Sciences], Vol. 309, Springer-Verlag, Berlin, 1994. MR MR1317096 (96f:20082)

[2] M. F. Atiyah, *Thom complexes*, Proc. London Math. Soc. (3) **11** (1961), 291–310. MR 0131880 (24 #A1727)

[3] D. Ayala and R. Hepworth, *Configurations spaces and theta-n*, arXiv:1202.2806.

[4] M. Barratt and S. Priddy, *On the homology of non-connected monoids and their associated groups*, Comment. Math. Helv. **47** (1972), 1–14. MR MR0314940 (47 #3489)

[5] C. Berger, *Opérades cellulaires et espaces de lacets itérés*, Ann. Inst. Fourier (Grenoble) **46** (1996), no. 4, 1125– 1157. MR 1415960 (98c:55011)

[6] C. Berger, *Combinatorial models for real configuration spaces and E_n-operads*, Operads: Proceedings of Renaissance Con- ferences (Hartford, CT/Luminy, 1995), Contemp. Math., Vol. 202, Amer. Math. Soc., Providence, RI, 1997, pp. 37–52. MR 1436916 (98j:18014)

[7] T. P. Bisson and A. Joyal, *Q-rings and the homology of the symmetric groups*, Operads: Proceedings of Renaissance Conferences (Hartford, CT/Luminy, 1995), Contemp. Math., vol. 202, Amer. Math. Soc., Providence, RI, 1997, pp. 235–286. MR MR1436923 (98e:55021)

[8] P. Blagojević and G. Ziegler, *Convex equipartitions via equivariant obstruction theory*, arXiv:1202.5504

[9] H. E. A. Campbell, F. R. Cohen, F. P. Peterson and P. S. Selick, Self-maps of loop spaces. II, Trans. Amer. Math. Soc. **293** (1986), no. 1, 41–51. MR 814910 (87e:55010a)

[10] H. E. A. Campbell, F. P. Peterson and P. S. Selick, *Self-maps of loop spaces*. I, Trans. Amer. Math. Soc. **293** (1986), no. 1, 1–39. MR 814910 (87e:55010a)

[11] E. Cheng and A. Lauda, *Higher-dimensional categories*, an illustrated guidebook., available at http://www.cheng.staff.shef.ac.uk/.

[12] F. R. Cohen, T. J. Lada and J. P. May, "The Homology of Iterated Loop Spaces", Springer-Verlag, Berlin, 1976. MR MR0436146 (55 #9096)

[13] M. Feshbach, *The mod 2 cohomology rings of the symmetric groups and invariants*, Topology **41** (2002), no. 1, 57–84. MR MR1871241 (2002h:20074)

[14] R. Fox and L. Neuwirth, The braid groups, Math. Scand. **10** (1962), 119–126. MR MR0150755 (27 #742)

[15] C. Giusti, P. Salvatore and Dev P. Sinha, *The mod 2 cohomology of symmetric groups as a hopf ring over the steenrod algebra*, arXiv:0909.3292.

[16] N. H. V. Hung, *The action of the Steenrod squares on the modular invariants of linear groups*, Proc. Amer. Math. Soc. **113** (1991), no. 4, 1097–1104. MR 1064904 (92c:55018)

[17] N. H. V. HU'NG and F. P. PETERSON, *Spherical classes and the Dickson algebra*, Math. Proc. Cambridge Philos. Soc. **124** (1998), no. 2, 253–264. MR MR1631123 (99i:55021)

[18] A. JOYAL, *Duality and θ-categories.*

[19] I. MADSEN, *On the action of the Dyer-Lashof algebra in $H_*(G)$*, Pacific J. Math. **60** (1975), no. 1, 235–275. MR 0388392 (52 #9228)

[20] R. J. MILGRAM, *Iterated loop spaces*, Ann. of Math. (2) **84** (1966), 386–403. MR MR0206951 (34 #6767)

[21] R. J. MILGRAM, *The mod 2 spherical characteristic classes*, Ann. of Math. (2) **92** (1970), 238–261. MR MR0263100 (41 #7705)

[22] M. NAKAOKA, *Homology of the infinite symmetric group*, Ann. of Math. (2) **73** (1961), 229–257. MR MR0131874 (24 #A1721)

[23] M. SALVETTI, *Topology of the complement of real hyperplanes in C^N*, Invent. Math. **88** (1987), no. 3, 603–618. MR 884802 (88k:32038)

[24] P. SELICK, *On the indecomposability of certain sphere-related spaces*, Current trends in algebraic topology, Part 1 (London, Ont., 1981), CMS Conf. Proc., Vol. 2, Amer. Math. Soc., Providence, R.I., 1982, pp. 359–372. MR 686125 (84b:55012)

[25] N. P. STRICKLAND and P. R. TURNER, *Rational Morava E-theory and DS^0*, Topology **36** (1997), no. 1, 137–151. MR MR1410468 (97g:55005)

[26] D. TAMAKI, *Cellular stratified spaces I : Face categories and classifying spaces*, arXiv:1106.3772.

[27] D. TAMAKI, *Cellular stratified spaces II: Basic constructions.* arXiv:1111.4774.

[28] D. TAMAKI, *The Salvetti complex and the little cubes.* math/0602085.

[29] D. TAMAKI, *A dual Rothenberg-Steenrod spectral sequence*, Topology **33** (1994), no. 4, 631–662. MR 1293304 (95f:55019)

[30] V. A. VASSILIEV, "Complements of Discriminants of Smooth Maps: Topology and Applications", Translations of Mathematical Monographs, Vol. 98, American Mathematical Society, Providence, RI, 1992, Translated from the Russian by B. Goldfarb. MR MR1168473 (94i:57020)

[31] R. J. WELLINGTON, *The unstable Adams spectral sequence for free iterated loop spaces*, Mem. Amer. Math. Soc. **36** (1982), no. 258, viii+225. MR 646741 (83c:55028)

[32] H. ZARE, *On spherical classes in $H_*(QX)$*, arXiv:1101.1215.

Basic questions on Artin-Tits groups

Eddy Godelle and Luis Paris

Abstract. This paper is a short survey on four basic questions on Artin-Tits groups: the torsion, the center, the word problem, and the cohomology ($K(\pi, 1)$ problem). It is also an opportunity to prove three new results concerning these questions: (1) if all free of infinity Artin-Tits groups are torsion free, then all Artin-Tits groups will be torsion free; (2) If all free of infinity irreducible non-spherical type Artin-Tits groups have a trivial center then all irreducible non-spherical type Artin-Tits groups will have a trivial center; (3) if all free of infinity Artin-Tits groups have solutions to the word problem, then all Artin-Tits groups will have solutions to the word problem. Recall that an Artin-Tits group is free of infinity if its Coxeter graph has no edge labeled by ∞.

1 Introduction

Let S be a finite set. A *Coxeter matrix* over S is a square matrix $M = (m_{s,t})_{s,t \in S}$ indexed by the elements of S such that $m_{s,s} = 1$ for all $s \in S$, and $m_{s,t} = m_{t,s} \in \{2, 3, 4, \dots, \infty\}$ for all $s, t \in S$, $s \neq t$. A Coxeter matrix is usually represented by its *Coxeter graph*, Γ. This is a labeled graph whose set of vertices is S, such that two vertices $s, t \in S$ are joined by an edge if $m_{s,t} \geq 3$, and such that this edge is labeled by $m_{s,t}$ if $m_{s,t} \geq 4$.

The *Coxeter system* associated with Γ is the pair (W, S) where $W = W_\Gamma$ is the group presented by

$$W = \left\langle S \;\middle|\; \begin{array}{ll} s^2 = 1 & \text{for } s \in S \\ (st)^{m_{s,t}} = 1 & \text{for } s, t \in S, s \neq t \text{ and } m_{s,t} \neq \infty \end{array} \right\rangle.$$

The group W is called *Coxeter group* of type Γ.

Both authors are partially supported by the *Agence Nationale de la Recherche (projet Théorie de Garside*, ANR-08-BLAN-0269-03).

If a, b are two letters and m is an integer greater or equal to 2, we set $\Pi(a, b : m) = (ab)^{\frac{m}{2}}$ if m is even, and $\Pi(a, b : m) = (ab)^{\frac{m-1}{2}} a$ if m is odd. Let $\Sigma = \{\sigma_s;\, s \in S\}$ be an abstract set in one-to-one correspondence with S. The *Artin-Tits system* associated with Γ is the pair (A, Σ), where $A = A_\Gamma$ is the group presented by

$$\begin{aligned} A = \langle \Sigma \mid \Pi(\sigma_s, \sigma_t : m_{s,t}) \\ = \Pi(\sigma_t, \sigma_s : m_{s,t}) \text{ for } s, t \in S, s \neq t \text{ and } m_{s,t} \neq \infty \rangle. \end{aligned}$$

The group A is called *Artin-Tits group* of type Γ.

Coxeter groups were introduced by Tits in his manuscript [28], which was used by Bourbaky as a basis for writing his seminal book "Lie groups and Lie algebras, Chapters 4, 5, and 6" [5]. They are involved in several areas of mathematics such as group theory, Lie theory, or hyperbolic geometry, and are in some sense fairly well understood. We recommend [13, 22] and [5] for a detailed study on these groups.

Artin-Tits groups were also introduced by Tits [29], as extensions of Coxeter groups. There is no general result on these groups, and the theory consists on the study of more or less extended families. In particular, the following basic questions are still open.

(1) Do Artin-Tits groups have torsion?
(2) How is the center of Artin-Tits groups?
(3) Do Artin-Tits groups have solutions to the word problem?
(4) What can we say about the cohomology of Artin-Tits groups?

Nevertheless, there are fairly precise conjectures on these four questions, and the history of the theory of Artin-Tits groups is intimately related to them. They will be presented within they historical context in Section 2.

A Coxeter graph Γ is called *free of infinity* if $m_{s,t} \neq \infty$ for all $s, t \in S$, $s \neq t$. In Section 3 we prove the following principle on the conjectures presented in Section 2 (except for the one on the cohomology, which has been already proved in [17] and [20]).

Principle. *If a property is true for all free of infinity Artin-Tits groups, then it will be true for all Artin-Tits groups.*

A careful reader may point out to the authors that this principle is false for the property "to be free of infinity", but we are sure that he is clever enough to understand that this principle is a general idea and not a theorem, and it may have exceptions.

Finally, note that the four questions mentioned above are not the only questions discussed in the theory of Artin-Tits groups. There are many others, more or less important, such as the conjugacy problem, the automaticity, the linearity, or the orderability.

2 Conjectures

The statement of the conjecture on the torsion is quite simple:

Conjecture A. *Artin-Tits groups are torsion free.*

However, in order to state the conjecture on the center, we need some preliminaries.

The center of a group G will be denoted by $Z(G)$. Let Γ be a Coxeter graph, and let $\Gamma_1, \dots, \Gamma_l$ be the connected components of Γ. It is easily seen that

$$A_\Gamma = A_{\Gamma_1} \times \cdots \times A_{\Gamma_l},$$

hence

$$Z(A_\Gamma) = Z(A_{\Gamma_1}) \times \cdots \times Z(A_{\Gamma_l}).$$

So, in order to understand the centers of the Artin-Tits groups, it suffices to consider connected Coxeter graphs.

We say that a Coxeter graph Γ is of *spherical type* if W_Γ is finite. It is known that $Z(A_\Gamma) \simeq \mathbb{Z}$ if Γ is connected and of spherical type (see [6, 14]), and we believe that the center of any other Artin-Tits group associated to a connected Coxeter graph is trivial. In other words:

Conjecture B. *Let Γ be a non-spherical connected Coxeter graph. Then* $\mathbb{Z}(A_\Gamma) = \{1\}$.

The third conjecture can be stated as easily as the first one:

Conjecture C. *Artin-Tits groups have solutions to the word problem.*

The statement of the last conjecture, known as *the $K(\pi, 1)$ conjecture*, is more sophisticated than the previous ones.

Let Γ be a Coxeter graph. The Coxeter group W of Γ has a faithful representation $W \hookrightarrow GL(V)$, called *canonical representation*, where V is a real vector space of dimension $|S|$. The group W, viewed as a subgroup of $GL(V)$, is generated by reflections and acts properly discontinuously on a non-empty open cone I, called *Tits cone* (see [5]). The set of reflections in W is $R = \{wsw^{-1}; s \in S \text{ and } w \in W\}$, and W acts freely and properly discontinuously on $I \setminus (\cup_{r \in R} H_r)$, where, for $r \in R$, H_r denotes the hyperplane of V fixed by r. Set

$$M = M_\Gamma = (I \times V) \setminus \left(\bigcup_{r \in R} (H_r \times H_r) \right).$$

This is a connected manifold of dimension $2|S|$ on which the group W acts freely and properly discontinuously. A key result in the domain is the following.

Theorem 2.1 (Van der Lek [23]). *The fundamental group of M_Γ / W is isomorphic to the Artin-Tits group A_Γ.*

Recall that a CW-complex X is a *a classifying space* for a (discrete) group G if $\pi_1(X) = G$ and the universal cover of X is contractible. These spaces play a prominent role in the calculation of the cohomology of groups (see [7]).

Conjecture D ($K(\pi, 1)$ conjecture). *Let Γ be a Coxeter graph. Then M_Γ / W is a classifying space for A_Γ.*

Observe that, if Conjecture D holds for a Coxeter graph Γ, then the cohomological dimension of A_Γ is finite, thus A_Γ is torsion free (see [7]). In other words, Conjecture D implies Conjecture A.

Curiously, for each of the conjectures A, B, C, or D, the set of Artin-Tits groups for which a proof of the conjecture is known coincides (up to some exceptions) with the set of Artin-Tits groups for which proofs for the other three conjectures are known.

The first family of Artin-Tits groups which has been studied was the family of spherical type Artin-Tits groups (recall that A_Γ is of spherical type if W_Γ is finite). Conjectures A, B and C were proved for these groups in [6, 14], and Conjecture D was proved in [14].

We say that Γ is of *large type* if $m_{s,t} \geq 3$ for all $s, t \in S$, $s \neq t$, and that Γ is of *extra-large type* if $m_{s,t} \geq 4$ for all $s, t \in S$, $s \neq t$. Conjectures A and C were proved in [4] for extra-large type Artin-Tits groups, and Conjecture B can be easily proved with the same techniques. Conjecture D for large type Artin-Tits groups is a straightforward consequence of [21].

Recall that a Coxeter graph Γ is *free of infinity* if $m_{s,t} \neq \infty$ for all $s, t \in S$. We say that Γ is of *FC type* if every free of infinity full subgraph of Γ is of spherical type. On the other hand, we say that Γ is of *dimension 2* if no full subgraph of Γ with three or more vertices is of spherical type.

After [4] and [21] the next significant step in the study of Artin-Tits groups and, more specifically, on the four conjectures stated above, was [10]. In this paper Conjecture D (and hence Conjecture A) was proved for FC type Artin-Tits groups and 2-dimensional Artin-Tits groups (that include the Artin-Tits groups of large type). Later, Conjecture B was proved for Artin-Tits groups of FC type in [18], and for 2-dimensional Artin-Tits groups in [19]. On the other hand, Conjecture C was proved for Artin-Tits groups of FC type in [2,3], and for 2-dimensional Artin-Tits groups in [12]. Note also that a new solution to the word problem for FC type Artin-Tits groups will follow from Theorem C stated in Section 3.

A challenging question in the domain is to prove Conjectures A, B, C, and D for the so-called Artin-Tits groups of affine type, that is, those Artin-Tits groups for which the associated Coxeter group is affine. As far as we know there is no known explicit algorithm for solving the word problem for these groups, except for the groups of type $\tilde{A}_n$ and $\tilde{C}_n$ (see [15, 16]). The techniques introduced in [1] may be used to solve the question for the groups of type $\tilde{B}_n$ and $\tilde{D}_n$, but we have no idea on how to treat the groups of type $\tilde{E}_k$, $k = 6, 7, 8$. The $K(\pi, 1)$ conjecture was proved for the groups of type $\tilde{A}_n$ and $\tilde{C}_n$ in [24] (see also [11]) and for the groups of type $\tilde{B}_n$ in [8]. The other cases are open. On the other hand, none of these papers addresses Conjecture B.

3 From free of infinity Artin-Tits groups to Artin-Tits groups

The aim of this section is to prove the following three theorems.

Theorem A. *If all free of infinity Artin-Tits groups are torsion free, then all Artin-Tits groups will be torsion free.*

Theorem B. *If, for every non-spherical connected free of infinity Coxeter graph Γ, we have $Z(A_\Gamma) = \{1\}$, then, for every non-spherical connected Coxeter graph Γ, we will have $Z(A_\Gamma) = \{1\}$.*

Note that, as pointed out before, if Γ is a spherical type connected Coxeter graph, then $Z(A_\Gamma) \simeq \mathbb{Z}$ (see [6, 14]).

Theorem C. *If all free of infinity Artin-Tits groups have solutions to the word problem, then all Artin-Tits groups will have solutions to the word problem.*

In order to complete our discussion, we state the following theorem without proof. This is already proved in [17] and [20].

Theorem D (Ellis, Sköldberg [17]). *If, for every free of infinity Coxeter graph Γ, the space M_Γ / W_Γ is a classifying space for A_Γ, then, for every Coxeter graph Γ, the space M_Γ / W_Γ will be a classifying space for A_Γ.*

The proofs of Theorems A, B and C use the notion of a parabolic subgroup defined as follows. For $X \subset S$, we set $M_X = (m_{s,t})_{s,t \in X}$, we denote by Γ_X the Coxeter graph of M_X, we denote by W_X the subgroup of $W = W_\Gamma$ generated by X, we set $\Sigma_X = \{\sigma_s;\, s \in X\}$, and we denote by A_X the subgroup of $A = A_\Gamma$ generated by Σ_X. It is known that (W_X, X) is the Coxeter system of Γ_X (see [5]), and that (A_X, Σ_X) is the Artin-Tits system of Γ_X (see [20, 23, 25]). The subgroup W_X is called *standard parabolic subgroup* of W, and A_X is called *standard parabolic subgroup* of A.

The proof of Theorem A is elementary. That of Theorem B is more complicated and requires some deeper knowledge on Artin-Tits groups. The proof of Theorem C is based on a combinatorial study of Artin-Tits groups made in [20]. In all three cases we use the following observation.

Observation. Let Γ be a Coxeter graph which is not free of infinity. Let $s, t \in S$, $s \neq t$, such that $m_{s,t} = \infty$. Set $X = S \setminus \{s\}$, $Y = S \setminus \{t\}$, and $Z = S \setminus \{s, t\}$. Recall that (A_X, Σ_X) (resp. (A_Y, Σ_Y), (A_Z, Σ_Z)) is the Artin-Tits system of type Γ_X (resp. Γ_Y, Γ_Z), and that the natural homomorphisms $A_Z \to A_X$ and $A_Z \to A_Y$ are injective. Then the following equality is a direct consequence of the presentation of A.

$$A = A_X *_{A_Z} A_Y\,.$$

Proof of Theorem A. We assume that all free of infinity Artin-Tits groups are torsion free. Let Γ be a Coxeter graph, S its set of vertices, and $A = A_\Gamma$ its associated Artin-Tits group. We prove by induction on $|S|$ that A is torsion free. If $|S| = 1$, then A is free of infinity, thus is torsion free. More generally, if A is free of infinity then it is torsion free by hypothesis. Assume that $|S| \geq 2$ and A is not free of infinity, plus the inductive hypothesis. Choose $s, t \in S$ such that $m_{s,t} = \infty$, and set $X = S \setminus \{s\}$, $Y = S \setminus \{t\}$ and $Z = S \setminus \{s, t\}$. Recall that $A = A_X *_{A_Z} A_Y$, hence, if α is a finite order element in A, then α is conjugate to an element in either A_X or A_Y (see [27]). By the inductive hypothesis A_X and A_Y are torsion free, thus α must be trivial. □

The following two lemmas are preliminaries to the proof of Theorem B.

Lemma 3.1. *Let Γ be a connected spherical type Coxeter graph, let S be its set of vertices, and let $X \subsetneq S$ be a proper subset of S. Then $Z(A) \cap A_X = \{1\}$.*

Proof. We assume first that Γ is a spherical type Coxeter graph, but not necessarily connected. The *Artin-Tits monoid* of Γ is defined by the monoid presentation

$$\begin{aligned} A^+ &= \langle \Sigma \mid \Pi(\sigma_s, \sigma_t : m_{s,t}) \\ &= \Pi(\sigma_t, \sigma_s : m_{s,t}),\ s, t \in S,\ s \neq t \text{ and } m_{s,t} \neq \infty \rangle^+ . \end{aligned}$$

By [6] the natural homomorphism $A^+ \to A$ is injective (see also [26]). Moreover, by [23], we have $A^+ \cap A_X = A_X^+$ (this equality is also a direct consequence of the normal forms defined in [9]). On the other hand, it is easily deduced from the presentation of A^+ that, for $a \in A^+$, we have $a \in A_X^+$ if and only if any expression of a in the elements of Σ is actually a word in the elements of Σ_X.

Now, we assume that Γ is connected. Let $s_1, \dots, s_n$ be the vertices of Γ, and, for $i \in \{1, \dots, n\}$, set $\sigma_i = \sigma_{s_i}$. By [6] the subgroup $Z(A)$ is the infinite cyclic group generated by

$$\delta = (\sigma_1 \cdots \sigma_n)^h ,$$

where h is some positive number. Furthermore, δ does not depend on the choice of the ordering of S. Now, the previous observations clearly imply that $\delta^k \notin A_X$ unless $k = 0$, thus $Z(A) \cap A_X = \{1\}$. □

The next lemma is well-known and can be easily proved using normal forms in free products with amalgamation (see [27]).

Lemma 3.2. *Let $G = H_1 *_K H_2$ be the free product with amalgamation of two groups H_1, H_2 along a common subgroup K. Then $Z(G) = Z(H_1) \cap Z(H_2) \subset K$.*

Proof of Theorem B. We take a connected Coxeter graph Γ, and we prove Theorem B by induction on the number $|S|$ of vertices of Γ. If $|S| = 1$, then $Z(A_\Gamma) = A_\Gamma \simeq \mathbb{Z}$. More generally, if Γ is free of infinity, then, by the starting hypothesis, $Z(A_\Gamma) \simeq \mathbb{Z}$ if Γ is of spherical type, and $Z(A_\Gamma) = \{1\}$ otherwise.

Suppose $|S| \geq 2$ and Γ is not free of infinity, plus the inductive hypothesis. Note that, in this case, Γ is not of spherical type, thus we should prove that $Z(A) = \{1\}$. Let $s, t \in S$ be such that $m_{s,t} = \infty$. Set $X = S \setminus \{s\}$, $Y = S \setminus \{t\}$ and $Z = S \setminus \{s, t\}$. We have $A = A_X *_{A_Z} A_Y$ thus, by Lemma 3.2,

$$Z(A) = Z(A_X) \cap Z(A_Y) \subset A_Z\,.$$

Let $X_1 \subset X$ be such that Γ_{X_1} is the connected component of Γ_X which contains t, let $X_2 = X \setminus X_1$, let $Y_1 \subset Y$ be such that Γ_{Y_1} is the connected component of Γ_Y which contains s, and let $Y_2 = Y \setminus Y_1$. First, we prove that $X_2 \subset Y_1$ and $Y_2 \subset X_1$.

Indeed, let $X_2' \subset X_2$ be such that $\Gamma_{X_2'}$ is a connected component of Γ_{X_2}. We have $X_2' \subset Y$, since $t \notin X_2'$. Moreover Γ_{Y_1} is a connected component of Γ_Y, thus either $X_2' \subset Y_1$, or $X_2' \cap Y_1 = \emptyset$. Suppose $X_2' \cap Y_1 = \emptyset$, that is $X_2' \subset Y_2$. Let $x \in S \setminus X_2'$. If $x \neq s$, then $x \in X \setminus X_2'$, thus $m_{x,y} = 2$ for all $y \in X_2'$, because $\Gamma_{X_2'}$ is a connected component of Γ_X. If $x = s$, then $m_{x,y} = 2$ for all $y \in X_2'$, because $X_2' \subset Y_2$. This implies that $\Gamma_{X_2'}$ is a connected component of Γ: a contradiction. So, $X_2' \subset Y_1$. Hense, $X_2 \subset Y_1$ and, by similarity, $Y_2 \subset X_1$.

Set $Z_1 = X_1 \setminus \{t\}$. We have $A_Z = A_{Z_1} \times A_{X_2}$ and $A_{Z_1} \subset A_{X_1}$, thus

$$\begin{aligned} Z(A_X) \cap A_Z &= (Z(A_{X_1}) \cap A_{Z_1}) \times (Z(A_{X_2}) \cap A_{X_2}) \\ &= (Z(A_{X_1}) \cap A_{Z_1}) \times Z(A_{X_2})\,. \end{aligned}$$

If Γ_{X_1} is not of spherical type, then, by the inductive hypothesis, $Z(A_{X_1}) = \{1\}$. If Γ_{X_1} is of spherical type, then, by Lemma 3.2, $Z(A_{X_1}) \cap A_{Z_1} = \{1\}$. So, $Z(A_X) \cap A_Z = Z(A_{X_2}) \subset A_{X_2}$. Now, since $X_2 \subset Y_1$, we have

$$Z(A_Y) \cap A_{X_2} = (Z(A_{Y_1}) \cap A_{X_2}) \times (Z(A_{Y_2}) \cap \{1\}) = Z(A_{Y_1}) \cap A_{X_2}\,.$$

If Γ_{Y_1} is not of spherical type, then, by the inductive hypothesis, $Z(A_{Y_1}) = \{1\}$. If Γ_{Y_1} is of spherical type, then, since $X_2 \subsetneq Y_1$, we have $Z(A_{Y_1}) \cap A_{X_2} = \{1\}$ by Lemma 3.2. So, $Z(A_Y) \cap A_{X_2} = \{1\}$. Finally,

$$Z(A) = Z(A_X) \cap Z(A_Y) = Z(A_X) \cap A_Z \cap Z(A_Y) \subset A_{X_2} \cap Z(A_Y) = \{1\}\,. \square$$

We turn now to the proof of Theorem C. We start with some definitions. Let $G = H_1 *_K H_2$ be an amalgamated free product. A *syllabic expression* of an element $\alpha \in G$ is a sequence $w = (\beta_1, \dots, \beta_l)$ of elements of $H_1 \cup H_2$ such that $\alpha = \beta_1 \beta_2 \cdots \beta_l$. We say that a syllabic expression w' is an *elementary reduction of type I* of w if there exists $i \in \{1, \dots, l\}$ such that $\beta_i = 1$ and

$$w' = (\beta_1, \dots, \beta_{i-1}, \beta_{i+1}, \dots, \beta_l) .$$

We say that a syllabic expression w' is an *elementary reduction of type II* of w if there exists $i \in \{1, \dots, l-1\}$ and $j \in \{1, 2\}$ such that $\beta_i, \beta_{i+1} \in H_j$ and

$$w' = (\beta_1, \dots, \beta_{i-1}, \beta_i \beta_{i+1}, \beta_{i+2}, \dots, \beta_l) .$$

Implicit in this definition there is the assumption that every element of K belongs to both, H_1 and H_2. Let w, w' be two syllabic expressions of α. If there is a finite sequence $w_0, w_1, \dots, w_n$ of syllabic expressions such that $w_0 = w$, $w_n = w'$ and w_i is an elementary reduction of w_{i-1} for all $i \in \{1, \dots, n\}$, then we say that w' is a *reduction* of w. A *reduced expression* is a syllabic expression which admits no elementary reduction. Clearly, a syllabic expression $w = (\beta_1, \dots, \beta_l)$ is reduced if and only if either $l = 0$, or $l = 1$ and $\beta_1 \neq 1$, or there exists a sequence $(k_1, \dots, k_l)$ in $\{1, 2\}$ such that $k_i \neq k_{i+1}$ for all $i \in \{1, \dots, l-1\}$, and $\beta_i \in H_{k_i} \setminus K$ for all $i \in \{1, \dots, l\}$. The following theorem is well-known and widely used in the study of amalgamated free products of groups (see [27]).

Theorem 3.3. *Let $G = H_1 *_K H_2$ be an amalgamated free product. Let $\alpha \in G$, and let $w = (\beta_1, \dots, \beta_l)$, $w' = (\beta'_1, \dots, \beta'_k)$ be two reduced syllabic expressions of α. Then $k = l$ and there exist $\gamma_0, \gamma_1, \dots, \gamma_l \in K$ such that*

$$\gamma_0 = \gamma_l = 1, \quad \textit{and } \beta'_i \gamma_i = \gamma_{i-1} \beta_i \textit{ for all } i \in \{1, \dots, l\} .$$

In particular, $\alpha = 1$ if and only if its (unique) reduced expression is the empty sequence of length 0. □

For a given set X we denote by X^* the free monoid generated by X. If G is a group generated by a family S and $w \in (S \sqcup S^{-1})^*$, we denote by $[w]_S$ the element of G represented by w. On the other hand, we will assume for the next corollary that H_1, H_2, and K are finitely generated, and we take finite generating sets S_1, S_2, and T, for H_1, H_2, and K, respectively.

Corollary 3.4. *Assume that, for each $j \in \{1, 2\}$,*

(a) *H_j has a solution to the word problem; and*
(b) *there exists an algorithm which, given a word $w \in (S_j \sqcup S_j^{-1})^*$, decides whether $\beta = [w]_{S_j}$ belongs to K and, if yes, determines a word $u \in (T \sqcup T^{-1})^*$ which represents β.*

Then $G = H_1 *_K H_2$ has a solution to the word problem.

Proof. Let $\alpha \in G$. A *syllabic pre-expression* of α is defined to be a pair of sequences of equal length,

$$U = ((w_1, \dots, w_l), (j_1, \dots, j_l)),$$

such that $j_i \in \{1, 2\}$, and $w_i \in (S_{j_i} \sqcup S_{j_i}^{-1})^*$, for all $i \in \{1, \dots, l\}$, and $\alpha = [w_1 \cdots w_l]_{S_1 \cup S_2}$. The *syllabic realization* of U is the syllabic expression

$$[U] = (\beta_1, \dots, \beta_l),$$

where $\beta_i = [w_i]_{S_{j_i}}$ for all $i \in \{1, \dots, l\}$. We will say that U is *reduced* if its realization $[U]$ is reduced.

Let $U = ((w_1, \dots, w_l), (j_1, \dots, j_l))$ be a syllabic pre-expression of α. Suppose there exists $i \in \{1, \dots, l\}$ such that $[w_i]_{S_{j_i}} = 1$. Note that, Since H_{j_i} has a solution to the word problem, there is an algorithm which decides whether $[w_i]_{S_{j_i}}$ is 1 or not. Set

$$U' = ((w_1, \dots, w_{i-1}, w_{i+1}, \dots, w_l), (j_1, \dots, j_{i-1}, j_{i+1}, \dots, j_l)).$$

Then U' is a syllabic pre-expression of α and $[U']$ is an elementary reduction of type I of $[U]$. Suppose there exists $i \in \{1, \dots, l-1\}$ such that $j_i = j_{i+1}$. Set

$$U' = ((w_1, \dots, w_{i-1}, w_i w_{i+1}, w_{i+2}, \dots, w_l), (j_1, \dots, j_{i-1}, j_i, j_{i+2}, \dots, j_l)).$$

Then U' is a syllabic pre-expression of α and $[U']$ is an elementary reduction of type II of $[U]$. Suppose there exists $i \in \{1, \dots, l-1\}$ such that $j_i \neq j_{i+1}$, but $\beta_i = [w_i]_{S_{j_i}} \in K$. Recall that, by hypothesis, there is an algorithm which decides whether this is true. Again by hypothesis, there is an algorithm which determines a word $u_i \in (T \sqcup T^{-1})^*$ which represents β_i. Writing the elements of T in the generators $S_{j_{i+1}} \sqcup S_{j_{i+1}}^{-1}$ of $H_{j_{i+1}}$, we get from u_i a word $w_i' \in (S_{j_{i+1}} \sqcup S_{j_{i+1}}^{-1})^*$ which represents β_i. Set

$$U' = ((w_1, \dots, w_{i-1}, w_i' w_{i+1}, w_{i+2}, \dots, w_l), (j_1, \dots, j_{i-1}, j_{i+1}, j_{i+2}, \dots, j_l)).$$

Then U' is a syllabic pre-expression of α and $[U']$ is an elementary reduction of type II of $[U]$. Suppose there exists $i \in \{1, \dots, l-1\}$ such that $j_i \neq j_{i+1}$, but $\beta_{i+1} = [w_{i+1}]_{S_{j_{i+1}}} \in K$. Then, in the same manner as in the previous case, one can explicitely calculate an expression w'_{i+1} for β_{i+1} in $(S_{j_i} \sqcup S_{j_i}^{-1})^*$,

$$U' = ((w_1, ..., w_{i-1}, w_i w'_{i+1}, w_{i+2}, ..., w_l), (j_1, ..., j_{i-1}, j_i, j_{i+2}, ..., j_l))$$

is a syllabic pre-expression of α, and $[U']$ is an elementary reduction of type II of $[U]$. It is easily seen that all elementary reductions of $[U]$ are of these forms. Hence, from a given syllabic pre-expression U of α one can effectively calculate a reduced syllabic pre-expression U' of α. Moreover, by Theorem 3.3, we have $\alpha = 1$ if and only if U' is a pair of empty sequences (of length 0).

Let $w = s_1^{\varepsilon_1} \cdots s_l^{\varepsilon_l} \in (S_1 \sqcup S_2 \sqcup S_1^{-1} \sqcup S_2^{-1})^*$, and let $\alpha = [w]_{S_1 \cup S_2} \in G$. For $i \in \{1, \dots, l\}$ we set $j_i = 1$ if $s_i \in S_1$ and $j_i = 2$ if $s_i \in S_2$. Then

$$U = ((s_1^{\varepsilon_1}, \dots, s_l^{\varepsilon_l}), (j_1, \dots, j_l))$$

is a syllabic pre-expression of α, and, by the above, one can decide from U if $\alpha = 1$. □

Now, the key in the proof of Theorem C is the following result proved in [20].

Theorem 3.5 (Godelle, Paris [20]). *Let Γ be a Coxeter graph, let S be its set of vertices, let $X \subset S$, and let (A, Σ) be the Artin-Tits system of Γ. If A has a solution to the word problem, then there is an algorithm which, given a word $w \in (\Sigma \sqcup \Sigma^{-1})^*$, decides whether $\alpha = [w]_\Sigma$ belongs to A_X, and, if yes, determines a word $w' \in (\Sigma_X \sqcup \Sigma_X^{-1})^*$ which represents α.*

Proof of Theorem C. We assume that all free of infinity Artin-Tits groups have solutions to the word problem. Let Γ be a Coxeter graph, let S be its set of vertices, and let (A, Σ) be its associated Artin-Tits system. We prove by induction on $|S|$ that A has a solution to the word problem.

If $|S| = 1$ or, more generally, if Γ is free of infinity, then, by hypothesis, A has a solution to the word problem. Assume that Γ is not free of infinity, plus the inductive hypothesis. Let $s, t \in S$, $s \neq t$, such that $m_{s,t} = \infty$. Set $X = S \setminus \{s\}$, $Y = S \setminus \{t\}$, and $Z = S \setminus \{s, t\}$. We have

$A = A_X *_{A_Z} A_Y$ and, by the inductive hypothesis, A_X, A_Y have solutions to the word problem. Moreover, by Theorem 3.5, for $T = X$ or Y, there is an algorithm which, given a word $w \in (\Sigma_T \sqcup \Sigma_T^{-1})^*$, decides whether $\alpha = [w]_{\Sigma_T}$ lies in A_Z and, if yes, determines a word $w' \in (\Sigma_Z \sqcup \Sigma_Z^{-1})^*$ which represents α. We conclude by Corollary 3.4 that A has a solution to the word problem. □

References

[1] D. Allcock, *Braid pictures for Artin groups*, Trans. Amer. Math. Soc. **354** (2002), no. 9, 3455–3474.

[2] J. Altobelli, *The word problem for Artin groups of FC type*, J. Pure Appl. Algebra **129** (1998), no. 1, 1–22.

[3] J. Altobelli and R. Charney, *A geometric rational form for Artin groups of FC type*, Geom. Dedicata **79** (2000), no. 3, 277–289.

[4] K. I. Appel and P. E. Schupp, *Artin groups and infinite Coxeter groups*, Invent. Math. **72** (1983), no. 2, 201–220.

[5] N. Bourbaki, "Eléments de mathématique", Fasc. XXXIV, Groupes et algèbres de Lie. Chapitre IV: Groupes de Coxeter et systèmes de Tits. Chapitre V: Groupes engendrés par des réflexions. Chapitre VI: Systèmes de racines, Actualités Scientifiques et Industrielles, No. 1337, Hermann, Paris, 1968.

[6] E. Brieskorn and K. Saito, *Artin-Gruppen und Coxeter-Gruppen.* Invent. Math. **17** (1972), 245–271.

[7] K. S. Brown, "Buildings", Springer-Verlag, New York, 1989.

[8] F. Callegaro, D. Moroni and M. Salvetti, *The $K(\pi, 1)$ problem for the affine Artin group of type $\tilde{B}_n$ and its cohomology*, J. Eur. Math. Soc. (JEMS) **12** (2010), no. 1, 1–22.

[9] R. Charney, *Geodesic automation and growth functions for Artin groups of finite type*, Math. Ann. **301** (1995), no. 2, 307–324.

[10] R. Charney and M. W. Davis, *The $K(\pi, 1)$-problem for hyperplane complements associated to infinite reflection groups.* J. Amer. Math. Soc. **8** (1995), no. 3, 597–627.

[11] R. Charney and D. Peifer, *The $K(\pi, 1)$-conjecture for the affine braid groups*, Comment. Math. Helv. **78** (2003), no. 3, 584–600.

[12] A. Chermak, *Locally non-spherical Artin groups*, J. Algebra **200** (1998), no. 1, 56–98.

[13] M. W. Davis, "The Geometry and Topology of Coxeter Groups", London Mathematical Society Monographs Series, 32. Princeton University Press, Princeton, NJ, 2008.

[14] P. DELIGNE, *Les immeubles des groupes de tresses généralisés*, Invent. Math. **17** (1972), 273–302.

[15] F. DIGNE, *Présentations duales des groupes de tresses de type affine $\widetilde{A}$*, Comment. Math. Helv. **81** (2006), no. 1, 23–47.

[16] F. DIGNE, *A Garside presentation for Artin-Tits groups of type $\tilde{C}_n$*, Preprint, arXiv: 1002.4320.

[17] G. ELLIS and E. SKÖLDBERG, *The $K(\pi, 1)$ conjecture for a class of Artin groups*, Comment. Math. Helv. **85** (2010), no. 2, 409–415.

[18] E. GODELLE, *Parabolic subgroups of Artin groups of type FC*, Pacific J. Math. **208** (2003), no. 2, 243–254.

[19] E. GODELLE, *Artin-Tits groups with CAT(0) Deligne complex*, J. Pure Appl. Algebra **208** (2007), no. 1, 39–52.

[20] E. GODELLE, L. PARIS, *$K(\pi, 1)$ and word problems for infinite type Artin-Tits groups, and applications to virtual braid groups*, Math. Z., to appear.

[21] H. HENDRIKS, *Hyperplane complements of large type*, Invent. Math. **79** (1985), no. 2, 375–381.

[22] J.E. HUMPHREYS, *Reflection groups and Coxeter groups*, Cambridge Studies in Advanced Mathematics, 29. Cambridge University Press, Cambridge, 1990.

[23] H. VAN DER LEK, *The homotopy type of complex hyperplane complements*, Ph. D. Thesis, Nijmegen, 1983.

[24] C. OKONEK, *Das $K(\pi, 1)$-Problem für die affinen Wurzelsysteme vom Typ A_n, C_n*, Math. Z. **168** (1979), no. 2, 143–148.

[25] L. PARIS, *Universal cover of Salvetti's complex and topology of simplicial arrangements of hyperplanes*, Trans. Amer. Math. Soc. **340** (1993), no. 1, 149–178.

[26] L. PARIS, *Artin monoids inject in their groups*, Comment. Math. Helv. **77** (2002), no. 3, 609–637.

[27] J.-P. SERRE, *Arbres, amalgames,* SL_2, Astérisque, No. 46. Société Mathématique de France, Paris, 1977.

[28] J. TITS, *Groupes et géométries de Coxeter*, Institut des Hautes Etudes Scientifiques, Paris, 1961.

[29] J. TITS, *Normalisateurs de tores. I. Groupes de Coxeter étendus*, J. Algebra **4** (1966), 96–116.

Rational cohomology of the real Coxeter toric variety of type A

Anthony Henderson

Abstract. The toric variety corresponding to the Coxeter fan of type A can also be described as a De Concini–Procesi wonderful model. Using a general result of Rains which relates cohomology of real De Concini–Procesi models to poset homology, we give formulas for the Betti numbers of the real toric variety, and the symmetric group representations on the rational cohomologies. We also show that the rational cohomology ring is not generated in degree 1.

1 Introduction

Fix a positive integer n. Let N denote the quotient $\mathbb{Z}^n/\mathbb{Z}(1,1,\cdots,1)$, which is a free $\mathbb{Z}$-module of rank $n-1$. We have a representation of the symmetric group S_n on $N \otimes \mathbb{R} = \mathbb{R}^n/\mathbb{R}(1,1,\cdots,1)$ by permuting coordinates, which is generated by reflections in the hyperplanes

$$\{(a_1, a_2, \cdots, a_n) \in \mathbb{R}^n \,|\, a_i = a_j\}/\mathbb{R}(1,1,\cdots,1), \text{ for } 1 \leq i \neq j \leq n.$$

Let Δ be the complete fan of rational strongly convex polyhedral cones in $N \otimes \mathbb{R}$ defined by this hyperplane arrangement (that is, the maximal cones of Δ are the closures of the chambers of the arrangement). Then to N and Δ we can associate an $(n-1)$-dimensional nonsingular projective toric variety $\mathcal{T}_n$ as in [4]. In terms of simplicial complexes, $\mathcal{T}_n$ is the toric variety associated to the Coxeter complex of S_n. Clearly S_n acts on $\mathcal{T}_n$ by variety automorphisms.

This is the $W = S_n$ case of a construction which can be carried out for any Weyl group W, producing the Coxeter toric variety $\mathcal{T}_W$. In [11], Procesi gave a formula for the rational cohomology $H^*(\mathcal{T}_W, \mathbb{Q})$ as a graded $\mathbb{Q}W$-module. Further results about the representation of W on

The author's research was supported by the Australian Research Council grant DP0985184.

$H^*(\mathcal{T}_W, \mathbb{Q})$ were obtained by Stembridge in [13, 14] and Lehrer in [10]. These results make use of the well-developed general theory of the cohomology of complex toric varieties.

The rational cohomology of the real variety $\mathcal{T}_W(\mathbb{R})$, which is not covered by any such general theory, was considered by the author and Lehrer in [8]. The main result of that paper was a formula for the alternating sum $\sum_i (-1)^i H^i(\mathcal{T}_W(\mathbb{R}), \mathbb{Q})$ as a virtual $\mathbb{Q}W$-module. The methods used there did not allow us to isolate an individual $H^i(\mathcal{T}_W(\mathbb{R}), \mathbb{Q})$.

The present paper restricts to the case $W = S_n$. As Procesi observed in [11, Section 3], there is an alternative construction of the variety $\mathcal{T}_n$, as a special case of what he and De Concini later called the wonderful model of a subspace arrangement. In the terminology of [1], if $\mathcal{G}$ denotes the building set in $(\mathbb{C}^n)^*$ which consists of all subspaces $\mathbb{C}\{x_i \mid i \in I\}$ where I is a nonempty subset of $\{1, 2, \cdots, n\}$ and $x_1, x_2, \cdots, x_n$ are the coordinate functions on $\mathbb{C}^n$, then the projective wonderful model $Y_n = \overline{Y}_{\mathcal{G}}$ is isomorphic to $\mathcal{T}_n$. We recall the definition of Y_n and explain this isomorphism further in Section 2.

Now the rational cohomology of $\overline{Y}_{\mathcal{G}}(\mathbb{R})$ may be described using the result of Rains [12, Theorem 3.7] (or rather its dual form, [9, Theorem 2.1]). This description in general involves the homology of the poset $\Pi_{\mathcal{G}}^{(2)}$, consisting of all direct sums of even-dimensional elements of $\mathcal{G}$; in our case, this poset is clearly isomorphic to the poset of even-size subsets of $\{1, 2, \cdots, n\}$, whose homology is easy to determine.

In Section 3 we carry out the calculation and derive the following formula. Here we let $R(S_n)$ denote the Grothendieck group of $\mathbb{Q}S_n$-modules and define the $\mathbb{N}$-graded ring $\bigoplus_{n \geq 0} R(S_n)$ using the induction product, with identity element $1 = 1_{S_0}$ (the trivial representation of S_0). We then complete this to a power series ring $R = \widehat{\bigoplus}_{n \geq 0} R(S_n)$ and adjoin an indeterminate t.

Theorem 1.1. *We have the following equality in* $R[t]$*:*

$$1 + \sum_{n \geq 1} \sum_i H^i(\mathcal{T}_n(\mathbb{R}), \mathbb{Q})\,(-t)^i = \left(\sum_{n \geq 0} 1_{S_n}\right) \left(1 + \sum_{\substack{n \geq 2 \\ n \text{ even}}} \varepsilon_{S_n} t^{n/2}\right)^{-1},$$

where 1_{S_n} *and* ε_{S_n} *denote the trivial and sign representations of* S_n.

Note that setting $t = 1$ in Theorem 1.1 recovers [8, (8)].

An alternative way to express this formula is as follows.

Corollary 1.2. *We have the following equality in $R(S_n)$:*

$$H^i(\mathcal{T}_n(\mathbb{R}), \mathbb{Q}) = \sum_{\substack{n_1, n_2, \cdots, n_m \geq 2 \\ n_1, n_2, \cdots, n_m \text{ even} \\ n_1 + n_2 + \cdots + n_m = 2i \leq n}} (-1)^{i+m} \operatorname{Ind}_{S_{n-2i} \times S_{n_1} \times \cdots \times S_{n_m}}^{S_n} (\varepsilon_{n_1, \cdots, n_m}),$$

where $\varepsilon_{n_1,\cdots,n_m}$ is the linear character whose restriction to the S_{n-2i} factor is trivial and whose restriction to each S_{n_j} factor is the sign character.

Corollary 1.2 answers the question posed after [8, Theorem 6] in the negative. Another consequence of Theorem 1.1 is a formula for the Betti numbers of $\mathcal{T}_n(\mathbb{R})$.

Corollary 1.3. *We have*

$$\dim H^i(\mathcal{T}_n(\mathbb{R}), \mathbb{Q}) = A_{2i} \binom{n}{2i},$$

where A_{2i} denotes the Euler secant number, i.e. *the coefficient of $\frac{x^{2i}}{(2i)!}$ in the power series* $\sec(x)$. *In particular, $H^i(\mathcal{T}_n(\mathbb{R}), \mathbb{Q}) = 0$ if $i > \frac{n}{2}$.*

From the De Concini–Procesi description one sees that there is a natural morphism $\overline{\mathcal{M}_{0,n+2}} \to \mathcal{T}_n$, where $\overline{\mathcal{M}_{0,n+2}}$ denotes the moduli space of stable genus 0 curves with $n+2$ marked points. In Section 4, we use this morphism and the results of Etingof *et al.* [3] on the rational cohomology of $\overline{\mathcal{M}_{0,n+2}}(\mathbb{R})$ to show that the cup products of elements of $H^1(\mathcal{T}_n(\mathbb{R}), \mathbb{Q})$ do not span $H^2(\mathcal{T}_n(\mathbb{R}), \mathbb{Q})$, for $n \geq 4$. Thus the cohomology ring $H^*(\mathcal{T}_n(\mathbb{R}), \mathbb{Q})$ is not generated in degree 1, in contrast to [3, Theorem 2.9]. It would be interesting to find a presentation for it.

ACKNOWLEDGEMENTS. This paper was conceived in June 2010 during the program 'Configuration Spaces: Geometry, Combinatorics and Topology' at the Centro di Ricerca Matematica Ennio De Giorgi in Pisa. I thank the organizers for their generous support. I am also indebted to Gus Lehrer for helpful conversations about $\mathcal{T}_n$, and to Michelle Wachs for the question which motivated Corollary 4.4.

2 The De Concini–Procesi model

Let V be a finite-dimensional complex vector space, and $\mathcal{G}$ a collection of nonzero subspaces of the dual space V^*, including V^* itself, which satisfies the definition [1, Section 2.3] of a building set. For any $A \in \mathcal{G}$, let $A^\perp$ denote the orthogonal subspace of V. Let $\mathcal{M}_\mathcal{G}$ be the complement in $\mathbb{P}(V)$ of the union of all $\mathbb{P}(A^\perp)$ for $A \in \mathcal{G}$, and let

$$\rho : \mathcal{M}_\mathcal{G} \to \prod_{A\in\mathcal{G}} \mathbb{P}(V/A^\perp)$$

be the morphism whose component in the factor labelled by A is the restriction to $\mathcal{M}_\mathcal{G}$ of the obvious map $\mathbb{P}(V)\setminus\mathbb{P}(A^\perp) \to \mathbb{P}(V/A^\perp)$. Since the factor labelled by $A = V^*$ is $\mathbb{P}(V)$ itself, ρ is injective. The projective De Concini–Procesi model $\overline{Y}_\mathcal{G}$ is defined in [1, Section 4.1] to be the closure of $\rho(\mathcal{M}_\mathcal{G})$ in $\prod_{A\in\mathcal{G}} \mathbb{P}(V/A^\perp)$.

In this paper, we take $V = \mathbb{C}^n$ for a positive integer n, let $x_1, x_2, \cdots, x_n$ be the coordinate functions in $(\mathbb{C}^n)^*$, and set

$$\mathcal{G} = \{\mathbb{C}\{x_i \mid i \in I\} \mid \emptyset \neq I \subseteq [n]\}. \tag{2.1}$$

Here $[n]$ denotes $\{1, 2, \cdots, n\}$. Then

$$\mathcal{M}_\mathcal{G} = \{[a_1 : a_2 : \cdots : a_n] \in \mathbb{P}^{n-1} \mid a_i \neq 0 \text{ for all } i\} = T_n \text{ say}, \tag{2.2}$$

the complement of the coordinate hyperplanes in $\mathbb{P}^{n-1}$. (Note that $\mathcal{G}$ is the maximal building set for this particular complement.) We may obviously identify T_n with the $(n-1)$-dimensional algebraic torus with cocharacter lattice $N = \mathbb{Z}^n/\mathbb{Z}(1, 1, \cdots, 1)$.

For any nonempty subset $I \subseteq [n]$, we set

$$\mathbb{P}^I = \mathbb{P}(\mathbb{C}^n/\{(a_1, \cdots, a_n) \in \mathbb{C}^n \mid a_i = 0 \text{ for all } i \in I\}) = \{[a_i^I]_{i\in I}\},$$

where $[a_i^I]_{i\in I}$ denotes an I-tuple of homogeneous coordinates. (We identify $\mathbb{P}^{[n]}$ with $\mathbb{P}^{n-1}$ and drop the superscript $[n]$ in $a_i^{[n]}$.) Then the morphism ρ becomes

$$\rho : T_n \to \prod_{\emptyset\neq I\subseteq[n]} \mathbb{P}^I, \tag{2.3}$$

where the I-component of $\rho([a_1 : a_2 : \cdots : a_n])$ is $[a_i]_{i\in I}$. The De Concini–Procesi model $Y_n = \overline{Y}_\mathcal{G}$ is the closure of $\rho(T_n)$ as above. Notice that the factors $\mathbb{P}^I$ where $|I| = 1$ are points, and make no difference to the definition of Y_n, but it is sometimes notationally convenient to keep them.

Example 2.1. Clearly $Y_1 \cong \mathbb{P}^0$ (a point), and $Y_2 \cong \mathbb{P}^1$.

Example 2.2. When $n = 3$, we can identify ρ with the map

$$T_3 \to \mathbb{P}^2\times\mathbb{P}^1\times\mathbb{P}^1\times\mathbb{P}^1 : [a_1{:}a_2{:}a_3] \mapsto ([a_1{:}a_2{:}a_3], [a_1{:}a_2], [a_1{:}a_3], [a_2{:}a_3]),$$

where the $\mathbb{P}^1$ factors are, respectively, $\mathbb{P}^{\{1,2\}}$, $\mathbb{P}^{\{1,3\}}$, and $\mathbb{P}^{\{2,3\}}$. Thus Y_3 is the blow-up of $\mathbb{P}^2$ at the three points $[0 : 0 : 1]$, $[0 : 1 : 0]$, and $[1 : 0 : 0]$.

Remark 2.3. For $n \geq 4$, Y_n can be obtained by an iterated blow-up procedure: first one blows up $\mathbb{P}^{n-1}$ at the n coordinate points, then one blows up again along the proper transforms of the $\binom{n}{2}$ coordinate lines, then one blows up again along the proper transforms of the $\binom{n}{3}$ coordinate planes, and so forth. This is the description given in [11, Section 3].

Note that the symmetric group S_n acts on $\prod_{\emptyset \neq I \subseteq [n]} \mathbb{P}^I$ in a natural way: the action of $w \in S_n$ uses the isomorphisms

$$\mathbb{P}^I \xrightarrow{\sim} \mathbb{P}^{w(I)} : [a_i^I]_{i \in I} \mapsto [a^I_{w^{-1}(j)}]_{j \in w(I)}. \tag{2.4}$$

Clearly ρ is S_n-equivariant, where S_n acts on T_n by permuting coordinates. Thus S_n acts on Y_n. There is also a natural action of the torus T_n on each $\mathbb{P}^I$ by

$$[a_1 : a_2 : \cdots : a_n].[a_i^I]_{i \in I} = [a_i a_i^I]_{i \in I}. \tag{2.5}$$

Clearly ρ is T_n-equivariant, so T_n acts on Y_n.

Proposition 2.4. *Y_n is the closed subvariety of $\prod_{\emptyset \neq I \subseteq [n]} \mathbb{P}^I$ defined by the condition that $(a_i^I)_{i \in I}$ and $(a_i^J)_{i \in I}$ are linearly dependent for all $\emptyset \neq I \subseteq J \subseteq [n]$.*

Note that $[a_j^J]_{j \in J} \in \mathbb{P}^J$ is a J-tuple of homogeneous coordinates, so by definition $a_j^J \neq 0$ for some $j \in J$. However, it is possible that $a_i^J = 0$ for all i in the smaller subset I, in which case the condition in Proposition 2.4 is automatically satisfied; otherwise, the condition is equivalent to saying that $[a_i^I]_{i \in I} = [a_i^J]_{i \in I}$.

Example 2.5. When $n = 3$, Proposition 2.4 asserts that

$$([a_1 : a_2 : a_3], [b_1 : b_2], [c_1 : c_3], [d_2 : d_3]) \in \mathbb{P}^2 \times \mathbb{P}^1 \times \mathbb{P}^1 \times \mathbb{P}^1$$

lies in Y_3 exactly when $a_1 b_2 = a_2 b_1$, $a_1 c_3 = a_3 c_1$, and $a_2 d_3 = a_3 d_2$.

Proof. It is clear that the condition in Proposition 2.4 can be rephrased in terms of equations in the coordinates a_i^I as in Example 2.5, so it does define a closed subvariety Z of $\prod_{\emptyset \neq I \subseteq [n]} \mathbb{P}^I$. It is also clear that $\rho(T_n) \subseteq Z$. Hence $Y_n \subseteq Z$. To prove the reverse inclusion, suppose that $p \in Z$ has I-component $[a_i^I]_{i \in I}$. Let $K_1 = [n]$, and define a chain of subsets $K_1 \supset K_2 \supset K_3 \supset \cdots \supset K_m \supset K_{m+1} = \emptyset$ by the recursive rule

$$K_{\ell+1} = \{k \in K_\ell \mid a_k^{K_\ell} = 0\} \text{ for } 1 \leq \ell \leq m, \tag{2.6}$$

where m is minimal such that $a_k^{K_m} \neq 0$ for all $k \in K_m$. For any nonempty $I \subseteq [n]$, there is some $s \leq m$ such that $I \subseteq K_s$, $I \not\subseteq K_{s+1}$. By definition of Z, we must have $[a_i^I]_{i \in I} = [a_i^{K_s}]_{i \in I}$. So p is determined by its K_s-components for $1 \leq s \leq m$, which are constrained only by the vanishing conditions (2.6). For any $t \in \mathbb{C}^\times$, define $q(t) = [a_1(t) : a_2(t) : \cdots : a_n(t)] \in T_n$ by the rule that $a_i(t) = t^s a_i^{K_s}$, if $i \in K_s$ and $i \notin K_{s+1}$. Then it is easy to see that $\lim_{t \to 0} \rho(q(t)) = p$, in the sense that we have a morphism

$$\sigma : \mathbb{C} \to Z : t \mapsto \begin{cases} \rho(q(t)), & \text{if } t \neq 0, \\ p, & \text{if } t = 0. \end{cases}$$

This proves that $p \in \overline{\rho(T_n)} = Y_n$. □

We can now decompose Y_n as the disjoint union of locally closed subvarieties $\mathcal{O}_{K_1, \cdots, K_{m+1}}$, one for each chain of subsets $[n] = K_1 \supset K_2 \supset \cdots \supset K_{m+1} = \emptyset$. Namely, a point of Y_n with I-component $[a_i^I]_{i \in I}$ lies in $\mathcal{O}_{K_1, \cdots, K_{m+1}}$ if and only if $K_1, K_2, \cdots, K_{m+1}$ are the subsets defined recursively by (2.6).

Example 2.6. For a point of Y_3 with notation as in Example 2.5, the conditions for belonging to certain pieces of the disjoint union are as follows:

$$\begin{aligned} \mathcal{O}_{[3], \emptyset} &: a_1, a_2, a_3 \neq 0, \\ \mathcal{O}_{[3], \{1\}, \emptyset} &: a_1 = 0,\ a_2, a_3 \neq 0, \\ \mathcal{O}_{[3], \{1,2\}, \emptyset} &: a_1 = a_2 = 0,\ b_1, b_2 \neq 0, \\ \mathcal{O}_{[3], \{1,2\}, \{1\}, \emptyset} &: a_1 = a_2 = 0,\ b_1 = 0. \end{aligned}$$

Proposition 2.7. *The subvarieties $\mathcal{O}_{K_1, \cdots, K_{m+1}}$ are exactly the T_n-orbits in Y_n.*

Proof. It is clear that each $\mathcal{O}_{K_1, \cdots, K_{m+1}}$ is T_n-stable. As observed in the proof of Proposition 2.4, a point of $\mathcal{O}_{K_1, \cdots, K_{m+1}}$ is determined by its K_s-components, and the coordinates of the K_s-component indexed by K_{s+1} must vanish by definition. From this it is immediate that T_n acts transitively on $\mathcal{O}_{K_1, \cdots, K_{m+1}}$. □

For example, $\mathcal{O}_{[n], \emptyset} = \rho(T_n)$. In general,

$$\mathcal{O}_{K_1, \cdots, K_{m+1}} \cong T_{|K_1| - |K_2|} \times T_{|K_2| - |K_3|} \times \cdots \times T_{|K_m| - |K_{m+1}|} \tag{2.7}$$

is a torus of dimension $n - m$.

Remark 2.8. The chains $[n] = K_1 \supset K_2 \supset \cdots \supset K_{m+1} = \emptyset$ are in obvious bijection with the nested subsets of $\mathcal{G}$, in the terminology of [1]. By an argument similar to the proof of Proposition 2.4, one can show that the closure $\overline{\mathcal{O}_{K_1,\cdots,K_{m+1}}}$ is the subvariety of Y_n defined by the condition that $a_k^{K_\ell} = 0$ for $k \in K_{\ell+1}$, for $1 \leq \ell \leq m$. This is the closed subvariety $D_{K_1,\cdots,K_{m+1}}$ of [1, Section 4.3]. It is clear that $D_{K_1,\cdots,K_{m+1}}$ is the union of all orbits $\mathcal{O}_{L_1,\cdots,L_{m'+1}}$ such that the chain (L_s) refines the chain (K_s). We have an obvious isomorphism

$$D_{K_1,\cdots,K_{m+1}} \cong Y_{|K_1|-|K_2|} \times Y_{|K_2|-|K_3|} \times \cdots \times Y_{|K_m|-|K_{m+1}|}, \tag{2.8}$$

which is merely what [1, Theorem 4.3] says for our special $\mathcal{G}$.

Now let $\mathcal{T}_n$ be the toric variety associated to the lattice $N = \mathbb{Z}^n/\mathbb{Z}(1, 1, \cdots, 1)$ and the Coxeter fan Δ, as in the introduction. The torus which naturally acts on $\mathcal{T}_n$ is the one with cocharacter lattice N, in other words the torus T_n. By [4, Section 3.1], the T_n-orbits in $\mathcal{T}_n$ are naturally in bijection with the cones of Δ, and these in turn are in bijection with the chains $[n] = K_1 \supset K_2 \supset \cdots \supset K_{m+1} = \emptyset$. So the above description of T_n-orbits in Y_n immediately suggests the following result.

Proposition 2.9 (De Concini–Procesi). *There is an isomorphism $\mathcal{T}_n \xrightarrow{\sim} Y_n$ which respects the actions of S_n and T_n on both varieties.*

Proof. The proof is omitted from [11, Section 3], but an argument similar to that in [2, Section IV] works. Namely, Y_n is nonsingular by [1, Theorem 4.2] and contains an open subvariety $\rho(T_n)$ on which T_n acts simply transitively, so Y_n is isomorphic to the toric variety associated to N and some complete fan Δ' in $N \otimes \mathbb{R}$. Since S_n acts on Y_n in a way compatible with its action on T_n, the fan Δ' must be S_n-stable. The maximal cones in Δ' correspond to the T_n-fixed points in Y_n, which are exactly the 0-dimensional orbits $\mathcal{O}_{K_1,\cdots,K_{n+1}}$, where $|K_i| = n + 1 - i$ for all i. Hence S_n permutes the maximal cones of Δ' simply transitively, which means that no reflecting hyperplane for the action of S_n can intersect the interior of a maximal cone. Hence the interior of each maximal cone of Δ' is contained in a single chamber of the hyperplane complement, forcing $\Delta' = \Delta$. □

3 Poset homology and the proof of Theorem 1.1

It is clear that the isomorphism of Proposition 2.9 respects the obvious real structures of $\mathcal{T}_n$ and Y_n, so the real variety $\mathcal{T}_n(\mathbb{R})$ is isomorphic to $Y_n(\mathbb{R})$. This allows us to use Rains' theorem [12, Theorem 3.7], or rather its dual form [9, Theorem 2.1], which expresses the rational cohomology of $\overline{Y}_{\mathcal{G}}(\mathbb{R})$ in terms of poset homology. By [5, Theorem 4.1], when $\mathcal{G}$ is the complexification of a real building set as in the present case, there is no difference between the real locus of the complex variety $\overline{Y}_{\mathcal{G}}$ and the real variety as defined in [12].

The general statement of Rains' theorem involves the poset $\Pi_{\mathcal{G}}^{(2)}$ of direct sums of even-dimensional elements of $\mathcal{G}$. In our case, this is clearly the same as the poset of even-dimensional elements of $\mathcal{G}$ with the minimal element $\{0\}$ added, which in turn is isomorphic to the poset B_n^{ev} of even-size subsets of $[n]$. Hence [9, Theorem 2.1] gives us an isomorphism of $\mathbb{Q}S_n$-modules

$$H^i(Y_n(\mathbb{R}), \mathbb{Q}) \cong \bigoplus_{I \in B_n^{\mathrm{ev}}} H_{|I|-i}(\emptyset, I) \otimes \varepsilon_I. \tag{3.1}$$

Here $H_m(\emptyset, I)$ denotes a poset homology group with $\mathbb{Q}$-coefficients of the open interval $(\emptyset, I)$ in the poset B_n^{ev}. Our degree convention follows [7, 9]: thus $H_m(\emptyset, I)$ is what would be called $\tilde{H}_{m-2}((\emptyset, I), \mathbb{Q})$ in texts such as [15]. By definition, $H_m(\emptyset, \emptyset)$ is $\mathbb{Q}$ if $m = 0$ and 0 otherwise, and if I is an atom of B_n^{ev} (that is, $|I| = 2$), then $H_m(\emptyset, I)$ is $\mathbb{Q}$ if $m = 1$ and 0 otherwise. The tensoring with ε_I also requires explanation: this is to indicate that if $w \in S_n$ preserves I, we want not its usual action on the homology group $H_{|I|-i}(\emptyset, I)$, but that action multiplied by $\varepsilon(w)$.

Now B_n^{ev} is a rank-selected subposet of the Boolean lattice B_n, and is itself a ranked poset with $\mathrm{rk}(I) = |I|/2$. By [15, Theorem 3.4.1], B_n^{ev} is Cohen–Macaulay. In particular, $H_m(\emptyset, I)$ is nonzero if and only if $m = \mathrm{rk}(I)$, or in other words $|I| = 2m$. So (3.1) can be simplified to

$$H^i(Y_n(\mathbb{R}), \mathbb{Q}) \cong \bigoplus_{\substack{I \in B_n^{\mathrm{ev}} \\ |I|=2i}} H_i(\emptyset, I) \otimes \varepsilon_I. \tag{3.2}$$

Here the right-hand side is just a sign-twisted version of the Whitney homology

$$WH_i(B_n^{\mathrm{ev}}) = \bigoplus_{\substack{I \in B_n^{\mathrm{ev}} \\ |I|=2i}} H_i(\emptyset, I).$$

Note that if $|I| = 2i$, then the closed interval $[\emptyset, I]$ is isomorphic to B_{2i}^{ev}.

The homology of B_n^{ev} is expressed as a representation of S_n by [15, Theorem 3.4.4]. Here we derive a different expression via the method of [15, Section 4.4] and [7], using the ring $R = \widehat{\bigoplus}_{n \geq 0} R(S_n)$ from the introduction.

Proposition 3.1. *We have the following equality in R:*

$$1+\sum_{\substack{n\geq 2\\ n \text{ even}}}(-1)^{n/2}H_{n/2}(\emptyset,[n])=\left(1+\sum_{\substack{n\geq 2\\ n \text{ even}}}1_{S_n}\right)^{-1}.$$

Proof. If we multiply $1+\sum_{\substack{n\geq 2\\ n \text{ even}}}(-1)^{n/2}H_{n/2}(\emptyset,[n])$ and $1+\sum_{\substack{n\geq 2\\ n \text{ even}}}1_{S_n}$ in R, the result clearly has terms only in even degrees. We must show that if $n\geq 2$ is even, then the degree-n term of this product vanishes. By definition, this term is

$$\sum_{\substack{0\leq m\leq n\\ m \text{ even}}}(-1)^{m/2}\,\mathrm{Ind}_{S_m\times S_{n-m}}^{S_n}(H_{m/2}(\emptyset,[m])\boxtimes 1_{S_{n-m}}),$$

which is clearly equal to

$$\sum_{\substack{0\leq m\leq n\\ m \text{ even}}}(-1)^{m/2}\bigoplus_{\substack{I\in B_n^{\mathrm{ev}}\\ |I|=m}}H_{m/2}(\emptyset,I)=\sum_{\substack{0\leq m\leq n\\ m \text{ even}}}(-1)^{m/2}\,WH_{m/2}(B_n^{\mathrm{ev}}).$$

This vanishes by a well-known general principle, for which one reference is [15, Theorem 4.4.1] (see also [7, Theorem 4.1]). □

We can now give the proof of Theorem 1.1.

Proof. Proposition 3.1 implies the following equality in $R[t]$:

$$\left(1+\sum_{\substack{n\geq 2\\ n \text{ even}}}\varepsilon_{S_n}t^{n/2}\right)^{-1}=\sum_{\substack{n\geq 0\\ n \text{ even}}}(H_{n/2}(\emptyset,[n])\otimes\varepsilon_{S_n})\,(-t)^{n/2}. \tag{3.3}$$

Hence for $n\geq 1$, the degree-n term of the right-hand side of Theorem 1.1 equals

$$\sum_{\substack{0\leq m\leq n\\ m \text{ even}}}\mathrm{Ind}_{S_m\times S_{n-m}}^{S_n}((H_{m/2}(\emptyset,[m])\otimes\varepsilon_{S_m})\boxtimes 1_{S_{n-m}})\,(-t)^{m/2},$$

which in turn equals

$$\sum_{\substack{0\leq m\leq n\\ m \text{ even}}}\left(\bigoplus_{\substack{I\in B_n^{\mathrm{ev}}\\ |I|=m}}H_{m/2}(\emptyset,I)\otimes\varepsilon_I\right)(-t)^{m/2}.$$

By (3.2) and Proposition 2.9, this equals the degree-n term of the left-hand side of Theorem 1.1. □

To deduce Corollary 1.2 from Theorem 1.1, we simply observe that

$$(1+\sum_{\substack{n\geq 2\\ n \text{ even}}} \varepsilon_{S_n} t^{n/2})^{-1}$$

$$=1+\sum_{m\geq 1}(-1)^m \left(\sum_{\substack{n\geq 2\\ n \text{ even}}} \varepsilon_{S_n} t^{n/2}\right)^m$$

$$=1+\sum_{m\geq 1}(-1)^m \sum_{\substack{n_1,n_2,\cdots,n_m\geq 2\\ n_1,n_2,\cdots,n_m \text{ even}}} \mathrm{Ind}_{S_{n_1}\times\cdots\times S_{n_m}}^{S_{n_1+\cdots+n_m}} (\varepsilon_{S_{n_1}} \boxtimes \cdots \boxtimes \varepsilon_{S_{n_m}})\, t^{(n_1+\cdots+n_m)/2}.$$

Finally, if we apply to Theorem 1.1 the ring homomorphism $R \to \mathbb{Q}[[x]]$ which sends $V \in R(S_n)$ to $(\dim V)\frac{x^n}{n!}$, we obtain

$$1+\sum_{n\geq 1}\sum_i \dim H^i(\mathcal{T}_n(\mathbb{R}),\mathbb{Q})\,(-t)^i \frac{x^n}{n!} = \exp(x)\,\mathrm{sech}(t^{1/2}x), \qquad (3.4)$$

from which Corollary 1.3 follows.

4 Comparison with the moduli space of stable genus-zero curves

In Section 2, we showed that $\mathcal{T}_n$ is isomorphic to the De Concini–Procesi model $\overline{Y}_{\mathcal{G}}$ associated to the building set $\mathcal{G} = \{\mathbb{C}\{x_i \mid i \in I\} \mid \emptyset \neq I \subseteq [n]\}$ in the dual of the vector space $V = \mathbb{C}^n$. A better-known example of the De Concini–Procesi construction is the moduli space of stable genus-0 curves with marked points.

Specifically, let $W = \mathbb{C}^{n+1}/\mathbb{C}(1,1,\cdots,1)$, and consider the building set

$$\mathcal{H} = \{\mathbb{C}\{y_j - y_{j'} \mid j, j' \in J\} \mid J \subseteq [n+1], |J| \geq 2\} \text{ in } W^*, \qquad (4.1)$$

where $y_1,\cdots,y_{n+1}$ are the coordinate functions on $\mathbb{C}^{n+1}$, so that $y_j - y_{j'}$ is a well-defined element of W^*. Then $\mathcal{M}_{\mathcal{H}}$ is the complement in $\mathbb{P}(W)$ of the hyperplanes $y_i = y_j$ generating the reflection representation of S_{n+1}. (Note that $\mathcal{H}$ is the minimal building set for this particular complement.) One can identify $\mathcal{M}_{\mathcal{H}}$ with the moduli space $\mathcal{M}_{0,n+2}$ of ordered configurations of $n+2$ points in $\mathbb{P}^1$, since the $(n+2)$th point can be assumed to be the point at infinity, leaving an $(n+1)$-tuple of distinct numbers modulo translation and scaling.

As observed in [1, Section 4.3, Remark (3)], the De Concini–Procesi model $\overline{Y}_{\mathcal{H}}$ is isomorphic to the moduli space $\overline{\mathcal{M}_{0,n+2}}$ of stable genus-0 curves with $n+2$ marked points, in such a way that the S_{n+1}-action on $\overline{Y}_{\mathcal{H}}$ is identified with the restriction to S_{n+1} of the S_{n+2}-action on $\overline{\mathcal{M}_{0,n+2}}$. (Some of the details of this identification are in [6] and [12].) By definition, we have a projection morphism from $\overline{Y}_{\mathcal{H}}$ to

$$\begin{aligned}&\mathbb{P}(W/\mathbb{C}\{y_j - y_{j'} \mid j, j' \in J\}^{\perp})\\&= \mathbb{P}(\mathbb{C}^{n+1}/\{(a_1, \cdots, a_{n+1}) \in \mathbb{C}^{n+1} \mid a_j = a_{j'} \text{ for all } j, j' \in J\})\end{aligned}$$

for all subsets $J \subseteq [n+1]$ with $|J| \geq 2$. A point p in $\overline{Y}_{\mathcal{H}}$ is determined by its images under these morphisms, which we will refer to as the J-components of p.

Example 4.1. When $|J| = 3$, the J-component of p may be thought of as a point of $\mathbb{P}^1 \cong \overline{\mathcal{M}_{0,4}}$. In the terminology of stable curves as found in [3, Section 2], the resulting projection morphism $\overline{\mathcal{M}_{0,n+2}} \to \overline{\mathcal{M}_{0,4}}$ is the map of 'stably forgetting' all marked points except those labelled by $J \cup \{n+2\}$.

Let $f : W \to V$ be the S_n-equivariant vector space isomorphism whose transpose map $f^* : V^* \to W^*$ sends x_i to $y_i - y_{n+1}$ for all $1 \leq i \leq n$. If $A \in \mathcal{G}$, then clearly $f^*(A) \in \mathcal{H}$. Thus, in the terminology of Rains, $f : (W, \mathcal{H}) \to (V, \mathcal{G})$ is a morphism of building sets. He observes in [12, Proposition 2.4] that the De Concini–Procesi construction is functorial, so we have a morphism $\overline{Y}_f : \overline{Y}_{\mathcal{H}} \to \overline{Y}_{\mathcal{G}} = Y_n$, which is surjective and birational by [12, Proposition 2.7]. By definition, for any $p \in \overline{Y}_{\mathcal{H}}$ and any nonempty subset $I \subseteq [n]$, the I-component of $\overline{Y}_f(p)$ is the image of the $(I \cup \{n+1\})$-component of p under the map

$$\begin{aligned}&\mathbb{C}^{n+1}/\{(a_1, \cdots, a_{n+1}) \in \mathbb{C}^{n+1} \mid a_i = a_{n+1} \text{ for all } i \in I\}\\&\qquad\to \mathbb{C}^n/\{(a_1, \cdots, a_n) \in \mathbb{C}^n \mid a_i = 0 \text{ for all } i \in I\} : \quad (4.2)\\&(a_1, \cdots, a_{n+1}) \mapsto (a_1 - a_{n+1}, \cdots, a_n - a_{n+1}).\end{aligned}$$

Composing the morphism $\overline{Y}_f$ with the isomorphism $\overline{\mathcal{M}_{0,n+2}} \xrightarrow{\sim} \overline{Y}_{\mathcal{H}}$, we obtain an S_n-equivariant surjective birational morphism $\tau : \overline{\mathcal{M}_{0,n+2}} \to Y_n$, which respects the real structures. We now aim to use τ to compare these two varieties.

For any ordered m-element subset $\{s_1, s_2, \cdots, s_m\}$ of $[n]$, we have a morphism $\phi_{s_1,\cdots,s_m} : Y_n \to Y_m$, analogous to that considered in [3, Section 2.3], which is defined as follows. For any $p \in Y_n$ with I-component $[a_i^I]_{i\in I}$ and any nonempty subset $J \subseteq [m]$, the J-component of $\phi_{s_1,\cdots,s_m}(p)$ is $[a_{s_j}^{S_J}]_{j\in J}$ where $S_j = \{s_j \mid j \in J\}$. It is clear from Proposition 2.4 that this does define a point of Y_m.

As a consequence, we have ring homomorphisms $\phi^*_{s_1,\ldots,s_m}: H^*(Y_m(\mathbb{R}),\mathbb{Q}) \to H^*(Y_n(\mathbb{R}),\mathbb{Q})$. For any $i \neq j$ in $[n]$, let ν_{ij} be the image under $\phi^*_{i,j}$: $H^1(Y_2(\mathbb{R}),\mathbb{Q}) \to H^1(Y_n(\mathbb{R}),\mathbb{Q})$ of the fundamental class of $Y_2(\mathbb{R}) \cong \mathbb{P}^1(\mathbb{R})$. Since $\phi_{j,i} : Y_n \to \mathbb{P}^1$ is the composition of $\phi_{i,j} : Y_n \to \mathbb{P}^1$ with the inversion map on $\mathbb{P}^1$, we have $\nu_{ji} = -\nu_{ij}$. If $w \in S_n$, it is clear that $w.\nu_{ij} = \nu_{w(i)w(j)}$.

Proposition 4.2. *The elements* $\{\nu_{ij} \mid 1 \leq i < j \leq n\}$ *form a basis of* $H^1(Y_n(\mathbb{R}),\mathbb{Q})$.

Proof. Corollary 1.3 says that $\dim H^1(Y_n(\mathbb{R}),\mathbb{Q}) = \binom{n}{2}$, so it suffices to show that the elements ν_{ij} for $i < j$ are linearly independent. For this, it is enough to show that their images $\tau^*(\nu_{ij})$ under the map τ^* : $H^1(Y_n(\mathbb{R}),\mathbb{Q}) \to H^1(\overline{\mathcal{M}_{0,n+2}}(\mathbb{R}),\mathbb{Q})$ are linearly independent. But from the definitions, it follows that the composition $\phi_{i,j} \circ \tau : \overline{\mathcal{M}_{0,n+2}} \to \mathbb{P}^1$ is precisely the map denoted $\phi_{i,j,n+1,n+2}$ in [3, Section 2.3]. Hence $\tau^*(\nu_{ij})$ is the element denoted $\omega_{i,j,n+1,n+2}$ there. Since the elements $\omega_{i,j,k,n+2}$ for $i < j < k < n+2$ form a basis of $H^1(\overline{\mathcal{M}_{0,n+2}}(\mathbb{R}),\mathbb{Q})$ by [3, Proposition 2.3, Theorem 2.9], the desired linear independence holds. □

However, there is a major difference between the rational cohomology rings $H^*(\overline{\mathcal{M}_{0,n+2}}(\mathbb{R}),\mathbb{Q})$ and $H^*(Y_n(\mathbb{R}),\mathbb{Q})$: the former is generated in degree 1 by [3, Theorem 2.9], whereas the latter is not (for $n \geq 4$), by the following result.

Proposition 4.3. *Assume* $n \geq 4$. *Let* C *be the subspace of* $H^2(Y_n(\mathbb{R}),\mathbb{Q})$ *spanned by the cup products of elements of* $H^1(Y_n(\mathbb{R}),\mathbb{Q})$. *Then* C *has basis*

$$\{\nu_{ij}\nu_{kl},\ \nu_{ik}\nu_{jl},\ \nu_{il}\nu_{jk} \mid 1 \leq i < j < k < l \leq n\}.$$

In particular, $\dim C = 3\binom{n}{4} < 5\binom{n}{4} = \dim H^2(Y_n(\mathbb{R}),\mathbb{Q})$.

Proof. Note first that $H^2(Y_3(\mathbb{R}),\mathbb{Q}) = 0$ by Corollary 1.3 (or since $Y_3(\mathbb{R})$ is a non-orientable surface). So in $H^*(Y_3(\mathbb{R}),\mathbb{Q})$ we have the relation $\nu_{12}\nu_{13} = 0$. Applying the map $\phi^*_{i,j,k}$ to this relation, we deduce that $\nu_{ij}\nu_{ik} = 0$ in $H^*(Y_n(\mathbb{R}),\mathbb{Q})$ for any distinct i, j, k in $[n]$ (and, of course, we also have $\nu_{ij}\nu_{ij} = 0$ by skew-symmetry of the cup product). Hence C is spanned by the given set. To show the linear independence, we can argue as in Proposition 4.2: it is enough to show that

$$\{\omega_{i,j,n+1,n+2}\,\omega_{k,l,n+1,n+2},\ \omega_{i,k,n+1,n+2}\,\omega_{j,l,n+1,n+2},\ \omega_{i,l,n+1,n+2}\,\omega_{j,k,n+1,n+2} \mid 1 \leq i < j < k < l \leq n\}$$

is linearly independent in $H^2(\overline{\mathcal{M}_{0,n+2}}(\mathbb{R}), \mathbb{Q})$, and this follows from [3, Proposition 2.3, Theorem 2.9]. □

In the case of $\overline{Y}_{\mathcal{H}} \cong \overline{\mathcal{M}_{0,n+2}}$, the S_{n+1}-action which arises naturally from the De Concini–Procesi construction can be extended to an S_{n+2}-action. As a final note, we can deduce from Proposition 4.3 that no such extension is possible for Y_n (at least, in a way that respects the real structure). This answers a question posed to the author by M. Wachs.

Corollary 4.4. *For $n \geq 4$, the S_n-action on $Y_n(\mathbb{R})$ cannot be extended to S_{n+1}.*

Proof. It suffices to show that the representation of S_n on the subspace C in Proposition 4.3 cannot be extended to S_{n+1}. We use the standard parametrization of irreducible $\mathbb{Q}S_n$-modules by partitions of n. It is immediate from the given basis that $C \cong \mathrm{Ind}_{S_4 \times S_{n-4}}^{S_n}(V_{(2,1,1)} \boxtimes 1_{S_{n-4}})$. Hence by the Pieri rule,

$$C \cong \begin{cases} V_{(2,1,1)}, & \text{if } n=4, \\ V_{(3,1,1)} \oplus V_{(2,2,1)} \oplus V_{(2,1,1,1)}, & \text{if } n=5, \\ V_{(n-2,1,1)} \oplus V_{(n-3,2,1)} \oplus V_{(n-3,1,1,1)} \oplus V_{(n-4,2,1,1)}, & \text{if } n \geq 6. \end{cases} \quad (4.3)$$

Now the $\mathbb{Q}S_{n+1}$-module V_λ, for λ a partition of $n+1$, restricts to give the $\mathbb{Q}S_n$-module $\bigoplus_\mu V_\mu$ where μ runs over all partitions of n whose diagram is contained in that of λ. It is easy to see from (4.3) that no sum of $\mathbb{Q}S_{n+1}$-modules V_λ can possibly restrict to give C. □

References

[1] C. De Concini and C. Procesi, *Wonderful models of subspace arrangements*, Selecta Math. (N.S.) **1** (1995), 459–494.

[2] F. De Mari, C. Procesi and M. A. Shayman, *Hessenberg varieties*, Trans. Amer. Math. Soc. **332** (1992), no. 2, 529–534.

[3] P. Etingof, A. Henriques, J. Kamnitzer and E. M. Rains, *The cohomology ring of the real locus of the moduli space of stable genus 0 curves with marked points*, Ann. of Math. (2) **171** (2010), no. 2, 731–777.

[4] W. Fulton, "Introduction to Toric Varieties", Annals of Mathematics Studies 131, Princeton University Press, Princeton, NJ, 1993.

[5] G. Gaiffi, *Real structures of models of arrangements*, Int. Math. Res. Not. **2004** (2004), no. 64, 3439–3467.

[6] A. HENDERSON, *Representations of wreath products on cohomology of De Concini–Procesi compactifications*, Int. Math. Res. Not. **2004** (2004), no. 20, 983–1021.

[7] A. HENDERSON, *Plethysm for wreath products and homology of sub-posets of Dowling lattices*, Electron. J. Combin. **13** (2006), no. 1, Research Paper 87, 25 pp.

[8] A. HENDERSON and G. LEHRER, *The equivariant Euler characteristic of real Coxeter toric varieties*, Bull. London Math. Soc. **41** (2009), no. 1, 515–523.

[9] A. HENDERSON and E. RAINS, *The cohomology of real De Concini–Procesi models*, Int. Math. Res. Not. IMRN **2008** (2008), no. 7, Art. ID rnn001, 29 pp.

[10] G. I. LEHRER, *Rational points and Coxeter group actions on the cohomology of toric varieties*, Ann. Inst. Fourier (Grenoble) **58** (2008), no. 2, 671–688.

[11] C. PROCESI, *The toric variety associated to Weyl chambers*, in *Mots*, Lang. Raison. Calc., Hermès, Paris, 1990, 153–161.

[12] E. M. RAINS, *The homology of real subspace arrangements*, J. Topol. **3** (2010), no. 4, 786–818.

[13] J. R. STEMBRIDGE, *Eulerian numbers, tableaux, and the Betti numbers of a toric variety*, Discrete Math. **99** (1992), 307–320.

[14] J. R. STEMBRIDGE, *Some permutation representations of Weyl groups associated with the cohomology of toric varieties*, Adv. Math. **106** (1994), no. 2, 244–301.

[15] M. L. WACHS, *Poset topology: tools and applications*, in *Geometric Combinatorics*, IAS/Park City Mathematics Series Vol. 13, American Mathematical Society, 2007, 497–615.

Arrangements stable under the Coxeter groups

Hidehiko Kamiya*, Akimichi Takemura† and Hiroaki Terao‡

Abstract. Let $\mathcal{B}$ be a real hyperplane arrangement which is stable under the action of a Coxeter group W. Then W acts naturally on the set of chambers of $\mathcal{B}$. We assume that $\mathcal{B}$ is disjoint from the Coxeter arrangement $\mathcal{A} = \mathcal{A}(W)$ of W. In this paper, we show that the W-orbits of the set of chambers of $\mathcal{B}$ are in one-to-one correspondence with the chambers of $\mathcal{C} = \mathcal{A} \cup \mathcal{B}$ which are contained in an arbitrarily fixed chamber of $\mathcal{A}$. From this fact, we find that the number of W-orbits of the set of chambers of $\mathcal{B}$ is given by the number of chambers of $\mathcal{C}$ divided by the order of W. We will also study the set of chambers of $\mathcal{C}$ which are contained in a chamber b of $\mathcal{B}$. We prove that the cardinality of this set is equal to the order of the isotropy subgroup W_b of b. We illustrate these results with some examples, and solve an open problem in [H. Kamiya, A. Takemura, H. Terao, Ranking patterns of unfolding models of codimension one, Adv. in Appl. Math. 47 (2011) 379–400] by using our results.

1 Introduction

Let $\mathcal{B}$ be a real hyperplane arrangement which is stable under the action of a Coxeter group W. Then W acts naturally on the set $\mathbf{Ch}(\mathcal{B})$ of chambers of $\mathcal{B}$. We want to find the number of W-orbits of $\mathbf{Ch}(\mathcal{B})$. A particular case of this problem was considered in the authors' previous paper (Kamiya, Takemura and Terao [11]) and the present paper is motivated by an open problem left in Section 6 of [11]. By the general results of the present paper, we give the affirmative answer to the open problem in Theorem 3.2.

Suppose throughout that $\mathcal{B} \cap \mathcal{A} = \emptyset$, where $\mathcal{A} = \mathcal{A}(W)$ is the Coxeter arrangement of W. In this paper, we will show that the orbit space of $\mathbf{Ch}(\mathcal{B})$ is in one-to-one correspondence with the set of chambers c of

* This work was partially supported by JSPS KAKENHI (22540134).

† This research was supported by JST CREST.

‡ This work was partially supported by JSPS KAKENHI (21340001).

$\mathcal{C} = \mathcal{A} \cup \mathcal{B}$ which are contained in a, $\{c \in \mathbf{Ch}(\mathcal{C}) \mid c \subseteq a\}$, where $a \in \mathbf{Ch}(\mathcal{A})$ is an arbitrary chamber of $\mathcal{A}$. From this fact, we find that the number of W-orbits of $\mathbf{Ch}(\mathcal{B})$ is given by $|\mathbf{Ch}(\mathcal{C})|/|W|$.

On the other hand, we will also study the set of chambers $c \in \mathbf{Ch}(\mathcal{C})$ which are contained in a chamber $b \in \mathbf{Ch}(\mathcal{B})$ of $\mathcal{B}$, $\{c \in \mathbf{Ch}(\mathcal{C}) \mid c \subseteq b\}$. We will prove that the cardinality of this set is equal to the order of the isotropy subgroup W_b of b. Moreover, we will investigate the structure of W_b.

Kamiya, Takemura and Terao [11] tried to find the number of "inequivalent ranking patterns generated by unfolding models of codimension one" in psychometrics, and obtained an upper bound for this number. It was left open to determine whether this upper bound is actually the exact number. The problem boils down to proving (or disproving) that the orbit space of the chambers of the restricted all-subset arrangement ([11]) $\mathcal{B}$ under the action of the symmetric group $\mathfrak{S}_m$ is in one-to-one correspondence with $\{c \in \mathbf{Ch}(\mathcal{A}(\mathfrak{S}_m) \cup \mathcal{B}) \mid c \subseteq a\}$ for a chamber $a \in \mathbf{Ch}(\mathcal{A}(\mathfrak{S}_m))$ of the braid arrangement $\mathcal{A}(\mathfrak{S}_m)$. The results of the present paper establish the one-to-one correspondence.

The paper is organized as follows. In Section 2, we verify our main results. Next, in Section 3, we illustrate our general results with five examples, some of which are taken from the authors' previous studies of unfolding models in psychometrics ([7, 11]). In Section 3, we also solve the open problem of [11] (Theorem 3.2) using our general results in Section 2 applied to one of our examples.

ACKNOWLEDGEMENTS. The authors are very grateful to an anonymous referee for valuable comments on an earlier version of this paper. The example of the Catalan arrangement in Subsection 3.1 was suggested by the referee.

2 Main results

In this section, we state and prove our main results.

Let V be a Euclidean space. Consider a Coxeter group W acting on V. Then the Coxeter arrangement $\mathcal{A} = \mathcal{A}(W)$ is the set of all reflecting hyperplanes of W. Suppose that $\mathcal{B}$ is a hyperplane arrangement which is stable under the natural action of W. We assume $\mathcal{A} \cap \mathcal{B} = \emptyset$ and define

$$\mathcal{C} := \mathcal{A} \cup \mathcal{B}.$$

Let $\mathbf{Ch}(\mathcal{A})$, $\mathbf{Ch}(\mathcal{B})$ and $\mathbf{Ch}(\mathcal{C})$ denote the set of chambers of $\mathcal{A}$, $\mathcal{B}$ and $\mathcal{C}$, respectively. Define

$$\varphi_{\mathcal{A}} : \mathbf{Ch}(\mathcal{C}) \to \mathbf{Ch}(\mathcal{A}), \quad \varphi_{\mathcal{B}} : \mathbf{Ch}(\mathcal{C}) \to \mathbf{Ch}(\mathcal{B})$$

by

$$\varphi_{\mathcal{A}}(c) := \text{the chamber of } \mathcal{A} \text{ containing } c,$$
$$\varphi_{\mathcal{B}}(c) := \text{the chamber of } \mathcal{B} \text{ containing } c$$

for $c \in \mathbf{Ch}(\mathcal{C})$. Note that the Coxeter group W naturally acts on $\mathbf{Ch}(\mathcal{A})$, $\mathbf{Ch}(\mathcal{B})$ and $\mathbf{Ch}(\mathcal{C})$.

Lemma 2.1. *$\varphi_{\mathcal{A}}$ and $\varphi_{\mathcal{B}}$ are both W-equivariant,* i.e.,

$$\varphi_{\mathcal{A}}(wc) = w(\varphi_{\mathcal{A}}(c)), \quad \varphi_{\mathcal{B}}(wc) = w(\varphi_{\mathcal{B}}(c))$$

for any $w \in W$ and $c \in \mathbf{Ch}(\mathcal{C})$.

The proof is easy and omitted.

The following result is classical (see, *e.g.*, [3, Chapter V, Section 3. 2. Theorem 1 (iii)]):

Theorem 2.2. *The group W acts on $\mathbf{Ch}(\mathcal{A})$ effectively and transitively. In particular, $|W| = |\mathbf{Ch}(\mathcal{A})|$.*

Using Theorem 2.2, we can prove the following lemma.

Lemma 2.3. *The group W acts on $\mathbf{Ch}(\mathcal{C})$ effectively. In particular, each W-orbit of $\mathbf{Ch}(\mathcal{C})$ is of size $|W|$.*

Proof. If $wc = c$ for $w \in W$ and $c \in \mathbf{Ch}(\mathcal{C})$, then we have $\varphi_{\mathcal{A}}(c) = w\varphi_{\mathcal{A}}(c)$, which implies $w = 1$ by Theorem 2.2. □

For $b \in \mathbf{Ch}(\mathcal{B})$, define the isotropy subgroup $W_b := \{w \in W \mid wb = b\}$. Then we have the next lemma.

Lemma 2.4. *For $b \in \mathbf{Ch}(\mathcal{B})$, the group W_b acts on $\varphi_{\mathcal{B}}^{-1}(b)$ effectively and transitively.*

Proof. The effective part follows from Lemma 2.3, so let us prove the transitivity. Let $c_1, c_2 \in \varphi_{\mathcal{B}}^{-1}(b)$. Define

$$\mathcal{A}(c_1, c_2) := \{H \in \mathcal{A} \mid c_1 \text{ and } c_2 \text{ are on different sides of } H\}.$$

Let us prove that there exists $w \in W$ such that $wc_1 = c_2$ by an induction on $|\mathcal{A}(c_1, c_2)|$. When $|\mathcal{A}(c_1, c_2)| = 0$, we have $\mathcal{A}(c_1, c_2) = \emptyset$ and $c_1 = c_2$. Thus we may choose $w = 1$. If $\mathcal{A}(c_1, c_2)$ is non-empty, then there exists $H_1 \in \mathcal{A}(c_1, c_2)$ such that H_1 contains a wall of c_1. Let s_1 denote the reflection with respect to H_1. Then

$$\mathcal{A}(s_1 c_1, c_2) = \mathcal{A}(c_1, c_2) \setminus \{H_1\}.$$

By the induction assumption, there exists $w_1 \in W$ with $w_1 s_1 c_1 = c_2$. Set $w := w_1 s_1$. Then $wc_1 = c_2$ and $c_2 = (wc_1) \cap c_2 \subseteq (wb) \cap b$, which implies that $(wb) \cap b$ is not empty. Thus $wb = b$ and $w \in W_b$. □

The following lemma states that the W-orbits of $\mathbf{Ch}(\mathcal{C})$ and those of $\mathbf{Ch}(\mathcal{B})$ are in one-to-one correspondence.

Lemma 2.5. *The map $\varphi_{\mathcal{B}} : \mathbf{Ch}(\mathcal{C}) \to \mathbf{Ch}(\mathcal{B})$ induces a bijection from the set of W-orbits of $\mathbf{Ch}(\mathcal{C})$ to the set of W-orbits of $\mathbf{Ch}(\mathcal{B})$.*

Proof. For $b \in \mathbf{Ch}(\mathcal{B})$ and $c \in \mathbf{Ch}(\mathcal{C})$, we denote the W-orbit of b and the W-orbit of c by $\mathcal{O}(b)$ and by $\mathcal{O}(c)$, respectively. It is easy to see that

$$\varphi_{\mathcal{B}}(\mathcal{O}(c)) = \mathcal{O}(\varphi_{\mathcal{B}}(c)), \quad c \in \mathbf{Ch}(\mathcal{C}),$$

by Lemma 2.1. Thus $\varphi_{\mathcal{B}}$ induces a map from the set of W-orbits of $\mathbf{Ch}(\mathcal{C})$ to the set of W-orbits of $\mathbf{Ch}(\mathcal{B})$. We will show the map is bijective.

Surjectivity: Let $\mathcal{O}(b)$ be an arbitrary orbit of $\mathbf{Ch}(\mathcal{B})$ with a representative point $b \in \mathbf{Ch}(\mathcal{B})$. Take an arbitrary $c \in \varphi_{\mathcal{B}}^{-1}(b)$. Then

$$\varphi_{\mathcal{B}}(\mathcal{O}(c)) = \mathcal{O}(\varphi_{\mathcal{B}}(c)) = \mathcal{O}(b),$$

which shows the surjectivity.

Injectivity: Suppose $\varphi_{\mathcal{B}}(\mathcal{O}(c_1)) = \varphi_{\mathcal{B}}(\mathcal{O}(c_2))$ $(c_1, c_2 \in \mathbf{Ch}(\mathcal{C}))$. Set $b_i := \varphi_{\mathcal{B}}(c_i)$ for $i = 1, 2$. We have

$$\mathcal{O}(b_1) = \mathcal{O}(\varphi_{\mathcal{B}}(c_1)) = \varphi_{\mathcal{B}}(\mathcal{O}(c_1)) = \varphi_{\mathcal{B}}(\mathcal{O}(c_2)) = \mathcal{O}(\varphi_{\mathcal{B}}(c_2)) = \mathcal{O}(b_2),$$

so we can pick $w \in W$ such that $wb_2 = b_1$. Then

$$\varphi_{\mathcal{B}}(wc_2) = w(\varphi_{\mathcal{B}}(c_2)) = wb_2 = b_1.$$

Therefore, both c_1 and wc_2 lie in $\varphi_{\mathcal{B}}^{-1}(b_1)$. By Lemma 2.4, we have $\mathcal{O}(c_1) = \mathcal{O}(wc_2) = \mathcal{O}(c_2)$. □

We are now in a position to state the main results of this paper.

Theorem 2.6. *The cardinalities of $\varphi_{\mathcal{A}}^{-1}(a)$, $\varphi_{\mathcal{B}}^{-1}(b)$ for $a \in \mathbf{Ch}(\mathcal{A})$, $b \in \mathbf{Ch}(\mathcal{B})$ are given as follows:*

(1) *For* $a \in \mathbf{Ch}(\mathcal{A})$, *we have*

$$|\varphi_{\mathcal{A}}^{-1}(a)| = \frac{|\mathbf{Ch}(\mathcal{C})|}{|\mathbf{Ch}(\mathcal{A})|} = \frac{|\mathbf{Ch}(\mathcal{C})|}{|W|} = |\{W\text{-orbits of } \mathbf{Ch}(\mathcal{C})\}|$$
$$= |\{W\text{-orbits of } \mathbf{Ch}(\mathcal{B})\}|.$$

(2) *For* $b \in \mathbf{Ch}(\mathcal{B})$, *we have* $|\varphi_{\mathcal{B}}^{-1}(b)| = |W_b|$.

Proof. Part 2 follows from Lemma 2.4, so we will prove Part 1. Since the map $\varphi_{\mathcal{A}} : \mathbf{Ch}(\mathcal{C}) \to \mathbf{Ch}(\mathcal{A})$ is W-equivariant (Lemma 2.1), we have for each $w \in W$ a bijection

$$\varphi_{\mathcal{A}}^{-1}(a) \to \varphi_{\mathcal{A}}^{-1}(wa)$$

sending $c \in \varphi_{\mathcal{A}}^{-1}(a)$ to wc. Thus every fiber of $\varphi_{\mathcal{A}}$ has the same cardinality because W acts transitively on $\mathbf{Ch}(\mathcal{A})$ (Theorem 2.2). The cardinality is equal to

$$\frac{|\mathbf{Ch}(\mathcal{C})|}{|\mathbf{Ch}(\mathcal{A})|} = \frac{|\mathbf{Ch}(\mathcal{C})|}{|W|}.$$

By Lemma 2.3, we have

$$|\{W\text{-orbits of } \mathbf{Ch}(\mathcal{C})\}| = \frac{|\mathbf{Ch}(\mathcal{C})|}{|W|}.$$

Finally, Lemma 2.5 proves the last equality. □

By Part 2 of Theorem 2.6, we can write $|\mathbf{Ch}(\mathcal{C})|$ as

$$\begin{aligned} |\mathbf{Ch}(\mathcal{C})| &= \sum_{b \in \mathbf{Ch}(\mathcal{B})} |\varphi_{\mathcal{B}}^{-1}(b)| = \sum_{b \in \mathbf{Ch}(\mathcal{B})} |W_b| \\ &= \sum_{w \in W} \mathrm{Fix}(w, \mathbf{Ch}(\mathcal{B})), \end{aligned} \tag{2.1}$$

where $\mathrm{Fix}(w, \mathbf{Ch}(\mathcal{B}))$ denotes the number of elements of $\mathbf{Ch}(\mathcal{B})$ fixed by w. In the case of the Catalan arrangement, (2.1) is stated in [18, page 561].

Next, let $x \in V \setminus \bigcup_{H \in \mathcal{B}} H$. Let $b \in \mathbf{Ch}(\mathcal{B})$ denote the unique chamber that contains x. Define the average $z(x)$ of x over the action of W_b:

$$z(x) = \frac{1}{|W_b|} \sum_{w \in W_b} wx.$$

Then it is easily seen that $z(x)$ lies in b because of the convexity of b, and that the map z is W-equivariant. Concerning the structure of W_b, we have the next proposition.

Proposition 2.7. *The following statements hold true:*

1. *For any $b \in \mathbf{Ch}(\mathcal{B})$, the set $\{z(x) \mid x \in b\}$ is equal to the set of all W_b-invariant points of b.*
2. *For any $x \in V \setminus \bigcup_{H \in \mathcal{B}} H$, the isotropy subgroup $W_{z(x)}$ of $z(x)$ is equal to W_b, where $b \in \mathbf{Ch}(\mathcal{B})$ is the unique chamber that contains x. In particular, $W_{z(x)}$ depends only on the chamber $b \in \mathbf{Ch}(\mathcal{B})$ containing x.*

Proof.
1. By the linearity of the action of W on V, the average $z(x) \in b$ of $x \in b$ is W_b-invariant: $wz(x) = z(x),\ w \in W_b$. Conversely, the average of any W_b-invariant point of b is the point itself.

2. Assume $w \in W_{z(x)}$. Then $z(x) = wz(x) \in b \cap wb$, which implies $b \cap wb \neq \emptyset$. Since b and wb are both chambers, they coincide: $wb = b$. Thus $w \in W_b$ and we obtain $W_{z(x)} \subseteq W_b$. We also have the reverse inclusion because the average $z(x)$ is W_b-invariant by the statement 1. □

When W is the symmetric group $\mathfrak{S}_m = \mathfrak{S}_{\{1,\dots,m\}}$, we have the following obvious fact (Proposition 2.8). Define

$$H_0 := \{x = (x_1, \dots, x_m)^T \in \mathbb{R}^m \mid x_1 + \dots + x_m = 0\}.$$

The group $W = \mathfrak{S}_m$ acts on $V = \mathbb{R}^m$ or $V = H_0$ by permuting coordinates. When $W = \mathfrak{S}_m$ and $V = \mathbb{R}^m$ or $V = H_0$, we agree that this action is considered.

Proposition 2.8. *Let $W = \mathfrak{S}_m$ and $V = \mathbb{R}^m$ or $V = H_0$. Then we have*

$$W_b = \mathfrak{S}_{k_1} \times \mathfrak{S}_{k_2} \times \dots \times \mathfrak{S}_{k_\ell}, \quad b \in \mathbf{Ch}(\mathcal{B}),$$

where $k_1, \dots, k_\ell$ $(k_1 + \dots + k_\ell = m,\ 1 \leq \ell \leq m)$ are defined by

$$\begin{aligned} z_{\sigma(1)} &= \dots = z_{\sigma(k_1)} > z_{\sigma(k_1+1)} \\ &= \dots = z_{\sigma(k_1+k_2)} > \dots > z_{\sigma(k_1+\dots+k_{\ell-1}+1)} = \dots = z_{\sigma(m)} \end{aligned}$$

for $z = (z_1, \dots, z_m)^T = z(x),\ x \in b$, and a permutation $\sigma \in \mathfrak{S}_{\{1,\dots,m\}}$.

Remark 2.9. Consider the map $b \mapsto W_b$ from $\mathbf{Ch}(\mathcal{B})$ to the set of subgroups of W. Since $W_{wb} = wW_bw^{-1},\ w \in W$, this induces a map τ from the set of W-orbits of $\mathbf{Ch}(\mathcal{B})$ to the set of conjugacy classes of subgroups of W:

$$\tau(\mathcal{O}(b)) = [W_b], \quad b \in \mathbf{Ch}(\mathcal{B}), \tag{2.2}$$

where $[W_b] := \{wW_bw^{-1} \mid w \in W\}$. This map τ is not injective in general. See Remarks 3.1 and 3.4.

3 Examples

In this section, we examine five examples. The first example (Subsection 3.1) is the Catalan arrangement, which has been well studied as a deformation of the braid arrangement. The Catalan arrangement is related to the semiorder introduced by Luce [13] in economics and mathematical psychology as a preference order that accounts for intransitive indifference. The next three examples are taken from problems in psychometrics—the arrangements in Subsections 3.2 and 3.3 (the braid arrangement in conjunction with the all-subset arrangement) appear naturally in the study of ranking patterns of unfolding models of codimension one (Kamiya, Takemura and Terao [11]), while the mid-hyperplane arrangement in Subsection 3.4 is needed in examining ranking patterns of unidimensional unfolding models (Kamiya, Orlik, Takemura and Terao [7]). In all four examples in Subsections 3.1–3.4, the Coxeter group W is of type A_{m-1}. In Subsection 3.5, we provide an illustration with the Coxeter group of type B_m. We also solve the open problem of [11] in Subsection 3.2.

3.1 Catalan arrangement

Let $W = \mathfrak{S}_m$ and $V = H_0$. Then $\mathcal{A} = \mathcal{A}(W)$ is the braid arrangement in H_0, consisting of the hyperplanes defined by $x_i = x_j,\ 1 \le i < j \le m$. All the $|\mathcal{A}| = m(m-1)/2$ hyperplanes form one orbit under the action of W on $\mathcal{A}$.

Let

$$\mathcal{B} = \{H_{ij} \mid i \neq j,\ 1 \le i \le m,\ 1 \le j \le m\},$$

where

$$H_{ij} := \{x = (x_1, \dots, x_m)^T \in H_0 \mid x_i = x_j + 1\}.$$

That is, $\mathcal{B}$ is an essentialization of the semiorder arrangement. We have $|\mathcal{B}| = m(m-1)$, and the action of W on $\mathcal{B}$ is transitive.

With these $\mathcal{A}$ and $\mathcal{B}$, the union $\mathcal{C} = \mathcal{A} \cup \mathcal{B}$ is an essentialization of the Catalan arrangement.

Now, as a chamber of $\mathcal{A}$, let us take $a \in \mathbf{Ch}(\mathcal{A})$ defined by $x_1 > \cdots > x_m$:

$$a : x_1 > \cdots > x_m. \tag{3.1}$$

For this a, we have always $x_i - x_j < 1\ (i > j)$ for $(x_1, \dots, x_m)^T \in a$, so the elements of $\varphi_{\mathcal{A}}^{-1}(a)$ are determined by the sets of pairs $i, j\ (i < j)$ such that $x_i - x_j < 1$, or equivalently the sets of maximal intervals $[i, j] := \{i, i+1, \dots, j\}\ (i < j)$ such that $x_i - x_j < 1$, for

$(x_1, \dots, x_m)^T \in a$. It is well known ([23, (1.1)], [6]) that the number $|\varphi_{\mathcal{A}}^{-1}(a)|$ is equal to the Catalan number

$$C_m := \frac{1}{m+1}\binom{2m}{m}.$$

Thus

$$|\mathbf{Ch}(\mathcal{C})| = |\varphi_{\mathcal{A}}^{-1}(a)| \times |W| = m!\, C_m = (2m)_{m-1} := (2m)(2m-1)\cdots(m+2).$$

Besides, from $|\mathbf{Ch}(\mathcal{C})| = m!\, C_m$, the characteristic polynomial $\chi(\mathcal{C}, t)$ of $\mathcal{C}$ ([15, Definition 2.52]) can be calculated as

$$\chi(\mathcal{C}, t) = (t - m - 1)(t - m - 2)\cdots(t - 2m + 1)$$

([22, Theorem 5.18]).

The W-orbits of $\mathbf{Ch}(\mathcal{C})$ are given as follows. Let $c_i \in \mathbf{Ch}(\mathcal{C})$, $i = 1, \dots, C_m$, be the chambers of $\mathcal{C}$ which are contained in a in (3.1): $\varphi_{\mathcal{A}}^{-1}(a) = \{c_i \mid i = 1, \dots, C_m\}$. As was mentioned earlier, each c_i, $i = 1, \dots, C_m$, can be indexed by the set of maximal intervals $\lfloor i, j \rfloor \subseteq \{1, \dots, m\}$ $(i < j)$ such that $x_i - x_j < 1$. The set $\varphi_{\mathcal{A}}^{-1}(a) = \{c_i \mid i = 1, \dots, C_m\} \subset \mathbf{Ch}(\mathcal{C})$ is a complete set of representatives of the W-orbits of $\mathbf{Ch}(\mathcal{C})$, *i.e.*, $\mathbf{Ch}(\mathcal{C})$ has exactly C_m orbits $\mathcal{O}(c_i)$, $i = 1, \dots, C_m$, under the action of W.

Next, consider the chambers of $\mathcal{B}$. The elements of $\mathbf{Ch}(\mathcal{B})$ are in one-to-one correspondence with the set of semiorders on $\{1, \dots, m\}$. Recall that a partial order $\succeq$ on $\{1, \dots, m\}$ is called a *semiorder* (or a *unit interval order*) if and only if the poset $(\{1, \dots, m\}, \succeq)$ contains no induced subposet isomorphic to $\mathbf{2}+\mathbf{2}$ or $\mathbf{3}+\mathbf{1}$, where $\boldsymbol{i}+\boldsymbol{j}$ stands for the disjoint union of an i-element chain and a j-element chain. The Scott-Suppes Theorem ([19]) states that a partial order $\succeq$ on $\{1, \dots, m\}$ is a semiorder if and only if there exist $x_1, \dots, x_m \in \mathbb{R}$ such that $i \succeq j \iff i \succ j$ or $i = j$ $(i, j \in \{1, \dots, m\})$, where

$$i \succ j \iff x_i > x_j + 1. \tag{3.2}$$

Now, a bijection from $\mathbf{Ch}(\mathcal{B})$ to the set $\mathbf{S}(m)$ of semiorders on $\{1, \dots, m\}$ is given by $\mathbf{Ch}(\mathcal{B}) \ni b \mapsto \succeq \in \mathbf{S}(m)$, where $\succeq$ is the partial order determined by (3.2) with an arbitrary $(x_1, \dots, x_m)^T \in b$. (The surjectivity of this map follows from the "Distinguishing Property" of $\mathbf{S}(m)$ ([16, Lemma 5.1], [17, Lemma 7.35]).)

The number of chambers of $\mathcal{B}$ is obtained as follows. Let us denote (only in this paragraph) our $\mathcal{B}$ and $\mathcal{C}$ as $\mathcal{B}_m$ and $\mathcal{C}_m$, respectively.

Postnikov and Stanley [18, Lemma 7.6] proved that $\mathrm{Fix}(w, \mathbf{Ch}(\mathcal{B}_m)) = |\mathbf{Ch}(\mathcal{B}_k)|$, where k is the number of cycles of $w \in W = \mathfrak{S}_m$. Thus (2.1) is written as

$$|\mathbf{Ch}(\mathcal{C}_m)| = \sum_{k=1}^{m} c(m,k)|\mathbf{Ch}(\mathcal{B}_k)|, \quad m \geq 1, \tag{3.3}$$

in this case ([18, Theorem 7.1], [21, Solution to Exercise 6.30]), where $c(m,k)$ is the signless Stirling number of the first kind. Because of [20, Proposition 1.9.1], equation (3.3) is equivalent to

$$\begin{aligned} |\mathbf{Ch}(\mathcal{B}_m)| &= \sum_{k=1}^{m} (-1)^{m-k} S(m,k) |\mathbf{Ch}(\mathcal{C}_k)| \\ &= \sum_{k=1}^{m} (-1)^{m-k} S(m,k)(2k)_{k-1}, \quad m \geq 1, \end{aligned} \tag{3.4}$$

where $S(m,k)$ is the Stirling number of the second kind. The number of semiorders, $|\mathbf{S}(m)| = |\mathbf{Ch}(\mathcal{B}_m)|$, in (3.4) was first obtained by Chandon, Lemaire and Pouget [4, Proposition 12]. In addition, by (3.4) and $\sum_{m=1}^{\infty} S(m,k)x^m/(m!) = (e^x - 1)^k/(k!)$, $k \geq 1$ ([14, page 174]), we obtain

$$\begin{aligned} \sum_{m=1}^{\infty} |\mathbf{Ch}(\mathcal{B}_m)| \frac{x^m}{m!} &= \sum_{k=1}^{\infty} |\mathbf{Ch}(\mathcal{C}_k)| \frac{(1-e^{-x})^k}{k!} \\ &= C(1-e^{-x}) \\ &= 1 \cdot x + 3 \cdot \frac{x^2}{2!} + 19 \cdot \frac{x^3}{3!} + 183 \cdot \frac{x^4}{4!} \\ &\quad + 2371 \cdot \frac{x^5}{5!} + 38703 \cdot \frac{x^6}{6!} + 763099 \cdot \frac{x^7}{7!} + \cdots \end{aligned} \tag{3.5}$$

([18, Theorem 7.1], [22, Corollary 5.12], [21, Exercise 6.30]; *cf.* [4, Table on page 79]), where

$$C(t) := \sum_{k=1}^{\infty} C_k t^k = \frac{1-\sqrt{1-4t}}{2t} - 1 = t + 2t^2 + 5t^3 + \cdots$$

([21, page 178]). By (3.4) and $\sum_{m=1}^{\infty} S(m,k)x^m = \prod_{j=1}^{k}\{x/(1-jx)\}$, $k \geq 1$ ([14, Theorem 4.3.1]), we can also get

$$\sum_{m=1}^{\infty} |\mathbf{Ch}(\mathcal{B}_m)|\, x^m = \sum_{k=1}^{\infty} \frac{(2k)_{k-1}\, x^k}{(1+x)(1+2x)\cdots(1+kx)}.$$

As for the W-orbits of $\mathbf{Ch}(\mathcal{B})$, we have $\{W\text{-orbits of } \mathbf{Ch}(\mathcal{B})\} = \{\mathcal{O}(b_1), \ldots, \mathcal{O}(b_{C_m})\}$ with cardinality C_m, where $b_i := \varphi_{\mathcal{B}}(c_i) \in \mathbf{Ch}(\mathcal{B})$, $i = 1, \ldots, C_m$.

Now, let us investigate the case $m = 3$.

The arrangement $\mathcal{A}$ consists of three lines in $V = H_0$, $\dim H_0 = 2$, each of which is defined by one of the following equations:

$$x_1 = x_2, \quad x_1 = x_3, \quad x_2 = x_3, \tag{3.6}$$

and $\mathcal{B}$ comprises six lines:

$$x_1 = x_2 \pm 1, \quad x_1 = x_3 \pm 1, \quad x_2 = x_3 \pm 1.$$

Figure 3.1 displays $\mathcal{A}$ and $\mathcal{B}$ in $V = H_0$.

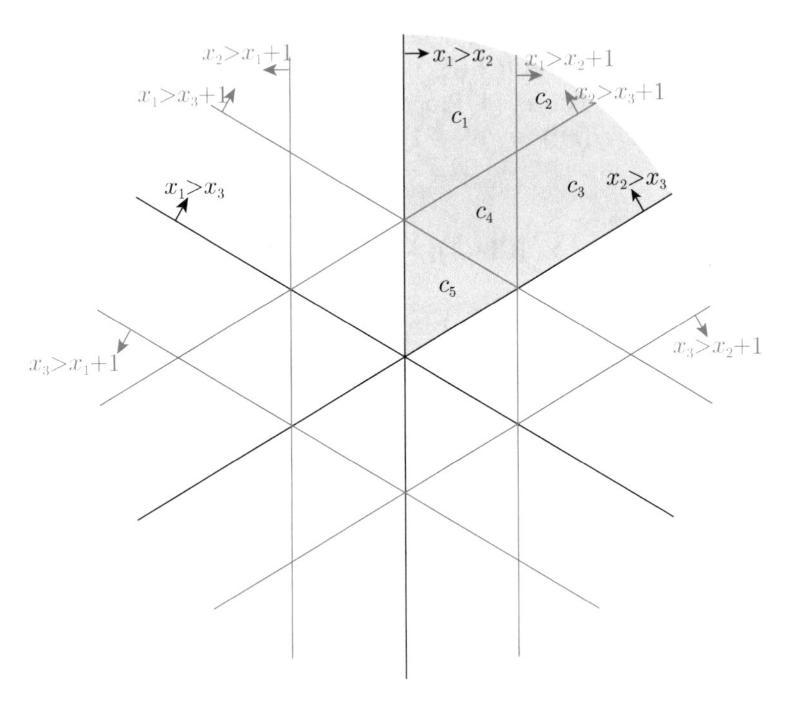

Figure 3.1. Essentialization of Catalan arrangement.

Let $a \in \mathbf{Ch}(\mathcal{A})$ be as in (3.1): $x_1 > x_2 > x_3$ (a is shaded in Figure 3.1). This chamber a of $\mathcal{A}$ contains exactly five ($= C_3$) chambers $c_1, \ldots, c_5$ of $\mathcal{C}$: $\varphi_{\mathcal{A}}^{-1}(a) = \{c_1, \ldots, c_5\}$. These five chambers $c_1, \ldots, c_5$ are indexed by the following sets of maximal intervals $[i, j]$ such that $x_i - x_j < 1$:

$$\begin{aligned} c_1 &: \{[1,2]\}, \\ c_2 &: \emptyset, \\ c_3 &: \{[2,3]\}, \\ c_4 &: \{[1,2],\ [2,3]\}, \\ c_5 &: \{[1,3]\}. \end{aligned}$$

Note that c_1 and c_3 are obtained from each other by changing (x_1, x_2, x_3) to $(-x_3, -x_2, -x_1)$. Since $a \in \mathbf{Ch}(\mathcal{A})$ consists of $|\varphi_{\mathcal{A}}^{-1}(a)| = 5$ chambers $c_1, \ldots, c_5$ of $\mathcal{C}$, we have $|\mathbf{Ch}(\mathcal{C})| = |\varphi_{\mathcal{A}}^{-1}(a)| \times |W| = 5 \times 3! = 30$, and $\mathbf{Ch}(\mathcal{C})$ has exactly five W-orbits $\mathcal{O}(c_1), \ldots, \mathcal{O}(c_5)$.

In $c_1 : \{[1,2]\}$, we have $x_1 - x_2 < 1,\ x_1 - x_3 > 1$ and $x_2 - x_3 > 1$, so the semiorder corresponding to $b_1 = \varphi_{\mathcal{B}}(c_1)$ is $1 \succ 3,\ 2 \succ 3$. (Recall that for the a in (3.1), and hence for any $c \in \mathbf{Ch}(\mathcal{C})$ contained in a, we have never $i \succ j$ for $i > j$.) By similar arguments, we can also obtain the semiorders corresponding to $b_i = \varphi_{\mathcal{B}}(c_i)$ for $i = 2, \ldots, 5$:

$$\begin{aligned} b_1 &: 1 \succ 3,\ 2 \succ 3, \\ b_2 &: 1 \succ 2 \succ 3, \\ b_3 &: 1 \succ 2,\ 1 \succ 3, \\ b_4 &: 1 \succ 3, \\ b_5 &: \text{none}. \end{aligned} \tag{3.7}$$

See Figure 3.2. Again, b_1 and b_3 are obtained from each other by the above-mentioned rule. The chamber $b_1 \in \mathbf{Ch}(\mathcal{B})$ is divided by the line of $\mathcal{A}$ defined by $x_1 = x_2$ into two chambers of $\mathcal{C}$, $|\varphi_{\mathcal{B}}^{-1}(b_1)| = 2$. The W_{b_1}-invariant points z of b_1 are $z = d(1, 1, -2)^T$, $d > 1/3$, so we have $W_{b_1} = \mathfrak{S}_{\{1,2\}}$. In a similar manner, b_2 is not divided by any line in $\mathcal{A}$; b_3 is divided by the line $x_2 = x_3$ into two; b_4 is not divided by any line; and b_5 is divided by the three lines $x_1 = x_2,\ x_1 = x_3,\ x_2 = x_3$ into six. The isotropy subgroups W_b and the W_b-invariant points z of b for $b = b_1, \ldots, b_5$ are given in Table 1.

We can confirm $|W_{b_1}| = 2! = |\varphi_{\mathcal{B}}^{-1}(b_1)|$, $|W_{b_2}| = 1 = |\varphi_{\mathcal{B}}^{-1}(b_2)|$, $|W_{b_3}| = 2! = |\varphi_{\mathcal{B}}^{-1}(b_3)|$, $|W_{b_4}| = 1 = |\varphi_{\mathcal{B}}^{-1}(b_4)|$, $|W_{b_5}| = 3! = |\varphi_{\mathcal{B}}^{-1}(b_5)|$ (Part 2 of Theorem 2.6).

We have $\{W\text{-orbits of } \mathbf{Ch}(\mathcal{B})\} = \{\mathcal{O}(b_1), \ldots, \mathcal{O}(b_5)\}$ and thus

$$|\{W\text{-orbits of } \mathbf{Ch}(\mathcal{B})\}| = 5.$$

We can see from (3.7) that $|\mathcal{O}(b_1)| = 3$, $|\mathcal{O}(b_2)| = 3!$, $|\mathcal{O}(b_3)| = 3$, $|\mathcal{O}(b_4)| = 3 \times 2 = 6$ and $|\mathcal{O}(b_5)| = 1$, coinciding with $|W|/|W_{b_i}|$, $i =$

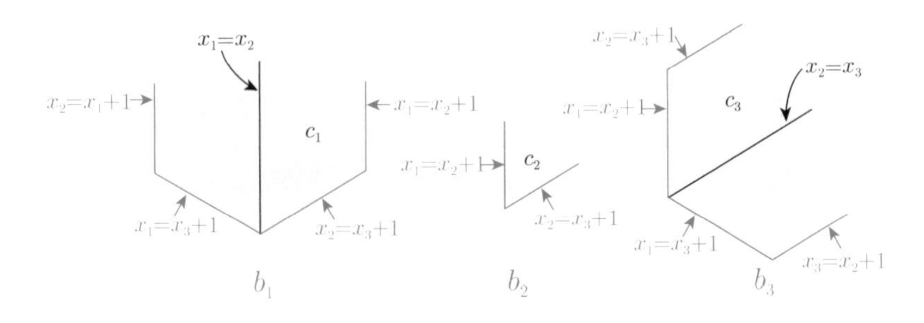

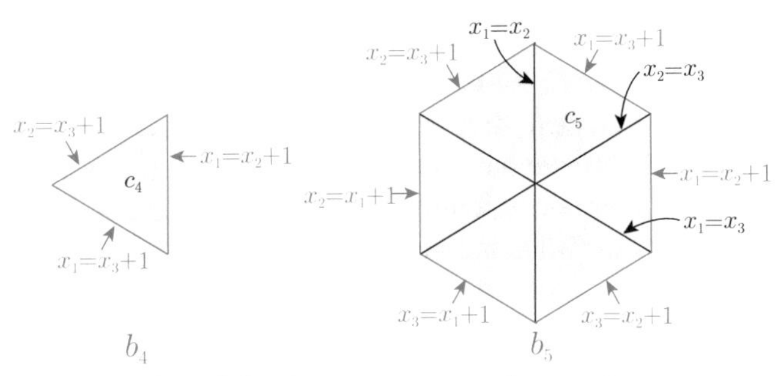

Figure 3.2. Essentialization of Catalan arrangement. $b_1, \ldots, b_5$.

Table 1. W_b and z for $b = b_1, \ldots, b_5$ in semiorder arrangement.

b	W_b	W_b-invariant points z of b
b_1	$\mathfrak{S}_{\{1,2\}}$	$d(1,1,-2)^T,\ d > \frac{1}{3}$
b_2	$\{1\}$	$(1,0,-1)^T + d_1(1,1,-2)^T + d_2(2,-1,-1)^T,\ d_1 > 0,\ d_2 > 0$
b_3	$\mathfrak{S}_{\{2,3\}}$	$d(2,-1,-1)^T,\ d > \frac{1}{3}$
b_4	$\{1\}$	$(1,0,-1)^T + d_1(-2,1,1)^T + d_2(-1,-1,2)^T, d_1 > 0, d_2 > 0, d_1 + d_2 < \frac{1}{3}$
b_5	$\mathfrak{S}_{\{1,2,3\}}$	$(0,0,0)^T$

$1, \ldots, 5$. With these values, $|\mathbf{Ch}(\mathcal{C})|$ can be computed also as

$$\begin{aligned}|\mathbf{Ch}(\mathcal{C})| &= \sum_{i=1}^{5} |\varphi_{\mathcal{B}}^{-1}(b_i)| \cdot |\mathcal{O}(b_i)| \\ &= 2 \cdot 3 + 1 \cdot 6 + 2 \cdot 3 + 1 \cdot 6 + 6 \cdot 1 = 30 \\ &= \sum_{i=1}^{5} \left(|W_{b_i}| \times \frac{|W|}{|W_{b_i}|} \right) = |W| \times |\{W\text{-orbits of } \mathbf{Ch}(\mathcal{B})\}|.\end{aligned}$$

In addition, $|\mathbf{Ch}(\mathcal{B})| = \sum_{i=1}^{5} |\mathcal{O}(b_i)| = 3 + 6 + 3 + 6 + 1 = 19$ in agreement with (3.5).

Remark 3.1. In Table 1, we find that $W_{b_2} = \{1\}$ and $W_{b_4} = \{1\}$ (respectively $W_{b_1} = \mathfrak{S}_{\{1,2\}}$ and $W_{b_3} = \mathfrak{S}_{\{2,3\}}$) are conjugate to each other, although b_2 and b_4 (respectively b_1 and b_3) are on different orbits. The map τ in (2.2) satisfies

$$\tau^{-1}([\{1\}]) = \{\mathcal{O}(b_2), \mathcal{O}(b_4)\}, \quad \tau^{-1}([\mathfrak{S}_{\{1,2\}}]) = \{\mathcal{O}(b_1), \mathcal{O}(b_3)\};$$

hence τ is not injective. It is evident that b_2 and b_4 are on different orbits, because b_2 is a cone and b_4 is a triangle. As for b_1 and b_3, we can confirm $\mathcal{O}(b_1) \neq \mathcal{O}(b_3)$ by looking at their corresponding semiorders in (3.7).

3.2 Coxeter group of type A_{m-1} and restricted all-subset arrangement

Let $W = \mathfrak{S}_m$ and $V = H_0$. Then $\mathcal{A} = \mathcal{A}(W)$ is the braid arrangement in H_0. Let $\mathcal{B}$ be the restricted all-subset arrangement (Kamiya, Takemura and Terao [11]):

$$\mathcal{B} = \{H_I^0 \mid \emptyset \neq I \subsetneq \{1, \ldots, m\}\},$$

where

$$H_I^0 := \{x = (x_1, \ldots, x_m)^T \in H_0 \mid \sum_{i \in I} x_i = 0\}.$$

Since $H_I^0 = H^0_{\{1,\ldots,m\}\setminus I}$ for $I \neq \emptyset, \{1, \ldots, m\}$, we have $|\mathcal{B}| = (2^m - 2)/2 = 2^{m-1} - 1$. The number of W-orbits of $\mathcal{B}$ is $(m-1)/2$ if m is odd and $m/2$ if m is even.

Theorem 2.6 applied to this case gives the affirmative answer to the open problem left in Section 6 of [11]. Using the terminology in Corollary 6.2 of [11], we state:

Theorem 3.2. *The number of inequivalent ranking patterns of unfolding models of codimension one is*

$$\frac{|\mathbf{Ch}(\mathcal{A} \cup \mathcal{B})|}{m!} - 1$$

for the braid arrangement $\mathcal{A}$ in H_0 and the restricted all-subset arrangement $\mathcal{B}$.

Proof. Part 1 of Theorem 2.6 implies that the number of W-orbits of $\mathbf{Ch}(\mathcal{B})$ for the restricted all-subset arrangement $\mathcal{B}$ is equal to $|\mathbf{Ch}(\mathcal{A} \cup \mathcal{B})|/(m!)$. This completes the proof because of the last sentence of [11, Corollary 6.2]. □

In the case $m = 3$, we can easily obtain $|\varphi_{\mathcal{A}}^{-1}(a)| = 2$ $(a \in \mathbf{Ch}(\mathcal{A}))$, $|\mathbf{Ch}(\mathcal{C})| = 12$, $|\mathbf{Ch}(\mathcal{B})| = 6$ and other quantities by direct observations.

Alternatively, $|\mathbf{Ch}(\mathcal{C})|$ can be computed by using the characteristic polynomial $\chi(\mathcal{C}, t) = (t-1)(t-5)$ ([11, Sec. 6.2.1]) of $\mathcal{C}$: Zaslavsky's result on the chamber-counting problem ([24, Theorem A], [15, Theorem 2.68]) yields $|\mathbf{Ch}(\mathcal{C})| = (-1)^2\chi(\mathcal{C}, -1) = 12$.

Let us investigate the case $m = 4$.

The elements of $\mathcal{A}$ are the six planes in $V = H_0$, $\dim H_0 = 3$, defined by the following equations:

$$x_1 = x_2, \quad x_1 = x_3, \quad x_1 = x_4, \quad x_2 = x_3, \quad x_2 = x_4, \quad x_3 = x_4, \tag{3.8}$$

whereas those of $\mathcal{B}$ are the seven planes below:

$$x_1 = 0, \quad x_2 = 0, \quad x_3 = 0, \quad x_4 = 0; \tag{3.9}$$

$$x_1 + x_2 = 0, \quad x_1 + x_3 = 0, \quad x_1 + x_4 = 0. \tag{3.10}$$

Figure 3.3 shows the intersection with the unit sphere $\mathbb{S}^2 = \{(x_1, ..., x_4)^T \in H_0 \mid x_1^2 + \cdots + x_4^2 = 1\}$ in H_0. Note that $\mathcal{B}$ has two orbits under the action of W—the four planes in (3.9) constitute one orbit, and the three in (3.10) form the other one. (In Figure 3.3, the planes in (3.9) are drawn in blue and those in (3.10) are sketched in red.)

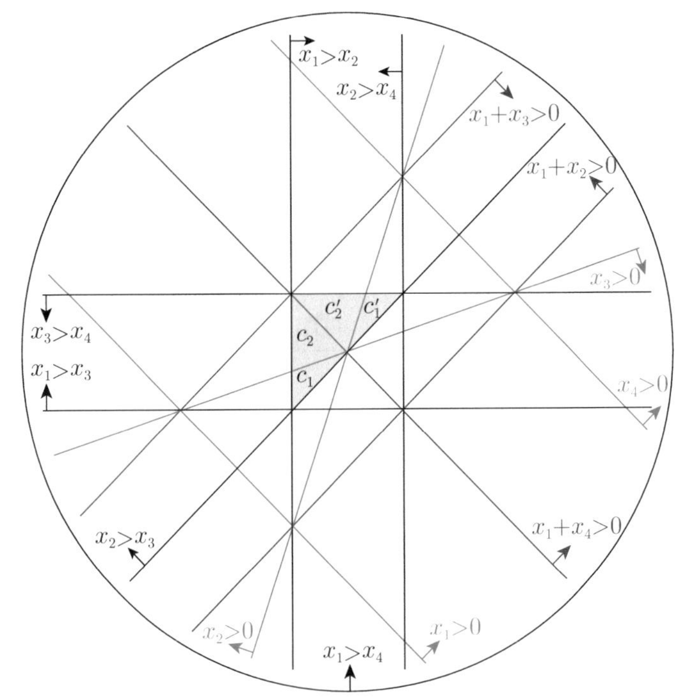

Figure 3.3. Braid plus restricted all-subset arrangement.

As a chamber of $\mathcal{A}$, let us take $a \in \mathbf{Ch}(\mathcal{A})$ defined by $x_1 > x_2 > x_3 > x_4$ (a is shaded in Figure 3.3). This chamber a of $\mathcal{A}$ contains exactly four chambers c_1, c_2, c_1', c_2' of $\mathcal{C}$: $\varphi_{\mathcal{A}}^{-1}(a) = \{c_1, c_2, c_1', c_2'\}$. These chambers have the following walls:

$$\text{Walls of } c_1 : x_1 = x_2,\ x_2 = x_3,\ x_3 = 0\ \ (x_1 > x_2 > x_3 > 0),$$
$$\text{Walls of } c_2 : x_1 = x_2,\ x_1 + x_4 = 0,\ x_3 = 0\ \ (-x_4 > x_1 > x_2,\ x_3 < 0),$$
$$\text{Walls of } c_1' : x_4 = x_3,\ x_3 = x_2,\ x_2 = 0\ \ (x_4 < x_3 < x_2 < 0),$$
$$\text{Walls of } c_2' : x_4 = x_3,\ x_4 + x_1 = 0,\ x_2 = 0\ \ (-x_1 < x_4 < x_3,\ x_2 > 0).$$

Note that c_1' (respectively c_2') is obtained from c_1 (respectively c_2) by changing (x_1, x_2, x_3, x_4) to $(-x_4, -x_3, -x_2, -x_1)$. Since $|\varphi_{\mathcal{A}}^{-1}(a)| = 4$, we have $|\mathbf{Ch}(\mathcal{C})| = |\varphi_{\mathcal{A}}^{-1}(a)| \times |W| = 4 \times 4! = 96$, and $\mathbf{Ch}(\mathcal{C})$ has exactly four orbits $\mathcal{O}(c_1), \mathcal{O}(c_2), \mathcal{O}(c_1'), \mathcal{O}(c_2')$.

The chambers $b_i := \varphi_{\mathcal{B}}(c_i),\ b_i' := \varphi_{\mathcal{B}}(c_i'),\ i = 1, 2$, of $\mathcal{B}$ containing c_1, c_2, c_1', c_2' have the following walls:

$$\text{Walls of } b_1 : x_1 = 0,\ x_2 = 0,\ x_3 = 0\ \ (x_1 > 0,\ x_2 > 0,\ x_3 > 0),$$
$$\text{Walls of } b_2 : x_1 + x_3 = 0,\ x_1 + x_4 = 0,\ x_3 = 0\ \ (-x_4 > x_1 > -x_3 > 0),$$
$$\text{Walls of } b_1' : x_4 = 0,\ x_3 = 0,\ x_2 = 0\ \ (x_4 < 0,\ x_3 < 0,\ x_2 < 0),$$
$$\text{Walls of } b_2' : x_4 + x_2 = 0,\ x_4 + x_1 = 0,\ x_2 = 0\ \ (-x_1 < x_4 < -x_2 < 0)$$

(Figure 3.4). The chamber $b_1 \in \mathbf{Ch}(\mathcal{B})$ is divided by the three planes $x_1 = x_2,\ x_1 = x_3,\ x_2 = x_3$ of $\mathcal{A}$ into six chambers of $\mathcal{C}$, whereas b_2 is divided by the plane $x_1 = x_2$ into two chambers of $\mathcal{C}$. For b_1, we have $W_{b_1} = \mathfrak{S}_{\{1,2,3\}}$ (the W_{b_1}-invariant points z of b_1 are $z = d(1, 1, 1, -3)^T$, $d > 0$), and for b_2, we find $W_{b_2} = \mathfrak{S}_{\{1,2\}}$ (the W_{b_2}-invariant points z of b_2 are $z = d_1(1, 1, -1, -1)^T + d_2(1, 1, 0, -2)^T$, $d_1, d_2 > 0$). So we see $|W_{b_1}| = |\varphi_{\mathcal{B}}^{-1}(b_1)|\ (= 6)$ and $|W_{b_2}| = |\varphi_{\mathcal{B}}^{-1}(b_2)|\ (= 2)$ hold true.

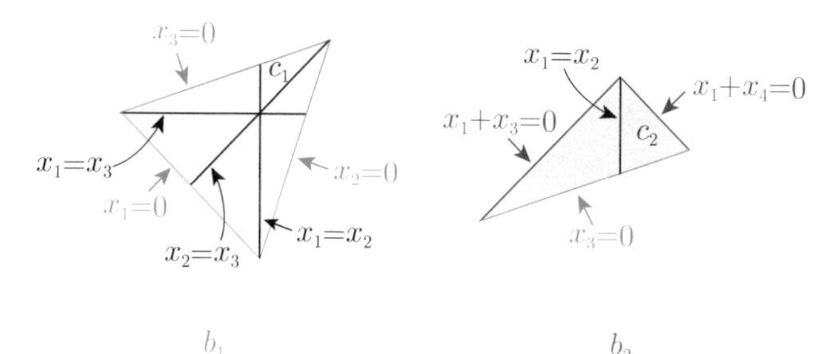

Figure 3.4. Braid plus restricted all-subset arrangement. b_1 and b_2.

We have $\{W\text{-orbits of } \mathbf{Ch}(\mathcal{B})\} = \{\mathcal{O}(b_1), \mathcal{O}(b_2), \mathcal{O}(b_1'), \mathcal{O}(b_2')\}$ and hence

$$|\{W\text{-orbits of } \mathbf{Ch}(\mathcal{B})\}| = 4.$$

The chambers $b \in \mathbf{Ch}(\mathcal{B})$ on the same W-orbit as b_1, $\mathcal{O}(b) = \mathcal{O}(b_1)$, have walls of the form $x_i > 0,\ x_j > 0,\ x_k > 0$ (i, j, k are all distinct), while $b \in \mathbf{Ch}(\mathcal{B})$ such that $\mathcal{O}(b) = \mathcal{O}(b_2)$ have walls of the form $x_i < 0,\ x_i + x_j > 0,\ x_i + x_k > 0$ (i, j, k are all distinct). Thus the orbit sizes are $|\mathcal{O}(b_1)| = \binom{4}{3} = 4$ $(= |W|/|W_{b_1}|)$ and $|\mathcal{O}(b_2)| = \binom{4}{3} \times 3 = 12$ $(= |W|/|W_{b_2}|)$. Accordingly, $|\mathbf{Ch}(\mathcal{C})|$ is again

$$\begin{aligned}|\mathbf{Ch}(\mathcal{C})| &= \left(|\varphi_{\mathcal{B}}^{-1}(b_1)| \cdot |\mathcal{O}(b_1)| + |\varphi_{\mathcal{B}}^{-1}(b_2)| \cdot |\mathcal{O}(b_2)|\right) \times 2 \\ &= (6 \times 4 + 2 \times 12) \times 2 = 96.\end{aligned}$$

Besides, $|\mathbf{Ch}(\mathcal{B})| = 2(|\mathcal{O}(b_1)| + |\mathcal{O}(b_2)|) = 2(4 + 12) = 32.$

The characteristic polynomial $\chi(\mathcal{C}, t)$ of $\mathcal{C}$ is

$$\chi(\mathcal{C}, t) = (t-1)(t-5)(t-7)$$

(Kamiya,Takemura and Terao [11, Sec. 6.2.2]). This polynomial yields

$$|\mathbf{Ch}(\mathcal{C})| = (-1)^3 \chi(\mathcal{C}, -1) = 96$$

in agreement with our observations above.

For $5 \leq m \leq 9$, we used the finite-field method (Athanasiadis [1, 2], Stanley [22, Lecture 5], Crapo and Rota [5], Kamiya, Takemura and Terao [8–10]) to calculate the characteristic polynomials $\chi(\mathcal{C}, t)$ of $\mathcal{C}$, and obtained the numbers of W-orbits of $\mathbf{Ch}(\mathcal{B})$ by using Zaslavsky's result [24, Theorem A] and Part 1 of Theorem 2.6: $|\{W\text{-orbits of } \mathbf{Ch}(\mathcal{B})\}| = |\varphi_{\mathcal{A}}^{-1}(a)| = |\mathbf{Ch}(\mathcal{C})|/|W|$ as follows.

$$\begin{aligned}
m = 5:\ & \chi(\mathcal{C}, t) = (t-1)(t-7)(t-8)(t-9), \\
& |\mathbf{Ch}(\mathcal{C})| = 1440, \quad |\varphi_{\mathcal{A}}^{-1}(a)| = 12, \\
m = 6:\ & \chi(\mathcal{C}, t) = (t-1)(t-7)(t-11)(t-13)(t-14), \\
& |\mathbf{Ch}(\mathcal{C})| = 40320, \quad |\varphi_{\mathcal{A}}^{-1}(a)| = 56, \\
m = 7:\ & \chi(\mathcal{C}, t) = (t-1)(t-11)(t-13)(t-17)(t-19)(t-23), \\
& |\mathbf{Ch}(\mathcal{C})| = 2903040, \quad |\varphi_{\mathcal{A}}^{-1}(a)| = 576, \\
m = 8:\ & \chi(\mathcal{C}, t) = (t-1)(t-19)(t-23)(t-25)(t-27)(t-29)(t-31), \\
& |\mathbf{Ch}(\mathcal{C})| = 670924800, \quad |\varphi_{\mathcal{A}}^{-1}(a)| = 16640, \\
m = 9:\ & \chi(\mathcal{C}, t) = (t-1)(t^7 - 290t^6 + 36456t^5 - 2573760t^4 \\
& \quad + 110142669t^3 - 2855339970t^2 + 41492561354t \\
& \quad - 260558129500), \\
& |\mathbf{Ch}(\mathcal{C})| = 610037568000, \quad |\varphi_{\mathcal{A}}^{-1}(a)| = 1681100.
\end{aligned}$$

Note that the characteristic polynomial $\chi(\mathcal{C}, t)$ factors into polynomials of degree one over $\mathbb{Z}$ for $m \leq 8$.

Remark 3.3. For $m = 5$ and $m = 6$, Kamiya, Takemura and Terao [11] identified all the elements c of $\varphi_{\mathcal{A}}^{-1}(a)$ for $a : x_1 > \cdots > x_m$ and gave an example of the W_b-invariant points z of $b = \varphi_{\mathcal{B}}(c)$ for each c. From those z, we immediately obtain W_b by Proposition 2.8.

3.3 Coxeter group of type A_{m-1} and unrestricted all-subset arrangement

Let $W = \mathfrak{S}_m$ and $V = \mathbb{R}^m$. Then $\mathcal{A} = \mathcal{A}(W)$ is the braid arrangement in $\mathbb{R}^m$. Let $\mathcal{B}$ be the (unrestricted) all-subset arrangement ([11]):

$$\mathcal{B} = \{H_I \mid \emptyset \neq I \subseteq \{1, \ldots, m\}\},$$

where

$$H_I := \{x = (x_1, \ldots, x_m)^T \in \mathbb{R}^m \mid \sum_{i \in I} x_i = 0\}.$$

Note $|\mathcal{B}| = 2^m - 1$. The number of orbits of $\mathcal{B}$ under the action of W is m.

We will examine the case $m = 3$.

The arrangement $\mathcal{A}$ has exactly the three planes in $V = \mathbb{R}^3$ defined by the same equations as those in (3.6). On the other hand, $\mathcal{B}$ consists of the seven planes defined by

$$\begin{gathered} x_1 = 0, \quad x_2 = 0, \quad x_3 = 0; \\ x_1 + x_2 = 0, \quad x_1 + x_3 = 0, \quad x_2 + x_3 = 0; \\ x_1 + x_2 + x_3 = 0 \end{gathered}$$

with each line corresponding to one orbit under the action of W on $\mathcal{B}$. Figure 3.5 exhibits the intersection with the unit sphere in $V = \mathbb{R}^3$.

Let us take $a \in \mathbf{Ch}(\mathcal{A})$ defined by $x_1 > x_2 > x_3$ (a with $x_1 + x_2 + x_3 > 0$ is shaded in Figure 3.5). Then $\varphi_{\mathcal{A}}^{-1}(a) = \{c_1, c_2, c_3, c_4, c_5, c_1', c_2', c_3', c_4', c_5'\}$, where

$$\begin{aligned} &c_1 : x_1 > x_2,\ x_1 + x_3 < 0,\ x_1 + x_2 + x_3 > 0, \\ &c_2 : x_2 > 0,\ x_1 + x_3 > 0,\ x_2 + x_3 < 0, \\ &c_3 : x_2 > x_3,\ x_2 < 0,\ x_1 + x_2 + x_3 > 0, \\ &c_4 : x_1 > x_2,\ x_3 < 0,\ x_2 + x_3 > 0, \\ &c_5 : x_1 > x_2,\ x_2 > x_3,\ x_3 > 0, \end{aligned}$$

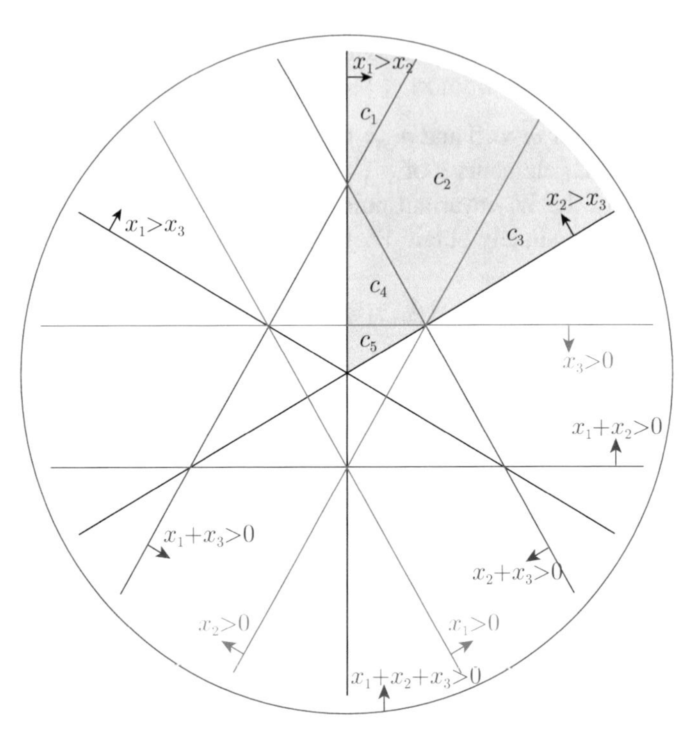

Figure 3.5. Braid plus unrestricted all-subset arrangement.

and c'_i, $i=1, ..., 5$, are the chambers obtained from c_i, $i = 1, \ldots, 5$, by changing (x_1, x_2, x_3) to $(-x_3, -x_2, -x_1)$. Thus $|\varphi_{\mathcal{A}}^{-1}(a)| = |\{W\text{-orbits of } \mathbf{Ch}(\mathcal{C})\}| = 10$ and $|\mathbf{Ch}(\mathcal{C})| = |\varphi_{\mathcal{A}}^{-1}(a)| \times |W| = 10 \times 3! = 60$.

The chambers $b_i := \varphi_{\mathcal{B}}(c_i) \in \mathbf{Ch}(\mathcal{B})$, $i = 1, \ldots, 5$, are

$$\begin{aligned} b_1 &: x_1 + x_3 < 0,\ x_2 + x_3 < 0,\ x_1 + x_2 + x_3 > 0, \\ b_2 &: x_2 > 0,\ x_1 + x_3 > 0,\ x_2 + x_3 < 0, \\ b_3 &: x_2 < 0,\ x_3 < 0,\ x_1 + x_2 + x_3 > 0, \\ b_4 &: x_3 < 0,\ x_1 + x_3 > 0,\ x_2 + x_3 > 0, \\ b_5 &: x_1 > 0,\ x_2 > 0,\ x_3 > 0, \end{aligned} \tag{3.11}$$

and $b'_i := \varphi_{\mathcal{B}}(c'_i) \in \mathbf{Ch}(\mathcal{B})$, $i = 1, \ldots, 5$, can be obtained from b_i, $i = 1, \ldots, 5$, by the above-mentioned rule. See Figure 3.6. The chamber b_1 is divided by the plane $x_1 = x_2$ into two chambers; b_2 is not divided by any plane in $\mathcal{A}$; b_3 is divided by $x_2 = x_3$ into two; b_4 is divided by $x_1 = x_2$ into two; and b_5 is divided by the three planes $x_1 = x_2$, $x_1 = x_3$, $x_2 = x_3$ into six. The isotropy subgroups W_b and the W_b-invariant points z of b for $b = b_1, \ldots, b_5$ are given in Table 2. We can confirm $|W_{b_1}| = 2! =$

Table 2. W_b and z for $b = b_1, \ldots, b_5$ in unrestricted all-subset arrangement.

b	W_b	W_b-invariant points z of b
b_1	$\mathfrak{S}_{\{1,2\}}$	$d_1(1,1,-1)^T + d_2(1,1,-2)^T,\ d_1, d_2 > 0$
b_2	$\{1\}$	$d_1(1,1,-1)^T + d_2(1,0,-1)^T + d_3(1,0,0)^T,\ d_1, d_2, d_3 > 0$
b_3	$\mathfrak{S}_{\{2,3\}}$	$d_1(1,0,0)^T + d_2(2,-1,-1)^T,\ d_1, d_2 > 0$
b_4	$\mathfrak{S}_{\{1,2\}}$	$d_1(1,1,-1)^T + d_2(1,1,0)^T,\ d_1, d_2 > 0$
b_5	$\mathfrak{S}_{\{1,2,3\}}$	$d(1,1,1)^T,\ d > 0$

$|\varphi_{\mathcal{B}}^{-1}(b_1)|$, $|W_{b_2}| = 1 = |\varphi_{\mathcal{B}}^{-1}(b_2)|$, $|W_{b_3}| = 2! = |\varphi_{\mathcal{B}}^{-1}(b_3)|$, $|W_{b_4}| = 2! = |\varphi_{\mathcal{B}}^{-1}(b_4)|$, $|W_{b_5}| = 3! = |\varphi_{\mathcal{B}}^{-1}(b_5)|$.

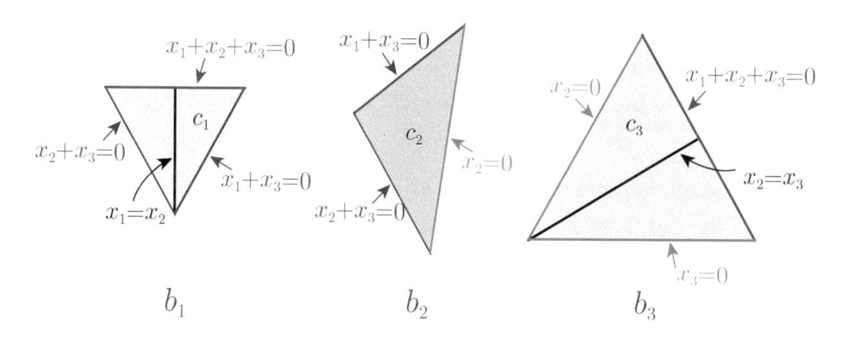

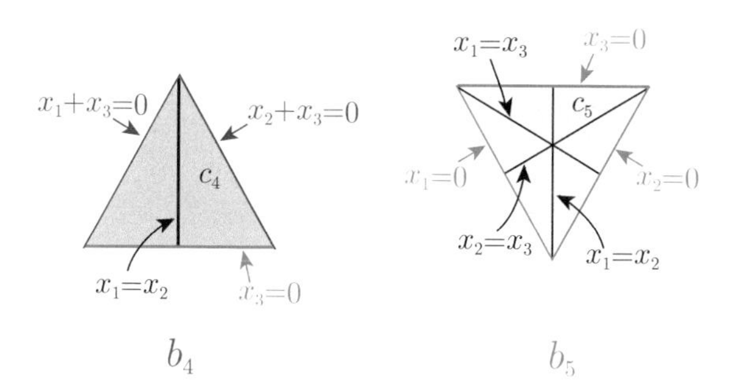

Figure 3.6. Braid plus unrestricted all-subset arrangement. $b_1, \ldots, b_5$.

We have $\{W\text{-orbits of } \mathbf{Ch}(\mathcal{B})\} = \{\mathcal{O}(b_1), \ldots, \mathcal{O}(b_5), \mathcal{O}(b_1'), \ldots, \mathcal{O}(b_5')\}$ and thus

$$|\{W\text{-orbits of } \mathbf{Ch}(\mathcal{B})\}| = 10.$$

From (3.11), we see $|\mathcal{O}(b_1)| = 3$, $|\mathcal{O}(b_2)| = 3 \times 2 = 6$, $|\mathcal{O}(b_3)| = 3$, $|\mathcal{O}(b_4)| = 3$, $|\mathcal{O}(b_5)| = 1$, which coincide with $|W|/|W_{b_i}|$, $i =$

$1, \dots, 5$. Hence, $|\mathbf{Ch}(\mathcal{C})| = 60$ can be obtained also from

$$|\mathbf{Ch}(\mathcal{C})| = 2\sum_{i=1}^{5} |\varphi_{\mathcal{B}}^{-1}(b_i)| \cdot |\mathcal{O}(b_i)| = 2\,(2\cdot 3 + 1\cdot 6 + 2\cdot 3 + 2\cdot 3 + 6\cdot 1) = 60.$$

We can also get $|\mathbf{Ch}(\mathcal{B})| = 2\sum_{i=1}^{5} |\mathcal{O}(b_i)| = 2\,(3+6+3+3+1) = 32$.

Remark 3.4. In Table 2, we find that $W_{b_1} = W_{b_4} = \mathfrak{S}_{\{1,2\}}$, $W_{b_3} = \mathfrak{S}_{\{2,3\}}$ are all conjugate to one another, although b_1, b_3, b_4 are on different orbits. We have

$$\tau^{-1}([\mathfrak{S}_{\{1,2\}}]) = \{\mathcal{O}(b_1), \mathcal{O}(b_3), \mathcal{O}(b_4)\},$$

so τ is not injective. The chambers b_1, b_3, b_4 are triangular cones (triangles in Figure 3.6) cut by a single plane (line) $x_i = x_j$ from the braid arrangement. However, these chambers are easily seen to be on different orbits, since their three walls (edges) are of different combinations of orbits of $\mathcal{B}$.

For $m \leq 7$, we computed $\chi(\mathcal{C}, t)$ using the finite-field method, and obtained the numbers of W-orbits of $\mathbf{Ch}(\mathcal{B})$ as follows:

$$\begin{aligned}
m = 3: \quad & \chi(\mathcal{C}, t) = (t-1)(t-4)(t-5), \\
& \qquad |\mathbf{Ch}(\mathcal{C})| = 60, \quad |\varphi_{\mathcal{A}}^{-1}(a)| = 10, \\
m = 4: \quad & \chi(\mathcal{C}, t) = (t-1)(t-5)(t-7)(t-8), \\
& \qquad |\mathbf{Ch}(\mathcal{C})| = 864, \quad |\varphi_{\mathcal{A}}^{-1}(a)| = 36, \\
m = 5: \quad & \chi(\mathcal{C}, t) = (t-1)(t-7)(t-9)(t-11)(t-13), \\
& \qquad |\mathbf{Ch}(\mathcal{C})| = 26880, \quad |\varphi_{\mathcal{A}}^{-1}(a)| = 224, \\
m = 6: \quad & \chi(\mathcal{C}, t) = (t-1)(t-11)(t-13)(t-17)^2(t-19), \\
& \qquad |\mathbf{Ch}(\mathcal{C})| = 2177280, \quad |\varphi_{\mathcal{A}}^{-1}(a)| = 3024, \\
m = 7: \quad & \chi(\mathcal{C}, t) = (t-1)(t-19)(t-23)(t^4 - 105t^3 + 4190t^2 \\
& \qquad - 75180t + 510834), \\
& \qquad |\mathbf{Ch}(\mathcal{C})| = 566697600, \quad |\varphi_{\mathcal{A}}^{-1}(a)| = 112440.
\end{aligned}$$

Note that the characteristic polynomial $\chi(\mathcal{C}, t)$ factors into polynomials of degree one over $\mathbb{Z}$ for $m \leq 6$.

3.4 Mid-hyperplane arrangement

Let $W = \mathfrak{S}_m$ and $V = H_0$, so $\mathcal{A} = \mathcal{A}(W)$ is the braid arrangement in H_0. We take

$$\mathcal{B} = \{H_{ijkl} \mid 1 \leq i < j \leq m,\ 1 \leq k < l \leq m,\ i < k,\ |\{i, j, k, l\}| = 4\},$$

where

$$H_{ijkl} := \{x = (x_1, \dots, x_m)^T \in H_0 \mid x_i + x_j = x_k + x_l\},$$

so that $\mathcal{C} = \mathcal{A} \cup \mathcal{B}$ is an essentialization of the mid-hyperplane arrangement (Kamiya, Orlik, Takemura and Terao [7]). We have $|\mathcal{B}| = 3\binom{m}{4}$, and the action of W on $\mathcal{B}$ is transitive.

Let us consider the case $m = 4$.

The elements of $\mathcal{A}$ are the six planes in $V = H_0$, $\dim H_0 = 3$, defined by the equations in (3.8), whereas those of $\mathcal{B}$ are the three planes defined by the following equations:

$$x_1 + x_2 = x_3 + x_4, \quad x_1 + x_3 = x_2 + x_4, \quad x_1 + x_4 = x_2 + x_3.$$

Figure 3.7 shows the intersection with the unit sphere $\mathbb{S}^2$ in H_0.

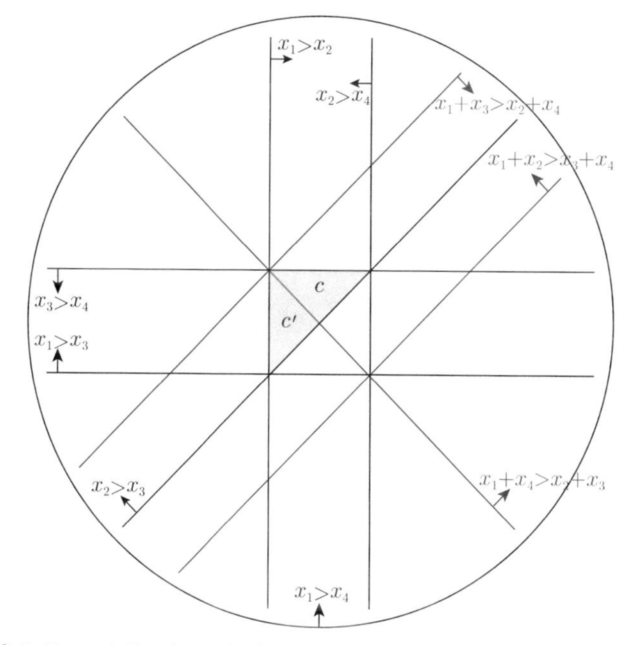

Figure 3.7. Essentialization of mid-hyperplane arrangement.

Let us take $a \in \mathbf{Ch}(\mathcal{A})$ defined by $x_1 > x_2 > x_3 > x_4$ (a is shaded in Figure 3.7). Then $\varphi_{\mathcal{A}}^{-1}(a) = \{c, c'\}$, where

$$c : x_2 > x_3,\ x_3 > x_4,\ x_1 + x_4 > x_2 + x_3,$$

$$c' : x_3 < x_2,\ x_2 < x_1,\ x_4 + x_1 < x_3 + x_2.$$

Note that c' is obtained from c by changing (x_1, x_2, x_3, x_4) to $(-x_4, -x_3, -x_2, -x_1)$. We have $|\varphi_{\mathcal{A}}^{-1}(a)| = |\{W\text{-orbits of } \mathbf{Ch}(\mathcal{C})\}| = 2$ and $|\mathbf{Ch}(\mathcal{C})| = |\varphi_{\mathcal{A}}^{-1}(a)| \times |W| = 2 \times 4! = 48$.

The chamber $b := \varphi_{\mathcal{B}}(c) \in \mathbf{Ch}(\mathcal{B})$ is

$$b : x_1 + x_2 > x_3 + x_4,\ x_1 + x_3 > x_2 + x_4,\ x_1 + x_4 > x_2 + x_3, \tag{3.12}$$

which is divided by the three planes $x_2 = x_3,\ x_2 = x_4,\ x_3 = x_4$ into six chambers (Figure 3.8). We find the W_b-invariant points z of b to be $z = d(3, -1, -1, -1)^T,\ d > 0$, so $W_b = \mathfrak{S}_{\{2,3,4\}}$ and $|W_b| = 3! = |\varphi_{\mathcal{B}}^{-1}(b)|$.

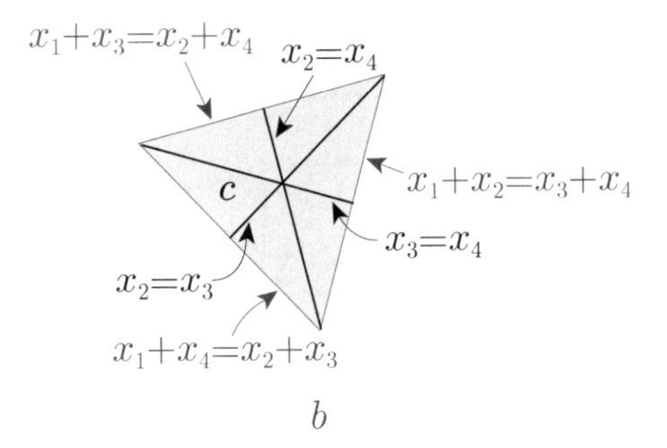

Figure 3.8. Essentialization of mid-hyperplane arrangement. b.

We have $\{W\text{-orbits of } \mathbf{Ch}(\mathcal{B})\} = \{\mathcal{O}(b), \mathcal{O}(b')\}$ with $b' := \varphi_{\mathcal{B}}(c')$. Thus

$$|\{W\text{-orbits of } \mathbf{Ch}(\mathcal{B})\}| = 2.$$

From (3.12), we see $|\mathcal{O}(b)| = 4\ (= |W|/|W_b| = 4!/(3!))$, so we can calculate $|\mathbf{Ch}(\mathcal{C})|$ alternatively as

$$|\mathbf{Ch}(\mathcal{C})| = 2(|\varphi_{\mathcal{B}}^{-1}(b)| \cdot |\mathcal{O}(b)|) = 2 \times 6 \cdot 4 = 48.$$

Moreover, we get $|\mathbf{Ch}(\mathcal{B})| = 2|\mathcal{O}(b)| = 2 \times 4 = 8$.

For $m \leq 10$, the characteristic polynomials of $\mathcal{C}$ are known ([7, 12]), so we can find the numbers of W-orbits of $\mathbf{Ch}(\mathcal{B})$:

$$
\begin{aligned}
m = 4 : \chi(\mathcal{C}, t) &= (t-1)(t-3)(t-5),\\
|\mathbf{Ch}(\mathcal{C})| &= 48, \quad |\varphi_{\mathcal{A}}^{-1}(a)| = 2,\\
m = 5 : \chi(\mathcal{C}, t) &= (t-1)(t-7)(t-8)(t-9),\\
|\mathbf{Ch}(\mathcal{C})| &= 1440, \quad |\varphi_{\mathcal{A}}^{-1}(a)| = 12,\\
m = 6 : \chi(\mathcal{C}, t) &= (t-1)(t-13)(t-14)(t-15)(t-17),\\
|\mathbf{Ch}(\mathcal{C})| &= 120960, \quad |\varphi_{\mathcal{A}}^{-1}(a)| = 168,\\
m = 7 : \chi(\mathcal{C}, t) &= (t-1)(t-23)(t-24)(t-25)(t-26)(t-27),\\
|\mathbf{Ch}(\mathcal{C})| &= 23587200, \quad |\varphi_{\mathcal{A}}^{-1}(a)| = 4680,\\
m = 8 : \chi(\mathcal{C}, t) &= (t-1)(t-35)(t-37)(t-39)(t-41)(t^2-85t+1926),\\
|\mathbf{Ch}(\mathcal{C})| &= 9248117760, \quad |\varphi_{\mathcal{A}}^{-1}(a)| = 229386,\\
m = 9 : \chi(\mathcal{C}, t) &= (t-1)(t^7 - 413t^6 + 73780t^5 - 7387310t^4\\
&\quad + 447514669t^3 - 16393719797t^2\\
&\quad + 336081719070t - 2972902161600),\\
|\mathbf{Ch}(\mathcal{C})| &= 6651665153280, \quad |\varphi_{\mathcal{A}}^{-1}(a)| = 18330206,\\
m = 10 : \chi(\mathcal{C}, t) &= (t-1)(t^8 - 674t^7 + 201481t^6 - 34896134t^5\\
&\quad + 3830348179t^4 - 272839984046t^3\\
&\quad + 12315189583899t^2 - 321989533359786t\\
&\quad + 3732690616086600),\\
|\mathbf{Ch}(\mathcal{C})| &= 8134544088921600,\\
|\varphi_{\mathcal{A}}^{-1}(a)| &= 2241662282.
\end{aligned}
$$

3.5 Signed all-subset arrangement

Let W be the Coxeter group of type B_m, *i.e.*, the semidirect product of $\mathfrak{S}_m$ by $(\mathbb{Z}/2\mathbb{Z})^m$: $W = (\mathbb{Z}/2\mathbb{Z})^m \rtimes \mathfrak{S}_m$, $|W| = 2^m \cdot m!$. Then W acts on $V = \mathbb{R}^m$ by permuting coordinates by $\mathfrak{S}_m$ and changing signs of coordinates by $(\mathbb{Z}/2\mathbb{Z})^m$. The Coxeter arrangement $\mathcal{A} = \mathcal{A}(W)$ consists of the hyperplanes defined by

$$x_i = 0, \quad 1 \leq i \leq m; \tag{3.13}$$

$$x_i + x_j = 0,\ x_i - x_j = 0, \quad 1 \leq i < j \leq m. \tag{3.14}$$

We have $|\mathcal{A}| = m + m(m-1) = m^2$. Moreover, the number of orbits of $\mathcal{A}$ under the action of W is two: one consisting of the m hyperplanes in (3.13) and the other made up of the $m(m-1)$ hyperplanes in (3.14).

Let

$$\mathcal{B} = \{H_{(\epsilon_1,\dots,\epsilon_m)} \mid \epsilon_1, \dots, \epsilon_m \in \{-1, 0, 1\}, \sum_{i=1}^{m} |\epsilon_i| \geq 3\},$$

where

$$H_{(\epsilon_1,\dots,\epsilon_m)} := \{x = (x_1, \dots, x_m)^T \in \mathbb{R}^m \mid \sum_{i=1}^{m} \epsilon_i x_i = 0\}.$$

Note $H_{(\epsilon_1,\dots,\epsilon_m)} = H_{(-\epsilon_1,\dots,-\epsilon_m)}$ so that $|\mathcal{B}| = \sum_{i=3}^{m} 2^{i-1}\binom{m}{i}$. The number of W-orbits of $\mathcal{B}$ is $m - 2$.

Let us study the case $m = 3$.
In this case, $\mathcal{A}$ comprises the nine planes in $V = \mathbb{R}^3$ defined by

$$x_1 = 0,\ x_2 = 0,\ x_3 = 0; \tag{3.15}$$

$$\begin{aligned} &x_1 + x_2 = 0, \quad x_1 - x_2 = 0, \quad x_1 + x_3 = 0, \\ &x_1 - x_3 = 0, \quad x_2 + x_3 = 0, \quad x_2 - x_3 = 0 \end{aligned} \tag{3.16}$$

with each of (3.15) and (3.16) corresponding to one orbit, and $\mathcal{B}$ consists of one orbit containing the four planes defined by

$$-x_1+x_2+x_3=0,\ x_1-x_2+x_3=0,\ x_1+x_2-x_3=0,\ x_1+x_2+x_3=0. \tag{3.17}$$

Figure 3.9 shows the intersection with the unit sphere in $V = \mathbb{R}^3$. (In Figure 3.9, the planes in (3.15), (3.16) and (3.17) are drawn in blue, black and purple, respectively.)

Let us take $a \in \mathbf{Ch}(\mathcal{A})$ defined by $x_1 > x_2 > x_3 > 0$ (this chamber is shaded in Figure 3.9). Then $\varphi_{\mathcal{A}}^{-1}(a) = \{c_1, c_2\}$, where

$$\begin{aligned} &c_1 : x_1 - x_2 > 0,\ x_2 - x_3 > 0,\ -x_1 + x_2 + x_3 > 0, \\ &c_2 : x_3 > 0,\ x_2 - x_3 > 0,\ -x_1 + x_2 + x_3 < 0. \end{aligned}$$

So $|\varphi_{\mathcal{A}}^{-1}(a)| = |\{W\text{-orbits of } \mathbf{Ch}(\mathcal{C})\}| = 2$ and $|\mathbf{Ch}(\mathcal{C})| = |\varphi_{\mathcal{A}}^{-1}(a)| \times |W| = 2 \times 2^3 \cdot 3! = 96$.

The chambers $b_i := \varphi_{\mathcal{B}}(c_i) \in \mathbf{Ch}(\mathcal{B}),\ i = 1, 2$, are

$$\begin{aligned} &b_1 : -x_1 + x_2 + x_3 > 0,\ x_1 - x_2 + x_3 > 0,\ x_1 + x_2 - x_3 > 0, \\ &b_2 : -x_1 + x_2 + x_3 < 0,\ x_1 - x_2 + x_3 > 0,\ x_1 + x_2 - x_3 > 0, \\ &\qquad x_1 + x_2 + x_3 > 0 \end{aligned}$$

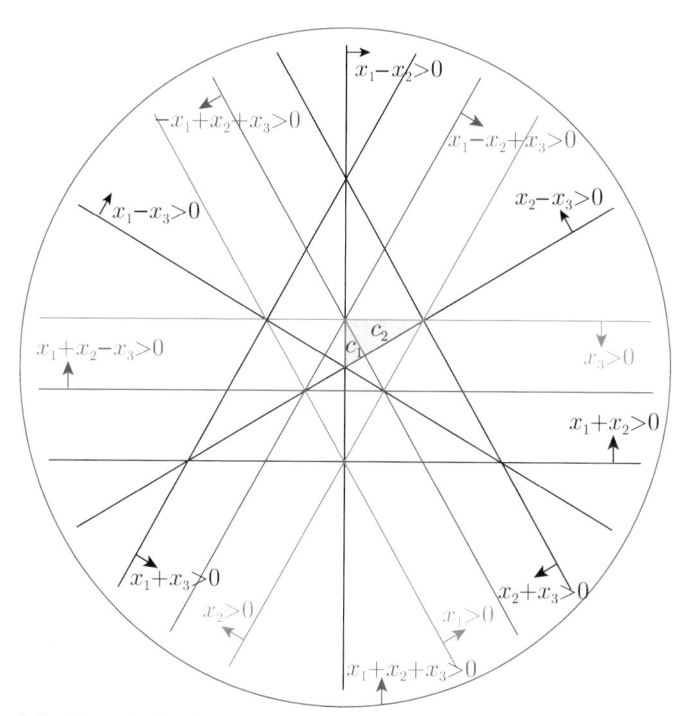

Figure 3.9. Signed all-subset arrangement.

(Figure 3.10). The chamber b_1 is divided by the three planes $x_1 - x_2 = 0,\ x_1 - x_3 = 0,\ x_2 - x_3 = 0$ into six chambers, and b_2 is divided by the four planes $x_2 = 0,\ x_3 = 0,\ x_2 + x_3 = 0,\ x_2 - x_3 = 0$ into eight. We see $W_{b_1} = \mathfrak{S}_{\{1,2,3\}}$ is the Coxeter group of type A_2 (the W_{b_1}-invariant points z of b_1 are $z = d(1,1,1)^T,\ d > 0$), and that $W_{b_2} = (\mathbb{Z}/2\mathbb{Z})^2 \rtimes \mathfrak{S}_{\{2,3\}}$ is the Coxeter group of type B_2 (the W_{b_2}-invariant points z of b_2 are $z = d(1,0,0)^T,\ d > 0$). Hence $|W_{b_1}| = 3! = |\varphi_{\mathcal{B}}^{-1}(b_1)|,\ |W_{b_2}| = 2^2 \cdot 2! = |\varphi_{\mathcal{B}}^{-1}(b_2)|$.

We have $\{W\text{-orbits of } \mathbf{Ch}(\mathcal{B})\} = \{\mathcal{O}(b_1), \mathcal{O}(b_2)\}$, so

$$|\{W\text{-orbits of } \mathbf{Ch}(\mathcal{B})\}| = 2.$$

From Figures 3.9 and 3.10, we see $|\mathcal{O}(b_1)| = 4 \times 2 = 8\ (= |W|/|W_{b_1}| = (2^3 \cdot 3!)/(3!))$, $|\mathcal{O}(b_2)| = 3 \times 2 = 6\ (= |W|/|W_{b_2}| = (2^3 \cdot 3!)/(2^2 \cdot 2!))$, so $|\mathbf{Ch}(\mathcal{C})|$ can be computed also as

$$|\mathbf{Ch}(\mathcal{C})| = |\varphi_{\mathcal{B}}^{-1}(b_1)| \cdot |\mathcal{O}(b_1)| + |\varphi_{\mathcal{B}}^{-1}(b_2)| \cdot |\mathcal{O}(b_2)| = 6 \cdot 8 + 8 \cdot 6 = 96.$$

Furthermore, we can get $|\mathbf{Ch}(\mathcal{B})| = |\mathcal{O}(b_1)| + |\mathcal{O}(b_2)| = 8 + 6 = 14$.

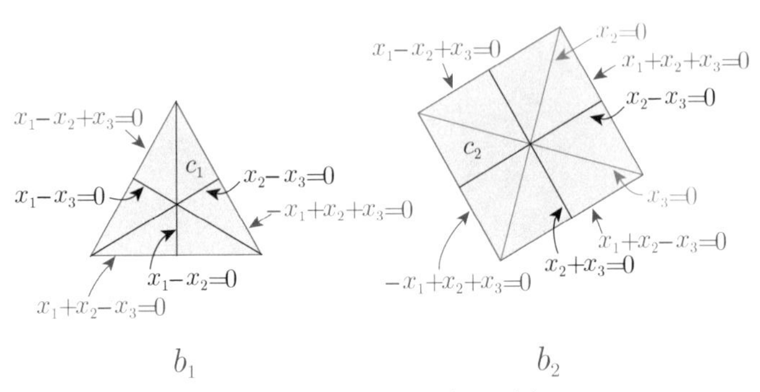

Figure 3.10. Signed all-subset arrangement. b_1 and b_2.

For $m \leq 6$, we computed $\chi(\mathcal{C}, t)$ and obtained the numbers of W-orbits of $\mathbf{Ch}(\mathcal{B})$ as follows:

$$
\begin{aligned}
m = 3: \quad & \chi(\mathcal{C}, t) = (t-1)(t-5)(t-7), \\
& \qquad |\mathbf{Ch}(\mathcal{C})| = 96, \quad |\varphi_{\mathcal{A}}^{-1}(a)| = 2, \\
m = 4: \quad & \chi(\mathcal{C}, t) = (t-1)(t-11)(t-13)(t-15), \\
& \qquad |\mathbf{Ch}(\mathcal{C})| = 5376, \quad |\varphi_{\mathcal{A}}^{-1}(a)| = 14, \\
m = 5: \quad & \chi(\mathcal{C}, t) = (t-1)(t-29)(t-31)(t^2-60t+971), \\
& \qquad |\mathbf{Ch}(\mathcal{C})| = 1981440, \quad |\varphi_{\mathcal{A}}^{-1}(a)| = 516, \\
m = 6: \quad & \chi(\mathcal{C}, t) = (t-1)(t^5-363t^4+54310t^3-4182690t^2 \\
& \qquad\qquad + 165591769t - 2691439347), \\
& \qquad |\mathbf{Ch}(\mathcal{C})| = 5722536960, \quad |\varphi_{\mathcal{A}}^{-1}(a)| = 124187.
\end{aligned}
$$

Note that the characteristic polynomial $\chi(\mathcal{C}, t)$ factors into polynomials of degree one over $\mathbb{Z}$ for $m \leq 4$.

References

[1] C. A. ATHANASIADIS, "Algebraic Combinatorics of Graph Spectra, Subspace Arrangements and Tutte Polynomials", Ph.D. thesis, MIT, 1996.

[2] C. A. ATHANASIADIS, *Characteristic polynomials of subspace arrangements and finite fields*, Adv. Math. **122** (1996) 193–233.

[3] N. BOURBAKI, "Groupes et Algèbres de Lie", Chapitres 4, 5 et 6, Hermann, Paris, 1968.

[4] J. L. CHANDON, J. LEMAIRE and J. POUGET, *Dénombrement des quasi-ordres sur un ensemble fini*, Math. Sci. Humaines **62** (1978), 61–80.

[5] H. CRAPO and G.-C. ROTA, "On the Foundations of Combinatorial Theory: Combinatorial Geometries", preliminary edition, MIT Press, Cambridge, MA, 1970.

[6] R. A. DEAN and G. KELLER, *Natural partial orders*, Canad. J. Math. **20** (1968), 535–554.

[7] H. KAMIYA, P. ORLIK, A. TAKEMURA and H. TERAO, *Arrangements and ranking patterns*, Ann. Comb. **10** (2006), 219–235.

[8] H. KAMIYA, A. TAKEMURA and H. TERAO, *Periodicity of hyperplane arrangements with integral coefficients modulo positive integers*, J. Algebraic Combin. **27** (2008) 317–330.

[9] H. KAMIYA, A. TAKEMURA and H. TERAO, *The characteristic quasi-polynomials of the arrangements of root systems and mid-hyperplane arrangements*, In: F. El Zein, A. I. Suciu, M. Tosun, A. M. Uludağ, S. Yuzvinsky, (Eds.), "Arrangements, Local Systems and Singularities", CIMPA Summer School, Galatasaray University, Istanbul, 2007, Progress in Mathematics 283, Birkhäuser Verlag, Basel, 2009, pp. 177–190.

[10] H. KAMIYA, A. TAKEMURA and H. TERAO, *Periodicity of non-central integral arrangements modulo positive integers*, Ann. Comb. **15** (2011), 449–464.

[11] H. KAMIYA, A. TAKEMURA and H. TERAO, *Ranking patterns of unfolding models of codimension one*, Adv. in Appl. Math. **47** (2011), 379–400.

[12] H. KAMIYA, A. TAKEMURA and N. TOKUSHIGE, *Application of arrangement theory to unfolding models*, In: "Arrangements of Hyperplanes (Sapporo 2009)", H. Terao, S. Yuzvinsky (eds.), Proceedings of the Second MSJ-SI, Hokkaido University, Sapporo, 2009, Advanced Studies in Pure Math., 62, Math. Soc. of Japan, Tokyo, 2012, 399–415.

[13] R. D. LUCE, *Semiorders and a theory of utility discrimination*, Econometrica **24** (1956), 178–191.

[14] D. R. MAZUR, *Combinatorics: A Guided Tour*, Mathematical Association of America, Washington, D.C., 2010.

[15] P. ORLIK and H. TERAO, "Arrangements of Hyperplanes", Springer-Verlag, Berlin, 1992.

[16] S. OVCHINNIKOV, *Hyperplane arrangements in preference modeling*, J. Math. Psych. **49** (2005), 481–488.

[17] S. OVCHINNIKOV, "Graphs and Cubes", Springer Science+Business Media, New York, NY, 2011.

[18] A. POSTNIKOV and R. P. STANLEY, *Deformations of Coxeter hyperplane arrangements*, J. Combin. Theory Ser. A **91** (2000), 544–597.

[19] D. Scott and P. Suppes, *Foundational aspects of theories of measurement*, J. Symbolic Logic **23** (1958), 113–128.
[20] R. P. Stanley, "Enumerative Combinatorics", Vol. 1, 2nd ed., Cambridge University Press, Cambridge, 2012.
[21] R. P. Stanley, "Enumerative Combinatorics", Vol. 2, Cambridge University Press, Cambridge, 1999.
[22] R. P. Stanley, *An introduction to hyperplane arrangements*, In: E. Miller, V. Reiner, B. Sturmfels, (Eds.), "Geometric Combinatorics, IAS/Park City Mathematics Series 13", American Mathematical Society, Providence, RI, 2007, pp. 389–496.
[23] R. L. Wine and J. E. Freund, *On the enumeration of decision patterns involving n means*, Ann. Math. Statist. **28** (1957), 256–259.
[24] T. Zaslavsky, *Facing up to arrangements: Face-count formulas for partitions of space by hyperplanes*, Mem. Amer. Math. Soc. 1, no. 154 (1975).

Quantum and homological representations of braid groups

Toshitake Kohno

Abstract. By means of a description of the solutions of the KZ equation using hypergeometric integrals we show that the homological representations of the braid groups studied by Lawrence, Krammer and Bigelow are equivalent at generic complex values to the monodromy of the KZ equation with values in the space of null vectors in the tensor product of Verma modules of $sl_2(\mathbf{C})$.

1 Introduction

The purpose of this paper is to clarify the relation between the Lawrence-Krammer-Bigelow (LKB) representations of the braid groups and the monodromy representations of the Knizhnik-Zamolodchikov (KZ) connection.

The LKB representations of the braid groups were studied by Lawrence [13] in relation with Hecke algebra representations of the braid groups and were extensively investigated by Bigelow [3] and Krammer [12].

On the other hand, it was shown by Schechtman-Varchenko [16] and others that the solutions of the KZ equation are expressed by hypergeometric integrals. From the expression of the integrals over homology cycles with coefficients in local systems it is clear that the LKB representation can be expressed as the monodromy representation of the KZ equation, however, I think it is worthwhile to state a precise relation between them.

There are two parameters λ and κ, which are related to the highest weight and the KZ connection respectively. We consider the KZ equation with values in the space of null vectors in the tensor product of Verma modules of $sl_2(\mathbf{C})$ and show that a specialization of the LKB representa-

The author is partially supported by Granr-in-Aid for Scientific Research, Japan Society of Promotion of Science and by World Premier Research Center Initiative, MEXT, Japan.

tion is equivalent to the monodromy representation of such KZ equation for generic parameters λ and κ. A complete statement is given in Theorem 6.1. We describe a sufficient condition for the parameters to be generic so that the statement of the theorem holds. This result was announced in [11] and the present paper is a more detailed account for this subject.

There is other approach due to Marin [14] expressing representations of the braid groups and their generalizations such as Artin groups as the monodromy of integrable connections by an infinitesimal method. Our approach depends on integral representations of the solutions of the KZ equation and is different from Marin's method.

In this article we will treat the case when the parameters are generic, but the case of special parameters are important from the viewpoint of conformal field theory (see [7, 17] and [19]). We will deal with this subject in a separate paper.

The paper is organized in the following way. In Section 2 we recall basic definitions for the LKB representations. In Section 3 we deal with the homology of local systems over the complement of a discriminantal arrangement. We recall the definition of the KZ equation in Section 4 and describe its solutions by hypergeometric integrals in Section 5. Section 6 is devoted to the statement of the main theorem and its proof.

ACKNOWLEDGEMENTS. The author would like to thank Ivan Marin for useful discussions.

2 Lawrence-Krammer-Bigelow representations

We denote by B_n the braid group with n strands. We fix a positive integer n and a set of distinct n points in $\mathbf{R}^2$ as

$$Q = \{(1,0), \cdots, (n,0)\},$$

where we set $p_\ell = (\ell, 0)$, $\ell = 1, \cdots, n$. We take a 2-dimensional disk in $\mathbf{R}^2$ containing Q in the interior. We fix a positive integer m and consider the configuration space of ordered distinct m points in $\Sigma = D \setminus Q$ defined by

$$\mathcal{F}_m(\Sigma) = \{(t_1, \cdots, t_m) \in \Sigma \ ; \ t_i \neq t_j \text{ if } i \neq j\},$$

which is also denoted by $\mathcal{F}_{n,m}(D)$. The symmetric group $\mathfrak{S}_m$ acts freely on $\mathcal{F}_m(\Sigma)$ by the permutations of distinct m points. The quotient space of $\mathcal{F}_m(\Sigma)$ by this action is by definition the configuration space of unordered distinct m points in Σ and is denoted by $\mathcal{C}_m(\Sigma)$. We also denote this configuration space by $\mathcal{C}_{n,m}(D)$.

In the original papers by Bigelow [3,4] and by Krammer [12] the case $m = 2$ was extensively studied, but for our purpose it is convenient to consider the case when m is an arbitrary positive integer such that $m \geq 2$.

We identify $\mathbf{R}^2$ with the complex plane $\mathbf{C}$. The quotient space $\mathbf{C}^m/\mathfrak{S}_m$ defined by the action of $\mathfrak{S}_m$ by the permutations of coordinates is analytically isomorphic to $\mathbf{C}^m$ by means of the elementary symmetric polynomials. Now the image of the hyperplanes defined by $t_i = p_\ell$, $\ell = 1, \cdots, n$, and the diagonal hyperplanes $t_i = t_j$, $1 \leq i \leq j \leq m$, are complex codimension one irreducible subvarieties of the quotient space $D^m/\mathfrak{S}_m$. This allows us to give a description of the first homology group of $C_{n,m}(D)$ as

$$H_1(C_{n,m}(D); \mathbf{Z}) \cong \mathbf{Z}^{\oplus n} \oplus \mathbf{Z} \tag{2.1}$$

where the first n components correspond to meridians of the images of hyperplanes $t_i = p_\ell$, $\ell = 1, \cdots, n$, and the last component corresponds to the meridian of the image of the diagonal hyperplanes $t_i = t_j$, $1 \leq i \leq j \leq m$, namely, the discriminant set. We consider the homomorphism

$$\alpha : H_1(C_{n,m}(D); \mathbf{Z}) \longrightarrow \mathbf{Z} \oplus \mathbf{Z} \tag{2.2}$$

defined by $\alpha(x_1, \cdots, x_n, y) = (x_1 + \cdots + x_n, y)$. Composing with the abelianization map $\pi_1(C_{n,m}(D), x_0) \rightarrow H_1(C_{n,m}(D); \mathbf{Z})$, we obtain the homomorphism

$$\beta : \pi_1(C_{n,m}(D), x_0) \longrightarrow \mathbf{Z} \oplus \mathbf{Z}. \tag{2.3}$$

Let $\pi : \widetilde{C}_{n,m}(D) \rightarrow C_{n,m}(D)$ be the covering corresponding to $\operatorname{Ker} \beta$. Now the group $\mathbf{Z} \oplus \mathbf{Z}$ acts as the deck transformations of the covering π and the homology group $H_*(\widetilde{C}_{n,m}(D); \mathbf{Z})$ is considered to be a $\mathbf{Z}[\mathbf{Z} \oplus \mathbf{Z}]$-module. Here $\mathbf{Z}[\mathbf{Z} \oplus \mathbf{Z}]$ stands for the group ring of $\mathbf{Z} \oplus \mathbf{Z}$. We express $\mathbf{Z}[\mathbf{Z} \oplus \mathbf{Z}]$ as the ring of Laurent polynomials $R = \mathbf{Z}[q^{\pm 1}, t^{\pm 1}]$. We consider the homology group

$$H_{n,m} = H_m(\widetilde{C}_{n,m}(D); \mathbf{Z})$$

as an R-module by the action of the deck transformations.

As is explained in the case of $m = 2$ in [3] it can be shown that $H_{n,m}$ is a free R-module of rank

$$d_{n,m} = \binom{m+n-2}{m}. \tag{2.4}$$

A basis of $H_{n,m}$ as a free R-module is discussed in relation with the homology of local systems in the next sections. Let $\mathcal{M}(D, Q)$ denote the mapping class group of the pair (D, Q), which consists of the isotopy classes of homeomorphisms of D fixing Q setwise and fixing the boundary ∂D pointwise. The braid group B_n is naturally isomorphic to the mapping class group $\mathcal{M}(D, Q)$. Now a homeomorphism f representing a class in $\mathcal{M}(D, Q)$ induces a homeomorphism $\tilde{f} : \mathcal{C}_{n,m}(D) \to \mathcal{C}_{n,m}(D)$, which is uniquely lifted to a homeomorphism of $\widetilde{\mathcal{C}}_{n,m}(D)$. This homeomorpshim commutes with the deck transformations.

Therefore, for $m \geq 2$ we obtain a representation of the braid group

$$\rho_{n,m} : B_n \longrightarrow \mathrm{Aut}_R\, H_{n,m} \tag{2.5}$$

which is called the homological representation of the braid group or the Lawrence-Krammer-Bigelow (LKB) representation. Let us remark that in the case $m = 1$ the above construction gives the reduced Burau representation over $\mathbf{Z}[q^{\pm 1}]$.

3 Discriminantal arrangements

First, we recall some basic definition for local systems. Let M be a smooth manifold and V a complex vector space. Given a linear representation of the fundamental group

$$r : \pi_1(M, x_0) \longrightarrow GL(V)$$

there is an associated flat vector bundle E over M. The local system $\mathcal{L}$ associated to the representation r is the sheaf of horizontal sections of the flat bundle E. Let $\pi : \widetilde{M} \to M$ be the universal covering. We denote by $\mathbf{Z}\pi_1$ the group ring of the fundamental group $\pi_1(M, x_0)$. We consider the chain complex

$$C_*(\widetilde{M}) \otimes_{\mathbf{Z}\pi_1} V$$

with the boundary map defined by $\partial(c \otimes v) = \partial c \otimes v$. Here $\mathbf{Z}\pi_1$ acts on $C_*(\widetilde{M})$ via the deck transformations and on V via the representation r. The homology of this chain complex is called the homology of M with coefficients in the local system $\mathcal{L}$ and is denoted by $H_*(M, \mathcal{L})$.

Let $\mathcal{A} = \{H_1, \cdots, H_N\}$ be a set of affine hyperplanes in the complex vector space $\mathbf{C}^n$. We call the set $\mathcal{A}$ a complex hyperplane arrangement. We consider the complement

$$M(\mathcal{A}) = \mathbf{C}^n \setminus \bigcup_{H \in \mathcal{A}} H.$$

Let $\mathcal{L}$ be a complex rank one local system over $M(\mathcal{A})$ associated with a representation of the fundamental group

$$r : \pi_1(M(\mathcal{A}), x_0) \longrightarrow \mathbf{C}^*.$$

We shall investigate the homology of $M(\mathcal{A})$ with coefficients in the local system $\mathcal{L}$. For our purpose the homology of locally finite chains $H_*^{lf}(M(\mathcal{A}), \mathcal{L})$ also plays an important role.

We briefly summarize basic properties of the above homology groups. For a complex hyperplane arrangement $\mathcal{A}$ we choose a smooth compactification $i : M(\mathcal{A}) \longrightarrow X$ with normal crossing divisors. We shall say that the local system $\mathcal{L}$ is generic if and only if there is an isomorphism

$$i_*\mathcal{L} \cong i_!\mathcal{L} \tag{3.1}$$

holds, where i_* is the direct image and $i_!$ is the extension by 0. This means that the monodromy of $\mathcal{L}$ along any divisor at infinity is not equal to 1. The following theorem was shown in [9].

Theorem 3.1. *If the local system $\mathcal{L}$ is generic in the above sense, then there is an isomorphism*

$$H_*(M(\mathcal{A}), \mathcal{L}) \cong H_*^{lf}(M(\mathcal{A}), \mathcal{L}).$$

Moreover, we have $H_k(M(\mathcal{A}), \mathcal{L}) = 0$ for any $k \neq n$.

Proof. In general we have isomorphisms

$$H^*(X, i_*\mathcal{L}) \cong H^*(M(\mathcal{A}), \mathcal{L}), \quad H^*(X, i_!\mathcal{L}) \cong H_c^*(M(\mathcal{A}), \mathcal{L})$$

where H_c denotes cohomology with compact supports.

There are Poincaré duality isomorphisms:

$$H_k^{lf}(M(\mathcal{A}), \mathcal{L}) \cong H^{2n-k}(M(\mathcal{A}), \mathcal{L})$$
$$H_k(M(\mathcal{A}), \mathcal{L}) \cong H_c^{2n-k}(M(\mathcal{A}), \mathcal{L}).$$

By the hypothesis $i_*\mathcal{L} \cong i_!\mathcal{L}$ we obtain an isomorphism

$$H_k^{lf}(M(\mathcal{A}), \mathcal{L}) \cong H_k(M(\mathcal{A}), \mathcal{L}).$$

It follows from the above Poincaré duality isomorpshims and the fact that $M(\mathcal{A})$ has a homotopy type of a CW complex of dimension at most n we have

$$H_k^{lf}(M(\mathcal{A}), \mathcal{L}) \cong 0, \quad k < n$$
$$H_k(M(\mathcal{A}), \mathcal{L}) \cong 0, \quad k > n.$$

Therefore we obtain $H_k(M(\mathcal{A}), \mathcal{L}) = 0$ for any $k \neq n$. □

Let us consider the configuration space of ordered distinct n points in the complex plane defined by

$$X_n = \{(z_1, \cdots, z_n) \in \mathbf{C}^n \ ; \ z_i \neq z_j \text{ if } i \neq j\}. \tag{3.2}$$

The fundamental group of X_n is the pure braid group with n strands denoted by P_n. For a positive integer m we consider the projection map

$$\pi_{n,m} : X_{n+m} \longrightarrow X_n \tag{3.3}$$

given by $\pi_{n,m}(z_1, \cdots, z_n, t_1, \cdots, t_m) = (z_1, \cdots, z_n)$, which defines a fiber bundle over X_n. For $p \in X_n$ the fiber $\pi_{n,m}^{-1}(p)$ is denoted by $X_{n,m}$. We denote by $(p_1, \cdots, p_n)$ the coordinates for p. Then, $X_{n,m}$ is the complement of hyperplanes defined by

$$t_i = p_\ell, \quad 1 \leq i \leq m, \ 1 \leq \ell \leq n, \quad t_i = t_j, \quad 1 \leq i < j \leq m. \tag{3.4}$$

Such arrangement of hyperplanes is called a discriminantal arrangement. The symmetric group $\mathfrak{S}_m$ acts on $X_{n,m}$ by the permutations of the coordinates functions $t_1, \cdots, t_m$. We put $Y_{n,m} = X_{n,m}/\mathfrak{S}_m$.

Identifying $\mathbf{R}^2$ with the complex plane $\mathbf{C}$, we have the inclusion map

$$\iota : \mathcal{F}_{n,m}(D) \longrightarrow X_{n,m}, \tag{3.5}$$

which is a homotopy equivalence. By taking the quotient by the action of the symmetric group $\mathfrak{S}_m$, we have the inclusion map

$$\bar{\iota} : \mathcal{C}_{n,m}(D) \longrightarrow Y_{n,m}, \tag{3.6}$$

which is also a homotopy equivalence.

We take $p = (1, 2, \cdots, n)$ as a base point. We consider a local system over $X_{n,m}$ defined in the following way. Let $\xi_{i\ell}$ and η_{ij} be normal loops around the hyperplanes $t_i = p_\ell$ and $t_i = t_j$ respectively. We fix complex numbers α_ℓ, $1 \leq \ell \leq n$, and γ and by the correspondence

$$\xi_{i\ell} \mapsto e^{2\pi\sqrt{-1}\alpha_\ell}, \quad \eta_{ij} \mapsto e^{4\pi\sqrt{-1}\gamma}$$

we obtain the representation

$$r : \pi_1(X_{n,m}, x_0) \longrightarrow \mathbf{C}^*.$$

We denote by $\mathcal{L}$ the associated rank one local system on $X_{n,m}$.

Let us consider the embedding

$$i_0 : X_{n,m} \longrightarrow (\mathbf{C}P^1)^m = \underbrace{\mathbf{C}P^1 \times \cdots \times \mathbf{C}P^1}_{m}. \tag{3.7}$$

Then we take blowing-ups at multiple points $\pi : \widehat{(\mathbf{C}P^1)^m} \longrightarrow (\mathbf{C}P^1)^m$ and obtain a smooth compactification $i : X_{n,m} \to \widehat{(\mathbf{C}P^1)^m}$ with normal crossing divisors. We are able to write down the condition $i_*\mathcal{L} \cong i_!\mathcal{L}$ explicitly by computing the monodromy of the local system $\mathcal{L}$ along divisors at infinity.

Examples. (1) In the case $m = 1$ the local system $\mathcal{L}$ is generic if and only if

$$\alpha_\ell \notin \mathbf{Z}, \quad 1 \leq \ell \leq n, \quad \alpha_1 + \cdots + \alpha_n \notin \mathbf{Z}.$$

(2) In the case $m = 2$ the local system $\mathcal{L}$ is generic if and only if

$$\begin{aligned} &\alpha_\ell \notin \mathbf{Z}, \quad 1 \leq \ell \leq n, \quad 2\gamma \notin \mathbf{Z}, \\ &2(\alpha_\ell + \gamma) \notin \mathbf{Z}, \quad 1 \leq \ell \leq n, \\ &2(\alpha_1 + \cdots + \alpha_n + \gamma) \notin \mathbf{Z}. \end{aligned}$$

The local system $\mathcal{L}$ on $X_{n,m}$ is invariant under the action of the symmetric group $\mathfrak{S}_m$ and induces the local system $\overline{\mathcal{L}}$ on $Y_{n,m}$.

We will deal with the case $\alpha_1 = \cdots = \alpha_\ell = \alpha$. In this case we have the following proposition.

Proposition 3.2. *There is an open dense subset V in $\mathbf{C}^2$ such that for $(\alpha, \gamma) \in V$ the associated local system $\overline{\mathcal{L}}$ on $Y_{n,m}$ satisfies*

$$H_*(Y_{n,m}, \overline{\mathcal{L}}) \cong H_*^{lf}(Y_{n,m}, \overline{\mathcal{L}})$$

and $H_k(Y_{n,m}, \overline{\mathcal{L}}) = 0$ for any $k \neq m$. Moreover, we have

$$\dim H_m(Y_{n,m}, \overline{\mathcal{L}}^*) = d_{n,m}, \tag{3.8}$$

where we use the same notation as in equation (2.4) for $d_{n,m}$.

Proof. We see that $Y_{n,m}$ is the complement of hypersurfaces in $\mathbf{C}^m$. We consider the embedding

$$i_0 : Y_{n,m} \longrightarrow \mathcal{S}^m \mathbf{C}P^1 \tag{3.9}$$

where $\mathcal{S}^m \mathbf{C}P^1$ is the symmetric product defined as $(\mathbf{C}P^1)^m / \mathfrak{S}_m$. We observe that $\mathcal{S}^m \mathbf{C}P^1$ is a smooth complex manifold. Now by taking blowing-ups we have a smooth compactification

$$i : Y_{n,m} \longrightarrow \widehat{\mathcal{S}^m \mathbf{C}P^1} \tag{3.10}$$

with normal crossing divisors. Let us remark that the argument of the proof of Theorem 3.1 can be applied to this situation and we have an isomorphism $H_*(Y_{n,m}, \overline{\mathcal{L}}) \cong H_*^{lf}(Y_{n,m}, \overline{\mathcal{L}})$ and the vanishing $H_k(Y_{n,m}, \mathcal{L}) = 0$ for $k \neq m$ if the condition $i_*\overline{\mathcal{L}} \cong i_!\overline{\mathcal{L}}$ is satisfied. Actually, by the Lefschetz hyperplane section theorem it is enough to verify the condition for a generic 2 dimensional section. In this case by expressing the monodromy along divisors with normal crossings at infinity by the parameter (α, γ) we can verify that the condition $i_*\overline{\mathcal{L}} \cong i_!\overline{\mathcal{L}}$ is satisfied for $(\alpha, \gamma) \in \mathbf{C}^2$ in an open dense subset of $\mathbf{C}^2$. The dimension formula for $H_m(Y_{n,m}, \overline{\mathcal{L}}^*)$ follows from the calculation of the Euler-Poincaré characteristic of $Y_{n,m}$. □

Remark. It was shown by Andreotti [1] that the above m-fold symmetric product $\mathcal{S}^m\mathbf{C}P^1$ is actually biholomorphically equivalent to $\mathbf{C}P^m$.

For the purpose of describing the homology group $H_m^{lf}(X_{n,m}, \mathcal{L})$ and $H_m^{lf}(Y_{n,m}, \overline{\mathcal{L}})$ we introduce the following notation. We fix the base point $p = (1, \cdots, n)$. For non-negative integers $m_1, \cdots, m_{n-1}$ satisfying

$$m_1 + \cdots + m_{n-1} = m \tag{3.11}$$

we define a bounded chamber $\Delta_{m_1,\cdots,m_{n-1}}$ in $\mathbf{R}^m$ by

$$\begin{aligned}
&1 < t_1 < \cdots < t_{m_1} < 2 \\
&2 < t_{m_1+1} < \cdots < t_{m_1+m_2} < 3 \\
&\cdots \\
&n-1 < t_{m_1+\cdots+m_{n-2}+1} + \cdots + t_m < n.
\end{aligned}$$

We put $M = (m_1, \cdots, m_{n-1})$ and we write Δ_M for $\Delta_{m_1,\cdots,m_{n-1}}$. We denote by $\overline{\Delta}_M$ the image of Δ_M by the projection map $\pi_{n,m}$. The bounded chamber Δ_M defines a homology class $[\Delta_M] \in H_m^{lf}(X_{n,m}, \mathcal{L})$ and its image $\overline{\Delta}_M$ defines a homology class $[\overline{\Delta}_M] \in H_m^{lf}(Y_{n,m}, \overline{\mathcal{L}})$. We shall show in Section 6 that under certain generic conditions $[\overline{\Delta}_M]$ for $M = (m_1, \cdots, m_{n-1})$ with $m_1 + \cdots + m_{n-1} = m$ form a basis of $H_m^{lf}(Y_{n,m}, \overline{\mathcal{L}})$. As we have shown in Theorem 3.1 there is an isomorphism $H_m(X_{n,m}, \mathcal{L}) \cong H_m^{lf}(X_{n,m}, \mathcal{L})$ if the condition $i_*\mathcal{L} \cong i_!\mathcal{L}$ is satisfied. In this situation we denote by $[\widetilde{\Delta}_M]$ the homology class in $H_m(X_{n,m}, \mathcal{L})$ corresponding to $[\Delta_M]$ in the above isomorphism and call $[\widetilde{\Delta}_M]$ the regularized cycle for $[\Delta_M]$.

Example. Let us consider the case $n = 2$, $m = 1$. The bounded chamber Δ_1 is the open unit interval $(0, 1)$. We suppose the condition $i_*\mathcal{L} \cong i_!\mathcal{L}$

is satisfied. The Pochhammer double loop Γ as depicted in Figure 3.1 is related to $[\widetilde{\Delta}_1]$ by

$$[\widetilde{\Delta}_1] = \frac{1}{(1 - e^{2\pi\sqrt{-1}\alpha_1})(1 - e^{2\pi\sqrt{-1}\alpha_2})}[\Gamma].$$

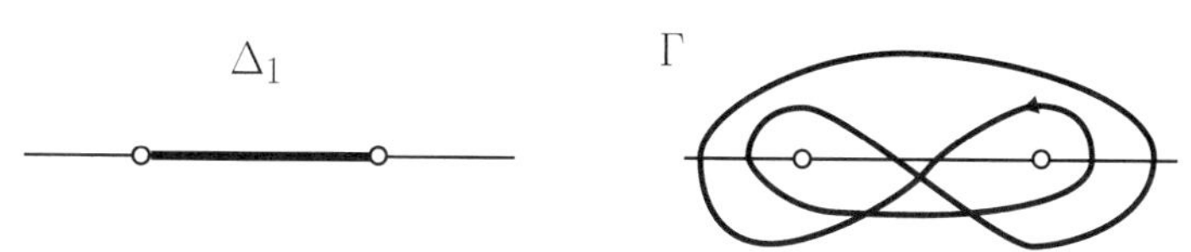

Figure 3.1. Pochhammer double loop.

In general regularized cycles can be constructed by means of the boundary of the tubular neighborhood of divisors at infinity. We refer the reader to [2] for more details about this subject.

4 KZ connection

Let $\mathfrak{g}$ be a complex semi-simple Lie algebra and $\{I_\mu\}$ be an orthonormal basis of $\mathfrak{g}$ with respect to the Cartan-Killing form. We set $\Omega = \sum_\mu I_\mu \otimes I_\mu$. Let $r_i : \mathfrak{g} \to \mathrm{End}(V_i)$, $1 \leq i \leq n$, be representations of the Lie algebra $\mathfrak{g}$. We denote by Ω_{ij} the action of Ω on the i-th and j-th components of the tensor product $V_1 \otimes \cdots \otimes V_n$. It is known that the Casimir element $c = \sum_\mu I_\mu \cdot I_\mu$ lies in the center of the universal enveloping algebra $U\mathfrak{g}$. Let us denote by $\Delta : U\mathfrak{g} \to U\mathfrak{g} \otimes U\mathfrak{g}$ be the coproduct, which is defined to be the algebra homomorphism determined by $\Delta(x) = x \otimes 1 + 1 \otimes x$ for $x \in \mathfrak{g}$. Since Ω is expressed as $\Omega = \frac{1}{2}(\Delta(c) - c \otimes 1 + 1 \otimes c)$ we have the relation

$$[\Omega, x \otimes 1 + 1 \otimes x] = 0 \tag{4.1}$$

for any $x \in \mathfrak{g}$ in the tensor product $U\mathfrak{g} \otimes U\mathfrak{g}$. By means of the above relation it can be shown that the infinitesimal pure braid relations:

$$[\Omega_{ik}, \Omega_{ij} + \Omega_{jk}] = 0, \quad (i, j, k \text{ distinct}), \tag{4.2}$$
$$[\Omega_{ij}, \Omega_{k\ell}] = 0, \quad (i, j, k, \ell \text{ distinct}) \tag{4.3}$$

hold. Let us briefly explain the reason why we have the above infinitesimal pure braid relations. For the first relation it is enough to show the case $i = 1, j = 3, k = 2$. Since we have

$$[\Omega \otimes 1, (I_\mu \otimes 1 + 1 \otimes I_\mu) \otimes I_\mu] = 0$$

by the equation (4.1) we obtained the desired relation. The equation (4.3) in the infinitesimal pure braid relations is clear from the definition of Ω on the tensor product.

We define the Knizhnik-Zamolodchikov (KZ) connection as the 1-form

$$\omega = \frac{1}{\kappa} \sum_{1 \le i < j \le n} \Omega_{ij} d \log(z_i - z_j) \tag{4.4}$$

with values in $\mathrm{End}(V_1 \otimes \cdots \otimes V_n)$ for a non-zero complex parameter κ.

We set $\omega_{ij} = d \log(z_i - z_j)$, $1 \le i, j \le n$. It follows from the above infinitesimal pure braid relations among Ω_{ij} together with Arnold's relation

$$\omega_{ij} \wedge \omega_{jk} + \omega_{jk} \wedge \omega_{k\ell} + \omega_{k\ell} \wedge \omega_{ij} = 0 \tag{4.5}$$

that $\omega \wedge \omega = 0$ holds. This implies that ω defines a flat connection for a trivial vector bundle over the configuration space $X_n = \{(z_1, \cdots, z_n) \in \mathbf{C}^n \ ; \ z_i \neq z_j \text{ if } i \neq j\}$ with fiber $V_1 \otimes \cdots \otimes V_n$. A horizontal section of the above flat bundle is a solution of the total differential equation

$$d\varphi = \omega\varphi \tag{4.6}$$

for a function $\varphi(z_1, \cdots, z_n)$ with values in $V_1 \otimes \cdots \otimes V_n$. This total differential equation can be expressed as a system of partial differential equations

$$\frac{\partial \varphi}{\partial z_i} = \frac{1}{\kappa} \sum_{j, j \neq i} \frac{\Omega_{ij}}{z_i - z_j} \varphi, \quad 1 \le i \le n, \tag{4.7}$$

which is called the KZ equation. The KZ equation was first introduced in [8] as the differential equation satisfied by n-point functions in Wess-Zumino-Witten conformal field theory.

Let $\phi(z_1, \cdots, z_n)$ be the matrix whose columns are linearly independent solutions of the KZ equation. By considering the analytic continuation of the solutions with respect to a loop γ in X_n with base point x_0 we obtain the matrix $\theta(\gamma)$ defined by

$$\phi(z_1, \cdots, z_n) \mapsto \phi(z_1, \cdots, z_n)\theta(\gamma). \tag{4.8}$$

Since the KZ connection ω is flat the matrix $\theta(\gamma)$ depends only on the homotopy class of γ. The fundamental group $\pi_1(X_n, x_0)$ is the pure braid group P_n. As the above holonomy of the connection ω we have a one-parameter family of linear representations of the pure braid group

$$\theta : P_n \to \mathrm{GL}(V_1 \otimes \cdots \otimes V_n). \tag{4.9}$$

The symmetric group $\mathfrak{S}_n$ acts on X_n by the permutations of coordinates. We denote the quotient space $X_n/\mathfrak{S}_n$ by Y_n. The fundamental group of Y_n is the braid group B_n. In the case $V_1 = \cdots = V_n = V$, the symmetric group $\mathfrak{S}_n$ acts diagonally on the trivial vector bundle over X_n with fiber $V^{\otimes n}$ and the connection ω is invariant by this action. Thus we have one-parameter family of linear representations of the braid group

$$\theta : B_n \to \mathrm{GL}(V^{\otimes n}). \tag{4.10}$$

It is known by [6] and [10] that this representation is described by means of quantum groups. We call θ the quantum representation of the braid group.

5 Solutions of KZ equation by hypergeometric integrals

In this section we describe solutions of the KZ equation for the case $\mathfrak{g} = sl_2(\mathbf{C})$ by means of hypergeometric integrals following Schechtman and Varchenko [16]. A description of the solutions of the KZ equation was also given by Date, Jimbo, Matsuo and Miwa [5]. We refer the reader to [2] and [15] for general treatments of hypergeometric integrals.

Let us recall basic facts about the Lie algebra $sl_2(\mathbf{C})$ and its Verma modules. As a complex vector space the Lie algebra $sl_2(\mathbf{C})$ has a basis H, E and F satisfying the relations:

$$[H, E] = 2E, \quad [H, F] = -2F, \quad [E, F] = H. \tag{5.1}$$

For a complex number λ we denote by M_λ the Verma module of $sl_2(\mathbf{C})$ with highest weight λ. Namely, there is a non-zero vector $v_\lambda \in M_\lambda$ called the highest weight vector satisfying

$$Hv_\lambda = \lambda v_\lambda, \quad Ev_\lambda = 0 \tag{5.2}$$

and M_λ is spanned by $F^j v_\lambda$, $j \geq 0$. The elements H, E and F act on this basis as

$$\begin{cases} H \cdot F^j v_\lambda = (\lambda - 2j) F^j v_\lambda \\ E \cdot F^j v_\lambda = j(\lambda - j + 1) F^{j-1} v_\lambda \\ F \cdot F^j v_\lambda = F^{j+1} v_\lambda. \end{cases} \tag{5.3}$$

It is known that if $\lambda \in \mathbf{C}$ is not a non-negative integer, then the Verma module M_λ is irreducible.

For $\Lambda = (\lambda_1, \cdots, \lambda_n) \in \mathbf{C}^n$ we put $|\Lambda| = \lambda_1 + \cdots + \lambda_n$ and consider the tensor product $M_{\lambda_1} \otimes \cdots \otimes M_{\lambda_n}$. For a non-negative integer m we define the space of weight vectors with weight $|\Lambda| - 2m$ by

$$W[|\Lambda| - 2m] = \{x \in M_{\lambda_1} \otimes \cdots \otimes M_{\lambda_n} \ ; \ Hx = (|\Lambda| - 2m)x\} \tag{5.4}$$

and consider the space of null vectors defined by

$$N[|\Lambda| - 2m] = \{x \in W[|\Lambda| - 2m]\ ;\ Ex = 0\}. \tag{5.5}$$

The KZ connection ω commutes with the diagonal action of $\mathfrak{g}$ on $V_{\lambda_1} \otimes \cdots \otimes V_{\lambda_n}$, hence it acts on the space of null vectors $N[|\Lambda| - 2m]$.

For parameters κ and λ we consider the multi-valued function

$$\Phi_{n,m} = \prod_{1\le i<j\le n} (z_i - z_j)^{\frac{\lambda_i\lambda_j}{\kappa}} \prod_{1\le i\le m, 1\le \ell\le n} (t_i - z_\ell)^{-\frac{\lambda_\ell}{\kappa}} \prod_{1\le i<j\le m} (t_i - t_j)^{\frac{2}{\kappa}} \tag{5.6}$$

defined over X_{n+m}. Let $\mathcal{L}$ denote the local system associated to the multi-valued function Φ. The restriction of $\mathcal{L}$ on the fiber $X_{n,m}$ is the local system associated with the parameters

$$\alpha_\ell = -\frac{\lambda_\ell}{\kappa}, \quad 1 \le \ell \le n, \quad \gamma = \frac{1}{\kappa} \tag{5.7}$$

in the notation of Section 3.

The symmetric group $\mathfrak{S}_m$ acts on $X_{n,m}$ by the permutations of the coordinate functions $t_1, \cdots, t_m$. The function $\Phi_{n,m}$ is invariant by the action of $\mathfrak{S}_m$. The local system $\mathcal{L}$ over $X_{n,m}$ defines a local system on $Y_{n,m}$, which we denote by $\overline{\mathcal{L}}$. The local system dual to $\mathcal{L}$ is denoted by $\mathcal{L}^*$.

We put $v = v_{\lambda_1} \otimes \cdots \otimes v_{\lambda_n}$ and for $J = (j_1, \cdots, j_n)$ set $F^J v = F^{j_1} v_{\lambda_1} \otimes \cdots \otimes F^{j_n} v_{\lambda_n}$, where $j_1, \cdots, j_n$ are non-negative integers. The weight space $W[|\Lambda| - 2m]$ has a basis $F^J v$ for each J with $|J| = j_1 + \cdots + j_n = m$. For the sequence of integers $(i_1, \cdots, i_m) = (\underbrace{1, \cdots, 1}_{j_1}, \cdots, \underbrace{n, \cdots, n}_{j_n})$ we set

$$S_J(z, t) = \frac{1}{(t_1 - z_{i_1}) \cdots (t_m - z_{i_m})} \tag{5.8}$$

and define the rational function $R_J(z, t)$ by

$$R_J(z, t) = \frac{1}{j_1! \cdots j_n!} \sum_{\sigma \in \mathfrak{S}_m} S_J(z_1, \cdots, z_n, t_{\sigma(1)}, \cdots, t_{\sigma(m)}). \tag{5.9}$$

For example, we have

$$R_{(1,0,\cdots,0)}(z, t) = \frac{1}{t_1 - z_1}, \quad R_{(2,0,\cdots,0)}(z, t) = \frac{1}{(t_1 - z_1)(t_2 - z_1)}$$

$$R_{(1,1,0,\cdots,0)}(z, t) = \frac{1}{(t_1 - z_1)(t_2 - z_2)} + \frac{1}{(t_2 - z_1)(t_1 - z_2)}$$

and so on.

Since $\pi_{n,m} : X_{m+n} \to X_m$ is a fiber bundle with fiber $X_{n,m}$ the fundamental group of the base space X_n acts naturally on the homology group $H_m(X_{n,m}, \mathcal{L}^*)$. Thus we obtain a representation of the pure braid group

$$r_{n,m} : P_n \longrightarrow \mathrm{Aut}\, H_m(X_{n,m}, \mathcal{L}^*) \tag{5.10}$$

which defines a local system on X_n denoted by $\mathcal{H}_{n,m}$. In the case $\lambda_1 = \cdots = \lambda_n$ there is a representation of the braid group

$$r_{n,m} : B_n \longrightarrow \mathrm{Aut}\, H_m(Y_{n,m}, \overline{\mathcal{L}}^*) \tag{5.11}$$

which defines a local system $\overline{\mathcal{H}}_{n,m}$ on $Y_{n,m}$. For any horizontal section $c(z)$ of the local system $\mathcal{H}_{n,m}$ we consider the hypergeometric type integral

$$\int_{c(z)} \Phi_{n,m} R_J(z,t)\, dt_1 \wedge \cdots \wedge dt_m \tag{5.12}$$

for the above rational function $R_J(z,t)$.

According to Schechtman and Varchenko, solutions of the KZ equation are described in the following way.

Theorem 5.1 (Schechtman and Varchenko [16]). *The integral*

$$\sum_{|J|=m} \left(\int_{c(z)} \Phi_{n,m} R_J(z,t)\, dt_1 \wedge \cdots \wedge dt_m \right) F^J v$$

lies in the space of null vectors $N[|\Lambda| - 2m]$ and is a solution of the KZ equation.

6 Relation between LKB representation and KZ connection

We fix a complex number λ and consider the space of null vectors

$$N[n\lambda - 2m] \subset M_\lambda^{\otimes n}$$

by putting $\lambda_1 = \cdots = \lambda_n = \lambda$ in the definition of Section 5. As the monodromy of the KZ connection

$$\omega = \frac{1}{\kappa} \sum_{1 \le i < j \le n} \Omega_{ij} d \log(z_i - z_j)$$

with values in $N[n\lambda - 2m]$ we obtain the linear representation of the braid group

$$\theta_{\lambda,\kappa} : B_n \longrightarrow \mathrm{Aut}\, N[n\lambda - 2m].$$

The next theorem describes a relationship between a specialization of the Lawrence-Krammer-Bigelow representation $\rho_{n,m}$ and the representation $\theta_{\lambda,\kappa}$.

Theorem 6.1. *There exists an open dense subset U in $(\mathbf{C}^*)^2$ such that for $(\lambda, \kappa) \in U$ the Lawrence-Krammer-Bigelow representation $\rho_{n,m}$ with the specialization*

$$q = e^{-2\pi\sqrt{-1}\lambda/\kappa}, \quad t = e^{2\pi\sqrt{-1}/\kappa}$$

is equivalent to the monodromy representation of the KZ connection $\theta_{\lambda,\kappa}$ with values in the space of null vectors

$$N[n\lambda - 2m] \subset M_\lambda^{\otimes n}.$$

We assume the conditions $i_*\mathcal{L} \cong i_!\mathcal{L}$ and $i_*\overline{\mathcal{L}} \cong i_!\overline{\mathcal{L}}$ in the following. By means of the argument in Section 3 these conditions are satisfied for (λ, κ) in an open dense subset in $(\mathbf{C}^*)^2$. By the assumption we have an isomorphism $H_m(X_{n,m}, \mathcal{L}) \cong H_m^{lf}(X_{n,m}, \mathcal{L})$ and we can take the regularized cycles $[\widetilde{\Delta}_M] \in H_m(X_{n,m}, \mathcal{L})$ for the bounded chamber Δ_M.

We will consider the integral

$$\sum_{|J|=m} \left(\int_{\Delta_M} \Phi_{n,m} R_J(z,t)\, dt_1 \wedge \cdots \wedge dt_m \right) F^J v$$

in the space of null vectors $N[|\Lambda| - 2m]$. In general the above integral is divergent. We replace the integration cycle by the regularized cycle $[\widetilde{\Delta}_M]$ to obtain the convergent integral. This is called the regularized integral. We refer the reader to [2] for details on this aspect.

The rest of this section is devoted to the proof of the above theorem. We first show the following proposition.

Proposition 6.2. *There exists an open dense subset U in $(\mathbf{C}^*)^2$ such that for $(\lambda, \kappa) \in U$ the following properties* (1) *and* (2) *are satisfied.*

(1) *The integrals in Theorem 5.1 over $[\widetilde{\Delta}_M]$ for $M = (m_1, \cdots, m_{n-1})$ with $m_1 + \cdots + m_{n-1} = m$ are linearly independent.*
(2) *The homology classes $[\overline{\Delta}_M]$ for $M = (m_1, \cdots, m_{n-1})$ with $m_1 + \cdots + m_{n-1} = m$ form a basis of $H_m^{lf}(Y_{n,m}, \overline{\mathcal{L}}^*) \cong H_m(Y_{n,m}, \overline{\mathcal{L}}^*)$.*

Here $m_1, \cdots, m_{n-1}$ are non-negative integers.

Proof. We prepare notation for a basis of $N[|\Lambda| - 2m]$. We suppose that λ_1 is not a non-negative integer. Let us observe that for $\Lambda = (\lambda_1, \cdots, \lambda_n)$ the space of null vectors $N[|\Lambda| - 2m]$ has dimension $d_{n,m}$. This can be shown as follows. First, let us consider the weight space

$$\begin{aligned} &M_{\lambda_2} \otimes \cdots \otimes M_{\lambda_n}[\lambda_2 + \cdots + \lambda_n - 2m] \\ &= \{x \in M_{\lambda_2} \otimes \cdots \otimes M_{\lambda_n} \ ; \ Hx = (\lambda_2 + \cdots + \lambda_n - 2m)x\}. \end{aligned}$$

There is an isomorphism

$$\xi : M_{\lambda_2} \otimes \cdots \otimes M_{\lambda_n}[\lambda_2 + \cdots + \lambda_n - 2m] \longrightarrow N[|\Lambda| - 2m]$$

defined by

$$u \mapsto v_{\lambda_1} \otimes u - \frac{1}{\lambda_1} F v_{\lambda_1} \otimes Eu + \frac{1}{\lambda_1(\lambda_1 - 1)} F^2 v_{\lambda_1} \otimes E^2 u - \cdots$$

This shows that $N[|\Lambda| - 2m]$ has a basis indexed by $J' = (j_1, j_2, \cdots, j_n)$ with $j_1 = 0$ and $j_2 + \cdots + j_n = m$, where $j_2, \cdots, j_n$ are non-negative integers. Let us denote by $S_{n,m}$ the set of such indices J'. The above weight space has a basis $u_{J'}$ indexed by $J' \in S_{n,m}$. We have the corresponding basis $\xi(u_{J'})$ of $N[|\Lambda| - 2m]$.

We set $\alpha_1, \cdots, \alpha_n$ and γ as in (5.7). We put

$$\widetilde{\Phi}_{n,m} = \prod_{1 \le i \le m, 1 \le \ell \le n} (t_i - z_\ell)^{\alpha_\ell} \prod_{1 \le i < j \le m} (t_i - t_j)^{2\gamma} \tag{6.1}$$

and for $J' \in S_{n,m}$ put

$$\alpha'_J = \prod_{k=2}^{n} (j_k)! \alpha_k (\alpha_k + \gamma) \cdots (\alpha_k + (j_k - 1)\gamma). \tag{6.2}$$

We assume that $\alpha_1, \cdots, \alpha_n$ and γ are positive. We express the integral in Theorem 5.1 over the cycle Δ_M in the linear combination for the basis $\xi(u_{J'})$ of $N[|\Lambda| - 2m]$ and we donote by $\widetilde{R}_{J'}(z, t)$ the corresponding rational function. In [18] Varchenko gave a formula for the determinant

$$\det_{M, J'} \left(\alpha_{J'} \int_{\Delta_M} \widetilde{\Phi}_{n,m} \widetilde{R}_{J'}(z, t) dt_1 \wedge \cdots \wedge dt_m \right), \tag{6.3}$$

where $M = (m_1, \cdots, m_{n-1})$ with $m_1 + \cdots + m_{n-1} = m$ and $J' \in S_{n,m}$. According to Varchenko's formula the above determinant is expressed as a non-zero constant times the gamma factor given by

$$\prod_{i=0}^{m-1} \left(\frac{\Gamma((i+1)\gamma + 1)^{n-1}}{\Gamma(\gamma + 1)^{n-1}} \frac{\Gamma(\alpha_1 + i\gamma + 1) \cdots (\alpha_n + i\gamma + 1)}{\Gamma(\alpha_1 + \cdots + \alpha_n + (2m - 2 - i)\gamma + 1)} \right)^{\nu_i} \tag{6.4}$$

where ν_i is defined by

$$\nu_i = \binom{m + n - i - 3}{m - i - 1}. \tag{6.5}$$

Since the gamma function does not have zeros and has only poles of order one at non-positive integers, it is clear that the determinant is zero only when the denominator of the gamma factor has a pole. Considering the regularized integrals over the cycles $[\widetilde{\Delta}_M]$ we can analytically continue the determinant formula to complex numbers $\alpha_1, \cdots, \alpha_n$ and γ.

Let us recall that we deal with the case

$$\alpha_\ell = -\frac{\lambda}{\kappa}, \quad 1 \leq \ell \leq n, \quad \gamma = \frac{1}{\kappa}.$$

From the determinant formula we observe that the linearly independence for the solutions of the KZ equation in (1) in the statement of the proposition is satisfied for (λ, κ) in an open dense subset in $(\mathbf{C}^*)^2$. Under the same condition we have the linear independence for the homology classes $[\overline{\Delta}_M]$ for $M = (m_1, \cdots, m_{n-1})$ with $m_1 + \cdots + m_{n-1} = m$. Since we have $\dim H_m^{lf}(Y_{n,m}, \overline{\mathcal{L}}^*) = d_{m,n}$ we obtain the property (2). This completes the proof of our proposition. □

Let us consider the specialization map

$$s : R = \mathbf{Z}[q^{\pm 1}, t^{\pm 1}] \longrightarrow \mathbf{C} \tag{6.6}$$

defined by the substitutions $q \mapsto e^{-2\pi\sqrt{-1}\lambda/\kappa}$ and $t \mapsto e^{2\pi\sqrt{-1}/\kappa}$. This induces in a natural way a homomorphism

$$H_m(\widetilde{C}_{n,m}(D); \mathbf{Z}) \longrightarrow H_m(Y_{n,m}, \overline{\mathcal{L}}^*). \tag{6.7}$$

We take a basis $[c_M]$ of $H_m(\widetilde{C}_{n,m}(D); \mathbf{Z})$ as the R-module for $M = (m_1, \cdots, m_{n-1})$ with $m_1 + \cdots + m_{n-1} = m$ in such a way that $[c_M]$ maps to the regularized cycle for $[\widetilde{\Delta}_M]$ by the above specialization map. We observe that the LKB representation specialized at $q \mapsto e^{-2\pi\sqrt{-1}\lambda/\kappa}$ and $t \mapsto e^{2\pi\sqrt{-1}/\kappa}$ is identified with the linear representation of the braid group $r_{n,m} : B_n \to \mathrm{Aut}\, H_m(Y_{n,m}, \overline{\mathcal{L}}^*)$.

Since the basis of $N[n\lambda - 2m]$ is indexed by the set $S_{n,m}$ we have an isomorphism

$$H_m(Y_{n,m}, \overline{\mathcal{L}}^*) \cong N[n\lambda - 2m].$$

Now the fundamental solutions of the KZ equation with values in $N[n\lambda - 2m]$ is given by the matrix of the form

$$\left(\int_{\widetilde{\Delta}_M} \omega_{M'} \right)_{M,M'}$$

with $M = (m_1, \cdots, m_{n-1})$ and $M' = (m'_1, \cdots, m'_{n-1})$ such that $m_1 + \cdots + m_{n-1} = m$ and $m'_1 + \cdots + m'_{n-1} = m$. Here $\omega_{M'}$ is a multivalued

m-form on $X_{n,m}$. The column vectors of the above matrix form a basis of the solutions of the KZ equation with values in $N[n\lambda - 2m]$. Thus the representation $r_{n,m} : B_n \to \text{Aut}\, H_m(Y_{n,m}, \overline{\mathcal{L}}^*)$ is equivalent to the action of B_n on the solutions of the KZ equation with values in $N[n\lambda - 2m]$. This completes the proof of Theorem 6.1.

References

[1] A. ANDREOTTI, On a theorem of Torelli, Amer. J. Math. **80** (1958), 801–828.

[2] K. AOMOTO and M. KITA, "Theory of Hypergeometric Functions", Springer Monographs in Mathematics, 2011.

[3] S. BIGELOW, *Braid groups are linear*, J. Amer. Math. Soc. **14** (2001), 471–486.

[4] S. BIGELOW, *The Lawrence-Krammer representation*, Topology and geometry of manifolds (Athens, GA, 2001), 51–68, Proc. Sympos. Pure Math. **71** Amer. Math. Soc., 2003.

[5] E. DATE, M JIMBO, A. MATSUO and T. MIWA, *Hypergeometric type integrals and the $sl(2, \mathbf{C})$ Knizhnik-Zamolodchikov equations*, Intern. J. Modern Phys. **B4** (1990), 1049–1057.

[6] V. G. DRINFEL'D, *Quasi-Hopf algebras*, Leningrad Math. J. **1** (1990), 1419–1457.

[7] B. FEIGIN, V. SCHECHTMAN and A. VARCHENKO, *On algebraic equations satisfied by hyergeometric correlators in WZW models*, I, Comm. Math. Phys. **163** (1994), 173–184.

[8] V. G. KNIZHNIK and A. B. ZAMOLODCHIKOV, *Current algebra and Wess-Zumino models in two dimensions*, Nuclear Phys. **B247** (1984), 83–103.

[9] T. KOHNO, *Homology of a local system on the complement of hyperplanes*, Proc. Japan Acad. Ser. A **62** (1986), 144–147.

[10] T. KOHNO, *Monodromy representations of braid groups and Yang-Baxter equations*, Ann. Inst. Fourier **37** (1987), 139–160.

[11] T. KOHNO, *Hyperplane arrangements, local system homology and iterated integrals*, Adv. Stud. Pure Math. **62** (2012), 157–174.

[12] D. KRAMMER, *Braid groups are linear*, Ann. of Math. **155** (2002), 131–156.

[13] R. J. LAWRENCE, *Homological representations of the Hecke algebra*, Comm. Math. Phys. **135** (1990), 141 – 191.

[14] I. MARIN, *Sur les représentations de Krammer génériques*, Ann. Inst. Fourier **57** (2007), 1883 – 1925.

[15] P. ORLIK and H. TERAO, "Arrangements and Hypergeometric Integrals", MSJ Memoirs **9**, Mathematical Society of Japan, 2001.

[16] V. SCHECHTMAN and A. VARCHENKO, *Hypergeometric solutions of the Knizhnik-Zamolodchikov equation*, Lett. in Math. Phys. **20** (1990), 93–102.
[17] R. SILVOTTI, *Local systems on the complement of hyperplanes and fusion rules in conformal field theory*, Internat. Math. Res. Notices 1994, no. 3, 111 ff., approx. 17 pp. (electronic).
[18] A. VARCHENKO, *Determinant formula for Selberg type integrals*, Funct. Analysis and Its Appl. **4** (1991), 65–66.
[19] A. VARCHENKO, *Multidimensional hypergeometric functions and representation theory of Lie algebras and quantum groups*, Advances in Math. Phys. **21**, World Scientific, 1995.

Cohomology of the complement to an elliptic arrangement

Andrey Levin* and Alexander Varchenko◇

Abstract. We consider the complement to an arrangement of hyperplanes in a cartesian power of an elliptic curve and describe its cohomology with coefficients in a nontrivial rank one local system.

1 Introduction

We start with a cartesian power E^k of an elliptic curve E and a nontrivial rank one local system on E^k. We consider an arrangement of elliptic hyperplanes in E^k and describe the cohomology of its complement with coefficients in the local system.We show that the cohomology is nontrivial only in degree k. We present each cohomology class by a unique closed holomorphic differential form. Our forms are elliptic analogs of the Arnold-Brieskorn-Orlik-Solomon logarithmic differential forms representing cohomology classes of the complement to an arrangement of hyperplanes in an affine space. For the elliptic discriminantal arrangement our forms are the forms considered in [2–4] to solve the KZB equations in hypergeometric integrals and to construct Bethe eigenfunctions to the elliptic Calogero-Moser operators.

To simplify the exposition, we first consider in Sections 2 and 3 the case of an elliptic discriminantal arrangement, then in Sections 4 and 5 we consider arbitrary elliptic arrangements.

ACKNOWLEDGEMENTS. The authors thank the Max Planck Institute for Mathematics in Bonn for hospitality.

* Supported in part by AG Laboratory GU-HSE, RF government grant, ag. 11 11.G34.31.0023.

◇ Supported in part by NSF grant DMS-1101508.

2 Cohomology of an elliptic discriminantal arrangement

Fix a natural number k and $\tau \in \mathbb{C}$, $\operatorname{Im}\tau > 0$. Denote $\Lambda = \tau\mathbb{Z} + \mathbb{Z} \subset \mathbb{C}$. The group $\Gamma = \mathbb{Z} \oplus \mathbb{Z}$ acts on $\mathbb{C}$ by transformations $(l, m) : t \mapsto t + l\tau + m$. The action on each factor gives an action of Γ^k on $\mathbb{C}^k$. Denote by $p : \mathbb{C}^k \to \mathbb{C}^k / \Gamma^k$ the canonical projection onto the space of orbits. We have $\mathbb{C}^k / \Gamma^k = E^k$, where E is the elliptic curve $\mathbb{C}/\Gamma$.

For each representation ρ of Γ^k on a vector space W we get a vector bundle over E^k with a flat connection, which is $(\mathbb{C}^k \times W)/\Gamma^k \to \mathbb{C}^k/\Gamma^k$. In particular, we may fix complex numbers $\boldsymbol{w} = (w_1, \ldots, w_k)$, take $W = \mathbb{C}$, and $\rho_{\boldsymbol{w}}(\gamma) = e^{2\pi\sqrt{-1}(w_1 l_1 + \cdots + w_k l_k)}$ for $\gamma = (l_1, m_1) \times \cdots \times (l_k, m_k)$. This line bundle over E^k with the flat connection will be denoted by $\mathcal{L}_{\boldsymbol{w}}$.

We say that the numbers $\boldsymbol{w} = (w_1, \ldots, w_k)$ are *discriminantal convenient* if for any subset $I \subset \{1, \ldots, k\}$ the sum $\sum_{i \in I} w_i$ is not in Λ.

Fix distinct complex numbers $z = (z_1, \ldots, z_n)$. The *discriminantal* arrangement $\mathcal{C}_z$ in $\mathbb{C}^k$ with parameters z is the arrangement of hyperplanes:

$$\begin{aligned} H_i^a &: t_i - z_a = 0, && i = 1, \ldots, k,\ a = 1, \ldots, n;\\ H_{ij} &: t_i - t_j = 0, && 1 \leqslant i < j \leqslant k. \end{aligned}$$

Let M_z denote its complement $\mathbb{C}^k - \cup_{H \in \mathcal{C}_z} H$.

The Γ^k-orbit of $\mathcal{C}_z$ is the infinite arrangement $\mathcal{C}_{z,\Gamma^k} = \{\gamma(H) \mid \gamma \in \Gamma^k, H \in \mathcal{C}_z\}$. Denote by M_{z,Γ^k} its complement $\mathbb{C}^k - \cup_{\gamma \in \Gamma^k, H \in \mathcal{C}_z} \gamma(H)$. Denote by $\tilde{M}_{z,\tau} \subset E^k$ the image of M_{z,Γ^k} under the projection p.

Theorem 2.1. *Assume that the numbers $\boldsymbol{w}$ are discriminantal convenient and $z_1, \ldots, z_n$ project to distinct points of E. Then $H^\ell(\tilde{M}_{z,\tau}; \mathcal{L}_{\boldsymbol{w}}) = 0$ for $\ell \neq k$ and $H^k(\tilde{M}_{z,\tau}; \mathcal{L}_{\boldsymbol{w}})$ is canonically isomorphic to $H^k(M_z; \mathbb{C})$, the k-th cohomology group with trivial coefficients of the complement in $\mathbb{C}^k$ to the discriminantal arrangement.*

Here $H^*(\tilde{M}_{z,\tau}; \mathcal{L}_{\boldsymbol{w}})$ denotes the cohomology of $\tilde{M}_{z,\tau}$ with coefficients in the local system of horizontal sections of $\mathcal{L}_{\boldsymbol{w}}$. Theorem 2.1 is proved in Section 5.

The space $\tilde{M}_{z,\tau}$ is a $K(\pi, 1)$-space and our theorem describes the cohomology of the fundamental group of $\tilde{M}_{z,\tau}$ with coefficients in $\mathcal{L}_{\boldsymbol{w}}$. Notice that the fundamental group of $\tilde{M}_{z,\tau}$ is a subgroup of the pure elliptic braid group with $n + k$ strings.

The cohomology $H^k(M_z; \mathbb{C})$ of the complement to the discriminantal arrangement in $\mathbb{C}^k$ are presented be explicit logarithmic forms by the Arnold-Brieskorn-Orlik-Solomon theory. Below we describe logarithmic differential forms representing elements of $H^k(\tilde{M}_{z,\tau}; \mathcal{L}_{\boldsymbol{w}})$. Those forms were used in [2–4] to give integral hypergeometric representations for solutions of the KZB equations with values in a tensor product of highest weight representations of a simple Lie algebra and to construct Bethe eigenfunctions of elliptic Calogero-Moser operators.

3 Differential forms of a discriminantal arrangement

In this section we follow [4]. Theorem 3.3 is new.

3.1 Combinatorial space

An *ordered k-forest* is a graph with no cycles, with k edges, and a numbering of its edges by the numbers $1, 2, \dots, k$. We consider the ordered k-forests on the vertex set of symbols $\{z_1, \dots, z_n, t_1, \dots, t_k\}$. An ordered k-forest T is *admissible* if all $t_1, \dots, t_k$ are among the vertices of T and each connected component of T has exactly one vertex from the subset $\{z_1, \dots, z_n\}$.

Let $\mathcal{A}_n^k$ be the complex vector space generated by the admissible ordered k-forests, modulo the following relations:

R1 $T_1 = -T_2$ if T_1 and T_2 have the same underlying graph, and the order of their edges differ by a transposition;

R2

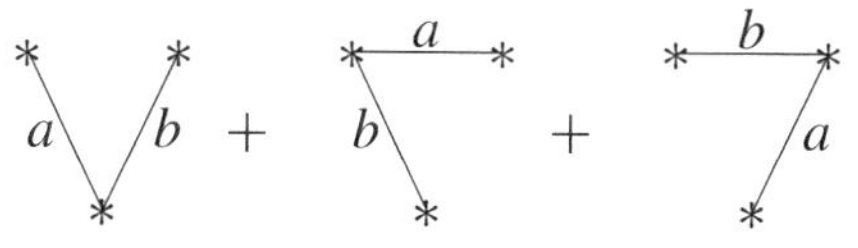

$$= 0 \qquad (a, b \in \{1, \dots, k\}),$$

that is, the sum of three k-forests that locally (i.e. their subgraphs spanned by 3 vertices) differ as above, but are otherwise identical, is 0.

A linear map ϕ of $\mathcal{A}_n^k$ to a vector space W is called a *representation* of $\mathcal{A}_n^k$. Suppose we are given a vector space W and a vector $\phi(T) \in W$ for every admissible k-forest T. This data induces a representation if the assignment $T \mapsto \phi(T)$ respects relations R1 and R2.

3.2 Rational representation

Let e be an edge of an admissible forest T. The connected component of T, containing e, has exactly one vertex, say z_a, from the set $\{z_1, \dots, z_n\}$. Denote by $h(e)$ and $t(e)$ the head and tail of the edge e, i.e. the vertices adjacent to e, farther resp. closer to the vertex z_a.

Fix distinct complex numbers $z = (z_1, \dots, z_n)$. To an admissible forest T with ordered edges $e_1, \dots, e_k$, we assign a closed holomorphic differential k-form $\phi_{rat}(T)$ on M_z by the formula

$$\phi_{rat}(T) = \wedge_{i=1}^{k} d\log(h(e_i) - t(e_i)).$$

This assignment defines a representation of $\mathcal{A}_n^k$ on the space of k-forms on M_z, see [1, 5]. By [1, 5], the representation is an isomorphism onto its image, see Proposition 2.1 in [4]. We denoted the image by $\mathcal{A}_z^k$. The assignment to a form of its cohomology class gives a linear map $\mathcal{A}_n^k \to H^k(M_z; \mathbb{C})$.

Theorem 3.1 ([1,5]). *The map $\mathcal{A}_n^k \to H^k(M_z; \mathbb{C})$ is an isomorphism.*

3.3 Theta representation.

For $z, \tau \in \mathbb{C}$, $\operatorname{Im} \tau > 0$, the first Jacobi theta function is defined by the infinite product

$$\theta(z) = \theta(z, \tau) = \sqrt{-1} e^{\pi\sqrt{-1}(\tau/4 - z)} (x; q) \left(\frac{q}{x}; q\right) (q; q),$$

$$q = e^{2\pi\sqrt{-1}\tau}, \; x = e^{2\pi\sqrt{-1}z}, \qquad (y; q) = \prod_{j=0}^{\infty} (1 - yq^j),$$

[9]. It is an entire holomorphic function of z satisfying

$$\theta(z+1, \tau) = -\theta(z, \tau), \quad \theta(z+\tau, \tau) = -e^{-\pi\sqrt{-1}\tau - 2\pi\sqrt{-1}z}\theta(z, \tau),$$
$$\theta(-z, \tau) = -\theta(z, \tau).$$

By $\theta'(z, \tau)$ we will mean the derivative in the z variable. Define

$$\sigma_w(t) = \sigma_w(t, \tau) = \frac{\theta(w - t, \tau)}{\theta(w, \tau)\theta(t, \tau)} \cdot \theta'(0, \tau).$$

The listed properties of the theta function yield that the function σ – viewed as a function of t – has simple poles at the points of $\Lambda \subset \mathbb{C}$, as well as the properties

$$\sigma_w(t+1, \tau) = \sigma_w(t, \tau), \qquad \sigma_w(t+\tau, \tau) = e^{2\pi\sqrt{-1}w}\sigma(t, \tau),$$
$$\operatorname{Res}_{t=0} \sigma_w(t, \tau) = 1.$$

We also have

$$\sigma_{w_1+w_2}(t-u)\sigma_{w_2}(s-t) - \sigma_{w_2}(s-u)\sigma_{w_1}(t-u) + \sigma_{w_1}(t-s)\sigma_{w_1+w_2}(s-u) = 0,$$

see for example, [4].

Fix discriminantal convenient complex numbers $\boldsymbol{w} = (w_1, \ldots, w_k)$ and distinct complex numbers $z = (z_1, \ldots, z_n)$. For $i = 1, \ldots, k$, we say that t_i has *weight* w_i.

Let T be an admissible forest and v a vertex of T. The connected component of T, containing v, has exactly one vertex, say z_a, from the set $\{z_1, \ldots, z_n\}$. We define the *branch* $B(v)$ of v to be the collection of those vertices u for which the unique path connecting u with z_a contains v. By definition $v \in B(v)$. The *load* $L(v)$ of a vertex v in the forest T is defined to be the sum of the weights of the vertices in $B(v)$.

To an admissible forest T with ordered edges $e_1, \ldots, e_k$, we assign a closed holomorphic differential k-form $\phi_\theta(T)$ on M_{z,Γ^k} by the formula

$$\phi_\theta(T) = \wedge_{i=1}^{k} \sigma_{L(h(e_i))}(h(e_i) - t(e_i), \tau)\, d(h(e_i) - t(e_i)).$$

Notice that if $\boldsymbol{w}$ are discriminantal convenient, then the load of each vertex $h(e_i)$ does not lie in Λ and the form is well-defined.

Theorem 3.2 ([4]). *Assume that* $\boldsymbol{w}$ *is discriminantal convenient and* $z_1, \ldots, z_n$ *project to distinct points of* E*. Then the assignment* $T \mapsto \phi_\theta(T)$ *defines a representation of* $\mathcal{A}_n^k$ *on the space of* k*-forms on* M_{z,Γ^k}*. The representation is an isomorphism onto its image, denoted by* $\mathcal{A}_{z,\Gamma^k}^k$*. Each element of* $\mathcal{A}_{z,\Gamma^k}^k$ *descends to a closed holomorphic differential form on* $\tilde{M}_{z,\tau}$ *with values in* $\mathcal{L}_{\boldsymbol{w}}$*.*

The assignment to a form of its cohomology class defines a linear map $\mathcal{A}_n^k \to H^k(\tilde{M}_{z,\tau}; \mathcal{L}_{\boldsymbol{w}})$. In Section 5 the following theorem will be proved.

Theorem 3.3. *Assume that the numbers* $\boldsymbol{w}$ *are discriminantal convenient and* $z_1, \ldots, z_n$ *project to distinct points of* E*. Then the map* $\mathcal{A}_n^k \to H^k(\tilde{M}_{z,\tau}; \mathcal{L}_{\boldsymbol{w}})$ *is an isomorphism.*

Theorems 3.1 and 3.3 imply the second statement of Theorem 2.1.

According to [7], $\dim \mathcal{A}_n^k = \sum_{m_1+\cdots+m_n=k} m_1! \ldots m_n!$.

4 Transversal elliptic hyperplanes

4.1 Elliptic hyperplanes in E^k

Denote $\mathcal{E} = E^k$. Any $k \times k$-matrix $C \in GL(k, \mathbb{Z})$ defines an isomorphism $\mathcal{E} \to \mathcal{E}$, $(\tilde{t}_1, \ldots, \tilde{t}_k) \mapsto (\sum_j \tilde{t}_j c_{j1}, \ldots, \sum_j \tilde{t}_j c_{jk})$. The collection $\sum_j \tilde{t}_j c_{j1}, \ldots, \sum_j \tilde{t}_j c_{jk}$ will be called *coordinates* on $\mathcal{E}$.

Let $\tilde{t}_1', \ldots, \tilde{t}_k'$ be coordinates on $\mathcal{E}$. Fibers of the projection $\mathcal{E} \to E^\ell$ along the last $k - \ell$ coordinates will be called *elliptic* $k - \ell$*-planes* in $\mathcal{E}$, in particular, elliptic $k-1$-planes are *elliptic hyperplanes*. The fibers of the same projection will be called *parallel* $k-\ell$-planes.

An elliptic $k-\ell$-plane is defined by equations $\tilde{t}'_i = \tilde{z}_i$, $i = 1, \dots, \ell$, for suitable $\tilde{z}_i \in E$. Each elliptic $k-\ell$-plane is isomorphic to $E^{k-\ell}$ as an algebraic variety.

Lemma 4.1. *The normal bundle of an elliptic $k-\ell$-plane in $\mathcal{E}$ is trivial.*

4.2 Intersection of $\ell \leqslant k$ transversal elliptic hyperplanes

We say that $\ell \leqslant k$ elliptic hyperplanes $\tilde{H}_1, \dots, \tilde{H}_\ell$ intersect transversally, if they are defined by equations

$$\tilde{t}_1 a_{1j} + \cdots + \tilde{t}_k a_{kj} - \tilde{z}_j = 0, \qquad j = 1, \dots, \ell, \tag{4.1}$$

and the rank of the $\ell \times k$-matrix $a = (a_{ij})$ equals ℓ.

By standard theorems, see for example [8], there are coordinates $\tilde{t}'_1, \dots, \tilde{t}'_k$ on $\mathcal{E}$ such that system (4.1) is equivalent to a system

$$d_j \tilde{t}'_j - \tilde{z}'_j = 0, \qquad j = 1, \dots, \ell, \tag{4.2}$$

where $\tilde{z}'_j \in E$, $d_j \in \mathbb{Z}_{>0}$ and $d_j | d_{j+1}$ for $j = 1, \dots, \ell - 1$. Therefore, the intersection X of ℓ transversal hyperplanes in $\mathcal{E}$ consists of $(d_1 \dots d_\ell)^2$ parallel elliptic $k-\ell$-planes.

Let $\boldsymbol{w} = (w_1, \dots, w_k)$ be complex numbers and $\mathcal{L}_{\boldsymbol{w}}$ the line bundle over $\mathcal{E}$ with a flat connection defined in Section 2. We say that $\boldsymbol{w}$ are *convenient* for $\mathcal{E}$, if there are no nonzero $\mathcal{L}_{\boldsymbol{w}}$-valued holomorphic differential k-forms on $\mathcal{E}$.

Lemma 4.2. *Complex numbers $\boldsymbol{w}$ are convenient for $\mathcal{E}$ if and only if $\boldsymbol{w} \notin \Lambda^k$.*

Proof. It is enough to prove the lemma for $k = 1$. If $k = 1$ and ω is an $\mathcal{L}_{\boldsymbol{w}}$-valued holomorphic 1-form, then it is 1-periodic and has the Fourier series expansion. The expansion easily implies the required statement. □

Similarly, we say that $\boldsymbol{w}$ are *convenient* for the transversal intersection X with $\dim X = k - \ell > 0$, if there are no nonzero $\mathcal{L}_{\boldsymbol{w}}$-valued holomorphic differential $k-\ell$-forms on any of the parallel $k-\ell$-planes composing X. Let X be defined by equations (4.1).

Lemma 4.3. *Complex numbers $\boldsymbol{w}$ are convenient for the transversal intersection X,* $\dim X > 0$, *if and only if there exist integers $l_1, \dots, l_k$ such that $\sum_{i=1} l_i a_{ij} = 0$ for $j = 1, \dots, \ell$, and $l_1 w_1 + \cdots + l_k w_k \notin \Lambda$.*

Proof. The proof of Lemma 4.3 is the same as the proof of Lemma 4.2. □

Lemma 4.4. *If the numbers $\boldsymbol{w}$ are convenient for $\mathcal{E}$, then $H^*(\mathcal{E}, \mathcal{L}_{\boldsymbol{w}}) = 0$. If $\boldsymbol{w}$ are convenient for the transversal intersection X,* $\dim X > 0$, *then $H^*(X, \mathcal{L}_{\boldsymbol{w}}|_X) = 0$.*

Proof. The lemma follows from the Kunneth formula and the fact that the cohomology of a circle with coefficients in a nontrivial local system is zero. □

4.3 Differential forms of k transversal hyperplanes in $\mathcal{E}$

This section contains the main construction of the paper.

Let k transversal elliptic hyperplanes $\tilde{H}_1, \dots, \tilde{H}_k$ in $\mathcal{E}$ be given by equations

$$\sum_{i=1}^{k} \tilde{t}_i a_{ij} = \tilde{z}_j \qquad j = 1, \dots, k, \tag{4.3}$$

where $\tilde{t}_1, \dots, \tilde{t}_k$ are coordinates on $\mathcal{E}$, $a = (a_{ij})$ is an integer matrix (with nonzero determinant) and $\tilde{z}_1, \dots, \tilde{z}_k$ are some points of E.

For a complex number c, we denote by $\tilde{c}$ its projection to E. In particular, $\tilde{0} \in E$ is the projection of 0. For given complex numbers $\boldsymbol{w} = (w_1, \dots, w_k)$, we consider the system of equations

$$\sum_{j=1}^{k} a_{ij} \tilde{v}_j = \tilde{w}_i \qquad i = 1, \dots, k, \tag{4.4}$$

with respect to the unknown $\tilde{v}_1, \dots, \tilde{v}_k \in E$. We say that $\boldsymbol{w}$ is *admissible* for $\tilde{H}_1, \dots, \tilde{H}_k$ if any coordinate $\tilde{v}_j$ of any solution of (4.4) is not equal to $\tilde{0}$.

Lemma 4.5. *Assume that the numbers $\boldsymbol{w}$ are convenient for each of the transversal intersections X_j, $j = 1, \dots, k$, where X_j is the intersection of the elliptic hyperplanes $\tilde{H}_1, \dots, \tilde{H}_{j-1}, \tilde{H}_{j+1}, \dots, \tilde{H}_k$, then $\boldsymbol{w}$ are admissible for $\tilde{H}_1, \dots, \tilde{H}_k$.*

Proof. Let $\tilde{v}_1, \dots, \tilde{v}_k$ be any solution of system (4.4). Let $l_1, \dots, l_k$ be integers such that $\sum_{i=1} l_i a_{ij} = 0$ for $j = 2, \dots, k$, and $l_1 w_1 + \dots + l_k w_k \notin \Lambda$. Then $\tilde{0} \neq \sum_i l_i \tilde{w}_i = \sum_{ij} l_i a_{ij} \tilde{v}_j = \sum_i l_i a_{i1} \tilde{v}_1$. Hence $\tilde{v}_1 \neq \tilde{0}$. Similarly we prove that $\tilde{v}_2, \dots, \tilde{v}_k$ are not equal to $\tilde{0}$. □

Let $\boldsymbol{w}$ be admissible for $\tilde{H}_1, \ldots, \tilde{H}_k$. Fix complex numbers $z_1, \ldots, z_k$ whose projections to E are $\tilde{z}_1, \ldots, \tilde{z}_k$. For any integers A_i, B_i, C_i, D_i with $i = 1, \ldots, k$, we consider two systems of equations:

$$\sum_{i=1}^{k} u_i a_{ij} = A_j \tau + B_j + z_j, \qquad j = 1, \ldots, k, \tag{4.5}$$

and

$$\sum_{j=1}^{k} a_{ij} v_j = C_i \tau + D_i + w_i, \qquad i = 1, \ldots, k. \tag{4.6}$$

The first system is with respect to complex numbers $\boldsymbol{u} = (u_1, ..., u_k)$ and the second system is with respect to complex numbers $\boldsymbol{v} = (v_1, ..., v_k)$.

To the solution $\boldsymbol{v} = (v_1, \ldots, v_k)$ of (4.6), we assign the meromorphic k-form on $\mathbb{C}^k$,

$$\begin{aligned} \omega_{\boldsymbol{v}}(\boldsymbol{t},\tau) &= \omega_{\boldsymbol{v}}(t_1, \ldots, t_k, \tau) \\ &= \det a \; e^{-2\pi\sqrt{-1}\sum_{i=1}^{k} C_i t_i} \prod_{j=1}^{k} \sigma_{v_j}\left(\sum_{i=1}^{k} t_i a_{ij} - z_j, \tau\right) dt_1 \wedge \cdots \wedge dt_k. \end{aligned} \tag{4.7}$$

The form is well-defined since the numbers $\boldsymbol{w}$ are admissible for $\tilde{H}_1, \ldots, \tilde{H}_k$.

Lemma 4.6. *The form $\omega_{\boldsymbol{v}}(\boldsymbol{t}, \tau)$ descends to an $\mathcal{L}_{\boldsymbol{w}}$-valued meromorphic form on $\mathcal{E}$, i.e. $\omega_{\boldsymbol{v}}(\boldsymbol{t} + \gamma, \tau) = \rho_{\boldsymbol{w}}(\gamma)\omega_{\boldsymbol{v}}(\boldsymbol{t}, \tau)$ for $\gamma \in \Gamma^k$.*

Lemma 4.7. *The form $\omega_{\boldsymbol{v}}(\boldsymbol{t}, \tau)$ does not change if $\boldsymbol{v}$ is changed by an element of Λ^k.*

Lemma 4.8. *Let $\boldsymbol{u} = (u_1, \ldots, u_k)$ be the solution of system* (4.5). *Then*

$$\begin{aligned} &\omega_{\boldsymbol{v}}(t_1 + u_1, \ldots, t_k + u_k, \tau) \\ &= M(\boldsymbol{u},\boldsymbol{v}) \; \det a \; e^{-2\pi\sqrt{-1}\sum_{i=1}^{k} C_i t_i} \prod_{j=1}^{k} \sigma_{v_j}\left(\sum_{i=1}^{k} t_i a_{ij}, \tau\right) dt_1 \wedge \cdots \wedge dt_k, \end{aligned}$$

where

$$M(\boldsymbol{u},\boldsymbol{v}) = e^{2\pi\sqrt{-1}\sum_{i=1}^{k}(A_i v_i - C_i u_i)}.$$

For a complex number c, we shall write $c = c_{\mathbb{R}} + \tau c_\tau$ with $c_{\mathbb{R}}, c_\tau \in \mathbb{R}$.

Lemma 4.9. *We have*

$$\sum_{i=1}^{k}(A_i v_i - C_i u_i) = \sum_{i=1}^{k}(A_i v_{i,\mathbb{R}} - B_i v_{i,\tau}) + \sum_{i=1}^{k}(u_i w_{i,\tau} - z_i v_{i,\tau}).$$

Proof. $\sum_i A_i v_i = \sum_i (A_i v_{i,\mathbb{R}} + \tau A_i v_{i,\tau})$ and $\sum_i C_i u_i = \sum_i (\sum_j a_{ij} v_{j,\tau} - w_{i,\tau}) u_i = \sum_{ij} u_{i,\mathbb{R}} a_{ij} v_{j,\tau} + \tau \sum_{ij} u_{i,\tau} a_{ij} v_{j,\tau} - \sum_i w_{i,\tau} u_i = \sum_j (B_j + z_{j,\mathbb{R}}) v_{j,\tau} + \tau \sum_j (A_j + z_{j,\tau}) v_{j,\tau} - \sum_i w_{i,\tau} u_i$. These equalities give the lemma. □

If $\boldsymbol{u}$ is a solution of (4.5), then $p(\boldsymbol{u})$ is a solution (4.3). All solutions of (4.3) have this form. Similar relations hold for systems (4.6) and (4.4).

Each of the systems (4.3) and (4.4) has $(\det a)^2$ solutions. For each solution $\tilde{\boldsymbol{u}}$ of (4.3) we fix a solution $\boldsymbol{u}$ of (4.5) such that $p(\boldsymbol{u}) = \tilde{\boldsymbol{u}}$. We denote by $\mathcal{U}$ the constructed set of $(\det a)^2$ points $\boldsymbol{u} \in \mathbb{C}^k$. For each solution $\tilde{\boldsymbol{v}}$ of (4.4) we fix a solution $\boldsymbol{v}$ of (4.6) such that $p(\boldsymbol{v}) = \tilde{\boldsymbol{v}}$. We denote by $\mathcal{V}$ the constructed set of $(\det a)^2$ points $\boldsymbol{v} \in \mathbb{C}^k$.

Theorem 4.10. *The matrix $M = (M(\boldsymbol{u}, \boldsymbol{v}))_{\boldsymbol{u}\in\mathcal{U}, \boldsymbol{v}\in\mathcal{V}}$ is nondegenerate.*

Proof. Let $M_1(\boldsymbol{u}, \boldsymbol{v}) = e^{2\pi\sqrt{-1}\sum_{i=1}^{k}(A_i v_{i,\mathbb{R}} - B_i v_{i,\tau})}$.
The matrix $M_1 = (M_1(\boldsymbol{u}, \boldsymbol{v}))_{\boldsymbol{u}\in\mathcal{U}, \boldsymbol{v}\in\mathcal{V}}$ is obtained from M by multiplication by nondegenerate diagonal matrices. Thus, it is enough to prove that M_1 is nondegenerate.

The nondegeneracy of M_1 follows from the nondegeneracy of M_1 for $\boldsymbol{w} = 0$ and $\boldsymbol{z} = 0$, since the matrix M_1 for $\boldsymbol{w}$, $\boldsymbol{z}$ not necessarily equal to zero is obtained from the matrix M_1 with $\boldsymbol{w} = 0$, $\boldsymbol{z} = 0$ by multiplication by nondegenerate diagonal matrices.

By elementary row and column transformations, the pair of systems (4.5) and (4.6) can be reduced to the case of a diagonal matrix a. For a diagonal a and $\boldsymbol{w} = 0$, $\boldsymbol{z} = 0$ the nondegeneracy of M_1 is obvious. □

Theorem 4.11. *Let $\boldsymbol{w}$ be admissible for $\tilde{H}_1, \ldots, \tilde{H}_k$. Let $\mathcal{U}$ be a set as above. Then there exist the unique differential k-forms $\omega_{\boldsymbol{u},\tilde{H}_1,\ldots,\tilde{H}_k}(\boldsymbol{t}, \tau)$, $\boldsymbol{u} \in \mathcal{U}$, such that each $\omega_{\boldsymbol{u},\tilde{H}_1,\ldots,\tilde{H}_k}(\boldsymbol{t}, \tau)$ is a $\mathbb{C}$-linear combination of forms $\omega_{\boldsymbol{v}}(\boldsymbol{t}, \tau)$, $\boldsymbol{v} \in \mathcal{V}$, and for any $\boldsymbol{u}, \boldsymbol{u}' \in \mathcal{U}$ we have the followings expansion,*

$$\omega_{\boldsymbol{u},\tilde{H}_1,\ldots,\tilde{H}_k}(\boldsymbol{t} + \boldsymbol{u}', \tau)$$
$$= (\delta_{\boldsymbol{u},\boldsymbol{u}'} + \mathcal{O}(\boldsymbol{t}))\, d\log\left(\sum_{i=1}^{k} t_i a_{i1}\right) \wedge \cdots \wedge d\log\left(\sum_{i=1}^{k} t_i a_{ik}\right),$$

where $\mathcal{O}(\boldsymbol{t})$ is a function holomorphic at $\boldsymbol{t} = 0$ and $\mathcal{O}(\boldsymbol{0}) = 0$.

Proof. The theorem is a direct corollary of Theorem 4.10. □

Given transversal $\tilde{H}_1, \ldots, \tilde{H}_k$, the set $\mathcal{U}$ is not unique, each point $\boldsymbol{u}' \in \mathcal{U}$ can be shifted by any element $\gamma = (l_1\tau + m_1, \ldots, l_k\tau + m_k)$ of Λ^k.

Lemma 4.12. *Assume that exactly one point $\boldsymbol{u}'$ of the set $\mathcal{U}$ is replaced with a point $\boldsymbol{u}'' = \boldsymbol{u}' + \gamma$. Consider the set of differential forms assigned to the new set $\mathcal{U}$ by Theorem 4.11. Then $\omega_{\boldsymbol{u}'', \tilde{H}_1, \ldots, \tilde{H}_k}(\boldsymbol{t}, \tau) = e^{2\pi\sqrt{-1}(w_1 l_1 + \ldots + w_k l_k)} \omega_{\boldsymbol{u}', \tilde{H}_1, \ldots, \tilde{H}_k}(\boldsymbol{t}, \tau)$ and all other differential forms $\omega_{\boldsymbol{u}, \tilde{H}_1, \ldots, \tilde{H}_k}(\boldsymbol{t}, \tau)$, $\boldsymbol{u} \in \mathcal{U}$, remain unchanged.*

4.3.1 The residue of $\omega_{\boldsymbol{u}, \tilde{H}_1, \ldots, \tilde{H}_k}$ Let H be a hyperplane in $\mathbb{C}^k$ defined by an equation $t_1 a_{1j} + \cdots + t_k a_{kj} = A_j\tau + B_j + z_j$, where $j \in \{1, \ldots, k\}$ and A_j, B_j are some integers, *cf.* (4.5) and (4.7). We have $p(H) = \tilde{H}_j$. Let $\boldsymbol{u}$ be a point of $\mathcal{U}$ and $\omega_{\boldsymbol{u}, \tilde{H}_1, \ldots, \tilde{H}_k}$ the corresponding differential form. We denote by $\eta_{\boldsymbol{u}}$ the residue of $\omega_{\boldsymbol{u}, \tilde{H}_1, \ldots, \tilde{H}_k}$ at H.

Lemma 4.13. *Assume that a vector $\gamma = (l_1\tau + m_1, \ldots, l_k\tau + m_k) \in \Lambda^k$ is tangent to H, i.e. $\sum_i (l_i\tau + m_i) a_{ij} = 0$. Then for all $\boldsymbol{t} \in H$, we have $\eta_{\boldsymbol{u}}(\boldsymbol{t} + \gamma) = e^{2\pi\sqrt{-1}(w_1 l_1 + \cdots + w_k l_k)} \eta_{\boldsymbol{u}}(\boldsymbol{t})$. That is, the form $\eta_{\boldsymbol{u}}$ defines an $\mathcal{L}_{\boldsymbol{w}}$-valued differential form over the elliptic hyperplane $p(H) = \tilde{H}_j \subset \mathcal{E}$.*

Now we choose H in Lemma 4.13 so that $\boldsymbol{u} \in H$.

For $i \neq j$, the intersection $\tilde{H}_i \cap \tilde{H}_j$ is a collection of parallel elliptic $k-2$-planes. We denote by $\tilde{H}_i^{(j)}$ that elliptic $k-2$-plane which contains $\tilde{\boldsymbol{u}} = p(\boldsymbol{u})$. Then $\tilde{H}_1^{(j)}, \ldots, \tilde{H}_{j-1}^{(j)}, \tilde{H}_{j+1}^{(j)}, \ldots, \tilde{H}_k^{(j)}$ are transversal elliptic hyperplanes in $\tilde{H}_j$.

Theorem 4.14. *We have $\eta_{\boldsymbol{u}}(\boldsymbol{t}) = (-1)^{j-1} \omega_{\boldsymbol{u}, \tilde{H}_1^{(j)}, \ldots, \tilde{H}_{j-1}^{(j)}, \tilde{H}_{j+1}^{(j)}, \ldots, \tilde{H}_k^{(j)}}$.*

Proof. The difference of the right hand side and the left hand side defines an $\mathcal{L}_{\boldsymbol{w}}$-valued form on $\tilde{H}_j$ with logarithmic singularities along $\tilde{H}_1 \cap \tilde{H}_j, \ldots, \tilde{H}_{j-1} \cap \tilde{H}_j, \tilde{H}_{j+1} \cap \tilde{H}_j, \ldots, \tilde{H}_k \cap \tilde{H}_j$. The difference has zero $k-1$-iterated residues at all points. Therefore, the difference vanishes due to the following lemma. □

Lemma 4.15. *Assume that for every $i = 1, \ldots, k$, we have a finite set of parallel hyperplanes $\{\tilde{H}_i^{l_i} \mid l_i \in L_i\}$ in $\mathcal{E}$. Assume that the hyperplanes $\tilde{H}_1^{l_1}, \ldots, \tilde{H}_k^{l_k}$ intersect transversally. Assume that numbers $\boldsymbol{w}$ are convenient for the transversal intersection of $\tilde{H}_1^{l_1}, \ldots, \tilde{H}_k^{l_k}$.*

Let Ω be an $\mathcal{L}_{\boldsymbol{w}}$-valued meromorphic differential k-form on $\mathcal{E}$ with logarithmic singularities at the union of all hyperplanes $\{\tilde{H}_i^{l_i} | i = 1, \ldots, k, l_i \in L_i\}$. Assume that Ω has zero k-iterated residues at all points of $\mathcal{E}$. Then Ω is the zero form.

Proof. The proof is by induction on k. If $k = 1$, then Ω is regular on E. Since $\boldsymbol{w}$ are convenient, Ω vanishes.

Step of the induction. The residue of Ω at any hyperplane $\tilde{H}_j^{l_j}$ has the same properties as Ω: the residue has logarithmic singularities at the union of all intersections $\tilde{H}_i^{l_i} \cap \tilde{H}_j^{l_j}$, the residue has zero $k-2$-iterated residue at any point. By the induction assumption, the residue of Ω at $\tilde{H}_j^{l_j}$ vanishes, hence, Ω is regular on $\mathcal{E}$ and Ω is the zero form due to the convenience of $\boldsymbol{w}$. □

4.3.2 Example Here is an example illustrating Theorem 4.11 for $k = 1$. Consider an analog of the pair of systems (4.3) and (4.4): $2\tilde{t} = 0$ and $2\tilde{v} = \tilde{w}$, where $w \notin \Lambda$. We can choose $\mathcal{U} = \{0, 1/2, \tau/2, 1/2 + \tau/2\}$ and $\mathcal{V} = \{w/2, w/2+1/2, w/2+\tau/2, w/2+1/2+\tau/2\}$. The differential forms $\omega_{\boldsymbol{v}}$, $\boldsymbol{v} \in \mathcal{V}$, given by formula (4.7), are

$$\omega_1 = 2\sigma_{w/2}(2t, \tau)dt, \quad \omega_2 = 2\sigma_{w/2+1/2}(2t, \tau)dt,$$
$$\omega_3 = 2e^{-2\pi\sqrt{-1}t}\sigma_{w/2+\tau/2}(2t,\tau)dt, \quad \omega_4 = 2e^{-2\pi\sqrt{-1}t}\sigma_{w/2+1/2+\tau/2}(2t,\tau)dt$$

Denote $\gamma = e^{-\pi\sqrt{-1}w}$. Then the differential forms $\omega_{\boldsymbol{u}}$, $\boldsymbol{u} \in \mathcal{U}$, given by Theorem 4.11, are

$$\tilde{\omega}_1 = \frac{1}{4}(\omega_1 + \omega_2 + \gamma\omega_3 + \gamma\omega_4), \quad \tilde{\omega}_2 = \frac{1}{4}(\omega_1 + \omega_2 - \gamma\omega_3 - \gamma\omega_4),$$

$$\tilde{\omega}_3 = \frac{1}{4}(\omega_1 - \omega_2 + \gamma\omega_3 - \gamma\omega_4), \quad \tilde{\omega}_4 = \frac{1}{4}(\omega_1 - \omega_2 - \gamma\omega_3 + \gamma\omega_4).$$

The forms $\tilde{\omega}_i$, $i = 1, \ldots, 4$, define meromorphic $\mathcal{L}_{\boldsymbol{w}}$-valued differential forms on E. The form $\tilde{\omega}_1$ is regular on $\mathbb{C} - \Lambda$, has simple poles at Λ, has residue 1 at $t = 0$. The forms $\tilde{\omega}_2, \tilde{\omega}_3, \tilde{\omega}_4$ have similar properties with respect to the sets $\mathbb{C} - (1/2 + \Lambda)$, $\mathbb{C} - (\tau/2 + \Lambda)$, $\mathbb{C} - (\tau/2 + 1/2 + \Lambda)$ and points $1/2$, $\tau/2$, $\tau/2 + 1/2$, respectively. These properties imply that

$$\tilde{\omega}_1 = \sigma_w(t, \tau)dt, \qquad \tilde{\omega}_2 = \sigma_w(t - 1/2, \tau)dt,$$
$$\tilde{\omega}_3 = \sigma_w(t - \tau/2, \tau)dt, \qquad \tilde{\omega}_4 = \sigma_w(t - \tau/2 - 1/2, \tau)dt.$$

5 Arbitrary elliptic arrangement

5.1 An elliptic arrangement

An elliptic arrangement in $\mathcal{E} = E^k$ is a finite collection $\mathcal{C} = \{\tilde{H}_j\}_{j\in J}$ of elliptic hyperplanes. We fix coordinates $\tilde{t}_1, \ldots, \tilde{t}_k$ on $\mathcal{E}$ and for every

$j \in J$ we fix an equation $\tilde{t}_1 a_{1j} + \cdots + \tilde{t}_k a_{kj} - \tilde{z}_j = 0$ defining the hyperplane $\tilde{H}_j$.

We denote by

$$\tilde{M}_{\mathcal{C}} = \mathcal{E} - \cup_{j \in J} \tilde{H}_j \ ,$$

the complement of the arrangement.

Consider the intersection of any $\ell \leqslant k$ transversal hyperplanes of $\mathcal{C}$. The intersection consists of a finite set of parallel elliptic $k - \ell$ planes. Each of these $k - \ell$-planes will be called an edge of $\mathcal{E}$. In particular, if $\ell = k$, then the 0-planes will be called *vertices* of $\mathcal{E}$.

For an edge X we denote $J_X = \{j \in\subset J \mid X \subset \tilde{H}_j\}$.

We denote by $\tilde{\mathcal{U}}$ the set of all vertices of $\mathcal{C}$. For every vertex $\tilde{\boldsymbol{u}} \in \tilde{\mathcal{U}}$ we choose a point $\boldsymbol{u} \in \mathbb{C}^k$ such that $p(\boldsymbol{u}) = \tilde{\boldsymbol{u}}$. The set of all chosen points in $\mathbb{C}^k$ is denoted by $\mathcal{U}$.

We say that complex numbers $\boldsymbol{w} = (w_1, \ldots, w_k)$ are *convenient* for the elliptic arrangement $\mathcal{C}$, if $\boldsymbol{w}$ are convenient for the intersection of every $\ell < k$ transversal hyperplanes $\tilde{H}_{j_1}, \ldots, \tilde{H}_{j_\ell}$ of $\mathcal{C}$ (in the sense of Section 4.2).

5.2 Differential k-forms of an elliptic arrangement

For a vertex $\tilde{\boldsymbol{u}} \in \tilde{\mathcal{U}}$, we denote by $\mathcal{C}_{\tilde{\boldsymbol{u}}} = \{\tilde{H}_j\}_{j \in I_{\tilde{\boldsymbol{u}}}}$ the subarrangement of all hyperplanes of $\mathcal{C}$ containing $\tilde{\boldsymbol{u}}$. In a small neighborhood of $\tilde{\boldsymbol{u}}$ the arrangement $\mathcal{C}_{\tilde{\boldsymbol{u}}}$ is isomorphic to a central arrangement of affine hyperplanes. We denote by $\mathcal{A}^k_{\tilde{\boldsymbol{u}}}$ the k-th graded component of the Orlik-Solomon algebra of that arrangement. More precisely, let $\mathcal{A}^k_{\tilde{\boldsymbol{u}}}$ be the complex vector space generated by symbols $(\tilde{H}_{j_1}, ..., \tilde{H}_{j_k})$ with $j_i \in J_{\tilde{\boldsymbol{u}}}$, subject to the relations:

(i) $(\tilde{H}_{j_1}, ..., \tilde{H}_{j_k}) = 0$ if $\tilde{H}_{j_1},...,\tilde{H}_{j_k}$ are not transversal;
(ii) $(\tilde{H}_{j_{\sigma(1)}}, ..., \tilde{H}_{j_{\sigma(k)}}) = (-1)^{|\sigma|}(\tilde{H}_{j_1}, ..., \tilde{H}_{j_k})$ for any $\sigma \in S_k$;
(iii) $\sum_{i=1}^{k+1} (-1)^i (\tilde{H}_{j_1}, ..., \widehat{\tilde{H}_{j_i}}, ..., \tilde{H}_{j_{k+1}}) = 0$ for any $k + 1$ elliptic hyperplanes of $\mathcal{C}_{\tilde{\boldsymbol{u}}}$.

We set

$$\mathcal{A}^k_{\mathcal{C}} = \oplus_{\tilde{\boldsymbol{u}} \in \tilde{\mathcal{U}}} \mathcal{A}^k_{\tilde{\boldsymbol{u}}}.$$

Let us fix $\boldsymbol{w} = (w_1, \ldots, w_k)$ convenient for $\mathcal{E}$. Let $\tilde{\boldsymbol{u}} \in \tilde{\mathcal{U}}$ and $\boldsymbol{u} \in \mathcal{U}$ be such that $p(\boldsymbol{u}) = \tilde{\boldsymbol{u}}$. Let $\tilde{H}_{j_l}, \ldots, \tilde{H}_{j_k}$ be any k transversal hyperplanes in $\mathcal{C}_{\tilde{\boldsymbol{u}}}$. Denote by $\omega_{\boldsymbol{u}, \tilde{H}_{j_l}, \ldots, \tilde{H}_{j_k}}(\boldsymbol{t}, \tau)$ the differential meromorphic k-form on $\mathbb{C}^k$ assigned by Theorem 4.11 to these k transversal hyperplanes and denoted by $\omega_{\boldsymbol{u},\tilde{H}_1,\ldots,\tilde{H}_k}(\boldsymbol{t},\tau)$ in Theorem 4.11. We denote by $A^k_{\tilde{\boldsymbol{u}}}$ the complex vector space generated by the forms $\omega_{\boldsymbol{u};\tilde{H}_{j_1},\ldots,\tilde{H}_{j_k}}(\boldsymbol{t},\tau)$. Notice that by Lemma 4.12, the space $A^k_{\tilde{\boldsymbol{u}}}$ does not depend on the choice of $\boldsymbol{u}$ such $p(\boldsymbol{u}) = \tilde{\boldsymbol{u}}$.

We denote by $A^k_{\mathcal{C}}$ the sum of vector spaces $A^k_{\tilde{\boldsymbol{u}}}$, $\tilde{\boldsymbol{u}} \in \tilde{\mathcal{U}}$.

Theorem 5.1.

(i) *The map* $\mathcal{A}^k_{\tilde{\boldsymbol{u}}} \to A^k_{\tilde{\boldsymbol{u}}}$, $(\tilde{H}_{j_1},\ldots,\tilde{H}_{j_k}) \mapsto \omega_{\boldsymbol{u};\tilde{H}_{j_1},\ldots,\tilde{H}_{j_k}}(\boldsymbol{t},\tau)$, *is an isomorphism of vector spaces.*

(ii) *We have* $A^k_{\mathcal{C}} = \oplus_{\tilde{\boldsymbol{u}}\in\tilde{\mathcal{U}}} A^k_{\tilde{\boldsymbol{u}}}$.

Proof. It is enough to prove that for any $k+1$ elliptic hyperplanes of $\mathcal{C}_{\tilde{\boldsymbol{u}}}$, we have the elliptic Orlik-Solomon relation

$$\sum_{i=1}^{k+1}(-1)^i \omega_{\boldsymbol{u},\tilde{H}_{j_1},\ldots,\widehat{\tilde{H}_{j_i}},\ldots,\tilde{H}_{j_{k+1}}} = 0. \tag{5.1}$$

The proof is by induction on k. If $k = 1$, the difference $\omega_{\boldsymbol{u},\tilde{H}_1} - \omega_{\boldsymbol{u},\tilde{H}_2}$ is regular on E and is the zero 1-form due to the convenience of $\boldsymbol{w}$.

Step of the induction. For every $i = 1,\ldots,k+1$, the residue at $\tilde{H}_{j_i}$ of the left hand side in (5.1) is the left hand side of an elliptic Orlik-Solomon relation for an arrangement in $\tilde{H}_{j_i}$, see Theorem 4.14. By the induction assumption, the residue of the left hand side at $\tilde{H}_{j_i}$ is the zero $k-1$-form. Hence, the left hand side in (5.1) is regular on $\mathcal{E}$ and vanishes due to the convenience of $\boldsymbol{w}$. □

5.3 Cohomology of the complement

Every form $\omega \in A^k_{\mathcal{C}}$ induces a holomorphic $\mathcal{L}_{\boldsymbol{w}}$-valued k-form $p_*(\omega)$ on the complement $\tilde{M}_{\mathcal{C}}$ of the elliptic arrangement $\mathcal{C}$. The image of $A^k_{\mathcal{C}}$ will be denoted by $p_*(A^k_{\mathcal{C}})$. The assignment to $p_*(\omega)$ its cohomology class $[p_*(\omega)]$ defines a linear map $\iota : A^k_{\mathcal{C}} \to H^k(\tilde{M}_{\mathcal{C}};\mathcal{L}_w)$. Here $H^*(\tilde{M}_{\mathcal{C}};\mathcal{L}_{\boldsymbol{w}})$ denotes the cohomology of $\tilde{M}_{\mathcal{C}}$ with coefficients in the local system of horizontal sections of $\mathcal{L}_{\boldsymbol{w}}$.

Theorem 5.2. *Assume that* $\boldsymbol{w}$ *are convenient for* $\mathcal{C}$. *Then* $H^\ell(\tilde{M}_{\mathcal{C}};\mathcal{L}_{\boldsymbol{w}}) = 0$ *for* $\ell \neq k$ *and* $\iota_{\mathcal{C}} : A^k_{\mathcal{C}} \to H^k(\tilde{M}_{\mathcal{C}};\mathcal{L}_{\boldsymbol{w}})$ *is an isomorphism.*

Proof. We need the following lemmas. □

Lemma 5.3. *The map* $\iota_{\mathcal{C}} : A^k_{\mathcal{C}} \to H^k(\tilde{M}_{\mathcal{C}};\mathcal{L}_{\boldsymbol{w}})$ *is a monomorphism.*

Proof. For a central affine arrangement of hyperplanes in $\mathbb{C}^k$, the k-th homology group of the complement with trivial coefficients is generated by k-dimensional tori located near the vertex of the arrangement and corresponding to the k-flags of the arrangement, see Section 4.4 in [7]. The nondegenerate pairing between the top degree cohomology of the complement with trivial coefficients and the top degree homology is given by the integrals of the Orlik-Solomon differential forms over the tori. The integrals are nothing else but the multiple residues of the differential forms at the flags of the arrangement. Locally at $\tilde{\boldsymbol{u}} \in \tilde{\mathcal{U}}$, the arrangement $\mathcal{C}_{\tilde{\boldsymbol{u}}}$ is isomorphic to a central affine arrangement. The k-dimensional tori of that central arrangement, considered as k-dimensional tori in a small neighborhood of $\tilde{\boldsymbol{u}}$ in $\tilde{M}$ induces a vector subspace $H_{\tilde{\boldsymbol{u}},k} \subset H_k(\tilde{M}; \mathcal{L}_{\boldsymbol{w}})$. The pairing between $H_{\tilde{\boldsymbol{u}},k}$ and $A_{\tilde{\boldsymbol{u}}'}$ is zero if $\tilde{\boldsymbol{u}} \neq \tilde{\boldsymbol{u}}'$ and the pairing is nondegenerate if $\tilde{\boldsymbol{u}} = \tilde{\boldsymbol{u}}'$. □

Lemma 5.4. *Theorem* 5.2 *is true for* $k = 1$.

Proof. The lemma follows from the convenience of $\boldsymbol{w}$ and the exact sequence for the pair $\tilde{M}_{\mathcal{C}} \subset E$. □

Let j_0 be an element of J. We consider the following three elliptic arrangements: $\mathcal{C}, \mathcal{C}', \mathcal{C}''$, where $\mathcal{C}' = \{\tilde{H}_j\}_{j\in J-\{j_0\}}$ and $\mathcal{C}''$ is the elliptic arrangement induced by $\mathcal{C}$ on $\tilde{H}_{j_0}$.

Lemma 5.5. *We have an exact sequence*

$$0 \to A^k_{\mathcal{C}'} \to A^k_{\mathcal{C}} \to A^{k-1}_{\mathcal{C}''} \to 0, \tag{5.2}$$

where the second map is the residue at $\tilde{H}_{j_0}$.

Proof. The lemma follows from the fact that $A^k_{\mathcal{C}}, A^k_{\mathcal{C}'}, A^{k-1}_{\mathcal{C}''}$ are isomorphic to the top degree components of the Orlik-Solomon algebras of central arrangements. □

Lemma 5.6.

(i) *For* $\ell \neq k$ *we have* $H^{\ell}(\tilde{M}_{\mathcal{C}'};\mathcal{L}_{\boldsymbol{w}}) = H^{\ell}(\tilde{M}_{\mathcal{C}}; \mathcal{L}_{\boldsymbol{w}}) = H^{\ell-1}(\tilde{M}_{\mathcal{C}''};\mathcal{L}_{\boldsymbol{w}}) = 0$.

(ii) *Consider the following diagram*

$$\begin{array}{ccccccccc}
0 & \longrightarrow & A^k_{\mathcal{C}'} & \longrightarrow & A^k_{\mathcal{C}} & \longrightarrow & A^{k-1}_{\mathcal{C}''} & \longrightarrow & 0 \\
 & & \downarrow{\scriptstyle \iota_{\mathcal{C}'}} & & \downarrow{\scriptstyle \iota_{\mathcal{C}}} & & \downarrow{\scriptstyle \iota_{\mathcal{C}''}} & & \\
0 & \longrightarrow & H^k(\tilde{M}_{\mathcal{C}'}; \mathcal{L}_{\boldsymbol{w}}) & \longrightarrow & H^k(\tilde{M}_{\mathcal{C}}; \mathcal{L}_{\boldsymbol{w}}) & \longrightarrow & H^{k-1}(\tilde{M}_{\mathcal{C}''}; \mathcal{L}_{\boldsymbol{w}}) & \longrightarrow & 0
\end{array}$$

where the top horizontal sequence is the sequence (5.2), *the homomorphisms of the bottom horizontal sequence are the homomorphisms of the exact sequence of the pair* $\tilde{M}_C \subset \tilde{M}_{C'}$, *cf. Lemma* 4.1. *Then the diagram is commutative, the horizontal sequence is exact and the vertical homomorphisms are isomorphisms.*

Proof. The proof of this lemma is similar to the corresponding proofs in Section 5.4 of [6]. Namely, using Lemma 4.1, one proves that there is a cohomology long exact sequence

$$\cdots \to H^{\ell}(\tilde{M}_{C'}; \mathcal{L}_w) \to H^{\ell}(\tilde{M}_C; \mathcal{L}_w) \\ \to H^{\ell-1}(\tilde{M}_{C''}; \mathcal{L}_w) \to H^{\ell+1}(\tilde{M}_{C'}; \mathcal{L}_w) \to \dots,$$

cf. Corollary 5.81 in [6]. Using the induction on k, one proves that $H^{\ell}(\tilde{M}_{C'}; \mathcal{L}_w) \simeq H^{\ell}(\tilde{M}_C; \mathcal{L}_w)$ if $\ell \neq k$. Using the induction on the number of hyperplanes in C, one concludes that $H^{\ell}(\tilde{M}_C; \mathcal{L}_w) = 0$ if $\ell \neq k$ and one gets an exact sequence $0 \to H^k(\tilde{M}_{C'}; \mathcal{L}_w) \to H^k(\tilde{M}_C; \mathcal{L}_w) \to H^{k-1}(\tilde{M}_{C''}; \mathcal{L}_w) \to 0$. Then using the double induction on k and the number of hyperplanes in C one gets the second statement of Lemma 5.6.

Lemma 5.6 implies Theorem 5.2.

Theorem 5.2 implies Theorems 3.3 and 2.1. □

References

[1] V. I. ARNOLD, *The cohomology ring of the pure braid group*, Mat. Zametki **5** (1969), 227–231, Math. Notes **5** (1969) 138–140.

[2] G. FELDER and A. VARCHENKO, *Integral representation of solutions of the elliptic Knizhnik-Zamolodchikov-Bernard equations*, Int. Math. Res. Notices n. 5 (1995), 221–233.

[3] G. FELDER and A. VARCHENKO *Three formulae for eigenfunctions of integrable Schrödinger operators*, Compositio Math. **107** (1997), 143–175.

[4] G. FELDER, R. RIMÁNYI and A. VARCHENKO, *Poincaré-Birkhoff-Witt expansions of the canonical elliptic differential form*, In: "Quantum Groups", 191–208, Contemp. Math., Vol. 433, Amer. Math. Soc., Providence, RI, 2007.

[5] P. ORLIK and L. SOLOMON, *Combinatorics and topology of complements of hyperplanes*, Invent. Math. **56** (1980), 167–189.

[6] P. ORLIK and H. TERAO, "Arrangements of Hyperplanes" Springer-Verlag, Berlin-Heidelberg-New York, 1992.

[7] V. SCHECHTMAN and A. VARCHENKO, *Arrangements of hyperplanes and Lie algebra homology*, Invent. Math., (1) **106** (1991), 139–194.
[8] E. B. VINBERG, "A Course in Algebra", AMS, 2003
[9] E. T. WHITTAKER and G. N. WATSON, "A Course of Modern Analysis", Reprint of the fourth (1927) edition, Cambridge University Press (September 1996).

Residual nilpotence for generalizations of pure braid groups

Ivan Marin

Abstract. It is known that the pure braid groups are residually torsion-free nilpotent. This property is however widely open for the most obvious generalizations of these groups, like pure Artin groups and like fundamental groups of hyperplane complements (even reflection ones). In this paper we relate this problem to the faithfulness of linear representations, and prove the residual torsion-free nilpotence for a few other groups.

1 Introduction

It has been known for a long time (see [10, 11]) that the pure braid groups are residually nilpotent, meaning that they have 'enough' nilpotent quotients to distinguish their elements, or equivalently that the intersection of their descending central series is trivial. Recall that a group G is called residually $\mathcal{F}$ for $\mathcal{F}$ a class of groups if for all $g \in G \setminus \{1\}$ there exists $\pi : G \twoheadrightarrow Q$ with $Q \in \mathcal{F}$ such that $\pi(g) \neq 1$. It is also known that they have the far stronger property of being residually torsion-free nilpotent. The strongness of this latter assumption is illustrated by the following implications (where 'residually p' corresponds to the class of p-groups).

$$\text{residually free} \Rightarrow \text{residually torsion-free nilpotent} \Rightarrow \text{residually } p \text{ for all } p$$

$$\Rightarrow \text{residually } p \text{ for some } p \Rightarrow \text{residually nilpotent} \Rightarrow \text{residually finite}$$

Pure braid groups are not residually free. The following proof of this fact has been communicated to me several years ago by Luis Paris (note however that the pure braid group on 3 strands $P_3 \simeq F_2 \times \mathbb{Z}$ is residually free; it has been announced this year that P_4 is not residually free, see [4]).

Proposition 1.1. *The pure braid group P_n is not residually free for $n \geq 5$.*

Proof. It is sufficient to show that P_5 is not residually free. Letting $\sigma_1, \ldots, \sigma_4$ denote the Artin generators of the braid group B_5, P_5 contains the subgroup H generated by $a = \sigma_1^2, b = \sigma_2^2, c = \sigma_3^2, d = \sigma_4^2$. As shown in [9] (see also [5]) this group is a right-angled Artin group, which contains a subgroup $H_0 =< a, b, d >$ isomorphic to $F_2 \times \mathbb{Z}$, which is well-known to be residually free but not fully residually free (see [1]).

If we can exhibit $x \in P_5$ such that the subgroup generated by x and H_0 is a free product $\mathbb{Z} * H_0$, then, by a result of [1] which states that the free product of two non-trivial groups can be residually free only if the two of them are fully residually free, this proves that P_5 is not residually free.

One can take $x = cbc^{-1}$. Indeed, if $< x, H_0 >$ were not a free product, then it would exist a word with trivial image of the form $cb^{u_1}c^{-1}y_1cb^{u_2}c^{-1}y_2 \ldots cb^{u_r}c^{-1}y_r$ with $y_i \in< a, b, d >$, $y_i \neq 1$ for $i < r$, $u_i \neq 0$ for $i \leq r$, and $r \geq 1$. But in a right-angled Artin group generated by a set X of letters, an expression can be reduced if and only if it contains a word of the form $x \ldots x^{-1}$ or $x^{-1} \ldots x$ with $x \in X$, such that all the letters in $\ldots$ commute with x (see *e.g.* [21]). From this it is straightforward to check that the former expression cannot be reduced, and this proves the claim. □

The original approach for proving this property of residual torsion-free nilpotence seems to fail for most of the usual generalizations of pure braid groups. Another approach has been used in [15, 16], using faithful linear representations, thus relating the linearity problem with this one. The main lemma is the following one.

Lemma 1.2. *Let $N \geq 1$, $\mathbf{k}$ a field of characteristic 0 and $A = \mathbf{k}[[h]]$ the ring of formal power series.Then the group $\mathrm{GL}_N^0(A) = \{X \in \mathrm{GL}_N(A) | X \equiv 1 \mod h\} = 1 + h\mathrm{Mat}_N(A)$ is residually torsion-free nilpotent.*

Proof. Let $G = \mathrm{GL}_N^0(A)$, $G_r = \{g \in G \mid g \equiv 1 \mod h^r\}$,

$$\begin{aligned} G^{(r)} &= \{X \in \mathrm{GL}_N(\mathbf{k}[h]/h^r) \mid X \equiv 1 \mod h\} \\ &= 1 + h\mathrm{Mat}_N(\mathbf{k}[h]/h^r) \subset \mathrm{Mat}_N(\mathbf{k}[h]/h^r) \end{aligned}$$

Clearly $G_r \lhd G$ and the natural map $G \to G^{(r)}$ has for kernel G_r, hence G/G_r is isomorphic to a subgroup of $G^{(r)}$. This latter group is clearly nilpotent, as $(1 + h^u x, 1 + h^v y) \equiv 1 + h^{uv}(xy - yx) \mod h^{uv+1}$ (where $(a, b) = aba^{-1}b^{-1}$), and torsion-free as $(1 + hx)^n \equiv 1 + nhx \mod h^2$ and $\mathbf{k}$ has characteristic 0. Thus all the G/G_r are torsion-free nilpotent, and since clearly $\bigcap_r G_r = \{1\}$ we get that G is residually torsion-free nilpotent. □

Usually, linear representations have their image in such a group when they appear as the monodromy of a flat connection on a *trivial* vector bundle (see [16]). However, we show how to (partly conjecturally) use this approach in situation where this geometric motivation is far less obvious. In particular, we prove the following.

Theorem 1.3. *If B is an Artin group for which the Paris representation is faithful, then its pure subgroup P is residually torsion-free nilpotent.*

So far, this Paris representation, which is a generalization of the Krammer representation of [12], has been shown to be faithful only for the case where W is a finite Coxeter group. By contrast, in the case of the pure braid groups of complex reflexion groups, which are other natural generalization of pure braid groups, and for which a natural and possibly faithful monodromy representation has been constructed in [17], we get the following more modest but unconditional result.

Theorem 1.4. *If B is the braid group of a complex reflection group of type G_{25}, G_{26}, G_{32}, G_{31}, then its pure braid group P is residually torsion-free nilpotent.*

The pure braid groups involved in the latter statement are equivalently described as the fundamental groups of complements of remarkable configurations of hyperplanes : the groups G_{25}, G_{26} are related to the symmetry group of the so-called Hessian configuration of the nine inflection points of nonsingular cubic curves, while G_{32} acts by automorphisms on the configuration of 27 lines on a nonsingular cubic surface. These groups belong to the special case of so-called 'Shephard groups', namely the symmetry groups of regular complex polytopes. The group G_{31}, introduced by H. Maschke in his first paper [19], is not a Shephard group, has all its reflections of order 2, and is connected to the theory of hyperelliptic functions. It is the only 'exceptional' reflection group in dimension $n \geq 3$ which cannot be generated by n reflections.

2 Artin groups and Paris representation

2.1 Preliminaries on Artin groups

Let S be a finite set. Recall that a *Coxeter matrix* based on S is a matrix $M = (m_{s,t})_{s,t \in S}$ indexed by elements of S such that

- $m_{ss} = 1$ for all $s \in S$
- $m_{st} = m_{ts} \in \{2, 3, \ldots, \infty\}$ for all $s, t \in S$, $s \neq t$.

and that the Coxeter system associated to M is the couple (W, S), with W the group presented by $< S \mid \forall s \in S\ s^2 = 1, \forall s, t \in S\ (st)^{m_{st}} = 1 >$. Let $\Sigma = \{\sigma_s, s \in S\}$ be a set in natural bijection with S. The *Artin system* associated to M is the pair (B, Σ) where B is the group presented by $< \Sigma \mid \forall s, t \in S\ \underbrace{\sigma_s\sigma_t\sigma_s\dots}_{m_{s,t}\text{ terms}} = \underbrace{\sigma_t\sigma_s\sigma_t\dots}_{m_{s,t}\text{ terms}} >$, and called the Artin group associated to M. The Artin monoid B^+ is the monoid with the same presentation. According to [20], the natural monoid morphism $\sigma_s \mapsto \sigma_s$, $B^+ \to B$, is an embedding. There is a natural morphism $B \twoheadrightarrow W$ given by $\sigma_s \mapsto s$, whose kernel is known as the pure Artin group P.

For the sequel we will need a slightly more specialized vocabulary, borrowed from [20]. A Coxeter matrix is said to be *small* if $m_{s,t} \in \{2, 3\}$ for all $s \neq t$, and it is called *triangle-free* if there is no triple (s, t, r) in S such that $m_{s,t}$, $m_{t,r}$ and $m_{r,s}$ are all greater than 2.

2.2 Paris representation

To a Coxeter system as above is naturally associated a linear representation of W, known as the reflection representation. We briefly recall its construction. Let $\Pi = \{\alpha_s; s \in S\}$ denote a set in natural bijection with S, called the set of simple roots. Let U denote the $\mathbf{R}$-vector space with basis Π, and $< , >: U \times U \to \mathbf{R}$ the symmetric bilinear form defined by

$$< \alpha_s, \alpha_t >= \begin{cases} -2\cos\left(\dfrac{\pi}{m_{st}}\right) & \text{if} \quad m_{st} < \infty \\ -2 & \text{otherwise} \end{cases}$$

In particular $< \alpha_s, \alpha_s >= 2$. There is a faithful representation $W \to \mathrm{GL}(U)$ defined by $s(x) = x - < \alpha_s, x > \alpha_s$ for $x \in U$, $s \in S$, which preserves the bilinear form $< , >$. Let $\Phi = \{w\alpha_s; s \in S, w \in W\}$ be the root system associated to W, $\Phi^+ = \{\sum_{s\in S} \lambda_s\alpha_s \in \Phi; \forall s \in S\ \lambda_s \geq 0\}$, and $\Phi^- = -\Phi^+$. We let ℓ denote the length function on W (resp. B^+) with respect to S (resp. Σ). The *depth* of $\beta \in \Phi^+$ is

$$dp(\beta) = \min\{m \in \mathbf{N} \mid \exists w \in W\ \ w.\beta \in \Phi^- \text{ and } \ell(w) = m\}.$$

We have (see [20, Lemma 2.5])

$$dp(\beta) = \min\{m \in \mathbf{N} \mid \exists w \in W, s \in S\ \ \beta = w^{-1}.\alpha_s \text{ and } \ell(w) + 1 = m\}.$$

When $s \in S$ and $\beta \in \Phi^+ \setminus \{\alpha_s\}$, we have

$$dp(s.\beta) = \begin{cases} dp(\beta) - 1 & \text{if } \langle\alpha_s, \beta\rangle > 0 \\ dp(\beta) & \text{if } \langle\alpha_s, \beta\rangle = 0 \\ dp(\beta) + 1 & \text{if } \langle\alpha_s, \beta\rangle < 0 \end{cases}$$

In [20], polynomials $T(s,\beta) \in \mathbf{Q}[y]$ are defined for $s \in S$ and $\beta \in \Phi^+$. They are constructed by induction on $dp(\beta)$, by the following formulas. When $dp(\beta) = 1$, that is $\beta = \alpha_t$ for some $t \in S$, then

$$\text{(D1)} \quad T(s,\alpha_t) = y^2 \quad \text{if} \quad t = s$$
$$\text{(D2)} \quad T(s,\alpha_t) = 0 \quad \text{if} \quad t \neq s$$

When $dp(\beta) \geq 2$, then there exists $t \in S$ such that $dp(t.\beta) = dp(\beta) - 1$, and we necessarily have $b = \langle \alpha_t, \beta \rangle > 0$. In case $\langle \alpha_s, \beta \rangle > 0$, we have

$$\text{(D3)} \quad T(s,\beta) = y^{dp(\beta)}(y-1);$$

in case $\langle \alpha_s, \beta \rangle = 0$, we have

$$\text{(D4)} \quad T(s,\beta) = yT(s,\beta - b\alpha_t) \qquad \text{if } \langle \alpha_s, \alpha_t \rangle = 0$$
$$\text{(D5)} \quad T(s,\beta) = (y-1)T(s,\beta - b\alpha_t) + yT(t,\beta - b\alpha_s - b\alpha_t) \qquad \text{if } \langle \alpha_s, \alpha_t \rangle = -1;$$

and, in case $\langle \alpha_s, \beta \rangle = -a < 0$, we have

$$\text{(D6)} \quad T(s,\beta) = yT(s,\beta - b\alpha_t) \qquad \text{if } \langle \alpha_s, \alpha_t \rangle = 0$$

$$\text{(D7)} \quad T(s,\beta) = (y-1)T(s,\beta - b\alpha_t) + yT(t,\beta - (b-a)\alpha_s - b\alpha_t) \quad \text{if } \langle \alpha_s, \alpha_t \rangle = -1 \text{ and } b > a$$

$$\text{(D8)} \quad T(s,\beta) = T(t,\beta - b\alpha_t) + (y-1)T(s,\beta - b\alpha_t) \quad \text{if } \langle \alpha_s, \alpha_t \rangle = -1 \text{ and } b = a$$

$$\text{(D9)} \quad T(s,\beta) = yT(s,\beta - b\alpha_t) + T(t,\beta - b\alpha_t) + y^{dp(\beta)-1}(1-y) \quad \text{if } \langle \alpha_s, \alpha_t \rangle = -1 \text{ and } b < a.$$

Now introduce $\mathcal{E} = \{e_\beta;\, \beta \in \Phi^+\}$ a set in natural bijection with Φ^+, and let V denote the free $\mathbf{Q}[x, y, x^{-1}, y^{-1}]$-module with basis $\mathcal{E}$. For $s \in S$, one defines a linear map $\varphi_s : V \to V$ by

$$\varphi_s(e_\beta) = \begin{cases} 0 & \text{if } \beta = \alpha_s \\ e_\beta & \text{if } \langle \alpha_s, \beta \rangle = 0 \\ ye_{\beta - a\alpha_s} & \text{if } \langle \alpha_s, \beta \rangle = a > 0 \text{ and } \beta \neq \alpha_s \\ (1-y)e_\beta + e_{\beta + a\alpha_s} & \text{if } \langle \alpha_s, \beta \rangle = -a < 0 \end{cases}$$

We have $\varphi_s\varphi_t = \varphi_t\varphi_s$ if $m_{s,t} = 2$, $\varphi_s\varphi_t\varphi_s = \varphi_t\varphi_s\varphi_t$ if $m_{s,t} = 3$. Now the Paris representation $\Psi : B \to \mathrm{GL}(V)$ is defined by $\Psi : \sigma_s \mapsto \psi_s$, with

$$\psi_s(e_\beta) = \varphi_s(e_\beta) + xT(s,\beta)e_{\alpha_s}.$$

2.3 Reduction modulo h

We embed $\mathbf{Q}[y]$ inside $\mathbf{Q}[[h]]$ under $y \mapsto e^h$ and consider congruences $\equiv$ modulo h.

Using the formulas of [20], we deduce the main technical step of our proof.

Proposition 2.1. *Let $s \in S$ and $\beta \in \Phi^+$. Then $T(s,\beta) \equiv 1$ if $\beta = \alpha_s$ and $T(s,\beta) \equiv 0$ otherwise.*

Proof. The case $dp(\beta) = 1$ is a consequence of (D1),(D2), as $y \equiv 1 \mod h$. We thus browse through the various cases when $dp(\beta) \geq 2$, and use induction on the depth. As in the definition of the polynomials, let $t \in S$ such that $dp(\gamma) = dp(\beta) - 1$ for $\gamma = t.\beta$, and recall that necessarily $b = \langle \alpha_t, \beta \rangle > 0$. In case $\langle \alpha_s, \beta \rangle > 0$ then (D3) implies $T(s,\beta) \equiv 0$. If $\langle \alpha_s, \beta \rangle = 0$, we have several subcases. If $\langle \alpha_s, \alpha_t \rangle = 0$, then (D4) implies $T(s,\beta) \equiv T(s,\gamma)$ with $\gamma = \beta - b\alpha_t = t.\beta$ hence $dp(\gamma) < dp(\beta)$ and $T(s,\gamma) \equiv 0$ by induction, unless $\gamma = \alpha_s$, that is $\alpha_s = \beta - b\alpha_t$, hence taking the scalar product by α_s we would get $2 = 0$, a contradiction. Otherwise, we have $\langle \alpha_s, \alpha_t \rangle = -1$. In that case, (D5) implies $T(s,\beta) \equiv T(t, \beta - b\alpha_s - b\alpha_t)$. Note that $\beta - b\alpha_t = \gamma = t.\beta$, $\langle \alpha_s, \gamma \rangle = 0 - b\langle \alpha_t, \alpha_s \rangle = b$ and $s.\gamma = \gamma - b\alpha_s$. Thus $T(s,\beta) \equiv T(t, st.\beta)$. Now $\langle \alpha_s, \gamma \rangle = b > 0$ hence $dp(s.\gamma) = dp(\gamma) - 1 < dp(\beta)$, unless $\gamma = \alpha_s$; but the case $\gamma = \alpha_s$ cannot occur here, as it would imply

$$2 = \langle \alpha_s, \alpha_s \rangle = \langle \alpha_s, \gamma \rangle = \langle \alpha_s, t.\beta \rangle = \langle \alpha_s, \beta - b\alpha_t \rangle = -b\langle \alpha_s, \alpha_t \rangle = b$$

hence $\alpha_s = \gamma = t.\beta = \beta - 2\alpha_t$, whence $-1 = \langle \alpha_s, \alpha_t \rangle = \langle \beta, \alpha_t \rangle - 2\langle \alpha_t, \alpha_t \rangle = b - 4 = -2$, a contradiction.

Finally, $T(t, st.\beta) \equiv 0$ by induction unless $\beta - b\alpha_s - b\alpha_t = \alpha_t$, in which case scalar product by α_t leads to the contradiction $2 = 0$.

The last case is when $\langle \alpha_s, \beta \rangle = -a < 0$, which is subdivided in 4 subcases. Either $\langle \alpha_s, \alpha_t \rangle = 0$, and then (D6) implies $T(s,\beta) \equiv T(s, \beta - b\alpha_t) \equiv 0$, as $\beta - b\alpha_t = \alpha_s$ cannot occur (scalar product with α_s yields $2 = -a < 0$). Or $\langle \alpha_s, \alpha_t \rangle = -1$ and $b > a$, then (D7) implies $T(s,\beta) \equiv T(t, \beta - (b-a)\alpha_s - b\alpha_t) \equiv 0$ unless $\alpha_t = \beta - (b-a)\alpha_s - b\alpha_t$, which cannot occur for the same reason as before (take the scalar product with α_t). Or, $\langle \alpha_s, \alpha_t \rangle = -1$ and $b = a$, in which case (D8) implies $T(s,\beta) \equiv T(t, \beta - b\alpha_t) \equiv 0$, as $\alpha_t \neq \beta - b\alpha_t$ (take the scalar product with α_t). Finally, the last subcase is $\langle \alpha_s, \alpha_t \rangle = -1$ and $b < a$, then (D9) implies $T(s,\beta) \equiv T(s, \beta - b\alpha_t) + T(t, \beta - b\alpha_t) \equiv 0$, unless $\alpha_s = \beta - b\alpha_t$, which leads to the contradiction $2 = b - a < 0$ under $\langle \alpha_s, \cdot \rangle$, or $\alpha_t = \beta - b\alpha_t$, which leads to the contradiction $2 = -b < 0$ under $\langle \alpha_t, \cdot \rangle$. □

We now embed $\mathbf{Q}[x^{\pm 1}, y^{\pm 1}]$ into $\mathbf{Q}(\sqrt{2})[[h]]$ under $y \mapsto e^h$, $x \mapsto e^{\sqrt{2}h}$ (any other irrational than $\sqrt{2}$ would also do), and define $\tilde{V} = V \otimes_\iota \mathbf{Q}(\sqrt{2})[[h]]$ where ι is the chosen embedding; that is, $\tilde{V}$ is the free $\mathbf{Q}(\sqrt{2})[[h]]$-module with basis $\mathcal{E}$, and clearly $V \subset \tilde{V}$. We similarly introduce the $\mathbf{Q}(\sqrt{2})$-vector space V_0 with basis $\mathcal{E}$. One has $\mathrm{GL}(V) \subset \mathrm{GL}(\tilde{V})$, and a reduction morphism $\mathrm{End}(\tilde{V}) \to \mathrm{End}(V_0)$. Composing both we get elements $\overline{\psi_s}, \overline{\varphi_s} \in \mathrm{End}(V_0)$ associated to the $\psi_s \in \mathrm{GL}(V)$, $\varphi_s \in \mathrm{End}(V)$. Because $x \equiv 1 \mod h$ and because of Proposition 2.1 one gets from the definition of ψ_s that

$$\begin{aligned} \overline{\psi}_s(e_\beta) &= \overline{\varphi_s}(e_\beta) & \text{if } \beta \neq \alpha_s \\ &= \overline{\varphi_s}(e_{\alpha_s}) + e_{\alpha_s} = e_{\alpha_s} & \text{if } \beta = \alpha_s. \end{aligned}$$

We denote $(w, \beta) \mapsto w \star \beta$ the natural action of W on Φ^+, that is $w \star \beta = \beta$ if $w.\beta \in \Phi^+$, $w \star \beta = -\beta \in \Phi^+$ if $w.\beta \in \Phi^-$. The previous equalities imply

$$\forall s \in S \ \ \forall \beta \in \Phi^+ \ \ \overline{\psi_s}(e_\beta) = e_{s\star\beta}$$

From this we deduce the following.

Proposition 2.2. *For all $g \in P$, $\overline{\Psi(g)} = \mathrm{Id}_{V_0}$.*

Proof. Recall that P is defined as $\mathrm{Ker}(\pi : B \twoheadrightarrow W)$. From $\overline{\psi_s}(e_\beta) = e_{s\star\beta}$ one gets $\overline{\Psi(g)}(e_\beta) = e_{\pi(g)\star\beta}$ for all $g \in B$ and the conclusion. □

As a consequence $\Psi(P) \subset \{\varphi \in \mathrm{GL}(V) \mid \overline{\varphi} = \mathrm{Id}_{V_0}\}$. The group $\{\varphi \in \mathrm{GL}(V) \mid \overline{\varphi} = \mathrm{Id}_{V_0}\}$ is a subgroup of $G = \{\varphi \in \mathrm{GL}(\tilde{V}) \mid \overline{\varphi} = \mathrm{Id}_{V_0}\}$, which we now prove to be residually torsion-free nilpotent. We adapt the argument of Lemma 1.2 to the infinite-dimensional case. Let $\mathbf{k} = \mathbf{Q}(\sqrt{2})$. The canonical projection $\mathbf{k}[[h]] \twoheadrightarrow \mathbf{k}[h]/h^r$ extends to a morphism $\pi_r : \mathrm{End}(\tilde{V}) \to \mathrm{End}(V_0) \otimes_{\mathbf{k}} \mathbf{k}[h]/h^r$ with clearly $\pi_1(\varphi) = \overline{\varphi}$. Let $G_r = \{\varphi \in \mathrm{GL}(\tilde{V}) \mid \pi_r(\varphi) = \mathrm{Id}\}$. Then G/G_r is identified to $\pi_r(G)$ which is a subgroup of $\{\mathrm{Id}_{V_0} + hu \mid u \in \mathrm{End}(V_0) \otimes_{\mathbf{k}} \mathbf{k}[h]/h^r\}$, which is clearly torsion-free and nilpotent. Since $\bigcap_r G_r = \{1\}$ this proves the residual torsion-free nilpotence of G and Theorem 1.3.

3 Braid groups of complex reflection groups

A pseudo-reflection in $\mathbf{C}^n$ is an endomorphism which fixes an hyperplane. For W a finite subgroup of $\mathrm{GL}_n(\mathbf{C})$ generated by pseudo-reflections (so-called complex reflection group), we have a *reflection arrangement* $\mathcal{A} = \{\mathrm{Ker}(s-1) | s \in \mathcal{R}\}$, where $\mathcal{R}$ is the set of reflections of W. Letting X denote the hyperplane complement $X = \mathbf{C}^n \setminus \bigcup \mathcal{A}$, the fundamental groups $P = \pi_1(X)$ and $B = \pi_1(X/W)$ are called the pure braid group and braid group associated to W. The case of spherical type Artin groups corresponds to the case where W is a finite Coxeter group.

It is conjectured that P is always residually torsion-free nilpotent. For this one can assume that W is irreducible. According to [22], such a W belongs either to an infinite series $G(de, e, n)$ depending on three integer parameters d, e, n, or to a finite set of 34 exceptions, denoted $G_4, \dots, G_{37}$. The fiber-type argument of [10, 11] to prove the residual torsion-free nilpotence only works for the groups $G(d, 1, n)$, and when $n = 2$.

For the case of W a finite Coxeter group, we used the Krammer representation to prove that P is residually torsion-free nilpotent in [15, 16]. The exceptional groups of rank $n > 2$ which are not Coxeter groups are the 9 groups $G_{24}, G_{25}, G_{26}, G_{27}, G_{29}, G_{31}, G_{32}, G_{33}, G_{34}$.

We show here that this argument can be adjusted to prove the residual torsion-free nilpotence for a few of them.

We begin with the Shephard groups G_{25}, G_{26}, G_{32}. The Coxeter-like diagrams of these groups are the following ones.

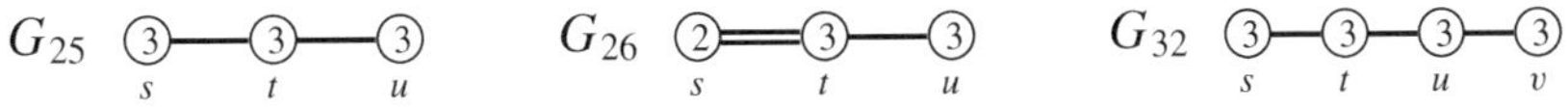

G_{25} ③—③—③ (s, t, u) G_{26} ②═③—③ (s, t, u) G_{32} ③—③—③—③ (s, t, u, v)

It is known (see [3]) that removing the conditions on the order of the generators gives a (diagrammatic) presentation of the corresponding braid group. In particular, these have for braid groups the Artin groups of Coxeter type A_3, B_3 and A_4, respectively.

We recall a matrix expression of the Krammer representation for B of Coxeter type A_{n-1}, namely for the classical braid group on n strands. Letting $\sigma_1, \dots, \sigma_{n-1}$ denote its Artin generators with relations $\sigma_i \sigma_j = \sigma_j \sigma_i$ if $|j - i| \geq 2$, $\sigma_i \sigma_{i+1} \sigma_i = \sigma_{i+1} \sigma_i \sigma_{i+1}$, their action on a specific basis x_{ij} $(1 \leq i < j \leq n)$ is given by the following formulas (see [12])

$$\left\{ \begin{array}{ll} \sigma_k x_{k,k+1} = tq^2 x_{k,k+1} & \\ \sigma_k x_{i,k} = (1-q) x_{i,k} + q x_{i,k+1} & i < k \\ \sigma_k x_{i,k+1} = x_{i,k} + tq^{k-i+1}(q-1) x_{k,k+1} & i < k \\ \sigma_k x_{k,j} = tq(q-1) x_{k,k+1} + q x_{k+1,j} & k+1 < j \\ \sigma_k x_{k+1,j} = x_{k,j} + (1-q) x_{k+1,j} & k+1 < j \\ \sigma_k x_{i,j} = x_{i,j} & i < j < k \text{ or } k+1 < i < j \\ \sigma_k x_{i,j} = x_{i,j} + tq^{k-i}(q-1)^2 x_{k,k+1} & i < k < k+1 < j \end{array} \right.$$

where t and q denote algebraically independent parameters. We embed the field $\mathbf{Q}(q, t)$ of rational fractions in q, t into $K = \mathbf{C}((h))$ by $q \mapsto -\zeta_3 e^h$ and $t \mapsto e^{\sqrt{2}h}$, where ζ_3 denotes a primitive 3-root of 1. We then check by an easy calculation that $\sigma_k^3 \equiv 1$ modulo h. Since the quotients of the braid group on n strands by the relations $\sigma_k^3 = 1$ are, for $n = 3, 4, 5$, the Shephard group of types G_4, G_{25} and G_{32}, respectively, it follows that the pure braid groups of these types embed in $\mathrm{GL}_N^0(A)$ with $N = n(n-1)/2$ and $\mathbf{k} = \mathbf{C}$, and this proves their residual torsion-free nilpotence by Lemma 1.2.

We now turn to type G_{26}. Types G_{25} and G_{26} are symmetry groups of regular complex polytopes which are known to be closely connected (for instance they both appear in the study of the Hessian configuration, see *e.g.* [6, Section 12.4] and [18] example 6.30). The hyperplane arrangement of type G_{26} contains the 12 hyperplanes of type G_{25} plus 9 additional ones. The natural inclusion induces morphisms between the corresponding pure braid groups, which cannot be injective, since a loop around one of the extra hyperplanes is non trivial in type G_{26}. However we will prove the following, which proves the residual torsion-free nilpotence in type G_{26}.

Proposition 3.1. *The pure braid group of type G_{26} embeds into the pure braid group of type G_{25}.*

More precisely, letting B_i, P_i, W_i denote the braid group, pure braid group and pseudo-reflection group of type G_i, respectively, we construct morphisms $B_{26} \hookrightarrow B_{25}$ and $W_{26} \twoheadrightarrow W_{25}$ such that the following diagram commutes, where the vertical arrows are the natural projections.

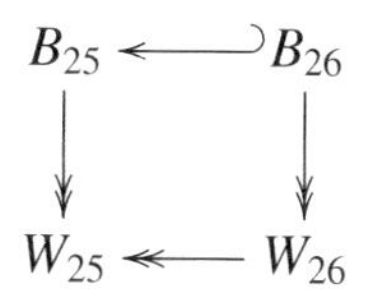

Both horizontal morphisms are given by the formula $(s,t,u) \mapsto ((tu)^3, s, t)$, where s, t, u denote the generators of the corresponding groups according to the above diagrams. The morphism between the pseudo-reflection groups is surjective because it is a retraction of an embedding $W_{25} \hookrightarrow W_{26}$ mapping (s, t, u) to $(t, u, t^{sut^{-1}u})$. The kernel of this projection is the subgroup of order 2 in the center of W_{26} (which has order 6).

We now consider the morphism between braid groups and prove that it is injective. First recall that the braid group of type G_{26} can be identified with the Artin group of type B_3. On the other hand, Artin groups of type B_n are isomorphic to the semidirect product of the Artin group of type A_{n-1}, that we denote $\mathcal{B}_n$ to avoid confusions, with a free group F_n on n generators $g_1, \ldots, g_n$, where the action (so-called 'Artin action') is given (on the left) by

$$\sigma_i : \begin{cases} g_i & \mapsto g_{i+1} \\ g_{i+1} & \mapsto g_{i+1}^{-1} g_i g_{i+1} \\ g_j & \mapsto g_j & \text{if } j \notin \{i, i+1\} \end{cases}$$

If $\tau, \sigma_1, \dots, \sigma_{n-1}$ are the standard generators of the Artin group of type B_n, with $\tau\sigma_1\tau\sigma_1 = \sigma_1\tau\sigma_1\tau$, $\tau\sigma_i = \sigma_i\tau$ for $i > 1$, and usual braid relations between the σ_i, then this isomorphism is given by $\tau \mapsto g_1, \sigma_i \mapsto \sigma_i$ (see [7, Proposition 2.1 (2)] for more details). Finally, there exists an embedding of this semidirect product into the Artin group $\mathcal{B}_{n+1}$ of type A_n which satisfies $g_1 \mapsto (\sigma_2 \dots \sigma_n)^n$, and $\sigma_i \mapsto \sigma_i$ $(i \le n-1)$. By composing both, we get an embedding which makes the square commute. This proves Proposition 3.1.

This embedding of type B_n into type A_n, different from the more standard one $\tau \mapsto \sigma_1^2, \sigma_i \mapsto \sigma_{i+1}$, has been considered in [13]. The algebraic proof given there being somewhat sketchy, we provide the details here. This embedding comes from the following construction.

Consider the (faithful) Artin action as a morphism $\mathcal{B}_{n+1} \to \mathrm{Aut}(F_{n+1})$, and the free subgroup $F_n = \langle g_1, \dots, g_n \rangle$ of F_{n+1}. The action of $\mathcal{B}_{n+1}$ preserves the product $g_1 g_2 \dots g_{n+1}$, and there is a natural retraction $F_{n+1} \twoheadrightarrow F_n$ which sends g_{n+1} to $(g_1 \dots g_n)^{-1}$. This induces a map $\Psi : \mathcal{B}_{n+1} \to \mathrm{Aut}(F_n)$, whose kernel is the center of $\mathcal{B}_{n+1}$ by a theorem of Magnus (see [14]). We claim that its image contains the group $\mathrm{Inn}(F_n)$ of inner automorphisms of F_n, which is naturally isomorphic to F_n.

Indeed, it is straightforward to check that $b_1 = (\sigma_2 \dots \sigma_n)^n$ is mapped to $\mathrm{Ad}(g_1) = x \mapsto g_1 x g_1^{-1}$. Defining $b_{i+1} = \sigma_i b_i \sigma_i^{-1}$, we get that b_i is mapped to $\mathrm{Ad}(g_i)$. In particular the subgroup $\mathcal{F}_n = \langle b_1, \dots, b_n \rangle$ of $\mathcal{B}_{n+1}$ is free and there is a natural isomorphism $\varphi : b_i \mapsto g_i$ to F_n characterized by the property $b.g = \mathrm{Ad}(\varphi(b))(g)$ for all $g \in F_n$ and $b \in \mathcal{F}_n$, that is $\mathrm{Ad}(\varphi(b)) = \Psi(b)$ for all $b \in \mathcal{F}_n$.

Now, let $\mathcal{B}_n \subset \mathcal{B}_{n+1}$ be generated by $\sigma_i, i \le n-1$. Its action on F_n is the usual Artin action recalled above. For $\sigma \in \mathcal{B}_n$ and $b \in \mathcal{F}_n$ we know that

$$\forall x \in F_n \quad \sigma b \sigma^{-1}.x = \sigma.\left(\varphi(b)(\sigma^{-1}.x)\varphi(b)^{-1}\right) = (\sigma.\varphi(b))x(\sigma.\varphi(b))^{-1}$$

that is $\sigma b \sigma^{-1}$ is mapped to $\mathrm{Ad}(\sigma.\varphi(b))$ in $\mathrm{Aut}(F_n)$, hence $\sigma b \sigma^{-1}$ and $\varphi^{-1}(\sigma.\varphi(b)) \in \mathcal{F}_n$ have the same image under Ψ. Since the kernel of Ψ is $Z(\mathcal{B}_{n+1})$, this proves that they may differ only by an element of the center $Z(\mathcal{B}_{n+1})$ of $\mathcal{B}_{n+1}$. On the other hand, $\varphi : \mathcal{F}_n \to F_n$ commutes with the maps $F_n \to \mathbf{Z}$ and $\eta : \mathcal{F}_n \to \mathbf{Z}$ which map every generator to 1. Likewise, the Artin action commutes with $F_n \to \mathbf{Z}$ hence

$\eta(\varphi^{-1}(\sigma.\varphi(b))) = \eta(b)$. We denote $\ell : \mathcal{B}_{n+1} \to \mathbf{Z}$ the abelianization map. We have $\ell(b_i) = n(n-1)$ for all i, hence $\ell(b) = n(n-1)\eta(b)$ for all $b \in \mathcal{F}_n$. Since $\ell(b) = \ell(\sigma b \sigma^{-1})$ it follows that $\sigma b \sigma^{-1}$ and $\varphi^{-1}(\sigma.\varphi(b)) \in \mathcal{F}_n$ differ by an element in $Z(\mathcal{B}_{n+1}) \cap (\mathcal{B}_{n+1}, \mathcal{B}_{n+1})$, where $(\mathcal{B}_{n+1}, \mathcal{B}_{n+1})$ denotes the commutators subgroup. But $Z(\mathcal{B}_{n+1})$ is generated by $(\sigma_1 \dots \sigma_n)^{n+1} \notin (\mathcal{B}_{n+1}, \mathcal{B}_{n+1})$ hence $\sigma b \sigma^{-1} = \varphi^{-1}(\sigma.\varphi(b)) \in \mathcal{F}_n$.

In particular $\mathcal{F}_n$ is stable under the action by conjugation of $\mathcal{B}_n$, which coincides with the Artin action. This is the embedding $\mathcal{B}_n \ltimes F_n \hookrightarrow \mathcal{B}_{n+1}$ that is needed to make the square commute. It remains to prove that we indeed have a semidirect product, namely that $\mathcal{B}_n \cap \mathcal{F}_n = \{1\}$. First notice that $\mathcal{F}_n$ is mapped to $\mathrm{Inn}(F_n)$ and recall that the outer Artin action $\mathcal{B}_n \to \mathrm{Out}(F_n)$ has for kernel $Z(\mathcal{B}_n)$, hence $\mathcal{F}_n \cap \mathcal{B}_n \subset Z(\mathcal{B}_n)$. Then $x \in \mathcal{F}_n \cap \mathcal{B}_n$ can be written $x = z^k$ for some $k \in \mathbf{Z}$ with $z = (\sigma_1 \dots \sigma_{n-1})^n$. It is classical and easy to check that the action of z on F_n is given by $\mathrm{Ad}((g_1 \dots g_n)^{-1})$, hence $\varphi(z^k) = \varphi((b_1 \dots b_n)^{-k})$ and $x = z^k = (b_1 \dots b_n)^{-k}$. Thus $\ell(x) = kn(n-1) = -kn^2(n-1)$ hence $k = 0$ and $x = 1$.

This concludes the case of G_{26}. The case of G_{31} is a consequence of the lifting of Springer's theory of 'regular elements' for complex reflection groups to their associated braid group. By Springer theory (see [23]), W_{31} appears as the centralizer of a regular element c of order 4 in W_{37}, which is the Coxeter group of type E_8, and, as a consequence of [2, Theorem 12.5 (iii)], B_{31} can be identified with the centralizer of a lift $\tilde{c} \in B_{37}$ of c, in such a way that the natural diagram

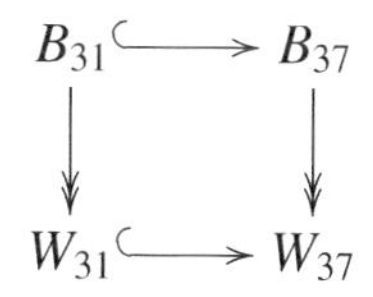

commutes. This embedding $B_{31} \hookrightarrow B_{37}$ is explicitly described in [8], to which we refer for more details. By commutation of the above diagram it induces an embedding $P_{31} \hookrightarrow P_{37}$. Since P_{37} is known to be residually torsion-free nilpotent by [15, 16], this concludes the proof of Theorem 1.4.

References

[1] B. BAUMSLAG, *Residually free groups*, Proc. London Math. Soc. **17** (1967), 402–418.

[2] D. BESSIS, *Finite complex reflection arrangements are* $K(\pi, 1)$, arXiv math/0610777 v3 (2007).

[3] M. BROUÉ, G. MALLE and R. ROUQUIER, *Complex reflection groups, braid groups, Hecke algebras*, J. Reine Angew. Math. 500 (1998), 127–190.

[4] D. COHEN, M. FALK and R. RANDELL, *Pure braid groups are not residually free*, In: "Configuration Spaces", Proceedings, A. Bjorner, F. Cohen, C. De Concini, C. Procesi and M. Salvetti (eds.), Edizioni della Normale, Pisa, 2012, 213–230.

[5] D. J. COLLINS, *Relations among the squares of the generators of the braid groups*, Invent. Math. **117** (1994), 525–529.

[6] H. S. M. COXETER, "Regular Complex Polytopes", 2nd edition, Cambridge University Press, 1991.

[7] J. CRISP and L. PARIS, *Artin groups of types B and D*, Adv. Geom. **5** (2005), 607–636.

[8] F. DIGNE, I. MARIN and J. MICHEL, *The center of pure complex braid groups*, J. Algebra **347** (2011), 206–213.

[9] C. DROMS, J. LEIN and H. SERVATIUS, *The Tits conjecture and the five string braid groups*, In: "Topology and Combinatorial Group Theory", LNM 1440, Springer-Verlag, 1990, 48–51.

[10] M. FALK and R. RANDELL, *The lower central series of a fiber-type arrangement*, Invent. Math. **82** (1985), 77–88.

[11] M. FALK and R. RANDELL, *Pure braid groups and products of free groups*, In: "Braids", Contemporary Mathematics, Vol. 78, 217-228, A.M.S., Providence, 1988.

[12] D. KRAMMER, *Braid groups are linear*, Annals of Math. **155** (2002), 131–156.

[13] D. LONG, *Constructing representations of braid groups*, Comm. Anal. Geom. **2** (1994), 217–238.

[14] W. MAGNUS, *Über Automorphismen von Fundamentalgruppen berandeter Flächen*, Math. Ann. **109** (1934) 617–646.

[15] I. MARIN, *On the residual nilpotence of pure Artin group*, J. Group Theory **9** (2006), 483–485.

[16] I. MARIN, *Sur les représentations de Krammer génériques*, Ann. Inst. Fourier (Grenoble) **57** (2007), 1883–1925.

[17] I. MARIN, *Krammer representation for complex braid groups*, preprint, arxiv:0711.3096v3 (2008).

[18] P. ORLIK and H. TERAO, "Arrangements of Hyperplanes", Springer, Berlin, 1992.

[19] H. MASCHKE, *Über die quaternäre, endliche, lineare Substitution's gruppe der Borchart'schen Moduln*, Math. Ann. **30** (1887), 496–515.

[20] L. PARIS, *Artin monoids inject in their group*, Comment. Math. Helv. **77** (2002), no. 3, 609–637.

[21] H. SERVATIUS, *Automorphisms of Graph Groups*, J. Algebra **126** (1987), 34–60.
[22] G. C. SHEPHARD and J. A. TODD, *Finite unitary reflection groups*, Canad. J. Math. **6** (1954), 274–304.
[23] T. SPRINGER, *Regular elements of finite reflection groups*, Inventiones Math. **25** (1974), 159–198.

Some topological problems on the configuration spaces of Artin and Coxeter groups

Davide Moroni, Mario Salvetti and Andrea Villa

Abstract. In the first part we review some topological and algebraic aspects in the theory of Artin and Coxeter groups, both in the finite and infinite case (but still, finitely generated). In the following parts, among other things, we compute the Schwartz genus of the covering associated to the orbit space for all affine Artin groups. We also give a partial computation of the cohomology of the braid group with non-abelian coefficients coming from geometric representations. We introduce an interesting class of "sheaves over posets", which we call "weighted sheaves over posets", and use them for explicit computations.

1 Introduction

This paper contains some results regarding some particular topological aspects in the theory of Artin and Coxeter groups. Among other things, we compute the Schwartz genus of the covering associated to the orbit space for all affine Artin groups. This result generalizes [15, 18].

Our paper contains also a brief review of some results, concerning the topology of Artin and Coxeter groups, both in the finite and infinite case (but still, finitely generated) which are essential to our computations. Other reviews, even if considering some interesting aspects of the theory, are not very satisfying about the topological underlying structure. Our review will still be very partial: the more than thirty years old literature on the subject would require a much longer paper. We concentrate essentially on a single line of research, which (in our opinion) gives the possibility to produce a neat picture of some basic topological situation underlying all the theory in very few pages. Such picture is based essentially on [27] (and [17]) and [18] (both papers are based on the construction [26]). Some of the several computations which use these constructions are cited below.

For some literature containing many works related to Artin groups the reader can see the paper [21] contained in this book. It should be added that almost all people who worked in the theory of Hyperplane Arrangements have given some contributions to the theory.

The new results that we present here concern some computations of:

- the cohomology of the braid groups with non-abelian coefficients, coming from geometric representations of the braid groups into the homology of an orientable surfaces (see part 3.1);
- the computation of the Schwartz genus of the covering associated to the orbit space of an affine Artin group (Theorem 3.15).

We skip the details of the computations of the first application, while we give details for the second one. We introduce here a particularly interesting class of "sheaves over posets", which we call "weighted sheaves over posets". We use them for some explicit computations for the top cohomology of the affine group of type $\tilde{A}_n$ (the whole cohomology was completely computed with different methods in [9, 10]). A natural spectral sequence is associated to such sheaves. We are going to exploit this general construction in future works.

Some of the ideas which we use here were first explained in [28] and formalized in [24].

2 General pictures

We will consider finitely generated *Coxeter systems* $(\mathbf{W}, S)$ (S finite), so

$$\mathbf{W} = < s \in S \mid (ss')^{m(s,s')} = 1 > \tag{2.1}$$

where $m(s, s') \in \mathbb{N} \cup \{\infty\}$, $m(s, s') = m(s', s)$, $m(s, s) = 1$ (for general reference see [4, 23]).

2.1 Case W finite

The group $\mathbf{W}$ can be realized as a group generated by reflections in $\mathbb{R}^n$ ($n = |S|$). Let $\mathcal{A}$ be the *reflection arrangement, i.e.*

$$\mathcal{A} = \{H \subset \mathbb{R}^n \mid H \text{ is fixed by some reflection in } \mathbf{W}\}.$$

Consider also the stratification into *facets* $\Phi := \{F\}$ of $\mathbb{R}^n$ induced by $\mathcal{A}$. The codimension-0 facets are called *chambers*. They are the connected components of the complement to the arrangement. All the chambers are simplicial cones, the group acting transitively over the set of all them. The Coxeter generator set S corresponds to the set of reflections with respect to the *walls* of a fixed base-chamber C_0 (see Figure 2.1).
Let $H_{\mathbb{C}} := H + iH \subset \mathbb{C}^n$ be the *complexification* of the hyperplane H, and set

$$\mathbf{Y} := \mathcal{M}(\mathcal{A}) := \mathbb{C}^n \setminus \cup_{H \in \mathcal{A}} H_{\mathbb{C}}$$

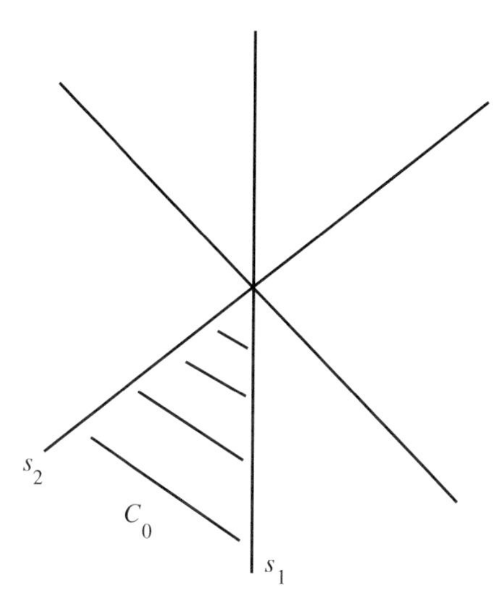

Figure 2.1. Case W finite.

be the complement to the complexified arrangement. The group W acts freely over Y so we can form the *orbit space*

$$\mathbf{Y_W} := \mathbf{Y/W}$$

which is a complex manifold, actually an affine manifold, by classical results.

Define

$$\mathbf{G_W} := \pi_1(\mathbf{Y}_W)$$

as the *Artin group* of type $\mathbf{W}$ (somebody calls it the *Artin-Brieskorn* group, somebody else calls it the *Artin-Tits* group: for brevity, we just take the intersection of them) (see [5, 13, 21]). A presentation of $\mathbf{G_W}$ is obtained by that of $\mathbf{W}$ by removing the relations $s^2 = 1$

$$\mathbf{G_W} = < g_s,\ s \in S \mid g_s g_{s'} g_s \cdots = g_{s'} g_s g_{s'} \cdots > \tag{2.2}$$

(same number $m(s, s')$ of factors on each side).

From Deligne's theorem ([13]) the orbit space $\mathbf{Y_W}$ is a space of type $k(\pi, 1)$, so we have

$$H^*(\mathbf{G_W}; L) = H^*(\mathbf{Y_W}; \mathbf{L})$$

where L is any G_W-module and $\mathbf{L}$ is the corresponding local system over $\mathbf{Y_W}$.

We also recall ([26, 27]):

Theorem 2.1. *The orbit space* $\mathbf{Y_W}$ *contracts over a cell complex* $\mathbf{X_W}$ *obtained from a convex polyhedron* Q *by explicit identifications over its faces.*

More precisely, one takes one point $x_0 \in C_0$, and set

$$Q := \text{ convex hull } [W.x_0]$$

a polyhedron obtained by the convex hull of the orbit of one point x_0 in the base-chamber.

For example, in the case of the *braid arrangement* one obtains the so called *permutohedron*.

One verifies the following facts.

The k-faces of Q are also polyhedra, each of them corresponding to a coset of a *parabolic* subgroup $\mathbf{W}_\Gamma$, where $\Gamma \subset S$ is a k-subset of S. The correspondence

$$\{\text{faces of } Q\} \leftrightarrow \{w.\mathbf{W}_\Gamma,\ \Gamma \subset S\}$$

is obtained by taking the polyhedron given by the convex-hull of the orbit $\mathbf{W}_\Gamma.x_0$ and translating it by w.

One has also ([4, 23]):

Proposition 2.2. *Inside each coset* $w.\mathbf{W}_\Gamma$ *there exists a unique element of* minimal length.

Here the length is the minimal number of letters (coming from S) in a reduced expression.

For every face e of Q, which corresponds to a coset $w.\mathbf{W}_\Gamma$, let $\beta(e) \in w.\mathbf{W}_\Gamma$ be the element of minimal length. Notice that $\mathbf{W}$ permutes faces of the same dimension. Then each pair of faces $e,\ e'$ belonging to the same orbit is identified by using the homeomorphism $\beta(e)\beta(e')^{-1}$.

We give in Figure 2.2, 2.3 the example of the group $\mathbf{W}$ of type A_2, so $\mathbf{G_W}$ is the braid group in three strands. The orbit space turns out to have the homotopy type of an hexagon whose edges are identified according to the given arrows.

2.2 Case W infinite

When $\mathbf{W}$ is infinite (but still S is finite) the theory is analogue with the following changes (see [4, 35]). One can still realize $\mathbf{W}$ as a group of (non-orthogonal) reflections in $\mathbb{R}^n$, $(n = |S|)$ starting from a base chamber C_0. In [4] the standard first octant is considered, (so $C_0 = \{x_i > 0,\ i = 1, \ldots, n\}$) but one can start from a more general open cone with

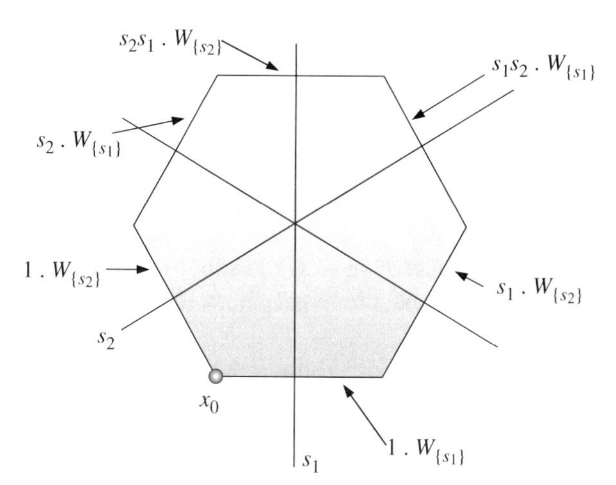

Figure 2.2. For each 1-cell e we indicate the corresponding coset $\beta(e).\mathbf{W}_\Gamma$.

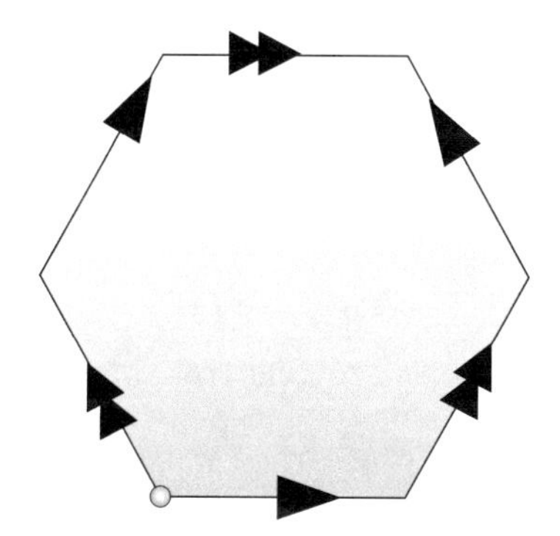

Figure 2.3. Glue 1-cells with the same type of arrows.

vertex 0 (see [35]). Again, S corresponds to the set of reflections with respect to the walls of C_0.

We recall here the main points which we need.

Let $U := \mathbf{W}.\overline{C}_0$ be the orbit of the closure of the base chamber. U is called the *Tits cone* of the Coxeter system.

Notice that the closure of the chamber $\overline{C}_0$ is endowed with a natural stratification into facets (which are still relatively open cones with vertex 0). When C_0 is the standard positive octant, each facet is given by imposing some coordinates equal to 0, and the remaining coordinates positive.

Each reflection in $\mathbf{W}$ is conjugated to a reflection with respect to a wall of C_0. So, the arrangement $\mathcal{A}$ of reflection hyperplanes is just the orbit of the walls of C_0. Each connected component of the complement inside U of the arrangement (again called a chamber) is of the shape $w.C_0$ for a unique $w \in \mathbf{W}$. Of course $\mathcal{A}$ is not locally finite (*e.g.* 0 is contained in all the hyperplanes). The orbits of the facets of C_0 give a "stratification" of U into relatively open cells, also called facets (in general, U is neither open nor closed in $\mathbb{R}^n$).

Recall also (see [4, 35]):

1. U is a convex cone in $\mathbb{R}^n$ with vertex 0.
2. $U = \mathbb{R}^n$ iff $\mathbf{W}$ is finite
3. The stabilizer of a facet F in U is the subgroup $\mathbf{W}_F$ generated by all the reflections with respect to hyperplanes (in $\mathcal{A}$) containing F. So, in general $\mathbf{W}_F$ is not finite.
4. $U^0 := int(U)$ is open in $\mathbb{R}^n$ and a (relatively open) facet $F \subset \overline{C}_0$ is contained in U^0 iff the stabilizer $\mathbf{W}_F$ is finite.

By property 4 the arrangement is locally finite in the interior part U^0.

So we take in this case

$$\mathbf{Y} := [U^0 + i\mathbb{R}^n] \setminus \cup_{H \in \mathcal{A}} H_{\mathbb{C}}$$

which corresponds to complexifying only the interior part of the Tits cone. The group $\mathbf{W}$ acts (as before) diagonally onto $\mathbf{Y}$, and one shows easily (exactly as in the finite case) that the action is free. Therefore, one has an orbit space

$$\mathbf{Y_W} := \mathbf{Y}/\mathbf{W}$$

which is still a manifold, and a regular covering

$$\mathbf{Y} \to \mathbf{Y_W}$$

with group $\mathbf{W}$.

We still define $\mathbf{G_W} := \pi_1(\mathbf{Y_W})$ as the Artin group of type $\mathbf{W}$. We will see in a moment that for $\mathbf{G_W}$ one has a presentation similar to (2.2).

In fact, we have very similar constructions.

Take $x_0 \in C_0$ and let Q be the *finite* CW-complex constructed as follows.

For all subsets $\Gamma \subset S$ such that $\mathbf{W}_\Gamma$ is finite, construct a $|\Gamma|$-cell Q_Γ in U^0 as the convex hull of the $\mathbf{W}_\Gamma$ orbit of x_0. Each Q_Γ is a finite convex polyhedron which contains the point x_0.

Let $\mathbf{X}_{\mathbf{W}_\Gamma}$ be obtained from Q_Γ by identifications on its faces defined as in the finite case (relative to the finite group $\mathbf{W}_\Gamma$). Define

$$Q := \cup\, Q_\Gamma \tag{2.3}$$

(a finite union of convex polyhedra) where the union is taken on all the above Γ for which $\mathbf{W}_\Gamma$ is finite. Define also

$$\mathbf{X_W} := \cup_\Gamma \mathbf{X}_{\mathbf{W}_\Gamma}. \tag{2.4}$$

Remark 2.3. The definition of $\mathbf{X_W}$ makes sense because of the following easy fact:

for any common cell $e \subset Q_\Gamma \cap Q_{\Gamma'}$ the minimal element $\beta(e)$ is the same when computed in $\mathbf{W}_\Gamma$ and in $\mathbf{W}_{\Gamma'}$.

Moreover, we have the following generalization of the finite case (see [27])

Theorem 2.4. *The CW-complex $\mathbf{X_W}$ is deformation retract of the orbit space $\mathbf{Y_W}$.*

Proof. Notice that $\mathbf{X_W}$ is a finite complex in all cases.

First, there exists a regular CW-complex $\mathbf{X} \subset \mathbf{Y}$ which is deformation retract of $\mathbf{Y}$, and $\mathbf{X}$ is constructed as in [26]. That paper already worked for the affine cases; however, the same procedure works in general because we reduce to the locally finite case around faces with finite stabilizer.

The construction of $\mathbf{X}$ can be chosen invariantly with respect to the action of $\mathbf{W}$, which permutes cells of the same dimension.

The action on $\mathbf{X}$ being free, we look at the orbit space $\mathbf{X}/\mathbf{W}$. By Remark 2.3, this reduces to finite cases. □

(A similar proof can be obtained by generalizing "combinatorial stratifications" to this situation, see [1].)

Below we give a picture for the case $\tilde{A}_2$ (Figure 2.4).

Conjecture 2.5. The space $\mathbf{Y}$ is a $k(\pi, 1)$ space.

Of course, if one of the spaces $\mathbf{Y}$, $\mathbf{Y_W}$, $\mathbf{X}$, $\mathbf{X_W}$ is a $k(\pi, 1)$, so are all the other spaces. This conjecture is known, besides the finite case, for some affine groups: $\tilde{A}_n$, $\tilde{C}_n$ ([25]; in case $\tilde{A}_n$ a different proof is given in [7]); case $\tilde{B}_n$ was solved in ([11]). For other known cases see [6, 22].

As an immediate corollary to Theorem 2.4 we re-find in a very short way a presentation of the fundamental group (see [34]).

Theorem 2.6. *A presentation for the Artin group $\mathbf{G_W}$ is similar to 2.2, where we have to consider only pairs s, s' such that $m(s, s')$ is finite.*

Proof. We have just to look at the 2-skeleton of $\mathbf{X_W}$, which is given as follows. For each pair s, s' such that $m(s, s')$ is finite, one has a $2m(s, s')-$gon with two orbits of edges, which glue in a similar way as Figure 2.3. This gives the relation

$$g_s g_{s'} g_s \ldots = g_{s'} g_s g_{s'} \ldots \quad (m(s, s') \text{ factors}).$$ □

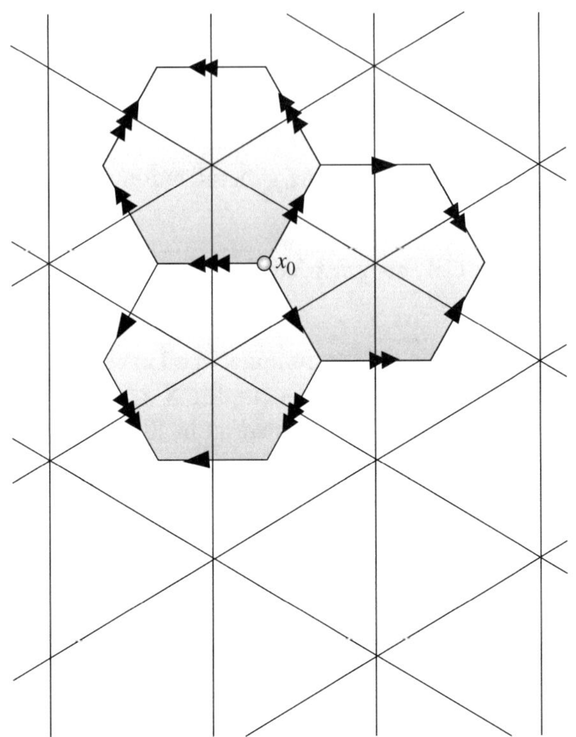

Figure 2.4. Q is the union of three hexagons. $\mathbf{X_W}$ is obtained by gluing 1-cells with same type of arrow.

2.3 Algebraic complexes for Artin groups

We refer here mainly to [17, 27].

We consider the algebraic complex related to the cell structure of $\mathbf{X_W}$. It is given in the following way.

Let $\mathbb{Z}[\mathbf{G_W}]$ be the group algebra of G_W. Let (C_*, ∂_*) be the algebraic complex of free $\mathbb{Z}[\mathbf{G_W}]$-modules such that in degree k it is free with basis e_J corresponding to subsets $J \subset S$ such that $\mathbf{W}_J$ is finite:

$$C_k := \bigoplus_{\substack{J \subset S \\ |J| = k \\ \mathbf{W}_J \text{ finite}}} \mathbb{Z}[\mathbf{G_W}]\, e_J \tag{2.5}$$

Let

$$\partial(e_J) := \sum_{\substack{I \subset J \\ |I| = k-1}} [I : J]\, T_I^J . e_I \tag{2.6}$$

where $[I : J]$ is the incidence number ($= 0,\ 1$ or -1) of the cells in $\mathbf{X_W}$ and

$$T_I^J := \sum_{\beta \in W_I^J} (-1)^{\ell(\beta)}\ g_\beta$$

where

1. $\mathbf{W}_I^J := \{\beta \in \mathbf{W}_J : \ell(\beta\ s) > \ell(\beta),\ \forall s \in \mathbf{W}_I\}$ is the set of elements of minimal length for the cosets $\frac{\mathbf{W}_J}{\mathbf{W}_I}$ (prop. 2.2);
2. if $\beta \in \mathbf{W}_I^J$ and $\beta = s_{i_1} \dots s_{i_k}$ is a reduced expression then $\ell(\beta) = k$;
3. if β is as in 2 then $g_\beta := g_{s_{i_1}} \dots g_{s_{i_k}}$. One shows that this map

$$\psi : \mathbf{W} \to \mathbf{G_W} \tag{2.7}$$

is a well-defined section (not a homomorphism) of the standard surjection $\mathbf{G_W} \to \mathbf{W}$.

Remark 2.7. The orbit space $\mathbf{X_W}$ is a $k(\pi, 1)$ iff (C_*, ∂_*) is acyclic (in this case the augmentation gives a resolution of $\mathbb{Z}$ into free $\mathbb{Z}[\mathbf{G_W}]$-modules).

In papers [17, 27] we used the knowledge of the fact that $\mathbf{X_W}$ is a $k(\pi, 1)$ in the finite case to deduce that (2.5), (2.6) is acyclic. One could try the converse: proof algebraically that the algebraic complex is exact and conclude that $\mathbf{X_W}$ is a $k(\pi, 1)$.

It is interesting to consider the following abelian representation ([19, 26, 29]).

Let $R := A[q, q^{-1}]$ be the ring of Laurent polynomials over a ring A. One can represent $\mathbf{G_W}$ by

$$g_s \mapsto [\text{multiplication by } -q] \quad , \forall s \in S \tag{2.8}$$

($\in \mathrm{Aut}(R)$). This representation, which coincides with the determinant of the Burau representation, has a very interesting meaning in the finite case: the cohomology of $\mathbf{G_W}$ with coefficient in this representation equals the trivial cohomology of the Milnor fibre of the associated discriminant bundle. In other words, the orbit space has a Milnor fibration over S^1, with Milnor fibre F, and the above twisted cohomology of $\mathbf{G_W}$ gives the trivial cohomology of F as a module over R, the $(-q)$-multiplication corresponding to the monodromy of the bundle.

The tensor product $C_* \otimes R$ has boundary

$$\partial(e_J) = \sum_{\substack{I \subset J \\ |I|=|J|-1}} [I : J]\ \frac{\mathbf{W}_J(q)}{\mathbf{W}_I(q)}\ e_I \tag{2.9}$$

where

$$\mathbf{W}_J(q) := \sum_{w \in W_J} q^{l(w)}$$

is the Poincaré series of the group $\mathbf{W}_J$ (here, a polynomial since the stabilizers are finite). The denominator $\mathbf{W}_I(q)$ divides the numerator $\mathbf{W}_J(q)$, so the quotient is still a polynomial in q.

Example 2.8. *Case* $\mathbf{A}_n$. We have Dynkin graph

$$\underset{1}{\circ} — \underset{2}{\circ} — \cdots — \underset{\text{n-1}}{\circ} — \underset{\text{n}}{\circ}$$

Let

$$[k] := \frac{q^k - 1}{q - 1}; \quad [k]! := \prod_{i=1}^{k} [i]; \quad \begin{bmatrix} k \\ h \end{bmatrix} := \frac{[k]!}{[h]![k-h]!} \tag{2.10}$$

For $J \subset S \cong \{1, \ldots, n\}$ one has

$$\mathbf{W}_J(q) = \prod_{i=1}^{m} [|\Gamma_i(J)| + 1]!$$

where $\Gamma_1(J),\ \Gamma_2(J), \ldots$ are the connected components of the subgraph of $\mathbf{A}_n$ generated by J (Figure 2.5).

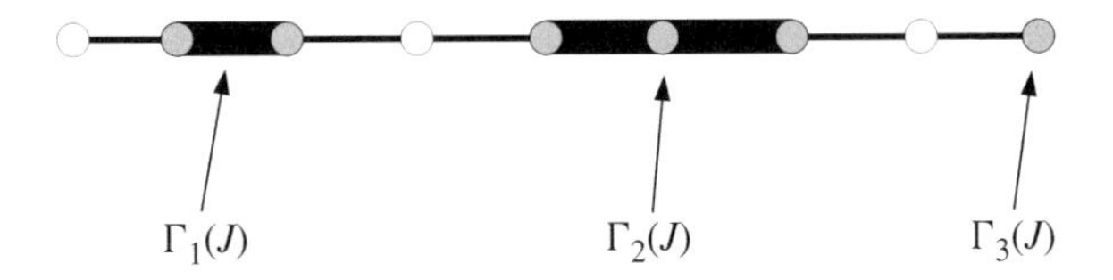

Figure 2.5. Here $\mathbf{W}_J(q) = [3]![4]![2]!$.

So, all coefficients are of the shape $\begin{bmatrix} k \\ h \end{bmatrix}$ $(k,\ h$ depending on $J,\ I)$.

For some computations using these methods for the finite case:

- for $H^*(\mathbf{G_W}, \mathbf{Q}[q, q^{-1}])$ [14,20] (case A_n) [16] (all other cases);
- for the top cohomology of $H^*(\mathbf{G_W}, \mathbb{Z}[q, q^{-1}])$ in all cases [19];
- for the cohomology $H^*(\mathbf{G_W}, \mathbb{Z}[q, q^{-1}])$ for all exceptional cases see [12]; the same for case A_n: see [8];

In the affine cases: see [10] (cohomology in case $\tilde{A}_n$), [9] ($k(\pi, 1)$ problem and cohomology for $\tilde{B}_n$).

2.4 CW-complexes for Coxeter groups

We refer here essentially to [18].

If $(\mathbf{W}, S)$ is a finite Coxeter system, which is realized as a reflection group in $\mathbb{R}^n$, with reflection arrangement $\mathcal{A}$, consider the subspace arrangement in $\mathbb{R}^{nd} \cong (\mathbb{R}^n)^d$ given by

$$\mathcal{A}^{(d)} := \{H^{(d)}\}$$

where $H^{(d)}$ is the codimensional-d subspace given by "d-complexification" of the hyperplane $H \in \mathcal{A}$:

$$H^{(d)} := \{(X_1, \ldots, X_d) \; : \; X_i \in \mathbb{R}^n, \; X_i \in H\}.$$

For $d = 2$ we have the standard complexification of the hyperplanes. Let

$$\mathbf{Y}^{(d)} := \mathbb{R}^{nd} \setminus \cup_{H \in \mathcal{A}} H^{(d)}.$$

As before, the group W acts freely on $Y^{(d)}$ and we consider the orbit space

$$\mathbf{Y}_{\mathbf{W}}^{(d)} := \mathbf{Y}^{(d)}/\mathbf{W}.$$

Recall:

Theorem 2.9. *The space*

$$\mathbf{Y}_{\mathbf{W}}^{(\infty)} := \left[\lim_{\overrightarrow{d}} \mathbf{Y}^{(d)}\right]/W = \left[\lim_{\overrightarrow{d}} \mathbf{Y}_{\mathbf{W}}^{(d)}\right]$$

is a space of type $k(W, 1)$.

In case (W, S) is infinite (still, S finite) one has to substitute $\mathbb{R}^n \times \cdots \times \mathbb{R}^n$ (d factors) with the space

$$\mathcal{U}_W^{(d)} := U_0 \times \mathbb{R}^n \times \cdots \times \mathbb{R}^n$$

($d - 1$ factors equal to $\mathbb{R}^n$), and the space $Y^{(d)}$ becomes

$$\mathbf{Y}^{(d)} := \mathcal{U}_W^{(d)} \setminus \cup_{H \in \mathcal{A}} H^{(d)}. \tag{2.11}$$

Theorem 2.10. *For finitely generated* $\mathbf{W}$, *the same conclusion as in Theorem* 2.9 *holds, by taking definition* (2.11) *for* $Y^{(d)}$.

So, different from the case of Artin groups, we always get a $k(\pi, 1)$ space here.

Recall also the construction of a CW-complex which generalizes that given for Artin groups.

Theorem 2.11. *When* $\mathbf{W}$ *is finite, the space* $\mathbf{Y}_{\mathbf{W}}^{(d)}$ *contracts over a* CW*-complex* $\mathbf{X}_{\mathbf{W}}^{(d)}$ *such that*

$$\{k\text{-cells of } \mathbf{X}_{\mathbf{W}}^{(d)}\} \longleftrightarrow \left\{\text{flags } \Gamma := (\Gamma_1 \supset \ldots \supset \Gamma_d) : \Gamma_1 \subset S, \sum_{i=1}^{d} |\Gamma_i| = k\right\}$$

Passing to the limit, $\mathbf{Y}_{\mathbf{W}}^{(\infty)} = k(W,1)$ *contracts over a* CW*-complex* $\mathbf{X}_{\mathbf{W}}^{(\infty)}$ *such that*

$$\{k\text{-cells of } \mathbf{X}_{\mathbf{W}}^{(\infty)}\} \longleftrightarrow \left\{\text{flags } \Gamma := (\Gamma_1 \supset \Gamma_2 \supset \ldots) : \Gamma_1 \subset S, \sum_{i \geq 1} |\Gamma_i| = k\right\}$$

Notice that $\mathbf{X}_{\mathbf{W}}^{(\infty)}$ does not have finite dimension but the number of k-cells is finite, given by $\binom{n+k-1}{k}$.

The case $\mathbf{W}$ infinite has to be modified as in case of Artin groups, by considering flags composed with subsets $\Gamma \subset S$ such that $\mathbf{W}_\Gamma$ is finite. In the limit, the theorem is

Theorem 2.12. *For* S *finite, the space* $\mathbf{Y}_{\mathbf{W}}^{(\infty)} = k(W,1)$ *contracts over a* CW*-complex* $\mathbf{X}_{\mathbf{W}}^{(\infty)}$ *such that the set of* k*-cells of* $\mathbf{X}_{\mathbf{W}}^{(\infty)}$ *corresponds to*

$$\left\{\text{flags } \Gamma := (\Gamma_1 \supset \Gamma_2 \supset \ldots) : \ \Gamma_1 \subset S, \ \mathbf{W}_{\Gamma_1} \text{ finite}, \ \sum_{i \geq 1} |\Gamma_i| = k\right\}$$

2.5 Algebraic complexes for Coxeter groups

Consider the algebraic complex $(C_*^{(d)}, \partial)$ of free $\mathbb{Z}[\mathbf{W}]$-modules, where

$$C_k^{(d)} := \bigoplus_{\substack{\Gamma : \sum_1^d |\Gamma_i| = k \\ |W_{\Gamma_1}| < \infty}} \mathbb{Z}[\mathbf{W}] e(\Gamma)$$

The generators of C_* are in one to one correspondence with the cells of $\mathbf{X}_{\mathbf{W}}^{(d)}$. The expression of the boundary is the following:

$$\partial e(\Gamma) = \sum_{\substack{1 \leq i \leq d \\ |\Gamma_i| > |\Gamma_{i+1}|}} \sum_{\tau \in \Gamma_i} \sum_{\substack{\beta \in \mathbf{W}_{\Gamma_i}^{\Gamma_i \setminus \{\tau\}} \\ \beta^{-1}\Gamma_{i+1}\beta \subset \Gamma_i \setminus \{\tau\}}} (-1)^{\alpha(\Gamma, i, \tau, \beta)} \beta e(\Gamma')$$

where

$$\Gamma' = (\Gamma_1 \supset \ldots \supset \Gamma_{i-1} \supset \Gamma_i \setminus \{\tau\} \supset \beta^{-1}\Gamma_{i+1}\beta \supset \ldots \supset \beta^{-1}\Gamma_d \beta)$$

and $(-1)^{\alpha(\Gamma,i,\tau,\beta)}$ is an incidence index. To get a precise expression for $\alpha(\Gamma, i, \tau, \beta)$, fix a linear order on S and let

$$\mu(\Gamma_i, \tau) := |j \in \Gamma \text{ s.t. } j \le \tau|$$
$$\sigma(\beta, \Gamma_j) := |(a, b) \in \Gamma_j \times \Gamma_j \text{ s.t. } a < b \text{ and } \beta(a) > \beta(b)|$$

in other words, $\mu(\Gamma_i, \tau)$ is the number of reflections in Γ_i less or equal to τ and $\sigma(\beta, \Gamma_j)$ is the number of inversions operated by β on Γ_j. Then we define:

$$\alpha(\Gamma, i, \tau, \beta) = i\ell(\beta) + \sum_{j=1}^{i-1} |\Gamma_j| + \mu(\Gamma_i, \tau) + \sum_{j=i+1}^{d} \sigma(\beta, \Gamma_j)$$

where ℓ is the length function in the Coxeter group.

Let now $C_* =: \lim_{d\to\infty} C_*^{(d)}$. The flags in C_k are the (infinite) sequences

$$\Gamma = (\Gamma_1 \supset \Gamma_2 \supset \ldots)$$

such that $\sum_{i\ge1} |\Gamma_i| = k$, so they still have a finite number of nonempty Γ_i.

Theorem 2.13. *For any finitely generated* $\mathbf{W}$, *the algebraic complex* (C_*, ∂_*) *gives a free resolution of the trivial* $\mathbb{Z}[\mathbf{W}]$*-module* $\mathbb{Z}$.

The proof follows straightforward from the remark that $\mathbf{Y}_{\mathbf{W}}^{(\infty)}$, so $\mathbf{X}_{\mathbf{W}}^{(\infty)}$, is a space of type $k(\pi, 1)$.

3 Applications

We give here some applications (see the introduction).

3.1 Cohomology of Artin groups over non-abelian representations

Let $M_{g,n}$ be an oriented surface of genus g with n boundary components. In what follows we fix the case $n = 1$, but what we say can be clearly generalized.

We construct a non-abelian representation of the braid group $Br_{2g+1} = \mathbf{G}_{\mathbf{A}_{2g}}$ as follows. Consider $2g$ simple curves $c_1, \ldots, c_{2g}$ inside $M := M_{2g,1}$, such that

$$|c_i \cap c_{i+1}| = 1, i = 1, \ldots, 2g-1; \;\; |c_i \cap c_j| = 0 \text{ otherwise.}$$

There are several ways to choose these curves, but we fix the standard one of Figure 3.1 (where we draw the case $g = 3$).

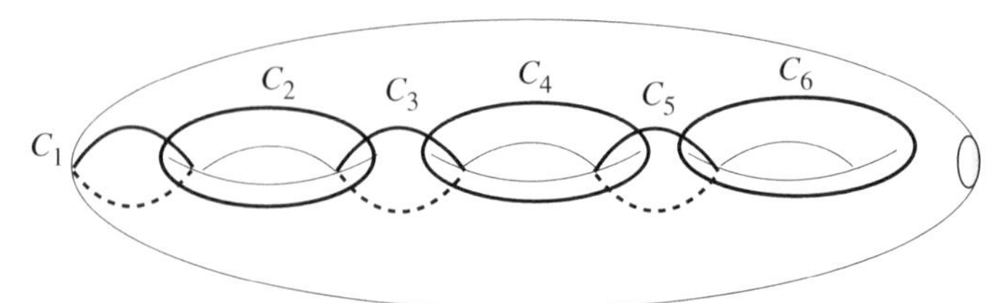

Figure 3.1. The simple curves c_i's for the surface $M_{3,1}$.

The Dehn twist t_{c_i} around the curve c_i belongs to the group $\mathrm{Homeo}^+(M_{g,1})$ of all orientation preserving homeomorphisms which restrict to the identity at the boundary, considered up to homotopy. So t_{c_i} induces an automorphism

$$(t_{c_i})_* \in \mathrm{Aut}(H_1(M_{g,1};\mathbb{Z}); <,>) \cong Sp(2g,\mathbb{Z})$$

which preserves the intersection form $<,>$. It is well known that Dehn twists $t_a,\ t_b$ around simple curves $a,\ b$ which intersect transversally in a single point satisfy the braid relation

$$t_a t_b t_a \ = \ t_b t_a t_b.$$

Therefore, the map (defined on the standard presentation of the braid group)

$$\sigma_i \to t_{c_i} \tag{3.1}$$

takes Br_{2g+1} homomorphically into $Sp(2g,\mathbb{Z})$.

Let M be the Br_{2g+1}-module which is given by $\mathbb{Z}^{2g}$ ($\cong H_1(M_{2g};\mathbb{Z})$) endowed with the action induced by the map (3.1). It is interesting to compute the cohomology

$$H^*(Br_{2g+1}; M).$$

We give here some calculation of this cohomology, up to genus 5, which is done by using the algebraic complex of part 2.3.

We add some explicit generators of these groups: we have all of them, but for lack of space we write only up to Br_7. A generator of degree k for Br_{2g+1} is written as a combination

$$c_k := \sum_{|J|=k} a_J \, e_J$$

	H^0	H^1	H^2	H^3	H^4	H^5	H^6	H^7	H^8	H^9	H^{10}
Br_3	0	0	$\mathbb{Z}_2$	0	0	0	0	0	0	0	0
Br_5	0	0	$\mathbb{Z}_2$	$\mathbb{Z}_2$	$\mathbb{Z}_2$	0	0	0	0	0	0
Br_7	0	0	$\mathbb{Z}_2$	$\mathbb{Z}_2$	$\mathbb{Z}_2^2$	$\mathbb{Z}_2^2$	$\mathbb{Z}_2$	0	0	0	0
Br_9	0	0	$\mathbb{Z}_2$	$\mathbb{Z}_2$	$\mathbb{Z}_2^2$	$\mathbb{Z}_2^3$	$\mathbb{Z}_2^3$	$\mathbb{Z}_2^2$	$\mathbb{Z}_2$	0	0
Br_{11}	0	0	$\mathbb{Z}_2$	$\mathbb{Z}_2$	$\mathbb{Z}_2^2$	$\mathbb{Z}_2^3$	$\mathbb{Z}_2^4$	$\mathbb{Z}_2^4$	$\mathbb{Z}_2^4$	$\mathbb{Z}_2^3$	$\mathbb{Z}_2$

Table 1. Some calculations.

for $J \subset S = \{1, \ldots, 2g\}$. For notational convenience, we write the coefficient $a_J \in \mathbb{Z}^{2g}$ as a degree-1 polynomial in the variables $x_1, \ldots, x_{2g}$.

Br$_3$.

$H^2 : c_2 := x_2\, e_{12}$

Br$_5$.

$H^2 : c_2 := -x_2 e_{12} + x_1 e_{23} + (x_1 - x_3) e_{34}$

$H^3 : c_3 := -x_1 e_{124} + x_3 e_{134} + (x_2 + x_3) e_{234}$

$H^4 : c_4 := x_3 e_{1234}$

Br$_7$.

$H^2 : c_2 := -x_2 e_{12} + x_1 e_{23} + (x_2 - x_4) e_{34} + x_3 e_{45} + (x_4 - x_6) e_{56}$

$$\begin{aligned} H^3 : c_3 := {} & x_2 e_{124} - x_1 e_{125} + x_2 e_{126} + x_4 e_{134} - x_3 e_{145} + (-x_4 + x_6) e_{156} \\ & - x_3 e_{235} - x_1 e_{236} + (x_2 - x_3) e_{245} + (-x_2 - x_4 + x_6) e_{256} \\ & + 2x_6 e_{345} + (-x_2 + x_4) e_{346} - (x_3 + x_4) e_{356} \end{aligned}$$

$$\begin{aligned} H^4 : c_{4,1} := {} & 2x_4 e_{1234} + (-x_5 + x_6) e_{1235} - x_6 e_{1236} + x_1 e_{1246} \\ & + (-x_1 - x_4 + x_5) e_{1345} + x_3 e_{1346} + x_6 e_{1356} - x_1 e_{1456} \\ & + 2x_5 e_{2345} - x_6 e_{2346} + x_2 e_{2456} \\ & + (-x_1 - 4x_2 + 3x_4 + 4x_6) e_{3456} \\ c_{4,2} := {} & 3x_4 e_{1234} + x_6 e_{1345} + 3x_5 e_{2345} \\ & + (-x_1 - 3x_2 + x_4 + 3x_5) e_{3456} \end{aligned}$$

$$\begin{aligned} H^5 : c_{5,1} := {} & -x_6 e_{12356} + x_1 e_{13456} - (x_2 + x_4 + x_6) e_{23456} \\ c_{5,2} := {} & x_3 c_{12346} + x_1 e_{13456} - (x_2 + x_4 + x_6) e_{23456} \end{aligned}$$

$H^6 : c_6 := x_6 e_{123456}$

3.2 The genus problem for infinite type Artin groups

Our main application here is the extension of some of the results found in [15, 18] about the genus of the covering associated to an Artin group. We first introduce some interesting machinery.

3.2.1 Some general remarks Let us consider the representation $\rho : \mathbf{G_W} \to \mathrm{Aut}(R)$ of (2.8), obtained by sending the standard generators of $\mathbf{G_W}$ into $(-q)$-multiplication. Let R_q be the the ring R with the prescribed structure of $\mathbf{G_W}$-module. It is convenient to indicate here by $D_*(\mathbf{W})$ the algebraic complex of part 2.3, so

$$D_k(\mathbf{W}) := \bigoplus_{\substack{J \subseteq S \\ |J|=k \\ |\mathbf{W}_J|<\infty}} R \cdot e_J \tag{3.2}$$

and the boundary is given by (2.9).

To motivate our next construction, notice that we can formally rewrite the boundary map in (2.9) as:

$$\partial\left(\frac{1}{\mathbf{W}_J(q)} e_J\right) = \sum_{\substack{I \subset J \\ |J|=|I|+1}} [I : J] \frac{1}{\mathbf{W}_I(q)} \cdot e_I \tag{3.3}$$

In particular, the fractions $e_I/\mathbf{W}_I(q)$ behave like the cells of a simplicial scheme.

We may therefore consider the complex:

$$D_k^0(\mathbf{W}) := \bigoplus_{\substack{J \subseteq S \\ |J|=k \\ |W_J|<\infty}} R \cdot e_J^0$$

with boundary:

$$\partial^0(e_J^0) = \sum_{I \subset J} [I : J] e_I^0 \tag{3.4}$$

Consider the diagonal map:

$$\Delta : D_*(\mathbf{W}) \to D_*^0(\mathbf{W}), \quad e_J \mapsto \mathbf{W}_J(q) e_J^0. \tag{3.5}$$

It is clear by the previous discussion that Δ is an injective chain-complex homomorphism, so there is an exact sequence of complexes:

$$0 \longrightarrow D_*(\mathbf{W}) \xrightarrow{\Delta} D_*^0(\mathbf{W}) \xrightarrow{\pi} L_*(\mathbf{W}) \longrightarrow 0 \tag{3.6}$$

where

$$L_k(\mathbf{W}) := \bigoplus_{\substack{J \subseteq S \\ |J|=k \\ |\mathbf{W}_J|<\infty}} \frac{R}{(\mathbf{W}_J(q))} \cdot \bar{e}_J$$

is the quotient complex.

Remark 3.1. Recall that for an affine Coxeter system $(\mathbf{W}, S)$ of rank $n+1$, a parabolic subgroup $\mathbf{W}_I$ is finite if and only if I is a proper subset of S. In particular the poset of finite parabolic subgroups is isomorphic to the poset of proper subsets of $I_{n+1} = \{1, \ldots, n+1\}$ (that is the boolean lattice minus its maximum). (For the classification of such systems, see again [4, 23].)

We assume hereinafter that $(\mathbf{W}, S)$ is an affine Coxeter system of rank $n+1$ Then, by Remark 3.1, the homology of $D^0_*(\mathbf{W})$ is the reduced homology of a $(n-1)$-sphere modulo a degree shift:

$$H_k(D^0_*(\mathbf{W})) \cong \tilde{H}_{k-1}(S^{n-1}; R) \cong \begin{cases} 0 & \text{if } k \neq n \\ R & \text{if } k = n \end{cases}$$

Using the long exact sequence associated to the short exact sequence of complexes 3.6:

$$\xrightarrow{\pi_*} H_{k+1}(L_*) \xrightarrow{\delta} H_k(D_*) \xrightarrow{\Delta_*} H_k(D^0_*) \xrightarrow{\pi_*} H_k(L_*) \xrightarrow{\delta} H_{k-1}(D_*) \xrightarrow{\Delta_*} \tag{3.7}$$

we have then isomorphisms for $0 \leq k \leq n-2$:

$$H_k(D_*(\mathbf{W})) \cong H_{k+1}(L_*(\mathbf{W}))$$

whereas the top and top-1 homology groups of $D_*(\mathbf{W})$ fit into:

$$0 \longrightarrow H_n(D_*(\mathbf{W})) \longrightarrow R \longrightarrow H_n(L_*(\mathbf{W})) \xrightarrow{\delta} H_{n-1}(D_*(\mathbf{W})) \longrightarrow 0 \tag{3.8}$$

In particular for the top homology we have

$$H_n(\mathbf{X_W}; R_q) \cong R \tag{3.9}$$

In general, one can try to determine $H_*(X(\mathbf{W}); R_q)$ by analyzing the homology $H_*(L_*(\mathbf{W}))$. The latter is a special case of the homology of a sheaf over a poset which we introduce in the next section.

3.2.2 Weighted Sheaves over posets Let $(P, \prec)$ be a poset.

Definition 3.2. (See also [3, 36].) Define a sheaf of rings over P as a collection

$$\{A_x,\ x \in P\}$$

of commutative rings together with a collection of ring homomorphisms

$$\{\rho_{x,y} :\ A_y \to A_x, \quad x \preceq y\}$$

satisfying

$$\rho_{x,x} = id_{A_x}\ ;$$

$$x \preceq y \preceq z\ \Rightarrow\ \rho_{x,z} = \rho_{x,y}\ \rho_{y,z}$$

Let $\mathcal{C}_P$ be the small category associated to P, with

$$Ob(\mathcal{C}_P) = P$$

and

$$\begin{aligned} \mathrm{Hom}(x, y) &= \{(x, y)\} \qquad \text{if } x \preceq y \\ &= \quad \emptyset \qquad\quad \text{otherwise} \end{aligned}$$

So, a sheaf of rings may be understood as a contravariant functor from $\mathcal{C}_P$ to the category of rings. As such it can be equivalently called a *diagram of rings*.

We will use a particular class of sheaves. From now on, we fix $A = \mathbf{Q}$, so $R = \mathbf{Q}[q^{\pm}1]$ is the ring of Laurent polynomials over the rational numbers; all we say generalizes to the class of principal ideal domains, in particular A may be any other field. We give R the structure of a poset, by considering as partial ordering the relation of divisibility in R. In particular any poset homomorphism

$$\psi :\ (P, \prec)\ \to\ (R, |)\ :\ x \longrightarrow \psi(x) = p_x\ \in\ R$$

defines a sheaf over P by the collections

$$\{\ R/(p_x)\ ,\ x \in P\}$$

and

$$\{\ i_{x,y} :\ R/(p_y)\ \to\ R/(p_x)\ \}$$

where $i_{x,y}$ is the map induced by the identity of R, well defined since by definition $(p_y) \subset (p_x)$.

Definition 3.3. We call the triple (P, R, ψ) a *weighted sheaf* over P and the coefficients p_x the *weights* of the sheaf.

Of special interest will be posets coming from simplicial schemes. Any such K, defined over a finite set

$$I_{n+1} := \{1, ..., n+1\},$$

is a poset with partial ordering

$$\sigma \prec \tau \Leftrightarrow \sigma \subset \tau.$$

It is convenient to denote by $C^0_*(K)$ the algebraic complex for the reduced simplicial R–homology of K, shifted in dimension by one, so that in dimension k one has

$$C^0_k(K) = \bigoplus_{\substack{\sigma \in K \\ |\sigma|=k}} R \cdot e^0_\sigma$$

where e^0_σ is a generator associated to a given orientation of σ ($C^0_0(K) = \mathbb{Z}\, e^0_\emptyset$); the boundary is given by

$$\partial^0(e^0_\sigma) = \sum_{|\tau|=k-1} [\tau : \sigma]\, e^0_\tau$$

where $[\tau : \sigma]$ denotes the incidence number holding ± 1 if $\tau \prec \sigma$ and vanishing otherwise.

We want to define a modified complex for the homology of weighted sheaves over a simplicial scheme K.

Definition 3.4. The *weighted complex* associated to the weighted sheaf (K, R, ψ) is the algebraic complex

$$L_* := L_*(K)$$

defined by

$$L_k := \bigoplus_{|\sigma|=k} \frac{R}{(p_\sigma)}\, \overline{e}_\sigma$$

and boundary

$$\partial : L_k \rightarrow L_{k-1}$$

induced by ∂^0 :

$$\partial(a_\sigma\, \overline{e}_\sigma) = \sum_{\tau \prec \sigma} [\tau : \sigma]\, i_{\tau,\sigma}(a_\sigma)\, \overline{e}_\tau$$

(The condition $\partial^2 = 0$ is straightforward from the definitions.)

Remark 3.5 (Main example). Let $(\mathbf{W}, S)$ be a Coxeter system and consider the simplicial scheme K (defined over the finite set S) associated to all finite parabolic subgroups of $\mathbf{W}$ (see also Remark 3.1). Then clearly the complexes $D_*^0(\mathbf{W})$, $L_*(\mathbf{W})$ introduced in the previous section 3.2.1 arise respectively as the complex $D_*^0(K)$ and as the weighted complex $L_*(K)$ associated to the homomorphism $\psi : K \to R$ taking $I \in K$ into $\mathbf{W}_I(q) \in R$.

3.2.3 Decomposition and filtration of $L_*(K)$ Let $L := L_*(K)$ be the algebraic complex associated to (K, R, ψ). Each single factor $\frac{R}{(p_\sigma)}$ of L decomposes into cyclic modules according to the irreducible factorization of p_σ :

$$p_\sigma = \prod_i \varphi_i^{m_i} \Rightarrow \frac{R}{(p_\sigma)} \cong \bigoplus_i \frac{R}{(\varphi_i^{m_i})}$$

Given any irreducible $\varphi \in R$, we define the φ-primary component of the weighted sheaf as follows. For $\sigma \in K$, let

$$m(\varphi, \sigma) := \text{max power of } \varphi \text{ dividing } p_\sigma$$

and let $\psi_{(\varphi)} : K \to R$ be defined by

$$\psi_{(\varphi)}(\sigma) := \varphi^{m(\varphi,\sigma)}.$$

Then $(K, R, \psi_{(\varphi)})$ defines a weighted sheaf which we call the φ-primary component of the given sheaf, and denote also by $(K, R, \psi)_{(\varphi)}$. The associated complex is $L_{(\varphi)}$, defined in degree k by:

$$(L_{(\varphi)})_k := \bigoplus_{|\sigma|=k} \frac{R}{(\varphi^{m(\varphi,\sigma)})} \bar{e}_\sigma .$$

There is a natural increasing filtration into subcomplexes of $L_{(\varphi)}$ by the powers of φ:

$$F^s(L_{(\varphi)}) := \bigoplus_{m(\varphi,\sigma)\leq s} \frac{R}{(\varphi^{m(\varphi,\sigma)})} \bar{e}_\sigma$$

The latter is associated to an increasing filtration of the simplicial complex K into subcomplexes:

$$K_{(\varphi),s} := \{\sigma \in K \mid m(\varphi, \sigma) \leq s\}.$$

Then $F^s(L_{(\varphi)})$ is the algebraic complex associated to the weighted sheaf $(K_{(\varphi),s}, R, \psi_{|K_{(\varphi),s}})$.

We get the following theorem.

Theorem 3.6. *Let (K, R, ψ) be a weighted sheaf, with associated algebraic complex L_*. For any irreducible $\varphi \in R$, there exists a spectral sequence*

$$E^0_{p,q} \quad \Rightarrow \quad H_*(L_{(\varphi)})$$

that abuts to the homology of the φ-primary component of the associated algebraic complex L_*.
Moreover the E^1-term:

$$E^1_{p,q} = H_{p+q}(F^p/F^{p-1}) \cong H_{p+q}(K_{(\varphi),p}, K_{(\varphi),p-1}; R/(\varphi^p)) \quad (3.10)$$

is isomorphic to the relative homology with trivial coefficients of the simplicial complexes pair $(K_{(\varphi),p}, K_{(\varphi),p-1})$.

The proof is standard. We stress that the second member of (3.10) is the homology with non-twisted coefficients, as the boundary is induced by that in D^0_* by (3.6).

3.2.4 Top and Top-1 homology of $\mathbf{G}_{\tilde{A}_n}$ We apply the results of the previous part to the determination of the groups $H_k(\mathbf{G}_\mathbf{W}; R_q)$ for $\mathbf{W}$ of type $\tilde{A}_n$, and $k = n-1, n$, *i.e.* in dimension top and top-1. This computation is just an example of the use of the methods developed before, since the cohomology of $\mathbf{G}_{\tilde{A}_n}$ with coefficients in R_q was completely determined in [9] (see also [10]). New cases are presently worked and will be published elsewhere.

Let $S := I_{n+1}$. For any subset $J \subset S$, let Γ be the subgraph of the Coxeter diagram for $\tilde{A}_n$ spanned by the vertices in J. Let also $\Gamma_1, \dots, \Gamma_k$ be the connected components of Γ and set $a_i := |\Gamma_i| + 1$. Since every proper connected component Γ_i gives rise to a Coxeter graph of type $A_{|\Gamma_i|}$, the Poincaré series for the parabolic subgroup $(\tilde{A}_n)_J$ is $\psi(J) := [a_1]![a_2]! \cdots [a_k]!$.

Let K be the simplicial scheme of finite parabolic subgroups of $\tilde{A}_n$ (see Remark 3.1) and consider the weighted sheaf (K, R, ψ). We start analyzing the associated algebraic complex L_*.

Let $d > 1$ be an integer and consider the φ_d-primary component of L_*, where we set as usual φ_d as the d-th cyclotomic polynomial. It is convenient to write for brevity $L^{(d)} := L_{(\varphi_d)}$.

Lemma 3.7.

$$L^{(d)} = \bigoplus_{J \subsetneq S} \frac{R}{\left(\varphi_d^{\lfloor a_1/d \rfloor + \ldots + \lfloor a_k/d \rfloor}\right)} \cdot \bar{e}_J$$

where $\lfloor x \rfloor$ is the integral part of the number x.

Proof. It is enough to observe that the maximal power of d dividing $a!$ is $\lfloor a/d \rfloor$. □

We may now determine the homology $H_k(\mathbf{G}_{\tilde{A}_n})$ for $k = n-1,\ n$. We have:

Theorem 3.8.

$$H_n(\mathbf{G}_{\tilde{A}_n}; R_q) = R$$

$$H_{n-1}(\mathbf{G}_{\tilde{A}_n}; R_q) = \bigoplus_{\substack{d|n+1 \\ d\geq 2}} \left(\frac{R}{(\varphi_d)}\right)^{d-1}$$

Proof. We can re-write (3.8) as:

$$0 \longrightarrow H_n(\mathbf{G}_{\tilde{A}_n}; R_q) \xrightarrow{\Delta_*} H_n(D^0_*) \xrightarrow{\pi_*} H_n(L_*) \xrightarrow{\delta} H_{n-1}(\mathbf{G}_{\tilde{A}_n}; R_q) \longrightarrow 0 \tag{3.11}$$

and observe that $H_n(D^0_*) \cong R$ is generated by $z^0 = \sum_{i=1}^{n+1}(-1)^i e^0_{S\setminus\{i\}}$. By construction for any n-subset $S \setminus \{i\} \subset S$ we have $\Delta(e_{S\setminus\{i\}}) = [n+1]! e^0_{S\setminus\{i\}}$. It is then clear that $H_n(\mathbf{G}_{\tilde{A}_n}; R_q) \cong R$ with generator $z = \sum_{i=1}^{n+1}(-1)^i e_{S\setminus\{i\}}$. Since $\Delta_*(z) = [n+1]! z^0$, we are left to analyze the sequence:

$$0 \longrightarrow \frac{R}{[n+1]!} \xrightarrow{\pi_*} H_n(L_*) \xrightarrow{\delta} H_{n-1}(\mathbf{G}_{\tilde{A}_n}; R_q) \longrightarrow 0 \tag{3.12}$$

We determine $H_n(L_*)$ using the decomposition in primary components and the related filtration. Fix thus an integer $d \geq 2$ and let $m =: \lfloor n + 1/d \rfloor$. We will consider in particular the short exact sequence:

$$0 \longrightarrow F^{m-1}L^{(d)}_* \longrightarrow F^m L^{(d)}_* \longrightarrow F^m L^{(d)}_*/F^{m-1}L^{(d)}_* \longrightarrow 0 \tag{3.13}$$

Since the Poincaré series of any n-subset $S \setminus \{i\} \subset S$ is $\psi(S \setminus \{i\}) = [n+1]!$, by Lemma 3.7 the maximal power of φ_d dividing $\psi(S \setminus \{i\})$ is precisely m. Thus, $F^m L^{(d)}_n = L^{(d)}_n$ whereas $F^{m-1}L^{(d)}_n = 0$.

We study similarly how the filtration behaves on $L^{(d)}_{n-1}$. Notice that $L^{(d)}_{n-1}$ is generated by the elements $\{\overline{e}_{S\setminus\{i,j\}}\}_{1\leq i<j\leq n+1}$ and that $\psi(S \setminus \{i,j\}) = [j-i]![n+1-j+i]!$. Thus the maximal power of φ_d dividing $\psi(S \setminus \{i,j\})$ is $\lfloor (j-i)/d \rfloor + \lfloor (n+1-j+i)/d \rfloor$. Let's write:

$$n+1 = \lfloor (n+1)/d \rfloor \cdot d + \langle n+1 \rangle_d$$

$$j-i = \lfloor (j-i)/d \rfloor \cdot d + \langle j-i \rangle_d$$

where we denote by $\langle a\rangle_d$ the remainder class of a modulo d with values in $0, 1, \dots, d-1$.
Then:

$$\begin{aligned} n+1-j+i = (\lfloor (n+1)/d\rfloor - \lfloor (j-i)/d\rfloor)\cdot d \\ + (\langle n+1\rangle_d - \langle j-i\rangle_d) \end{aligned} \tag{3.14}$$

If $\langle j-i\rangle_d \le \langle n+1\rangle_d$, it results from Equation (3.14) that $\lfloor (n+1-j+i)/d\rfloor = \lfloor (n+1)/d\rfloor - \lfloor (j-i)/d\rfloor$ and thus the maximal power of φ_d dividing $\psi(S\setminus\{i,j\})$ is m. Otherwise, if $\langle j-i\rangle_d > \langle n+1\rangle_d$, we have $\lfloor (n+1-j+i)/d\rfloor = \lfloor (n+1)/d\rfloor - \lfloor (j-i)/d\rfloor - 1$ and the maximal power of φ_d dividing $\psi(S\setminus\{i,j\})$ is $m-1$. This shows that $F^{m-2}L^{(d)}_{n-1} = 0$, whereas:

$$\frac{F^m L^{(d)}_{n-1}}{F^{m-1}L^{(d)}_{n-1}} = \bigoplus_{\substack{1\le i<j\le n+1\\ \langle j-i\rangle_d\le\langle n+1\rangle_d}} \left(\frac{R}{(\varphi_d)^m}\right) p(\overline{e}_{S\setminus\{i,j\}})$$

$$\frac{F^{m-1} L^{(d)}_{n-1}}{F^{m-2}L^{(d)}_{n-1}} = \bigoplus_{\substack{1\le i<j\le n+1\\ \langle j-i\rangle_d>\langle n+1\rangle_d}} \left(\frac{R}{(\varphi_d)^{m-1}}\right) p(\overline{e}_{S\setminus\{i,j\}})$$

where p denotes the projection in the suitable quotient space. We now claim that:

$$H_n\left(\frac{F^m L^{(d)}_*}{F^{m-1}L^{(d)}_*}\right) \cong \begin{cases} \left(\dfrac{R}{(\varphi_d)^m}\right)\cdot p(\overline{z}) & \text{for } d\nmid n+1 \\ \bigoplus_{\lambda=1}^d \left(\dfrac{R}{(\varphi_d)^m}\right)\cdot p(\overline{z}_\lambda) & \text{for } d\mid n+1 \end{cases} \tag{3.15}$$

where $\overline{z}$ and $\overline{z}_\lambda$ are defined by:

$$\overline{z} =: \sum_{i=1}^{n+1} (-1)^i \overline{e}_{S\setminus\{i\}}$$

$$\overline{z}_\lambda =: \sum_{s=0}^{m-1} (-1)^{\lambda+sd}\overline{e}_{S\setminus\{\lambda+sd\}}$$

Notice that $\pi_*(z^0) = \overline{z}$; in addition, for a divisor $d\mid n+1$, we have $\overline{z} = \sum_{\lambda=1}^d \overline{z}_\lambda$. To prove Equation (3.15), let $\overline{c} = \sum_{i=1}^{n+1} a_i\overline{e}_{S\setminus\{i\}}$ be an n-cycle. In case $d\nmid n+1$, observe that the cells of type $p(\overline{e}_{S\setminus\{i,i+1\}})$ where $1\le i\le n$ belong to $F^m L^{(d)}_{n-1}/F^{m-1}L^{(d)}_{n-1}$ and that the only n-cells incident with $p(\overline{e}_{S\setminus\{i,i+1\}})$ are $\overline{e}_{S\setminus\{i\}}$ and $\overline{e}_{S\setminus\{i+1\}}$. If $\overline{c}$ is a cycle, it is thus necessary that $a_i = -a_{i+1}$ for $1 \leq i \leq n$. This shows that $\overline{c}$ must be a multiple of $\overline{z}$. Since $\overline{z}$ is clearly a n-cycle, it follows that it is a generator of $H_n\left(F^m L_*^{(d)}/F^{m-1}L_*^{(d)}\right)$. In case $d \mid n+1$, the only $(n-1)$-cells in $F^m L_{n-1}^{(d)}/F^{m-1}L_{n-1}^{(d)}$ are of the form $p(\overline{e}_{S\setminus\{h,k\}})$ with $d \mid (k-h)$. From this observation it follows easily that the $\overline{z}_\lambda$ $(1 \leq \lambda \leq d)$ are n-cycles. In addition, since the only n-cells incident with $p(\overline{e}_{S\setminus\{\lambda,\lambda+sd\}})$ (for $1 \leq \lambda \leq d$ and $1 \leq s < m$) are $\overline{e}_{S\setminus\{\lambda\}}$ and $\overline{e}_{S\setminus\{\lambda+sd\}}$, if $\overline{c} = \sum_{i=1}^{n+1} a_i \overline{e}_{S\setminus\{i\}}$ is a cycle, then we must have $a_\lambda = (-1)^{sd} a_{\lambda+sd}$ for $1 \leq \lambda \leq d$ and $1 \leq s < m$. This shows that $\overline{c}$ must be a linear combination of the $\overline{z}_\lambda$ $(1 \leq \lambda \leq d)$. This concludes the proof of Equation (3.15).

Using the long exact sequence associate to the short exact sequence in Equation (3.13), we may identify $H_n(F^m)$ with the kernel of the connecting homomorphism $H_n(F^m/F^{m-1}) \to H_{n-1}(F^{m-1})$. When $d \nmid n+1$, the generator $p(\overline{z})$ is obviously in the kernel. When $d \mid n+1$, we should check whether the boundary of a linear combination $\sum a_\lambda \overline{z}_\lambda$ is zero modulo φ_d^{m-1}. By an argument similar to the one used for proving Equation (3.15), it turns out that the kernel is generated by $\overline{z} = \sum_\lambda \overline{z}_\lambda$ and by the multiples $(\varphi_d)^{m-1}\overline{z}_\lambda$ for $\lambda = 1, \ldots, d-1$.

Since $H_n(F^m) = H_n(L_*^{(d)})$, the stated result follows now easily from the exact sequence (3.12). Indeed, when $d \nmid n+1$ the generator $\overline{z}$ of $H_n(F^m) = H_n(L_*^{(d)})$ is in the image of π_*, since $\overline{z} = \pi_*(z^0)$. Therefore, there is no φ_d-primary component in $H_{n-1}(\mathbf{G}_{\tilde{A}_n}; R_q)$ for $d \nmid n+1$. When $d \mid n+1$, we have instead:

$$\begin{aligned} H_{n-1}^{(d)}(\mathbf{G}_{\tilde{A}_n}; R_q) &\cong H_n(L_*^{(d)})/\mathrm{Im}(\pi_*) \\ &\cong < \overline{z}, (\varphi_d)^{m-1}\overline{z}_1, \ldots, (\varphi_d)^{m-1}\overline{z}_d >_{R/(\varphi_d)^m} / < \overline{z} >_{R/(\varphi_d)^m} \\ &\cong \left(\frac{R}{(\varphi_d)}\right)^{d-1}. \end{aligned}$$
□

3.2.5 Cohomological version Since R is a PID it follows

$$H^n(\mathbf{G}_{\tilde{A}_n}; R_q) \cong H_n(\mathbf{G}_{\tilde{A}_n}; R_q) \oplus H_{n-1}(\mathbf{G}_{\tilde{A}_n}; R_q).$$

By using calculations similar to those in the previous parts (obtained by using the dual to the diagonal map Δ) we can prove

Theorem 3.9. *Let* $\mathbf{W}$ *be an affine Weyl group of rank* $n+1$ *and let* $M = \mathbb{Z}[q^{\pm}1]$ *or* $M = \mathbb{Z}[-1]$. *Then* $H^n(\mathbf{G}_\mathbf{W}; M)$ *has rank* 1 *as* M*-module.*

Proof. Notice that the entries of the diagonal map Δ are non-zero in M. In fact, the sign representation is obtained from (2.8) by specializing to $q = 1$. Then both results follow straightforwardly from suitable analogue of the exact sequence in Equation (3.8). □

3.2.6 Schwarz and homological genera We start recalling the definition of Schwarz genus and discussing briefly some of its properties (we refer to [33] for details; the original exposition is in [30]).

Definition 3.10. For a locally trivial fibration $f : \mathbf{Y} \to \mathbf{X}$, the Schwarz genus $g(f)$ is the minimal cardinality of an open cover $\mathcal{U}$ of $\mathbf{X}$ such that f admits a section over each set $U \in \mathcal{U}$.

Remark. The Schwarz genus is the extension to fibrations of the Lusternik-Schnirelmann category of a topological space; indeed the category of a path connected topological space coincides with the Schwarz genus of its Serre fibration.

When $\mathbf{X}$ has the homotopy type of a finite dimensional CW complex, we have an upper bound for the genus of whatever fibration:

Theorem 3.11 ([30]). *If X has the homotopy type of a CW complex of dimension N, then $g(f) \leq N + 1$.*

Let now $f : \mathbf{Y} \to \mathbf{X}$ be a regular G-covering. Consider the classifying space BG for G and the universal G-bundle $EG \to BG$ (see *e.g.* [32]). Then there is a unique (up to homotopy) classifying map $a : \mathbf{X} \to BG$ such that the covering f is induced as pull-back of the universal G-bundle $EG \to BG$.
Let M be an arbitrary G-module and a^*M be the local system on X induced by the map a.

Definition 3.12. The homological M-genus of $f : \mathbf{Y} \to \mathbf{X}$ is the smallest integer $h_M(f)$ such that the induced map in cohomology:

$$a^* : H^j(BG; M) \to H^j(\mathbf{X}; a^*M)$$

is zero in degree j for $j \geq h_M(f)$.
The homological genus is defined as the maximum $h(f) = \max_M h_M(f)$ of the homological M-genera.

Homological genus provides a lower bound for Schwarz genus:

Theorem 3.13 ([30]). *For any regular covering $f : \mathbf{Y} \to \mathbf{X}$, we have $g(f) \geq h(f)$.*

3.2.7 The genus problem and Coxeter groups Let $\mathbf{W}$ be a Coxeter group and consider the regular covering $f_{\mathbf{W}} : \mathbf{Y}_{\mathbf{W}} \to \mathbf{X}_{\mathbf{W}}$. We are interested in the genus $g(f_{\mathbf{W}})$ of $f_{\mathbf{W}}$.

When $\mathbf{W} = A_n$, this number has an interesting interpretation. Recall that in this case, $\mathbf{X}_{\mathbf{W}}$ is the space of monic polynomials of degree $n + 1$ with distinct roots and $\mathbf{Y}_{\mathbf{W}}$ is the space of such roots. Suppose that f

admits a section s on an open set $U \subset \mathbf{X_W}$. The section s may be then understood as a function that "computes" the roots of a polynomial $P \in U$. The Schwarz genus corresponds then to the minimal number of such functions needed to compute the roots of any polynomial in $\mathbf{X_W}$.

Smale [31] links this problem to the topological complexity $\tau(n,\epsilon)$ of an algorithm that computes the roots of a polynomial of degree $n+1$ with a given precision ϵ. In particular for small ϵ, one has $\tau(n,\epsilon) \geq g(f_{A_n}) - 1$. Further it is known that the previous inequality becomes an equality when $n+1$ is a prime power.

For a finite Coxeter group $\mathbf{W}$ of rank n, it was shown in [18] that the Schwarz genus reaches the upper bound in Theorem 3.11 (*i.e.* $g(f_{\mathbf{W}}) = n+1$), except when $\mathbf{W} = A_n$ and $n+1$ is not a prime power. In this case, the exact determination of the genus is still open; for the first non prime power $n+1=6$, it is known that $g(f_{A_5}) = 5$ (and so the genus does not reach the upper bound of Theorem 3.11) [15]. Next, Arone [2] showed by different methods that $g(f_{A_n}) < n+1$ when $n+1 \neq p^k, 2p^k$ for p prime.

Recall that we have an inclusion $i : \mathbf{X_W} \hookrightarrow \mathbf{X_W^{(\infty)}}$ and that $\mathbf{X_W}$ may be identified with the subcomplex of $\mathbf{X_W^{(\infty)}}$ consisting of cells of type $\Gamma = (\Gamma_1 \supseteq \emptyset \supseteq \emptyset \supseteq \ldots)$.

Let M be a $\mathbf{W}$-module and M' the local coefficient system on $\mathbf{X_W}$ induced by M via i. The associated map of cochains

$$i^* : C^*(\mathbf{W}; M) \to C^*(\mathbf{X_W}; M')$$

may then be easily described as the restriction of $c \in C^*(\mathbf{W}; M)$ to the chains for $\mathbf{X_W}$. Let n be the maximal cardinality of a subset $I \subset S$ s.t. $|\mathbf{W}_I| < \infty$. Then in degree n we have:

$$\begin{array}{ccccccc}
\longrightarrow C^{n-1}(\mathbf{W}; M) & \longrightarrow & C^n(\mathbf{W}; M) & \longrightarrow & C^{n+1}(\mathbf{W}; M) & \longrightarrow \\
\downarrow & & \downarrow & & \downarrow & \\
\longrightarrow C^{n-1}(\mathbf{X}(\mathbf{W}); M') & \longrightarrow & C^n(\mathbf{X_W}; M') & \longrightarrow & 0 & \longrightarrow
\end{array} \tag{3.16}$$

The following proposition is useful in the computation of the homological genus. The proof comes from the explicit form of the boundaries in the complexes $C^*(\mathbf{W}),\ C^*(\mathbf{X_W})$.

Proposition 3.14 ([18]). *Let $M = \mathbb{Z}[-1]$ be the sign representation. Then the map $i^* : H^n(\mathbf{W}; M) \to H^n(\mathbf{X_W}; M')$ is an epimorphism.*

Let now $\mathbf{W}_a$ be an affine Weyl group. From Theorem 3.9, we know that the top-cohomology of $\mathbf{X}_{\mathbf{W}_a}$ with coefficients in the sign representation does not vanish. Using Proposition 3.14, the homological genus $h(f_{\mathbf{W}_a})$ is greater than $n+1$. Since $\mathbf{X}_{\mathbf{W}_a}$ has dimension n, using Theorems 3.13 and 3.11, we get:

Theorem 3.15. *Let $\mathbf{W}_a$ be an affine Weyl group of rank $n+1$. Then the Schwarz genus of the fibration $\mathbf{Y}_{\mathbf{W}_a} \to \mathbf{X}_{\mathbf{W}_a}$ is precisely $n+1$.*

Remark. It should be observed that in the previous corollary the machinery of Theorem 3.9 is only needed to deal with the affine Weyl group of type $\tilde{A}_n$, while in the other cases the result follows readily from the finite type case. Indeed any affine group $\mathbf{W}_a \neq \tilde{A}_n$ has a finite parabolic subgroup H of rank n of type $\neq A_n$. Since $H^n(\mathbf{X}_{\mathbf{W}_a}; \mathbb{Z}[-1])$ is onto $H^n(\mathbf{X}_H; \mathbb{Z}[-1])$, the latter being non trivial (see [18]), the top cohomology $H^n(\mathbf{X}_{\mathbf{W}_a}; \mathbb{Z}[-1])$ is non-trivial as well.

References

[1] A. Bjorner and G. Ziegler , *Combinatorial stratifications of complex arrangements*, Jour. Amer. Math. Soc. **5** (1992), 105–149.

[2] G. Arone, *A note on the homology of Σ_n, the Schwartz genus, and solving polynomial equations*, 2005.

[3] K. Baclawski, *Whitney numbers of geometric lattices*, Advances in Mathematics **16** (1975), 125–138.

[4] N. Bourbaki, "Groupes et algebrès de Lie", Vol. Chapters IV-VI, Hermann, 1968.

[5] E. Brieskorn, *Die Fundamentalgruppe des Raumes der regulären Orbits einer endlichen komplexen Spiegelunsgruppe*, Invent. Math. **12** (1971), 57–61.

[6] R. Charney and M. W. Davis, *The $k(\pi, 1)$-problem for hyperplane complements associated to infinite reflection groups*, J. of AMS **8** (1995), 597–627.

[7] R. Charney, *Geodesic automation and growth functions for Artin groups of finite type*, Math. Ann. **301** (1995), 307–324.

[8] F. Callegaro, *The homology of the Milnor fiber for classical braid groups*, Alg. Geom. Top. **6** (2006), 1903–1923.

[9] F. Callegaro, D. Moroni and M. Salvetti, *Cohomology of affine Artin groups and applications*, Transactions of the American Mathematical Society **360** (2008), 4169–4188.

[10] F. Callegaro, D. Moroni and M. Salvetti, *Cohomology of Artin groups of type $\tilde{A}_n$, b_n and applications*, Geometry and Topology Monographs **13** (2008), 85–104.

[11] F. Callegaro, D. Moroni and M. Salvetti, *The $k(\pi,1)$ problem for the affine Artin group of type $\tilde{B}_n$ and its cohomology*, J. European Mathematical Society **12** (2010), 1–22.

[12] F. CALLEGARO and M. SALVETTI, *Integral cohomology of the Milnor fibre of the discriminant bundle associated with a finite Coxeter group*, C. R. Acad. Sci. Paris, Ser. I **339** (2004), 573–578.
[13] P. DELIGNE, "Equation Différentielles à Points Singuliers Réguliers", Lecture Notes in Mathematics, Vol. 163, Springer-Verlag, 1970.
[14] C. DE CONCINI, C. PROCESI and M. SALVETTI, *Arithmetic properties of the cohomology of braid groups*, Topology **40** (2001), no. 4, 739–751.
[15] C. DE CONCINI, C. PROCESI and M. SALVETTI, *On the equation of degree* 6, Comm. Math. Helv. **79** (2004), 605–617.
[16] C. DE CONCINI, C. PROCESI, M. SALVETTI and F. STUMBO, *Arithmetic properties of the cohomology of Artin groups*, Ann. Scuola Norm. Sup. Pisa Cl. Sci. **28** (1999), no. 4, 695–717.
[17] C. DE CONCINI and M. SALVETTI, *Cohomology of Artin groups*, Math. Res. Lett. **3** (1996), 293–297.
[18] C. DE CONCINI and M. SALVETTI, *Cohomology of Artin groups and Coxeter groups*, Math. Res. Lett. **7** (2000), 213–232.
[19] C. DE CONCINI, M. SALVETTI and F. STUMBO, *The top-cohomology of Artin groups with coefficients in rank-1 local systems over* $\mathbb{Z}$, Topology and its Applications **78** (1997), 5–20.
[20] E. V. FRENKEL, *Cohomology of the commutator subgroup of the braids group*, Func. Anal. Appl. **22** (1988), no. 3, 248–250.
[21] E. GODELLE and L. PARIS, *Basic questions on Artin-Tits groups*, In: "Configuration Spaces", Proceedings, A. Bjorner, F. Cohen, C. De Concini, C. Procesi and M. Salvetti (eds.), Edizioni della Normale, Pisa, 2012, 299–311.
[22] H. HENDRIKS, *Hyperplane complements of large type*, Invent. Math. **79** (1985), 375–381.
[23] J. E. HUMPHREYS,"Reflection Groups and Coxeter Groups", Cambridge University Press, 1990.
[24] D. MORONI, "Finite and Infinite Type artin Groups: Topological Aspects and Cohomological Computations", PhD thesis, 2006.
[25] C. OKONEK, *Das* $k(\pi, 1)$*-Problem für die affinen Wurzelsysteme vom typ* A_n, C_n, Mathematische Zeitschrift **168** (1979), 143–148.
[26] M. SALVETTI, *Topology of the complement of real hyperplanes in* $\mathbb{C}^n$, Invent. Math. **88** (1987), no. 3, 603–618.
[27] M. SALVETTI, *The homotopy type of Artin groups*, Math. Res. Lett. **1** (1994), 567–577.
[28] M. SALVETTI, *On the cohomology and topology of artin and coxeter groups*, Pubblicazioni Dipartimento di Matematica L.Tonelli, Pisa (2005).

[29] M. SALVETTI and F. STUMBO, *Artin groups associated to infinite Coxeter groups*, Discrete Mathematics **163** (1997), 129–138.
[30] A. S. SCHWARZ, *Genus of a fibre bundle*, Trudy Moscow. Math. Obshch **10** (1961), 217–272.
[31] S. SMALE, *On the topology of algorithms*, Int. J. Complexity **3** (1987), 81–89.
[32] N. STEENROD, "The Topology of Fibre Bundles", Princeton University Press, Princeton, N.J., 1951.
[33] V. A. VASSILIEV, "Complements of Discriminants of Smooth Maps: Topology and Applications", Translations of Mathematical Monographs, Vol. 98, AMS, 1992.
[34] H. VAN DER LEK, "The Homotopy Type of Complex Hyperplane Complements", Ph.D. thesis, University of Nijmegan, 1983.
[35] E. B. VINBERG, *Discrete linear groups generated by reflections*, Math. USSR Izvestija **5** (1971), no. 5, 1083–1119.
[36] S. YUZVINSKY, *Cohen-Macaulay rings of sections*, Advances in Mathematics **63** (1987), 172–195.

Chromatic quasisymmetric functions and Hessenberg varieties

John Shareshian[1] and Michelle L. Wachs[2]

Abstract. We discuss three distinct topics of independent interest; one in enumerative combinatorics, one in symmetric function theory, and one in algebraic geometry. The topic in enumerative combinatorics concerns a q-analog of a generalization of the Eulerian polynomials, the one in symmetric function theory deals with a refinement of the chromatic symmetric functions of Stanley, and the one in algebraic geometry deals with Tymoczko's representation of the symmetric group on the cohomology of the regular semisimple Hessenberg variety of type A. Our purpose is to explore some remarkable connections between these topics.

1 Introduction

Let $H(z) := \sum_{n\geq 0} h_n z^n$, where h_n is the complete homogeneous symmetric function of degree n. The formal power series

$$\frac{(1-t)H(z)}{H(tz)-tH(z)}, \tag{1.1}$$

has occurred in various contexts in the literature, including

- the study of a class of q-Eulerian polynomials
- the enumeration of Smirnov words by descent number
- the study of the toric variety associated with the Coxeter complex of the symmetric group.

We briefly describe these contexts now and give more detail in Section 2 with definitions in Section 1.1. In [22,23] Shareshian and Wachs initiated a study of the joint distribution of the Mahonian permutation statistic, major index, and the Eulerian permutation statistic, excedance number. They showed that the q-Eulerian polynomial defined by

$$\sum_{\sigma\in\mathfrak{S}_n} q^{\mathrm{maj}(\sigma)} t^{\mathrm{exc}(\sigma)}$$

[1] Supported in part by NSF Grant DMS 0902142.

[2] Supported in part by NSF Grant DMS 0902323.

can be obtained by taking a certain specialization of the coefficient of z^n in (1.1) known as the stable principal specialization. This enabled them to derive the q-analog of a formula of Euler that is given in (2.9).

The work in [22] led Stanley to derive a refinement of a formula of Carlitz, Scoville and Vaughan [1] for the enumerator of words over the alphabet $\mathbb{P}$ with no adjacent repeats (see [23]). These are known as Smirnov words. In Stanley's refinement the Smirnov words are enumerated according to their number of descents and the enumerator is equal to the symmetric function obtained by applying the standard involution ω on the ring of symmetric functions to (1.1).

In another direction, Stanley [27, Proposition 7.7] used a formula of Procesi [18] to show that (1.1) is the generating function for the Frobenius characteristic of the representation of the symmetric group $\mathfrak{S}_n$ on the cohomology of the toric variety associated with the Coxeter complex of type A_{n-1}.

The q-Eulerian polynomial, the Smirnov word enumerator, and the $\mathfrak{S}_n$-representation on the cohomology of the toric variety all have very nice generalizations. In this paper we discuss these generalizations and explore the relationships among these generalizations and the ramifications of the relationships. The proofs of the results surveyed here either have appeared in [23] or will appear in [25].

The generalization of the q-Eulerian numbers, as defined in [25], is obtained by considering the joint distribution of a Mahonian statistic of Rawlings [19] that interpolates between inversion number and major index, and a generalized Eulerian statistic of De Mari and Shayman [4]. We discuss these generalized q-Eulerian numbers and a more general version of them associated with posets in Section 3.

The generalization of the Smirnov word enumerator, as defined in [25], is a refinement of Stanley's well known chromatic symmetric function [30], which we call a chromatic quasisymmetric function. In Section 3 we report on results on the chromatic quasisymmetric functions, which are proved in [25]. In particular we present a refinement of Gasharov's [10] expansion of certain chromatic symmetric functions in the Schur basis and a refinement of Chow's [2] expansion in the basis of fundamental quasisymmetric functions. We also present a conjectured refinement of Stanley's [30] expansion in the power-sum basis and a refinement of the conjecture of Stanley and Stembridge (see [32, Conjecture 5.5] and [30, Conjecture 5.1]) on e-positivity.

The generalization of the toric variety associated with the type A Coxeter complex is the regular semisimple Hessenberg variety of type A first studied by De Mari and Shayman [4] and further studied by De Mari, Procesi and Shayman [3]. Tymoczko [38] defined a representation of the symmetric group on the cohomology of this Hessenberg variety, which generalizes the representation studied by Procesi and Stanley. This generalization is described in more detail in Section 5.

In Section 4 we present the following extension of the relation between q-Eulerian polynomials and Smirnov words. As noted above, we can associate to a finite poset P a generalized q-Eulerian polynomial, which will be denoted by $A_P(q,t)$. Let $X_G(\mathbf{x},t)$ denote the chromatic quasisymmetric function of a graph G.

Theorem 1.1 (Corollary 4.18). *Let G be the incomparability graph of a poset P on [n]. Then*

$$A_P(q,t) = (q;q)_n \, \mathbf{ps}(\omega X_G(\mathbf{x},t)),$$

where **ps** *denotes stable principal specialization.*

In Section 5 we discuss the following conjectured generalization of relationship between Smirnov words and the representation of the symmetric group on the cohomology of the toric variety.

Conjecture 1.2 (Conjecture 5.3). Let G be the incomparability graph of a natural unit interval order P. Then

$$\omega X_G(\mathbf{x},t) = \sum_{j=0}^{|E(G)|} \operatorname{ch} H^{2j}(\mathcal{H}(P))t^j,$$

where ch denotes the Frobenius characteristic and $H^{2j}(\mathcal{H}(P))$ denotes Tymoczko's representation of $\mathfrak{S}_n$ on the degree $2j$ cohomology of the Hessenberg variety $\mathcal{H}(P)$ associated with P.

This conjecture has the potential of providing solutions to several open problems such as,

1. *Our conjecture that the generalized q-Eulerian polynomials are unimodal (Conjecture* 3.3*).* This would follow from Theorem 1.1 and the hard Lefschetz theorem applied to Tymoczko's representation on the cohomology of the Hessenberg variety.
2. *Tymoczko's problem of finding a decomposition of her representation into irreducibles* [38]. Such a decomposition would be provided by the expansion of the chromatic quasisymmetric functions in the Schur function basis given in Theorem 4.13.
3. *Stanley and Stembridge's well-known conjecture that the chromatic symmetric functions (for unit interval orders) are e-positive* [30, 32]. This would be equivalent to Conjecture 5.4 that Tymoczko's representation is a permutation representation in which each point stabilizer is a Young subgroup.

1.1 Preliminaries

We assume that the reader is familiar with some basic notions from combinatorics and symmetric function theory. All terms used but not defined in this paper are defined in [28] or [29].

Let $\sigma \in \mathfrak{S}_n$, where $\mathfrak{S}_n$ denotes the symmetric group on $[n] := \{1,2,\dots,n\}$. The excedance number of σ is given by

$$\operatorname{exc}(\sigma) := |\{i \in [n-1] : \sigma(i) > i\}|.$$

The descent set of σ is given by

$$\operatorname{DES}(\sigma) := \{i \in [n-1] : \sigma(i) > \sigma(i+1)\}$$

and the descent number and major index are

$$\operatorname{des}(\sigma) := |\operatorname{DES}(\sigma)| \quad \text{and} \quad \operatorname{maj}(\sigma) := \sum_{i \in \operatorname{DES}(\sigma)} i.$$

It is well-known that the permutation statistics exc and des are equidistributed on $\mathfrak{S}_n$. The common generating functions for these statistics are called Eulerian polynomials. That is, the Eulerian polynomials are defined as

$$A_n(t) := \sum_{\sigma \in \mathfrak{S}_n} t^{\operatorname{des}(\sigma)} = \sum_{\sigma \in \mathfrak{S}_n} t^{\operatorname{exc}(\sigma)}.$$

Any permutation statistic with generating function $A_n(t)$ is called an Eulerian statistic. The Eulerian numbers are the coefficients of the Eulerian polynomials; they are defined as

$$a_{n,j} := |\{\sigma \in \mathfrak{S}_n : \operatorname{des}(\sigma) = j\}| = |\{\sigma \in \mathfrak{S}_n : \operatorname{exc}(\sigma) = j\}|,$$

for $0 \le j \le n-1$.

It is also well-known that the major index is equidistributed with the inversion index defined as

$$\operatorname{inv}(\sigma) = |\{(i,j) \in [n] \times [n] : i < j \text{ and } \sigma(i) > \sigma(j)\}|$$

and that

$$\sum_{\sigma \in \mathfrak{S}_n} q^{\operatorname{maj}(\sigma)} = \sum_{\sigma \in \mathfrak{S}_n} q^{\operatorname{inv}(\sigma)} = [n]_q!,$$

where

$$[n]_q := 1 + q + \cdots + q^{n-1} \text{ and } [n]_q! := [n]_q[n-1]_q \dots [1]_q.$$

Any permutation statistic equidistributed with maj and inv is called a Mahonian permutation statistic.

Let R be a ring with a partial order relation, e.g., $\mathbb{Q}$ with its usual total order. Given a sequence $(a_0, a_1, \ldots, a_n)$ of elements of R we say that the sequence is *palindromic* with center of symmetry $\frac{n}{2}$ if $a_j = a_{n-j}$ for $0 \leq j \leq n$. The sequence is said to be *unimodal* if

$$a_0 \leq a_1 \leq \cdots \leq a_c \geq a_{c+1} \geq a_{c+2} \geq \cdots \geq a_n$$

for some c. The sequence is said to be *positive* if $a_i \geq 0$ for each i. We say that the polynomial $A(t) := a_0 + a_1 t + \cdots + a_n t^n \in R[t] - \{0\}$ is positive, palindromic and unimodal with center of symmetry $\frac{n}{2}$ if $(a_0, a_1, \ldots, a_n)$ has these properties.

Let $\Lambda_{\mathbb{Q}}$ be the ring of symmetric functions in variables $x_1, x_2, \ldots$ with coefficients in $\mathbb{Q}$ and let b be any basis for $\Lambda_{\mathbb{Q}}$. If f is a symmetric function whose expansion in b has nonnegative coefficients, we say that f is b-positive. This induces a partial order relation on the ring $\Lambda_{\mathbb{Q}}$ given by $f \leq_b g$ if $g - f$ is b-positive. We use this partial order to define b-unimodality of sequences of symmetric functions in $\Lambda_{\mathbb{Q}}$ and polynomials in $\Lambda_{\mathbb{Q}}[t]$. Similarly, when we say that a sequence of polynomials in $\mathbb{Q}[q_1, \ldots, q_m]$ or a polynomial in $\mathbb{Q}[q_1, \ldots, q_m][t]$ is unimodal, we are using the partial order relation on $\mathbb{Q}[q_1, \ldots, q_m]$ defined by $f(\mathbf{q}) \leq g(\mathbf{q})$ if $g(\mathbf{q}) - f(\mathbf{q})$ has nonnegative coefficients.

The bases for $\Lambda_{\mathbb{Q}}$ that are relevant here are the Schur basis $(s_\lambda)_{\lambda \in \mathrm{Par}}$, the elementary symmetric function basis $(e_\lambda)_{\lambda \in \mathrm{Par}}$, the complete homogeneous symmetric function basis $(h_\lambda)_{\lambda \in \mathrm{Par}}$, and the power-sum symmetric function basis $(z_\lambda^{-1} p_\lambda)_{\lambda \in \mathrm{Par}}$, where Par is the set of all integer partitions,

$$z_\lambda := \prod_i m_i(\lambda)!\, i^{m_i(\lambda)}$$

and $m_i(\lambda)$ is the number of parts of λ equal to i. These are all integral bases in that every symmetric function in the ring $\Lambda_{\mathbb{Z}}$ of symmetric functions over $\mathbb{Z}$ is an integral linear combination of the basis elements.

The next two propositions give basic tools for establishing unimodality.

Proposition 1.3 (see [27, Proposition 1]). *Let $A(t)$ and $B(t)$ be positive, palindromic and unimodal polynomials in $\mathbb{Q}[t]$ with respective centers of symmetry c_A and c_B. Then*

1. *$A(t)B(t)$ is positive, palindromic and unimodal with center of symmetry $c_A + c_B$.*

2. *If* $c_A = c_B$ *then* $A(t) + B(t)$ *is positive, palindromic and unimodal with center of symmetry* c_A.

Proposition 1.4. *Let* $b := (b_\lambda)_\lambda$ *be a basis for* $\Lambda_{\mathbb{Q}}$. *The polynomial* $f(t) \in \Lambda_{\mathbb{Q}}[t]$ *is b-positive, palindromic and b-unimodal with center of symmetry c if and only if each coefficient* $a_\lambda(t)$ *in the expansion* $f(t) = \sum_\lambda a_\lambda(t) b_\lambda$ *is positive, palindromic and unimodal with center of symmetry c.*

For example, from Proposition 1.3 (1) we see that $[n]_t!$ is positive, palindromic and unimodal with center of symmetry $\frac{n(n+1)}{4}$. From Propositions 1.3 and 1.4 we have that $[5]_t[2]_t h_{(5,2)} + [4]_t[3]_t h_{(4,3)}$ is h-positive, palindromic and h-unimodal with center of symmetry $5/2$.

Recall that the Frobenius characteristic ch is a ring isomorphism from the ring of virtual representations of the symmetric groups to $\Lambda_{\mathbb{Z}}$. It takes the irreducible Specht module S^λ to the Schur function s_λ. As is customary, we use ω to denote the involution on $\Lambda_{\mathbb{Q}}$ that takes h_i to e_i. Recall $\omega(s_\lambda) = s_{\lambda'}$ for all $\lambda \in \mathrm{Par}$, where λ' denotes the conjugate shape.

For any ring R, let $\mathcal{Q}_R$ be the ring of quasisymmetric functions in variables $x_1, x_2, \ldots$ with coefficients in R. For $n \in \mathbb{P}$ and $S \subseteq [n-1]$, let $D(S)$ be the set of all functions $f : [n] \to \mathbb{P}$ such that

- $f(i) \geq f(i+1)$ for all $i \in [n-1]$, and
- $f(i) > f(i+1)$ for all $i \in S$.

Recall Gessel's *fundamental quasisymmetric function*

$$F_{n,S} := \sum_{f \in D(S)} \mathbf{x}_f,$$

where $\mathbf{x}_f := x_{f(1)} x_{f(2)} \cdots x_{f(n)}$. It is straightforward to confirm that $F_{n,S} \in \mathcal{Q}_{\mathbb{Z}}$. In fact (see [29, Proposition 7.19.1]), $\{F_{n,S} : S \subseteq [n-1]\}$ is a basis for the free $\mathbb{Z}$-module of homogeneous degree n quasisymmetric functions with coefficients in $\mathbb{Z}$.

See [29] for further information on symmetric functions, quasisymmetric functions and representations.

ACKNOWLEDGEMENTS. We are grateful to Richard Stanley for informing us of Theorem 2.12, which led us to introduce the chromatic quasisymmetric functions. We also thank Julianna Tymoczko for useful discussions and the anonymous referee for helpful comments.

2 q-Eulerian numbers and toric varieties

It is well known that the Eulerian numbers $(a_{n,0}, a_{n,1}, \ldots, a_{n,n-1})$ form the h-vector of the Coxeter complex Δ_n of the symmetric group $\mathfrak{S}_n$. Danilov and Jurciewicz (see [27, Equation (26)]) showed that the jth entry of the h-vector of any convex simplicial polytope P is equal to $\dim H^{2j}(\mathcal{V}(P))$ where $\mathcal{V}(P)$ is the toric variety associated with P and $H^i(\mathcal{V}(P))$ is the degree i singular cohomology of $\mathcal{V}(P)$ over $\mathbb{C}$. (In odd degrees cohomology vanishes.) It follows that

$$a_{n,j} = \dim H^{2j}(\mathcal{V}(\Delta_n)), \tag{2.1}$$

for all $j = 0, \ldots, n-1$.

It is also well known that the Eulerian numbers are palindromic and unimodal. Although there are elementary ways to prove this result, palindromicity and unimodality can be viewed as consequences of the hard Lefschetz theorem applied to the toric variety $\mathcal{V}(\Delta_n)$.

The classical hard Lefschetz theorem states that if $\mathcal{V}$ is a projective smooth variety of complex dimension n then there is a linear operator ϕ on $H^*(\mathcal{V})$ sending each $H^i(\mathcal{V})$ to $H^{i+2}(\mathcal{V})$ such that for $0 \le i \le 2n-2$, the map $\phi^{n-i} : H^i(\mathcal{V}) \to H^{2n-i}(\mathcal{V})$ is a bijection. It follows from this that the sequence of even degree Betti numbers and the sequence of odd degree Betti numbers are palindromic and unimodal. Since $\mathcal{V}(\Delta_n)$ is a smooth variety, it follows that the Eulerian numbers form a palindromic and unimodal sequence. See [26, 27] for further information on the use of the hard Lefschetz theorem in combinatorics.

2.1 Action of the symmetric group

The symmetric group $\mathfrak{S}_n$ acts naturally on the Coxeter complex Δ_n and this induces a representation of $\mathfrak{S}_n$ on cohomology of the toric variety $\mathcal{V}(\Delta_n)$. Procesi [18] derived a recursive formula for this representation, which Stanley [27, see Proposition 12] used to obtain the following formula for the Frobenius characteristic of the representation,

$$1 + \sum_{n\ge1}\sum_{j=0}^{n-1} \operatorname{ch} H^{2j}(\mathcal{V}(\Delta_n))t^j z^n = \frac{(1-t)H(z)}{H(tz) - tH(z)}. \tag{2.2}$$

(See [6, 14, 33, 34] for related work on this representation.) This is a symmetric function analog of the classical formula of Euler,

$$1 + \sum_{n\ge1}\sum_{j=0}^{n-1} a_{n,j} t^j \frac{z^n}{n!} = \frac{(1-t)e^z}{e^{tz} - te^z}. \tag{2.3}$$

An immediate consequence of (2.2) is that ch $H^{2j}(\mathcal{V}(\Delta_n))$ is h-positive, which is equivalent to saying that the linear representation of $\mathfrak{S}_n$ on $H^{2j}(\mathcal{V}(\Delta_n))$ is obtained from an action on some set in which each point stabilizer is a Young subgroup. Formula (2.2) has a number of additional interesting, but not so obvious, consequences. We discuss some of them below.

Stembridge [33] uses (2.2) to characterize the multiplicity of each irreducible $\mathfrak{S}_n$-module in $H^{2j}(\mathcal{V}(\Delta_n))$ in terms of marked tableaux. A *marked tableau* of shape λ is a semistandard tableau of shape λ with entries in $\{0, 1, \dots, k\}$ for some $k \in \mathbb{N}$, such that each $a \in [k]$ occurs at least twice in the tableau, together with a mark on one occurrence of each $a \in [k]$; the marked occurrence cannot be the leftmost occurrence. The index of a marked a is the number of occurrences of a to the left of the marked a. The index of a marked tableau is the sum of the indices of the marked entries of the tableau.

Theorem 2.1 (Stembridge [33]). *Let $\lambda \vdash n$. The multiplicity of the irreducible Specht module S^λ in $H^{2j}(\mathcal{V}(\Delta_n))$ is the number of marked tableaux of shape λ and index j.*

Recently, Shareshian and Wachs [25] obtained a different formula for the multiplicity of S^λ in terms of a different type of tableau (see Corollary 4.14).

The character of $H^{2j}(\mathcal{V}(\Delta_n))$ can be obtained from the following expansion in the p-basis.

Theorem 2.2 (Stembridge [33], Dolgachev and Lunts [6]).
For all $n \geq 1$,

$$\sum_{j=0}^{n-1} \mathrm{ch}\, H^{2j}(\mathcal{V}(\Delta_n))t^j = \sum_{\lambda \vdash n} A_{l(\lambda)}(t) \prod_i [\lambda_i]_t \, z_\lambda^{-1} p_\lambda .$$

2.2 Expansion in the basis of fundamental quasisymmetric functions

In their study of q-Eulerian numbers [22, 23], Shareshian and Wachs obtain a formula for the expansion of the right hand side of (2.2) in the basis of fundamental quasisymmetric functions. In order to describe this decomposition the notion of DEX set of a permutation is needed. For $n \geq 1$, we set

$$[\overline{n}] := \{\overline{1}, \dots, \overline{n}\}$$

and totally order the alphabet $[n] \cup [\overline{n}]$ by

$$\overline{1} < \dots < \overline{n} < 1 < \dots < n. \tag{2.4}$$

For a permutation $\sigma = \sigma_1 \dots \sigma_n \in \mathfrak{S}_n$, we define $\overline{\sigma}$ to be the word over alphabet $[n] \cup [\overline{n}]$ obtained from σ by replacing σ_i with $\overline{\sigma_i}$ whenever $i \in \mathrm{EXC}(\sigma)$. For example, if $\sigma = 531462$ then $\overline{\sigma} = \overline{5}\overline{3}14\overline{6}2$. We define a descent in a word $w = w_1 \dots w_n$ over any totally ordered alphabet to be any $i \in [n-1]$ such that $w_i > w_{i+1}$ and let $\mathrm{DES}(w)$ be the set of all descents of w. Now, for $\sigma \in \mathfrak{S}_n$, we define

$$\mathrm{DEX}(\sigma) := \mathrm{DES}(\overline{\sigma}).$$

For example, $\mathrm{DEX}(531462) = \mathrm{DES}(\overline{5}\overline{3}14\overline{6}2) = \{1, 4\}$.

Theorem 2.3 (Shareshian and Wachs [23, Theorem 1.2]).

$$1 + \sum_{n \geq 1} \sum_{\sigma \in \mathfrak{S}_n} F_{n,\mathrm{DEX}(\sigma)} t^{\mathrm{exc}(\sigma)} z^n = \frac{(1-t)H(z)}{H(tz) - tH(z)}. \tag{2.5}$$

Corollary 2.4 ([23, Theorem 7.4]). *For* $0 \leq j \leq n-1$,

$$\mathrm{ch}\, H^{2j}(\mathcal{V}(\Delta_n)) = \sum_{\substack{\sigma \in \mathfrak{S}_n \\ \mathrm{exc}(\sigma)=j}} F_{n,\mathrm{DEX}(\sigma)}\,.$$

Expansion in the basis of fundamental quasisymmetric functions is useful because it can yield interesting results about permutation statistics via specialization. For a quasisymmetric function $Q(x_1, x_2, \dots) \in \mathcal{Q}_R$, the stable principal specialization $\mathbf{ps}(Q) \in R[[q]]$ is, by definition, obtained from Q by substituting q^{i-1} for x_i for each $i \in \mathbb{P}$. Now (see for example [29, Lemma 7.19.10]), for any $n \in \mathbb{P}$ and $S \subseteq [n-1]$, we have

$$\mathbf{ps}(F_{n,S}) = (q;q)_n^{-1}\, q^{\sum_{i \in S} i}, \tag{2.6}$$

where

$$(p;q)_n = \prod_{j=1}^{n} (1 - pq^{j-1}).$$

From Corollary 2.4 and the fact that that $\sum_{i \in \mathrm{DEX}(\sigma)} i = \mathrm{maj}(\sigma) - \mathrm{exc}(\sigma)$ (see [23, Lemma 2.2]) we obtain the following q-analog of (2.1).

Corollary 2.5. *For* $0 \leq j \leq n-1$,

$$(q;q)_n\, \mathbf{ps}(\mathrm{ch}\, H^{2j}(\mathcal{V}(\Delta_n))) = a_{n,j}(q),$$

where

$$a_{n,j}(q) := \sum_{\substack{\sigma \in \mathfrak{S}_n \\ \mathrm{exc}(\sigma)=j}} q^{\mathrm{maj}(\sigma)-\mathrm{exc}(\sigma)}. \tag{2.7}$$

The q-Eulerian numbers $a_{n,j}(q)$ defined in Corollary 2.5 were initially studied in [22,23] and have been further studied in [8,9,12,20,24]. Define the q-Eulerian polynomials,

$$A_n(q,t) := \sum_{j=1}^{n-1} a_{n,j}(q) t^j = \sum_{\sigma \in \mathfrak{S}_n} q^{\mathrm{maj}(\sigma)-\mathrm{exc}(\sigma)} t^{\mathrm{exc}(\sigma)}. \tag{2.8}$$

In [22, 23], the following q-analog of Euler's formula (2.3) was obtained via specialization of (2.5),

$$1 + \sum_{n \geq 1} A_n(q,t) \frac{z^n}{[n]_q!} = \frac{(1-t)\exp_q(z)}{\exp_q(tz) - t\exp_q(z)}, \tag{2.9}$$

where

$$\exp_q(z) := \sum_{n \geq 0} \frac{z^n}{[n]_q!}.$$

It is surprising that when one evaluates $a_{n,j}(q)$ at any nth root of unity, one always gets a positive integer.

Theorem 2.6 (Sagan, Shareshian and Wachs [20, Corollary 6.2]). *Let $dm = n$ and let ξ_d be any primitive dth root of unity. Then*

$$A_n(\xi_d, t) = A_m(t)[d]_t^m.$$

Consequently $A_n(\xi_d, t)$ is a positive, palindromic, unimodal polynomial in $\mathbb{Z}[t]$.

Theorem 2.2, Corollary 2.5 and the following result, which is implicit in [5] and stated explicitly in [20], can be used to give a proof of Theorem 2.6.

Lemma 2.7. *Suppose $u(q) \in \mathbb{Z}[q]$ and there exists a homogeneous symmetric function U of degree n with coefficients in $\mathbb{Z}$ such that*

$$u(q) = (q;q)_n \, \mathbf{ps}(U).$$

If $dm = n$ then $u(\xi_d)$ is the coefficient of $z_{d^m}^{-1} p_{d^m}$ in the expansion of U in the power-sum basis.

2.3 Unimodality

Let us now consider what the hard Lefschetz theorem tells us about the sequence $(\mathrm{ch}\, H^{2j}(\mathcal{V}(\Delta_n)))_{0\leq j\leq n-1}$. Since the action of the symmetric group $\mathfrak{S}_n$ commutes with the hard Lefschetz map ϕ (see [27, p. 528]), we can conclude that for $0 \leq i \leq n$ the map $\phi : H^i(\mathcal{V}(\Delta_n)) \to H^{i+2}(\mathcal{V}(\Delta_n))$ is an $\mathfrak{S}_n$-module monomorphism. Hence, by Schur's lemma, for each $\lambda \vdash n$, the multiplicity of the Specht module S^λ in $H^i(\mathcal{V}(\Delta_n))$ is less than or equal to the multiplicity in $H^{i+2}(\mathcal{V}(\Delta_n))$. Equivalently, the coefficient of the Schur function s_λ in the expansion of $\mathrm{ch}\, H^i(\mathcal{V}(\Delta_n))$ in the Schur basis is less than or equal to the coefficient in $\mathrm{ch}\, H^{i+2}(\mathcal{V}(\Delta_n))$. Hence, it follows from the hard Lefschetz theorem that the sequence $(\mathrm{ch}\, H^{2j}(\mathcal{V}(\Delta_n)))_{0\leq j\leq n-1}$ is palindromic and Schur-unimodal.

The following lemma is useful for establishing unimodality of polynomials in $\mathbb{Q}[q][t]$.

Lemma 2.8 (see for example [23, Lemma 5.2]). *If U is a Schur-positive homogeneous symmetric function of degree n then*

$$(q;q)_n \mathbf{ps}(U)$$

is a polynomial in q with nonnegative coefficients.

It follows from Lemma 2.8 and Corollary 2.5 that the palindromicity and Schur-unimodality of $(\mathrm{ch}\, H^{2j}(\mathcal{V}(\Delta_n)))_{0\leq j\leq n-1}$ can be specialized to yield the following q-analog of the palindromicity and unimodality of the Eulerian numbers.

Theorem 2.9 (Shareshian and Wachs [23, Theorem 5.3]). *For any $n \in \mathbb{P}$, the sequence $(a_{n,j}(q))_{0\leq j\leq n-1}$ is palindromic and unimodal.*

We remark that it is not really necessary to use the hard Lefschetz theorem to establish the above mentioned unimodality results. In fact, by manipulating the right hand side of (2.2) we obtain (2.10) below[1]. The palindromicity and h-unimodality consequences follow by Propositions 1.3 and 1.4.

[1] This formula is different from a similar looking formula given in [23, Corollary 4.2].

Theorem 2.10 (Shareshian and Wachs [25]). *For all $n \geq 1$,*

$$\sum_{j=0}^{n-1} \operatorname{ch} H^{2j}(\mathcal{V}(\Delta_n))t^j = \sum_{m=1}^{\lfloor \frac{n+1}{2} \rfloor} t^{m-1} \sum_{\substack{k_1,\dots,k_m \geq 2 \\ \sum k_i = n+1}} \prod_{i-1}^{m} [k_i - 1]_t \, h_{k_1-1} h_{k_2} \cdots h_{k_m}. \tag{2.10}$$

Consequently, the sequence $(\operatorname{ch} H^{2j}(\mathcal{V}(\Delta_n)))_{0 \leq j \leq n-1}$ *is h-positive, palindromic and h-unimodal.*

By specializing (2.10) we obtain the following result. Recall that the q-analog of the multinomial coefficients is defined by

$$\begin{bmatrix} n \\ k_1, \dots, k_m \end{bmatrix}_q := \frac{[n]_q!}{[k_1]_q! \cdots [k_m]_q!}$$

for all $k_1, \dots, k_m \in \mathbb{N}$ such that $\sum_{i=1}^{n} k_i = n$.

Corollary 2.11. *For all $n \geq 1$,*

$$A_n(q,t) = \sum_{m=1}^{\lfloor \frac{n+1}{2} \rfloor} t^{m-1} \sum_{\substack{k_1,\dots,k_m \geq 2 \\ \sum k_i = n+1}} \prod_{i=1}^{m} [k_i - 1]_t \begin{bmatrix} n \\ k_1 - 1, k_2, \dots, k_m \end{bmatrix}_q .$$

Consequently, $A_n(q,t)$ is a palindromic and unimodal polynomial in t.

2.4 Smirnov words

The expression on the right hand side of (2.2) has appeared in several other contexts (see [23, Section 7]). We mention just one of these contexts here.

A Smirnov word is a word with no equal adjacent letters. Let W_n be the set of all Smirnov words of length n over alphabet $\mathbb{P}$. Define the enumerator

$$Y_{n,j}(x_1, x_2, \dots) := \sum_{\substack{w \in W_n \\ \operatorname{des}(w) = j}} \mathbf{x}_w,$$

where $\mathbf{x}_w := x_{w(1)} \cdots x_{w(n)}$. (Here we are calculating $\operatorname{des}(w)$ using the standard total order on $\mathbb{P}$.) A formula for the generating function of $Y_n := \sum_{j=0}^{n-1} Y_{n,j}$ was initially obtained by Carlitz, Scoville and Vaughan [1] and further studied by Dollhopf, Goulden and Greene [7] and Stanley [30]. Stanley pointed out to us the following refinement of this formula. This refinement follows from results in [23, Section 3.3] by P-partition reciprocity.

Theorem 2.12 (Stanley (see [23, Theorem 7.2])).

$$\sum_{n,j\geq 0} Y_{n,j}(\mathbf{x})t^j z^n = \frac{(1-t)E(z)}{E(zt)-tE(z)}, \tag{2.11}$$

where $E(z) := \sum_{n\geq 0} e_n z^n$ *and* e_n *is the elementary symmetric function of degree* n.

Corollary 2.13. *For* $0 \leq j \leq n-1$,

$$\operatorname{ch} H^{2j}(\mathcal{V}(\Delta_n)) = \omega Y_{n,j}.$$

3 Rawlings major index and generalized Eulerian numbers

In [19], Rawlings studies Mahonian permutation statistics that interpolate between the major index and the inversion index. Fix $n \in \mathbb{P}$ and $k \in [n]$. For $\sigma \in \mathfrak{S}_n$, set

$$\begin{aligned}
\mathrm{DES}_{\geq k}(\sigma) &:= \{i \in [n-1] : \sigma(i) - \sigma(i+1) \geq k\},\\
\mathrm{maj}_{\geq k}(\sigma) &:= \sum_{i\in \mathrm{DES}_{\geq k}} i,\\
\mathrm{INV}_{<k}(\sigma) &:= \{(i,j) \in [n]\times[n] : i < j \text{ and } 0 < \sigma(i)-\sigma(j) < k\},\\
\mathrm{inv}_{<k}(\sigma) &:= |\mathrm{INV}_{<k}(\sigma)|.
\end{aligned}$$

Now define the *Rawlings major index* to be

$$r\mathrm{maj}_k(\sigma) := \mathrm{inv}_{<k}(\sigma) + \mathrm{maj}_{\geq k}(\sigma).$$

Note that $r\mathrm{maj}_1$ and $r\mathrm{maj}_n$ are, respectively, the well studied major index maj and inversion number inv. Rawlings shows in [19] that each $r\mathrm{maj}_k$ is a Mahonian statistic, that is,

$$\sum_{\sigma\in\mathfrak{S}_n} q^{r\mathrm{maj}_k(\sigma)} = [n]_q!. \tag{3.1}$$

For a proof of (3.1) that is different from that of Rawlings and a generalization of (3.1) from permutations to labeled trees, see [15].

In [4] De Mari and Shayman introduce a class of numbers, closely related to the Rawlings major index, which they call generalized Eulerian numbers. For $k \in [n]$ and $0 \leq j$ define the De Mari-Shayman generalized Eulerian numbers to be

$$a_{n,j}^{(k)} := |\{\sigma \in \mathfrak{S}_n : \mathrm{inv}_{<k}(\sigma) = j\}|.$$

Note that $\mathrm{inv}_{<2}(\sigma) = \mathrm{des}(\sigma^{-1})$, which implies that $a_{n,j}^{(2)} = a_{n,j}$, justifying the name "generalized Eulerian numbers". De Mari and Shayman introduce the generalized Eulerian numbers in connection with their study of Hessenberg varieties. It follows from their work and the hard Lefschetz theorem that for each fixed $k \in [n]$, the generalized Eulerian numbers $(a_{n,j}^{(k)})_{j\geq 0}$ form a palindromic unimodal sequence of numbers.

In [25] Shareshian and Wachs consider a q-analog of the De Mari-Shayman generalized Eulerian numbers defined for $k \in [n]$ and $0 \leq j$ by

$$a_{n,j}^{(k)}(q) := \sum_{\substack{\sigma \in S_n \\ \mathrm{inv}_{<k}(\sigma)=j}} q^{\mathrm{maj}_{\geq k}(\sigma)}.$$

Similarly, the generalized q-Eulerian polynomials are defined by

$$A_n^{(k)}(q,t) = \sum_{\sigma \in \mathfrak{S}_n} q^{\mathrm{maj}_{\geq k}(\sigma)} t^{\mathrm{inv}_{<k}(\sigma)}.$$

Now (3.1) is equivalent to,

$$A_n^{(k)}(q,q) = [n]_q!.$$

We now consider the question of whether unimodality and other known properties of the generalized Eulerian numbers $a_{n,j}^{(k)}$ extend to the generalized q-Eulerian numbers $a_{n,j}^{(k)}(q)$. Although it is not at all obvious, it turns out that in the case $k = 2$, the generalized q-Eulerian numbers are equal to the q-Eulerian numbers defined in (2.7),

$$a_{n,j}^{(2)}(q) = a_{n,j}(q)$$

(see Theorem 4.19). Hence by Theorem 2.9, unimodality holds when $k = 2$. We consider this question in a more general setting, which we now describe.

Let $E(G)$ denote the edge set of a graph G. For $\sigma \in \mathfrak{S}_n$ and G a graph with vertex set $[n]$, the *G-inversion set* of σ is

$$\mathrm{INV}_G(\sigma) := \{(i,j) : i < j,\ \sigma(i) > \sigma(j) \text{ and } \{\sigma(i), \sigma(j)\} \in E(G)\}$$

and the *G-inversion number* is

$$\mathrm{inv}_G(\sigma) := |\,\mathrm{INV}_G(\sigma)|.$$

For $\sigma \in \mathfrak{S}_n$ and P a poset on $[n]$, the *P-descent set* of σ is

$$\mathrm{DES}_P(\sigma) := \{i \in [n-1] : \sigma(i) >_P \sigma(i+1)\},$$

and the *P-major index* is

$$\mathrm{maj}_P(\sigma) := \sum_{i \in \mathrm{DES}_P(\sigma)} i.$$

Define the *incomparability graph* $\mathrm{inc}(P)$ of a poset P on $[n]$ to be the graph with vertex set $[n]$ and edge set $\{\{a, b\} : a \not\leq_P b \text{ and } b \not\leq_P a\}$. For $0 \leq j \leq |E(\mathrm{inc}(P))|$ define the (q, P)-Eulerian numbers,

$$a_{P,j}(q) := \sum_{\substack{\sigma \in \mathfrak{S}_n \\ \mathrm{inv}_{\mathrm{inc}(P)}(\sigma) = j}} q^{\mathrm{maj}_P(\sigma)},$$

and the (q, P)-Eulerian polynomials

$$A_P(q, t) := \sum_{j=0}^{|E(\mathrm{inc}(P))|} a_{P,j}(q) t^j = \sum_{\sigma \in \mathfrak{S}_n} q^{\mathrm{maj}_P(\sigma)} t^{\mathrm{inv}_{\mathrm{inc}(P)}(\sigma)}.$$

Define $P_{n,k}$ to be the poset on vertex set $[n]$ such that $i < j$ in $P_{n,k}$ if and only if $j - i \geq k$ and let $G_{n,k}$ be the incomparability graph of $P_{n,k}$. Then

$$\mathrm{DES}_{\geq k}(\sigma) = \mathrm{DES}_{P_{n,k}}(\sigma)$$

and

$$\mathrm{INV}_{<k}(\sigma) = \mathrm{INV}_{G_{n,k}}(\sigma).$$

Hence if $P = P_{n,k}$, we have $a_{P,j}(q) = a_{n,j}^{(k)}(q)$ and $A_P(q, t) = A_n^{(k)}(q.t)$.

The unimodality property that holds for the (q, P)-Eulerian numbers in case $P = P_{n,2}$ does not hold for general P. Indeed, consider the poset P on $[3]$ whose only relation is $1 < 2$. We have

$$A_P(q, t) := (1 + q) + 2t + (1 + q^2)t^2, \tag{3.2}$$

which is neither palindromic as a polynomial in t, nor unimodal. The property $A_P(q, q) = [n]_q!$ fails as well. However there is a very nice class of posets for which unimodality seems to hold.

A *unit interval order* is a poset that is isomorphic to a finite collection $\mathcal{I}$ of intervals $[a, a + 1]$ on the real line, partially ordered by the relation $[a, a + 1] <_{\mathcal{I}} [b, b + 1]$ if $a + 1 < b$. Define a *natural unit interval order* to be a poset P on $[n]$ that satisfies (1) $x <_P y$ implies $x < y$ in the natural order on $[n]$ and (2) if the direct sum $\{x <_P z\} + \{y\}$ is an induced subposet of P then $x < y < z$ in the natural order on $[n]$. It is not difficult to see that every natural unit interval order is a unit interval order and that every unit interval order is isomorphic to a unique natural unit interval order. The poset $P_{n,k}$ is an example of a natural unit interval order.

It is a consequence of a result of Kasraoui [13, Theorem 1.8] that if P is a natural unit interval order then $\mathrm{inv}_{\mathrm{inc}(P)} + \mathrm{maj}_P$ is Mahonian. In other words, if P is a natural unit interval order on $[n]$ then

$$A_P(q,q) = [n]_q!\,.$$

From the work of De Mari, Procesi and Shayman [3] we have the following result. The proof is discussed in Section 5.

Theorem 3.1 (see [28, Exercise 1.50 (f)]). *Let P be a natural unit interval order. Then the P-Eulerian polynomial $A_P(1,t)$ is palindromic and unimodal.*

Palindromicity of the q-analog $A_P(q,t)$ follows from results discussed in Section 4 and unimodality is implied by a conjecture discussed in Section 4.

Theorem 3.2 (Shareshian and Wachs [25]). *Let P be a natural unit interval order on $[n]$. Then the sequence $(a_{P,j}(q))_{0\le j\le |E(\mathrm{inc}(P))|}$ is palindromic.*

Conjecture 3.3. Let P be a natural unit interval order on $[n]$. Then the palindromic sequence $(a_{P,j}(q))_{0\le j\le |E(\mathrm{inc}(P))|}$ is unimodal.

The following conjecture generalizes Theorem 2.6.

Conjecture 3.4. Let P be a natural unit interval order on $[n]$. If $dm = n$ then there is a polynomial $B_{P,d}(t) \in \mathbb{N}[t]$ such that

$$A_P(\xi_d, t) = B_{P,d}(t)[d]_t^m.$$

Moreover, $(a_{P,j}(\xi_d))_{0\le j\le |E(\mathrm{inc}(P))|}$ is a palindromic unimodal sequence of positive integers.

In addition to the case $P = P_{n,2}$ (Theorems 2.9 and 2.6), these conjectures have been verified for $P_{n,k}$ when $k = 1, n-2, n-1, n$, and by computer for all k when $n \le 8$. An approach to proving them in general will be presented in Sections 4 and 5.

4 Chromatic quasisymmetric functions

4.1 Stanley's chromatic symmetric functions

Let G be a graph with vertex set $[n]$ and edge set $E = E(G) \subseteq \binom{[n]}{2}$. A *proper $\mathbb{P}$-coloring* of G is a function c from $[n]$ to the set $\mathbb{P}$ of positive integers such that whenever $\{i, j\} \in E$ we have $c(i) \neq c(j)$. Given any function $c : [n] \to \mathbb{P}$, set

$$\mathbf{x}_c := \prod_{i=1}^{n} x_{c(i)}.$$

Let $C(G)$ be the set of proper $\mathbb{P}$-colorings of G. In [30], Stanley defined the *chromatic symmetric function* of G,

$$X_G(\mathbf{x}) := \sum_{c \in C(G)} \mathbf{x}_c.$$

It is straightforward to confirm that $X_G \in \Lambda_{\mathbb{Z}}$. The chromatic symmetric function is a generalization of the chromatic polynomial $\chi_G : \mathbb{P} \to \mathbb{P}$, where $\chi_G(n)$ is the number of proper colorings of G with n colors. Indeed, $X_G(1^n) = \chi_G(n)$, where $X_G(1^n)$ is the specialization of $X_G(\mathbf{x})$ obtained by setting $x_i = 1$ for $1 \leq i \leq n$ and $x_i = 0$ for $i > n$. Chromatic symmetric functions are studied in various papers, including [2, 10, 17, 30, 31].

We recall Stanley's description of the power sum decomposition of $X_G(\mathbf{x})$, for arbitrary G with vertex set $[n]$. We call a partition $\pi = \pi_1|\ldots|\pi_l$ of $[n]$ into nonempty subsets *G-connected* if the subgraph of G induced on each block π_i of π is connected. The set $\Pi_{G,n}$ of all G-connected partitions of $[n]$ is partially ordered by refinement (that is, $\pi \leq \theta$ if each block of π is contained in some block of θ). We write μ_G and $\hat{0}$, respectively, for the Möbius function on $\Pi_{G,n}$, and the minimum element $1|\ldots|n$ of $\Pi_{G,n}$. For $\pi \in \Pi_{G,n}$, we write $\mathrm{par}(\pi)$ for the partition of n whose parts are the sizes of the blocks of π.

Theorem 4.1 (Stanley [30, Theorem 2.6]). *Let G be a graph with vertex set $[n]$. Then*

$$\omega X_G(\mathbf{x}) = \sum_{\pi \in \Pi_{G,n}} |\mu_G(\hat{0}, \pi)| p_{\mathrm{par}(\pi)}. \tag{4.1}$$

Consequently $\omega X_G(\mathbf{x})$ is p-positive.

A poset P is called *$(r + s)$-free* if it contains no induced subposet isomorphic to the direct sum of an r element chain and an s element chain. A classical result (see [21]) says that a poset is a unit interval order if and only if it is both $(3 + 1)$-free and $(2 + 2)$-free.

Conjecture 4.2. (Stanley and Stembridge [32, Conjecture 5.5], [30, Conjecture 5.1). Let G be the incomparability graph of a $(3 + 1)$-free poset. Then $X_G(\mathbf{x})$ is e-positive.

This conjecture is still open even for unit interval orders. Gasharov [10] obtains the weaker, but still very interesting, result that, under the assumptions of Conjecture 4.2, $X_G(\mathbf{x})$ is Schur-positive. Let P be a poset on n and let λ be a partition if n. Gasharov defines a *P-tableau of shape λ* to be a filling of a Young diagram of shape λ (in English notation) with elements of P such that

- each element of P appears exactly once,
- if $y \in P$ appears immediately to the right of $x \in P$, then $y >_P x$, and
- if $y \in P$ appears immediately below $x \in P$, then $y \not<_P x$.

Given a P-tableau T, let $\lambda(T)$ be the shape of T. Let $\mathcal{T}_P$ be the set of all P-tableaux.

Theorem 4.3 (Gasharov [10]). *Let P be a $(3+1)$-free poset. Then*

$$X_{\mathrm{inc}(P)}(\mathbf{x}) = \sum_{T \in \mathcal{T}_P} s_{\lambda(T)}.$$

4.2 A quasisymmetric refinement

We define the *chromatic quasisymmetric function* of G as

$$X_G(\mathbf{x}, t) := \sum_{c \in C(G)} t^{\mathrm{asc}(c)} \mathbf{x}_c,$$

where

$$\mathrm{asc}(c) := |\{\{i, j\} \in E(G) : i < j \text{ and } c(i) < c(j)\}|.$$

It is straightforward to confirm that $X_G(\mathbf{x}, t) \in \mathcal{Q}_{\mathbb{Z}}[t]$.

While the chromatic quasisymmetric function is defined for an arbitrary graph on vertex set $[n]$, our results will concern incomparability graphs of partially ordered sets. The natural unit interval orders are particularly significant in our theory of chromatic quasisymmetric functions because they yield symmetric functions.

Proposition 4.4 (Shareshian and Wachs [25]). *Let P be a natural unit interval order. Then*

$$X_{\mathrm{inc}(P)}(\mathbf{x}, t) \in \Lambda_{\mathbb{Z}}[t].$$

Before looking at some examples of $X_{\mathrm{inc}(P)}(\mathbf{x}, t)$ for natural unit interval orders, let us consider the poset P on vertex set $[3]$ whose only relation is $1 < 2$. Clearly P is not a natural unit interval order. (Recall $A_P(q, t)$ is given in (3.2).) Using Theorem 4.17 below we compute

$$X_{\mathrm{inc}(P)}(\mathbf{x}, t) = (e_3 + F_{3,\{2\}}) + 2e_3 t + (e_3 + F_{3,\{1\}})t^2,$$

which is clearly not in $\Lambda_{\mathbb{Z}}[t]$.

Example 4.5. $P = P_{n,1}$. The incomparability graph has no edges. Hence

$$X_{\mathrm{inc}(P_{n,1})}(\mathbf{x}, t) = e_1^n. \tag{4.2}$$

Example 4.6. $P = P_{n,n}$. The incomparability graph is the complete graph. Each proper coloring c is an injective map which can be associated with a permutation $\sigma \in \mathfrak{S}_n$ for which $\mathrm{inv}(\sigma) = \mathrm{asc}(c)$. It follows that

$$X_{\mathrm{inc}(P_{n,n})}(\mathbf{x}, t) = e_n \sum_{\sigma \in \mathfrak{S}_n} t^{\mathrm{inv}(\sigma)} = [n]_t!\, e_n. \tag{4.3}$$

Example 4.7. $P = P_{n,2}$. Recall that the incomparability graph of $P_{n,2}$ is a path. To each proper coloring c of the path $\mathrm{inc}(P_{n,2})$ one can associate the word $w(c) := c(n), c(n-1), \ldots, c(1)$. This word is clearly a Smirnov word of length n (c.f. Section 2.4) and $\mathrm{des}(w(c)) = \mathrm{asc}(c)$. Since w is a bijection from $C(\mathrm{inc}(P_{n,2}))$ to W_n, we have

$$X_{\mathrm{inc}(P_{n,2})}(\mathbf{x}, t) = \sum_{j=0}^{n-1} Y_{n,j}(\mathbf{x}) t^j.$$

It therefore follows from Corollary 2.13 that

$$\omega X_{\mathrm{inc}(P_{n,2})}(\mathbf{x}, t) = \sum_{j=0}^{n-1} \mathrm{ch}\, H^{2j}(\mathcal{V}(\Delta_n)) t^j. \tag{4.4}$$

The following result is a consequence of Proposition 4.4.

Theorem 4.8 (Shareshian and Wachs [25]). *Let P be a natural unit interval order. Then $X_{\mathrm{inc}(P)}(\mathbf{x}, t)$ is a palindromic polynomial in t.*

From (4.2), (4.3), (4.4) and Theorem 2.10 we see that the following conjectured refinement of the unit interval order case of the Stanley-Stembridge conjecture (Conjecture 4.2) is true for $P = P_{n,k}$ when $k = 1, 2, n$.

Conjecture 4.9. Let P be a natural unit interval order. Then $X_{\mathrm{inc}(P)}(\mathbf{x}, t)$ is an e-positive and e-unimodal polynomial in t.

We have verified e-positivity and e-unimodality for several other cases including $P = P_{n,k}$ when $k = n-1, n-2$, and by computer for all $n \leq 8$. Our computation yields,

$$X_{\mathrm{inc}(P_{n,n-1})}(\mathbf{x}, t) = [n-2]_t! \left([n]_t[n-2]_t e_n + t^{n-2} e_{n-1,1}\right) \tag{4.5}$$

$$\begin{aligned} X_{\mathrm{inc}(P_{n,n-2})}(\mathbf{x}, t) = [n-4]_t!\, ([n]_t[n-3]_t^3 e_n + [n-2]_t t^{n-3}([n-3]_t \\ + [2]_t[n-4]_t) e_{n-1,1} + t^{2n-7}[2]_t e_{n-2,2}), \end{aligned} \tag{4.6}$$

from which the conjecture is easily verified using Propositions 1.3 and 1.4.

In [30] Stanley proves that for any graph G on $[n]$, the number of acyclic orientations of G with j sinks is equal to $\sum_{\lambda \in \mathrm{Par}(n,j)} c_\lambda^G$, where $\mathrm{Par}(n, j)$ is the set of partitions of n into j parts and c_λ^G is the coefficient of e_λ in the e-basis expansion of the chromatic symmetric function $X_G(\mathbf{x})$. We obtain the following refinement of this result using essentially the same proof as Stanley's, thereby providing a bit of further evidence for e-positivity of $X_G(\mathbf{x}, t)$.

Theorem 4.10 (Shareshian and Wachs [25]). *Let G be the incomparability graph of a natural unit interval order on $[n]$. For each $\lambda \vdash n$, let $c_\lambda^G(t)$ be the coefficient of e_λ in the e-basis expansion of $X_G(\mathbf{x}, t)$. Then*

$$\sum_{\lambda \in \mathrm{Par}(n,j)} c_\lambda^G(t) = \sum_{o \in \mathcal{O}(G,j)} t^{\mathrm{asc}(o)},$$

where $\mathcal{O}(G, j)$ is the set of acyclic orientations of G with j sinks and $\mathrm{asc}(o)$ is the number of directed edges (i, j) of o for which $i < j$.

Theorem 4.10 gives a combinatorial description of the coefficient of e_n in the e-basis expansion of $X_G(\mathbf{x}, t)$. In Theorem 4.11 and Conjecture 4.12 below we give alternative descriptions. Let P be a poset on $[n]$. We say that $\sigma \in \mathfrak{S}_n$ has a left-to-right P-maximum at $r \in [n]$ if $\sigma(r) >_P \sigma(s)$ for $1 \leq s < r$. The left-to-right P-maximum at 1 will be referred to as trivial. (The notion of (trivial) left-to-right P-maximum can be extended in an obvious way to permutations of any subset of P.) Let

$$c_P(t) := \sum_{\sigma} t^{\mathrm{inv}_{\mathrm{inc}(P)}(\sigma)}, \tag{4.7}$$

where σ ranges over the permutations in $\mathfrak{S}_n$ with no P-descents and no nontrivial left-to-right P-maxima.

Theorem 4.11 (Shareshian and Wachs [25]). *Let P be a natural unit interval order on $[n]$. The coefficient of e_n in the expansion of $X_{\mathrm{inc}(P)}(\mathbf{x}, t)$ in the e-basis of $\Lambda_{\mathbb{Z}[t]}$ is equal to $c_P(t)$.*

Conjecture 4.12. Let P be a natural unit interval order on $[n]$. Then

$$c_P(t) = [n]_t \prod_{i=2}^{n} [a_i]_t,$$

where $a_i = |\{j \in [i-1] : \{i, j\} \in E(\mathrm{inc}(P))\}|$. Consequently $c_P(t)$ is palindromic and unimodal.

In the case that $P = P_{n,2}$, this conjecture is true by Theorem 4.11 and equations (2.10) and (4.4). It is also true for $P_{n,k}$ when $k = 1, n, n-1, n-2$ by Theorem 4.11 and equations (4.2), (4.3), (4.5) and (4.6), respectively.

4.3 Schur and power sum decompositions

When P is a natural unit interval order, we have the following refinement of Gasharov's Schur positivity result (Theorem 4.3). For a P-tableau T and a graph G with vertex set $[n]$, let $\mathrm{inv}_G(T)$ be the number of edges $\{i, j\} \in E(G)$ such that $i < j$ and i appears to the south of j in T.

Theorem 4.13 (Shareshian and Wachs [25]). *Let P be a natural unit interval order poset on $[n]$ and let G be the incomparability graph of P. Then*

$$X_G(\mathbf{x}, t) = \sum_{T \in \mathcal{T}_P} t^{\mathrm{inv}_G(T)} s_{\lambda(T)}.$$

Now by (4.4) we have the following.

Corollary 4.14. *Let $\lambda \vdash n$. The multiplicity of the irreducible Specht module S^λ in $H^{2j}(\mathcal{V}(\Delta_n))$ is equal to the number of $P_{n,2}$-tableaux of shape λ' with* $\mathrm{inv}_{\mathrm{inc}(P_{n,2})}(T) = j$.

By comparing this decomposition to Stembridge's decomposition (Theorem 2.1) we see that the number of marked tableaux of shape λ and index j equals the number of $P_{n,2}$-tableaux of shape λ' and $\mathrm{inv}_{\mathrm{inc}(P_{n,2})}(T) = j$. It would be interesting to find a bijective proof of this fact.

Theorem 4.13 shows that when G is the incomparability graph of a natural unit interval order, the coefficient of each power of t in $\omega X_G(\mathbf{x}, t)$ is a nonnegative integer combination of Schur functions and therefore the Frobenius characteristic of an actual representation of $\mathfrak{S}_n$. Conjecture 4.9 says that this linear representation arises from a permutation representation in which each point stabilizer is a Young subgroup. In Section 5 we present a very promising concrete candidate for the desired permutation representation.

Next we attempt to refine Stanley's p-basis decomposition of the chromatic symmetric functions (Theorem 4.1). In Conjectures 4.15 and 4.16 below, we provide two proposed formulae for this power sum decomposition. With $\mu = (\mu_1 \geq \mu_2 \geq \cdots \geq \mu_l)$ a partition of n and P a natural unit interval order on $[n]$, we call a permutation $\sigma \in \mathfrak{S}_n$, $(P, \mu, 1)$-*compatible* if, when we break σ (in one line notation) into consecutive segments of lengths $\mu_1, \dots, \mu_l$, the segments have no P-descents and no nontrivial left-to-right P-maxima. We call $\sigma \in \mathfrak{S}_n$, $(P, \mu, 2)$-*compatible* if, when we break σ into consecutive segments of lengths $\mu_1, \dots, \mu_l$, the segments have no P-ascents and they begin with the numerically smallest letter of the segment. We write $\mathfrak{S}_{P,\mu,i}$, where $i = 1, 2$, for the set of all (P, μ, i)-compatible elements of $\mathfrak{S}_n$.

Conjecture 4.15. Let P be a natural unit interval order on $[n]$ with incomparability graph G. Then

$$\omega X_G(\mathbf{x},t) = \sum_{\mu \vdash n} z_\mu^{-1} p_\mu \sum_{\sigma \in \mathfrak{S}_{P,\mu,1}} t^{\mathrm{inv}_G(\sigma)}.$$

Conjecture 4.16. Let P be a natural unit interval order on $[n]$ with incomparability graph G. Then

$$\omega X_G(\mathbf{x},t) = \sum_{\mu \vdash n} z_\mu^{-1} p_\mu \prod_{i=1}^{l(\mu)} [\mu_i]_t \sum_{\sigma \in \mathfrak{S}_{P,\mu,2}} t^{\mathrm{inv}_G(\sigma)}.$$

It can be shown that both conjectures reduce to the formula in Theorem 4.1 when $t = 1$. Conjecture 4.15 says that the coefficient of $z_{(n)}^{-1} p_{(n)}$ is equal to $c_P(t)$ defined in (4.7). Since the coefficient of $h_{(n)}$ in the h-basis decomposition must equal the coefficient of $z_{(n)}^{-1} p_{(n)}$ in the p-basis decomposition, by Theorem 4.11, Conjecture 4.15 gives the correct coefficient of $z_{(n)}^{-1} p_{(n)}$. When $P = P_{n,2}$ it is not difficult to see that both conjectures reduce to Theorem 2.2.

4.4 Fundamental quasisymmetric function basis decomposition

Theorem 4.17 (Shareshian and Wachs [25]). *Let G be the incomparability graph of a poset P on $[n]$. Then*

$$\omega X_G(\mathbf{x},t) = \sum_{\sigma \in \mathfrak{S}_n} t^{\mathrm{inv}_G(\sigma)} F_{n,\mathrm{DES}_P(\sigma)},$$

where ω is the involution on $\mathcal{Q}_{\mathbb{Z}}$ that maps $F_{n,S}$ to $F_{n,[n-1]\setminus S}$ for each $n \in \mathbb{N}$ and $S \subseteq [n-1]$. (This extends the involution ω on $\Lambda_{\mathbb{Z}}$ that maps h_n to e_n)

Theorem 4.17 refines Corollary 2 in Chow's paper [2]. Indeed, one obtains Chow's result by setting $t = 1$ in Theorem 4.17. Our proof of Theorem 4.17 follows the same path as Chow's proof of his Corollary 2.

We use Theorem 4.17 and (2.6) to compute the principal stable specialization of $\omega X_G(\mathbf{x},t)$.

Corollary 4.18. *Let G be the incomparability graph of a poset P on $[n]$. Then*

$$(q;q)_n \mathbf{ps}(\omega X_G(\mathbf{x},t)) = A_P(q,t).$$

Recall that for $P = P_{n,2}$ we have a formulation for the expansion of $\omega X_{\mathrm{inc}(P)}(\mathbf{x},t)$ in the fundamental quasisymmetric function basis that is different from that of Theorem 4.17. It is obtained by combining (2.5) and (2.11). By equating these formulations, we obtain the identity,

$$\sum_{\sigma \in \mathfrak{S}_n} t^{\mathrm{des}(\sigma^{-1})} F_{n,\mathrm{DES}_{\geq 2}(\sigma)} = \sum_{\sigma \in \mathfrak{S}_n} t^{\mathrm{exc}(\sigma)} F_{n,\mathrm{DEX}(\sigma)}. \tag{4.8}$$

Taking the stable principle specialization of both sides of (4.8) yields the following new Euler-Mahonian result.

Theorem 4.19 (Shareshian and Wachs [25]). *For all $n \in \mathbb{P}$,*

$$\sum_{\sigma \in \mathfrak{S}_n} q^{\mathrm{rmaj}_2(\sigma)} t^{\mathrm{des}(\sigma^{-1})} = \sum_{\sigma \in \mathfrak{S}_n} q^{\mathrm{maj}(\sigma)} t^{\mathrm{exc}(\sigma)}.$$

It would be interesting to find bijective proofs of (4.8) and Theorem 4.19.

Theorem 3.2, asserting palindromicity of $A_P(q,t)$, is proved by taking stable principal specializations in Theorem 4.8, by means of Corollary 4.18. Conjecture 3.3, asserting unimodality of $A_P(q,t)$, can be obtained by taking stable principal specializations in Conjecture 4.9 (see Lemma 2.8). By taking a specialization called (nonstable) principal specialization, a stronger unimodality conjecture follows from Conjecture 4.9. For a poset P on $[n]$ and $\sigma \in \mathfrak{S}_n$, define $\mathrm{des}_P(\sigma) := |\mathrm{DES}_P(\sigma)|$.

Conjecture 4.20. Let P a natural unit interval order on $[n]$. Then

$$A_P(q,p,t) := \sum_{\sigma \in \mathfrak{S}_n} q^{\mathrm{maj}_P(\sigma)} p^{\mathrm{des}_P(\sigma)} t^{\mathrm{inv}_{\mathrm{inc}(P)}(\sigma)}$$

is a palindromic unimodal polynomial in t.

It follows from Corollary 4.18 and Lemma 2.7 that $A_P(\xi_d, t)$ is the coefficient of $z_{d^m}^{-1} p_{d^m}$ in the p-basis expansion of $\omega X_{\mathrm{inc}(P)}(\mathbf{x}, t)$ when P is a natural unit interval order on $[dm]$. Hence Conjecture 3.4 on the evaluation of $A_P(q,t)$ at roots of unity is a consequence of Conjectures 4.9 and 4.16.

5 Hessenberg varieties

Let $G = GL_n(\mathbb{C})$ and let B be the set of upper triangular matrices in G. The (type A) *flag variety* is the quotient space G/B. Fix now a natural unit interval order P and let $M_{n,P}$ be the set of all $n \times n$ complex matrices (a_{ij}) such that $a_{ij} = 0$ whenever $i >_P j$. Fix a nonsingular diagonal $n \times n$ matrix s with n distinct eigenvalues. The *regular semisimple Hessenberg variety of type A* associated to P is

$$\mathcal{H}(P) := \{gB \in G/B : g^{-1}sg \in M_{n,P}\}.$$

(Note that $\mathcal{H}(P)$ is well defined, since the group B normalizes the set $M_{n,P}$.)

Certain regular semisimple Hessenberg varieties of type A were studied initially by De Mari and Shayman in [4]. Hessenberg varieties for other Lie types are defined and studied by De Mari, Procesi and Shayman in [3]. Such varieties are determined by certain subsets of a root system. In [3, Theorem 11] it is noted that, for arbitrary Lie type, the Hessenberg variety associated with the set of simple roots is precisely the toric variety associated with the corresponding Coxeter complex. In particular, in Lie type A, the poset giving rise to the regular semisimple Hessenberg variety associated with simple roots is $P_{n,2}$ since $E(\mathrm{inc}(P_{n,2})) = \{\{i, i+1\} : i \in [n-1]\}$. Thus

$$\mathcal{H}(P_{n,2}) = \mathcal{V}(\Delta_n). \tag{5.1}$$

De Mari, Procesi and Shayman also show that the cohomology of $\mathcal{H}(P)$ is concentrated in even degrees and that for $0 \leq j \leq |E(\mathrm{inc}(P))|$,

$$\dim H^{2j}(\mathcal{H}(P)) = a_{P,j}(1). \tag{5.2}$$

Hence unimodality of the P-Eulerian polynomials $A_P(1,t)$ (Theorem 3.1) follows from the hard Lefschetz theorem since $\mathcal{H}(P)$ is smooth. Our approach to establishing the unimodality of the (q, P)-Eulerian polynomial $A_P(q,t)$ is to find a representation of the symmetric group on $H^{2j}(\mathcal{H}(P))$ such that the stable principal specialization of its Frobenius characteristic yields $a_{P,j}(q)$. If the hard Lefschetz map commutes with the action of the symmetric group then it follows from Schur's lemma and Lemma 2.8 that $A_P(q,t)$ is unimodal.

For the cases $P = P_{n,k}$, where k=1,2, we already have the desired representations. Indeed, $\mathcal{H}(P_{n,1})$ consists of $n!$ isolated points and the representation of $\mathfrak{S}_n$ on $H^0(\mathcal{H}(P_{n,1}))$ is the regular representation. For $k = 2$, we can use the representation of $\mathfrak{S}_n$ on $H^*(\mathcal{H}(P_{n,2})) = H^*(\mathcal{V}(\Delta_n))$ discussed in Section 2.

For general natural unit interval orders P, Tymoczko [37], [38] has defined a representation of $\mathfrak{S}_n$ on $H^*(\mathcal{H}(P))$ via a theory of Goresky, Kottwitz and MacPherson known as GKM theory (see [11], and see [36] for an introductory description of GKM theory). The centralizer $T = C_G(s)$ consists of the diagonal matrices in $G := GL_n(\mathbb{C})$. It follows that the torus T acts (by left translation) on $\mathcal{H}(P)$. The technical conditions required for application of GKM theory are satisfied by this action, and it follows that one can describe the cohomology of $\mathcal{H}(P)$ using the moment graph associated to this action. The moment graph M is a subgraph of the Cayley graph of $\mathfrak{S}_n$ with generating set consisting of the transpositions. The vertex set of M is $\mathfrak{S}_n$ and the edges connect pairs of elements that differ by a transposition (i, j) such that $\{i, j\} \in E(\mathrm{inc}(P))$. Thus M admits an action of $\mathfrak{S}_n$, and this action can be used to define a linear representation of $\mathfrak{S}_n$ on the cohomology $H^*(\mathcal{H}(P))$.

Tymoczko's representation of $\mathfrak{S}_n$ on $H^*(\mathcal{H}(P))$ in the cases $P = P_{n,k}$, $k = 1, 2$, is the same as the respective representations of $\mathfrak{S}_n$ discussed above. In the case that $P = P_{n,n}$, it follows from [38, Proposition 4.2] that for all j, Tymoczko's representation is isomorphic to $a_{n,j}^{(n)}$ copies of the trivial representation.

By applying the hard Lefschetz theorem, MacPherson and Tymoczko obtain the following result.

Theorem 5.1 (MacPherson and Tymoczko [16]). *For all natural unit interval orders P, the sequence $(\mathrm{ch}\, H^{2j}(\mathcal{H}(P)))_{0\le j\le |E(\mathrm{inc}(P))|}$ is palindromic and Schur-unimodal.*

Tymoczko poses the following problem.

Problem 5.2 (Tymoczko [38]). Given any natural unit interval order P on vertex set $[n]$, describe the decomposition of the representation of $\mathfrak{S}_n$ on $H^{2j}(\mathcal{H}(P))$ into irreducibles.

We finally come to the conjecture that ties together the three topics of this paper.

Conjecture 5.3. Let P be a natural unit interval order on $[n]$. Then

$$\sum_{j=0}^{|E(\mathrm{inc}(P))|} \mathrm{ch}\, H^{2j}(\mathcal{H}(P))t^j = \omega X_{\mathrm{inc}(P)}(\mathbf{x}, t). \tag{5.3}$$

By (5.1) and (4.4) the conjecture is true in the case that $P = P_{n,2}$. It is straightforward to verify in the case $P = P_{n,k}$, when $k = 1, n-1, n$, and for all natural unit interval orders on $[n]$ when $n \le 4$. The conjecture is also true for the parabolic Hessenberg varieties studied by Teff [35]. It follows from (5.2) that the coefficient of the monomial symmetric function m_{1^n} in the expansion of the left hand side of (5.3) in the monomial symmetric function basis equals the coefficient of m_{1^n} in the expansion of the right hand side of (5.3).

Conjecture 5.3, if true, would have many important ramifications. It would allow us to transfer what we know about chromatic quasisymmetric functions to Tymoczko's representation and vice-versa. For instance Theorem 4.13 would provide a solution to Tymoczko's problem. Conjecture 5.3 provides an approach to proving the Stanley-Stembridge conjecture (Conjecture 4.2) for unit interval orders. Indeed one would only have to prove the following conjecture.

Conjecture 5.4. For all natural interval orders P, Tymoczko's representation of $\mathfrak{S}_n$ on $H^*(\mathcal{H}(P))$ is a permutation representation in which each point stabilizer is a Young subgroup.

Our refinement of the Stanley-Stembridge conjecture (Conjecture 4.9) would be equivalent to the following strengthening of the result of MacPherson and Tymoczko.

Conjecture 5.5. For all natural unit interval orders P, the palindromic sequence $(\mathrm{ch}\, H^{2j}(\mathcal{H}(P)))_{0 \leq j \leq |E(\mathrm{inc}(P))|}$ is h-unimodal.

To establish the unimodality of the (q, P)-Eulerian numbers (Conjecture 3.3) one would only need to prove Conjecture 5.3. Indeed Conjecture 5.3, Theorem 5.1, Corollary 4.18 and Lemma 2.8 together imply the unimodality result. The more general unimodality conjecture for $A_P(q, p, t)$ (Conjecture 4.20) is also a consequence of Conjecture 5.3.

References

[1] L. CARLITZ, R. SCOVILLE and T. VAUGHAN, *Enumeration of pairs of sequences by rises, falls and levels*, Manuscripta Math. **19** (1976), 211–243.

[2] T. CHOW, *Descents, quasi-symmetric functions, Robinson-Schensted for posets, and the chromatic symmetric function*, J. Algebraic Combin. **10** (1999), no. 3, 227–240.

[3] F. DE MARI, C. PROCESI and M. SHAYMAN, *Hessenberg varieties*, Trans. Amer. Math. Soc. **332** (1992), no. 2, 529–534.

[4] F. DE MARI and M. SHAYMAN, *Generalized Eulerian numbers and the topology of the Hessenberg variety of a matrix*, Acta Appl. Math. **12** (1988), no. 3, 213–235.

[5] J. DÉSARMÉNIEN, *Fonctions symétriques associées à des suites classiques de nombres*, Ann. scient. Éc. Norm. Sup. **16** (1983), 271–304.

[6] I. DOLGACHEV and V. LUNTS, *A character formula for the representation of a Weyl group in the cohomology of the associated toric variety*, J. Algebra **168** (1994), 741–772.

[7] J. DOLLHOPF, I. GOULDEN and C. GREENE, *Words avoiding a reflexive acyclic relation*, Electron. J. Combin. **11** (2006), #R28.

[8] D. FOATA and G.-N. HAN, *Fix Mahonian calculus III; A quadruple distribution*, Monatshefte für Mathematik **154** (2008), 177–197.

[9] D. FOATA and G.-N. HAN, *The q-tangent and q-secant numbers via basic Eulerian polynomials*, Proc. Amer. Math. Soc. **138** (2010), 385–393.

[10] V. Gasharov, *Incomparability graphs of* $(3+1)$*-free posets are s-positive*, Proceedings of the 6th Conference on Formal Power Series and Algebraic Combinatorics (New Brunswick, NJ, 1994), Discrete Math. **157** (1996), no. 1-3, 193–197.

[11] M. Goresky, R. Kottwitz and R. MacPherson, *Equivariant cohomology, Koszul duality, and the localization theorem*, Invent. Math. **131** (1998), no. 1, 25–83.

[12] A. Henderson and M. L. Wachs, *Unimodality of Eulerian quasisymmetric functions*, J. Combin. Theory Ser. A 119 (2012), no. 1, 135–145.

[13] A. Kasraoui, *A classification of Mahonian maj-inv statistics*, Adv. in Appl. Math. **42** (2009), no. 3, 342–357.

[14] G. I. Lehrer, *Rational points and Coxeter group actions on the cohomology of toric varieties*, Ann. Inst. Fourier (Grenoble) **58** (2008), no. 2, 671–688.

[15] K. Liang and M. L. Wachs, *Mahonian statistics on labeled forests*, Discrete Math. **99** (1992), 181–197.

[16] R. MacPherson and J. Tymoczko, personal communication.

[17] J. Martin, M. Morin and J. Wagner, *On distinguishing trees by their chromatic symmetric functions*, J. Combin. Theory Ser. A **115** (2008), no. 2, 237–253.

[18] C. Procesi, *The toric variety associated to Weyl chambers*, In: Mots, 153–161, Lang. Raison. Calc., Hermès, Paris, 1990.

[19] D. Rawlings, *The r-major index*, J. Combin. Theory Ser. A **31** (1981), no. 2, 175–183.

[20] B. Sagan, J. Shareshian and M. L. Wachs, *Eulerian quasisymmetric functions and cyclic sieving*, Adv. Applied Math. **46** (2011), 536–562.

[21] D. Scott and P. Suppes, *Foundational aspects of theories of measurement*, J. Symb. Logic **23** (1958), 113–128.

[22] J. Shareshian and M. L. Wachs, *q-Eulerian polynomials: excedance number and major index*, Electron. Res. Announc. Amer. Math. Soc. **13** (2007), 33–45.

[23] J. Shareshian and M. L. Wachs, *Eulerian quasisymmetric functions, Adv. Math.* **225** (2010), no. 6, 2921–2966.

[24] J. Shareshian and M. L. Wachs, *Poset homology of Rees products and q-Eulerian polynomials*, Elect. J. Combin. **16** (2009), R20.

[25] J. Shareshian and M. L. Wachs, *Chromatic quasisymmetric functions*, in preparation.

[26] R. P. Stanley, *Combinatorial Applications of the Hard Lefschetz Theorem*, Proc. Int. Congress of Mathematicians, 1983, Warszawa.

[27] R. P. STANLEY, *Log-concave and unimodal sequences in algebra, combinatorics, and geometry*, In: Graph theory and its applications: East and West (Jinan, 1986), 500–535, Ann. New York Acad. Sci., Vol. 576, New York Acad. Sci., New York, 1989.

[28] R. P. STANLEY, "Enumerative Combinatorics", Volume 1, 2nd ed., Cambridge Studies in Advanced Mathematics, Vol. 49, Cambridge University Press, Cambridge, 2011.

[29] R. P. STANLEY, "Enumerative Combinatorics", Volume 2, Cambridge Studies in Advanced Mathematics, Vol. 62, Cambridge University Press, Cambridge, 1999.

[30] R. P. STANLEY, *A symmetric function generalization of the chromatic polynomial of a graph*, Adv. Math. **111** (1995), no. 1, 166–194.

[31] R. P. STANLEY, *Graph colorings and related symmetric functions: ideas and applications: a description of results, interesting applications, & notable open problems*, Selected papers in honor of Adriano Garsia (Taormina, 1994), Discrete Math. **193** (1998), no. 1-3, 267–286.

[32] R. P. STANLEY and J. R. STEMBRIDGE, *On immanants of Jacobi Trudi matrices and permutations with restricted positions*, J. Combin. Theory Ser. A **62** (1993), 261–279.

[33] J. R. STEMBRIDGE, *Eulerian numbers, tableaux, and the Betti numbers of a toric variety*, Discrete Math. **99** (1992), 307–320.

[34] J. R. STEMBRIDGE, *Some permutation representations of Weyl groups associated with the cohomology of toric varieties*, Adv. Math. **106** (1994), 244–301.

[35] N. TEFF, "Representations on Hessenberg Varieties and Young's Rule", Proceedings of FPSAC 2011, Reykjavik.

[36] J. TYMOCZKO, *An introduction to equivariant cohomology and homology, following Goresky, Kottwitz, and MacPherson*, Snowbird lectures in algebraic geometry, 169–188, Contemp. Math. **388**, Amer. Math. Soc., Providence, RI, 2005.

[37] J. TYMOCZKO, *Permutation representations on Schubert varieties*, Amer. J. Math. **130** (2008), no. 5, 1171–1194.

[38] J. TYMOCZKO, *Permutation actions on equivariant cohomology*, Toric topology, 365–384, Contemp. Math. **460**, Amer. Math. Soc., Providence, RI, 2008.

Geometric and homological finiteness in free abelian covers

Alexander I. Suciu

Abstract. We describe some of the connections between the Bieri-Neumann-Strebel-Renz invariants, the Dwyer-Fried invariants, and the cohomology support loci of a space X. Under suitable hypotheses, the geometric and homological finiteness properties of regular, free abelian covers of X can be expressed in terms of the resonance varieties, extracted from the cohomology ring of X. In general, though, translated components in the characteristic varieties affect the answer. We illustrate this theory in the setting of toric complexes, as well as smooth, complex projective and quasi-projective varieties, with special emphasis on configuration spaces of Riemann surfaces and complements of hyperplane arrangements.

1 Introduction

1.1 Finiteness properties

This investigation is motivated by two seminal papers that appeared in 1987: one by Bieri, Neumann, and Strebel [7], and the other by Dwyer and Fried [17]. Both papers dealt with certain finiteness properties of normal subgroups of a group (or regular covers of a space), under the assumption that the factor group (or the group of deck transformations) is free abelian.

In [7], Bieri, Neumann, and Strebel associate to every finitely generated group G a subset $\Sigma^1(G)$ of the unit sphere $S(G)$ in the real vector space $\mathrm{Hom}(G, \mathbb{R})$. This "geometric" invariant of the group G is cut out of the sphere by open cones, and is independent of a finite generating set for G. In [8], Bieri and Renz introduced a nested family of higher-order invariants, $\{\Sigma^i(G, \mathbb{Z})\}_{i\geq 1}$, which record the finiteness properties of normal subgroups of G with abelian quotients.

*Partially supported by NSA grant H98230-09-1-0021 and NSF grant DMS–1010298

In a recent paper [21], Farber, Geoghegan and Schütz further extended these definitions. To each connected, finite-type CW-complex X, these authors assign a sequence of invariants, $\{\Sigma^i(X,\mathbb{Z})\}_{i\geq 1}$, living in the unit sphere $S(X)\subset H^1(X,\mathbb{R})$. The sphere $S(X)$ can be thought of as parametrizing all free abelian covers of X, while the Σ-invariants (which are again open subsets), keep track of the geometric finiteness properties of those covers.

Another tack was taken by Dwyer and Fried in [17]. Instead of looking at all free abelian covers of X at once, they fix the rank, say r, of the deck-transformation group, and view the resulting covers as being parametrized by the rational Grassmannian $\operatorname{Gr}_r(H^1(X,\mathbb{Q}))$. Inside this Grassmannian, they consider the subsets $\Omega^i_r(X)$, consisting of all covers for which the Betti numbers up to degree i are finite, and show how to determine these sets in terms of the support varieties of the relevant Alexander invariants of X. Unlike the Σ-invariants, though, the Ω-invariants need not be open subsets, see [17] and [38].

Our purpose in this note is to explore several connections between the geometric and homological invariants of a given space X, and use these connections to derive useful information about the rather mysterious Σ-invariants from concrete knowledge of the more accessible Ω-invariants.

1.2 Characteristic varieties and Ω-sets

Let $G=\pi_1(X,x_0)$ be the fundamental group of X, and let $\widehat{G}=\operatorname{Hom}(G,\mathbb{C}^\times)$ be the group of complex-valued characters on G, thought of as the parameter space for rank 1 local systems on X. The key tool for comparing the aforementioned invariants are the *characteristic varieties* $\mathcal{V}^i(X)$, consisting of those characters $\rho\in\widehat{G}$ for which $H_j(X,\mathbb{C}_\rho)\neq 0$, for some $j\leq i$.

Let X^{ab} be the universal abelian cover of X, with group of deck transformations G_{ab}. We may view each homology group $H_j(X^{\mathrm{ab}},\mathbb{C})$ as a finitely generated module over the Noetherian ring $\mathbb{C}G_{\mathrm{ab}}$. As shown in [33], the variety $\mathcal{V}^i(X)$ is the support locus for the direct sum of these modules, up to degree i. It follows then from the work of Dwyer and Fried [17], as further reinterpreted in [33, 38], that $\Omega^i_r(X)$ consists of those r-planes P inside $H^1(X,\mathbb{Q})$ for which the algebraic torus $\exp(P\otimes\mathbb{C})$ intersects the variety $\mathcal{V}^i(X)$ in only finitely many points.

Let $\mathcal{W}^i(X)$ be the intersection of $\mathcal{V}^i(X)$ with the identity component of $\widehat{G}$, and let $\tau_1(\mathcal{W}^i(X))$ be the set of points $z\in H^1(X,\mathbb{C})$ such that $\exp(\lambda z)$ belongs to $\mathcal{W}^i(X)$, for all $\lambda\in\mathbb{C}$. As noted in [16], this set is a finite union of rationally defined subspaces. For each $r\geq 1$, we then have

$$\Omega^i_r(X)\subseteq \operatorname{Gr}_r(H^1(X,\mathbb{Q}))\setminus \sigma_r(\tau_1^{\mathbb{Q}}(\mathcal{W}^i(X))), \tag{1.1}$$

where $\tau_1^{\mathbb{Q}}$ denotes the rational points on τ_1, and $\sigma_r(V)$ denotes the variety of incident r-planes to a homogeneous subvariety $V \subset H^1(X, \mathbb{Q})$, see [38]. There are many classes of spaces for which inclusion (1.1) holds as equality — for instance, toric complexes, or, more generally, the "straight spaces" studied in [37] — but, in general, the inclusion is strict.

1.3 Comparing the Ω-sets and the Σ-sets

A similar inclusion holds for the BNSR-invariants. As shown in [21], the set $\Sigma^i(X, \mathbb{Z})$ consists of those elements $\chi \in S(X)$ for which the homology of X with coefficients in the Novikov-Sikorav completion $\widehat{\mathbb{Z}G}_{-\chi}$ vanishes, up to degree i. Using this interpretation, we showed in [33] that the following inclusion holds:

$$\Sigma^i(X, \mathbb{Z}) \subseteq S(X) \setminus S(\tau_1^{\mathbb{R}}(\mathcal{W}^i(X))), \tag{1.2}$$

where $\tau_1^{\mathbb{R}}$ denotes the real points on τ_1, and $S(V)$ denotes the intersection of $S(X)$ with a homogeneous subvariety $V \subset H^1(X, \mathbb{R})$. Again, there are several classes of spaces for which inclusion (1.2) holds as equality — for instance, nilmanifolds, or compact Kähler manifolds without elliptic pencils with multiple fibers — but, in general, the inclusion is strict.

Clearly, formulas (1.1) and (1.2) hint at a connection between the Dwyer-Fried invariants and the Bieri-Neumann-Strebel-Renz invariants of a space X. We establish an explicit connection here, by comparing the conditions insuring those inclusions hold as equalities. Our main results reads as follows.

Theorem 1.1. *Let X be a connected CW-complex with finite k-skeleton. Suppose that, for some $i \leq k$,*

$$\Sigma^i(X, \mathbb{Z}) = S(X) \setminus S(\tau_1^{\mathbb{R}}(\mathcal{W}^i(X))).$$

Then, for all $r \geq 1$,

$$\Omega^i_r(X) = \mathrm{Gr}_r(H^1(X, \mathbb{Q})) \setminus \sigma_r(\tau_1^{\mathbb{Q}}(\mathcal{W}^i(X))).$$

Simple examples show that, in general, the above implication cannot be reversed.

The main usefulness of Theorem 1.1 resides in the fact that it allows one to show that the Σ-invariants of a space X are smaller than the upper bound given by (1.2), once one finds certain components in the characteristic varieties of X (*e.g.*, positive-dimensional translated subtori) insuring that the Ω-invariants are smaller than the upper bound given by (1.1).

1.4 Formality, straightness, and resonance

The above method can still be quite complicated, in that it requires computing cohomology with coefficients in rank 1 local systems on X. As noted in [33] and [37], though, in favorable situations the right-hand sides of (1.1) and (1.2) can be expressed in terms of ordinary cohomological data.

By definition, the i-th *resonance variety* of X, with coefficients in a field $\Bbbk$ of characteristic 0, is the set $\mathcal{R}^i(X, \Bbbk)$ of elements $a \in H^1(X, \Bbbk)$ for which the cochain complex whose terms are the cohomology groups $H^j(X, \Bbbk)$, and whose differentials are given by multiplication by a fails to be exact in some degree $j \leq i$. It is readily seen that each of these sets is a homogeneous subvariety of $H^1(X, \Bbbk)$.

If X is 1-formal (in the sense of rational homotopy theory), or, more generally, if X is locally 1-straight, then $\tau_1(\mathcal{W}^1(X)) = \mathrm{TC}_1(\mathcal{W}^1(X)) = \mathcal{R}^1(X)$. Thus, formulas (1.1) and (1.2) yield the following inclusions:

$$\Omega^1_r(X) \subseteq \mathrm{Gr}_r(H^1(X, \mathbb{Q})) \setminus \sigma_r(\mathcal{R}^1(X, \mathbb{Q})), \tag{1.3}$$

$$\Sigma^1(X, \mathbb{Z}) \subseteq S(X) \setminus S(\mathcal{R}^1(X, \mathbb{R})). \tag{1.4}$$

If X is locally k-straight, then the analogue of (1.3) holds in degrees $i \leq k$, with equality if X is k-straight. In general, though, neither (1.3) nor (1.4) is an equality.

Applying now Theorem 1.1, we obtain the following corollary.

Corollary 1.2. *Let X be a locally 1-straight space (for instance, a 1-formal space). Suppose $\Sigma^1(X, \mathbb{Z}) = S(X) \setminus S(\mathcal{R}^1(X, \mathbb{R}))$. Then, for all $r \geq 1$,*

$$\Omega^1_r(X) = \mathrm{Gr}_r(H^1(X, \mathbb{Q})) \setminus \sigma_r(\mathcal{R}^1(X, \mathbb{Q})).$$

1.5 Applications

We illustrate the theory outlined above with several classes of examples, coming from toric topology and algebraic geometry, as well as the study of configuration spaces and hyperplane arrangements.

1.5.1 Toric complexes Every simplicial complex L on n vertices determines a subcomplex T_L of the n-torus, with k-cells corresponding to the $(k-1)$-simplices of L. The fundamental group $\pi_1(T_L)$ is the right-angled Artin group G_L attached to the 1-skeleton of L, while a classifying space for G_L is the toric complex associated to the flag complex Δ_L.

It is known that all toric complexes are both straight and formal; their characteristic and resonance varieties were computed in [31], whereas the Σ-invariants of right-angled Artin groups were computed in [10, 29].

These computations, as well as work from [33, 37] show that $\Omega^1_r(T_L) = \sigma_r(\mathcal{R}^i(T_L, \mathbb{Q}))^{\complement}$ and $\Sigma^i(G_L, \mathbb{Z}) \subseteq S(\mathcal{R}^i(G_L, \mathbb{R}))^{\complement}$, though this last inclusion may be strict, unless a certain torsion-freeness assumption on the subcomplexes of Δ_L is satisfied.

1.5.2 Quasi-projective varieties The basic structure of the characteristic varieties of smooth, complex quasi-projective varieties was determined by Arapura [1], building on work of Beauville, Green and Lazarsfeld, and others. In particular, if X is such a variety, then all the components of $\mathcal{W}^1(X)$ are torsion-translated subtori.

If X is also 1-formal (*e.g.*, if X is compact), then inclusions (1.3) and (1.4) hold, but not always as equalities. For instance, if $\mathcal{W}^1(X)$ has a 1-dimensional component not passing through 1, and $\mathcal{R}^1(X, \mathbb{C})$ has no codimension-1 components, then, as shown in [37], inclusion (1.3) is strict for $r = 2$, and thus, by Corollary 1.2, inclusion (1.4) is also strict.

1.5.3 Configuration spaces An interesting class of quasi-projective varieties is provided by the configuration spaces $X = F(\Sigma_g, n)$ of n ordered points on a closed Riemann surface of genus g. If $g = 1$ and $n \geq 3$, then the resonance variety $\mathcal{R}^1(X, \mathbb{C})$ is irreducible and non-linear, and so X is not 1-formal, by [16]. To illustrate the computation of the Ω-sets and Σ-sets in such a non-formal setting, we work out the details for $F(\Sigma_1, 3)$.

1.5.4 Hyperplane arrangements Given a finite collection of hyperplanes, $\mathcal{A}$, in a complex vector space $\mathbb{C}^{\ell}$, the complement $X(\mathcal{A})$ is a smooth, quasi-projective variety; moreover, $X(\mathcal{A})$ is formal, locally straight, but not always straight. Arrangements of hyperplanes have been the main driving force behind the development of the theory of cohomology jump loci, and still provide a rich source of motivational examples for this theory. Much is known about the characteristic and resonance varieties of arrangement complements; in particular, $\mathcal{R}^1(X(\mathcal{A}), \mathbb{C})$ admits a purely combinatorial description, owing to the pioneering work of Falk [19], as sharpened by many others since then. We give here both lower bounds and upper bounds for the Ω-invariants and the Σ-invariants of arrangements.

In [37], we gave an example of an arrangement $\mathcal{A}$ for which the Dwyer-Fried set $\Omega^1_2(X(\mathcal{A}))$ is strictly contained in $\sigma_2(\mathcal{R}^1(X(\mathcal{A}), \mathbb{Q}))^{\complement}$. Using Corollary 1.2, we show here that the BNS invariant $\Sigma^1(X(\mathcal{A}), \mathbb{Z})$ is strictly contained in $S(\mathcal{R}^1(X(\mathcal{A}), \mathbb{R}))^{\complement}$. This answers a question first raised at an Oberwolfach Mini-Workshop [22], and revisited in [33, 36].

ACKNOWLEDGEMENTS. A preliminary version of this paper was presented at the Centro di Ricerca Matematica Ennio De Giorgi in Pisa, Italy, in May–June, 2010. I wish to thank the organizers of the Intensive Research Period on *Configuration Spaces: Geometry, Combinatorics and Topology* for their friendly hospitality.

Most of the work was done during the author's visit at the Université de Caen, France in June, 2011. Likewise, I wish to thank the Laboratoire de Mathématiques Nicolas Oresme for its support and hospitality.

Finally, I wish to thank the referee for helpful comments and suggestions.

2 Characteristic and resonance varieties

We start with a brief review of the characteristic varieties of a space, and their relation to the resonance varieties, via two kinds of tangent cone constructions.

2.1 Jump loci for twisted homology

Let X be a connected CW-complex with finite k-skeleton, for some $k \geq 1$. Without loss of generality, we may assume X has a single 0-cell, call it x_0. Let $G = \pi_1(X, x_0)$ be the fundamental group of X, and let $\widehat{G} = \operatorname{Hom}(G, \mathbb{C}^\times)$ be the group of complex characters of G. Clearly, $\widehat{G} = \widehat{G}_{\mathrm{ab}}$, where $G_{\mathrm{ab}} = H_1(X, \mathbb{Z})$ is the abelianization of G. The universal coefficient theorem allows us to identify $\widehat{G} = H^1(X, \mathbb{C}^\times)$.

Each character $\rho \colon G \to \mathbb{C}^\times$ determines a rank 1 local system, $\mathbb{C}_\rho$, on our space X. Computing the homology groups of X with coefficients in such local systems carves out some interesting subsets of the character group.

Definition 2.1. The *characteristic varieties* of X are the sets

$$\mathcal{V}^i_d(X) = \{\rho \in H^1(X, \mathbb{C}^\times) \mid \dim_{\mathbb{C}} H_i(X, \mathbb{C}_\rho) \geq d\}. \tag{2.1}$$

Clearly, $1 \in \mathcal{V}^i_d(X)$ if and only if $d \leq b_i(X)$. In degree 0, we have $\mathcal{V}^0_1(X) = \{1\}$ and $\mathcal{V}^0_d(X) = \emptyset$, for $d > 1$. In degree 1, the sets $\mathcal{V}^i_d(X)$ depend only on the group $G = \pi_1(X, x_0)$ — in fact, only on its maximal metabelian quotient, G/G''.

For the purpose of computing the characteristic varieties up to degree $i = k$, we may assume without loss of generality that X is a finite CW-complex of dimension $k + 1$, see [33]. With that in mind, it can be shown that the jump loci $\mathcal{V}^i_d(X)$ are Zariski closed subsets of the algebraic group $H^1(X, \mathbb{C}^\times)$, and that they depend only on the homotopy type of X. For details and further references, see [38].

One may extend the definition of characteristic varieties to arbitrary fields $\Bbbk$. The resulting varieties, $\mathcal{V}^i_d(X, \Bbbk)$, behave well under field extensions: if $\Bbbk \subseteq \mathbb{K}$, then $\mathcal{V}^i_d(X, \Bbbk) = \mathcal{V}^i_d(X, \mathbb{K}) \cap H^1(X, \Bbbk^\times)$.

Most important for us are the depth one characteristic varieties, $\mathcal{V}^i_1(X)$, and their unions up to a fixed degree, $\mathcal{V}^i(X) = \bigcup_{j=0}^{i} \mathcal{V}^j_1(X)$. These varieties yield an ascending filtration of the character group,

$$\{1\} = \mathcal{V}^0(X) \subseteq \mathcal{V}^1(X) \subseteq \cdots \subseteq \mathcal{V}^k(X) \subseteq \widehat{G}. \tag{2.2}$$

Now let $\widehat{G}^\circ$ be the identity component of the character group $\widehat{G}$. Writing $n = b_1(X)$, we may identify $\widehat{G}^\circ$ with the complex algebraic torus $(\mathbb{C}^\times)^n$. Set

$$\mathcal{W}^i(X) = \mathcal{V}^i(X) \cap \widehat{G}^\circ. \tag{2.3}$$

These varieties yield an ascending filtration of the complex algebraic torus $\widehat{G}^\circ$,

$$\{1\} = \mathcal{W}^0(X) \subseteq \mathcal{W}^1(X) \subseteq \cdots \subseteq \mathcal{W}^k(X) \subseteq (\mathbb{C}^\times)^n. \tag{2.4}$$

The characteristic varieties behave well with respect to direct products. For instance, suppose both X_1 and X_2 have finite k-skeleton. Then, by [33] (see also [38]), we have that

$$\mathcal{W}^i(X_1 \times X_2) = \bigcup_{p+q=i} \mathcal{W}^p(X_1) \times \mathcal{W}^q(X_2), \tag{2.5}$$

for all $i \leq k$.

2.2 Tangent cones and exponential tangent cones

Let $W \subset (\mathbb{C}^\times)^n$ be a Zariski closed subset, defined by an ideal I in the Laurent polynomial ring $\mathbb{C}[t_1^{\pm 1}, \ldots, t_n^{\pm 1}]$. Picking a finite generating set for I, and multiplying these generators with suitable monomials if necessary, we see that W may also be defined by the ideal $I \cap R$ in the polynomial ring $R = \mathbb{C}[t_1, \ldots, t_n]$. Finally, let J be the ideal in the polynomial ring $S = \mathbb{C}[z_1, \ldots, z_n]$, generated by the polynomials $g(z_1, \ldots, z_n) = f(z_1 + 1, \ldots, z_n + 1)$, for all $f \in I \cap R$.

Definition 2.2. The *tangent cone* of W at 1 is the algebraic subset $\mathrm{TC}_1(W) \subset \mathbb{C}^n$ defined by the ideal $\mathrm{in}(J) \subset S$ generated by the initial forms of all non-zero elements from J.

The tangent cone $\mathrm{TC}_1(W)$ is a homogeneous subvariety of $\mathbb{C}^n$, which depends only on the analytic germ of W at the identity. In particular, $\mathrm{TC}_1(W) \neq \emptyset$ if and only if $1 \in W$. Moreover, TC_1 commutes with finite unions.

The following, related notion was introduced in [16], and further studied in [33] and [38].

Definition 2.3. The *exponential tangent cone* of W at 1 is the homogeneous subvariety $\tau_1(W)$ of $\mathbb{C}^n$, defined by

$$\tau_1(W) = \{z \in \mathbb{C}^n \mid \exp(\lambda z) \in W, \text{ for all } \lambda \in \mathbb{C}\}. \tag{2.6}$$

Again, $\tau_1(W)$ depends only on the analytic germ of W at the identity, and so $\tau_1(W) \neq \emptyset$ if and only if $1 \in W$. Moreover, τ_1 commutes with finite unions, as well as arbitrary intersections. The most important properties of this construction are summarized in the following result from [16] (see also [33] and [38]).

Theorem 2.4. *For every Zariski closed subset $W \subset (\mathbb{C}^\times)^n$, the following hold*:

(1) *$\tau_1(W)$ is a finite union of rationally defined linear subspaces of $\mathbb{C}^n$.*
(2) $\tau_1(W) \subseteq \mathrm{TC}_1(W)$.

If W is an algebraic subtorus of $(\mathbb{C}^\times)^n$, then $\tau_1(W) = \mathrm{TC}_1(W)$, and both types of tangent cones coincide with the tangent space at the origin, $T_1(W)$; moreover, $W = \exp(\tau_1(W))$ in this case. More generally, if all positive-dimensional components of W are algebraic subtori, then $\tau_1(W) = \mathrm{TC}_1(W)$. In general, though, the inclusion from Theorem 2.4 (2) can be strict.

For brevity, we shall write $\tau_1^{\mathbb{Q}}(W) = \mathbb{Q}^n \cap \tau_1(W)$ for the rational points on the exponential tangent cone, and $\tau_1^{\mathbb{R}}(W) = \mathbb{R}^n \cap \tau_1(W)$ for the real points.

The main example we have in mind is that of the characteristic varieties $\mathcal{W}^i(X)$, viewed as Zariski closed subsets of the algebraic torus $H^1(X, \mathbb{C}^\times)^\circ = (\mathbb{C}^\times)^n$, where $n = b_1(X)$. By Theorem 2.4, the exponential tangent cone to $\mathcal{W}^i(X)$ can be written as a union of rationally defined linear subspaces,

$$\tau_1(\mathcal{W}^i(X)) = \bigcup_{L \in \mathcal{C}_i(X)} L \otimes \mathbb{C}. \tag{2.7}$$

We call the resulting rational subspace arrangement, $\mathcal{C}_i(X)$, the *i-th characteristic arrangement* of X; evidently, $\tau_1^{\mathbb{Q}}(\mathcal{W}^i(X))$ is the union of this arrangement.

2.3 Resonance varieties

Now consider the cohomology algebra $A = H^*(X, \mathbb{C})$, with graded ranks the Betti numbers $b_i = \dim_{\mathbb{C}} A^i$. For each $a \in A^1$, we have $a^2 = 0$, by graded-commutativity of the cup product. Thus, right-multiplication by a defines a cochain complex,

$$(A, \cdot a)\colon\ A^0 \xrightarrow{\ a\ } A^1 \xrightarrow{\ a\ } A^2 \longrightarrow \cdots, \tag{2.8}$$

The jump loci for the cohomology of this complex define a natural filtration of the affine space $A^1 = H^1(X, \mathbb{C})$.

Definition 2.5. The *resonance varieties* of X are the sets

$$\mathcal{R}^i_d(X) = \{a \in A^1 \mid \dim_{\mathbb{C}} H^i(A, a) \geq d\}. \tag{2.9}$$

For the purpose of computing the resonance varieties in degrees $i \leq k$, we may assume without loss of generality that X is a finite CW-complex of dimension $k + 1$. The sets $\mathcal{R}^i_d(X)$, then, are homogeneous, Zariski closed subsets of $A^1 = \mathbb{C}^n$, where $n = b_1$. In each degree $i \leq k$, the resonance varieties provide a descending filtration,

$$H^1(X, \mathbb{C}) \supseteq \mathcal{R}^i_1(X) \supseteq \cdots \supseteq \mathcal{R}^i_{b_i}(X) \supseteq \mathcal{R}^i_{b_i+1}(X) = \emptyset. \tag{2.10}$$

Note that, if $A^i = 0$, then $\mathcal{R}^i_d(X) = \emptyset$, for all $d > 0$. In degree 0, we have $\mathcal{R}^0_1(X) = \{0\}$, and $\mathcal{R}^0_d(X) = \emptyset$, for $d > 1$. In degree 1, the varieties $\mathcal{R}^1_d(X)$ depend only on the fundamental group $G = \pi_1(X, x_0)$ — in fact, only on the cup-product map $\cup\colon H^1(G, \mathbb{C}) \wedge H^1(G, \mathbb{C}) \to H^2(G, \mathbb{C})$.

One may extend the definition of resonance varieties to arbitrary fields $\Bbbk$, with the proviso that $H_1(X, \mathbb{Z})$ should be torsion-free, if $\Bbbk$ has characteristic 2. The resulting varieties, $\mathcal{R}^i_d(X, \Bbbk)$, behave well under field extensions: if $\Bbbk \subseteq \mathbb{K}$, then $\mathcal{R}^i_d(X, \Bbbk) = \mathcal{R}^i_d(X, \mathbb{K}) \cap H^1(X, \Bbbk)$. In particular, $\mathcal{R}^i_d(X, \mathbb{Q})$ is just the set of rational points on the integrally defined variety $\mathcal{R}^i_d(X) = \mathcal{R}^i_d(X, \mathbb{C})$.

Most important for us are the depth-1 resonance varieties, $\mathcal{R}^i_1(X)$, and their unions up to a fixed degree, $\mathcal{R}^i(X) = \bigcup_{j=0}^{i} \mathcal{R}^j_1(X)$. The latter varieties can be written as

$$\mathcal{R}^i(X) = \{a \in A^1 \mid H^j(A, \cdot a) \neq 0, \text{ for some } j \leq i\}. \tag{2.11}$$

These algebraic sets provide an ascending filtration of the first cohomology group,

$$\{0\} = \mathcal{R}^0(X) \subseteq \mathcal{R}^1(X) \subseteq \cdots \subseteq \mathcal{R}^k(X) \subseteq H^1(X, \mathbb{C}) = \mathbb{C}^n. \tag{2.12}$$

As noted in [33], the resonance varieties also behave well with respect to direct products: if both X_1 and X_2 have finite k-skeleton, then, for all $i \leq k$,

$$\mathcal{R}^i(X_1 \times X_2) = \bigcup_{p+q=i} \mathcal{R}^p(X_1) \times \mathcal{R}^q(X_2). \tag{2.13}$$

2.4 Tangent cone and resonance

An important feature of the theory of cohomology jumping loci is the relationship between characteristic and resonance varieties, based on the tangent cone construction. A foundational result in this direction is the following theorem of Libgober [27], which generalizes an earlier result of Green and Lazarsfeld [24].

Theorem 2.6. *Let X be a connected CW-complex with finite k-skeleton. Then, for all $i \leq k$ and $d > 0$, the tangent cone at 1 to $\mathcal{W}^i_d(X)$ is included in $\mathcal{R}^i_d(X)$.*

Putting together Theorems 2.4 (2) and 2.6, and using the fact that both types of tangent cone constructions commute with finite unions, we obtain an immediate corollary.

Corollary 2.7. *For each $i \leq k$, the following inclusions hold:*

(1) $\tau_1(\mathcal{W}^i_d(X)) \subseteq \mathrm{TC}_1(\mathcal{W}^i_d(X)) \subseteq \mathcal{R}^i_d(X)$, *for all $d > 0$.*
(2) $\tau_1(\mathcal{W}^i(X)) \subseteq \mathrm{TC}_1(\mathcal{W}^i(X)) \subseteq \mathcal{R}^i(X)$.

In general, the above inclusions may very well be strict. In the presence of formality, though, they become equalities, at least in degree $i = 1$.

2.5 Formality

Let us now collect some known facts on various formality notions. For further details and references, we refer to the recent survey [32].

Let X be a connected CW-complex with finite 1-skeleton. The space X is *formal* if there is a zig-zag of commutative, differential graded algebra quasi-isomorphisms connecting Sullivan's algebra of polynomial differential forms, $A_{\mathrm{PL}}(X, \mathbb{Q})$, to the rational cohomology algebra, $H^*(X, \mathbb{Q})$, endowed with the zero differential. The space X is merely k-*formal* (for some $k \geq 1$) if each of these morphisms induces an isomorphism in degrees up to k, and a monomorphism in degree $k + 1$.

Examples of formal spaces include rational cohomology tori, Riemann surfaces, compact connected Lie groups, as well as their classifying spaces. On the other hand, a nilmanifold is formal if and only if it is a torus. Formality is preserved under wedges and products of spaces, and connected sums of manifolds.

The 1-minimality property of a space depends only on its fundamental group. A finitely generated group G is said to be 1-formal if it admits a classifying space $K(G, 1)$ which is 1-formal, or, equivalently, if the Malcev Lie algebra $\mathfrak{m}(G)$ (that is, the Lie algebra of the rational, prounipotent completion of G) admits a quadratic presentation. Examples of 1-formal groups include free groups and free abelian groups of finite rank, surface groups, and groups with first Betti number equal to 0 or 1. The 1-formality property is preserved under free products and direct products.

Theorem 2.8 ([16]). *Let X be a 1-formal space. Then, for each $d > 0$,*

$$\tau_1(\mathcal{V}^1_d(X)) = \mathrm{TC}_1(\mathcal{V}^1_d(X)) = \mathcal{R}^1_d(X). \tag{2.14}$$

In particular, the first resonance variety, $\mathcal{R}^1(X)$, of a 1-formal space X is a finite union of rationally defined linear subspaces.

2.6 Straightness

In [37], we delineate another class of spaces for which the resonance and characteristic varieties are intimately related to each other via the tangent cone constructions.

Definition 2.9. We say that X is *locally k-straight* if, for each $i \leq k$, all components of $\mathcal{W}^i(X)$ passing through the origin 1 are algebraic subtori, and the tangent cone at 1 to $\mathcal{W}^i(X)$ equals $\mathcal{R}^i(X)$. If, moreover, all positive-dimensional components of $\mathcal{W}^i(X)$ contain the origin, we say X is *k-straight*. If these conditions hold for all $k \geq 1$, we say X is *(locally) straight*.

Examples of straight spaces include Riemann surfaces, tori, and knot complements. Under some further assumptions, the straightness properties behave well with respect to finite direct products and wedges.

It follows from [16] that every 1-formal space is locally 1-straight. In general, though, examples from [37] show that 1-formal spaces need not be 1-straight, and 1-straight spaces need not be 1-formal.

Theorem 2.10 ([37]). *Let X be a locally k-straight space. Then, for all $i \leq k$,*

(1) $\tau_1(\mathcal{W}^i(X)) = \mathrm{TC}_1(\mathcal{W}^i(X)) = \mathcal{R}^i(X)$.

(2) $\mathcal{R}^i(X, \mathbb{Q}) = \bigcup_{L \in \mathcal{C}_i(X)} L$.

In particular, all the resonance varieties $\mathcal{R}^i(X)$ of a locally straight space X are finite unions of rationally defined linear subspaces.

3 The Dwyer–Fried invariants

In this section, we recall the definition of the Dwyer–Fried sets, and the way these sets relate to the (co)homology jump loci of a space.

3.1 Betti numbers of free abelian covers

As before, let X be a connected CW-complex with finite 1-skeleton, and let $G = \pi_1(X, x_0)$. Denote by $n = b_1(X)$ the first Betti number of X. Fix an integer r between 1 and n, and consider the (connected) regular covers of X, with group of deck-transformations $\mathbb{Z}^r$.

Each such cover, $X^\nu \to X$, is determined by an epimorphism $\nu\colon G \twoheadrightarrow \mathbb{Z}^r$. The induced homomorphism in rational cohomology, $\nu^*\colon H^1(\mathbb{Z}^r, \mathbb{Q}) \hookrightarrow H^1(G, \mathbb{Q})$, defines an r-dimensional subspace, $P_\nu = \operatorname{im}(\nu^*)$, in the n-dimensional $\mathbb{Q}$-vector space $H^1(G, \mathbb{Q}) = H^1(X, \mathbb{Q})$. Conversely, each r-dimensional subspace $P \subset H^1(X, \mathbb{Q})$ can be written as $P = P_\nu$, for some epimorphism $\nu\colon G \twoheadrightarrow \mathbb{Z}^r$, and thus defines a regular $\mathbb{Z}^r$-cover of X.

In summary, the regular $\mathbb{Z}^r$-covers of X are parametrized by the Grassmannian of r-planes in $H^1(X, \mathbb{Q})$, via the correspondence

$$\begin{array}{ccc} \{\text{regular } \mathbb{Z}^r\text{-covers of } X\} & \longleftrightarrow & \{r\text{-planes in } H^1(X, \mathbb{Q})\} \\ X^\nu \to X & \longleftrightarrow & P_\nu = \operatorname{im}(\nu^*). \end{array} \tag{3.1}$$

Moving about the rational Grassmannian and recording how the Betti numbers of the corresponding covers vary leads to the following definition.

Definition 3.1. The *Dwyer–Fried invariants* of the space X are the subsets

$$\Omega^i_r(X) = \left\{ P_\nu \in \operatorname{Gr}_r(H^1(X, \mathbb{Q})) \;\middle|\; b_j(X^\nu) < \infty \text{ for } j \le i \right\}.$$

For a fixed integer r between 1 and n, these sets form a descending filtration of the Grassmannian of r-planes in $H^1(X, \mathbb{Q}) = \mathbb{Q}^n$,

$$\operatorname{Gr}_r(\mathbb{Q}^n) = \Omega^0_r(X) \supseteq \Omega^1_r(X) \supseteq \Omega^2_r(X) \supseteq \cdots . \tag{3.2}$$

If $r > n$, we adopt the convention that $\operatorname{Gr}_r(\mathbb{Q}^n) = \emptyset$, and define $\Omega^i_r(X) = \emptyset$.

As noted in [38], the Ω-sets are invariants of homotopy-type: if $f\colon X \to Y$ is a homotopy equivalence, then the induced isomorphism in cohomology, $f^*\colon H^1(Y, \mathbb{Q}) \to H^1(X, \mathbb{Q})$, defines isomorphisms $f^*_r\colon \operatorname{Gr}_r(H^1(Y, \mathbb{Q})) \to \operatorname{Gr}_r(H^1(X, \mathbb{Q}))$, which send each subset $\Omega^i_r(Y)$ bijectively onto $\Omega^i_r(X)$.

Particularly simple is the situation when $n = b_1(X) > 0$ and $r = n$. In this case, $\operatorname{Gr}_n(H^1(X, \mathbb{Q})) = \{\text{pt}\}$. Under the correspondence from (3.1), this single point is realized by the maximal free abelian cover, $X^\alpha \to X$, where $\alpha\colon G \twoheadrightarrow G_{\text{ab}}/\operatorname{Tors}(G_{\text{ab}}) = \mathbb{Z}^n$ is the canonical projection. The sets $\Omega^i_n(X)$ are then given by

$$\Omega^i_n(X) = \begin{cases} \{\text{pt}\} & \text{if } b_j(X^\alpha) < \infty \text{ for all } j \le i, \\ \emptyset & \text{otherwise.} \end{cases} \tag{3.3}$$

3.2 Dwyer-Fried invariants and characteristic varieties

The next theorem reduces the computation of the Ω-sets to a more standard computation in algebraic geometry. The theorem was proved by Dwyer and Fried in [17], using the support loci for the Alexander invariants, and was recast in a slightly more general context by Papadima and Suciu in [33], using the characteristic varieties. We state this result in the form established in [38].

Theorem 3.2 ([17,33,38]). *Suppose X has finite k-skeleton, for some $k \geq 1$. Then, for all $i \leq k$ and $r \geq 1$,*

$$\Omega_r^i(X) = \left\{ P \in \mathrm{Gr}_r(\mathbb{Q}^n) \,\middle|\, \dim\left(\exp(P \otimes \mathbb{C}) \cap \mathcal{W}^i(X)\right) = 0 \right\}. \tag{3.4}$$

In other words, an r-plane $P \subset \mathbb{Q}^n$ belongs to the set $\Omega_r^i(X)$ if and only if the algebraic torus $T = \exp(P \otimes \mathbb{C})$ intersects the characteristic variety $W = \mathcal{W}^i(X)$ only in finitely many points. When this happens, the exponential tangent cone $\tau_1(T \cap W)$ equals $\{0\}$, forcing $P \cap L = \{0\}$, for every subspace $L \subset \mathbb{Q}^n$ in the characteristic subspace arrangement $\mathcal{C}_i(X)$. Consequently,

$$\Omega_r^i(X) \subseteq \left(\bigcup_{L \in \mathcal{C}_i(X)} \left\{ P \in \mathrm{Gr}_r(H^1(X,\mathbb{Q})) \,\middle|\, P \cap L \neq \{0\} \right\} \right)^{\complement}. \tag{3.5}$$

As noted in [38], this inclusion may be reinterpreted in terms of the classical incidence correspondence from algebraic geometry. Let V be a homogeneous variety in $\Bbbk^n$. The locus of r-planes in $\Bbbk^n$ that meet V,

$$\sigma_r(V) = \left\{ P \in \mathrm{Gr}_r(\Bbbk^n) \,\middle|\, P \cap V \neq \{0\} \right\}, \tag{3.6}$$

is a Zariski closed subset of the Grassmannian $\mathrm{Gr}_r(\Bbbk^n)$. In the case when V is a linear subspace $L \subset \Bbbk^n$, the incidence variety $\sigma_r(L)$ is known as the *special Schubert variety* defined by L. If L has codimension d in $\Bbbk^n$, then $\sigma_r(L)$ has codimension $d - r + 1$ in $\mathrm{Gr}_r(\Bbbk^n)$.

Applying this discussion to the homogeneous variety $\tau_1^{\mathbb{Q}}(\mathcal{W}^i(X)) = \bigcup_{L \in \mathcal{C}_i(X)} L$ lying inside $H^1(X,\mathbb{Q}) = \mathbb{Q}^n$, and using formula (3.5), we obtain the following corollary.

Corollary 3.3 ([38]). *Let X be a connected CW-complex with finite k-skeleton. For all $i \leq k$ and $r \geq 1$,*

$$\begin{aligned} \Omega_r^i(X) &\subseteq \mathrm{Gr}_r(H^1(X,\mathbb{Q})) \setminus \sigma_r\left(\tau_1^{\mathbb{Q}}(\mathcal{W}^i(X))\right) \\ &= \mathrm{Gr}_r(H^1(X,\mathbb{Q})) \setminus \bigcup_{L \in \mathcal{C}_i(X)} \sigma_r(L). \end{aligned} \tag{3.7}$$

In other words, each set $\Omega^i_r(X)$ is contained in the complement of a Zariski closed subset of the Grassmanian $\mathrm{Gr}_r(H^1(X,\mathbb{Q}))$, namely, the union of the special Schubert varieties $\sigma_r(L)$ corresponding to the subspaces L in $\mathcal{C}_i(X)$.

Under appropriate hypothesis, the inclusion from Corollary 3.3 holds as equality. The next two propositions illustrate this point.

Proposition 3.4 ([38]). *Let X be a connected CW-complex with finite k-skeleton. Suppose that, for some $i \leq k$, all positive-dimensional components of $\mathcal{W}^i(X)$ are algebraic subtori. Then, for all $r \geq 1$,*

$$\Omega^i_r(X) = \mathrm{Gr}_r(H^1(X,\mathbb{Q})) \setminus \bigcup_{L \in \mathcal{C}_i(X)} \sigma_r(L). \tag{3.8}$$

Proposition 3.5 ([17,33,38]). *Let X be a CW-complex with finite k-skeleton. Then, for all $i \leq k$,*

$$\Omega^i_1(X) = \mathbb{P}(H^1(X,\mathbb{Q})) \setminus \bigcup_{L \in \mathcal{C}_i(X)} \mathbb{P}(L), \tag{3.9}$$

where $\mathbb{P}(V)$ denotes the projectivization of a homogeneous subvariety $V \subseteq H^1(X,\mathbb{Q})$.

In either of these two situations, the sets $\Omega^i_r(X)$ are Zariski open subsets of $\mathrm{Gr}_r(H^1(X,\mathbb{Q}))$. In general, though, the sets $\Omega^i_r(X)$ need not be open, not even in the usual topology on the rational Grassmanian. This phenomenon was first noticed by Dwyer and Fried, who constructed in [17] a 3-dimensional cell complex X for which $\Omega^2_2(X)$ is a finite set (see Example 5.12). In [38], we provide examples of finitely presented groups G for which $\Omega^1_2(G)$ is not open.

3.3 Straightness and the Ω-sets

Under appropriate straightness assumptions, the upper bounds for the Dwyer–Fried sets can be expressed in terms of the resonance varieties associated to the cohomology ring $H^*(X,\mathbb{Q})$.

Theorem 3.6 ([37]). *Let X be a connected CW-complex.*

(1) *If X is locally k-straight, then $\Omega^i_r(X) \subseteq \mathrm{Gr}_r(H^1(X,\mathbb{Q})) \setminus \sigma_r(\mathcal{R}^i(X,\mathbb{Q}))$, for all $i \leq k$ and $r \geq 1$.*
(2) *If X is k-straight, then $\Omega^i_r(X) = \mathrm{Gr}_r(H^1(X,\mathbb{Q})) \setminus \sigma_r(\mathcal{R}^i(X,\mathbb{Q}))$, for all $i \leq k$ and $r \geq 1$.*

If X is locally k-straight, we also know from Theorem 2.10 (2) that $\mathcal{R}^i(X, \mathbb{Q})$ is the union of the linear subspaces comprising $C_i(X)$, for all $i \leq k$. Thus, if X is k-straight, then $\Omega^i_r(X)$ is the complement of a finite union of special Schubert varieties in the Grassmannian of r-planes in $H^1(X, \mathbb{Q})$; in particular, $\Omega^i_r(X)$ is a Zariski open set in $\mathrm{Gr}_r(H^1(X, \mathbb{Q}))$.

The straightness hypothesis is crucial for the equality in Theorem 3.6 (2) to hold.

Example 3.7. Let $G = \langle x_1, x_2 \mid x_1^2 x_2 = x_2 x_1^2 \rangle$. Then $\mathcal{W}^1(G) = \{1\} \cup \{t \in (\mathbb{C}^\times)^2 \mid t_1 = -1\}$, while $\mathcal{R}^1(G) = \{0\}$. Thus, G is locally 1-straight, but not 1-straight. Moreover, $\Omega^1_2(G) = \emptyset$, yet $\sigma_2(\mathcal{R}^1(G, \mathbb{Q}))^{\complement} = \{\mathrm{pt}\}$.

4 The Bieri-Neumann-Strebel-Renz invariants

We now go over the several definitions of the Σ-invariants of a space X (and, in particular, of a group G), and discuss the way these invariants relate to the (co)homology jumping loci.

4.1 A finite type property

We start with a finiteness condition for chain complexes, following the approach of Farber, Geoghegan, and Schütz [21]. Let $C = (C_i, \partial_i)_{i \geq 0}$ be a non-negatively graded chain complex over a ring R, and let k be a non-negative integer.

Definition 4.1. We say C is of *finite k-type* if there is a chain complex C' of finitely generated, projective R-modules and a k-equivalence between C' and C, *i.e.*, a chain map $C' \to C$ inducing isomorphisms $H_i(C') \to H_i(C)$ for $i < k$ and an epimorphism $H_k(C') \to H_k(C)$.

Equivalently, there is a chain complex of free R-modules, call it D, such that D_i is finitely generated for all $i \leq k$, and there is a chain map $D \to C$ inducing an isomorphism $H_*(D) \to H_*(C)$. When C itself is free, we have the following alternate characterization from [21].

Lemma 4.2. *Let C be a free chain complex over a ring R. Then C is of finite k-type if and only if C is chain-homotopy equivalent to a chain complex D of free R-modules, such that D_i is finitely generated for all $i \leq k$.*

Remark 4.3. Suppose $\rho\colon R \to S$ is a ring morphism, and C is a chain complex over R. Then $C \otimes_R S$ naturally acquires the structure of an S-chain complex via extension of scalars. Now, if C is free, and of finite k-type over R, it is readily seen that $C \otimes_R S$ is of finite k-type over S.

4.2 The Σ-invariants of a chain complex

Let G be a finitely generated group, and denote by $\mathrm{Hom}(G, \mathbb{R})$ the set of homomorphisms from G to the additive group of the reals. Clearly, this is a finite-dimensional $\mathbb{R}$-vector space, on which the multiplicative group of positive reals naturally acts. After fixing an inner product on $\mathrm{Hom}(G, \mathbb{R})$, the quotient space,

$$S(G) = (\mathrm{Hom}(G, \mathbb{R}) \setminus \{0\})/\mathbb{R}^+, \tag{4.1}$$

may be identified with the unit sphere in $\mathrm{Hom}(G, \mathbb{R})$. Up to homeomorphism, this sphere is determined by the first Betti number of G. Indeed, if $b_1(G) = n$, then $S(G) = S^{n-1}$; in particular, if $b_1(G) = 0$, then $S(G) = \emptyset$. To simplify notation, we will denote both a non-zero homomorphism $G \to \mathbb{R}$ and its equivalence class in $S(G)$ by the same symbol, and we will routinely view $S(G)$ as embedded in $\mathrm{Hom}(G, \mathbb{R})$.

Given a homomorphism $\chi\colon G \to \mathbb{R}$, consider the set $G_\chi = \{g \in G \mid \chi(g) \geq 0\}$. Clearly, G_χ is a submonoid of G, and the monoid ring $\mathbb{Z}G_\chi$ is a subring of the group ring $\mathbb{Z}G$. Thus, any $\mathbb{Z}G$-module naturally acquires the structure of a $\mathbb{Z}G_\chi$-module, by restriction of scalars.

Definition 4.4 ([21]). Let C be a chain complex over $\mathbb{Z}G$. For each integer $k \geq 0$, the *k-th Bieri–Neumann–Strebel–Renz invariant* of C is the set

$$\Sigma^k(C) = \{\chi \in S(G) \mid C \text{ is of finite } k\text{-type over } \mathbb{Z}G_\chi\}. \tag{4.2}$$

Note that $\mathbb{Z}G$ is a flat $\mathbb{Z}G_\chi$-module, and $\mathbb{Z}G \otimes_{\mathbb{Z}G_\chi} M \cong M$ for every $\mathbb{Z}G$-module M. Thus, if $\Sigma^k(C)$ is non-empty, then C must be of finite k-type over $\mathbb{Z}G$.

To a large extent, the importance of the Σ-invariants lies in the fact that they control the finiteness properties of kernels of projections to abelian quotients. More precisely, let N be a normal subgroup of G, with G/N abelian. Define

$$S(G, N) = \{\chi \in S(G) \mid N \leq \ker(\chi)\}. \tag{4.3}$$

It is readily seen that $S(G, N)$ is the great subsphere obtained by intersecting the unit sphere $S(G) \subset H^1(G, \mathbb{R})$ with the linear subspace $P_\nu \otimes \mathbb{R}$, where $\nu\colon G \twoheadrightarrow G/N$ is the canonical projection, and $P_\nu = \mathrm{im}(\nu^*\colon H^1(G/N, \mathbb{Q}) \hookrightarrow H^1(G, \mathbb{Q}))$. Notice also that every $\mathbb{Z}G$-module acquires the structure of a $\mathbb{Z}N$-module by restricting scalars.

Theorem 4.5 ([21]). *Let C be a chain complex of free $\mathbb{Z}G$-modules, with C_i finitely generated for $i \leq k$, and let N be a normal subgroup of G, with G/N is abelian. Then C is of finite k-type over $\mathbb{Z}N$ if and only if $S(G, N) \subset \Sigma^k(C)$.*

4.3 The Σ-invariants of a CW-complex

Let X be a connected CW-complex with finite 1-skeleton, and let $G = \pi_1(X, x_0)$ be its fundamental group. Picking a classifying map $X \to K(G,1)$, we obtain an induced isomorphism, $H^1(G,\mathbb{R}) \xrightarrow{\simeq} H^1(X,\mathbb{R})$, which identifies the respective unit spheres, $S(G)$ and $S(X)$.

The cell structure on X lifts to a cell structure on the universal cover $\widetilde{X}$. Clearly, this lifted cell structure is invariant under the action of G by deck transformations. Thus, the cellular chain complex $C_*(\widetilde{X},\mathbb{Z})$ is a chain complex of (free) $\mathbb{Z}G$-modules.

Definition 4.6. For each $k \geq 0$, the *k-th Bieri-Neumann-Strebel-Renz invariant* of X is the subset of $S(X)$ given by

$$\Sigma^k(X,\mathbb{Z}) = \Sigma^k(C_*(\widetilde{X},\mathbb{Z})). \tag{4.4}$$

It is shown in [21] that $\Sigma^k(X,\mathbb{Z})$ is an open subset of $S(X)$, which depends only on the homotopy type of the space X. Clearly, $\Sigma^0(X,\mathbb{Z}) = S(X)$.

Now let G be a finitely generated group, and pick a classifying space $K(G,1)$. In view of the above discussion, the sets

$$\Sigma^k(G,\mathbb{Z}) := \Sigma^k(K(G,1),\mathbb{Z}), \tag{4.5}$$

are well-defined invariants of the group G. These sets, which live inside the unit sphere $S(G) \subset \mathrm{Hom}(G,\mathbb{R})$, coincide with the classical geometric invariants of Bieri, Neumann and Strebel [7] and Bieri and Renz [8].

The Σ-invariants of a CW-complex and those of its fundamental group are related, as follows.

Proposition 4.7 ([21]). *Let X be a connected CW-complex with finite 1-skeleton. If $\widetilde{X}$ is k-connected, then*

$$\Sigma^k(X) = \Sigma^k(\pi_1(X)), \text{ and } \Sigma^{k+1}(X) \subseteq \Sigma^{k+1}(\pi_1(X)). \tag{4.6}$$

The last inclusion can of course be strict. For instance, if $X = S^1 \vee S^{k+1}$, with $k \geq 1$, then $\Sigma^{k+1}(X) = \emptyset$, though $\Sigma^{k+1}(\pi_1(X)) = S^0$. We shall see a more subtle occurrence of this phenomenon in Example 5.12.

4.4 Generalizations and discussion

More generally, if M is a $\mathbb{Z}G$-module, the invariants $\Sigma^k(G,M)$ of Bieri and Renz [8] are given by $\Sigma^k(G,M) = \Sigma^i(F_\bullet)$, where $F_\bullet \to M$ is a projective $\mathbb{Z}G$-resolution of M. In particular, we have the invariants $\Sigma^k(G,\Bbbk)$, where $\Bbbk$ is a field, viewed as a trivial $\mathbb{Z}G$-module. There is always an inclusion $\Sigma^k(G, \mathbb{Z}) \subseteq \Sigma^i(G, \Bbbk)$, but this inclusion may be strict, as we shall see in Example 6.5 below.

An alternate definition of the Σ-invariants of a group is as follows. Recall that a monoid (in particular, a group) G is of type FP_k if there is a projective $\mathbb{Z}G$-resolution $F_\bullet \to \mathbb{Z}$, with F_i finitely generated, for all $i \le k$. In particular, G is of type FP_1 if and only if G is finitely generated. We then have

$$\Sigma^k(G, \mathbb{Z}) = \{\chi \in S(G) \mid \text{the monoid } G_\chi \text{ is of type } \mathrm{FP}_k\}. \tag{4.7}$$

These sets form a descending chain of open subsets of $S(G)$, starting at $\Sigma^1(G) = \Sigma^1(G, \mathbb{Z})$. Moreover, $\Sigma^k(G, \mathbb{Z})$ is non-empty only if G is of type FP_k.

If N is a normal subgroup of G, with G/N abelian, then Theorem 4.5 implies the following result from [7,8]: The group N is of type FP_k if and only if $S(G, N) \subseteq \Sigma^k(G, \mathbb{Z})$. In particular, the kernel of an epimorphism $\chi : G \twoheadrightarrow \mathbb{Z}$ is finitely generated if and only if both χ and $-\chi$ belong to $\Sigma^1(G)$.

Onc class of groups for which the BNSR invariants can be computed explicitly is that of one-relator groups. K. Brown gave in [9] an algorithm for computing $\Sigma^1(G)$ for groups G in this class, while Bieri and Renz in [8] reinterpreted this algorithm in terms of Fox calculus, and showed that, for 1-relator groups, $\Sigma^k(G, \mathbb{Z}) = \Sigma^1(G)$, for all $k \ge 2$.

Another class of groups for which the Σ-invariants can be completely determined is that of non-trivial free products. Indeed, if G_1 and G_2 are two non-trivial, finitely generated groups, then $\Sigma^k(G_1 * G_2, \mathbb{Z}) = \emptyset$, for all $k \ge 1$ (see, for instance, [33]).

Finally, it should be noted that the BNSR invariants obey some very nice product formulas. For instance, let G_1 and G_2 be two groups of type F_k, for some $k \ge 1$. Identify the sphere $S(G_1 \times G_2)$ with the join $S(G_1) * S(G_2)$. Then, for all $i \le k$,

$$\Sigma^i(G_1 \times G_2, \mathbb{Z})^{\complement} \subseteq \bigcup_{p+q=i} \Sigma^p(G_1, \mathbb{Z})^{\complement} * \Sigma^q(G_2, \mathbb{Z})^{\complement}, \tag{4.8}$$

where again $A * B$ denotes the join of two spaces, with the convention that $A * \emptyset = A$. As shown in [7, Theorem 7.4], the above inclusion holds as equality for $i = 1$, *i.e.*,

$$\Sigma^1(G_1 \times G_2, \mathbb{Z})^{\complement} = \Sigma^1(G_1, \mathbb{Z})^{\complement} \cup \Sigma^1(G_2, \mathbb{Z})^{\complement}. \tag{4.9}$$

The general formula (4.8) was established by Meinert (unpublished) and Gehrke [23]. Recently, it was shown by Schütz [34] and Bieri-Geoghegan

[6] that equality holds in (4.8) for all $i \leq 3$, although equality may fail for $i \geq 4$. Furthermore, it was shown in [6] that the analogous product formula for the Σ-invariants with coefficients in a field $\Bbbk$ holds as an equality, for all $i \leq k$.

4.5 Novikov homology

In his 1987 thesis, J.-Cl. Sikorav reinterpreted the BNS invariant of a finitely generated group in terms of Novikov homology. This interpretation was extended to all BNSR invariants by Bieri [5], and later to the BNSR invariants of CW-complexes by Farber, Geoghegan and Schütz [21].

The *Novikov–Sikorav completion* of the group ring $\mathbb{Z}G$ with respect to a homomorphism $\chi : G \to \mathbb{R}$ consists of all formal sums $\sum_j n_j g_j$, with $n_j \in \mathbb{Z}$ and $g_j \in G$, having the property that, for each $c \in \mathbb{R}$, the set of indices j for which $n_j \neq 0$ and $\chi(g_j) \geq c$ is finite. With the obvious addition and multiplication, the Novikov–Sikorav completion, $\widehat{\mathbb{Z}G}_\chi$, is a ring, containing $\mathbb{Z}G$ as a subring; in particular, $\widehat{\mathbb{Z}G}_\chi$ carries a natural G-module structure. We refer to M. Farber's book [20] for a comprehensive treatment, and to R. Bieri [5] for further details.

Let X be a connected CW-complex with finite k-skeleton, and let $G = \pi_1(X, x_0)$ be its fundamental group.

Theorem 4.8 ([21]). *With notation as above,*

$$\Sigma^k(X, \mathbb{Z}) = \{\chi \in S(X) \mid H_i(X, \widehat{\mathbb{Z}G}_{-\chi}) = 0, \text{ for all } i \leq k\}. \quad (4.10)$$

Now, every non-zero homomorphism $\chi : G \to \mathbb{R}$ factors as $\chi = \iota \circ \xi$, where $\xi : G \twoheadrightarrow \Gamma$ is a surjection onto a lattice $\Gamma \cong \mathbb{Z}^r$ in $\mathbb{R}$, and $\iota : \Gamma \hookrightarrow \mathbb{R}$ is the inclusion map. A Laurent polynomial $p = \sum_\gamma n_\gamma \gamma \in \mathbb{Z}\Gamma$ is said to be ι-monic if the greatest element in $\iota(\operatorname{supp}(p))$ is 0, and $n_0 = 1$; every such polynomial is invertible in the completion $\widehat{\mathbb{Z}\Gamma}_\iota$. We denote by $\mathcal{R}\Gamma_\iota$ the localization of $\mathbb{Z}\Gamma$ at the multiplicative subset of all ι-monic polynomials. Using the known fact that $\mathcal{R}\Gamma_\iota$ is both a G-module and a PID, one may define for each $i \leq k$ the *i-th Novikov Betti number*, $b_i(X, \chi)$, as the rank of the finitely generated $\mathcal{R}\Gamma_\iota$-module $H_i(X, \mathcal{R}\Gamma_\iota)$.

4.6 Σ-invariants and characteristic varieties

The following result from [33] creates a bridge between the Σ-invariants of a space and the real points on the exponential tangent cones to the respective characteristic varieties.

Theorem 4.9 ([33]). *Let X be a CW-complex with finite k-skeleton, for some $k \geq 1$, and let $\chi \in S(X)$. The following then hold.*

(1) *If $-\chi$ belongs to $\Sigma^k(X, \mathbb{Z})$, then $H_i(X, \mathcal{R}\Gamma_\iota) = 0$, and so $b_i(X, \chi) = 0$, for all $i \leq k$.*
(2) *χ does not belong to $\tau_1^{\mathbb{R}}(\mathcal{W}^k(X))$ if and only if $b_i(X, \chi) = 0$, for all $i \leq k$.*

Recall that $S(V)$ denotes the intersection of the unit sphere $S(X)$ with a homogeneous subvariety $V \subset H^1(X, \mathbb{R})$. In particular, if $V = \{0\}$, then $S(V) = \emptyset$.

Corollary 4.10 ([33]). *Let X be a CW-complex with finite k-skeleton. Then, for all $i \leq k$,*

$$\Sigma^i(X, \mathbb{Z}) \subseteq S(X) \setminus S(\tau_1^{\mathbb{R}}(\mathcal{W}^i(X))). \tag{4.11}$$

Qualitatively, Corollary 4.10 says that each BNSR set $\Sigma^i(X, \mathbb{Z})$ is contained in the complement of a union of rationally defined great subspheres.

As noted in [33], the above bound is sharp. For example, if X is a nilmanifold, then $\Sigma^i(X, \mathbb{Z}) = S(X)$, while $\mathcal{W}^i(X, \mathbb{C}) = \{1\}$, and so $\tau_1^{\mathbb{R}}(\mathcal{W}^i(X)) = \{0\}$, for all $i \geq 1$. Thus, the inclusion from Corollary 4.10 holds as an equality in this case.

If the space X is (locally) straight, Theorem 2.10 allows us to replace the exponential tangent cone by the corresponding resonance variety.

Corollary 4.11. *If X is locally k-straight, then, for all $i \leq k$,*

$$\Sigma^i(X, \mathbb{Z}) \subseteq S(X) \setminus S(\mathcal{R}^i(X, \mathbb{R})). \tag{4.12}$$

Corollary 4.12 ([33]). *If G is a 1-formal group, then*

$$\Sigma^1(G) \subseteq S(G) \setminus S(\mathcal{R}^1(G, \mathbb{R})). \tag{4.13}$$

5 Relating the Ω-invariants and the Σ-invariants

In this section, we prove Theorem 1.1 from the Introduction, which essentially says the following: if inclusion (3.7) is strict, then inclusion (4.11) is also strict.

5.1 Finiteness properties of abelian covers

Let X be a connected CW-complex with finite 1-skeleton, and let $\widetilde{X}$ be the universal cover, with group of deck transformation $G = \pi_1(X, x_0)$.

Let $p\colon X^\nu \to X$ be a (connected) regular, free abelian cover, associated to an epimorphism $\nu\colon G \twoheadrightarrow \mathbb{Z}^r$. Fix a basepoint $\tilde{x}_0 \in p^{-1}(x_0)$, and identify the fundamental group $\pi_1(X^\nu, \tilde{x}_0)$ with $N = \ker(\nu)$. Note that the universal cover $\widetilde{X^\nu}$ is homeomorphic to $\widetilde{X}$.

Finally, set

$$S(X, X^\nu) = \{\chi \in S(X) \mid p^*(\chi) = 0\}. \tag{5.1}$$

Then $S(X, X^\nu)$ is a great sphere of dimension $r - 1$. In fact,

$$S(X, X^\nu) = S(X) \cap (P_\nu \otimes \mathbb{R}), \tag{5.2}$$

where recall P_ν is the r-plane in $H^1(X, \mathbb{Q})$ determined by ν via the correspondence from (3.1). Moreover, under the identification $S(X) = S(G)$, the subsphere $S(X, X^\nu)$ corresponds to $S(G, N)$.

Proposition 5.1. *Let X be a connected CW-complex with finite k-skeleton, and let $p\colon X^\nu \to X$ be a regular, free abelian cover. Then the chain complex $C_*(\widetilde{X^\nu}, \mathbb{Z})$ is of finite k-type over $\mathbb{Z}N$ if and only if $S(X, X^\nu) \subseteq \Sigma^k(X, \mathbb{Z})$.*

Proof. Consider the free $\mathbb{Z}G$-chain complex $C = C_*(\widetilde{X}, \mathbb{Z})$. Upon restricting scalars to the subring $\mathbb{Z}N \subset \mathbb{Z}G$, the resulting $\mathbb{Z}N$-chain complex may be identified with $C_*(\widetilde{X^\nu}, \mathbb{Z})$. The desired conclusion follows from Theorem 4.5. □

5.2 Upper bounds for the Σ- and Ω-invariants

Now assume X has finite k-skeleton, for some $k \geq 1$. As we just saw, great subspheres in the Σ-invariants indicate directions in which the corresponding covers have good finiteness properties. The next result compares the rational points on such subspheres with the corresponding Ω-invariants.

Proposition 5.2. *Let $P \subseteq H^1(X, \mathbb{Q})$ be an r-dimensional linear subspace. If the unit sphere in P is included in $\Sigma^k(X, \mathbb{Z})$, then P belongs to $\Omega^k_r(X)$.*

Proof. Realize $P = P_\nu$, for some epimorphism $\nu\colon G \twoheadrightarrow \mathbb{Z}^r$, and let $X^\nu \to X$ be the corresponding $\mathbb{Z}^r$-cover. By assumption, $S(X, X^\nu) \subseteq \Sigma^k(X, \mathbb{Z})$. Thus, by Proposition 5.1, the chain complex $C_*(\widetilde{X^\nu}, \mathbb{Z})$ is of finite k-type over $\mathbb{Z}N$.

Now, in view of Remark 4.3, the chain complex $C_*(X^\nu, \mathbb{Z}) = C_*(\widetilde{X^\nu}, \mathbb{Z}) \otimes_{\mathbb{Z}N} \mathbb{Z}$ is of finite k-type over $\mathbb{Z}$. Therefore, $b_i(X^\nu) < \infty$, for all $i \leq k$. Hence, $P \in \Omega^k_r(X)$. □

Corollary 5.3. *Let $P \subseteq H^1(X, \mathbb{Q})$ be a linear subspace. If $S(P) \subseteq \Sigma^k(X, \mathbb{Z})$, then $P \cap \tau_1^{\mathbb{Q}}(\mathcal{W}^k(X)) = \{0\}$.*

Proof. Set $r = \dim_{\mathbb{Q}} P$. From Corollary 3.3, we know that $\Omega^k_r(X)$ is contained in the complement in $\mathrm{Gr}_r(H^1(X, \mathbb{Q}))$ to the incidence variety $\sigma_r(\tau_1^{\mathbb{Q}}(\mathcal{W}^k(X)))$. The conclusion follows from Proposition 5.2. □

We are now in a position to state and prove the main result of this section, which is simply a restatement of Theorem 1.1 from the Introduction.

Theorem 5.4. *If X is a connected CW-complex with finite k-skeleton, then, for all $r \geq 1$,*

$$\Sigma^k(X, \mathbb{Z}) = S(\tau_1^{\mathbb{R}}(\mathcal{W}^k(X)))^{\complement} \implies \Omega^k_r(X) = \sigma_r(\tau_1^{\mathbb{Q}}(\mathcal{W}^k(X)))^{\complement}. \quad (5.3)$$

Proof. From Corollary 3.3, we know that $\Omega^k_r(X) \subseteq \sigma_r(\tau_1^{\mathbb{Q}}(\mathcal{W}^k(X)))^{\complement}$. Suppose this inclusion is strict. There is then an r-dimensional linear subspace $P \subseteq H^1(X, \mathbb{Q})$ such that

(1) $P \cap \tau_1^{\mathbb{Q}}(\mathcal{W}^k(X)) = \{0\}$, and
(2) $P \notin \Omega^k_r(X)$.

By supposition (2) and Proposition 5.2, we must have $S(P) \not\subseteq \Sigma^k(X, \mathbb{Z})$. Thus, there exists an element $\chi \in S(P)$ such that $\chi \notin \Sigma^k(X, \mathbb{Z})$.

Now, supposition (1) and the fact that $\chi \in P$ imply that $\chi \notin \tau_1^{\mathbb{Q}}(\mathcal{W}^k(X))$. Since χ belongs to $H^1(X, \mathbb{Q})$, we infer that $\chi \notin \tau_1^{\mathbb{R}}(\mathcal{W}^k(X))$.

We have shown that $\Sigma^k(X, \mathbb{Z}) \subsetneq S(\tau_1^{\mathbb{R}}(\mathcal{W}^k(X)))^{\complement}$, a contradiction. □

Recall now that the incidence variety $\sigma_r(\tau_1^{\mathbb{Q}}(\mathcal{W}^k(X)))$ is a Zariski closed subset of the Grassmannian $\mathrm{Gr}_r(H^1(X, \mathbb{Q}))$. Recall also that the Dwyer-Fried sets $\Omega^k_1(X)$ are Zariski open, but that $\Omega^k_r(X)$ is not necessarily open, if $1 < r < b_1(X)$. We thus have the following immediate corollary.

Corollary 5.5. *Suppose there is an integer $r \geq 2$ such that $\Omega^k_r(X)$ is* not *Zariski open in* $\mathrm{Gr}_r(H^1(X, \mathbb{Q}))$. *Then $\Sigma^k(X, \mathbb{Z}) \neq S(\tau_1^{\mathbb{R}}(\mathcal{W}^k(X)))^{\complement}$.*

5.3 The straight and formal settings

When the space X is locally k-straight, we may replace in the above the exponential tangent cone to the k-th characteristic variety of X by the corresponding resonance variety.

Corollary 5.6. *Let X be a locally k-straight space. Then, for all $r \geq 1$,*

$$\begin{aligned}\Sigma^k(X,\mathbb{Z}) &= S(X) \setminus S(\mathcal{R}^k(X,\mathbb{R}))\\ &\Longrightarrow \Omega_r^k(X) = \mathrm{Gr}_r(H^1(X,\mathbb{Q})) \setminus \sigma_r(\mathcal{R}^k(X,\mathbb{Q})).\end{aligned}$$

Proof. Follows at once from Theorems 5.4 and 2.10. □

Recalling now that every 1-formal space is locally 1-straight (*cf.* Theorem 2.8), we derive the following corollary.

Corollary 5.7. *Let X be a 1-formal space. Then, for all $r \geq 1$,*

$$\begin{aligned}\Sigma^1(X,\mathbb{Z}) &= S(X) \setminus S(\mathcal{R}^1(X,\mathbb{R}))\\ &\Longrightarrow \Omega_r^1(X) = \mathrm{Gr}_r(H^1(X,\mathbb{Q})) \setminus \sigma_r(\mathcal{R}^1(X,\mathbb{Q})).\end{aligned}$$

Using Corollary 5.3 and Theorem 2.8, we obtain the following consequences, which partially recover Corollary 4.12.

Corollary 5.8. *Let X be a 1-formal space, and $P \subseteq H^1(X,\mathbb{Q})$ a linear subspace. If the unit sphere in P is included in $\Sigma^1(X)$, then $P \cap \mathcal{R}^1(X,\mathbb{Q}) = \{0\}$.*

Corollary 5.9. *Let G be a 1-formal group, and let $\chi : G \to \mathbb{Z}$ be a non-zero homomorphism. If $\{\pm\chi\} \subseteq \Sigma^1(G)$, then $\chi \notin \mathcal{R}^1(X,\mathbb{Q})$.*

The formality assumption is really necessary in the previous two corollaries. For instance, let X be the Heisenberg nilmanifold, *i.e.*, the S^1-bundle over $S^1 \times S^1$ with Euler number 1. Then $\Sigma^1(X) = S(X) = S^1$, yet $\mathcal{R}^1(X,\mathbb{Q}) = H^1(X,\mathbb{Q}) = \mathbb{Q}^2$.

5.4 Discussion and examples

As we shall see in the last few sections, there are several interesting classes of spaces for which the implication from Theorem 5.4 holds as an equivalence. Nevertheless, as the next two examples show, neither the implication from Theorem 5.4, nor the one from Corollary 5.7 can be reversed, in general. We will come back to this point in Example 6.5.

Example 5.10. Consider the 1-relator group $G = \langle x_1, x_2 \mid x_1 x_2 x_1^{-1} = x_2^2 \rangle$. Clearly, $G_{\mathrm{ab}} = \mathbb{Z}$, and so G is 1-formal and $\mathcal{R}^1(G) = \{0\} \subset \mathbb{C}$. A Fox calculus computation shows that $\mathcal{W}^1(G) = \{1, 2\} \subset \mathbb{C}^\times$; thus, $\Omega_1^1(G) = \{\mathrm{pt}\}$, and so $\Omega_1^1(G) = \sigma_1(\mathcal{R}^1(G, \mathbb{Q}))^{\complement}$. On the other hand, algorithms from [8,9] show that $\Sigma^1(G) = \{-1\}$, whereas $S(\mathcal{R}^1(G, \mathbb{R}))^{\complement} = \{\pm 1\}$.

To see why this is the case, consider the abelianization map, $\mathrm{ab}\colon G \twoheadrightarrow \mathbb{Z}$. Then $G' = \ker(\mathrm{ab})$ is isomorphic to $\mathbb{Z}[1/2]$. Hence, $H_1(G', \mathbb{Q}) = \mathbb{Z}[1/2] \otimes \mathbb{Q} = \mathbb{Q}$, which explains why the character ab belongs to $\Omega_1^1(G)$. On the other hand, the group $\mathbb{Z}[1/2]$ is not finitely generated, which explains why $\{\pm\, \mathrm{ab}\} \not\subseteq \Sigma^1(G)$, although $-\,\mathrm{ab} \in \Sigma^1(G)$.

Example 5.11. Consider the space $X = S^1 \vee \mathbb{RP}^2$, with fundamental group $G = \mathbb{Z} * \mathbb{Z}_2$. As before, X is 1-formal and $\mathcal{R}^1(X) = \{0\} \subset \mathbb{C}$. The maximal free abelian cover X^α, corresponding to the projection $\alpha\colon G \twoheadrightarrow \mathbb{Z}$, is homotopy equivalent to a countably infinite wedge of projective planes. Thus, $b_1(X^\alpha) = 0$, and so $\Omega_1^1(X) = \{\mathrm{pt}\}$, which equals $\sigma_1(\mathcal{R}^1(X, \mathbb{Q}))^{\complement}$. On the other hand, since G splits as a non-trivial free product, $\Sigma^1(X, \mathbb{Z}) = \emptyset$, which does not equal $S(\mathcal{R}^1(X, \mathbb{R}))^{\complement} = S^0$.

Finally, here is an example showing how Corollary 5.5 can be used to prove that the inclusions from (4.11) and (4.6) are proper, in general. The construction is based on an example of Dwyer and Fried [17], as revisited in more detail in [36].

Example 5.12. Let $Y = T^3 \vee S^2$. Then $\pi_1(Y) = \mathbb{Z}^3$, a free abelian group on generators x_1, x_2, x_3, and $\pi_2(Y) = \mathbb{Z}\mathbb{Z}^3$, a free module generated by the inclusion $S^2 \hookrightarrow Y$. Attaching a 3-cell to Y along a map $S^2 \to Y$ representing the element $x_1 - x_2 + 1$ in $\pi_2(Y)$, we obtain a CW-complex X, with $\pi_1(X) = \mathbb{Z}^3$ and $\pi_2(X) = \mathbb{Z}\mathbb{Z}^3/(x_1 - x_2 + 1)$. Identifying $\widehat{\mathbb{Z}^3} = (\mathbb{C}^\times)^3$, we have that $\mathcal{W}^2(X) = \{t \in (\mathbb{C}^\times)^3 \mid t_1 - t_2 + 1 = 0\}$, and thus $\tau_1(\mathcal{W}^2(X)) = \{0\}$.

Making use of Theorem 3.2, we see that $\Omega_2^2(X)$ consists of precisely two points in $\mathrm{Gr}_2(\mathbb{Q}^3) = \mathbb{QP}^2$; in particular, $\Omega_2^2(X)$ is not a Zariski open subset. Corollary 5.5 now shows that $\Sigma^2(X, \mathbb{Z}) \subsetneqq S(\tau_1(\mathcal{W}^2(X)))^{\complement} = S^2$. On the other hand, $\Sigma^2(\mathbb{Z}^3, \mathbb{Z}) = S^2$; thus, $\Sigma^2(X, \mathbb{Z}) \subsetneqq \Sigma^2(\pi_1(X), \mathbb{Z})$.

6 Toric complexes

In this section, we illustrate our techniques on a class of spaces that arise in toric topology, as a basic example of polyhedral products. These "toric complexes" are both straight and formal, so it comes as no surprise that both their Ω-invariants and their Σ-invariants are closely related to the resonance varieties.

6.1 Toric complexes and right-angled Artin groups

Let L be a simplicial complex with n vertices, and let T^n be the n-torus, with the standard cell decomposition, and with basepoint $*$ at the unique 0-cell.

The *toric complex* associated to L, denoted T_L, is the union of all subcomplexes of the form $T^\sigma = \{x \in T^n \mid x_i = * \text{ if } i \notin \sigma\}$, where σ runs through the simplices of L. Clearly, T_L is a connected CW-complex; its k-cells are in one-to-one correspondence with the $(k-1)$-simplices of L.

Denote by V the set of 0-cells of L, and by E the set of 1-cells of L. The fundamental group, $G_L = \pi_1(T_L)$, is the right-angled Artin group associated to the graph $\Gamma = (\mathsf{V}, \mathsf{E})$, with a generator v for each vertex $v \in \mathsf{V}$, and a commutation relation $vw = wv$ for each edge $\{v, w\} \in \mathsf{E}$.

Much is known about toric complexes and their fundamental groups. For instance, the group G_L has as classifying space the toric complex T_Δ, where $\Delta = \Delta_L$ is the flag complex of L, *i.e.*, the maximal simplicial complex with 1-skeleton equal to that of L. Moreover, the homology groups of T_L are torsion-free, while the cohomology ring $H^*(T_L, \mathbb{Z})$ is isomorphic to the exterior Stanley-Reisner ring of L, with generators the dual classes $v^* \in H^1(T_L, \mathbb{Z})$, and relations the monomials corresponding to the missing faces of L. Finally, all toric complexes are formal spaces.

For more details and references on all this, we refer to [30, 31, 33, 37].

6.2 Jump loci and Ω-invariants

The resonance and characteristic varieties of right-angled Artin groups and toric complexes were studied in [30] and [16], with the complete computation achieved in [31]. We recall here those results, in a form suited for our purposes.

Let $\Bbbk$ be a coefficient field. Fixing an ordering on the vertex set V allows us to identify $H^1(T_L, \Bbbk^\times)$ with the algebraic torus $(\Bbbk^\times)^{\mathsf{V}} = (\Bbbk^\times)^n$ and $H^1(T_L, \Bbbk)$ with the vector space $\Bbbk^{\mathsf{V}} = \Bbbk^n$. Each subset $\mathsf{W} \subseteq \mathsf{V}$ gives rise to an algebraic subtorus $(\Bbbk^\times)^{\mathsf{W}} \subset (\Bbbk^\times)^{\mathsf{V}}$ and a coordinate subspace $\Bbbk^{\mathsf{W}} \subset \Bbbk^{\mathsf{V}}$.

In what follows, we denote by L_{W} the subcomplex induced by L on W, and by $\operatorname{lk}_K(\sigma)$ the link of a simplex $\sigma \in L$ in a subcomplex $K \subseteq L$.

Theorem 6.1 ([31]). *Let L be a simplicial complex on vertex set V. Then, for all $i \geq 1$,*

$$\mathcal{V}^i(T_L, \Bbbk) = \bigcup_{\mathsf{W}} (\Bbbk^\times)^{\mathsf{W}} \quad \textit{and} \quad \mathcal{R}^i(T_L, \Bbbk) = \bigcup_{\mathsf{W}} \Bbbk^{\mathsf{W}}, \tag{6.1}$$

where, in both cases, the union is taken over all subsets $\mathsf{W} \subseteq \mathsf{V}$ *for which there is a simplex* $\sigma \in L_{\mathsf{V}\setminus\mathsf{W}}$ *and an index* $j \leq i$ *such that* $\widetilde{H}_{j-1-|\sigma|}(\mathrm{lk}_{L_{\mathsf{W}}}(\sigma), \Bbbk) \neq 0$.

In degree 1, the resonance formula takes a simpler form, already noted in [30]. Clearly, $\mathcal{R}^1(T_L, \Bbbk)$ depends only on the 1-skeleton $\Gamma = L^{(1)}$; moreover, $\mathcal{R}^1(T_L, \Bbbk) = \bigcup_{\mathsf{W}} \Bbbk^{\mathsf{W}}$, where the union is taken over all maximal subsets $\mathsf{W} \subseteq \mathsf{V}$ for which the induced graph Γ_{W} is disconnected.

As a consequence of Theorem 6.1, we see that every toric complex is a straight space. Theorem 3.6 (2), then, allows us to determine the Dwyer–Fried invariants of such spaces.

Corollary 6.2 ([33, 37]). *Let L be a simplicial complex on vertex set* V. *Then, for all $i, r \geq 1$,*

$$\Omega^i_r(T_L) = \mathrm{Gr}_r(\mathbb{Q}^{\mathsf{V}}) \setminus \sigma_r(\mathcal{R}^i(T_L, \mathbb{Q})). \tag{6.2}$$

6.3 Σ-invariants

In [3], Bestvina and Brady considered the "diagonal" homomorphism $\nu\colon G_L \to \mathbb{Z}$, $v \mapsto 1$, and the finiteness properties of the corresponding subgroup, $N_L = \ker(\nu)$. One of the main results of [3] determines the maximal integer k for which ν belongs to $\Sigma^k(G_L, \mathbb{Z})$.

The picture was completed by Meier, Meinert, and VanWyk [29] and by Bux and Gonzalez [10], who computed explicitly the Bieri-Neumann-Strebel-Renz invariants of right-angled Artin groups.

Theorem 6.3 ([10, 29]). *Let L be a simplicial complex, and let Δ be the associated flag complex. Let $\chi \in S(G_L)$ be a non-zero homomorphism, with support* $\mathsf{W} = \{v \in \mathsf{V} \mid \chi(v) \neq 0\}$, *and let $\Bbbk = \mathbb{Z}$ or a field. Then, $\chi \in \Sigma^i(G_L, \Bbbk)$ if and only if*

$$\widetilde{H}_j(\mathrm{lk}_{\Delta_{\mathsf{W}}}(\sigma), \Bbbk) = 0, \tag{6.3}$$

for all $\sigma \in \Delta_{\mathsf{V}\setminus\mathsf{W}}$ *and* $-1 \leq j \leq i - \dim(\sigma) - 2$.

We would like now to compare the BNSR invariants of a toric complex T_L to the resonance varieties of T_L. Using the fact that toric complexes are straight spaces, Corollary 4.11 gives

$$\Sigma^i(T_L, \mathbb{Z}) \subseteq S(T_L) \setminus S(\mathcal{R}^i(T_L, \mathbb{R})). \tag{6.4}$$

For right-angled Artin groups, we can say more. Comparing the description of the Σ-invariants of the group G_L given in Theorem 6.3 to that of the resonance varieties of the space $T_\Delta = K(G_L, 1)$ given in Theorem 6.1, yields the following result.

Corollary 6.4 ([33]). *Let G_L be a right-angled Artin group. For each $i \geq 0$, the following hold.*

(1) $\Sigma^i(G_L, \mathbb{R}) = S(\mathcal{R}^i(G_L, \mathbb{R}))^{\complement}$.
(2) $\Sigma^i(G_L, \mathbb{Z}) = S(\mathcal{R}^i(G_L, \mathbb{R}))^{\complement}$, *provided that, for every $\sigma \in \Delta$, and every $\mathsf{W} \subseteq \mathsf{V}$ with $\sigma \cap W = \emptyset$, the groups $\widetilde{H}_j(\mathrm{lk}_{\Delta_{\mathsf{W}}}(\sigma), \mathbb{Z})$ are torsion-free, for all $j \leq i - \dim(\sigma) - 2$.*

The torsion-freeness condition from Corollary 6.4 (2) is always satisfied in degree $i = 1$. Thus,

$$\Sigma^1(G_L, \mathbb{Z}) = S(\mathcal{R}^1(G_L, \mathbb{R}))^{\complement}, \tag{6.5}$$

an equality already proved (by different methods) in [30]. Nevertheless, the condition is not always satisfied in higher degrees, thus leading to situations where the equality from Corollary 6.4 (2) fails. The next example (extracted from [33]) illustrates this phenomenon, while also showing that the implication from Theorem 5.4 cannot always be reversed, even for right-angled Artin groups.

Example 6.5. Let Δ be a flag triangulation of the real projective plane, $\mathbb{RP}^2$, and let $\nu\colon G_\Delta \to \mathbb{Z}$ be the diagonal homomorphism. Then $\nu \notin \Sigma^2(G_\Delta, \mathbb{Z})$, even though $\nu \in \Sigma^2(G_\Delta, \mathbb{R})$. Consequently,

$$\Sigma^2(G_\Delta, \mathbb{Z}) \subsetneqq S(\mathcal{R}^2(G_\Delta, \mathbb{R}))^{\complement}, \tag{6.6}$$

although $\Omega^2_r(G_\Delta) = \sigma_r(\mathcal{R}^2(T_\Delta, \mathbb{Q}))^{\complement}$, for all $r \geq 1$.

7 Quasi-projective varieties

We now discuss the cohomology jumping loci, the Dwyer-Fried invariants, and the Bieri-Neumann-Strebel-Renz invariants of smooth, complex projective and quasi-projective varieties.

7.1 Complex algebraic varieties

A smooth, connected manifold X is said to be a (*smooth*) *quasi-projective variety* if there is a smooth, complex projective variety $\overline{X}$ and a normal-crossings divisor D such that $X = \overline{X} \setminus D$. By a well-known result of Deligne, each cohomology group of a quasi-projective variety X admits a mixed Hodge structure. This puts definite constraints on the topology of such varieties.

For instance, if X admits a non-singular compactification $\overline{X}$ with $b_1(\overline{X}) = 0$, the weight 1 filtration on $H^1(X, \mathbb{C})$ vanishes; in turn, by work of Morgan, this implies the 1-formality of X. Thus, as noted by

Kohno, if X is the complement of a hypersurface in $\mathbb{CP}^n$, then $\pi_1(X)$ is 1-formal. In general, though, quasi-projective varieties need not be 1-formal.

If X is actually a (compact, smooth) projective variety, then a stronger statement holds: as shown by Deligne, Griffiths, Morgan, and Sullivan, such a manifold (and, more generally, a compact Kähler manifold) is formal. In general, though, quasi-projective varieties are not formal, even if their fundamental groups are 1-formal.

Example 7.1. Let $T = E^n$ be the n-fold product of an elliptic curve $E = \mathbb{C}/\mathbb{Z}\oplus\mathbb{Z}$. The closed form $\frac{1}{2}\sqrt{-1}\sum_{i=1}^n dz_i \wedge d\bar{z}_i = \sum_{i=1}^n dx_i \wedge dy_i$ defines a cohomology class ω in $H^{1,1}(T) \cap H^2(T, \mathbb{Z})$. By the Lefschetz theorem on $(1,1)$-classes (see [25, page 163]), ω can be realized as the first Chern class of an algebraic line bundle $L \to T$.

Let X be the complement of the zero-section of L; then X is a connected, smooth, quasi-projective variety. Moreover, X deform-retracts onto N, the total space of the circle bundle over the torus $T = (S^1)^{2n}$ with Euler class ω. Clearly, the Heisenberg-type nilmanifold N is not a torus, and thus it is not formal. In fact, as shown by Măcinic in [28, Remark 5.4], the manifold N is $(n-1)$-formal, but not n-formal. Thus, if $n = 1$, the variety X is not 1-formal, whereas if $n > 1$, the variety X is 1-formal, but not formal.

7.2 Cohomology jump loci

The existence of mixed Hodge structures on the cohomology groups of connected, smooth, complex quasi-projective varieties also puts definite constraints on the nature of their cohomology support loci.

The structure of the characteristic varieties of such spaces (and, more generally, Kähler and quasi-Kähler manifolds) was determined through the work of Beauville, Green and Lazarsfeld, Simpson, Campana, and Arapura in the 1990s. Further improvements and refinements have come through the recent work of Budur, Libgober, Dimca, Artal-Bartolo, Cogolludo, and Matei. We summarize these results, essentially in the form proved by Arapura, but in the simplified (and slightly updated) form we need them here.

Theorem 7.2 ([1]). *Let $X = \overline{X} \setminus D$, where $\overline{X}$ is a smooth, projective variety and D is a normal-crossings divisor.*

(1) *If either $D = \emptyset$ or $b_1(\overline{X}) = 0$, then each characteristic variety $\mathcal{V}^i(X)$ is a finite union of unitary translates of algebraic subtori of $H^1(X, \mathbb{C}^\times)$.*

(2) *In degree $i = 1$, the condition that $b_1(\overline{X}) = 0$ if $D \neq \emptyset$ may be lifted. Furthermore, each positive-dimensional component of $\mathcal{V}^1(X)$ is of the form $\rho \cdot T$, with T an algebraic subtorus, and ρ a* torsion *character.*

For instance, if C is a connected, smooth complex curve with $\chi(C) < 0$, then $\mathcal{V}^1(C) = H^1(C, \mathbb{C}^\times)$. More generally, if X is a smooth, quasi-projective variety, then every positive-dimensional component of $\mathcal{V}^1(X)$ arises by pullback along a suitable pencil. More precisely, if $\rho \cdot T$ is such a component, then $T = f^*(H^1(C, \mathbb{C}^\times))$, for some curve C, and some holomorphic, surjective map $f : X \to C$ with connected generic fiber.

In the presence of 1-formality, the quasi-projectivity of X also imposes stringent conditions on the degree 1 resonance varieties. Theorems 7.2 and 2.8 yield the following characterization of these varieties.

Corollary 7.3 ([16]). *Let X be a 1-formal, smooth, quasi-projective variety. Then $\mathcal{R}^1(X)$ is a finite union of rationally defined linear subspaces of $H^1(X, \mathbb{C})$.*

In fact, much more is proved in [16] about those subspaces. For instance, any two of them intersect only at 0, and the restriction of the cup-product map $H^1(X, \mathbb{C}) \wedge H^1(X, \mathbb{C}) \to H^2(X, \mathbb{C})$ to any one of them has rank equal to either 0 or 1.

Corollary 7.4 ([37]). *If X is a 1-formal, smooth, quasi-projective variety, then X is locally 1-straight. Moreover, X is 1-straight if and only if $\mathcal{W}^1(X)$ contains no positive-dimensional translated subtori.*

7.3 Ω-invariants

The aforementioned structural results regarding the cohomology jump loci of smooth, quasi-projective varieties inform on the Dwyer-Fried sets of such varieties. For instance, Theorem 7.2 together with Proposition 3.4 yield the following corollary.

Corollary 7.5. *Let $X = \overline{X} \setminus D$ be a smooth, quasi-projective variety with $D = \emptyset$ or $b_1(\overline{X}) = 0$. If $\mathcal{W}^i(X)$ contains no positive-dimensional translated subtori, then $\Omega^i_r(X) = \sigma_r(\tau_1^{\mathbb{Q}}(\mathcal{W}^i(X)))^{\mathsf{C}}$, for all $r \geq 1$.*

Likewise, Corollary 7.4 together with Theorem 3.6 yield the following corollary.

Corollary 7.6 ([37]). *Let X be a 1-formal, smooth, quasi-projective variety. Then*:

(1) $\Omega^1_1(X) = \mathbb{P}(\mathcal{R}^1(X, \mathbb{Q}))^{\complement}$ *and* $\Omega^1_r(X) \subseteq \sigma_r(\mathcal{R}^1(X, \mathbb{Q}))^{\complement}$, *for* $r \geq 2$.
(2) *If* $\mathcal{W}^1(X)$ *contains no positive-dimensional translated subtori, then* $\Omega^1_r(X) = \sigma_r(\mathcal{R}^1(X, \mathbb{Q}))^{\complement}$, *for all* $r \geq 1$.

If the characteristic variety $\mathcal{W}^1(X)$ contains positive-dimensional translated components, the resonance variety $\mathcal{R}^1(X, \mathbb{Q})$ may fail to determine all the Dwyer-Fried sets $\Omega^1_r(X)$. This phenomenon is made concrete by the following result.

Theorem 7.7 ([37]). *Let X be a 1-formal, smooth, quasi-projective variety. Suppose $\mathcal{W}^1(X)$ has a 1-dimensional component not passing through 1, while $\mathcal{R}^1(X)$ has no codimension-1 components. Then $\Omega^1_2(X)$ is strictly contained in* $\mathrm{Gr}_2(H^1(X, \mathbb{Q})) \setminus \sigma_2(\mathcal{R}^1(X, \mathbb{Q}))$.

7.4 Σ-invariants

The cohomology jump loci of smooth, quasi-projective varieties also inform on the Bieri-Neumann-Strebel sets of such varieties. For instance, using Corollary 5.7, we may identify a class of 1-formal, quasi-projective varieties for which inclusion (4.13) is strict.

Corollary 7.8. *Let X be a 1-formal, smooth, quasi-projective variety. Suppose $\mathcal{W}^1(X)$ has a 1-dimensional component not passing through 1, while $\mathcal{R}^1(X)$ has no codimension-1 components. Then $\Sigma^1(X)$ is strictly contained in $S(X) \setminus S(\mathcal{R}^1(X, \mathbb{R}))$.*

We shall see in Example 9.11 a concrete variety to which this corollary applies.

In the case of smooth, complex projective varieties (or, more generally, compact Kähler manifolds), a different approach is needed in order to show that inclusion (4.13) may be strict. Indeed, by Theorem 7.2, all components of $\mathcal{W}^1(X)$ are even-dimensional, so Corollary 7.8 does not apply.

On the other hand, as shown by Delzant in [13], the BNS invariant of a compact Kähler manifold M is determined by the pencils supported by M.

Theorem 7.9 ([13]). *Let M be a compact Kähler manifold. Then*

$$\Sigma^1(M) = S(M) \setminus \bigcup\nolimits_{\alpha} S(f^*_{\alpha}(H^1(C_{\alpha}, \mathbb{R}))), \tag{7.1}$$

where the union is taken over those pencils $f_{\alpha}\colon M \to C_{\alpha}$ with the property that either $\chi(C_{\alpha}) < 0$, or $\chi(C_{\alpha}) = 0$ and f_{α} has some multiple fiber.

This theorem, together with results from [16], yields the following characterization of those compact Kähler manifolds M for which the inclusion from Corollary 4.12 holds as equality.

Theorem 7.10 ([33]). *Let M be a compact Kähler manifold. Then $\Sigma^1(M) = S(\mathcal{R}^1(M,\mathbb{R}))^{\complement}$ if and only if there is no pencil $f\colon M \to E$ onto an elliptic curve E such that f has multiple fibers.*

7.5 Examples and discussion

As noted in [33, Remark 16.6], a general construction due to Beauville [2] shows that equality does not always hold in Theorem 7.10. More precisely, if N is a compact Kähler manifold on which a finite group π acts freely, and $p\colon C \to E$ is a ramified, regular π-cover over an elliptic curve, with at least one ramification point, then the quotient $M = (C \times N)/\pi$ is a compact Kähler manifold admitting a pencil $f\colon M \to E$ with multiple fibers.

An example of this construction is given in [15]. Let C be a Fermat quartic in $\mathbb{CP}^2$, viewed as a 2-fold branched cover of E, and let N be a simply-connected compact Kähler manifold admitting a fixed-point free involution, for instance, a Fermat quartic in $\mathbb{CP}^3$, viewed as a 2-fold unramified cover of the Enriques surface. Then, the Kähler manifold $M = (C \times N)/\mathbb{Z}_2$ admits a pencil with base E, having four multiple fibers, each of multiplicity 2; thus, $\Sigma^1(M,\mathbb{Z}) = \emptyset$. Moreover, direct computation shows that $\mathcal{R}^1(M) = \{0\}$, and so $\Sigma^1(M,\mathbb{Z}) \subsetneqq S(\mathcal{R}^1(M,\mathbb{R}))^{\complement}$.

We provide here another example, in the lowest possible dimension, using a complex surface studied by Catanese, Ciliberto, and Mendes Lopes in [12]. For this manifold M, the resonance variety does not vanish, yet $\Sigma^1(M,\mathbb{Z})$ is strictly contained in the complement of $S(\mathcal{R}^1(M,\mathbb{R}))$. We give an alternate explanation of this fact which is independent of Theorems 7.9 and 7.10, but relies instead on results from [38] and on Corollary 5.7.

Example 7.11. Let C_1 be a (smooth, complex) curve of genus 2 with an elliptic involution σ_1 and let C_2 be a curve of genus 3 with a free involution σ_2. Then $\Sigma_1 = C_1/\sigma_1$ is a curve of genus 1, and $\Sigma_2 = C_2/\sigma_2$ is a curve of genus 2.

Now let $M = (C_1 \times C_2)/\mathbb{Z}_2$, where $\mathbb{Z}_2$ acts via the involution $\sigma_1 \times \sigma_2$. Then M is a smooth, complex projective surface. Projection onto the first coordinate yields a pencil $f_1\colon M \to \Sigma_1$ with two multiple fibers, each of multiplicity 2, while projection onto the second coordinate defines a smooth fibration $f_2\colon M \to \Sigma_2$. By Theorem 7.10, we have that $\Sigma^1(M,\mathbb{Z}) \subsetneqq S(\mathcal{R}^1(M,\mathbb{R}))^{\complement}$.

Here is an alternate explanation. Using the fact that $H_1(M,\mathbb{Z}) = \mathbb{Z}^6$, we may identify $H^1(M,\mathbb{C}^\times) = (\mathbb{C}^\times)^6$. In [38], we showed that

$$\mathcal{V}^1(M) = \{t_4 = t_5 = t_6 = 1,\ t_3 = -1\} \cup \{t_1 = t_2 = 1\}, \tag{7.2}$$

with the two components corresponding to the pencils f_1 and f_2, respectively, from which we inferred that the set $\Omega_2^1(M)$ is not open, not even in the usual topology on $\mathrm{Gr}_2(\mathbb{Q}^6)$. Now, from (7.2), we also see that $\mathcal{R}^1(M) = \{x_1 = x_2 = 0\}$. Since $\Omega_2^1(M)$ is not open, it must be a proper subset of $\sigma_2(\mathcal{R}^1(M,\mathbb{Q}))^{\complement}$. In view of Corollary 5.7, we conclude once again that $\Sigma^1(M,\mathbb{Z}) \subsetneq S(\mathcal{R}^1(M,\mathbb{R}))^{\complement}$.

8 Configuration spaces

We now consider in more detail a particularly interesting class of quasi-projective varieties, obtained by deleting the "fat diagonal" from the n-fold Cartesian product of a smooth, complex algebraic curve.

8.1 Ordered configurations on algebraic varieties

A construction due to Fadell and Neuwirth associates to a space X and a positive integer n the space of ordered configurations of n points in X,

$$F(X,n) = \{(x_1,\dots,x_n) \in X^n \mid x_i \neq x_j \text{ for } i \neq j\}. \tag{8.1}$$

The most basic example is the configuration space of n ordered points in $\mathbb{C}$, which is a classifying space for P_n, the pure braid group on n strings, whose cohomology ring was computed by Arnol'd in the late 1960s.

The E_2-term of the Leray spectral sequence for the inclusion $F(X,n) \hookrightarrow X^n$ was described concretely by Cohen and Taylor in the late 1970s. If X is a smooth, complex projective variety of dimension m, then, as shown by Totaro in [39], the Cohen-Taylor spectral sequence collapses at the E_{m+1}-term, and $H^*(F(X,n),\mathbb{C}) = E_{m+1}$, as graded algebras.

Particularly interesting is the case of a Riemann surface Σ_g. The ordered configuration space $F(\Sigma_g,n)$ is a classifying space for $P_n(\Sigma_g)$, the pure braid group on n strings of the underlying surface. In [4], Bezrukavnikov gave explicit presentations for the Malcev Lie algebras $\mathfrak{m}_{g,n} = \mathfrak{m}(P_n(\Sigma_g))$, from which he concluded that the pure braid groups on surfaces are 1-formal for $g > 1$ or $g = 1$ and $n \leq 2$, but not 1-formal for $g = 1$ and $n \geq 3$. The non-1-formality of the groups $P_n(\Sigma_1)$, $n \geq 3$, is also established in [16], by showing that the tangent cone formula (2.14) fails in this situation (see Example 8.2 below for the case $n = 3$).

Remark 8.1. In [11, Proposition 5], Calaque, Enriquez, and Etingof prove that $P_n(\Sigma_1)$ is formal, for all $n \geq 1$. But the notion of formality that these authors use is weaker than the usual notion of 1-formality: their result is that $\mathfrak{m}_{1,n}$ is isomorphic as a filtered Lie algebra with the completion (with respect to the bracket length filtration) of the associated graded Lie algebra, $\mathrm{gr}(\mathfrak{m}_{1,n})$. The failure of 1-formality comes from the fact that $\mathrm{gr}(\mathfrak{m}_{1,n})$ is *not* a quadratic Lie algebra, for $n \geq 3$.

8.2 Ordered configurations on the torus

For the reasons outlined above, it makes sense to look more carefully at the configuration spaces of an elliptic curve Σ_1. The resonance varieties $\mathcal{R}^1(F(\Sigma_1, n))$ were computed in [16], while the positive-dimensional components of $\mathcal{V}^1(F(\Sigma_1, n))$ were determined in [14].

Since $\Sigma_1 = S^1 \times S^1$ is a topological group, the space $F(\Sigma_1, n)$ splits up to homeomorphism as a direct product, $F(\Sigma_1', n-1) \times \Sigma_1$, where Σ_1' denotes Σ_1 with the identity removed. Thus, for all practical purposes, it is enough to consider the space $F(\Sigma_1', n-1)$. For the sake of concreteness, we will work out in detail the case $n = 3$; the general case may be treated similarly.

Example 8.2. Let $X = F(\Sigma_1', 2)$ be the configuration space of 2 labeled points on a punctured torus. The cohomology ring of X is the exterior algebra on generators a_1, a_2, b_1, b_2 in degree 1, modulo the ideal spanned by the forms $a_1b_2 + a_2b_1$, a_1b_1, and a_2b_2. The first resonance variety is an irreducible quadric hypersurface in $\mathbb{C}^4$, given by

$$\mathcal{R}^1(X) = \{x_1y_2 - x_2y_1 = 0\}.$$

Corollary 7.3, then, shows that X is not 1-formal.

The first characteristic variety of X consists of three 2-dimensional algebraic subtori of $(\mathbb{C}^\times)^4$:

$$\mathcal{V}^1(X) = \{t_1 = t_2 = 1\} \cup \{s_1 = s_2 = 1\} \cup \{t_1s_1 = t_2s_2 = 1\}.$$

These three subtori arise from the fibrations $F(\Sigma_1', 2) \to \Sigma_1'$ obtained by sending a point (z_1, z_2) to z_1, z_2, and $z_1z_2^{-1}$, respectively. It follows that $\tau_1(\mathcal{V}^1(X)) = \mathrm{TC}_1(\mathcal{V}^1(X))$, but both types of tangent cones are properly contained in the resonance variety $\mathcal{R}^1(X)$. Moreover, the characteristic subspace arrangement $\mathcal{C}_1(X)$ consists of three, pairwise transverse planes in $\mathbb{Q}^4$, namely,

$$L_1 = \{x_1 = x_2 = 0\},\ L_2 = \{y_1 = y_2 = 0\},\ L_3 = \{x_1 + y_1 = x_2 + y_2 = 0\}.$$

By Proposition 3.4, the Dwyer-Fried sets $\Omega^1_r(X)$ are obtained by removing from $\mathrm{Gr}_r(\mathbb{Q}^4)$ the Schubert varieties $\sigma_r(L_1)$, $\sigma_r(L_2)$, and $\sigma_r(L_3)$. We treat each rank r separately.

- When $r = 1$, the set $\Omega^1_1(X)$ is the complement in $\mathbb{QP}^3$ of the three projective lines defined by L_1, L_2, and L_3.
- When $r = 2$, the Grassmannian $\mathrm{Gr}_2(\mathbb{Q}^4)$ is the quadric hypersurface in $\mathbb{QP}^5$ given in Plücker coordinates by the equation $p_{12}p_{34} - p_{13}p_{24} + p_{23}p_{14} = 0$. The set $\Omega^1_2(X)$, then, is the complement in $\mathrm{Gr}_2(\mathbb{Q}^4)$ of the variety cut out by the hyperplanes $p_{12} = 0$, $p_{34} = 0$, and $p_{12} - p_{23} + p_{14} + p_{34} = 0$.
- When $r \geq 3$, the set $\Omega^1_r(X)$ is empty.

Finally, by Corollary 4.10, the BNS set $\Sigma^1(X, \mathbb{Z})$ is included in the complement in S^3 of the three great circles cut out by the real planes spanned by L_1, L_2, and L_3, respectively. It would be interesting to know whether this inclusion is actually an equality.

9 Hyperplane arrangements

We conclude with another interesting class of quasi-projective varieties, obtained by deleting finitely many hyperplanes from a complex affine space.

9.1 Complement and intersection lattice

A (central) hyperplane arrangement $\mathcal{A}$ is a finite collection of codimension 1 linear subspaces in a complex affine space $\mathbb{C}^\ell$. A defining polynomial for $\mathcal{A}$ is the product $Q(\mathcal{A}) = \prod_{H\in\mathcal{A}} \alpha_H$, where $\alpha_H\colon \mathbb{C}^\ell \to \mathbb{C}$ is a linear form whose kernel is H.

The main topological object associated to an arrangement is its complement, $X(\mathcal{A}) = \mathbb{C}^\ell \setminus \bigcup_{H\in\mathcal{A}} H$. This is a connected, smooth, quasi-projective variety, whose homotopy-type invariants are intimately tied to the combinatorics of the arrangement. The latter is encoded in the intersection lattice, $L(\mathcal{A})$, which is the poset of all (non-empty) intersections of $\mathcal{A}$, ordered by reverse inclusion. The rank of the arrangement, denoted $\mathrm{rk}(\mathcal{A})$, is the codimension of $\bigcap_{H\in\mathcal{A}} H$.

Example 9.1. A familiar example is the rank $\ell - 1$ braid arrangement, consisting of the diagonal hyperplanes $H_{ij} = \{z_i - z_j = 0\}$ in $\mathbb{C}^\ell$. The complement is the configuration space $F(\mathbb{C}, \ell)$, while the intersection lattice is the lattice of partitions of $\{1, \dots, \ell\}$, ordered by refinement.

For a general arrangement $\mathcal{A}$, the cohomology ring of the complement was computed by Brieskorn in the early 1970s, building on work of Arnol'd on the cohomology ring of the braid arrangement. It follows from Brieskorn's work that the space $X(\mathcal{A})$ is formal. In 1980, Orlik and Solomon gave a simple combinatorial description of the ring $H^*(X(\mathcal{A}), \mathbb{Z})$: it is the quotient of the exterior algebra on degree-one classes e_H dual to the meridians around the hyperplanes $H \in \mathcal{A}$, modulo a certain ideal (generated in degrees greater than one) determined by the intersection lattice.

Let $\overline{\mathcal{A}} = \{\mathbb{P}(H)\}_{H \in \mathcal{A}}$ be the projectivization of $\mathcal{A}$, and let $X(\overline{\mathcal{A}})$ be its complement in $\mathbb{CP}^{\ell-1}$. The standard $\mathbb{C}^{\times}$-action on $\mathbb{C}^{\ell}$ restricts to a free action on $X(\mathcal{A})$; the resulting fiber bundle, $\mathbb{C}^{\times} \to X(\mathcal{A}) \to X(\overline{\mathcal{A}})$, is readily seen to be trivial. Under the resulting identification, $X(\mathcal{A}) = X(\overline{\mathcal{A}}) \times \mathbb{C}^{\times}$, the group $H^1(\mathbb{C}^{\times}, \mathbb{Z}) = \mathbb{Z}$ is spanned by the vector $\sum_{H \in \mathcal{A}} e_H \in H^1(X(\mathcal{A}), \mathbb{Z})$.

9.2 Cohomology jump loci

The resonance varieties $\mathcal{R}^i(\mathcal{A}) := \mathcal{R}^i(X(\mathcal{A}), \mathbb{C})$ were first defined and studied by Falk in [19]. Clearly, these varieties depend only on the graded ring $H^*(X(\mathcal{A}), \mathbb{C})$, and thus, only on the intersection lattice $L(\mathcal{A})$.

Now fix a linear ordering on the hyperplanes of $\mathcal{A}$, and identify $H^1(X(\mathcal{A}), \mathbb{C}) = \mathbb{C}^n$, where $n = |\mathcal{A}|$. From the product formula (2.13) for resonance varieties (or from an old result of Yuzvinsky [40]), we see that $\mathcal{R}^i(\mathcal{A})$ is isomorphic to $\mathcal{R}^i(\overline{\mathcal{A}})$, and lies in the hyperplane $x_1 + \cdots + x_n = 0$ inside $\mathbb{C}^n$. Similarly, the characteristic varieties $\mathcal{V}^i(\mathcal{A}) := \mathcal{V}^i(X(\mathcal{A}), \mathbb{C})$ lie in the subtorus $t_1 \cdots t_n = 1$ inside the complex algebraic torus $H^1(X(\mathcal{A}), \mathbb{C}^{\times}) = (\mathbb{C}^{\times})^n$.

In view of the Lefschetz-type theorem of Hamm and Lê, taking a generic two-dimensional section does not change the fundamental group of the complement. Thus, in order to describe the variety $\mathcal{R}^1(\mathcal{A})$, we may assume $\mathcal{A}$ is a affine arrangement of n lines in $\mathbb{C}^2$, for which no two lines are parallel.

The structure of the first resonance variety of an arrangement was worked out in great detail in work of Cohen, Denham, Falk, Libgober, Pereira, Suciu, Yuzvinsky, and many others. It is known that each component of $\mathcal{R}^1(\mathcal{A})$ is a linear subspace in $\mathbb{C}^n$, while any two distinct components meet only at 0. The simplest components of the resonance variety are those corresponding to multiple points of $\mathcal{A}$: if m lines meet at a point, then $\mathcal{R}^1(\mathcal{A})$ acquires an $(m-1)$-dimensional linear subspace. The remaining components (of dimension either 2 or 3), correspond to certain "neighborly partitions" of sub-arrangements of $\mathcal{A}$.

Example 9.2. Let $\mathcal{A}$ be a generic 3-slice of the braid arrangement of rank 3, with defining polynomial $Q(\mathcal{A}) = z_0z_1z_2(z_0 - z_1)(z_0 - z_2)(z_1 - z_2)$. Take a generic plane section, and label the corresponding lines as $1, \dots, 6$. Then, the variety $\mathcal{R}^1(\mathcal{A}) \subset \mathbb{C}^6$ has 4 local components, corresponding to the triple points 124, 135, 236, 456, and one non-local component, corresponding to the neighborly partition $(16|25|34)$.

From Theorem 7.2, we know that $\mathcal{V}^1(\mathcal{A})$ consists of subtori in $(\mathbb{C}^\times)^n$, possibly translated by roots of unity, together with a finite number of torsion points. By Theorem 2.8 — first proved in the context of hyperplane arrangements by Cohen-Suciu and Libgober — we have that $\mathrm{TC}_1(\mathcal{V}^1(\mathcal{A})) = \mathcal{R}^1(\mathcal{A})$. Thus, the components of $\mathcal{V}^1(\mathcal{A})$ passing through the origin are completely determined by $\mathcal{R}^1(\mathcal{A})$, and hence, by $L(\mathcal{A})$: to each linear subspace L in $\mathcal{R}^1(\mathcal{A})$ there corresponds an algebraic subtorus, $T = \exp(L)$, in $\mathcal{V}^1(\mathcal{A})$.

As pointed out in [35], though, the characteristic variety $\mathcal{V}^1(\mathcal{A})$ may contain translated subtori — that is, components not passing through 1. Despite much work since then, it is still not known whether such components are combinatorially determined.

9.3 Upper bounds for the Ω- and Σ-sets

We are now ready to consider the Dwyer-Fried and the Bieri-Neumann-Strebel-Renz invariants associated to a hyperplane arrangement $\mathcal{A}$. For simplicity of notation, we will write

$$\Omega^i_r(\mathcal{A}) := \Omega^i_r(X(\mathcal{A})), \tag{9.1}$$

and view this set as lying in the Grassmannian $\mathrm{Gr}_r(\mathcal{A}) := \mathrm{Gr}_r(H^1(X(\mathcal{A}), \mathbb{Q}))$ of r-planes in a rational vector space of dimension $n = |\mathcal{A}|$, with a fixed basis given by the meridians around the hyperplanes. Similarly, we will write

$$\Sigma^i(\mathcal{A}) := \Sigma^i(X(\mathcal{A}), \mathbb{Z}), \tag{9.2}$$

and view this set as an open subset inside the $(n-1)$-dimensional sphere $S(\mathcal{A}) = S(H^1(X(\mathcal{A}), \mathbb{R}))$.

As noted in [33, 37], it follows from work of Arapura [1] and Esnault, Schechtman and Viehweg [18] that every arrangement complement is locally straight. In view of Theorem 3.6 (1) and Corollary 4.11, then, we have the following corollaries.

Corollary 9.3 ([37]). *For all $i \geq 1$ and $r \geq 1$,*

$$\Omega^i_r(\mathcal{A}) \subseteq \mathrm{Gr}_r(\mathcal{A}) \setminus \sigma_r(\mathcal{R}^i(X(\mathcal{A}), \mathbb{Q})). \tag{9.3}$$

Corollary 9.4 ([33]). *For all $i \geq 1$,*

$$\Sigma^i(\mathcal{A}) \subseteq S(\mathcal{A}) \setminus S(\mathcal{R}^i(X(\mathcal{A}), \mathbb{R})). \tag{9.4}$$

Since the resonance varieties of $\mathcal{A}$ depend only on its intersection lattice, these upper bounds for the Ω- and Σ-invariants are combinatorially determined. Furthermore, Corollary 5.6 yields the following.

Corollary 9.5. *Suppose $\Sigma^i(\mathcal{A}) = S(\mathcal{A}) \setminus S(\mathcal{R}^i(X(\mathcal{A}), \mathbb{R}))$. Then $\Omega^i_r(\mathcal{A}) = \mathrm{Gr}_r(\mathcal{A}) \setminus \sigma_r(\mathcal{R}^i(X(\mathcal{A}), \mathbb{Q}))$, for all $r \geq 1$.*

9.4 Lower bounds for the Σ-sets

In this context, the following recent result of Kohno and Pajitnov [26] is relevant.

Theorem 9.6 ([26]). *Let $\mathcal{A}$ be an arrangement of n hyperplanes, and let $\chi = (\chi_1, \dots, \chi_n)$ be a vector in $S^{n-1} = S(\mathcal{A})$, with all components χ_j strictly positive. Then $H_i(X, \widehat{\mathbb{Z}G}_\chi) = 0$, for all $i < \mathrm{rk}(\mathcal{A})$.*

Let χ be a positive vector as above. Making use of Theorem 4.8, we infer that $-\chi \in \Sigma^i(\mathcal{A})$, for all $i < \mathrm{rk}(\mathcal{A})$. On the other hand, since $\sum_{j=1}^n \chi_j \neq 0$, we also have that $-\chi \notin S(\mathcal{R}^i(X(\mathcal{A}), \mathbb{R}))$, a fact predicted by Corollary 9.4.

Denote by $S^-(\mathcal{A}) = S^{n-1} \cap (\mathbb{R}_{<0})^n$ the negative octant in the unit sphere $S(\mathcal{A})$. In view of the above discussion, Theorem 4.8 provides the following lower bound for the BNSR invariants of arrangements.

Corollary 9.7. *Let $\mathcal{A}$ be a (central) hyperplane arrangement. Then $S^-(\mathcal{A}) \subset \Sigma^i(\mathcal{A})$, for all $i < \mathrm{rk}(\mathcal{A})$. In particular, $S^-(\mathcal{A}) \subset \Sigma^1(\mathcal{A})$.*

The above lower bound for the BNS invariant of an arrangement $\mathcal{A}$ can be improved quite a bit, by considering the projectivized arrangement $\overline{\mathcal{A}}$. Set $n = |\mathcal{A}|$, and identify the unit sphere $S(\overline{\mathcal{A}}) = S(H^1(X(\overline{\mathcal{A}}), \mathbb{R}))$ with the great sphere $S^{n-2} = \{\chi \in S^{n-1} \mid \sum_{i=1}^n \chi_j = 0\}$ inside $S^{n-1} = S(\mathcal{A})$. Clearly, $S^-(\mathcal{A}) \subset S(\mathcal{A}) \setminus S(\overline{\mathcal{A}})$.

Proposition 9.8. *For any central arrangement $\mathcal{A}$,*

$$\Sigma^1(\mathcal{A}) = S(\mathcal{A}) \setminus (S(\overline{\mathcal{A}}) \setminus \Sigma^1(\overline{\mathcal{A}})). \tag{9.5}$$

In particular, $S(\mathcal{A}) \setminus S(\overline{\mathcal{A}}) \subseteq \Sigma^1(\mathcal{A})$.

Proof. By the product formula (4.9) for the BNS invariants, we have that $\Sigma^1(\mathcal{A})^\complement = \Sigma^1(\overline{\mathcal{A}})^\complement$. The desired conclusion follows. $\square$

9.5 Discussion and examples

In simple situations, the Dwyer-Fried and Bieri-Neumann-Strebel invariants of an arrangement can be computed explicitly, and the answers agree with those predicted by the upper bounds from Section 9.3.

Example 9.9. Let $\mathcal{A}$ be a pencil of $n \geq 3$ lines through the origin of $\mathbb{C}^2$. Then $X(\mathcal{A})$ is diffeomorphic to $\mathbb{C}^\times \times (\mathbb{C} \setminus \{n-1 \text{ points}\})$, which in turn is homotopy equivalent to the toric complex T_L, where $L = K_{1,n-1}$ is the bipartite graph obtained by coning a discrete set of $n-1$ points. Thus, the Ω- and Σ-invariants of $\mathcal{A}$ can be computed using the formulas from Section 6. For instance, $\mathcal{R}^1(\mathcal{A}) = \mathbb{C}^{n-1} \subset \mathbb{C}^n$. Therefore, $\Omega^1_1(\mathcal{A}) = \mathbb{QP}^{n-1} \setminus \mathbb{QP}^{n-2}$ and $\Omega^1_2(\mathcal{A}) = \emptyset$. Moreover, $\Sigma^1(\mathcal{A}) = S^{n-1} \setminus S^{n-2}$, which is the lower bound from Proposition 9.8.

For arbitrary arrangements, the computation of the Ω- and Σ-invariants is far from being done, even in degree $i = 1$. A more detailed analysis of the Ω-invariants of arrangements is given in [37, 38]. Here is a sample result.

Proposition 9.10 ([37]). *Let $\mathcal{A}$ be an arrangement of n lines in $\mathbb{C}^2$. Suppose $\mathcal{A}$ has 1 or 2 lines which contain all the intersection points of multiplicity 3 and higher. Then $\Omega^1_r(\mathcal{A}) = \operatorname{Gr}_r(\mathbb{Q}^n) \setminus \sigma_r(\mathcal{R}^1(X(\mathcal{A}), \mathbb{Q}))$, for all $r \geq 1$.*

The reason is that, by a result of Nazir and Raza, the first characteristic variety of such an arrangement has no translated components, and so $X(\mathcal{A})$ is 1-straight. In general, though, translated tori in the characteristic variety may affect both the Ω-sets and the Σ-sets of the arrangement.

Example 9.11. Let $\mathcal{A}$ be the deleted B_3 arrangement, with defining polynomial $Q(\mathcal{A}) = z_0 z_1 (z_0^2 - z_1^2)(z_0^2 - z_2^2)(z_1^2 - z_2^2)$. The jump loci of this arrangement were computed in [35]. The resonance variety $\mathcal{R}^1(\mathcal{A}) \subset \mathbb{C}^8$ contains 7 local components, corresponding to 6 triple points and one quadruple point, and 5 other components, corresponding to braid subarrangements. In particular, $\operatorname{codim} \mathcal{R}^1(\mathcal{A}) = 5$. In addition to the 12 subtori arising from the subspaces in $\mathcal{R}^1(\mathcal{A})$, the characteristic variety $\mathcal{V}^1(\mathcal{A}) \subset (\mathbb{C}^\times)^8$ also contains a component of the form $\rho \cdot T$, where T is a 1-dimensional algebraic subtorus, and ρ is a root of unity of order 2.

Of course, the complement of $\mathcal{A}$ is a formal, smooth, quasi-projective variety. From Theorem 7.7, we deduce that the Dwyer-Fried set $\Omega^1_2(\mathcal{A})$ is strictly contained in $\sigma_2(\mathcal{R}^1(X(\mathcal{A}), \mathbb{Q}))^{\complement}$. Using Corollary 9.5, we conclude that the BNS set $\Sigma^1(\mathcal{A})$ is strictly contained in $S(\mathcal{R}^1(X(\mathcal{A}), \mathbb{R}))^{\complement}$.

This example answers in the negative Question 9.18(ii) from [36]. It would be interesting to compute explicitly the Ω-invariants and Σ-

invariants of wider classes of arrangements, and see whether these invariants depend only on the intersection lattice, or also on other, more subtle data.

References

[1] D. ARAPURA, *Geometry of cohomology support loci for local systems.* I., J. Algebraic Geom. (3) **6** (1997), 563–597.

[2] A. BEAUVILLE, *Annulation du* H^1 *pour les fibrés en droites plats*, In: "Complex Algebraic Varieties" (Bayreuth, 1990), 1–15, Lecture Notes in Math., Vol. 1507, Springer, Berlin, 1992.

[3] M. BESTVINA and N. BRADY, *Morse theory and finiteness properties of groups*, Invent. Math. (3) **129** (1997), 445–470.

[4] R. BEZRUKAVNIKOV, *Koszul DG-algebras arising from configuration spaces*, Geom. Funct. Anal. (2) **4** (1994), 119–135.

[5] R. BIERI, *Deficienc y and the geometric invariants of a group*, J. Pure Appl. Alg. (3) **208** (2007), 951–959.

[6] R. BIERI and R. GEOGHEGAN, *Sigma invariants of direct products of groups*, Groups Geom. Dyn. (2) **4** (2010), 251–261.

[7] R. BIERI, W. NEUMANN and R. STREBEL, *A geometric invariant of discrete groups*, Invent. Math. (3) **90** (1987), 451–477.

[8] R. BIERI and B. RENZ, *Valuations on free resolutions and higher geometric invariants of groups*, Comment. Math. Helvetici (3) **63** (1988), 464–497.

[9] K. S. BROWN, *Trees, valuations, and the Bieri-Neumann-Strebel invariant*, Invent. Math. (3) **90** (1987), 479–504.

[10] K.-U. BUX and C. GONZALEZ, *The Bestvina-Brady construction revisited: geometric computation of* Σ*-invariants for right-angled Artin groups*, J. London Math. Society (3) **60** (1999), 793–801.

[11] D. CALAQUE, B. ENRIQUEZ and P. ETINGOF, *Universal KZB equations: the elliptic case*, In: "Algebra, Arithmetic, and Geometry: in Honor of Yu. I. Manin", Vol. I, 165–266, Progr. Math., Vol. 269, Birkhäuser, Boston, MA, 2009.

[12] F. CATANESE, C. CILIBERTO and M. MENDES LOPES, *On the classification of irregular surfaces of general type with nonbirational bicanonical map*, Trans. Amer. Math. Soc. (1) **350** (1998), 275–308.

[13] T. DELZANT, *L'invariant de Bieri Neumann Strebel des groupes fondamentaux des variétés kählériennes*, Math. Annalen (1) **348** (2010), 119–125.

[14] A. DIMCA, *Characteristic varieties and logarithmic differential* 1*-forms*, Compositio Math. (1) **146** (2010), 129–144.

[15] A. DIMCA and S. PAPADIMA, *Arithmetic group symmetry and finiteness properties of Torelli groups*, arXiv:1002.0673v3.

[16] A. DIMCA, S. PAPADIMA and A. SUCIU, *Topology and geometry of cohomology jump loci*, Duke Math. J. (3) **148** (2009), 405–457.

[17] W. G. DWYER and D. FRIED, *Homology of free abelian covers*. I, Bull. London Math. Soc. (4) **19** (1987), 350–352.

[18] H. ESNAULT, V. SCHECHTMAN and E. VIEHWEG, *Cohomology of local systems of the complement of hyperplanes*, Invent. Math. (3) **109** (1992), 557–561.

[19] M. FALK, *Arrangements and cohomology*, Ann. Combin. (2) **1** (1997), 135–157.

[20] M. FARBER, "Topology of Closed One-Forms", Math. Surveys Monogr., Vol. 108, Amer. Math. Soc., Providence, RI, 2004.

[21] M. FARBER, R. GEOGHEGAN and D. SCHÜTZ, *Closed 1-forms in topology and geometric group theory*, Russian Math. Surveys (1) **65** (2010), 143–172.

[22] M. FARBER, A. SUCIU and S. YUZVINSKY, *Mini-Workshop: Topology of closed one-forms and cohomology jumping loci*, Oberwolfach Reports (3) **4** (2007), 2321–2360.

[23] R. GEHRKE, *The higher geometric invariants for groups with sufficient commutativity*, Comm. Algebra (4) **26** (1998), 1097–1115.

[24] M. GREEN and R. LAZARSFELD, *Deformation theory, generic vanishing theorems and some conjectures of Enriques, Catanese and Beauville*, Invent. Math. (2) **90** (1987), 389–407.

[25] P. GRIFFITHS and J. HARRIS, "Principles of Algebraic Geometry", Pure and Applied Mathematics, John Wiley & Sons, New York, 1978.

[26] T. KOHNO and A. PAJITNOV, *Circle-valued Morse theory for complex hyperplane arrangements*, preprint arXiv:1101.0437v2.

[27] A. LIBGOBER, *First order deformations for rank one local systems with a non-vanishing cohomology*, Topology Appl. (1-2) **118** (2002), 159–168.

[28] A. MĂCINIC, *Cohomology rings and formality properties of nilpotent groups*, J. Pure Appl. Alg. (10) **214** (2010), 1818–1826.

[29] J. MEIER, H. MEINERT and L. VANWYK, *Higher generation subgroup sets and the Σ-invariants of graph groups*, Comment. Math. Helv. (1) **73** (1998), 22–44.

[30] S. PAPADIMA and A. SUCIU, *Algebraic invariants for right-angled Artin groups*, Math. Annalen (3) **334** (2006), 533–555.

[31] S. PAPADIMA and A. SUCIU, *Toric complexes and Artin kernels*, Advances in Math. (2) **220** (2009), 441–477.

[32] S. PAPADIMA and A. SUCIU, *Geometric and algebraic aspects of 1-formality*, Bull. Math. Soc. Sci. Math. Roumanie (3) **52** (2009), 355–375.

[33] S. PAPADIMA and A. SUCIU, *Bieri-Neumann-Strebel-Renz invariants and homology jumping loci*, Proc. London Math. Soc. (3) **100** (2010), 795–834.

[34] D. SCHÜTZ, *On the direct product conjecture for sigma invariants*, Bull. Lond. Math. Soc. (4) **40** (2008), 675–684.

[35] A. SUCIU, *Translated tori in the characteristic varieties of complex hyperplane arrangements*, Topology Appl. (1-2) **118** (2002), 209–223.

[36] A. SUCIU, *Fundamental groups, Alexander invariants, and cohomology jumping loci*, In: "Topology of algebraic varieties and singularities", 179–223, Contemp. Math., Vol. 538, Amer. Math. Soc., Providence, RI, 2011.

[37] A. SUCIU, *Resonance varieties and Dwyer–Fried invariants*, In: "Arrangements of Hyperplanes (Sapporo 2009)", Advanced Studies Pure Math., Vol. 62, Kinokuniya, Tokyo, 2012, 359–398.

[38] A. SUCIU, *Characteristic varieties and Betti numbers of free abelian covers*, preprint arXiv:1111.5803v2.

[39] B. TOTARO, *Configuration spaces of algebraic varieties*, Topology (6) **35** (1996), 1057–1067.

[40] S. YUZVINSKY, *Cohomology of the Brieskorn-Orlik-Solomon algebras*, Comm. Algebra (14) **23** (1995), 5339–5354.

Minimal stratifications for line arrangements and positive homogeneous presentations for fundamental groups

Masahiko Yoshinaga

Abstract. The complement of a complex hyperplane arrangement is known to be homotopic to a minimal CW complex. There are several approaches to the minimality. In this paper, we restrict our attention to real two dimensional cases, and introduce the "dual" objects so called minimal stratifications. The strata are explicitly described as semialgebraic sets. The stratification induces a partition of the complement into a disjoint union of contractible spaces, which is minimal in the sense that the number of codimension k pieces equals the k-th Betti number.

We also discuss presentations for the fundamental group associated to the minimal stratification. In particular, we show that the fundamental groups of complements of a real arrangements have positive homogeneous presentations.

1 Introduction

In 1980s Randell found an algorithm for presenting the fundamental group of the complement $M(\mathcal{A})$ of arrangement $\mathcal{A}$ of complexified lines in $\mathbb{C}^2$ ([8, 17]). Various algorithms for doing this were found subsequently ([1,3,13,20]). It was observed that these presentations are minimal in the sense that the numbers of generators and relations are equal to $b_1(\pi_1)$ and $b_2(\pi_1)$, respectively, (*c.f.* $b_i(M) = b_i(\pi_1(M(\mathcal{A})))$ for $i \leq 2$ [18]). Moreover these presentations are known to be homotopic to $M(\mathcal{A})$ ([8]).

These works have been generalized to higher dimensional cases. Let $\mathcal{A}$ be an arrangement of hyperplanes in $\mathbb{C}^\ell$. The complement $M(\mathcal{A}) = \mathbb{C}^\ell \setminus \mathcal{A}$ is proved to be homotopic to a minimal CW complex, that is, a finite CW complex in which the number of p-cells equals the p-th Betti number [5, 16, 19]. The minimality is expected to have applications to topological problems of arrangements. In order to apply, we need to make explicit how cells in the minimal CW complex are attached. There are two approaches to describe the minimal structure of $M(\mathcal{A})$, one is based on classical Morse theoretic study of Lefschetz's theorem on hyperplane section [23], the other is based on discrete Morse theory of Salvetti complex [4, 22]. There are also some applications to computations of local system (co-)homology groups [9, 24, 25].

The purpose of this paper is to describe the "dual" object to the minimal CW complex for $\ell = 2$. We introduce the minimal stratification $M(\mathcal{A}) = X_0 \supset X_1 \supset X_2$ for the complement $M(\mathcal{A})$ such that

- $X_0 \setminus X_1 = U$ is a contractible 4-manifolds,
- $X_1 \setminus X_2 = \bigsqcup_{i=1}^{b_1(M)} S_i^{\circ}$ is a disjoint union of contractible 3-manifolds, such that the number of pieces is equal to the 1st Betti number $b_1(M)$, and
- $X_2 = \bigsqcup_{\lambda=1}^{b_2(M)} C_\lambda$ is a disjoint union of contractible 2-manifolds (chambers), such that the number of pieces is equal to the 2nd Betti number $b_2(M)$.

(See Theorem 4.2 for details). We describe explicitly the strata as semialgebraic sets. By analyzing the incidence relation of strata, we obtain a presentation for π_1. The resulting presentation has only positive homogeneous relations.

This paper is organized as follows. In Section 2, as a motivating example, we compare the minimal stratification with Morse theoretic description of minimal CW complex for a very simple example: two points $\{0, 1\}$ in $\mathbb{R}$. In Section 3 we recall basic facts and introduce the sail $S(\alpha, \beta)$ bound to lines. The sail is a 3-dimensional semialgebraic submanifold of $M(\mathcal{A})$ which will be used to define the minimal stratification. Section 4 contains the main result. The proof will be given in Section 7. In Section 5 and Section 6, we discuss presentations for π_1 associated with the minimal stratification.

Remark 1.1. The input data used in the presentation (Section 6.1) look similar to those in braid monodromy presentation ([13, 20]). It seems natural to expect that these are Tietze-I equivalent. The author does not know whether or not all minimal presentations are Tietze-I equivalent (see [3, 12] for more backgrounds).

Remark 1.2. In [2, 21], stratifications of $M(\mathcal{A})$ which are not minimal are studied. The minimal stratification in this paper is apparently of a different nature, for a chamber itself is not a stratum of previous stratifications. However certain refinements and discrete Morse theoretic techniques ([4, 22]) might connect them. Also note that, recently in [11], a minimal partition of $M(\mathcal{A})$ into contractible submanifolds is constructed for any dimension. However it does not come from a stratification.

ACKNOWLEDGEMENTS. A part of this work was done while the author visited Università di Pisa and Centro di Ricerca Matematica Ennio De

Giorgi. The author appreciates their supports and hospitality especially to Prof. Mario Salvetti. The author greatefully acknowledges the financial support of JSPS.

2 A one-dimensional example

Example 2.1. Let $M = \mathbb{C} \setminus \{0, 1\}$ and $\varphi(z) := \frac{(z+1)^2}{\sqrt{z(z-1)}}$. We consider $|\varphi| : M \to \mathbb{R}$ as a Morse function which has three critical points $z = -1, \frac{5-\sqrt{17}}{4}, \frac{5+\sqrt{17}}{4}$ with index 0, 1, 1 respectively. Note that all critical points are real and $0 < \frac{5-\sqrt{17}}{4} < 1 < \frac{5+\sqrt{17}}{4}$. The unstable manifolds present a one-dimensional CW complex which is homotopic to M. Since $|\varphi(z)| \to \infty$ as $|z| \to \infty$, the unstable cells are as in Figure 2.1. It is not easy to describe the unstable manifolds explicitly even for one-dimensional cases. Nevertheless, the stable manifolds can be explicitly described: two open segments $(0, 1), (1, \infty)$ and the remainder $U = M \setminus ((0, 1) \cup (1, \infty))$.

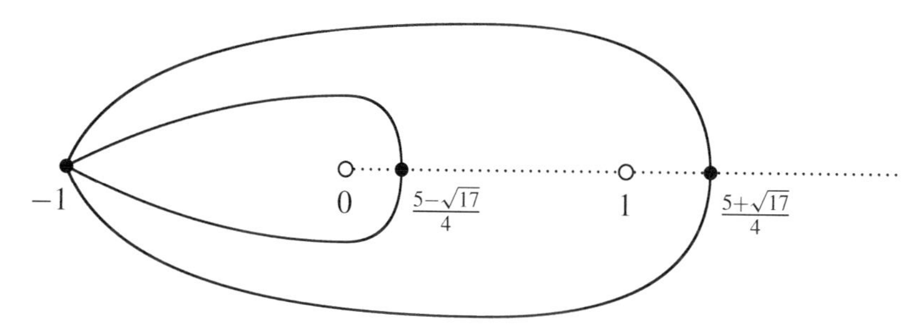

Figure 2.1. Unstable and stable manifolds (thick and dotted line, respectively).

We have a partition $U \sqcup (0, 1) \sqcup (1, \infty)$ of M by contractible pieces, and note that the number of codimension zero piece is equal to $b_0(M) = 1$ and that of codimension one is $b_1(M) = 2$. Also note that codimension one pieces $(0, 1)$ and $(1, \infty)$ are nothing but chambers of the real hyperplane arrangement $\{0, 1\}$. These pieces are expressed in terms of defining linear forms as follows,

$$\begin{aligned} (0,1) &= \left\{ z \in M \,\middle|\, \frac{z-1}{z} \in \mathbb{R}_{<0} \right\}, \\ (1,\infty) &= \left\{ z \in M \,\middle|\, \frac{-1}{z-1} \in \mathbb{R}_{<0} \right\}, \end{aligned} \tag{2.1}$$

where $\mathbb{R}_{<0}$ is the set of negative real numbers.

The homotopy types of the unstable cells for higher dimensional cases are discussed in [23]. The unstable cell itself is highly transcendental.

We will see that the submanifolds defined by formulae similar to (2.1) stratify the complement $\mathbb{C}^2$ minus lines. Also it gives a partition into the disjoint union of contractible manifolds.

3 Basic notation

3.1 Setting

A real arrangement $\mathcal{A} = \{H_1, \dots, H_n\}$ is a finite set of affine lines in the affine plane $\mathbb{R}^2$. Each line is defined by some affine linear form

$$\alpha_H(x_1, x_2) = ax_1 + bx_2 + c = 0, \tag{3.1}$$

with $a, b, c \in \mathbb{R}$ and $(a, b) \neq (0, 0)$. A connected component of $\mathbb{R}^2 \setminus \bigcup_{H\in\mathcal{A}} H$ is called a chamber. The set of all chambers is denoted by $\mathsf{ch}(\mathcal{A})$. The affine linear equation (3.1) defines a complex line $\{(z_1, z_2) \in \mathbb{C}^2 \mid az_1 + bz_2 + c = 0\}$ in $\mathbb{C}^2$. We denote the set of complexified lines by $\mathcal{A}_\mathbb{C} = \{H_\mathbb{C} = H \otimes \mathbb{C} \mid H \in \mathcal{A}\}$. The object of our interest is the complexified complement $M(\mathcal{A}) = \mathbb{C}^2 \setminus \bigcup_{H\in\mathcal{A}} H_\mathbb{C}$.

3.2 Generic flags and numbering of lines

Let $\mathcal{F}$ be a generic flag in $\mathbb{R}^2$

$$\mathcal{F} : \emptyset = \mathcal{F}^{-1} \subset \mathcal{F}^0 \subset \mathcal{F}^1 \subset \mathcal{F}^2 = \mathbb{R}^2,$$

where $\mathcal{F}^k$ is a generic k-dimensional affine subspace.

Definition 3.1. For $k = 0, 1, 2$, define the subset $\mathsf{ch}_k^\mathcal{F}(\mathcal{A}) \subset \mathsf{ch}(\mathcal{A})$ by

$$\mathsf{ch}_k^\mathcal{F}(\mathcal{A}) := \{C \in \mathsf{ch}(\mathcal{A}) \mid C \cap \mathcal{F}^k \neq \emptyset, C \cap \mathcal{F}^{k-1} = \emptyset\}.$$

The set of chambers decomposes into a disjoint union, $\mathsf{ch}(\mathcal{A}) = \mathsf{ch}_0^\mathcal{F}(\mathcal{A}) \sqcup \mathsf{ch}_1^\mathcal{F}(\mathcal{A}) \sqcup \mathsf{ch}_2^\mathcal{F}(\mathcal{A})$. The cardinality of $\mathsf{ch}_k^\mathcal{F}(\mathcal{A})$ is given as follows, which is an application of Zaslawsky's formula [26].

Proposition 3.2.

$$\begin{aligned}
\sharp\mathsf{ch}_0^\mathcal{F}(\mathcal{A}) &= b_0(M(\mathcal{A})) = 1,\\
\sharp\mathsf{ch}_1^\mathcal{F}(\mathcal{A}) &= b_1(M(\mathcal{A})) = n,\\
\sharp\mathsf{ch}_2^\mathcal{F}(\mathcal{A}) &= b_2(M(\mathcal{A})).
\end{aligned}$$

3.3 Assumptions on generic flag and numbering

Throughout this paper, we assume that the generic flag $\mathcal{F}$ satisfies the following conditions:

- $\mathcal{F}^1$ does not separate intersections of $\mathcal{A}$,
- $\mathcal{F}^0$ does not separate n-points $\mathcal{A} \cap \mathcal{F}^1$.

Then we can choose coordinates x_1, x_2 so that $\mathcal{F}^0$ is the origin, $\mathcal{F}^1$ is given by $x_2 = 0$, all intersections of $\mathcal{A}$ are contained in the upper-half plane $\{(x_1, x_2) \in \mathbb{R}^2 \mid x_2 > 0\}$ and $\mathcal{A} \cap \mathcal{F}^1$ is contained in the half line $\{(x_1, 0) \mid x_1 > 0\}$.

We set $H_i \cap \mathcal{F}^1$ has coordinates $(a_i, 0)$. By changing the numbering of lines and signs of the defining equation α_i of $H_i \in \mathcal{A}$ we may assume

- $0 < a_n < a_{n-1} < \cdots < a_1$, and
- the origin $\mathcal{F}^0$ is contained in the negative half plane $H_i^- = \{\alpha_i < 0\}$.

Remark 3.3. Sometimes it is convenient to consider 0-th line H_0 to be the line at infinity with defining equation $\alpha_0 = -1$ and $a_0 = +\infty$.

We also put $\mathsf{ch}_0^{\mathcal{F}}(\mathcal{A}) = \{C_0\}$ and $\mathsf{ch}_1^{\mathcal{F}}(\mathcal{A}) = \{C_1, \ldots, C_n\}$ so that $C_k \cap \mathcal{F}^1$ is equal to the interval (a_k, a_{k-1}). (We use the convention $a_0 = +\infty$.) It is easily seen that the chambers C_0 and C_k $(k = 1, \ldots, n)$ have the following expression.

$$
\begin{aligned}
C_0 &= \bigcap_{i=1}^{n} \{\alpha_i < 0\}, \\
C_k &= \bigcap_{i=0}^{k-1} \{\alpha_i < 0\} \cap \bigcap_{i=k}^{n} \{\alpha_i > 0\}, \quad (k = 1, \ldots, n).
\end{aligned}
\tag{3.2}
$$

(We consider $\alpha_0 < 0$ whole $\mathbb{R}^2$.) The notations introduced in this section are illustrated in Figure 3.1.

3.4 Sails bound to lines

Let $\alpha, \beta \in \mathbb{C}[z_1, z_2]$ be polynomials of deg ≤ 1. We assume that $\alpha \neq 0, \beta \neq 0$ and they are linearly independent over $\mathbb{C}$. (Note that we allow the situation that one of α or β is equal to a non-zero constant.)

Definition 3.4. For α and β as above, we define the *sail bound to α and β* by

$$
S(\alpha.\beta) = \left\{ z = (z_1, z_2) \in \mathbb{C}^2 \,\middle|\, \alpha(z)\beta(z) \neq 0, \ \frac{\alpha(z)}{\beta(z)} \in \mathbb{R}_{<0} \right\}.
$$

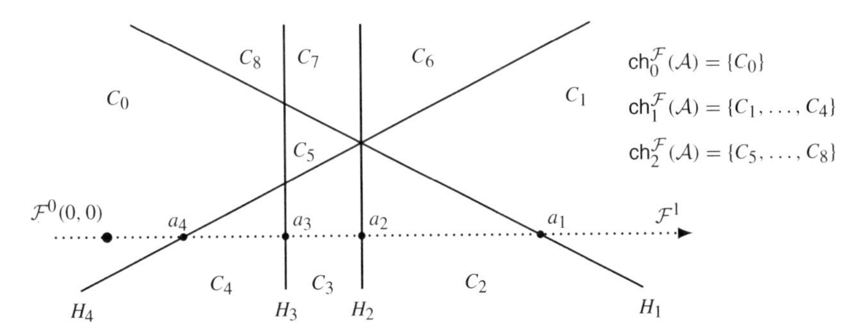

Figure 3.1. Numbering of lines and chambers.

The sail $S(\alpha, \beta)$ is a closed subset of $\mathbb{C}^2 \setminus \{\alpha\beta = 0\}$. Furthermore we have:

Lemma 3.5. *$S(\alpha, \beta)$ is an orientable 3-dimensional manifold. More precisely,*

(1) *if α and β determine intersecting lines, then $S(\alpha, \beta)$ is diffeomorphic to $\mathbb{C}^* \times \mathbb{R}_{<0}$.*
(2) *else, (i.e., either α and β determine parallel lines or one of α and β is a nonzero constant), then $S(\alpha, \beta)$ is diffeomorphic to $\mathbb{C} \times \mathbb{R}_{<0}$.*

Proof. Suppose that $\deg \alpha = \deg \beta = 1$ and two lines intersects. Then the map $(s, t) \longmapsto (t, \frac{s}{t})$ obviously gives an isomorphism $S(\alpha, \beta) \xrightarrow{\simeq} \mathbb{C}^* \times \mathbb{R}_{<0}$. Other cases are similar. $\square$

3.5 Orientations

For the purpose of obtaining a presentation for the fundamental group of $M(\mathcal{A})$, intersection numbers of loops and sails play crucial roles. It is necessary to specify the orientation of the sail $S(\alpha, \beta)$.

We first recall that the orientation of $\mathbb{C}^2$ is given by the identification

$$\begin{aligned} \mathbb{C}^2 &\xrightarrow{\sim} \mathbb{R}^4 \\ (z_1, z_2) &\longmapsto (x_1, y_1, x_2, y_2), \end{aligned}$$

where $z_i = x_i + \sqrt{-1}y_i$. Consider the map $\varphi = \frac{\alpha}{\beta} : \mathbb{C}^2 \setminus \{\alpha\beta = 0\} \to \mathbb{C}$. Since $S(\alpha, \beta)$ is connected, it is enough to specify an orientation of $T_pS(\alpha, \beta)$ for a point $p \in S(\alpha, \beta)$. The following two ordered direct sums determine an orientation of $S(\alpha, \beta)$:

$$\begin{aligned} T_pS(\alpha, \beta) \oplus N_p(S(\alpha, \beta), \mathbb{C}^2) &= T_p\mathbb{C}^2 \\ T_{\varphi(p)}R_{<0} \oplus \varphi_* N_p(S(\alpha, \beta), \mathbb{C}^2) &= T_{\varphi(p)}\mathbb{C}, \end{aligned}$$

where $N_p(S, \mathbb{C}^2)$ is a normal bundle. Note that we consider the orientation of $\mathbb{R}_{<0}$ induced from the inclusion $\mathbb{R}_{<0} \subset \mathbb{R}$.

Remark 3.6. $S(\alpha, \beta)$ and $S(\beta, \alpha)$ are the same as manifolds, but orientations are different.

The above definition is equivalent to saying as follows. Let $c:(-\varepsilon,\varepsilon) \longrightarrow \mathbb{C}^2 \setminus \{\alpha\beta = 0\}$ be a differentiable map transversal to $S(\alpha, \beta)$. Assume that $c^{-1}(S(\alpha, \beta)) = \{0\}$. Then c intersects $S(\alpha, \beta)$ positively (denoted by $I_{c(0)}(S(\alpha, \beta), c) = +1$) if and only if

$$\varphi_*(\dot{c}(0)) \in T_{\varphi(c(0))}\mathbb{C} \simeq \mathbb{C}$$

has positive imaginary part (Figure 3.2).

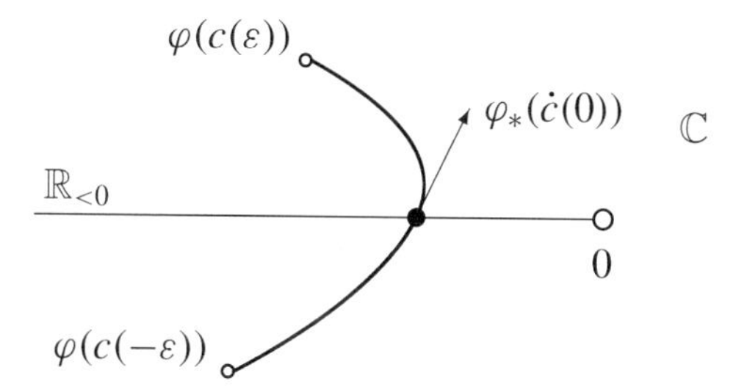

Figure 3.2. $\varphi \circ c : (-\varepsilon, \varepsilon) \longrightarrow \mathbb{C}$.

Let us look at an example showing how the intersection numbers are computed.

Example 3.7. Let $\varphi(z_2, z_1) = \frac{z_2}{z_1}$ and

$$S := S(z_2, z_1) = \left\{(z_1, z_2) \in (\mathbb{C}^*)^2 \mid \varphi(z_2, z_1) \in \mathbb{R}_{<0}\right\}.$$

Fix positive real numbers $r, \varepsilon > 0$ and an argument $0 \leq \theta_0 < 2\pi$. Consider the continuous map

$$\begin{aligned} \gamma : \mathbb{R}/2\pi\mathbb{Z} &\longrightarrow (\mathbb{C}^*)^2 \\ t &\longmapsto r(\cos\theta_0, \sin\theta_0) + \sqrt{-1}\varepsilon(\cos t, \sin t). \end{aligned}$$

Then $\gamma(t) \in S$ if and only if $\varphi(\gamma(t)) = \frac{r\sin\theta_0 + \sqrt{-1}\varepsilon\sin t}{r\cos\theta_0 + \sqrt{-1}\varepsilon\cos t}$ is a negative real number. Since

$$\begin{aligned} &\frac{r\sin\theta_0 + \sqrt{-1}\varepsilon\sin t}{r\cos\theta_0 + \sqrt{-1}\varepsilon\cos t} \\ &= \frac{r^2\sin\theta_0\cos\theta_0 + \varepsilon^2\sin t\cos t + \sqrt{-1}\cdot r\cdot\varepsilon\sin(t-\theta_0)}{r^2\cos^2\theta_0 + \varepsilon^2\cos^2 t}, \end{aligned}$$

it is contained in $\mathbb{R}_{<0}$ if and only if $t = \theta_0, \theta_0 + \pi$ and $\sin\theta_0 \cdot \cos\theta_0 < 0$ (equivalently either $\frac{\pi}{2} < \theta_0 < \pi$ or $\frac{3\pi}{2} < \theta_0 < 2\pi$). In such cases it is easily seen that $\mathfrak{Im}\,\varphi_*(\dot{\gamma}(\theta_0)) > 0$ and $\mathfrak{Im}\,\varphi_*(\dot{\gamma}(\theta_0+\pi)) < 0$. Hence we have

$$I_{\gamma(\theta_0)}(S,\gamma) = +1, \text{ and } I_{\gamma(\theta_0+\pi)}(S,\gamma) = -1.$$

4 Minimal stratification

4.1 Main result

In this section we shall give an explicit stratification of the complement $M(\mathcal{A})$ by using chambers and sails. As in Section 3.3, the sail defined by α_i and α_{i-1} is

$$S(\alpha_{i-1}, \alpha_i) = \left\{ z \in \mathbb{C}^2 \;\middle|\; \alpha_{i-1}(z)\cdot\alpha_i(z) \neq 0,\ \frac{\alpha_{i-1}(z)}{\alpha_i(z)} \in \mathbb{R}_{<0} \right\},$$

where we use the convention $\alpha_0 = -1$. Then $S_i := S(\alpha_{i-1}, \alpha_i) \cap M(\mathcal{A})$ is an oriented 3-dimensional closed submanifold of $M(\mathcal{A})$ for $i = 1, \dots, n$. These S_i's stratify the complement $M(\mathcal{A})$. The next proposition is easy.

Proposition 4.1. *Let $C \in \mathsf{ch}(\mathcal{A})$ and $i = 1, \dots, n$. The following are equivalent.*

(a) $C \subset S_i$.
(b) $C \cap S_i \neq \emptyset$.
(c) $\alpha_i(C)\cdot\alpha_{i-1}(C) < 0$. *(We use the convention $\alpha_0 = -1$.)*

Now we state the main result.

Theorem 4.2. *The closed submanifolds $S_1, \dots, S_n \subset M(\mathcal{A})$ satisfy the following.*

(i) *S_i and S_j $(i \neq j)$ intersect transversely, and $S_i \cap S_j = \bigsqcup C$, where C runs all chambers satisfying $\alpha_i(C)\alpha_{i-1}(C) < 0$ and $\alpha_j(C)\alpha_{j-1}(C) < 0$.*
(ii) *$S_i^\circ := S(\alpha_i, \alpha_{i-1}) \setminus \bigcup_{C \in \mathsf{ch}_2^{\mathcal{F}}(\mathcal{A})} C$ is a contractible 3-manifold.*
(iii) *$U := M(\mathcal{A}) \setminus \bigcup_{i=1}^n S_i$ is a contractible 4-manifold.*

The proof will be given in Section 7.

5 Dual presentation for the fundamental group

Using Theorem 4.2, we give a presentation for the fundamental group $\pi_1(M(\mathcal{A}))$. The idea is that we take the base point in U and transversal loop to each S_i as a generator, then relations are generated by loops around chambers $C \in \mathsf{ch}_2^{\mathcal{F}}(\mathcal{A})$.

5.1 Transversal generators

Fix a base point $* \in U$ and a point $p_i \in S_i^\circ$. There exists a continuous curve $\eta_i : [0, 1] \to M(\mathcal{A})$ such that

- $\eta_i(0) = \eta_i(1) = *$,
- $\eta_i(\frac{1}{2}) = p_i$ and $\eta_i^{-1}(S_i) = \{\frac{1}{2}\}$,
- η_i intersects S_i° transversely and positively, that is, $I_{p_i}(S_i^\circ, \eta_i) = 1$, and it does not intersect S_j for $j \neq i$.

Since U and S_i° are contractible, η_i is unique up to homotopy equivalence.

Let $\eta : [0, 1] \to M(\mathcal{A})$ be a continuous map with $\eta(0), \eta(1) \in U$ (not necessarily $\eta(0) = \eta(1) = *$). Since U is contractible, there exist paths c_1 from the base point $*$ to $\eta(0)$ and c_2 from $\eta(1)$ to $*$. Then $c_1\eta c_2$ is a loop. The homotopy type $[c_1\eta c_2] \in \pi_1(M(\mathcal{A}), *)$ is uniquely determined by η. We denote the class by $[\eta] \in \pi_1(M(\mathcal{A}), *)$ for simplicity.

Lemma 5.1. *With the notation above, $[\eta_1], \dots, [\eta_n]$ generate $\pi_1(M(\mathcal{A}), *)$.*

Proof. Let $\eta : [0, 1] \to M(\mathcal{A})$ be a continuous map such that $\eta(0) = \eta(1) = *$. By the transversality homotopy theorem (*e.g.*, [10, Chapter 2]), we can perturb η into a new loop such that the following hold:

- The image of η is disjoint from $\bigsqcup_{C \in \mathsf{ch}_2^{\mathcal{F}}(\mathcal{A})} C$.
- The image of η intersects $\bigsqcup_{i=1}^n S_i^\circ$ transversely.

Suppose that $\eta^{-1}(\bigsqcup_{i=1}^n S_i^\circ) = \{t_1, \dots, t_N\}$ with $0 < t_1 < \cdots < t_N < 1$ and $\eta(t_k) \in S_{m_k}^\circ$. From the transversality, the intersection number $\varepsilon_k := I_{\eta(t_k)}(S_{m_k}^\circ, \eta)$ is either $+1$ or -1 because of transversality. The class $[\eta] \in \pi_1(M(\mathcal{A}), *)$ is expressed as

$$[\eta] = [\eta_{m_1}]^{\varepsilon_1}[\eta_{m_2}]^{\varepsilon_2} \dots [\eta_{m_N}]^{\varepsilon_N}.$$

Thus any $[\eta] \in \pi_1(M)$ is generated by $[\eta_1], \dots, [\eta_n]$. □

Remark 5.2. If we fix the base point in $\mathcal{F}_{\mathbb{C}}^1 = \mathcal{F}^1 \otimes \mathbb{C}$, then we may choose transversal generators as in Figure 5.1.

5.2 Chamber relations

As we have seen in the previous section, the transversal generators determine a surjective homomorphism

$$G : F\langle \eta_1, \dots, \eta_n \rangle \longrightarrow \pi_1(M(\mathcal{A}), *),$$

from the free group generated by $\eta_1, \dots, \eta_n$ to $\pi_1(M(\mathcal{A}), *)$. We will prove that the kernel of the above map is generated by conjugacy classes of meridian loops around chambers $C \subset M(\mathcal{A})$, $C \in \mathsf{ch}_2^{\mathcal{F}}(\mathcal{A})$.

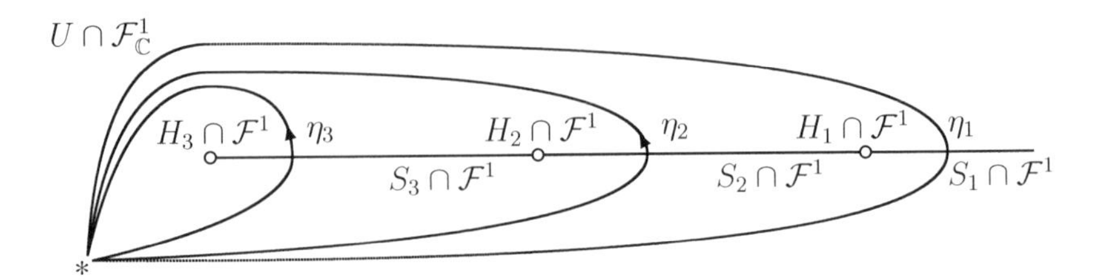

Figure 5.1. Transversal generators η_1, η_2, η_3.

Let $\eta : [0, 1] \to M(\mathcal{A})$ be a loop with $\eta(0) = \eta(1) = *$. Suppose that η represents an element of $\operatorname{Ker} G$. Then η is null-homotopic in $M(\mathcal{A})$, and hence there is a homotopy $\sigma : [0, 1]^2 \to M(\mathcal{A})$ such that $\sigma(t, 0) = \eta(t)$, $\sigma(t, 1) = \sigma(0, s) = \sigma(1, s) = *$. We can perturb σ in such a way that

- $\sigma(\partial[0, 1]^2) \cap \bigsqcup_{C \in \mathsf{ch}_2} C = \emptyset$.
- σ intersects $\bigsqcup_{C \in \mathsf{ch}_2} C$ transversely.

Let $\sigma^{-1}(\bigsqcup_{C \in \mathsf{ch}_2} C) = \{q_1, \dots, q_L\}$. We choose a meridian loop v_i in $[0, 1]^2$ around each point q_i with the base point $(0, 0)$. Let $\alpha : [0, 1] \to \partial([0, 1]^2)$ be the loop with the base point $(0, 0)$ that goes along the boundary in the counter clockwise direction. Then α is homotopically equivalent to a product of meridians $v_1, \dots, v_n$. Since η is homotopically equivalent to $\sigma \circ \alpha$, it is also homotopically equivalent to the product of meridian loops $\sigma \circ v_i$ that are meridian loops of chambers (Figure 5.2).

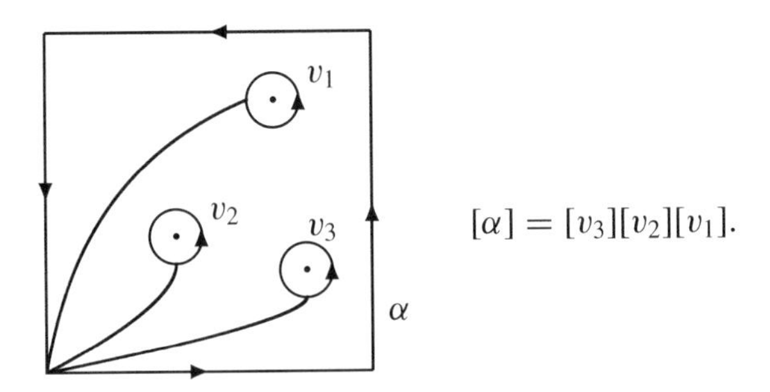

Figure 5.2. Inverse images of chambers.

We will describe the relations more explicitly in the next section.

5.3 Dual presentation

Let $i = 1, \ldots, n$ and $C \in \mathsf{ch}_2^{\mathcal{F}}(\mathcal{A})$. We define the i-th degree $d_i(C) \in \{-1, 0, +1\}$ by

$$d_i(C) := \begin{cases} -1 & \text{if } \alpha_{i-1}(C) < 0 < \alpha_i(C), \\ +1 & \text{if } \alpha_{i-1}(C) > 0 > \alpha_i(C), \\ 0 & \text{otherwise.} \end{cases} \tag{5.1}$$

(Here we use the convention $\alpha_0 = -1$, in particular, $\alpha_0(C) < 0$ for any chamber C. See Section 5.6 for examples.)

We will prove (in Section 5.5) that the meridian loop of $C \subset M(\mathcal{A})$ ($C \in \mathsf{ch}_2^{\mathcal{F}}(\mathcal{A})$) is conjugate to the word

$$E(C) := \eta_n^{d_n(C)}\eta_{n-1}^{d_{n-1}(C)}\cdots\eta_1^{d_1(C)}\cdot\eta_n^{-d_n(C)}\eta_{n-1}^{-d_{n-1}(C)}\cdots\eta_1^{-d_1(C)}. \tag{5.2}$$

Theorem 5.3. *With notation as above, the fundamental group $\pi_1(M(\mathcal{A}),*)$ is isomorphic to the group defined by the presentation*

$$\langle \eta_1, \ldots, \eta_n \mid E(C), C \in \mathsf{ch}_2^{\mathcal{F}}(\mathcal{A})\rangle.$$

Remark 5.4. The information about homotopy type of $M(\mathcal{A})$ is encoded in the degree map $d_i : \mathsf{ch}_2^{\mathcal{F}}(\mathcal{A}) \to \{0, \pm 1\}$. Indeed, it plays a role when we present cellular chain complex with coefficients in a local system (see Section 5.7).

Before proving Theorem 5.3 we introduce some terminology.

5.4 Pivotal argument

Let us denote the argument of the line H_i by θ_i, that is the angle of two positive half lines of $\mathcal{F}^1$ and H_i (see Figure 5.3). By the assumption on generic flag, arguments $\theta_1, \ldots, \theta_n$ satisfy

$$0 < \theta_n \leq \theta_{n-1} \leq \cdots \leq \theta_1 < \pi. \tag{5.3}$$

Remark 5.5. Sometimes it is convenient to define $\theta_0 := \theta_1$.

Definition 5.6. Let $p = (x_1, x_2) \in \mathbb{R}^2$ be a point different from $H_i \cap H_{i-1}$. For $i = 1, \ldots, n$, define the *i-th pivotal argument* $\mathrm{pvarg}_i(p) \in [0, 2\pi)$ by

$$\mathrm{pvarg}_i(p) = \begin{cases} \arg(\overrightarrow{qp}), & \text{if } i > 1 \text{ and } H_i \cap H_{i-1}(\neq \emptyset) = \{q\}, \\ \\ \theta_i + \pi, & \text{if } i = 1 \text{ or } i > 1,\ H_i \text{ is parallel to } H_{i-1}. \end{cases}$$

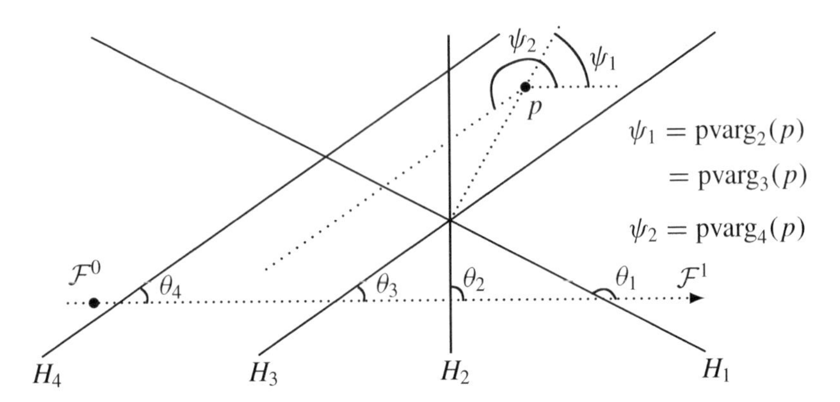

Figure 5.3. Pivotal arguments.

And also

$$|\,\mathrm{pvarg}_i(p)| = \begin{cases} \mathrm{pvarg}_i(p), & \text{if } 0 \le \mathrm{pvarg}_i(p) < \pi, \\ \mathrm{pvarg}_i(p) - \pi, & \text{if } \pi \le \mathrm{pvarg}_i(p) < 2\pi. \end{cases}$$

We have the following.

Proposition 5.7. *Let $p \in \mathbb{R}^2$. Suppose $\alpha_i(p) \cdot \alpha_{i-1}(p) < 0$ ($i > 1$).*

- *If H_{i-1} and H_i intersects, then $\theta_i < |\,\mathrm{pvarg}_i(p)| < \theta_{i-1}$.*
- *If H_{i-1} and H_i are parallel, then $\theta_i = |\,\mathrm{pvarg}_i(p)| = \theta_{i-1}$.*

Using pvarg, we can describe the intersection number of the sail $S_i = S(\alpha_{i-1}, \alpha_i) \cap M(\mathcal{A})$ and a curve, which is a generalization of Example 3.7.

Example 5.8. Let $p(x_1, x_2) \in \mathbb{R}^2 \setminus \bigcup_{H \in \mathcal{A}} H$ and $\varepsilon > 0$. Consider the loop

$$\begin{aligned} \gamma : \mathbb{R}/2\pi\mathbb{Z} &\longrightarrow M(\mathcal{A}) \\ t &\longmapsto (x_1, x_2) + \sqrt{-1}\varepsilon(\cos t, \sin t). \end{aligned}$$

If $\alpha_i(p)\cdot\alpha_{i-1}(p) > 0$, then γ does not intersect S_i. If $\alpha_i(p)\cdot\alpha_{i-1}(p) < 0$, then $\gamma^{-1}(S_i) = \{\mathrm{pvarg}_i(p), \mathrm{pvarg}_i(p) + \pi\}$. We have

$$\begin{aligned} I_{\gamma(\mathrm{pvarg}_i(p))}(S_i, \gamma) &= 1, \text{ and} \\ I_{\gamma(\mathrm{pvarg}_i(p)+\pi)}(S_i, \gamma) &= -1. \end{aligned}$$

Combining this with the degree d_i, we have the following.

Proposition 5.9. *Let $p(x_1, x_2) \in \mathbb{R}^2 \setminus \bigcup_{H \in \mathcal{A}} H$ and the loop γ be as in Example 5.8. Let us denote by C the chamber which contains p. We have*

$$\begin{aligned} I_{\gamma(|\,\mathrm{pvarg}_i(p)|)}(S_i, \gamma) &= d_i(C), \text{ and} \\ I_{\gamma(|\,\mathrm{pvarg}_i(p)|+\pi)}(S_i, \gamma) &= -d_i(C). \end{aligned}$$

5.5 Proof of Theorem 5.3

Now we prove Theorem 5.3. Let $C \in \mathsf{ch}_2^{\mathcal{F}}(\mathcal{A})$ and $p \in C$. We take a meridian loop $\gamma : \mathbb{R}/2\pi\mathbb{Z} \to M(\mathcal{A}), t \mapsto \gamma(t)$ as in Example 5.8. Then γ intersects S_i at $t = |\operatorname{pvarg}_i(p)|$ and $t = |\operatorname{pvarg}_i(p)| + \pi$ with intersection numbers $d_i(C)$ and $-d_i(C)$, respectively. (This logically includes that γ does not intersect C if and only if $d_i(C) = 0$.) In particular, from Proposition 5.7, $\theta_{i-1} \leq |\operatorname{pvarg}_i(p)| \leq \theta_i$ provided $d_i(C) \neq 0$. From Equation (5.3), the loop γ intersects $S_n, S_{n-1}, \ldots, S_1, S_n, S_{n-1}, \ldots, S_1$ in this order with intersection numbers $d_n(C), d_{n-1}(C), \ldots, d_1(C), -d_n(C), -d_{n-1}(C), \ldots, -d_1(C)$. Hence the loop γ is homotopic to the word $E(C)$ in Equation (5.2).

5.6 Examples

Example 5.10. Let $\mathcal{A} = \{H_1, \ldots, H_5\}$ be a line arrangement and $\mathcal{F}$ be a flag pictured in Figure 5.4. Then $\mathsf{ch}_2^{\mathcal{F}}(\mathcal{A}) = \{C_6, C_7, \ldots, C_{12}\}$ consists of 7 chambers. The degrees can be computed as follows.

	d_1	d_2	d_3	d_4	d_5
C_6	0	0	−1	1	−1
C_7	0	−1	0	1	−1
C_8	0	−1	1	0	−1
C_9	0	−1	1	0	0
C_{10}	−1	0	1	0	0
C_{11}	−1	0	0	1	0
C_{12}	−1	0	0	1	−1

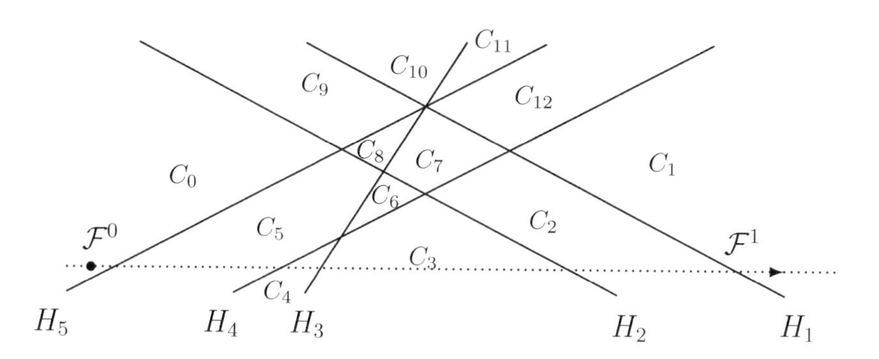

Figure 5.4. Example 5.10 and 6.7.

The fundamental group $\pi_1(M(\mathcal{A}), *)$ has the following presentation.

$$\begin{aligned}\pi_1(M(\mathcal{A}), *) = \langle \eta_1, \dots, \eta_5 \,| E(C_6) &: \eta_5^{-1}\eta_4\eta_3^{-1}\eta_5\eta_4^{-1}\eta_3 \\ E(C_7) &: \eta_5^{-1}\eta_4\eta_2^{-1}\eta_5\eta_4^{-1}\eta_2 \\ E(C_8) &: \eta_5^{-1}\eta_3\eta_2^{-1}\eta_5\eta_3^{-1}\eta_2 \\ E(C_9) &: \eta_3\eta_2^{-1}\eta_3^{-1}\eta_2 \\ E(C_{10}) &: \eta_3\eta_1^{-1}\eta_3^{-1}\eta_1 \\ E(C_{11}) &: \eta_4\eta_1^{-1}\eta_4^{-1}\eta_1 \\ E(C_{12}) &: \eta_5^{-1}\eta_4\eta_1^{-1}\eta_5\eta_4^{-1}\eta_1\rangle.\end{aligned}$$

Example 5.11. Let $\mathcal{A} = \{H_1, \dots, H_6\}$ be a line arrangement and $\mathcal{F}$ be a flag pictured in Figure 5.5. Then $\mathsf{ch}_2^{\mathcal{F}}(\mathcal{A}) = \{C_7, C_8, \dots, C_{17}\}$ consists of 11 chambers. The degrees can be computed as follows.

	d_1	d_2	d_3	d_4	d_5	d_6
C_7	-1	1	0	-1	0	0
C_8	-1	1	0	0	-1	0
C_9	-1	1	0	0	-1	1
C_{10}	-1	0	1	0	-1	1
C_{11}	-1	0	1	-1	0	1
C_{12}	-1	0	1	-1	0	0
C_{13}	-1	1	0	0	0	0
C_{14}	-1	0	1	0	0	0
C_{15}	-1	0	0	1	0	0
C_{16}	-1	0	0	0	1	0
C_{17}	-1	0	0	0	0	1

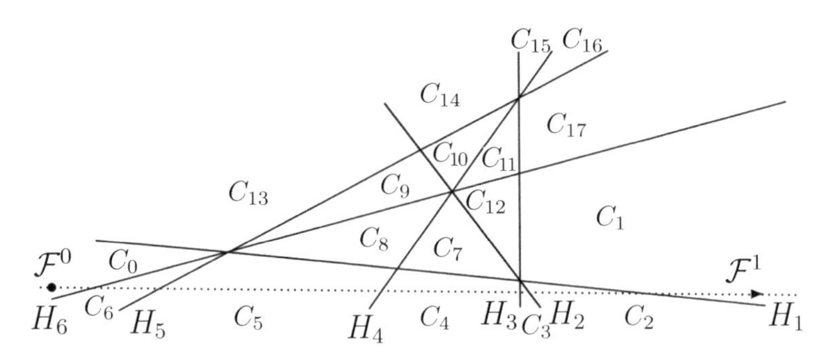

Figure 5.5. Example 5.11 and 6.8.

The fundamental group $\pi_1(M(\mathcal{A}), *)$ has the following presentation.

$$\begin{aligned}\pi_1(M(\mathcal{A}), *) = \langle \eta_1, \ldots, \eta_6 \mid & E(C_7) : \eta_4^{-1}\eta_2\eta_1^{-1}\eta_4\eta_2^{-1}\eta_1 \\ & E(C_8) : \eta_5^{-1}\eta_2\eta_1^{-1}\eta_5\eta_2^{-1}\eta_1 \\ & E(C_9) : \eta_6\eta_5^{-1}\eta_2\eta_1^{-1}\eta_6^{-1}\eta_5\eta_2^{-1}\eta_1 \\ & E(C_{10}) : \eta_6\eta_5^{-1}\eta_3\eta_1^{-1}\eta_6^{-1}\eta_5\eta_3^{-1}\eta_1 \\ & E(C_{11}) : \eta_6\eta_4^{-1}\eta_3\eta_1^{-1}\eta_6^{-1}\eta_4\eta_3^{-1}\eta_1 \\ & E(C_{12}) : \eta_4^{-1}\eta_3\eta_1^{-1}\eta_4\eta_3^{-1}\eta_1 \\ & E(C_{13}) : \eta_2\eta_1^{-1}\eta_2^{-1}\eta_1 \\ & E(C_{14}) : \eta_3\eta_1^{-1}\eta_3^{-1}\eta_1 \\ & E(C_{15}) : \eta_4\eta_1^{-1}\eta_4^{-1}\eta_1 \\ & E(C_{16}) : \eta_5\eta_1^{-1}\eta_5^{-1}\eta_1 \\ & E(C_{17}) : \eta_6\eta_1^{-1}\eta_6^{-1}\eta_1 \rangle.\end{aligned}$$

The relations $E(C_{13}), \ldots, E(C_{17})$, indicate that the large loop η_1 is contained in the center of the group.

Remark 5.12. Example 5.11 gives a presentation for the pure braid group with 4-strands. See also Example 6.8.

5.7 Twisted minimal chain complex

Let $\mathcal{L}$ be a complex rank one local system on $M(\mathcal{A})$. $\mathcal{L}$ is determined by nonzero complex numbers (monodromy around H_i) $q_i \in \mathbb{C}^*$, $i = 1, \ldots, n$. Fix a square root $q_i^{1/2} \in \mathbb{C}^*$ for each i. For given chambers C, C', let us define

$$\Delta(C, C') := \prod_{H_i \in \mathrm{Sep}(C,C')} q^{1/2} - \prod_{H_i \in \mathrm{Sep}(C,C')} q^{-1/2},$$

where $H_i \in \mathrm{Sep}(C, C')$ runs over all hyperplanes which separate C and C'. With these notation, we can describe a chain complex which computes homology groups with coefficients in $\mathcal{L}$.

Theorem 5.13. *Denote by* $\mathbb{C}[\mathsf{ch}_k^{\mathcal{F}}(\mathcal{A})] := \bigoplus_{C \in \mathsf{ch}_k^{\mathcal{F}}(\mathcal{A})} \mathbb{C} \cdot [C]$ *the vector space spanned by* $\mathsf{ch}_k^{\mathcal{F}}(\mathcal{A})$. *Recall that* $\mathsf{ch}_1^{\mathcal{F}}(\mathcal{A}) = \{C_1, C_2, \ldots, C_n\}$ *and* $\mathsf{ch}_0^{\mathcal{F}}(\mathcal{A}) = \{C_0\}$. *Then the linear maps*

$$\nabla : \mathsf{ch}_2^{\mathcal{F}}(\mathcal{A}) \longrightarrow \mathsf{ch}_1^{\mathcal{F}}(\mathcal{A}),\ [C] \longmapsto \sum_{i=1}^{n} d_i(C)\Delta(C, C_i)[C_i],$$

$$\nabla : \mathsf{ch}_1^{\mathcal{F}}(\mathcal{A}) \longrightarrow \mathsf{ch}_0^{\mathcal{F}}(\mathcal{A}),\ [C_i] \longmapsto \Delta(C_0, C_i)[C_0],$$

determines a chain complex $(\mathbb{C}[\mathsf{ch}_\bullet^{\mathcal{F}}(\mathcal{A})], \nabla)$ *which homology group is isomorphic to*

$$H_k(\mathbb{C}[\mathsf{ch}_\bullet^{\mathcal{F}}(\mathcal{A})], \nabla) \simeq H_k(M(\mathcal{A}), \mathcal{L}).$$

See [24, 25] for details and applications.

6 Positive homogeneous presentations

6.1 Left and right lines

In this section, we give an alternative presentation for the fundamental group $\pi_1(M(\mathcal{A}))$. It is presented with positive homogeneous relations as:

$$\begin{aligned} \text{Generators} &: \gamma_1, \gamma_2, \dots, \gamma_n, \\ \text{Relations}, R(C) &: \gamma_1\gamma_2\cdots\gamma_n = \gamma_{i_1(C)}\gamma_{i_2(C)}\cdots\gamma_{i_n(C)}, \end{aligned}$$

where C runs over all $\mathsf{ch}_2^{\mathcal{F}}(\mathcal{A})$ and $(i_1(C), \dots, i_n(C))$ is a permutation of $(1, \dots, n)$ associated to C.

Definition 6.1. Let $C \in \mathsf{ch}(\mathcal{A})$ be a chamber. The line $H_i \in \mathcal{A}$ is said to be *passing the left side of* C if $C \subset \{\alpha_i > 0\}$. Similarly, The line $H_i \in \mathcal{A}$ is said to be *passing the right side of* C if $C \subset \{\alpha_i < 0\}$.

Remark 6.2. Sometimes it is convenient to consider 0-th line H_0 is passing the right side of C for any chamber C. (Recall that $\alpha_0(C) = -1$ by our convention.)

Definition 6.3. For a chamber $C \in \mathsf{ch}(\mathcal{A})$, define the decomposition $\{1, \dots, n\} = I_R(C) \sqcup I_L(C)$ as follows.

$$\begin{aligned} I_R(C) =& \{i \mid H_i \text{ passes the right side of } C\}, \\ I_L(C) =& \{i \mid H_i \text{ passes the left side of } C\}. \end{aligned}$$

The notion right/left is related to the map d_i. The proof of the next proposition is straightforward.

Proposition 6.4. *Let* $C \in \mathsf{ch}_2^{\mathcal{F}}(\mathcal{A})$.

- *If* H_{i-1} *is passing right side of* C *and* H_i *is passing left side of* C, *then* $d_i(C) = -1$.
- *If* H_{i-1} *is passing left side of* C *and* H_i *is passing right side of* C, *then* $d_i(C) = 1$.
- *Otherwise,* $d_i(C) = 0$.

6.2 Positive homogeneous relations

For a chamber $C \in \mathrm{ch}_2^{\mathcal{F}}(\mathcal{A})$, arranging the right/left indices increasingly as

$$\begin{aligned} I_R(C) =& \{i_1(C) < i_2(C) < \cdots < i_k(C)\}, \\ I_L(C) =& \{i_{k+1}(C) < i_{k+2}(C) < \cdots < i_n(C)\}. \end{aligned}$$

Then we introduce the following homogeneous relation.

$$\Gamma(C) : \gamma_1\gamma_2 \dots \gamma_n = \gamma_{i_1(C)}\gamma_{i_2(C)} \cdots \gamma_{i_n(C)}. \tag{6.1}$$

Theorem 6.5. *With notation as above, the fundamental group $\pi_1(M(\mathcal{A}),*)$ is isomorphic to the group defined by the presentation*

$$\langle \gamma_1, \dots, \gamma_n \mid \Gamma(C), C \in \mathrm{ch}_2^{\mathcal{F}}(\mathcal{A})\rangle.$$

Remark 6.6. Note that all relations in the above presentation are positive homogeneous. It is similar to the "conjugation-free geometric presentation" introduced in [6, 7]. However they require stronger properties on relations. Indeed they prove that the fundamental group of Ceva arrangement (Figure 5.5) does not have conjugation-free geometric presentation.

Example 6.7. Let $\mathcal{A} = \{H_1, \dots, H_5\}$ be a line arrangement and $\mathcal{F}$ be a flag pictured in Figure 5.4.

	$I_R(C)$	$I_L(C)$
C_6	124	35
C_7	14	235
C_8	134	25
C_9	1345	2
C_{10}	345	12
C_{11}	45	123
C_{12}	4	1235

Hence the fundamental group has the following presentation.

$$\begin{aligned} \pi_1(M(\mathcal{A}), *) \simeq \langle \gamma_1, \dots, \gamma_5 \mid & 12345 = 12435 = 14235 = 13425 \\ & = 13452 = 34512 = 45123 = 41235\rangle. \end{aligned}$$

Here we denote 12345 instead of $\gamma_1\gamma_2\gamma_3\gamma_4\gamma_5$ for simplicity.

Example 6.8. Let $\mathcal{A} = \{H_1, \dots, H_6\}$ be a line arrangement and $\mathcal{F}$ be a flag pictured in Figure 5.5.

	$I_R(C)$	$I_L(C)$
C_7	23	1456
C_8	234	156
C_9	2346	15
C_{10}	346	125
C_{11}	36	1245
C_{12}	3	12456
C_{13}	23456	1
C_{14}	3456	12
C_{15}	456	123
C_{16}	56	1234
C_{17}	6	12345

Hence the fundamental group has the following presentation.

$$\begin{aligned}\pi_1(M(\mathcal{A}), *) \simeq \langle \gamma_1, \dots, \gamma_6 \mid & 123456 \\ &= 231456 = 234156 = 234615 = 346125 \\ &= 361245 = 312456 = 234561 = 345612 \\ &= 456123 = 561234 = 612345\rangle.\end{aligned}$$

6.3 Proof of Theorem 6.5

The new presentation in Theorem 6.5 is obtained by changing generators as $\eta_i = \gamma_i\gamma_{i+1}\dots\gamma_n$, or equivalently,

$$\begin{aligned}\gamma_1 &= \eta_1\eta_2^{-1} \\ \gamma_2 &= \eta_2\eta_3^{-1} \\ &\dots \\ \gamma_{n-1} &= \eta_{n-1}\eta_n^{-1} \\ \gamma_n &= \eta_n.\end{aligned} \tag{6.2}$$

Remark 6.9. If we fix the base point in $\mathcal{F}^1_{\mathbb{C}} = \mathcal{F}^1 \otimes \mathbb{C}$, then we may choose meridian generators $\gamma_1, \dots, \gamma_n$ as in Figure 6.1. (Compare Figure 5.1.)

Proposition 6.10. *By the change* (6.2), *the relation* $E(C) = 1$ (*Equation* (5.2)) *is equivalent to* $\Gamma(C)$ (*Equation* (6.1)).

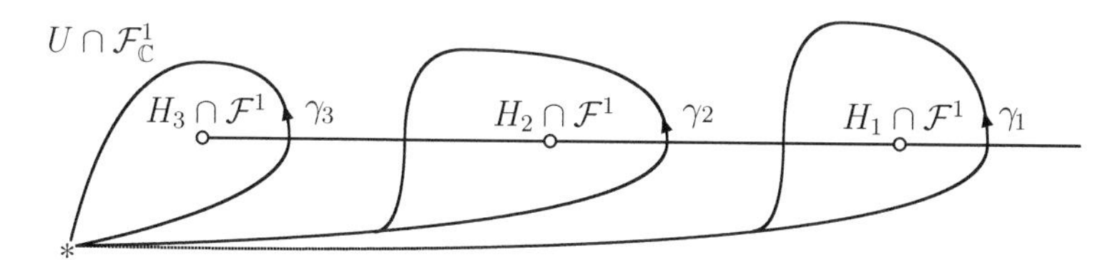

Figure 6.1. Meridian generators $\gamma_1, \gamma_2, \gamma_3$.

Proof. We distinguish four cases according to H_1 and H_n are passing right/left of C.

Case (1). Both H_1 and H_n are passing right side of C.

Case (2). H_1 is passing right and H_n is passing left side of C.

Case (3). Both H_1 and H_n are passing left side of C.

Case (4). H_1 is passing left and H_n is passing right side of C.

Case (1). We may take $1 < i_1 < \cdots < i_{2k} < n$ in such a way that

$$\overbrace{1, 2, \ldots,}^{\text{right}} \overbrace{i_1, i_1+1, \ldots,}^{\text{left}} \overbrace{i_2, i_2+1, \ldots,}^{\text{right}} \ldots, \overbrace{i_{2k}, i_{2k}+1, \ldots, n}^{\text{right}}.$$

In this case we have

$$I_R(C)=\{1, 2, \ldots, i_1-1, i_2, i_2+1, \ldots, i_3-1, \ldots, i_{2k}, i_{2k}+1 \ldots, n\},$$
$$I_L(C)=\{i_1, i_1+1, \ldots, i_2-1, i_3, \ldots, i_4-1, \ldots, i_{2k-1}, i_{2k-1}+1, \ldots, i_{2k}-1\}.$$

Then by Proposition 6.4, $d_{i_{2g-1}}(C) = -1, d_{i_{2g}}(C) = 1$ $(g = 1, \ldots, k)$ and otherwise, $d_i(C) = 0$. Hence the word $E(C)$ is equal to

$$E(C) = \eta_{i_1}^{-1} \eta_{i_2}^{1} \eta_{i_3}^{-1} \ldots \eta_{i_{2k}}^{1} \cdot \eta_{i_1}^{1} \eta_{i_2}^{-1} \eta_{i_3}^{1} \ldots \eta_{i_{2k}}^{-1}.$$

Using (6.2), we have

$$\begin{aligned} E(C) =& \eta_{i_1}^{-1}(\eta_{i_2}\eta_{i_3}^{-1}) \ldots (\eta_{i_{2k-2}}\eta_{i_{2k-1}}^{-1})\eta_{i_{2k}} \cdot (\eta_{i_1}\eta_{i_2}^{-1}) \ldots (\eta_{i_{2k-1}}\eta_{i_{2k}}^{-1}) \\ =& \eta_{i_1}^{-1} \cdot (\gamma_{i_2} \ldots \gamma_{i_3-1}) \ldots (\gamma_{i_{2k-2}} \ldots \gamma_{i_{2k-1}-1}) \cdot (\gamma_{i_{2k}} \ldots \gamma_n) \\ & \cdot (\gamma_{i_1} \ldots \gamma_{i_2-1}) \ldots (\gamma_{i_{2k-1}} \ldots \gamma_{i_{2k}-1}). \end{aligned}$$

Since the equality $E(C) = e$ holds, by multiplying $\gamma_1\gamma_2 \ldots \gamma_n$ from the left, we have (note that $\gamma_1\gamma_2 \ldots \gamma_n \eta_{i_1}^{-1} = \gamma_1\gamma_2 \ldots \gamma_{i_1-1}$)

$$\begin{aligned} \gamma_1\gamma_2 \ldots \gamma_n =& (\gamma_1\gamma_2 \ldots \gamma_{i_1-1})(\gamma_{i_2} \ldots \gamma_{i_3-1}) \ldots (\gamma_{i_{2k}} \ldots \gamma_n) \\ & \cdot (\gamma_{i_1}\gamma_{i_1+1} \ldots \gamma_{i_2-1})(\gamma_{i_3} \ldots \gamma_{i_4-1}) \ldots (\gamma_{i_{2k-1}} \ldots \gamma_{i_{2k}-1}), \end{aligned}$$

which is identical to the relation $\Gamma(C)$.

The remaining cases (2), (3) and (4) are handled in the same way. □

7 Proofs of main results

In this section, we prove Theorem 4.2. For this purposes, it is convenient to describe $M(\mathcal{A})$ in terms of tangent bundle of $\mathbb{R}^2$.

7.1 Tangent bundle description

We identify $\mathbb{C}^2$ with the total space $T\mathbb{R}^2$ of the tangent bundle of $\mathbb{R}^2$ via

$$\begin{aligned} T\mathbb{R}^2 &\longrightarrow \mathbb{C}^2 \\ (\boldsymbol{x}, \boldsymbol{y}) &\longmapsto \boldsymbol{x} + \sqrt{-1}\boldsymbol{y}, \end{aligned}$$

where $\boldsymbol{y} \in T_{\boldsymbol{x}}\mathbb{R}^2$ is a tangent vector of $\mathbb{R}^2$ at $\boldsymbol{x} \in \mathbb{R}^2$. Let $H \subset \mathbb{R}^2$ be a line and $H_{\mathbb{C}} \subset \mathbb{C}^2$ be its complexification. Then $H_{\mathbb{C}}$ is identified by the above map with

$$H_{\mathbb{C}} \simeq \{(\boldsymbol{y} \in T_{\boldsymbol{x}}\mathbb{R}^2) \mid \boldsymbol{x} \in H, \boldsymbol{y} \in T_{\boldsymbol{x}}H\}. \tag{7.1}$$

For $\boldsymbol{x} \in \mathbb{R}^2$, write $\mathcal{A}_{\boldsymbol{x}}$ the set of lines passing through $\boldsymbol{x}$. Then we have the following (see [23, Section 3.1]):

$$M(\mathcal{A}) \simeq \{(\boldsymbol{y} \in T_{\boldsymbol{x}}\mathbb{R}^2) \mid \boldsymbol{x} \in \mathbb{R}^2, \boldsymbol{y} \notin T_{\boldsymbol{x}}H, \text{ for } H \in \mathcal{A}_{\boldsymbol{x}}\}.$$

It is straightforward to check the following from (7.1).

Lemma 7.1. *If $\boldsymbol{x} + \sqrt{-1}\boldsymbol{y} \in M(\mathcal{A})$, then $(\boldsymbol{x} + t\boldsymbol{y}) + \sqrt{-1}\boldsymbol{y} \in M(\mathcal{A})$ for any $t \in \mathbb{R}$.*

Thus lines and the complement $M(\mathcal{A})$ are preserved under the linear uniform motion. The next lemma shows that the sail $S(\alpha, \beta)$ is also preserved under the linear uniform motion. The next lemma will be used repeatedly to construct deformation retractions for certain subsets of $M(\mathcal{A})$.

Lemma 7.2. *Let α, β be linear forms (as in Definition 3.4). Suppose $\boldsymbol{x} + \sqrt{-1}\boldsymbol{y} \in S(\alpha, \beta)$. Then $(\boldsymbol{x} + t\boldsymbol{y}) + \sqrt{-1}\boldsymbol{y} \in S(\alpha, \beta)$ for any $t \in \mathbb{R}$. Conversely, if $\boldsymbol{x} + \sqrt{-1}\boldsymbol{y} \notin S(\alpha, \beta)$, then $(\boldsymbol{x} + t\boldsymbol{y}) + \sqrt{-1}\boldsymbol{y} \notin S(\alpha, \beta)$ for any $t \in \mathbb{R}$.*

Proof. Set $\alpha(\boldsymbol{x}) = \boldsymbol{a} \cdot \boldsymbol{x} + b$ and $\beta(\boldsymbol{x}) = \boldsymbol{c} \cdot \boldsymbol{x} + d$, where $\boldsymbol{a}, \boldsymbol{c} \in (\mathbb{R}^2)^*$ and $b, d \in \mathbb{R}$. By assumption,

$$\frac{\alpha(\boldsymbol{x} + \sqrt{-1}\boldsymbol{y})}{\beta(\boldsymbol{x} + \sqrt{-1}\boldsymbol{y})} = \frac{\boldsymbol{a} \cdot \boldsymbol{x} + \sqrt{-1}\boldsymbol{a} \cdot \boldsymbol{y} + b}{\boldsymbol{c} \cdot \boldsymbol{x} + \sqrt{-1}\boldsymbol{c} \cdot \boldsymbol{y} + d} = \frac{\alpha(\boldsymbol{x}) + \sqrt{-1}\boldsymbol{a} \cdot \boldsymbol{y}}{\beta(\boldsymbol{x}) + \sqrt{-1}\boldsymbol{c} \cdot \boldsymbol{y}} = r \in \mathbb{R}_{<0}.$$

Hence

$$\alpha(\boldsymbol{x}) = r\beta(\boldsymbol{x}) \text{ and } \boldsymbol{a} \cdot \boldsymbol{y} = r\boldsymbol{c} \cdot \boldsymbol{y}. \tag{7.2}$$

The assertion follows from

$$\frac{\alpha(\boldsymbol{x}+t\boldsymbol{y}+\sqrt{-1}\boldsymbol{y})}{\beta(\boldsymbol{x}+t\boldsymbol{y}+\sqrt{-1}\boldsymbol{y})}=\frac{\alpha(\boldsymbol{x})+t\boldsymbol{a}\cdot\boldsymbol{y}+\sqrt{-1}\boldsymbol{a}\cdot\boldsymbol{y}}{\beta(\boldsymbol{x})+t\boldsymbol{c}\cdot\boldsymbol{y}+\sqrt{-1}\boldsymbol{c}\cdot\boldsymbol{y}}=r. \tag{7.3}$$

The second part follows immediately from the first part. □

Suppose that $\boldsymbol{x}+\sqrt{-1}\boldsymbol{y}\in S(\alpha,\beta)$ and $\boldsymbol{a}\cdot\boldsymbol{y}\neq 0$. Set $t=-\frac{\alpha(\boldsymbol{x})}{\boldsymbol{a}\cdot\boldsymbol{y}}$. Then by (7.3) above, $\alpha(\boldsymbol{x})+t\boldsymbol{a}\cdot\boldsymbol{y}=\alpha(\boldsymbol{x}+t\boldsymbol{y})=0$ and $\beta(\boldsymbol{x})+t\boldsymbol{c}\cdot\boldsymbol{y}=\beta(\boldsymbol{x}+t\boldsymbol{y})=0$, which implies that the line $\boldsymbol{x}+\mathbb{R}\cdot\boldsymbol{y}$ is passing through the intersection $H_\alpha\cap H_\beta$ of two lines $H_\alpha=\{\alpha=0\}$ and $H_\beta=\{\beta=0\}$. We obtain the following description of the sail.

Proposition 7.3. *Let α and β be as in Lemma* 7.2.

(i) *Suppose $H_\alpha=\{\alpha=0\}$ and $H_\beta=\{\beta=0\}$ are not parallel. Then $\boldsymbol{x}+\sqrt{-1}\boldsymbol{y}\in S(\alpha,\beta)$ if and only if either*
  - *$\alpha(\boldsymbol{x})\beta(\boldsymbol{x})<0$ and $\boldsymbol{y}$ is tangent to the line $\overline{\boldsymbol{x}\cdot(H_\alpha\cap H_\beta)}$ passing through $\boldsymbol{x}$ and the intersection $H_\alpha\cap H_\beta$, or*
  - *$\alpha(\boldsymbol{x})=\beta(\boldsymbol{x})=0$ (i.e., $\{\boldsymbol{x}\}=H_\alpha\cap H_\beta$) and $\boldsymbol{y}\neq 0$ such that the line $\boldsymbol{x}+\mathbb{R}\cdot\boldsymbol{y}$ is passing through the domain $\{\boldsymbol{x}\in\mathbb{R}^2\mid \alpha(\boldsymbol{x})\beta(\boldsymbol{x})<0\}$.*

(ii) *Suppose H_α and H_β are parallel. Then $\boldsymbol{x}+\sqrt{-1}\boldsymbol{y}\in S(\alpha,\beta)$ if and only if $\alpha(\boldsymbol{x})\beta(\boldsymbol{x})<0$ and $\boldsymbol{y}$ is either zero or parallel to H_α.*

(iii) *Suppose α is a nonzero constant. (In this case, β should be degree one.) Then $\boldsymbol{x}+\sqrt{-1}\boldsymbol{y}\in S(\alpha,\beta)$ if and only if $\alpha(\boldsymbol{x})\beta(\boldsymbol{x})<0$ and $\boldsymbol{y}$ is either zero or parallel to H_β.*

(See Figure 7.1.*)*

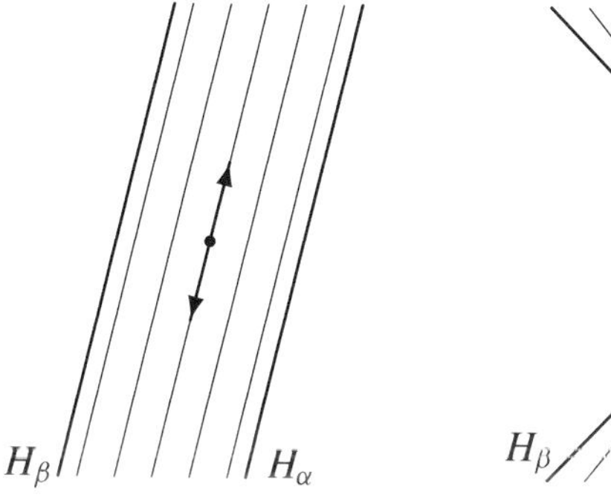

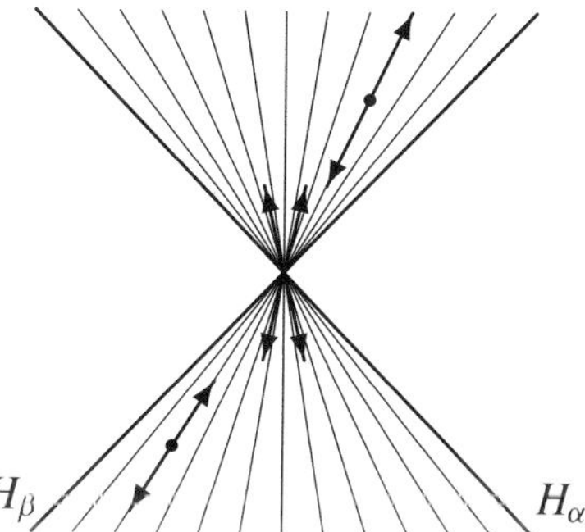

Figure 7.1. Sails $S(\alpha,\beta)$.

Define

$$|\arg(\boldsymbol{y})| := \begin{cases} \arg(\boldsymbol{y}), & \text{if } 0 \le \arg(\boldsymbol{y}) < \pi \\ \arg(\boldsymbol{y}) - \pi, & \text{if } \pi \le \arg(\boldsymbol{y}) < 2\pi. \end{cases}$$

Using the above and Proposition 5.7, we have

Proposition 7.4. *Let* $\boldsymbol{x} + \sqrt{-1}\boldsymbol{y} \in S_i = S(\alpha_{i-1}, \alpha_i) \cap M(\mathcal{A})$.

- *If* $\boldsymbol{x}$ *is the intersection* $H_{i-1} \cap H_i$, *then* $\boldsymbol{y} \neq \boldsymbol{0}$ *and* $\theta_{i-1} < |\arg(\boldsymbol{y})| < \theta_i$.
- *If* $\boldsymbol{x}$ *is not the intersection* $H_{i-1} \cap H_i$, *then* $\alpha_{i-1}(\boldsymbol{x})\alpha_i(\boldsymbol{x}) < 0$ *and* $\boldsymbol{y} \neq \boldsymbol{0}$ *with* $|\arg(\boldsymbol{y})| = |\operatorname{pvarg}_i(\boldsymbol{x})|$ *or* $\boldsymbol{y} = \boldsymbol{0}$.

Now we prove Theorem 4.2 (i):

$$S_i \cap S_j = \bigsqcup C,$$

where C runs all chambers satisfying $\alpha_{i-1}(C)\alpha_i(C)<0$, $\alpha_{j-1}(C)\alpha_j(C)< 0$ $(1 \le i < j \le n)$. Suppose that $\boldsymbol{x} + \sqrt{-1}\boldsymbol{y} \in S_i \cap S_j$. Then by Proposition 7.3 and 7.4, we have $\alpha_{i-1}(\boldsymbol{x})\alpha_i(\boldsymbol{x}) < 0 and \alpha_{j-1}(\boldsymbol{x})\alpha_j(\boldsymbol{x}) < 0$. Since the imaginary parts of S_i and S_j are linearly independent, we have $\boldsymbol{y} = \boldsymbol{0}$. (See Figure 7.2.)

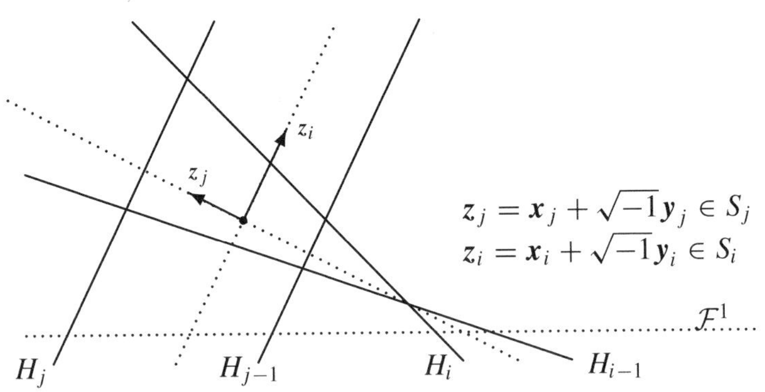

Figure 7.2. S_i and S_j intersect transversely.

7.2 Contractibility of S_i°

Now we prove that $S_i^\circ = S_i \setminus \bigcup_{C \in \mathrm{ch}_2^{\mathcal{F}}(\mathcal{A})} C$ is contractible. Let us denote $A_i := S(\alpha_{i-1}, \alpha_i) \cap \mathcal{F}^1$ the 1-dimensional segment. Since A_i is obviously contractible, therefore it suffices to construct a deformation retract onto A_i.

Define a continuous map $\rho : S_i^\circ \to A_i$, $\boldsymbol{x} + \sqrt{-1}\boldsymbol{y} \mapsto \rho(\boldsymbol{x} + \sqrt{-1}\boldsymbol{y})$ by

(1) if $\boldsymbol{y} \neq \boldsymbol{0}$, then $\rho(\boldsymbol{x} + \sqrt{-1}\boldsymbol{y}) = A_i \cap (\boldsymbol{x} + \mathbb{R} \cdot \boldsymbol{y})$,
(2) if $H_i \cap H_{i-1} \neq \emptyset$ and $\boldsymbol{y} = \boldsymbol{0}$, then $\rho(\boldsymbol{x} + \sqrt{-1}\boldsymbol{y}) = A_i \cap \overline{(\boldsymbol{x} \cdot H_i \cap H_{i-1})}$, where $\overline{(\boldsymbol{x} \cdot H_i \cap H_{i-1})}$ is the line passing through $\boldsymbol{x}$ and the intersection $H_i \cap H_{i-1}$,
(3) if $H_i \cap H_{i-1} = \emptyset$ and $\boldsymbol{y} = \boldsymbol{0}$, then $\rho(\boldsymbol{x} + \sqrt{-1}\boldsymbol{y}) = A_i \cap L_{\boldsymbol{x}}$, where $L_{\boldsymbol{x}}$ is the line passing through $\boldsymbol{x}$ and parallel to H_i.

By Proposition 7.4, ρ is a well-defined continuous map. Note that $\rho|_{A_i} = \mathrm{id}_{A_i}$.

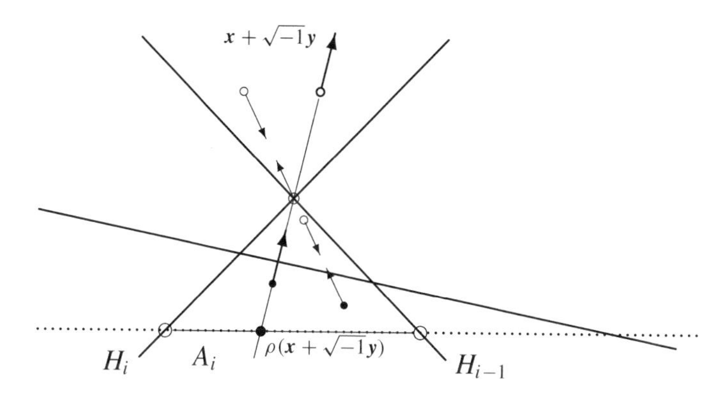

Figure 7.3. Deformation retract $\rho(\boldsymbol{x} + \sqrt{-1}\boldsymbol{y})$.

Define

$$f_t(\boldsymbol{x} + \sqrt{-1}\boldsymbol{y}) = ((1-t)\boldsymbol{x} + t\rho(\boldsymbol{x} + \sqrt{-1}\boldsymbol{y})) + \sqrt{-1}(1-t)\boldsymbol{y}.$$

If $\boldsymbol{y} \neq \boldsymbol{0}$, then the real part $((1-t)\boldsymbol{x} + t\rho(\boldsymbol{x} + \sqrt{-1}\boldsymbol{y}))$ is on the line $\boldsymbol{x} + \mathbb{R} \cdot \boldsymbol{y}$ and the imaginary part is nonzero provided $t \neq 1$. Hence $f_t(\boldsymbol{x} + \sqrt{-1}\boldsymbol{y}) \in S_i^\circ$ (see also Lemma 7.2). If $\boldsymbol{y} = \boldsymbol{0}$, then $\boldsymbol{x}$ is contained in the chamber C_i. (Otherwise, $\boldsymbol{x}$ is contained in some chamber $C \in \mathsf{ch}_2^{\mathcal{F}}(\mathcal{A})$ which does not intersects $\mathcal{F}^1$.) Hence $f_t(\boldsymbol{x}) \in C_i \subset S_i^\circ$. The map f_t determines a deformation contraction of S_i° onto A_i. (See Figure 7.3.)

7.3 Contractibility of U

We break the proof of the contractibility of $U = M(\mathcal{A}) \setminus \bigcup_{i=1}^n S_i$ up into n steps.

7.3.1 Filtration U_k

Definition 7.5. Define $U_0 = U$ and

$$U_k := \{z = \boldsymbol{x} + \sqrt{-1}\boldsymbol{y} \in U \mid \alpha_1(\boldsymbol{x}) \leq 0, \ldots, \alpha_k(\boldsymbol{x}) \leq 0\},$$

for $k = 1, \ldots, n$.

Obviously $U = U_0 \supset U_1 \supset \cdots \supset U_n$, and we have easily the following.

Proposition 7.6. *U_n is star-shaped. In particular, U_n is contractible.*

Therefore it is enough to construct a deformation retract $\rho_k : U_k \to U_{k+1}$ for $k = 0, \ldots, n-1$.

7.3.2 The case $k = 0$ We construct a deformation retraction $\rho_0 : U = U_0 \to U_1 = \{z = \boldsymbol{x} + \sqrt{-1}\boldsymbol{y} \mid \alpha_1(\boldsymbol{x}) \leq 0\}$ as follows (Figure 7.4). Let $z = \boldsymbol{x} + \sqrt{-1}\boldsymbol{y} \in U_0 \setminus U_1$. Then, by definition, $\alpha_1(\boldsymbol{x}) > 0$. Recall Proposition 7.3 that

$$S_1 = \{\boldsymbol{x} + \sqrt{-1}\boldsymbol{y} \mid \alpha(\boldsymbol{x}) > 0 \text{ and } \boldsymbol{y} \text{ is either zero or parallel to } H_1\}.$$

Therefore $z \notin S_1$ implies that the affine line $\boldsymbol{x} + \mathbb{R} \cdot \boldsymbol{y} \subset \mathbb{R}^2$ is not parallel to H_1, hence intersects H_1. Denote by $\tau(z) \in \mathbb{R}$ the unique real number satisfying $\boldsymbol{x} + \tau(z)\boldsymbol{y} \in H_1$. Define $\rho_0(z)$ by

$$\rho_0(z) = \begin{cases} (\boldsymbol{x} + t \cdot \tau(z)\boldsymbol{y}) + \sqrt{-1}\boldsymbol{y} & \text{if } z \in U_0 \setminus U_1 \\ \boldsymbol{x} + \sqrt{-1}\boldsymbol{y} & \text{if } z \in U_1. \end{cases}$$

Then by Lemma 7.2, ρ_0 is a deformation retraction.

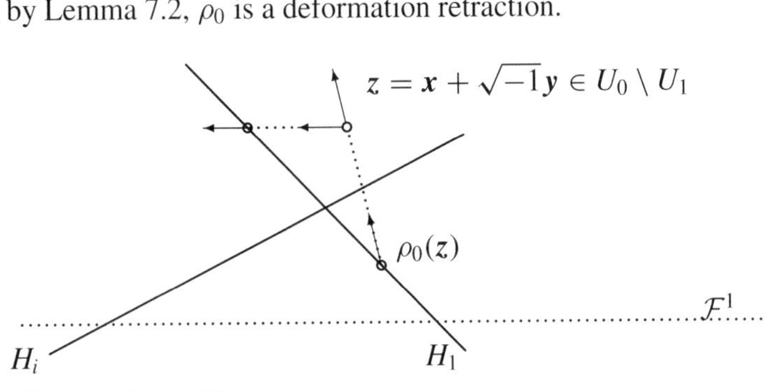

Figure 7.4. $\rho_0 : U_0 \to U_1$.

7.3.3 The case that H_k and H_{k+1} are parallel Here, we assume that H_k and H_{k+1} are parallel ($1 \leq k \leq n-1$).

Definition 7.7.

(1) Define the closed subset $D_k \subset \mathbb{R}^2$ by

$$D_k = \{\boldsymbol{x} \in \mathbb{R}^2 \mid \alpha_1(\boldsymbol{x}) \leq 0, \alpha_2(\boldsymbol{x}) \leq 0, \ldots, \alpha_k(\boldsymbol{x}) \leq 0, \text{ and } \alpha_{k+1}(\boldsymbol{x}) \geq 0\}.$$

(2) Denote the upper roof of D_k by R_k. More precisely, R_k is the closure of $\partial(D_k) \setminus (H_k \cup H_{k+1})$.

(3) Suppose $\alpha_k(\boldsymbol{x}) \leq 0$ and $\alpha_{k+1}(\boldsymbol{x}) \geq 0$. Then denote the line passing through $\boldsymbol{x}$ which is parallel to H_k by $L_{\boldsymbol{x}}$.

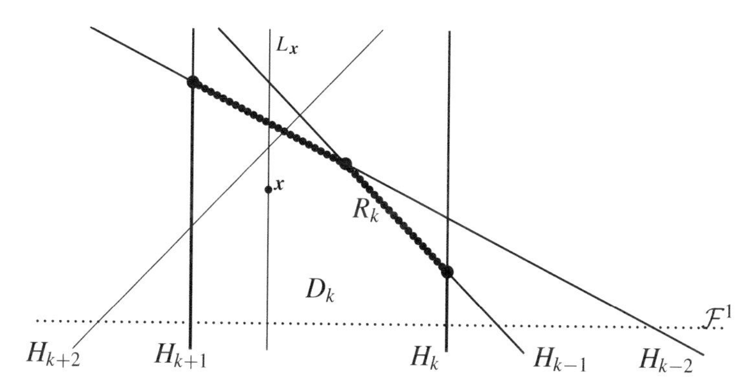

Figure 7.5. D_k and its roof R_k.

Remark 7.8. By definitions, if $\boldsymbol{x} + \sqrt{-1}\boldsymbol{y} \in U_k \setminus U_{k+1}$, then $\boldsymbol{x} \in D_k$.

The set $\{\boldsymbol{x} \in \mathbb{R}^2 \mid \alpha_k(\boldsymbol{x}) \leq 0 \leq \alpha_{k+1}(\boldsymbol{x})\}$ is a strip with boundaries H_k and H_{k+1}. We can define a deformation retract of this strip to D_k by

$$\mathrm{pr}_k(\boldsymbol{x}) = \begin{cases} R_k \cap L_{\boldsymbol{x}} & \text{if } \boldsymbol{x} \notin D_k, \\ \boldsymbol{x} & \text{if } \boldsymbol{x} \in D_k. \end{cases}$$

Suppose $z = \boldsymbol{x} + \sqrt{-1}\boldsymbol{y} \in U_k \setminus U_{k+1}$. Since $z \notin S_k$, $\boldsymbol{y}$ is neither zero nor parallel to H_{k+1}. Hence there exists a unique real number $\tau(\boldsymbol{x}, \boldsymbol{y}) \in \mathbb{R}$ such that $\boldsymbol{x} + \tau(\boldsymbol{x}, \boldsymbol{y}) \cdot \boldsymbol{y} \in H_{k+1}$.

Define the family of continuous map F_t ($0 \leq t \leq 1$) by

$$F_t(\boldsymbol{x} + \sqrt{-1}\boldsymbol{y}) = \mathrm{pr}_k(\boldsymbol{x} + t \cdot \tau(\boldsymbol{x}, \boldsymbol{y}) \cdot \boldsymbol{y}) + \sqrt{-1}\boldsymbol{y} \tag{7.4}$$

for $\boldsymbol{x} + \sqrt{-1}\boldsymbol{y} \in U_k$ with $\boldsymbol{x} \in D_k$ (Figure 7.6).

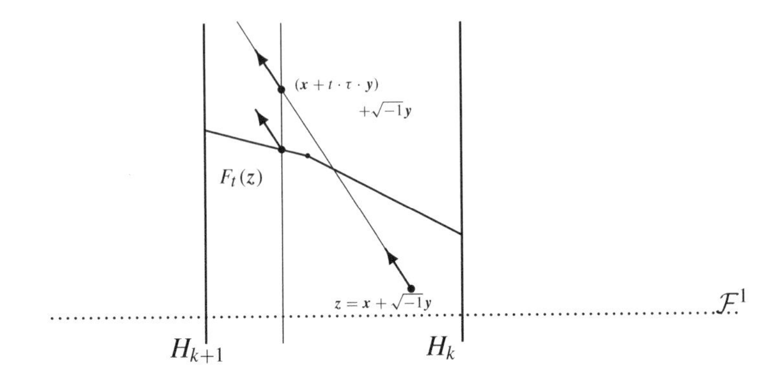

Figure 7.6. Retraction F_t.

Proposition 7.9. *Let us extend the above F_t by*

$$F_t(\boldsymbol{x} + \sqrt{-1}\boldsymbol{y}) = \begin{cases} F_t(z) \ (\text{as above}) & \text{if } z \in U_k \setminus U_{k+1}, \\ z & \text{if } z \in U_{k+1}. \end{cases}$$

Then $F_t(z) \in U_k$ for any $z \in U_k$ and hence F_1 determines a deformation retract $U_k \to U_{k+1}$.

Proof. Let $z = \boldsymbol{x} + \sqrt{-1}\boldsymbol{y} \in U_k$ and $F_t(z) = z' = \boldsymbol{x}' + \sqrt{-1}\boldsymbol{y}$. Suppose that $z' \notin U_k$. If $\boldsymbol{x} + t \cdot \tau \cdot \boldsymbol{y} \in D_k$, then $F_t(z) = (\boldsymbol{x} + t \cdot \tau \cdot \boldsymbol{y}) + \sqrt{-1}\boldsymbol{y}$. By Lemma 7.2 $F_t(z) \in M(\mathcal{A})$, hence contained in U_k. Thus we may assume that $\boldsymbol{x} + t \cdot \tau \cdot \boldsymbol{y} \notin D_k$ and $\boldsymbol{x}' \in R_k$. Furthermore, we may assume that $\boldsymbol{y} \in T_{\boldsymbol{x}'}\mathbb{R}^2$ is contained in a line $H_j \subset T_{\boldsymbol{x}'}$ with $\boldsymbol{x}' \in H_j$ for some $1 \leq j < k$. Then $\boldsymbol{x} + t \cdot \tau \cdot \boldsymbol{y}$ must be contained in the domain $\{\alpha_j > 0\}$. However, this contradicts $\boldsymbol{x} \in \{\alpha_j \leq 0\}$ and the fact that $\boldsymbol{y}$ is parallel to H_j. Hence $F_t(z) \in U_k$. □

7.3.4 LQ-curves The remaining case is the construction of deformation retract $U_k \to U_{k+1}$ when H_k and H_{k+1} are not parallel. The idea is similar to the previous case, however, it requires more technicality.

An Linear-Quadric curve on $\mathbb{R}^2$ is, roughly speaking, a C^1-curve which is linear when $x_1 \leq 1$ and quadric when $x_1 \geq 1$. The precise definition is as follows.

Definition 7.10. An *LQ-curve* (Linear-Quadric-curve) C on the real plane $\mathbb{R}^2$ is either a vertical line $C = \{(x_1, x_2) \in \mathbb{R}^2 \mid x_1 = t\}$ or the graph $\{(x, f(x)) \mid x \in \mathbb{R}\}$ of a C^1-function $f(x)$ such that

$$f(x_1) = \begin{cases} ax_1 + b & \text{for } x_1 \leq 1, \\ cx_1^2 + dx_1 & \text{for } x_1 \geq 1, \end{cases}$$

where $t, a, b, c, d \in \mathbb{R}$.

Remark 7.11.

(1) Since $f(x_1)$ is C^1 at $x_1 = 1$, $f(x_1)$ should have the following expression.

$$f(x_1) = \begin{cases} ax_1 + b & \text{for } x_1 \le 1, \\ -bx_1^2 + (a+2b)x_1 & \text{for } x_1 \ge 1. \end{cases} \tag{7.5}$$

(2) $f(x_1)$ and the derivative $f'(x_1)$ for some $x_1 \in \mathbb{R}$ determines the unique LQ-curve.

Let $\boldsymbol{x} \in \mathbb{R}^2$ be a point in the positive quadrant and $\boldsymbol{y} \in T_{\boldsymbol{x}}\mathbb{R}^2 \setminus \{\boldsymbol{0}\}$ a nonzero tangent vector. Then there exists a unique C^1-map $X_{\boldsymbol{x},\boldsymbol{y}} : \mathbb{R} \to \mathbb{R}^2$ such that

- $X(0) = \boldsymbol{x}$, $\dot{X}(0) = \boldsymbol{y}$,
- $\{X(t) \mid t \in \mathbb{R}\} \subset \mathbb{R}^2$ is an LQ-curve.
- $|\dot{X}(t)| = |\boldsymbol{y}|$.

Roughly speaking, $X(t)$ is a motion along an LQ-curve with constant velocity. $X_{\boldsymbol{x},\boldsymbol{y}}(t)$ is continuous with respect to $\boldsymbol{x}$, $\boldsymbol{y}$ and t.

In the remainder of this section, we assume $\boldsymbol{x} \in (\mathbb{R}_{\ge 0})^2$ and $\boldsymbol{x} \ne \boldsymbol{0}$. Then $0 \le \arg \boldsymbol{x} \le \frac{\pi}{2}$. We also assume that $\boldsymbol{y} \notin \mathbb{R} \cdot \boldsymbol{x}$. We call $\boldsymbol{y}$ positive (resp. negative) if $\arg \boldsymbol{x} < \arg \boldsymbol{y} < \arg \boldsymbol{x} + \pi$ (resp. $\arg \boldsymbol{x} - \pi < \arg \boldsymbol{y} < \arg \boldsymbol{x}$). It is easily seen if $\boldsymbol{y}$ is positive (resp. negative), then $\arg X_{\boldsymbol{x},\boldsymbol{y}}(t)$ is increasing (resp. decreasing) in t.

Lemma 7.12. *Let* $\boldsymbol{x}$ *and* $\boldsymbol{y}$ *as above. Then the LQ-curve* $X_{\boldsymbol{x},\boldsymbol{y}}(t)$ *intersects the positive* x_1*-axis* $\{(x_1, 0) \mid x_1 > 0\}$ *exactly once.*

Proof. If $\boldsymbol{y}$ is vertical, the assertion holds obviously. Assume that $\boldsymbol{y}$ is not vertical. We use the expression (7.5). From the assumption that $\boldsymbol{y} \notin \mathbb{R} \cdot \boldsymbol{x}$, $b \ne 0$. Suppose $b > 0$. The quadric equation $-bx^2 + (a+2b)x = 0$ has the solution $x = \frac{a+2b}{b}$ (and $x = 0$). If $\frac{a+2b}{b} > 1$, then we have $a + b = f(1) > 0$. Since $f(0) = b > 0$, $f(t) \ne 0$ for $0 \le t \le 1$. Hence $x = \frac{a+2b}{b}$ is the unique solution. $\frac{a+2b}{b} = 1$ is equivalent to say $a + b = 0$. Hence $x_1 = 1$ is the unique solution of $f(x_1) = 0$. $\frac{a+2b}{b} < 1$ implies that $a + b < 0$. Since $f(0) > 0 > f(1)$, there exists the unique solution $f(t) = 0$ with $0 \le t \le 1$. The case $b < 0$ is similar. □

Definition 7.13. Let $\boldsymbol{x}$ and $\boldsymbol{y}$ be as above. Denote by $\tau = \tau(\boldsymbol{x}, \boldsymbol{y})$ the unique real number such that $X_{\boldsymbol{x},\boldsymbol{y}}(\tau) \in \{(x_1, 0) \mid x_1 > 0\}$. (Figure 7.7.)

Remark 7.14. $\tau(\boldsymbol{x}, \boldsymbol{y})$ is continuous on $\{(\boldsymbol{x}, \boldsymbol{y}) \mid \boldsymbol{x} \in (\mathbb{R}_{\ge 0})^2 \setminus \{\boldsymbol{0}\}, \boldsymbol{y} \notin \mathbb{R} \cdot \boldsymbol{x}\}$.

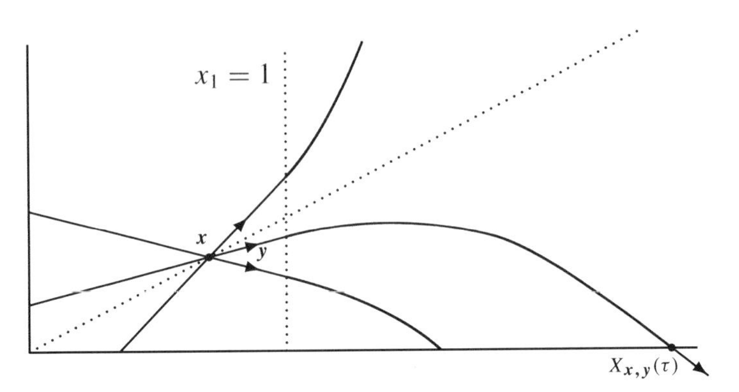

Figure 7.7. LQ-curves $X_{x,y}(t)$.

7.3.5 The case that H_k and H_{k+1} are not parallel Next we assume that H_k and H_{k+1} are not parallel, and constructing a deformation retraction $\rho_k : U_k \to U_{k+1}$. The idea is similar to the parallel case (Section 7.3.3). However we need LQ-curves to construct the retraction.

Here we choose coordinates x_1, x_2 such that $\alpha_k = -x_1, \alpha_{k+1} = x_2$ and $\mathcal{F}^1 = \{x_1 + x_2 = 1\}$. Recall Definition 7.7 that $D_k \subset \mathbb{R}^2$ is defined by

$$D_k = \{\boldsymbol{x} \in \mathbb{R}^2 \mid \alpha_1(\boldsymbol{x}) \leq 0, \alpha_2(\boldsymbol{x}) \leq 0, \ldots, \alpha_k(\boldsymbol{x}) \leq 0, \text{ and } \alpha_{k+1}(\boldsymbol{x}) \geq 0\},$$

and the roof R_k is defined as the closure of $\partial(D_k) \setminus (H_k \cup H_{k+1})$.

Definition 7.15. Suppose $\boldsymbol{x} \neq \boldsymbol{0}$. Then denote the line passing through $\boldsymbol{x}$ and the intersection $\{\boldsymbol{0}\} = H_k \cap H_{k+1}$ by $L_{\boldsymbol{x}}$.

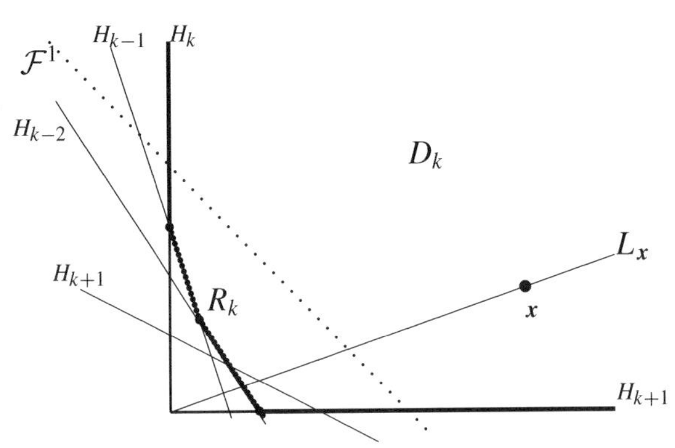

Figure 7.8. D_k, R_k and $L_{\boldsymbol{x}}$.

We can define a deformation retract of $(\mathbb{R}_{\geq 0})^2 \setminus \{\boldsymbol{0}\}$ to D_k by

$$\mathrm{pr}_k(\boldsymbol{x}) = \begin{cases} R_k \cap L_{\boldsymbol{x}} & \text{if } \boldsymbol{x} \notin D_k, \\ \boldsymbol{x} & \text{if } \boldsymbol{x} \in D_k. \end{cases}$$

Suppose $z = \boldsymbol{x} + \sqrt{-1}\boldsymbol{y} \in U_k \setminus U_{k+1}$. Since $z \notin S_k$, $\boldsymbol{y} \notin \mathbb{R} \cdot \boldsymbol{x}$ (Proposition 7.3). Hence there exists a unique real number $\tau = \tau(\boldsymbol{x}, \boldsymbol{y}) \in \mathbb{R}$ such that $X_{\boldsymbol{x},\boldsymbol{y}}(\tau) \in H_{k+1}$.

Define the family of continuous map F_t $(0 \leq t \leq 1)$ by

$$F_t(\boldsymbol{x} + \sqrt{-1}\boldsymbol{y}) = \mathrm{pr}_k(X_{\boldsymbol{x},\boldsymbol{y}}(t \cdot \tau(\boldsymbol{x}, \boldsymbol{y}))) + \sqrt{-1}\dot{X}_{\boldsymbol{x},\boldsymbol{y}}(t \cdot \tau(\boldsymbol{x}, \boldsymbol{y})), \quad (7.6)$$

for $\boldsymbol{x} + \sqrt{-1}\boldsymbol{y} \in U_k$ with $\boldsymbol{x} \in D_k$. (Figure 7.9.)

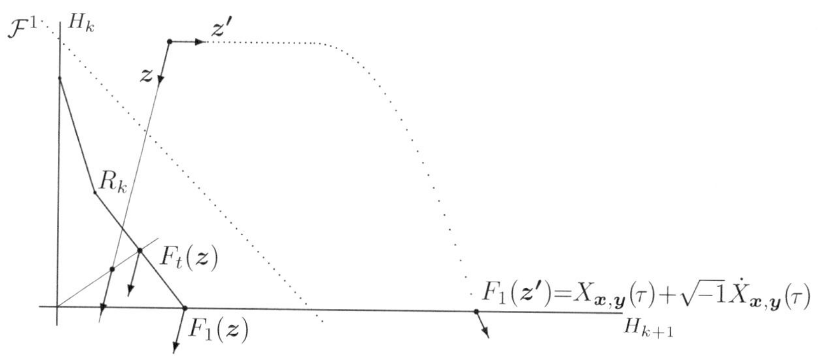

Figure 7.9. $F_t(z)$.

The next proposition completes the proof of the main result, which is proved in a similar way to the proof of Proposition 7.9.

Proposition 7.16. *Let us extend the above F_t by*

$$F_t(\boldsymbol{x} + \sqrt{-1}\boldsymbol{y}) = \begin{cases} F_t(z) \text{ (as above)} & \text{if } z \in U_k \setminus U_{k+1}, \\ z & \text{if } z \in U_{k+1}, \end{cases}$$

Then $F_t(z) \in U_k$ for any $z \in U_k$ and hence F_1 determines a deformation retract $U_k \to U_{k+1}$.

References

[1] W. ARVOLA, *The fundamental group of the complement of an arrangement of complex hyperplanes*, Topology **31** (1992), 757–765.

[2] A. BJÖRNER and G. ZIEGLER, *Combinatorial stratification of complex arrangements*, J. Amer. Math. Soc. (1) **5** (1992), 105–149.

[3] D. C. COHEN and A. SUCIU, *The braid monodromy of plane algebraic curves and hyperplane arrangements*, Comment. Math. Helvetici (2) **72** (1997), 285–315.

[4] E. DELUCCHI, "Shelling-type Orderings of Regular CW-complexes and Acyclic Matchings of the Salvetti complex", Int. Math. Res. Not. n. 6, 2008, pp. 39.

[5] A. DIMCA and S. PAPADIMA, *Hypersurface complements, Milnor fibers and higher homotopy groups of arrangements*, Ann. of Math. (2) **158** (2003), 473–507.

[6] M. ELIYAHU, D. GARBER and M. TEICHER, *Conjugation-free geometric presentations of fundamental groups of arrangements*, Manuscripta Math. (1-2) **133** (2010), 247–271.

[7] M. ELIYAHU, D. GARBER and M. TEICHER, *A conjugation-free geometric presentation of fundamental groups of arrangements II: expansion and some properties*, Internat. J. Algebra Comput. (5) **21** (2011), 775–792.

[8] M. FALK, *Homotopy types of line arrangements*, Invent. Math. (1) **111** (1993), 139–150.

[9] G. GAIFFI and M. SALVETTI *The Morse complex of a line arrangement*, J. of Algebra (1) **321** (2009), 316–337.

[10] V. GUILLEMIN and A. POLLACK, "Differential Topology", Prentice-Hall, Inc., Englewood Cliffs, N.J., 1974. xvi+222.

[11] K. ITO and M. YOSHINAGA, *Semi-algebraic partition and basis of Borel-Moore homology of hyperplane arrangements*, Proc. Amer. Math. Soc. **140** (2012), 2065–2074.

[12] A. LIBGOBER, *On the homotopy type of the complement to plane algebraic curves*, J. Reine Angew. Math. **367** (1986), 103–114.

[13] B. MOISHEZON and M. TEICHER, *Braid group technique in complex geometry I: line arrangements in* $\mathbb{CP}^2$, Contemp. Math. **78** (1988), 425–555.

[14] P. ORLIK and L. SOLOMON, *Combinatorics and topology of complements of hyperplanes*, Invent. Math. **56** (1980), 167–189.

[15] P. ORLIK and H. TERAO, "Arrangements of Hyperplanes", Grundlehren Math. Wiss. Vol. 300, Springer-Verlag, New York, 1992.

[16] S. PAPADIMA and A. SUCIU, *Higher homotopy groups of complements of complex hyperplane arrangements*, Adv. Math. (1) **165** (2002), 71–100.

[17] R. RANDELL, *The fundamental group of the complement of a union of complex hyperplanes*, Invent. Math. (1) **69** (1982), 103–108. Correction, Invent. Math. **80** (1985), 467–468.

[18] R. RANDELL, *Homotopy and group cohomology of arrangements*, Topology Appl. **78** (1997), 201–213.

[19] R. RANDELL, *Morse theory, Milnor fibers and minimality of hyperplane arrangements*, Proc. Amer. Math. Soc. (9) **130** (2002), 2737–2743.

[20] M. SALVETTI, *Arrangements of lines and monodromy of plane curves*, Compositio Math. **68** (1988), 103–122.

[21] M. SALVETTI, *Topology of the complement of real hyperplanes in* $\mathbb{C}^N$, Invent. Math. (3) **88** (1987), 603–618.

[22] M. SALVETTI and S. SETTEPANELLA, *Combinatorial Morse theory and minimality of hyperplane arrangements*, Geom. Topol. **11** (2007), 1733–1766.

[23] M. YOSHINAGA, *Hyperplane arrangements and Lefschetz's hyperplane section theorem*, Kodai Math. J. (2) **30** (2007), 157–194.

[24] M. YOSHINAGA, *The chamber basis of the Orlik-Solomon algebra and Aomoto complex*, Arkiv för Matematik **47** (2009), 393–407.

[25] M. YOSHINAGA, *Minimality of hyperplane arrangements and basis of local system cohomology*, In: "The Proceedings of the 5-th Franco-Japanese Symposium on Singularities", IRMA Lectures in Mathematics and Theoretical Physics, to appear.

[26] T. ZASLAVSKY, *Facing up to arrangements: Face-count formulas for partitions of space by hyperplanes*, Memoirs Amer. Math. Soc. **154** (1975).

CRM Series
Publications by the Ennio De Giorgi Mathematical Research Center Pisa

The Ennio De Giorgi Mathematical Research Center in Pisa, Italy, was established in 2001 and organizes research periods focusing on specific fields of current interest, including pure mathematics as well as applications in the natural and social sciences like physics, biology, finance and economics. The CRM series publishes volumes originating from these research periods, thus advancing particular areas of mathematics and their application to problems in the industrial and technological arena.

Published volumes

1. Matematica, cultura e società 2004 (2005). ISBN 88-7642-158-0
2. Matematica, cultura e società 2005 (2006). ISBN 88-7642-188-2
3. M. GIAQUINTA, D. MUCCI, *Maps into Manifolds and Currents: Area and* $W^{1,2}$*-*, $W^{1/2}$*-*, BV*-Energies*, 2006. ISBN 88-7642-200-5
4. U. ZANNIER (editor), *Diophantine Geometry*. Proceedings, 2005 (2007). ISBN 978-88-7642-206-5
5. G. MÉTIVIER, *Para-Differential Calculus and Applications to the Cauchy Problem for Nonlinear Systems*, 2008. ISBN 978-88-7642-329-1
6. F. GUERRA, N. ROBOTTI, *Ettore Majorana. Aspects of his Scientific and Academic Activity*, 2008. ISBN 978-88-7642-331-4
7. Y. CENSOR, M. JIANG, A. K. LOUISR (editors), *Mathematical Methods in Biomedical Imaging and Intensity-Modulated Radiation Therapy (IMRT)*, 2008. ISBN 978-88-7642-314-7
8. M. ERICSSON, S. MONTANGERO (editors), *Quantum Information and Many Body Quantum systems*. Proceedings, 2007 (2008). ISBN 978-88-7642-307-9
9. M. NOVAGA, G. ORLANDI (editors), *Singularities in Nonlinear Evolution Phenomena and Applications*. Proceedings, 2008 (2009). ISBN 978-88-7642-343-7

Matematica, cultura e società 2006 (2009). ISBN 88-7642-315-4

10. H. HOSNI, F. MONTAGNA (editors), *Probability, Uncertainty and Rationality*, 2010. ISBN 978-88-7642-347-5
11. L. AMBROSIO (editor), *Optimal Transportation, Geometry and Functional Inequalities*, 2010. ISBN 978-88-7642-373-4
12*. O. COSTIN, F. FAUVET, F. MENOUS, D. SAUZIN (editors), *Asymptotics in Dynamics, Geometry and PDEs; Generalized Borel Summation*, vol. I, 2011. ISBN 978-88-7642-374-1, e-ISBN 978-88-7642-379-6
12**. O. COSTIN, F. FAUVET, F. MENOUS, D. SAUZIN (editors), *Asymptotics in Dynamics, Geometry and PDEs; Generalized Borel Summation*, vol. II, 2011. ISBN 978-88-7642-376-5, e-ISBN 978-88-7642-377-2
13. G. MINGIONE (editor), *Topics in Modern Regularity Theory*, 2011. ISBN 978-88-7642-426-7, e-ISBN 978-88-7642-427-4
14. A. BJORNER, F. COHEN, C. DE CONCINI, C. PROCESI, M. SALVETTI (editors), *Configuration Spaces*, Geometry, Combinatorics and Topology, 2012. ISBN 978-88-7642-430-4, e-ISBN 978-88-7642-431-1

Volumes published earlier

Dynamical Systems. Proceedings, 2002 (2003)
Part I: *Hamiltonian Systems and Celestial Mechanics.*
ISBN 978-88-7642-259-1
Part II: *Topological, Geometrical and Ergodic Properties of Dynamics.*
ISBN 978-88-7642-260-1

Matematica, cultura e società 2003 (2004). ISBN 88-7642-129-7

Ricordando Franco Conti, 2004. ISBN 88-7642-137-8

N.V. KRYLOV, *Probabilistic Methods of Investigating Interior Smoothness of Harmonic Functions Associated with Degenerate Elliptic Operators,* 2004. ISBN 978-88-7642-261-1

Phase Space Analysis of Partial Differential Equations. Proceedings, vol. I, 2004 (2005). ISBN 978-88-7642-263-1

Phase Space Analysis of Partial Differential Equations. Proceedings, vol. II, 2004 (2005). ISBN 978-88-7642-263-1

Fotocomposizione "CompoMat" Loc. Braccone, 02040 Configni (RI) Italy
Finito di stampare nel mese di settembre 2012

ˈd States